To Luca,
and to Michele,
who knows about EEG and MEG

MS&A

Ana Alonso Rodríguez and Alberto Valli

Eddy Current Approximation of Maxwell Equations

Theory, algorithms and applications

Springer

Ana Alonso Rodríguez
Department of Mathematics
University of Trento
Trento, Italy

Alberto Valli
Department of Mathematics
University of Trento
Trento, Italy

ISBN 978-88-470-5584-1 ISBN 978-88-470-1506-7 (eBook)

DOI 10.1007/978-88-470-1506-7

Springer Milan Dordrecht Heidelberg London New York

Softcover re-print of the Hardcover 1st edition 2010

9 8 7 6 5 4 3 2 1

The image on the cover shows the eddy current in a trefoil knot (real and imaginary part)
Cover-Design: Beatrice 8, Milano

Typesetting with LaTeX: PTP-Berlin, Protago TeX-Production GmbH, Germany (www.ptp-berlin.eu)
Printing and Binding: Grafiche Porpora, Segrate (Mi)

Springer-Verlag Italia srl – Via Decembrio 28 – 20137 Milano
Springer is a part of Springer Science+Business Media (www.springer.com)

Preface

Continuamente nascono i fatti
a confusione delle teorie [1]
Carlo Dossi [2]

Electromagnetism is without any doubt a fascinating area of physics, engineering and mathematics. Since the early pioneering works of Ampère, Faraday, and Maxwell, the scientific literature on this subject has become immense, and books devoted to almost all of its aspects have been published in the meantime.

However, we believe that there is still some place for new books dealing with electromagnetism, particularly if they are focused on more specific models, or try to mix different levels of analysis: rigorous mathematical results, sound numerical approximation schemes, real-life examples from physics and engineering.

The complete mathematical description of electromagnetic problems is provided by the celebrated Maxwell equations, a system of partial differential equations expressed in terms of physical quantities like the electric field, the magnetic field and the current density. Maxwell's contribution to the formulation of these equations is related to the introduction of a specific term, called displacement current, that he proposed to add to the set of equations generally assumed to hold at that time, in order to ensure the conservation of the electric charge.

The presence of the displacement current permits to describe one of the most important phenomenon in electromagnetism, namely, wave propagation; however, in many interesting applications the propagation speed of the wave is very high with respect to the ratio of some typical length and time scale of the considered device, and therefore the dominant aspect becomes the diffusion of the electromagnetic fields.

When the focus is on diffusion instead of propagation, from the modeling point of view this corresponds to neglecting the time derivative of the electric induction (i.e., the displacement current introduced by Maxwell) or, alternatively, neglecting the time derivative of the magnetic induction.

[1] Constantly facts arise to muddle theories.
[2] Carlo Dossi, 1849–1910, Italian writer.

This book is devoted to the former model. The resulting equations are usually called magneto-quasistatic equations, or else *eddy current equations*, and can be seen as a low-frequency approximation of the full Maxwell system. In the following we are indeed concerned with the time-harmonic case, in which the data and the electromagnetic fields are assumed to be sinusoidal in time. This model is very often used in electrical engineering (for some examples, see Section 1.2 and Chapter 9). Indeed, for the typical problems in this field alternating currents are applied, the electromagnetic wave propagation can be neglected, but the variation of the magnetic field is still significative: in fact, in conducting media this variation generates current densities that have to be taken into account. Summing up, the term that can be dropped is the displacement current.

In our opinion, the reasons for the interest in the time-harmonic eddy current model are manifold. In fact, it is not only an important topic in electromagnetism, but also an intriguing mathematical problem in which one has to face some delicate aspects that can also be present in other situations. Therefore, the study of this problem can be useful for understanding general techniques that can be applied in other contexts, as well.

One of these peculiar aspects is that the time-harmonic eddy current problem presents differential constraints: the magnetic field is curl-free and the electric field is divergence-free in the insulating region, and the magnetic induction is divergence-free in the whole physical domain (insulator plus conductor).

There are several mathematical approaches that allow us to treat these constraints. In this book we refer to the following:

- saddle-point formulations with Lagrange multipliers;
- introduction of vector and scalar potentials;
- penalization methods.

Each of these approaches gives rise to different finite element approximations: mixed finite element methods are used when considering saddle-point formulations, and in these cases edge elements are needed for describing the discrete magnetic and electric fields; nodal vector elements and nodal scalar elements are used for approximating vector and scalar potentials, respectively; nodal vector elements are employed when dealing with penalization methods. Our aim is to give a presentation in which all these different approaches are considered and analyzed.

One could ask why it is necessary to introduce many different methods for solving the same problem. Let us quote from the well-known book by Silvester and Ferrari [227], p. 345: “In recent years, a considerable literature dealing with the numerical solution of problems relating to eddy currents has accumulated. Practical configurations are invariably irreducibly three-dimensional. No clear consensus appears to have emerged as to the best method of attack, although in many cases some finite element approach or other is used.”

In fact, as hopefully it will be clear by the end of the book, each method has assets and drawbacks:

- saddle-point and Lagrange multipliers. *Plus*: physical fields as principal unknowns; no difficulty with the topology of the conducting domain. *Minus*: many degrees of freedom; algebraic problem with a more complex structure;
- magnetic scalar potential. *Plus*: few degrees of freedom; "positive definite" algebraic problem. *Minus*: some difficulties coming from the topology of the computational domain, in particular of the conductor; need to compute in advance a vector potential of the current density;
- magnetic vector potential and penalization. *Plus*: standard nodal finite elements for all the unknowns; no difficulty with the topology of the conducting domain; "positive definite" algebraic problem. *Minus*: many degrees of freedom; lack of convergence for re-entrant corners of the computational domain.

Therefore, it is not an easy task to devise the best method for all seasons: this is also apparent looking at the literature, especially the part related mainly to engineering applications, in which new methods are proposed in each issue.

Nevertheless, let us note that, as far as we know, there are no books where eddy current problems are widely treated from both the mathematical and the engineering point of view. In fact, various monographs are devoted to modeling through partial differential equations and their numerical approximation (just to quote a couple of the most known, see Eriksson et al. [102] and Quarteroni [198]), but in general they do not cover electromagnetism and its mathematical theory.

On the other hand, among classical texts on electromagnetism only Silvester and Ferrari [227] and especially Bossavit [58], [59] devote some pages to this topic. The eddy current model is also briefly presented in Křížek and Neittaanmäki [158], though only for conductive media, and in Bondeson et al. [55]. Finally, a chapter in Gross and Kotiuga [115] is concerned with eddy current problems, but more specifically with those topological issues that are relevant for their numerical approximation.

In the engineering literature we recall the books by Tegopoulos and Kriezis [233] and Mayergoyz [173], where analytical methods are systematically employed for determining the explicit solution of eddy current problems, but only in simple geometrical configurations, the former for linear materials, the latter in the nonlinear case.

This book is the story of a falling in love. When in the mid 1990s we started to study eddy current problems, we even did not know the usual way these equations are referred to (indeed, we wrote a paper on "heterogeneous low-frequency Maxwell equations"). However, we were quickly attracted by their peculiar aspects:

- variational formulations set in somehow unusual spaces like $H(\text{div};\Omega)$ and $H(\text{curl};\Omega)$, for which some basic results were not completely clarified (for instance, the characterization of the space of tangential traces of functions belonging to $H(\text{curl};\Omega)$);
- the presence of differential constraints, which give rise to some difficulties in devising efficient finite element numerical approximation schemes;

- the strong interplay between the topological shape of the computational domain and the well-posedness of the problem, involving delicate arguments of algebraic topology not surprisingly already considered by Maxwell himself, but not always addressed in a correct way in the more recent literature;
- the problem of determining meaningful boundary conditions, or else realistic excitation terms associated to significative physical quantities such as voltage or current intensity;
- the breaking of the symmetry between the electric and the magnetic fields, which is specific in this context, and does not take place in the case of the full Maxwell equations;
- the unusually large number of different methods proposed for finding the approximate solution, some of them based on various choices of vector and scalar potentials, mainly already present in classical works in electromagnetism but not completely understood in the framework of eddy current problems;
- the loss of convergence of nodal finite element approximation schemes in the presence of re-entrant corners or edges.

This book is the story of an obsession. Having to face such a large number of different aspects, and their even larger possible interplays, our research work on eddy current problems has soon become a never-ending wandering among formulations, approximation methods, analyses of convergence, topological obstructions, choices of boundary conditions, and so on. Trying to write in a structured way all these topics has been a way to exit the labyrinth and to stop looking for a further result. (As a matter of fact, we have in mind another possible approach, but the margin of the page is too narrow for writing it here.[3]) We hope we succeeded in giving a map to people interested in the mathematical theory of low-frequency electromagnetism and the related numerical approximation schemes.

We have tried to write a self-contained book. Starting from the Maxwell equations we derive the eddy current model, and we make clear in which sense it is an approximation of the full Maxwell system. The existence and uniqueness of the solution are proved for all the described formulations, and stability and convergence of the finite element numerical schemes are presented. Some useful tools from functional analysis and finite element theory are collected in the Appendix.

Due to the structure described above, this monograph is addressed to researchers and Ph.D. students in mathematical electromagnetism, as well as to electrical engineers and practitioners, who can find here a sound mixture of theory, numerical approximation schemes and implementation issues, with a limited need of prerequisites.

The book is organized as follows.

In Chapter 1 we introduce the eddy current problem and we present its mathematical formulation, for the time-harmonic case and for three alternative sets of boundary conditions. Particular attention is devoted to the description of certain spaces of harmonic fields, which are related to the topological shape of the computational domain and must be taken into account in order to devise a well-posed problem.

[3] Hanc marginis exiguitas non caperet.

The second chapter deals with a mathematical justification of the eddy current model in a domain composed by a conductor and an insulator. It is obtained through two different asymptotic limits of the full Maxwell equations: in the first case the electric permittivity vanishes, and in the second case the frequency vanishes.

The analysis of well-posedness of eddy current problems is performed in Chapter 3: the existence and uniqueness of the solution is proved, and, moreover, an important remark is presented, concerning the verification of the Faraday equation on the so-called "cutting" surfaces contained in the insulator. This fact has been sometimes overlooked in the existing literature, leading to incorrect results for the numerical computations based on formulations where the principal unknown is the magnetic field.

In Chapter 4 we describe and analyze some coupled formulations that employ Lagrange multipliers for imposing the differential constraints on the magnetic and electric fields. The advantage of these approaches is that they involve no restrictions originating from the topology of the conductor, and that the used meshes do not need to match on the interface. To test the performance of the methods we present some numerical computations for domains of general shape, in particular some results for problem 7 of the TEAM workshop and for a conducting domain given by the trefoil knot.

Two formulations based on the introduction of a scalar magnetic potential in the insulator are illustrated in the fifth chapter: the unknown used in the conductor is the magnetic field in the first case, and the electric field in the second case. These methods use a small number of degrees of freedom (the unknowns are a vector function in the conductor and a scalar function in the insulator, plus a few degrees of freedom associated to the topological shape), but require some pre-processing, like the determination of the "cutting" surfaces and that of a vector potential of the applied current density.

The classical approaches using vector potentials are presented in Chapter 6, mainly for the case of a magnetic vector potential. The gauge conditions, needed for finding a unique potential, are analyzed in depth, in particular in the case of the Coulomb gauge and the Lorenz gauge. The advantage of these formulations lies in the fact that classical nodal finite elements are employed, so that the same discrete basis functions can be used for all the unknowns. Moreover, no difficulty comes from the topology of the conductor.

In Chapter 7 we set the problem in the whole space and we introduce some coupled finite element/boundary element methods, which, by using potential theory, allow to reduce the degrees of freedom in the insulator to degrees of freedom on the interface. In particular, we present in more detail the coupled approach based on the magnetic vector potential and the scalar electric potential in the conductor: this method has the characteristic of being stable with respect to the frequency, hence can be also used without modification for the static case.

The eighthğ¡ chapter deals with the case of excitation terms given by a voltage drop or a current intensity, a situation that can be interesting when the coupling with circuit problems has to be considered. In order to devise a well-posed problem it is necessary to choose suitable boundary conditions. For other boundary conditions the solution can be found only if the voltage or the current intensity are interpreted as an excitation term giving rise to a specific current density.

In Chapter 9 we present some real-life problems that are based on the eddy current equations. The description is not fully detailed, the aim being only to show the importance of eddy current problems in applications.

The book ends with an appendix, devoted to the functional framework, to nodal and edge finite elements, to some orthogonal decomposition results, and to a more complete characterization of the spaces of harmonic fields.

This book would not have been written without the help of some friends and colleagues. First of all, we want to thank Paolo Fernandes, Ralf Hiptmair, Oszkár Bíró and Rafael Vázquez Hernández, who worked with us on some eddy current problems, and with whom we had many enlightening and pleasant scientific conversations. Special thanks are due to Alfredo Bermúdez, Rodolfo Rodríguez, Pilar Salgado and Virginia Selgas, who provided us with many of their numerical results and figures, enriching the content and the final aspect of our book. We have learnt many interesting things about harmonic fields, homology theory and algebraic topology from our colleagues Domenico Luminati and, especially, Riccardo Ghiloni, and it is a pleasure to acknowledge their help.

We are grateful to Jarke J. van Wijk (Eindhoven University of Technology), Victor Valcarcel (Universidade de Santiago de Compostela) and Elekta, who have permitted us to reproduce and insert in the book some figures and photographs. We also thank Gianpaolo Demarchi and Elisa Leonardelli: the former showed us some of the technological devices installed at CIMeC (Centro Interdipartimentale Mente/Cervello, University of Trento), explaining their fascinating operation; the latter permitted us to take some pictures of the EEG recording cap system. Our official photographer has been Luca Manini: thanks a lot!

Finally, we express our gratitude to the Editors (Tom Hou, Claude Le Bris, Anthony T. Patera, and Enrique Zuazua) and to Alfio Quarteroni for having accepted to publish this monograph in the MS&A Series and for their several suggestions that have contributed to improve the final result; to Peter Laurence, who helped us with the English language; and to Francesca Bonadei from Springer, who with great expertise and attention has taken care of the realization of this book.

Povo (Trento), April 2010

Ana Alonso Rodríguez
Alberto Valli

Contents

1

Setting the problem

In this chapter, starting from the classical Maxwell equations, we describe and motivate the problem we are going to consider.

In particular, we derive the full Maxwell system, for both the time-dependent and the time-harmonic case, and we explain how eddy currents are generated and why they are the most relevant aspect in a series of engineering problems. Then we introduce the eddy current approximation of the Maxwell equations, often used in low-frequency electromagnetism, presenting the complete set of equations together with some suitable choices of boundary conditions.

It is worth noting that, in order to properly formulate the problem, we need to introduce certain spaces of vector fields: the so-called harmonic fields. These spaces are strongly related to the topological properties of the insulator, namely, the domain where the electric conductivity vanishes, and their characterization is an important tool for proving well-posedness of the problem and devising efficient numerical approximation schemes.

1.1 Maxwell equations and time-harmonic Maxwell equations

The study of the propagation and the diffusion of electromagnetic fields is an important topic in physical sciences. The first attempt to describe in a rigorous mathematical way these phenomena dates back to the beginning of the nineteenth century, when Ampère and Faraday, among others, started to make experiments on electricity and magnetism.

The physical quantities that have to be taken into account are the magnetic field $\mathcal{H}$, the electric field $\mathcal{E}$, the magnetic induction $\mathcal{B}$, the electric induction $\mathcal{D}$ and the electric current density $\mathcal{J}$. The electric field and the magnetic induction can be defined at the microscopic level, and at this level $\mathcal{D}$ and $\mathcal{H}$ are simply a multiple of $\mathcal{E}$ and $\mathcal{B}$, respectively. At the macroscopic level, where the properties of the material media play a significant role, all these fields are in some sense averaged quantities, related through some constitutive equations. A linear dependence of the form $\mathcal{D} = \varepsilon\mathcal{E}$, $\mathcal{B} = \mu\mathcal{H}$ is usually assumed; here ε and μ are called the electric permittivity and magnetic perme-

A. Alonso Rodríguez, A. Valli: Eddy Current Approximation of Maxwell Equations.

ability, respectively (for a complete presentation of the physics of electromagnetism, see, e.g., Jackson [137]).

In most interesting physical and engineering problems, the region of interest is composed of a non-homogeneous and non-isotropic medium: namely, ε and μ are not constant, but are symmetric and uniformly positive definite matrices (with entries that are bounded functions of the space variable $\mathbf{x}$). In general, a nonlinear dependence between $\mathcal{D}$ and $\mathcal{E}$, $\mathcal{B}$ and $\mathcal{H}$ can also be taken into account (for instance, for hysteresis problems). However, in this book we only consider the linear case.

The basic equations relating the electromagnetic fields are derived by some experimental results. The first one, that takes the name of Ampère, states that, in the steady case, the electric current I^0 passing through a surface is equal to the line integral (with the counterclockwise orientation) of the magnetic field $\mathcal{H}$ on the boundary of that surface. A second relation, which is due to Faraday, comes from the observation that a time-variation of the magnetic field generates an electric field: precisely, the time derivative of the flux of the magnetic induction through a given surface is equal to the line integral (with the clockwise orientation) of that induced electric field on the boundary of that surface.

These relations can be easily written in a differential form: first of all, the Ampère law reads

$$I^0 = \int_S \mathcal{J} \cdot \mathbf{n} = \int_{\partial S} \mathcal{H} \cdot \boldsymbol{\tau} \,,$$

where $\mathbf{n}$ is the unit normal vector on S and $\boldsymbol{\tau}$ is the unit tangent vector on ∂S (oriented counterclockwise with respect to $\mathbf{n}$). Therefore, from the Stokes theorem we find

$$\int_S \mathcal{J} \cdot \mathbf{n} = \int_S \operatorname{curl} \mathcal{H} \cdot \mathbf{n} \,.$$

Since the surface S is arbitrarily placed in the space, it follows that

$$\mathcal{J} = \operatorname{curl} \mathcal{H} \,.$$

On the other hand, the Faraday law can be written as

$$\frac{d}{dt} \int_S \mathcal{B} \cdot \mathbf{n} = - \int_{\partial S} \mathcal{E} \cdot \boldsymbol{\tau} \,,$$

hence by the Stokes theorem

$$\frac{d}{dt} \int_S \mathcal{B} \cdot \mathbf{n} = - \int_S \operatorname{curl} \mathcal{E} \cdot \mathbf{n} \,,$$

and thus

$$\frac{\partial \mathcal{B}}{\partial t} = - \operatorname{curl} \mathcal{E} \,.$$

The celebrated contribution of Maxwell was the observation that the Ampère law was not completely satisfactory in the time-dependent case, and that it has to be corrected by adding another term. It is possible to devise its form by taking into consideration two facts: the first is the Gauss electrical equation, stating that the total charge contained in a volume V is equal to the external flux of the electric induction through the boundary of that volume, namely,

$$\int_V \rho = \int_{\partial V} \mathcal{D} \cdot \mathbf{n} \,,$$

where ρ is the volume electric charge density (supposed to vanish in any non-conducting region) and $\mathbf{n}$ is the unit outward normal vector on ∂V; the second is the charge conservation law

$$\frac{d}{dt} \int_V \rho = - \int_{\partial V} \mathcal{J} \cdot \mathbf{n} \,,$$

similar to the mass conservation law in fluid dynamics. As a consequence one has

$$\frac{d}{dt} \int_{\partial V} \mathcal{D} \cdot \mathbf{n} = - \int_{\partial V} \mathcal{J} \cdot \mathbf{n} \,,$$

and then, by the divergence theorem and since the volume V is arbitrary,

$$\operatorname{div} \left(\frac{\partial \mathcal{D}}{\partial t} + \mathcal{J} \right) = 0 \,.$$

Being divergence-free, $\frac{\partial \mathcal{D}}{\partial t} + \mathcal{J}$ has to be equal to the curl of a suitable vector field: since in the time-independent case the Ampère law $\mathcal{J} = \operatorname{curl} \mathcal{H}$ holds, for time-dependent fields Maxwell proposed the following generalization of the Ampère law

$$\frac{\partial \mathcal{D}}{\partial t} + \mathcal{J} = \operatorname{curl} \mathcal{H} \,.$$

Maxwell himself called the added term $\frac{\partial \mathcal{D}}{\partial t}$ the displacement current.

Summing up, the complete *Maxwell system* of electromagnetism reads

$$\begin{cases} \dfrac{\partial \mathcal{D}}{\partial t} + \mathcal{J} = \operatorname{curl} \mathcal{H} & \text{Maxwell–Ampère equation} \\ \dfrac{\partial \mathcal{B}}{\partial t} + \operatorname{curl} \mathcal{E} = 0 & \text{Faraday equation} \\ \operatorname{div} \mathcal{D} = \rho & \text{Gauss electrical equation} \\ \operatorname{div} \mathcal{B} = 0 & \text{Gauss magnetic equation} \,, \end{cases} \tag{1.1}$$

where the Gauss magnetic equation is a consequence of the experimental fact that magnetic charges do not exist.

To close the system, another relation is introduced, which expresses the current density in a conductor in terms of the electric field: the classic Ohm law, based on physical observations about electrical circuits, states that $\mathcal{J} = \boldsymbol{\sigma} \mathcal{E}$, where $\boldsymbol{\sigma}$ is the electric conductivity, which, in conducting regions, is assumed to be a symmetric and uniformly positive definite matrix (with entries that are bounded functions of the space variable $\mathbf{x}$), while it is vanishing in insulators.

When the problem is driven by an applied current density $\mathcal{J}_e$, one needs to consider the generalized Ohm law $\mathcal{J} = \boldsymbol{\sigma} \mathcal{E} + \mathcal{J}_e$. Let us note that, as a consequence of Maxwell–Ampère and Gauss electrical equations, it is necessary to assume that $\operatorname{div} \mathcal{J}_e = 0$ in any non-conducting region.

Though more general situations are also of interest, in this book we focus on problems where the physical quantities vary periodically with time[1]: typically, this happens when the applied current density $\mathcal{J}_e$ is an alternating current, having the form

$$\mathcal{J}_e(\mathbf{x},t) = \mathbf{J}_*(\mathbf{x})\cos(\omega t + \phi) \ ,$$

where $\mathbf{J}_*(\mathbf{x})$ is a real-valued vector function, $\omega \neq 0$ is the angular frequency and ϕ is the phase angle. This is equivalent to the representation

$$\mathcal{J}_e(\mathbf{x},t) = \mathrm{Re}\left[\mathbf{J}_*(\mathbf{x})e^{i(\omega t+\phi)}\right] = \mathrm{Re}\left[\mathbf{J}_e(\mathbf{x})e^{i\omega t}\right] \ ,$$

where we have introduced the complex-valued vector function $\mathbf{J}_e(\mathbf{x}) := \mathbf{J}_*(\mathbf{x})e^{i\phi}$.

Accordingly, we look for a time-periodic (or else, time-harmonic) solution given by

$$\begin{aligned} \mathcal{E}(\mathbf{x},t) &= \mathrm{Re}\left[\mathbf{E}(\mathbf{x})e^{i\omega t}\right] \\ \mathcal{H}(\mathbf{x},t) &= \mathrm{Re}\left[\mathbf{H}(\mathbf{x})e^{i\omega t}\right] \ , \end{aligned}$$

where $\mathbf{E}$ and $\mathbf{H}$ are complex-valued vector functions (often called “phasors”).

The *time-harmonic Maxwell equations* are directly derived from the complete system under these assumptions, and read

$$\begin{cases} \mathrm{curl}\,\mathbf{H} - (i\omega\varepsilon + \boldsymbol{\sigma})\mathbf{E} = \mathbf{J}_e \\ \mathrm{curl}\,\mathbf{E} + i\omega\boldsymbol{\mu}\mathbf{H} = \mathbf{0} \ , \end{cases} \tag{1.2}$$

determining the electric charge density by setting, separately in the conducting and non-conducting regions,

$$\rho(\mathbf{x},t) = \mathrm{div}\left(\varepsilon(\mathbf{x})\mathcal{E}(\mathbf{x},t)\right) = \mathrm{div}\left(\mathrm{Re}[\varepsilon(\mathbf{x})\mathbf{E}(\mathbf{x})e^{i\omega t}]\right) .$$

Note that the Gauss magnetic equation $\mathrm{div}(\boldsymbol{\mu}\mathbf{H}) = 0$ is a consequence of the Faraday equation; moreover, the Maxwell–Ampère equation and the assumption that $\mathrm{div}\,\mathcal{J}_e$ is vanishing in any non-conducting region imply that the charge density is vanishing there, too.

1.2 Eddy currents and eddy current approximation

As observed in experiments and stated by the Faraday law, a time-variation of the magnetic field generates an electric field. Therefore, in each conductor a current density

[1] We believe that most of the methods proposed in this book, very likely all of them, can be adapted to the time-dependent case: for instance, it should be possible to prove existence and uniqueness of the solution, and stability and convergence of suitable numerical schemes. However in this treatise we limit ourselves to the important case of time-periodic models. For some additional issues on the time-dependent problem see, e.g., the books by Silvester and Ferrari [227], Bossavit [59] and van Rienen [238], the papers by Nicolet and Delincé [189], Clemens and Weiland [84] and Weiland [243].

$\mathbf{J}_{\text{eddy}} = \boldsymbol{\sigma}\mathbf{E}$ arises; this term expresses the presence in conducting media of the so-called *eddy currents*. This phenomenon, and the related heating of the conductor, was observed and studied by the French physicist L. Foucault in the mid of the nineteenth century, and in fact the generated currents are also known as Foucault currents.

The heat Q generated by the current density in a conductor is given by the Joule law

$$Q = \boldsymbol{\sigma}^{-1}\mathcal{J} \cdot \mathcal{J}\,.$$

Moreover, eddy currents also generate Lorentz forces

$$\mathbf{f}_l = \mathcal{J} \times \mathcal{B}\,.$$

Let us have a deeper look at these two aspects.

The Joule effect can have a good use, as it is the basis of induction furnaces, widely used in the metallurgic industry. Probably melting systems were the first industrial application of eddy currents. Basically a induction furnace for melting consists of a conducting crucible charged with the metal to be melted and of a helical coil, turning around the crucible, carrying an alternating current. This alternating current produces an oscillating magnetic field, which generates eddy currents in the crucible and in its load. These currents, due to the Joule effect, heat the metal until it melts. However, at the same time, they can also generate very high temperatures in the crucible, damaging it and reducing its lifetime. Some parameters, as the frequency and intensity of the applied current, the thermal and electrical conductivity of the crucible or its distance from the coil, affect the temperature profile in the furnace and must be taken into account in the construction of the melting system. Moreover, Lorentz forces act on the molten metal and cannot be ignored in melting processes, as the stirring effect modifies the properties of the final product. A more detailed description of this application of eddy currents is given in Section 9.1.1.

Joule heating can also produces undesirable power losses and overheating of electrical devices. For instance it is an important aspect in the design of power transformers. Transformers are used to produce an alternating current with low intensity and high voltage starting from another one with high intensity and low voltage, and vice-versa. They basically consist in two windings wrapped around an iron core. An alternating current passing through the primary winding generates a time-varying magnetic field in the core that induces a current in the secondary winding. The ratio between the voltages of the current in the primary and the secondary winding is proportional to the ratio between the number of their turns. In theory, a transformer would have no energy losses; in practice, energy is dissipated in the windings, the core, and the other metallic components of the transformer. Very soon it was observed that cores constructed from solid iron have extremely high eddy current losses, so later designs are based on a core made up of thin steel layers in order to reduce losses. The overheating of the clamping structure that maintains the core and the coils properly assembled can affect the reliability and the operating life of large power transformers. Numerical simulations are very useful for the optimal design of transformers; Section 9.4 includes a more extended presentation of this kind of simulations.

Lorentz forces can be used, for instance, for levitation or for the design of electromagnetic braking systems. A simple way to illustrate magnetic levitation principles is to consider a toroidal inductor carrying an alternating current $\mathcal{J}_e$ placed below a conducting sheet. By the Ampère law the tangential current in the inductor generates a time-varying magnetic flux. By the Faraday law this changing magnetic flux induces an electric field in the conducting plate. The dominant current component in the plate is along the direction of $\mathcal{J}_e$. This current interacts with the radial component of the magnetic field to generate (by the Lorentz law) a lift perpendicular to the plate. Section 9.3 is devoted to give a more precise description of magnetic levitation phenomena.

Eddy currents are also used as a non-destructive technique to detect the flaws in conductive objects: a coil fed by an alternating current is placed near the object, thus eddy currents arise inside, and flaws are located by a suitable measure of the variation of the impedance. A more detailed presentation of non-destructive techniques for defect detection is in Section 9.5.

Summing up, the computation of the eddy current distribution and of the related energy loss is an important task for engineering applications in electromagnetism.

In all these applications, it can be checked that the time of propagation of the electromagnetic wave is very small with respect to the inverse of the angular frequency ω, therefore one can think that the speed of propagation is infinite, and take into account only the diffusion of the electromagnetic fields: if one wants to express this fact with a mathematical recipe, one has not to face a "hyperbolic" problem but rather a "parabolic" problem.

Rephrasing this concept, one can also say that, when considering time-dependent problems in electromagnetism, one can distinguish between "fast" varying fields and "slowly" varying fields. In the latter case, one is led to simplify the set of equations, neglecting time derivatives, or, depending on the specific situation at hand, one time derivative, either $\frac{\partial \mathcal{D}}{\partial t}$ or $\frac{\partial \mathcal{B}}{\partial t}$. Typically, problems of this type arise in electrical engineering, where low frequencies are involved, but not in electronic engineering, where the frequency ranges in much larger bands.

When neglecting both the time derivative terms, one obtains the electro-magnetostatic model: an approximation of the Maxwell system for which diffusion of the electromagnetic fields is not considered and eddy currents and their effects are not taken into account.

If the time derivative of the magnetic induction is disregarded, the governing equations are called electro-quasistatic equations, and describe "slowly" varying fields for which the electric field is somehow independent of the magnetic field and the displacement current makes a significant contribution. These equations can be used for modeling problems in electrical engineering where the frequency is relatively low but the voltage is high (for a more detailed description, see, e.g., van Rienen [238]).

In this book we focus on the case in which the displacement current term $\frac{\partial \mathcal{D}}{\partial t}$ can be disregarded, while the time-variation of the magnetic induction is still important. In particular, as already noted, this means that the electromagnetic waves are neglected, as their time of propagation is very small with respect to $1/\omega$, or, equivalently, their wave length is much larger than the diameter of the physical domain.

Let us make more precise this statement, referring, e.g., to Haus and Melcher [119], Bossavit [59] and van Rienen [238] for a more detailed discussion concerning the physical validity of this assumption. Clearly, the point is that $\frac{\partial \mathcal{D}}{\partial t}$ should be small in comparison with $\operatorname{curl} \mathcal{H}$ and $\mathcal{J} = \boldsymbol{\sigma}\mathcal{E} + \mathcal{J}_e$. A thumb rule can be formulated as follows: if L is a typical length in Ω (say, its diameter), and we choose ω^{-1} as a typical time, it is possibile to disregard the displacement current term provided that

$$|\mathcal{D}||\omega| \ll |\mathcal{H}|L^{-1} \ , \ |\mathcal{D}||\omega| \ll |\boldsymbol{\sigma}\mathcal{E}| \, .$$

Using the Faraday equation, we can write $\mathcal{E}$ in terms of $\mathcal{H}$, finding

$$|\mathcal{E}|L^{-1} \approx |\omega||\boldsymbol{\mu}\mathcal{H}| \, .$$

Hence, recalling that $\mathcal{D} = \varepsilon\mathcal{E}$ and putting everything together, one should have

$$\mu_{\max}\varepsilon_{\max}\omega^2 L^2 \ll 1 \ , \ \sigma_{\min}^{-1}\varepsilon_{\max}|\omega| \ll 1 \, ,$$

where $\mu_{\max}$ and $\varepsilon_{\max}$ are uniform upper bounds in Ω for the maximum eigenvalues of $\boldsymbol{\mu}(\mathbf{x})$ and $\varepsilon(\mathbf{x})$, respectively, and $\sigma_{\min}$ denotes a uniform lower bound in Ω_C for the minimum eigenvalues of $\boldsymbol{\sigma}(\mathbf{x})$. Since the magnitude of the velocity of the electromagnetic wave can be estimated by $(\mu_{\max}\varepsilon_{\max})^{-1/2}$, the first relation is requiring that the wave length is large compared to L. Let us also note that for industrial electrical applications some typical values of the parameters involved are $\mu_0 = 4\pi \times 10^{-7}$ H/m, $\varepsilon_0 = 8.9 \times 10^{-12}$ F/m, $\sigma_{\text{copper}} = 5.7 \times 10^7$ S/m, $\omega = 2\pi \times 50$ rad/s (power frequency of 50 Hz), hence in that case

$$\frac{1}{\sqrt{\mu_0\varepsilon_0}|\omega|} \approx 10^6 \,\text{m} \ , \ \sigma_{\text{copper}}^{-1}\varepsilon_0|\omega| \approx 4.9 \times 10^{-17} \, ,$$

and dropping the displacement current term looks appropriate. Though less apparent, the same is true for a typical conductivity in a physiological problem, say, $\sigma_{\text{tissue}} \approx 10^{-1}$ S/m, for which $\sigma_{\text{tissue}}^{-1}\varepsilon_0|\omega| \approx 2.8 \times 10^{-8}$.

The system of equations obtained when the displacement current term $\frac{\partial \mathcal{D}}{\partial t}$ (or, equivalently, $i\omega\varepsilon\mathbf{E}$) is disregarded is called *eddy current approximation* (or magneto-quasistatic approximation) of the Maxwell equations. In the time-harmonic case, the resulting set of equations is therefore

$$\begin{cases} \operatorname{curl} \mathbf{H} - \boldsymbol{\sigma}\mathbf{E} = \mathbf{J}_e & \text{in } \Omega \\ \operatorname{curl} \mathbf{E} + i\omega\boldsymbol{\mu}\mathbf{H} = \mathbf{0} & \text{in } \Omega \\ \operatorname{div}(\varepsilon\mathbf{E}) = 0 & \text{in } \Omega_I \, , \end{cases} \tag{1.3}$$

where we have denoted by Ω the physical domain and by Ω_I the insulator.

A few remarks are in order: first, as in the case of the full Maxwell system we have to assume that the condition

$$\operatorname{div} \mathbf{J}_e = 0 \quad \text{in } \Omega_I \tag{1.4}$$

is satisfied. Then, note that again the constraint $\operatorname{div}(\boldsymbol{\mu}\mathbf{H}) = 0$ has been dropped from system (1.3), as it follows from the Faraday equation. Finally, note that in the eddy current approximation the equation $\operatorname{div}(\varepsilon\mathbf{E}) = 0$ in Ω_I, ensuring that the electric charge is vanishing in the insulator, is no longer a consequence of the Ampère equation and of the assumption (1.4). This is why we have kept it (1.3).

However, in the problem above something is still missing (clearly, beside the boundary conditions). In Alonso and Valli [7] it has been proved that other equations, related with the geometry of the domain Ω_I, have to be added in order to close the system. We present the complete model in Section 1.5; moreover, in Chapter 2 we give its rigorous mathematical justification, showing in particular that the difference between the solution of the full Maxwell system (1.2) and the solution of the complete eddy current model is vanishing as the angular frequency ω goes to 0.

1.3 Geometrical setting and boundary conditions

Let us make precise the geometrical context we consider in the sequel (with the exception of Chapter 8): the physical domain Ω is a bounded connected open set in $\mathbb{R}^3$, with a Lipschitz boundary $\partial\Omega$. We assume that an open subset Ω_C, the conductor, is strictly contained in Ω, namely, $\overline{\Omega_C} \subset \Omega$, and, as before, we denote by $\Omega_I := \Omega \setminus \overline{\Omega_C}$ the insulator (see Figure 1.1). For the sake of simplicity, we also suppose that Ω_I is connected: we believe that the interested reader will not find difficult to extend the results presented in this book to the case of a non-connected insulator Ω_I, though some formal changes are needed since in that case at least one connected component of Ω_I has empty intersection with the boundary $\partial\Omega$.

We denote by $\Gamma := \partial\Omega_I \cap \partial\Omega_C$ the interface between the two subdomains, and we assume that it is a Lipschitz surface; note that, in the present situation, $\partial\Omega_C = \Gamma$ and $\partial\Omega_I = \partial\Omega \cup \Gamma$.

The unit outward normal vector on $\partial\Omega$ is denoted by $\mathbf{n}$, while $\mathbf{n}_C = -\mathbf{n}_I$ denotes the unit normal vector on the interface Γ, pointing towards Ω_I.

Let us present now some suitable boundary conditions for the eddy current model. If the boundary $\partial\Omega$ can be considered as a perfect conductor, say, a fictitious medium

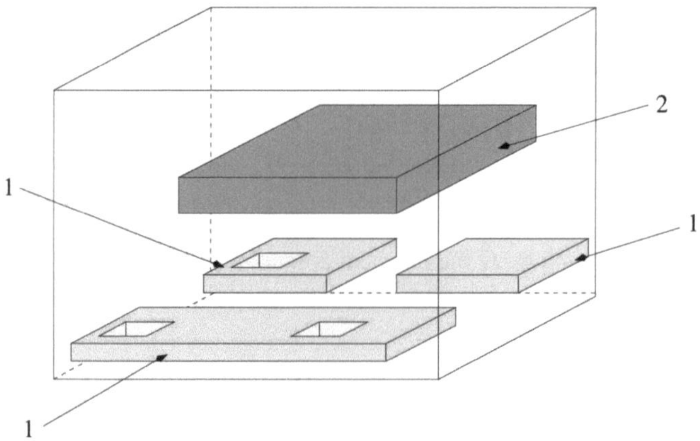

Fig. 1.1. The geometry of the problem: 1 conductors, 2 a region not included in the domain Ω

where the electric conductivity is infinite, then the boundary condition is the so-called *electric boundary condition*

$$\mathbf{E} \times \mathbf{n} = \mathbf{0} \quad \text{on } \partial\Omega \, . \tag{1.5}$$

It is easily checked that a boundary condition for the magnetic field follows from this: in fact, from the Faraday equation,

$$\boldsymbol{\mu}\mathbf{H} \cdot \mathbf{n} = -(i\omega)^{-1} \operatorname{curl} \mathbf{E} \cdot \mathbf{n} = -(i\omega)^{-1} \operatorname{div}_{\tau}(\mathbf{E} \times \mathbf{n}) = 0 \quad \text{on } \partial\Omega$$

(see Section A.1 for the definition and the properties of the tangential divergence operator $\operatorname{div}_{\tau}$).

If the boundary $\partial\Omega$ can be considered as an infinitely permeable medium (say, iron), then the so-called *magnetic boundary condition* can be imposed

$$\mathbf{H} \times \mathbf{n} = \mathbf{0} \quad \text{on } \partial\Omega \, . \tag{1.6}$$

Proceeding as above, and recalling that the conductivity $\boldsymbol{\sigma}$ vanishes near the boundary, from the Ampère equation this implies that the following compatibility condition has to be satisfied

$$\mathbf{J}_e \cdot \mathbf{n} = \operatorname{curl} \mathbf{H} \cdot \mathbf{n} = \operatorname{div}_{\tau}(\mathbf{H} \times \mathbf{n}) = 0 \quad \text{on } \partial\Omega \, . \tag{1.7}$$

However, the magnetic boundary condition is not enough for the determination of the electric field in the insulator. Recalling that for the solution of the full Maxwell system (1.2) one would have

$$0 = \operatorname{curl} \mathbf{H} \cdot \mathbf{n} = i\omega\varepsilon\mathbf{E} \cdot \mathbf{n} + \mathbf{J}_e \cdot \mathbf{n} \quad \text{on } \partial\Omega \, ,$$

one is led to require

$$\varepsilon\mathbf{E} \cdot \mathbf{n} = 0 \quad \text{on } \partial\Omega \, . \tag{1.8}$$

Summing up, when the magnetic boundary condition (1.6) is considered, one has also to impose (1.8) and to assume that (1.7) is satisfied.

A third set of boundary conditions has been proposed in the literature (see, e.g., Bossavit [61], Bermúdez et al. [43]), especially for voltage and current excitation problems (see Chapter 8). They are usually called *no-flux boundary conditions*, and look like a mixture of the preceding boundary conditions, namely,

$$\begin{cases} \boldsymbol{\mu}\mathbf{H} \cdot \mathbf{n} = 0 & \text{on } \partial\Omega \\ \varepsilon\mathbf{E} \cdot \mathbf{n} = 0 & \text{on } \partial\Omega \, . \end{cases} \tag{1.9}$$

In this book we mainly focus on the magnetic and the electric boundary conditions, and we present a more specific analysis of condition (1.9) in Section 3.5 and Chapter 8 only. Instead, we are not treating the impedance (or absorbing) condition

$$\mathbf{n} \times \mathbf{H} \times \mathbf{n} + \alpha \mathbf{E} \times \mathbf{n} = \mathbf{0} \; , \quad \alpha \in \mathbb{C} \, ,$$

which for eddy current problems has a correct physical meaning mainly as an interface condition on Γ (and not on $\partial\Omega$), provided that the penetration depth is small enough (see, e.g., MacCamy and Stephan [171], Ammari et al. [24], Sterz and Schwab [229]).

1.4 Harmonic fields in electromagnetism

Harmonic fields are those vector fields $\mathbf{v}$ satisfying $\operatorname{curl} \mathbf{v} = \mathbf{0}$ and $\operatorname{div} \mathbf{v} = 0$ (or, more generally, $\operatorname{div}(\eta \mathbf{v}) = 0$, where $\eta = \eta(\mathbf{x})$ is a symmetric and uniformly positive definite matrix, with bounded entries). In other words, if the physical domain under consideration is the entire space $\mathbb{R}^3$, they are the gradient of a harmonic function.

Suppose now that the physical domain is a bounded domain $\mathcal{O}$, and assume that its boundary is divided into two non-overlapping Lipschitz surfaces Γ_D and Γ_N (it is possible that one of the two could be empty).

A couple of questions are in order. If we also require that the boundary conditions $\mathbf{v} \times \mathbf{n} = \mathbf{0}$ on Γ_D and $\mathbf{v} \cdot \mathbf{n} = 0$ on Γ_N are satisfied, do non-trivial harmonic fields exist (here "non-trivial" means "not vanishing everywhere")? In that case, do harmonic fields appear in electromagnetism?

Both questions have an affermative answer. Let us start from the first question. If the domain $\mathcal{O}$ is homeomorphic to a three-dimensional ball, a curl-free vector field $\mathbf{v}$ must be a gradient of a scalar function ψ, that must be harmonic due to the constraint on the divergence. If the boundary condition is $\mathbf{v} \times \mathbf{n} = \mathbf{0}$ on $\partial\mathcal{O}$, which in this case is a connected surface, then it follows $\psi =$ const. on $\partial\mathcal{O}$, and therefore $\psi =$ const. in $\mathcal{O}$ and $\mathbf{v} = \mathbf{0}$ in $\mathcal{O}$. On the other hand, if the boundary condition is $\mathbf{v} \cdot \mathbf{n} = 0$ on $\partial\mathcal{O}$, then ψ satisfies a homogeneous Neumann boundary condition and thus $\psi =$ const. in $\mathcal{O}$ and $\mathbf{v} = \mathbf{0}$ in $\mathcal{O}$. The same result follows if the boundary conditions are $\mathbf{v} \times \mathbf{n} = \mathbf{0}$ on Γ_D and $\mathbf{v} \cdot \mathbf{n} = 0$ on Γ_N, and Γ_D is a connected surface: in fact, we still have $\psi =$ const. on Γ_D and $\operatorname{grad} \psi \cdot \mathbf{n} = 0$ on Γ_N, hence ψ satisfies a mixed boundary value problem and we obtain $\psi =$ const. in $\mathcal{O}$ and $\mathbf{v} = \mathbf{0}$ in $\mathcal{O}$.

However, the problem is different in a more general geometry. In fact, take the magnetic field generated in the vacuum by a current of constant intensity I^0 passing along the x_3-axis: as it is well-known, for $x_1^2 + x_2^2 > 0$ it is given by

$$\mathbf{H}(x_1, x_2, x_3) = \frac{I^0}{2\pi} \left(-\frac{x_2}{x_1^2 + x_2^2}, \frac{x_1}{x_1^2 + x_2^2}, 0 \right) .$$

It is easily checked that, as Maxwell equations require, $\operatorname{curl} \mathbf{H} = \mathbf{0}$ and $\operatorname{div} \mathbf{H} = 0$. Let us consider now the torus $\mathcal{T}$ obtained by rotating around the x_3-axis the disk of centre $(a, 0, 0)$ and radius b, with $0 < b < a$. One sees at once that $\mathbf{H} \cdot \mathbf{n} = 0$ on $\partial\mathcal{T}$; hence we have found a non-trivial harmonic field $\mathbf{H}$ in $\mathcal{T}$ satisfying $\mathbf{H} \cdot \mathbf{n} = 0$ on $\partial\mathcal{T}$.

On the other hand, consider now the electric field generated in the vacuum by a pointwise charge ρ_0 placed at the origin. For $\mathbf{x} \neq \mathbf{0}$ it is given by

$$\mathbf{E}(x_1, x_2, x_3) = \frac{\rho_0}{4\pi\varepsilon_0} \frac{\mathbf{x}}{|\mathbf{x}|^3} ,$$

where ε_0 is the electric permittivity of the vacuum. It satisfies $\operatorname{div} \mathbf{E} = 0$ and $\operatorname{curl} \mathbf{E} = \mathbf{0}$, and moreover $\mathbf{E} \times \mathbf{n} = \mathbf{0}$ on the boundary of $\mathcal{C} := B_{R_2} \setminus \overline{B_{R_1}}$, where $0 < R_1 < R_2$ and $B_R := \{\mathbf{x} \in \mathbb{R}^3 \mid |\mathbf{x}| < R\}$ is the ball of centre $\mathbf{0}$ and radius R. We have thus found a non-trivial harmonic field $\mathbf{E}$ in $\mathcal{C}$ satisfying $\mathbf{E} \times \mathbf{n} = \mathbf{0}$ on $\partial\mathcal{C}$.

These two examples show that the geometry of the domain and the type of boundary conditions play an essential role when considering harmonic fields. What are the relevant differences between the set $\mathcal{O}$, homeomorphic to a ball, and the sets $\mathcal{T}$ and $\mathcal{C}$? For the former, the point is that in $\mathcal{T}$ we have a *non-bounding* cycle, namely, a cycle that is not the boundary of a surface contained in $\mathcal{T}$ (take for instance the circle of centre $\mathbf{0}$ and radius a in the (x_1, x_2)-plane). In the latter case, the boundary of $\mathcal{C}$ is *not connected*.

In eddy current problems we have not only the constraint on the divergence of the electric and the magnetic fields, but also the one on the curl of the magnetic field in the insulator. As a consequence, we will see in the sequel that the formulation and the analysis of these problems require the introduction of several spaces of harmonic fields.

These spaces are presented, e.g., in Bossavit [59], Hiptmair [126], Cantarella et al. [73] and Gross and Kotiuga [115]; however, the most complete description and analysis is given by Ghiloni [110]. Here we introduce their basis functions; a more detailed description of them is given in Section A.4.

We need to make precise the geometry of the domains Ω, Ω_C and Ω_I (see Figures 1.2, 1.3, 1.4, 1.5 and 1.6). We indicate by Γ_j, $j = 1, \ldots, p_\Gamma + 1$, the connected components of Γ, and by $(\partial\Omega)_r$, $r = 0, 1, \ldots, p_{\partial\Omega}$, the connected components of $\partial\Omega$ (in particular, we have denoted by $(\partial\Omega)_0$ the external one).

We also denote by n_{Ω_I} the number of independent non-bounding cycles in Ω_I, and similarly by n_Ω the number of independent non-bounding cycles in Ω. Here, we say that a finite family $\mathcal{F}$ of disjoint cycles of Ω_I is formed by independent cycles if, for each non-empty sub-family $\mathcal{F}'$ of $\mathcal{F}$, the union of the cycles of $\mathcal{F}'$ cannot be equal to the boundary of a surface contained in Ω_I. A similar definition applies for cycles in Ω. We recall that n_{Ω_I} is a topological invariant of Ω_I, namely, using the terminology of algebraic topology, its first Betti number, or, equivalently, the dimension of the first homology space of Ω_I. One can also show that n_{Ω_I} is the number of "cutting" surfaces Ξ_α^* such that every curl-free vector in Ω_I has a global potential in $\widetilde{\Omega}_I := \Omega_I \setminus \cup_\alpha \Xi_\alpha^*$ (this does not mean that $\widetilde{\Omega}_I$ is simply-connected nor that it is homologically trivial: an example is furnished by $\Omega_I = \Omega \setminus \overline{\Omega_C}$, where Ω is a cube and Ω_C is the trefoil knot, see Benedetti et al. [36]).

Finally, n_Γ is the number of $\partial\Omega$-independent non-bounding cycles in Ω_I. Similarly, $n_{\partial\Omega}$ is the number of Γ-independent non-bounding cycles in Ω_I. Here, we say that a finite family $\mathcal{G}$ of disjoint cycles of Ω_I is formed by $\partial\Omega$-independent cycles (Γ-independent cycles, respectively) if, for each non-empty sub-family $\mathcal{G}'$ of $\mathcal{G}$, the union of the cycles of $\mathcal{G}'$ cannot be equal to $\partial S \setminus \gamma$, S being a surface contained in Ω_I and γ a disjoint union of cycles, possibly empty, contained in $\partial\Omega$ (in Γ, respectively). For instance, in Figure 1.3 we have two non-bounding cycles on Γ, but both of them can be brought on $\partial\Omega$, therefore they are not $\partial\Omega$-independent, hence $n_\Gamma = 0$. Similarly, there are two non-bounding cycles on $\partial\Omega$, but none of them is Γ-independent and $n_{\partial\Omega} = 0$.

In order to help the reader to become acquainted with these notations, let us refer to Figure 1.1: there one has $p_\Gamma = 2$, $p_{\partial\Omega} = 1$, $n_{\Omega_I} = 3$, $n_\Omega = 0$, $n_\Gamma = 3$, $n_{\partial\Omega} = 0$. For Figures 1.2, 1.3, 1.4, 1.5 and 1.6, see the captions there.

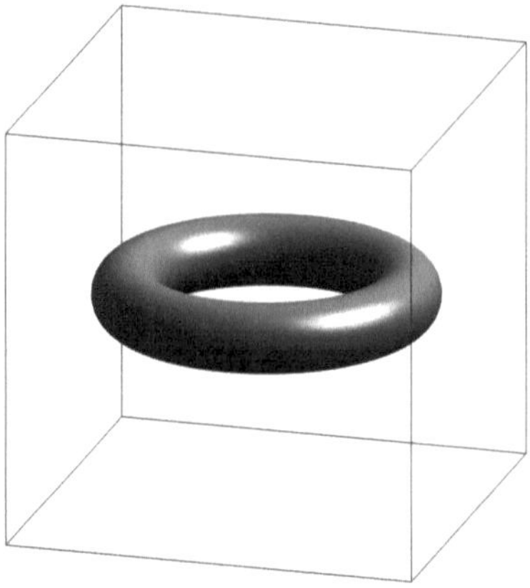

Fig. 1.2. An example of the geometry of the problem: the conductor is dark (here one has $p_\Gamma = 0$, $p_{\partial\Omega} = 0$, $n_{\Omega_I} = 1$, $n_\Omega = 0$, $n_\Gamma = 1$, $n_{\partial\Omega} = 0$)

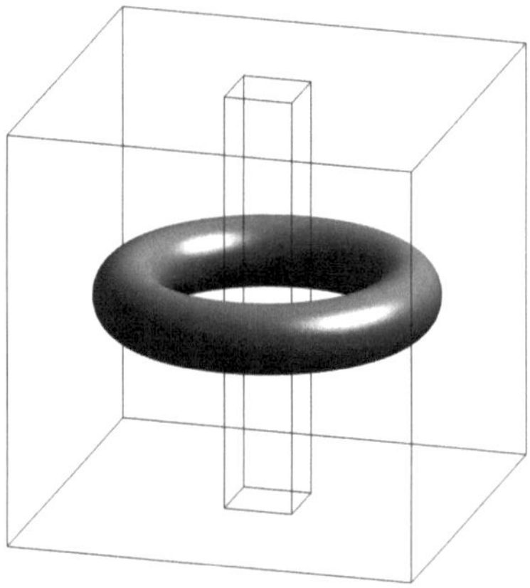

Fig. 1.3. An example of the geometry of the problem: the conductor is dark (here one has $p_\Gamma = 0$, $p_{\partial\Omega} = 0$, $n_{\Omega_I} = 2$, $n_\Omega = 1$, $n_\Gamma = 0$, $n_{\partial\Omega} = 0$)

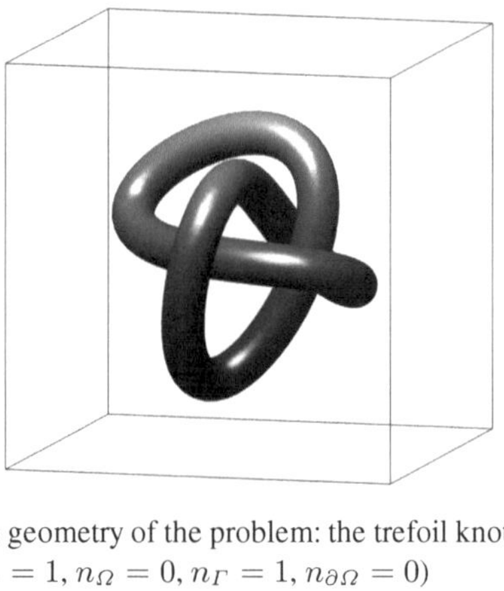

Fig. 1.4. An example of the geometry of the problem: the trefoil knot is the conductor (here one has $p_\Gamma = 0$, $p_{\partial\Omega} = 0$, $n_{\Omega_I} = 1$, $n_\Omega = 0$, $n_\Gamma = 1$, $n_{\partial\Omega} = 0$)

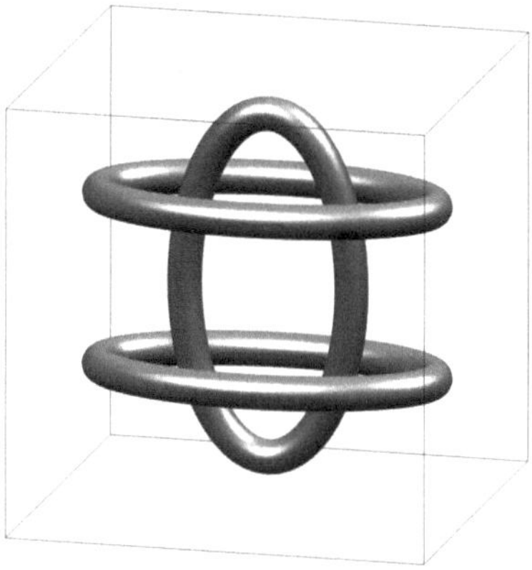

Fig. 1.5. An example of the geometry of the problem: the three rings are the conductor (here one has $p_\Gamma = 2, p_{\partial\Omega} = 0, n_{\Omega_I} = 3, n_\Omega = 0, n_\Gamma = 3, n_{\partial\Omega} = 0$)

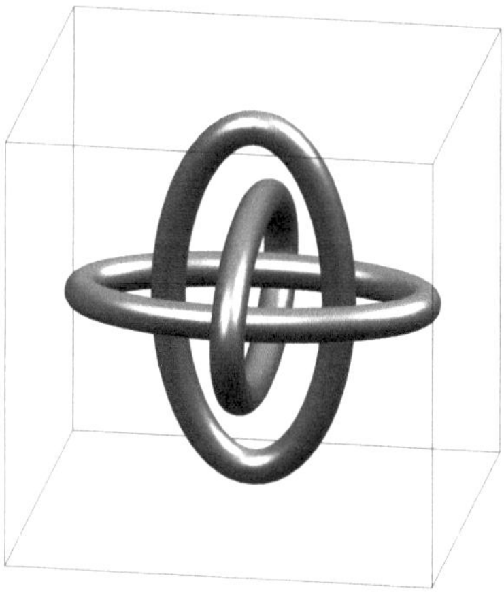

Fig. 1.6. An example of the geometry of the problem: the three Borromean rings are the conductor (here one has $p_\Gamma = 2, p_{\partial\Omega} = 0, n_{\Omega_I} = 3, n_\Omega = 0, n_\Gamma = 3, n_{\partial\Omega} = 0$)

We set $\mathbf{v}_I := \mathbf{v}_{|\Omega_I}$, $\mathbf{v}_C := \mathbf{v}_{|\Omega_C}$ and similary for all the other functions and matrices. The first space we introduce is

$$\begin{aligned}\mathcal{H}_{\varepsilon_I}(\Gamma, \partial\Omega; \Omega_I) := \{\mathbf{v}_I \in (L^2(\Omega_I))^3 \mid \operatorname{curl} \mathbf{v}_I = \mathbf{0}, \operatorname{div}(\varepsilon_I \mathbf{v}_I) = 0,\\ \mathbf{v}_I \times \mathbf{n}_I = \mathbf{0} \text{ on } \Gamma, \varepsilon_I \mathbf{v}_I \cdot \mathbf{n} = 0 \text{ on } \partial\Omega\}\,,\end{aligned} \tag{1.10}$$

whose dimension is equal to $n_{\partial\Omega} + p_\Gamma$. We denote a basis by $\boldsymbol{\pi}_{k,I}$ and $\operatorname{grad} w_{j,I}$, $k = 1, \ldots, n_{\partial\Omega}$, $j = 1, \ldots, p_\Gamma$. The fields $\boldsymbol{\pi}_{k,I}$ are more precisely described in Section A.4, and their construction requires the determination of a suitable set of "cutting" surfaces; the functions $w_{j,I}$ are the solutions of the elliptic problems

$$\begin{cases} \operatorname{div}(\varepsilon_I \operatorname{grad} w_{j,I}) = 0 & \text{in } \Omega_I \\ \varepsilon_I \operatorname{grad} w_{j,I} \cdot \mathbf{n} = 0 & \text{on } \partial\Omega \\ w_{j,I} = 0 & \text{on } \Gamma \setminus \Gamma_j \\ w_{j,I} = 1 & \text{on } \Gamma_j \, . \end{cases}$$

It is worth noting that the determination of $w_{j,I}$ is easier than that of $\boldsymbol{\pi}_{k,I}$, as the latter needs the identification of the "cutting" surface.

A second space is given by

$$\begin{aligned} \mathcal{H}_{\mu_I}(\partial\Omega, \Gamma; \Omega_I) := \{\mathbf{v}_I \in (L^2(\Omega_I))^3 \mid \operatorname{curl} \mathbf{v}_I = \mathbf{0}, \operatorname{div}(\boldsymbol{\mu}_I \mathbf{v}_I) = 0, \\ \mathbf{v}_I \times \mathbf{n} = \mathbf{0} \text{ on } \partial\Omega, \boldsymbol{\mu}_I \mathbf{v}_I \cdot \mathbf{n}_I = 0 \text{ on } \Gamma\} \, , \end{aligned} \tag{1.11}$$

and its dimension is equal to $n_\Gamma + p_{\partial\Omega}$. A basis is denoted by $\boldsymbol{\rho}_{l,I}$ and $\operatorname{grad} z_{r,I}$, $l = 1, \ldots, n_\Gamma$, $r = 1, \ldots, p_{\partial\Omega}$, where $\boldsymbol{\rho}_{l,I}$ are explicitly characterized in Section A.4, while $z_{r,I}$ is the solution of the elliptic problem

$$\begin{cases} \operatorname{div}(\boldsymbol{\mu}_I \operatorname{grad} z_{r,I}) = 0 & \text{in } \Omega_I \\ \boldsymbol{\mu}_I \operatorname{grad} z_{r,I} \cdot \mathbf{n}_I = 0 & \text{on } \Gamma \\ z_{r,I} = 0 & \text{on } \partial\Omega \setminus (\partial\Omega)_r \\ z_{r,I} = 1 & \text{on } (\partial\Omega)_r \, . \end{cases}$$

Note that the dimension of the space $\mathcal{H}_{\mu_I}(\partial\Omega, \Gamma; \Omega_I)$ is equal to 1 for both the examples shown in Figures 1.2 and 1.4. The difference resides in the basis function $\boldsymbol{\rho}_{1,I}$: as described in (A.34), it is associated to a "cutting" surface. For the torus in Figure 1.2 this surface is the one "filling" the "hole", for the trefoil knot in Figure 1.4 is the surface illustrated in Figure 4.2.

Another space is

$$\begin{aligned} \mathcal{H}_{\varepsilon_I}(e; \Omega_I) := \{\mathbf{v}_I \in (L^2(\Omega_I))^3 \mid \operatorname{curl} \mathbf{v}_I = \mathbf{0}, \operatorname{div}(\varepsilon_I \mathbf{v}_I) = 0, \\ \mathbf{v}_I \times \mathbf{n}_I = \mathbf{0} \text{ on } \Gamma \cup \partial\Omega\} \, , \end{aligned} \tag{1.12}$$

whose dimension is equal to $p_{\partial\Omega} + p_\Gamma + 1$, and which has the basis functions $\operatorname{grad} w^*_{\gamma,I}$, $\gamma = 0, \ldots, p_{\partial\Omega} + p_\Gamma$, where $w^*_{\gamma,I}$ is the solution of the elliptic problem

$$\begin{cases} \operatorname{div}(\varepsilon_I \operatorname{grad} w^*_{\gamma,I}) = 0 & \text{in } \Omega_I \\ w^*_{\gamma,I} = 0 & \text{on } (\partial\Omega \cup \Gamma) \setminus \Theta_\gamma \\ w^*_{\gamma,I} = 1 & \text{on } \Theta_\gamma \, , \end{cases}$$

having set $\Theta_\gamma := (\partial\Omega)_\gamma$ for $\gamma = 0, \ldots, p_{\partial\Omega}$ and $\Theta_\gamma := \Gamma_{\gamma - p_{\partial\Omega}}$ for $\gamma = p_{\partial\Omega} + 1, \ldots, p_{\partial\Omega} + p_\Gamma$. Note that the dimension of this space is one less than the number of connected components of $\Gamma \cup \partial\Omega$, the boundary of Ω_I.

A fourth space is defined by

$$\begin{aligned} \mathcal{H}_{\mu_I}(m; \Omega_I) := \{\mathbf{v}_I \in (L^2(\Omega_I))^3 \mid \operatorname{curl} \mathbf{v}_I = \mathbf{0}, \operatorname{div}(\boldsymbol{\mu}_I \mathbf{v}_I) = 0, \\ \boldsymbol{\mu}_I \mathbf{v}_I \cdot \mathbf{n} = 0 \text{ on } \Gamma \cup \partial\Omega\} \, , \end{aligned} \tag{1.13}$$

and its dimension is equal to n_{Ω_I}. A basis is given by $\boldsymbol{\rho}^*_{\alpha,I}$, $\alpha = 1, \ldots, n_{\Omega_I}$ (for a precise characterization see Section A.4). Note that its dimension is the number of independent non-bounding cycles of Ω_I; therefore, it is equal to 3 for both the examples shown in Figures 1.5 and 1.6. Again, the difference is in the basis functions $\boldsymbol{\rho}^*_{\alpha,I}$, $\alpha = 1, 2, 3$, which are associated to three "cutting" surfaces. For the three rings in Figure 1.5 these surfaces are disjoint, for the Borromean rings in Figure 1.6 they have non-empty intersection.

When $\varepsilon_I = \mathrm{Id}$ or $\boldsymbol{\mu}_I = \mathrm{Id}$, where Id is the identity matrix, we simply write $\mathcal{H}(\Gamma, \partial\Omega; \Omega_I)$, $\mathcal{H}(\partial\Omega, \Gamma; \Omega_I)$, $\mathcal{H}(e; \Omega_I)$ and $\mathcal{H}(m; \Omega_I)$, respectively.

Finally, we introduce two last spaces: the first is

$$\begin{aligned} \mathcal{H}(e; \Omega) := \{\mathbf{v} \in (L^2(\Omega))^3 \mid \operatorname{curl} \mathbf{v} = \mathbf{0}, \operatorname{div} \mathbf{v} = 0, \\ \mathbf{v} \times \mathbf{n} = \mathbf{0} \text{ on } \partial\Omega\} \ , \end{aligned} \tag{1.14}$$

whose dimension is equal to $p_{\partial\Omega}$, one less than the number of connected components of $\partial\Omega$, and that admits the basis functions $\operatorname{grad} \hat{z}_r$, $r = 1, \ldots, p_{\partial\Omega}$, where $\hat{z}_r$ is the solution of the elliptic problem

$$\begin{cases} \Delta \hat{z}_r = 0 & \text{in } \Omega \\ \hat{z}_r = 0 & \text{on } \partial\Omega \setminus (\partial\Omega)_r \\ \hat{z}_r = 1 & \text{on } (\partial\Omega)_r \ . \end{cases}$$

The second is

$$\begin{aligned} \mathcal{H}(m; \Omega) := \{\mathbf{v} \in (L^2(\Omega))^3 \mid \operatorname{curl} \mathbf{v} = \mathbf{0}, \operatorname{div} \mathbf{v} = 0, \\ \mathbf{v} \cdot \mathbf{n} = 0 \text{ on } \partial\Omega\} \ , \end{aligned} \tag{1.15}$$

whose dimension is equal to n_Ω, the number of independent non-bounding cycles of Ω, and for which a basis is denoted by $\hat{\boldsymbol{\pi}}_t$, $t = 1, \ldots, n_\Omega$.

Remark 1.1. We note that n_Ω, the number of independent non-bounding cycles of Ω or, equivalently, the Betti number of Ω, is equal to 0 if and only if the domain Ω is simply-connected. There appears to be some confusion concerning this point in the literature devoted to electromagnetism (see, e.g., the discussion in Bossavit et al. [64] and Kotiuga et al. [155]; see also Kettunen et al. [150]). Its proof can be found in Benedetti et al. [36]. □

1.5 The complete eddy current model

In this section we finally introduce the complete set of equations describing the eddy current problem. Beside the Ampère and Faraday equations, the vanishing of the electric charge in Ω_I and a suitable choice of the boundary conditions, we show that, in order to obtain a well-posed problem, other equations related to the specific geometry of Ω_I must be considered. In fact, if $\mathbf{E}$ is a solution of this set of equations, it is still a solution if we add to it in Ω_I a harmonic field $\mathbf{h}_I$ with $\mathbf{h}_I \times \mathbf{n}_I = \mathbf{0}$ on Γ and the same boundary condition of $\mathbf{E}_I$ on $\partial\Omega$.

The further conditions to impose can be determined in several ways. A heuristic argument suggests to devise these equations just checking which relations are satisfied by the solution of the full Maxwell system (1.2) but are not satisfied by the solution of the eddy current model (1.3).

From the Stokes theorem, a solution of the eddy current problem (1.3) must satisfy

$$0 = \int_S \operatorname{curl} \mathbf{H}_I \cdot \mathbf{n} = \int_S \mathbf{J}_{e,I} \cdot \mathbf{n} \,, \qquad (1.16)$$

for each connected components S of $\Gamma \cup \partial\Omega$, the boundary of Ω_I: this is a necessary condition to be verified by the current density.

Denoting by $\mathbf{E}^M$ and $\mathbf{H}^M$ the solutions of (1.2), in Ω_I one has

$$\operatorname{curl} \mathbf{H}_I^M = i\omega\varepsilon_I \mathbf{E}_I^M + \mathbf{J}_{e,I} \,,$$

thus from the Stokes theorem

$$\int_S (i\omega\varepsilon_I \mathbf{E}_I^M + \mathbf{J}_{e,I}) \cdot \mathbf{n} = 0 \,.$$

Therefore, it is natural to assume that the electric field $\mathbf{E}_I$, solution of the eddy current problem, satisfies

$$\int_S \varepsilon_I \mathbf{E}_I \cdot \mathbf{n} = 0 \,, \qquad (1.17)$$

as would be the case for the solution of the full Maxwell system (1.2) under the assumption (1.16).

For the electric boundary condition $\mathbf{E} \times \mathbf{n} = \mathbf{0}$ on $\partial\Omega$ these equations are enough. Instead, for the magnetic boundary conditions $\mathbf{H} \times \mathbf{n} = \mathbf{0}$ and $\varepsilon\mathbf{E} \cdot \mathbf{n} = 0$ on $\partial\Omega$ one has to proceed further. First of all, it is useful to observe that equations (1.17) reduce to those associated with the connected components Γ_j only, as on $\partial\Omega$ one has $\varepsilon\mathbf{E} \cdot \mathbf{n} = 0$. Moreover, considering the basis functions $\boldsymbol{\pi}_{k,I}$ of $\mathcal{H}_{\varepsilon_I}(\Gamma, \partial\Omega; \Omega_I)$, from (1.3) in Ω_I we have

$$\begin{aligned} \textstyle\int_{\Omega_I} \mathbf{J}_{e,I} \cdot \boldsymbol{\pi}_{k,I} &= \textstyle\int_{\Omega_I} \operatorname{curl} \mathbf{H}_I \cdot \boldsymbol{\pi}_{k,I} \\ &= \textstyle\int_{\Omega_I} \mathbf{H}_I \cdot \operatorname{curl} \boldsymbol{\pi}_{k,I} + \int_\Gamma \mathbf{n}_I \times \mathbf{H}_I \cdot \boldsymbol{\pi}_{k,I} \\ &\quad + \textstyle\int_{\partial\Omega} \mathbf{n} \times \mathbf{H}_I \cdot \boldsymbol{\pi}_{k,I} = 0 \,, \end{aligned} \qquad (1.18)$$

a new set of necessary conditions for the current density.

Similarly, from the Maxwell equations (1.2) in Ω_I we find

$$\textstyle\int_{\Omega_I} (i\omega\varepsilon_I \mathbf{E}_I^M + \mathbf{J}_{e,I}) \cdot \boldsymbol{\pi}_{k,I} = \int_{\Omega_I} \operatorname{curl} \mathbf{H}_I^M \cdot \boldsymbol{\pi}_{k,I} = 0 \,.$$

Thus, as in the case the solution of the full Maxwell problem (1.2) under the conditions (1.18), one is led to require

$$\textstyle\int_{\Omega_I} \varepsilon_I \mathbf{E}_I \cdot \boldsymbol{\pi}_{k,I} = 0 \,, \qquad (1.19)$$

for each $k = 1, \ldots, n_{\partial\Omega}$.

Summing up, in the case of the electric boundary condition the complete set of equations is

$$\begin{cases} \operatorname{curl}\mathbf{H} - \boldsymbol{\sigma}\mathbf{E} = \mathbf{J}_e & \text{in } \Omega \\ \operatorname{curl}\mathbf{E} + i\omega\boldsymbol{\mu}\mathbf{H} = \mathbf{0} & \text{in } \Omega \\ \operatorname{div}(\varepsilon_I \mathbf{E}_I) = 0 & \text{in } \Omega_I \\ \mathbf{E}_I \times \mathbf{n} = \mathbf{0} & \text{on } \partial\Omega \\ \int_{\Gamma_j} \varepsilon_I \mathbf{E}_I \cdot \mathbf{n}_I = 0 & \forall\, j = 1, \ldots, p_\Gamma \\ \int_{(\partial\Omega)_r} \varepsilon_I \mathbf{E}_I \cdot \mathbf{n} = 0 & \forall\, r = 0, 1, \ldots, p_{\partial\Omega}\,, \end{cases} \tag{1.20}$$

with the necessary assumptions

$$\begin{array}{ll} \operatorname{div}\mathbf{J}_{e,I} = 0 & \text{in } \Omega_I \\ \int_{\Gamma_j} \mathbf{J}_{e,I} \cdot \mathbf{n}_I = 0 & \forall\, j = 1, \ldots, p_\Gamma \\ \int_{(\partial\Omega)_r} \mathbf{J}_{e,I} \cdot \mathbf{n} = 0 & \forall\, r = 0, 1, \ldots, p_{\partial\Omega}\,. \end{array} \tag{1.21}$$

Instead, in the case of the magnetic boundary conditions the complete set of equations is

$$\begin{cases} \operatorname{curl}\mathbf{H} - \boldsymbol{\sigma}\mathbf{E} = \mathbf{J}_e & \text{in } \Omega \\ \operatorname{curl}\mathbf{E} + i\omega\boldsymbol{\mu}\mathbf{H} = \mathbf{0} & \text{in } \Omega \\ \operatorname{div}(\varepsilon_I \mathbf{E}_I) = 0 & \text{in } \Omega_I \\ \mathbf{H} \times \mathbf{n} = \mathbf{0} & \text{on } \partial\Omega \\ \varepsilon_I \mathbf{E}_I \cdot \mathbf{n} = 0 & \text{on } \partial\Omega \\ \int_{\Gamma_j} \varepsilon_I \mathbf{E}_I \cdot \mathbf{n}_I = 0 & \forall\, j = 1, \ldots, p_\Gamma \\ \int_{\Omega_I} \varepsilon_I \mathbf{E}_I \cdot \boldsymbol{\pi}_{k,I} = 0 & \forall\, k = 1, \ldots, n_{\partial\Omega}\,, \end{cases} \tag{1.22}$$

with the necessary assumptions

$$\begin{array}{ll} \operatorname{div}\mathbf{J}_{e,I} = 0 & \text{in } \Omega_I \\ \mathbf{J}_{e,I} \cdot \mathbf{n} = 0 & \text{on } \partial\Omega \\ \int_{\Gamma_j} \mathbf{J}_{e,I} \cdot \mathbf{n}_I = 0 & \forall\, j = 1, \ldots, p_\Gamma \\ \int_{\Omega_I} \mathbf{J}_{e,I} \cdot \boldsymbol{\pi}_{k,I} = 0 & \forall\, k = 1, \ldots, n_{\partial\Omega}\,. \end{array} \tag{1.23}$$

Note that, as a consequence of the Gauss divergence theorem, the solution to (1.20) also satisfies

$$\int_{\Gamma_{p_\Gamma+1}} \varepsilon_I \mathbf{E}_I \cdot \mathbf{n}_I = 0\,.$$

Therefore this equation could be added to (1.20). However, in general we have preferred to drop from the final problem all the equations that are not independent of the others. The same remark applies to the problem (1.22) or to problem (1.24) here below.

Remark 1.2. The conditions $\int_{\Omega_I} \varepsilon_I \mathbf{E}_I \cdot \boldsymbol{\pi}_{k,I} = 0$ (as well as $\int_{\Omega_I} \mathbf{J}_{e,I} \cdot \boldsymbol{\pi}_{k,I} = 0$) have a physical interpretation.

In fact, as explained in Section A.4, the basis functions $\boldsymbol{\pi}_{k,I}$ can be written in a more explicit way. Precisely, let us start recalling that in Ω_I there exist $n_{\partial\Omega}$ surfaces Σ_k, with $\partial\Sigma_k \subset \partial\Omega$, each one "cutting" a Γ-independent non-bounding cycle in Ω_I. The functions $\boldsymbol{\pi}_{k,I}$ are the $(L^2(\Omega_I))^3$-extension of $\operatorname{grad} q_{k,I}$, where $q_{k,I} \in H^1(\Omega_I \setminus \Sigma_k)$ are the solutions to

$$\begin{cases} \operatorname{div}(\varepsilon_I \operatorname{grad} q_{k,I}) = 0 & \text{in } \Omega_I \setminus \Sigma_k \\ \varepsilon_I \operatorname{grad} q_{k,I} \cdot \mathbf{n} = 0 & \text{on } \partial\Omega \setminus \partial\Sigma_k \\ q_{k,I} = 0 & \text{on } \Gamma \\ [\varepsilon_I \operatorname{grad} q_{k,I} \cdot \mathbf{n}_\Sigma]_{\Sigma_k} = 0 \\ [q_{k,I}]_{\Sigma_k} = 1 \,, \end{cases}$$

having denoted by $[\,\cdot\,]_{\Sigma_k}$ the jump across the surface Σ_k and by $\mathbf{n}_\Sigma$ the unit normal vector on Σ_k.

Integration by parts gives

$$\begin{aligned} \int_{\Omega_I} \varepsilon_I \mathbf{E}_I \cdot \boldsymbol{\pi}_{k,I} &= \int_{\Omega_I \setminus \Sigma_k} \varepsilon_I \mathbf{E}_I \cdot \operatorname{grad} q_{k,I} \\ &= -\int_{\Omega_I \setminus \Sigma_k} \operatorname{div}(\varepsilon_I \mathbf{E}_I)\, q_{k,I} + \int_{\partial\Omega \setminus \partial\Sigma_k} \varepsilon_I \mathbf{E}_I \cdot \mathbf{n}\, q_{k,I} \\ &\qquad + \int_\Gamma \varepsilon_I \mathbf{E}_I \cdot \mathbf{n}_I\, q_{k,I} + \int_{\Sigma_k} \varepsilon_I \mathbf{E}_I \cdot \mathbf{n}_\Sigma \, [q_{k,I}]_{\Sigma_k} \\ &= \int_{\Sigma_k} \varepsilon_I \mathbf{E}_I \cdot \mathbf{n}_\Sigma \,, \end{aligned}$$

thus conditions $(1.22)_7$ are equivalent to require that the flux of the electric induction is vanishing on each "cutting" surface. □

Though in this book we are mainly concerned with problems (1.20) and (1.22), for the sake of completeness we also present the eddy current problem with the no-flux boundary conditions (1.9).

The detailed procedure for devising this problem needs some preliminaries, and is fully described in Section 3.5: here we limit ourselves to present the final result.

The complete problem reads

$$\begin{cases} \operatorname{curl} \mathbf{H} - \boldsymbol{\sigma} \mathbf{E} = \mathbf{J}_e & \text{in } \Omega \\ \operatorname{curl} \mathbf{E} + i\omega \boldsymbol{\mu} \mathbf{H} = \mathbf{0} & \text{in } \Omega \\ \operatorname{div}(\varepsilon_I \mathbf{E}_I) = 0 & \text{in } \Omega_I \\ \boldsymbol{\mu} \mathbf{H} \cdot \mathbf{n} = 0 & \text{on } \partial\Omega \\ \varepsilon_I \mathbf{E}_I \cdot \mathbf{n} = 0 & \text{on } \partial\Omega \\ \int_{\Gamma_j} \varepsilon_I \mathbf{E}_I \cdot \mathbf{n}_I = 0 & \forall\, j = 1, \dots, p_\Gamma \\ \int_{\Omega_I} \varepsilon_I \mathbf{E}_I \cdot \boldsymbol{\pi}_{k,I} = 0 & \forall\, k = 1, \dots, n_{\partial\Omega} \\ \int_{\partial\Omega} \mathbf{E}_I \times \mathbf{n} \cdot \boldsymbol{\rho}^*_{\alpha,I} = 0 & \forall\, \alpha = 1, \dots, n^*_{\Omega_I} \,, \end{cases} \tag{1.24}$$

with the necessary assumptions (1.21). Here $\boldsymbol{\rho}^*_{\alpha,I}$, $\alpha = 1, \dots, n_{\Omega_I}$, are the basis functions of the space of harmonic fields $\mathcal{H}_{\mu_I}(m; \Omega_I)$; the number $n^*_{\Omega_I}$ is defined in Remark 1.3 here below and satisfies $0 \leq n^*_{\Omega_I} \leq n_{\Omega_I}$.

Remark 1.3. Conditions $(1.24)_8$ have a physical interpretation. In fact, as made explicit in Section A.4, we recall that in Ω_I we have a collection of "cutting" surfaces Ξ^*_α, $\alpha = 1, \dots, n_{\Omega_I}$, with $\partial\Xi^*_\alpha \subset \partial\Omega \cup \Gamma$. The basis functions $\boldsymbol{\rho}^*_{\alpha,I}$ are the $(L^2(\Omega_I))^3$-extension of $\operatorname{grad} p^*_{\alpha,I}$, where $p^*_{\alpha,I}$, defined in $\Omega_I \setminus \Xi^*_\alpha$, is the solution, determined up to an additive constant, to

$$\begin{cases} \operatorname{div}(\boldsymbol{\mu}_I \operatorname{grad} p^*_{\alpha,I}) = 0 & \text{in } \Omega_I \setminus \Xi^*_\alpha \\ \boldsymbol{\mu}_I \operatorname{grad} p^*_{\alpha,I} \cdot \mathbf{n}_I = 0 & \text{on } (\partial\Omega \cup \Gamma) \setminus \partial\Xi^*_\alpha \\ \left[\boldsymbol{\mu}_I \operatorname{grad} p^*_{\alpha,I} \cdot \mathbf{n}_{\Xi^*}\right]_{\Xi^*_\alpha} = 0 & \\ \left[p^*_{\alpha,I}\right]_{\Xi^*_\alpha} = 1\,, & \end{cases} \tag{1.25}$$

having denoted by $[\,\cdot\,]_{\Xi^*_\alpha}$ the jump across the surface Ξ^*_α and by $\mathbf{n}_{\Xi^*}$ the unit normal vector on Ξ^*_α.

Let us denote by $n^*_{\Omega_I}$ the number of "cutting" surfaces Ξ^*_α such that $\partial\Xi^*_\alpha \cap \partial\Omega \neq \emptyset$. Clearly one has $0 \leq n^*_{\Omega_I} \leq n_{\Omega_I}$. If $n^*_{\Omega_I} \geq 1$, for $\alpha = 1, \ldots, n^*_{\Omega_I}$ one has

$$\begin{aligned} \int_{\partial\Omega} \mathbf{E}_I \times \mathbf{n} \cdot \boldsymbol{\rho}^*_{\alpha,I} &= \int_{\partial\Omega\setminus\partial\Xi^*_\alpha} \mathbf{E}_I \times \mathbf{n} \cdot \operatorname{grad} p^*_{\alpha,I} \\ &= -\int_{\partial\Omega\setminus\partial\Xi^*_\alpha} \operatorname{div}_\tau(\mathbf{E}_I \times \mathbf{n})\, p^*_{\alpha,I} + \int_{\partial\Xi^*_\alpha} (\mathbf{E}_I \times \mathbf{n}) \cdot \mathbf{n}_{\Xi^*} \left[p^*_{\alpha,I}\right]_{\Xi^*_\alpha} \\ &= -\int_{\partial\Omega\setminus\partial\Xi^*_\alpha} \operatorname{curl} \mathbf{E}_I \cdot \mathbf{n}\, p^*_{\alpha,I} + \int_{\partial\Xi^*_\alpha} \mathbf{E}_I \cdot (\mathbf{n} \times \mathbf{n}_{\Xi^*}) \\ &= i\omega \int_{\partial\Omega\setminus\partial\Xi^*_\alpha} \boldsymbol{\mu}_I \mathbf{H}_I \cdot \mathbf{n}\, p^*_{\alpha,I} + \int_{\partial\Xi^*_\alpha} \mathbf{E}_I \cdot d\boldsymbol{\tau} \\ &= \int_{\partial\Xi^*_\alpha} \mathbf{E}_I \cdot d\boldsymbol{\tau}\,, \end{aligned}$$

hence with $(1.24)_8$ we are imposing that the line integral of the electric field is vanishing on the boundary of each "cutting" surface intersecting $\partial\Omega$. □

It is worth noting that the conditions stemming from the geometrical properties of the domain Ω_I (with the exception of $(1.24)_8$) are orthogonality conditions of $\mathbf{E}_I$ to a suitable space of harmonic fields. More precisely, the additional conditions for the magnetic boundary value problem (1.22) or for problem (1.24) are orthogonality conditions to $\mathcal{H}_{\varepsilon_I}(\Gamma, \partial\Omega; \Omega_I)$, whereas those appearing in the electric boundary value problem (1.20) are orthogonality conditions to $\mathcal{H}_{\varepsilon_I}(e; \Omega_I)$, in all cases with respect to the scalar product

$$(\mathbf{w}_I, \mathbf{z}_I)_{\varepsilon_I,\Omega_I} := \int_{\Omega_I} \varepsilon_I \mathbf{w}_I \cdot \overline{\mathbf{z}_I}\,,$$

where $\overline{\mathbf{z}_I}$ denotes the complex conjugate of $\mathbf{z}_I$. In fact,

$$\begin{aligned} &\int_{\Omega_I} \varepsilon_I \mathbf{E}_I \cdot \operatorname{grad} w_{j,I} \\ &\quad = -\int_{\Omega_I} \operatorname{div}(\varepsilon_I \mathbf{E}_I)\, w_{j,I} + \int_\Gamma \varepsilon_I \mathbf{E}_I \cdot \mathbf{n}_I\, w_{j,I} + \int_{\partial\Omega} \varepsilon_I \mathbf{E}_I \cdot \mathbf{n}\, w_{j,I} \\ &\quad = \int_{\Gamma_j} \varepsilon_I \mathbf{E}_I \cdot \mathbf{n}_I\,, \end{aligned}$$

as the function $w_{j,I}$ satisfies $w_{j,I} = 1$ on Γ_j and $w_{j,I} = 0$ on Γ_i, $i \neq j$, $j = 1, \ldots, p_\Gamma$ (see Section 1.4). Similarly,

$$\begin{aligned} &\int_{\Omega_I} \varepsilon_I \mathbf{E}_I \cdot \operatorname{grad} w^*_{\gamma,I} \\ &\quad = -\int_{\Omega_I} \operatorname{div}(\varepsilon_I \mathbf{E}_I)\, w^*_{\gamma,I} + \int_\Gamma \varepsilon_I \mathbf{E}_I \cdot \mathbf{n}_I\, w^*_{\gamma,I} + \int_{\partial\Omega} \varepsilon_I \mathbf{E}_I \cdot \mathbf{n}\, w^*_{\gamma,I}\,, \end{aligned}$$

and the result follows as, taking γ equal to a value from 0 to $p_{\partial\Omega} + p_\Gamma$, the function $w^*_{\gamma,I}$ has boundary value 0 on $\Gamma_{p_\Gamma+1}$, value 1 on only one connected component of $\Gamma \cup \partial\Omega$ and value 0 on all the remaining components.

2

A mathematical justification of the eddy current model

The aim of this chapter is to analyze in which sense the eddy current model is a proper approximation of the full Maxwell system. As explained in the previous chapter, the eddy current problem is a simplified model derived from the full system of Maxwell equations by neglecting the displacement currents, namely, the term $i\omega\varepsilon\mathbf{E}$. Therefore it can be seen either as the low electric permittivity limit or as the low-frequency limit of the full Maxwell system. The analysis is mainly based on the **E**-based formulation of Maxwell equations obtained by eliminating the magnetic field.

2.1 The E-based formulation of Maxwell equations

In this chapter we are not concerned with the problem of well-posedness of the eddy current model, an aspect that is dealt with in Chapter 3. We simply assume that a solution of the eddy current equations exists, and that a solution of the full Maxwell system exists as well, both solutions being smooth enough to justify all the computations we will perform. Moreover, we focus on the magnetic boundary value problem (1.22), leaving to the reader the modifications needed for treating the electric boundary value case (1.20).

The geometrical situation is that described in Section 1.3. Moreover, as already indicated, in agreement with well-known physical considerations we suppose that the matrix $\boldsymbol{\mu}$ is symmetric and uniformly positive definite in Ω, with entries in $L^\infty(\Omega)$, the matrix ε_I is symmetric and uniformly positive definite in Ω_I, with entries in $L^\infty(\Omega_I)$, and the matrix $\boldsymbol{\sigma}$ is symmetric and uniformly positive definite in Ω_C, with entries in $L^\infty(\Omega_C)$, whereas it is vanishing in Ω_I. Finally, the current density $\mathbf{J}_e \in (L^2(\Omega))^3$ satisfies the necessary conditions (1.23).

In the Maxwell system, and also in the eddy current model, it is possible to eliminate either the electric field (as it will be done in the first part of Chapter 3) or the magnetic field. For the full Maxwell system the two formulations are quite similar, but this is not the case for the eddy current model. In particular, in order to compare the two problems it is convenient to use the electric approach, eliminating the magnetic field.

A. Alonso Rodríguez, A. Valli: Eddy Current Approximation of Maxwell Equations.

From the Faraday law in (1.22) one has $\boldsymbol{\mu}^{-1} \operatorname{curl} \mathbf{E} = -i\omega\mathbf{H}$, then substituting in the Ampère law we obtain the $\mathbf{E}$-based formulation of the eddy current problem

$$\begin{cases} \operatorname{curl}(\boldsymbol{\mu}^{-1} \operatorname{curl} \mathbf{E}) + i\omega\boldsymbol{\sigma}\mathbf{E} = -i\omega\mathbf{J}_e & \text{in } \Omega \\ \operatorname{div}(\varepsilon_I \mathbf{E}_I) = 0 & \text{in } \Omega_I \\ \boldsymbol{\mu}^{-1} \operatorname{curl} \mathbf{E} \times \mathbf{n} = \mathbf{0} & \text{on } \partial\Omega \\ \varepsilon_I \mathbf{E}_I \cdot \mathbf{n} = 0 & \text{on } \partial\Omega \\ \int_{\Gamma_j} \varepsilon_I \mathbf{E}_I \cdot \mathbf{n}_I = 0 & \forall\, j = 1, \dots, p_\Gamma \\ \int_{\Omega_I} \varepsilon_I \mathbf{E}_I \cdot \boldsymbol{\pi}_{k,I} = 0 & \forall\, k = 1, \dots, n_{\partial\Omega} . \end{cases} \tag{2.1}$$

To help the reader, we remark that if the boundary of the conductor Ω_C is connected then $p_\Gamma = 0$, and that if Ω is simply-connected then $n_{\partial\Omega} = 0$. An example of this simplified geometry is that of a connected conductor (possibly with "handles") contained in a computational domain similar to a "box".

Using integration by parts it is easily seen that a solution $\mathbf{E}$ to (2.1) satisfies

$$\begin{aligned} \int_\Omega \boldsymbol{\mu}^{-1} \operatorname{curl} \mathbf{E} \cdot \operatorname{curl} \overline{\mathbf{z}} &+ i\omega \int_{\Omega_C} \boldsymbol{\sigma}\mathbf{E}_C \cdot \overline{\mathbf{z}_C} \\ &+ c_0^* \int_{\Omega_I} \operatorname{div}(\varepsilon_I \mathbf{E}_I) \operatorname{div}(\varepsilon_I \overline{\mathbf{z}_I}) = -i\omega \int_\Omega \mathbf{J}_e \cdot \overline{\mathbf{z}} \end{aligned} \tag{2.2}$$

for all $\mathbf{z} \in H(\operatorname{curl}; \Omega)$ with $\operatorname{div}(\varepsilon_I \mathbf{z}_I) \in L^2(\Omega_I)$, where $c_0^* > 0$ is an arbitrarily chosen dimensional constant.

Let us consider the space

$$\begin{aligned} W_{\varepsilon_I}(\Omega_I; \Omega) := \{\mathbf{z} \in H(\operatorname{curl}; \Omega) \,|\, &\mathbf{z}_I \in H_{0,\partial\Omega}(\varepsilon_I, \operatorname{div}; \Omega_I), \\ &\mathbf{z}_I \perp^{\varepsilon_I} \mathcal{H}_{\varepsilon_I}(\Gamma, \partial\Omega; \Omega_I)\} \end{aligned} \tag{2.3}$$

where the symbol $\perp^{\varepsilon_I}$ denotes the orthogonality with respect to the scalar product

$$(\mathbf{w}_I, \mathbf{z}_I)_{\varepsilon_I, \Omega_I} := \int_{\Omega_I} \varepsilon_I \mathbf{w}_I \cdot \overline{\mathbf{z}_I}$$

(for the other notations see Sections 1.4 and A.1). In $W_{\varepsilon_I}(\Omega_I; \Omega)$ we define the norm

$$\|\mathbf{z}\|_{W_{\varepsilon_I}(\Omega_I;\Omega)} := \left(\| \operatorname{curl} \mathbf{z}\|_{0,\Omega}^2 + \|\mathbf{z}\|_{0,\Omega}^2 + \| \operatorname{div}(\varepsilon_I \mathbf{z}_I)\|_{0,\Omega_I}^2\right)^{1/2} .$$

Recalling that, as shown in Section 1.5, the integral conditions in (2.1) are orthogonality conditions with respect to the space of harmonic fields $\mathcal{H}_{\varepsilon_I}(\Gamma, \partial\Omega; \Omega_I)$, with respect to the scalar product $(\cdot, \cdot)_{\varepsilon_I, \Omega_I}$, we have in particular that $\mathbf{E} \in W_{\varepsilon_I}(\Omega_I; \Omega)$.

In the space $W_{\varepsilon_I}(\Omega_I; \Omega)$ let us define the sesquilinear form

$$\begin{aligned} a_e^*(\mathbf{w}, \mathbf{z}) := \int_\Omega \boldsymbol{\mu}^{-1} \operatorname{curl} \mathbf{w} \cdot \operatorname{curl} \overline{\mathbf{z}} + i\omega \int_{\Omega_C} \boldsymbol{\sigma}\mathbf{w}_C \cdot \overline{\mathbf{z}_C} \\ + c_0^* \int_{\Omega_I} \operatorname{div}(\varepsilon_I \mathbf{w}_I) \operatorname{div}(\varepsilon_I \overline{\mathbf{z}_I}) . \end{aligned} \tag{2.4}$$

With the aim of analyzing the asymptotic behaviour of the solution $\mathbf{E}$ as the electric permittivity ε tends to 0 or the angular frequency ω tends to 0, an important point is to show that the sesquilinear form $a_e^*(\cdot, \cdot)$ is coercive in $W_{\varepsilon_I}(\Omega_I; \Omega)$.

We need some preliminary results. It is known that there are several ways of writing a vector function belonging to $(L^2(\Omega_I))^3$ as the sum of a curl, a gradient and a harmonic field. In particular, let us recall (see Theorem A.6) that $\mathbf{z}_I \in (L^2(\Omega_I))^3$ can be represented as

$$\mathbf{z}_I = \varepsilon_I^{-1} \operatorname{curl} \mathbf{q}_I + \operatorname{grad} \varphi_I + \mathbf{h}_I \ , \tag{2.5}$$

where $\mathbf{q}_I \in H_{0,\partial\Omega}(\mathrm{curl};\Omega_I) \cap H^0_{0,\Gamma}(\mathrm{div};\Omega_I) \cap \mathcal{H}(\partial\Omega,\Gamma;\Omega_I)^\perp$, $\varphi_I \in H^1_{0,\Gamma}(\Omega_I)$ and $\mathbf{h}_I \in \mathcal{H}_{\varepsilon_I}(\Gamma,\partial\Omega;\Omega_I)$. Moreover, we also know that if $\mathbf{z}_I \perp^{\varepsilon_I} \mathcal{H}_{\varepsilon_I}(\Gamma,\partial\Omega;\Omega_I)$ we have $\mathbf{h}_I = \mathbf{0}$.

A second useful result is the following one.

Lemma 2.1. *There exists a constant $C > 0$ such that*

$$\begin{aligned}\|\mathbf{z}_I\|_{0,\Omega_I} \le C\big(&\|\operatorname{curl} \mathbf{z}_I\|_{0,\Omega_I} + \|\operatorname{div}(\varepsilon_I \mathbf{z}_I)\|_{0,\Omega_I} \\ &+\|\mathbf{z}_I \times \mathbf{n}_I\|_{H^{-1/2}(\mathrm{div}_\tau;\Gamma)} + \|\varepsilon_I \mathbf{z}_I \cdot \mathbf{n}_I\|_{-1/2,\partial\Omega}\big)\end{aligned}$$

for all $\mathbf{z}_I \in H(\mathrm{curl};\Omega_I) \cap H(\varepsilon_I,\mathrm{div};\Omega_I)$ with $\mathbf{z}_I \perp^{\varepsilon_I} \mathcal{H}_{\varepsilon_I}(\Gamma,\partial\Omega;\Omega_I)$.

Proof. Since $\mathbf{z}_I \perp^{\varepsilon_I} \mathcal{H}_{\varepsilon_I}(\Gamma,\partial\Omega;\Omega_I)$, from (2.5) we can write

$$\mathbf{z}_I = \varepsilon_I^{-1} \operatorname{curl} \mathbf{q}_I + \operatorname{grad} \varphi_I \ .$$

Then we estimate the norm of the two terms on the right hand side. Looking at problem (A.14), we start by considering $\int_{\Omega_I} \mathbf{z}_I \cdot \operatorname{curl} \overline{\mathbf{q}_I}$. Integrating by parts we have

$$\begin{aligned}\left|\int_{\Omega_I} \mathbf{z}_I \cdot \operatorname{curl} \overline{\mathbf{q}_I}\right| &= \left|\int_{\Omega_I} \operatorname{curl} \mathbf{z}_I \cdot \overline{\mathbf{q}_I} + \int_\Gamma \mathbf{z}_I \times \mathbf{n}_I \cdot \overline{\mathbf{q}_I}\right| \\ &\le C(\|\operatorname{curl} \mathbf{z}_I\|_{0,\Omega_I} + \|\mathbf{z}_I \times \mathbf{n}_I\|_{H^{-1/2}(\mathrm{div}_\tau;\Gamma)})(\|\mathbf{q}_I\|_{0,\Omega_I} + \|\operatorname{curl} \mathbf{q}_I\|_{0,\Omega_I}) \ ,\end{aligned}$$

where we have used the duality estimate

$$\left|\int_\Gamma \mathbf{z}_I \times \mathbf{n}_I \cdot \overline{\mathbf{q}_I}\right| \le C\|\mathbf{z}_I \times \mathbf{n}_I\|_{H^{-1/2}(\mathrm{div}_\tau;\Gamma)}\|\mathbf{n}_I \times \mathbf{q}_I \times \mathbf{n}_I\|_{H^{-1/2}(\mathrm{curl}_\tau;\Gamma)}$$

(see Section A.1) and the tangential trace inequality (A.11)

$$\|\mathbf{n}_I \times \mathbf{q}_I \times \mathbf{n}_I\|_{H^{-1/2}(\mathrm{curl}_\tau;\Gamma)} \le C(\|\mathbf{q}_I\|_{0,\Omega_I} + \|\operatorname{curl} \mathbf{q}_I\|_{0,\Omega_I}) \ .$$

On the other hand, from the Poincaré-like inequality (A.15) and taking also into account that $\operatorname{div} \mathbf{q}_I = \mathbf{0}$ in Ω_I we find

$$\int_{\Omega_I} |\mathbf{q}_I|^2 \le C \int_{\Omega_I} (|\operatorname{curl} \mathbf{q}_I|^2 + |\operatorname{div} \mathbf{q}_I|^2) = C \int_{\Omega_I} |\operatorname{curl} \mathbf{q}_I|^2 \ .$$

Summing up, choosing $\mathbf{p}_I = \mathbf{q}_I$ in (A.14) gives

$$\|\varepsilon^{-1} \operatorname{curl} \mathbf{q}_I\|_{0,\Omega_I} \le C(\|\operatorname{curl} \mathbf{z}_I\|_{0,\Omega_I} + \|\mathbf{z}_I \times \mathbf{n}_I\|_{H^{-1/2}(\mathrm{div}_\tau;\Gamma)}) \ .$$

Another integration by parts in the right hand side of (A.17) furnishes

$$\begin{aligned}\left|\int_{\Omega_I} \varepsilon_I \mathbf{z}_I \cdot \operatorname{grad} \overline{\varphi_I}\right| &= \left|-\int_{\Omega_I} \operatorname{div}(\varepsilon_I \mathbf{z}_I)\overline{\varphi_I} + \int_{\partial\Omega} \varepsilon_I \mathbf{z}_I \cdot \mathbf{n}\,\overline{\varphi_I}\right| \\ &\le C(\|\operatorname{div}(\varepsilon_I \mathbf{z}_I)\|_{0,\Omega_I} + \|\varepsilon_I \mathbf{z}_I \cdot \mathbf{n}_I\|_{-1/2,\partial\Omega})(\|\varphi_I\|_{0,\Omega_I} + \|\operatorname{grad} \varphi_I\|_{0,\Omega_I}) \ ,\end{aligned}$$

having used the duality estimate

$$\left| \int_{\partial\Omega} \varepsilon_I \mathbf{z}_I \cdot \mathbf{n} \overline{\varphi_I} \right| \leq C \|\varepsilon_I \mathbf{z}_I \cdot \mathbf{n}\|_{-1/2,\partial\Omega} \|\varphi_{I|\partial\Omega}\|_{1/2,\partial\Omega}$$

(see Section A.1) and the trace inequality (A.8)

$$\|\varphi_{I|\partial\Omega}\|_{1/2,\partial\Omega} \leq C(\|\varphi_I\|_{0,\Omega_I} + \|\operatorname{grad} \varphi_I\|_{0,\Omega_I}) .$$

Since the Poincaré inequality (A.18)

$$\int_{\Omega_I} |\varphi_I|^2 \leq C \int_{\Omega_I} |\operatorname{grad} \varphi_I|^2$$

holds in $H^1_{0,\Gamma}(\Omega_I)$, choosing $\eta_I = \varphi_I$ in (A.17) we have

$$\|\operatorname{grad} \varphi_I\|_{0,\Omega_I} \leq C(\|\operatorname{div}(\varepsilon_I \mathbf{z}_I)\|_{0,\Omega_I} + \|\varepsilon_I \mathbf{z}_I \cdot \mathbf{n}_I\|_{-1/2,\partial\Omega}) ,$$

and the thesis follows. □

As a consequence we have the following lemma, that is the main point for proving the coerciveness of the sesquilinear form $a_e^*(\cdot,\cdot)$.

Lemma 2.2. *There exists a constant $C > 0$ such that for each $\mathbf{z} \in W_{\varepsilon_I}(\Omega_I;\Omega)$*

$$\|\mathbf{z}_I\|_{0,\Omega_I} \leq C\big(\|\operatorname{curl} \mathbf{z}\|_{0,\Omega} + \|\operatorname{div}(\varepsilon_I \mathbf{z}_I)\|_{0,\Omega_I} + \|\mathbf{z}_C\|_{0,\Omega_C}\big) .$$

Proof. First we recall that $\mathbf{z} \in H(\operatorname{curl};\Omega)$ if and only if $\mathbf{z}_C \in H(\operatorname{curl};\Omega_C)$, $\mathbf{z}_I \in H(\operatorname{curl};\Omega_I)$ and $\mathbf{z}_C \times \mathbf{n}_C = -\mathbf{z}_I \times \mathbf{n}_I$ on Γ.

From Lemma 2.1 we have

$$\begin{aligned} \|\mathbf{z}_I\|_{0,\Omega_I} &\leq C\left(\|\operatorname{curl} \mathbf{z}_I\|_{0,\Omega_I} + \|\operatorname{div}(\varepsilon_I \mathbf{z}_I)\|_{0,\Omega_I} + \|\mathbf{z}_I \times \mathbf{n}_I\|_{H^{-1/2}(\operatorname{div}_\tau;\Gamma)}\right) \\ &= C\left(\|\operatorname{curl} \mathbf{z}_I\|_{0,\Omega_I} + \|\operatorname{div}(\varepsilon_I \mathbf{z}_I)\|_{0,\Omega_I} + \|\mathbf{z}_C \times \mathbf{n}_C\|_{H^{-1/2}(\operatorname{div}_\tau;\Gamma)}\right) . \end{aligned}$$

Taking into account the tangential trace inequality (A.10), namely,

$$\|\mathbf{z}_C \times \mathbf{n}_C\|_{H^{-1/2}(\operatorname{div}_\tau;\Gamma)} \leq \kappa \|\mathbf{z}_C\|_{H(\operatorname{curl};\Omega_C)} ,$$

the proof is complete. □

Now we are in condition to prove the main result on this section.

Theorem 2.3. *The sesquilinear form $a_e^*(\cdot,\cdot)$ is coercive in $W_{\varepsilon_I}(\Omega_I;\Omega)$, i.e., there exists a constant $C_0 > 0$ such that*

$$|a_e^*(\mathbf{z},\mathbf{z})| \geq C_0 \|\mathbf{z}\|^2_{W_{\varepsilon_I}(\Omega_I;\Omega)} \qquad \textit{for all } \mathbf{z} \in W_{\varepsilon_I}(\Omega_I;\Omega) . \tag{2.6}$$

Proof. As a consequence of Lemma 2.2 we have, for all $\mathbf{z} \in W_{\varepsilon_I}(\Omega_I;\Omega)$

$$\|\mathbf{z}\|^2_{W_{\varepsilon_I}(\Omega_I;\Omega)} \leq C_1 \left(\| \operatorname{curl} \mathbf{z}\|^2_{0,\Omega} + \|\mathbf{z}_C\|^2_{0,\Omega_C} + \| \operatorname{div}(\varepsilon_I \mathbf{z}_I)\|^2_{0,\Omega_I}\right) , \tag{2.7}$$

for some positive constant C_1. Since $\boldsymbol{\nu} := \boldsymbol{\mu}^{-1}$ and $\boldsymbol{\sigma}$ are symmetric and uniformly positive definite in Ω and Ω_C, respectively, we have

$$\begin{aligned}
|a_e^*(\mathbf{z},\mathbf{z})|^2 &= \left(\int_\Omega \boldsymbol{\mu}^{-1} \operatorname{curl} \mathbf{z} \cdot \operatorname{curl} \overline{\mathbf{z}} + c_0^* \int_{\Omega_I} \operatorname{div}(\varepsilon_I \mathbf{z}_I) \operatorname{div}(\varepsilon_I \overline{\mathbf{z}_I})\right)^2 \\
&\qquad\qquad\qquad\qquad + \left(\omega \int_{\Omega_C} \boldsymbol{\sigma} \mathbf{w}_C \cdot \overline{\mathbf{z}_C}\right)^2 \\
&\geq \left(\nu_{\min} \| \operatorname{curl} \mathbf{z}\|^2_{0,\Omega} + c_0^* \| \operatorname{div}(\varepsilon_I \mathbf{z}_I)\|^2_{0,\Omega_I}\right)^2 + \left(\omega \sigma_{\min} \|\mathbf{z}_C\|^2_{0,\Omega_C}\right)^2 \\
&\geq C_2 \left(\| \operatorname{curl} \mathbf{z}\|^2_{0,\Omega} + \| \operatorname{div}(\varepsilon_I \mathbf{z}_I)\|^2_{0,\Omega_I} + \|\mathbf{z}_C\|^2_{0,\Omega_C}\right)^2 \\
&\geq C_2 C_1^{-2} \|\mathbf{z}\|^4_{W_{\varepsilon_I}(\Omega_I;\Omega)} ,
\end{aligned}$$

where $\nu_{\min}$ is a uniform lower bound in Ω for the minimum eigenvalues of $\boldsymbol{\nu}(\mathbf{x})$, $\sigma_{\min}$ is a uniform lower bound in Ω_C for the minimum eigenvalues of $\boldsymbol{\sigma}(\mathbf{x})$, and $C_2 = \frac{1}{2} \min(\nu^2_{\min}, c_0^{*2}, \omega^2 \sigma^2_{\min})$. □

Remark 2.4. The proof of the coerciveness of the sesquilinear form $a_e^*(\cdot,\cdot)$ is the crucial point in showing, via the Lax–Milgram lemma, that the weak problem

$$\begin{aligned}
&\text{Find } \mathbf{E} \in W_{\varepsilon_I}(\Omega_I;\Omega) \text{ such that} \\
&\quad \int_\Omega \boldsymbol{\mu}^{-1} \operatorname{curl} \mathbf{E} \cdot \operatorname{curl} \overline{\mathbf{z}} + i\omega \int_{\Omega_C} \boldsymbol{\sigma} \mathbf{E}_C \cdot \overline{\mathbf{z}_C} \\
&\qquad\qquad + c_0^* \int_{\Omega_I} \operatorname{div}(\varepsilon_I \mathbf{E}_I) \operatorname{div}(\varepsilon_I \overline{\mathbf{z}_I}) = -i\omega \int_\Omega \mathbf{J}_e \cdot \overline{\mathbf{z}} \\
&\text{for each } \mathbf{z} \in W_{\varepsilon_I}(\Omega_I;\Omega)
\end{aligned}$$

is well-posed.

Starting from this result, some additional work gives that also the solution to (2.1) exists and is unique. However, we do not want to consider this aspect here, and we refer to Chapter 3 for a systematic approach to the existence and uniqueness theory for eddy current problems, where the result is based on a simpler weak formulation in terms of the magnetic field $\mathbf{H}$ only, and to Section 6.1.5 for a complete analysis of the $\mathbf{E}$-based formulation (2.1). □

2.2 The eddy current model as the low electric permittivity limit

By eliminating the magnetic field in the time-harmonic Maxwell equations we obtain the following boundary value problem for the electric field $\mathbf{E}^M$ (with the magnetic boundary condition)

$$\begin{cases}
\operatorname{curl}(\boldsymbol{\mu}^{-1} \operatorname{curl} \mathbf{E}^M) - \omega^2 \varepsilon \mathbf{E}^M + i\omega \boldsymbol{\sigma} \mathbf{E}^M = -i\omega \mathbf{J}_e & \text{in } \Omega \\
\boldsymbol{\mu}^{-1} \operatorname{curl} \mathbf{E}^M \times \mathbf{n} = \mathbf{0} & \text{on } \partial\Omega .
\end{cases} \tag{2.8}$$

Here ε is a symmetric tensor, uniformly positive definite in Ω, with coefficients in $L^\infty(\Omega)$ and, as usual, $\varepsilon_{|\Omega_I} = \varepsilon_I$. As we have already explained in Section 1.2, the eddy current model is obtained by neglecting the term $\omega^2\varepsilon\mathbf{E}^M$, that corresponds to the displacement currents.

In this section we consider an electric permittivity of the form $\varepsilon_\delta := \delta\varepsilon$, where $\delta > 0$ is a real number. Clearly, the physical problem described by the full Maxwell system corresponds to the case $\delta = 1$. We want to show that the eddy current model is the limit as δ tends to 0 of the problem with electric permittivity ε_δ. This is the notion of eddy current limit presented in Bossavit [58], Chap. 4.

We will show that the norm in $H(\text{curl};\Omega)$ of the difference between the electric field solution of the full Maxwell system and the electric field solution of the eddy current problem is of order δ. This result has been proved in Costabel et al. [90] who impose the electric boundary condition $\mathbf{E} \times \mathbf{n} = \mathbf{0}$ on $\partial\Omega$. Here we report a proof of this result in the case of the magnetic boundary condition $\mathbf{H} \times \mathbf{n} = \mathbf{0}$ on $\partial\Omega$, that, in terms of the electric field, is the boundary condition considered in (2.1) and (2.8).

Let us denote by $\mathbf{E}$ the solution of the eddy current problem (2.1) and by $\mathbf{E}_\delta^M$ the solution of the full Maxwell system

$$\begin{cases} \operatorname{curl}(\boldsymbol{\mu}^{-1} \operatorname{curl} \mathbf{E}_\delta^M) - \omega^2\delta\varepsilon\mathbf{E}_\delta^M + i\omega\boldsymbol{\sigma}\mathbf{E}_\delta^M = -i\omega\mathbf{J}_e & \text{in } \Omega \\ \boldsymbol{\mu}^{-1} \operatorname{curl} \mathbf{E}_\delta^M \times \mathbf{n} = \mathbf{0} & \text{on } \partial\Omega\,. \end{cases} \tag{2.9}$$

For the existence and uniqueness of solution of (2.9) see, for instance, Alonso and Valli [8], Alonso and Raffetto [15].

As shown in Section 1.5, the current density $\mathbf{J}_e$ has to satisfy conditions (1.23), that are necessary conditions for the solvability of the eddy current problem. Let us rewrite them here

$$\begin{array}{ll} \operatorname{div} \mathbf{J}_{e,I} = 0 & \text{in } \Omega_I \\ \mathbf{J}_{e,I} \cdot \mathbf{n} = 0 & \text{on } \partial\Omega \\ \int_{\Gamma_j} \mathbf{J}_{e,I} \cdot \mathbf{n}_I = 0 & \forall j = 1, \ldots, p_\Gamma \\ \int_{\Omega_I} \mathbf{J}_{e,I} \cdot \boldsymbol{\pi}_{k,I} = 0 & \forall k = 1, \ldots, n_{\partial\Omega}\,. \end{array} \tag{2.10}$$

We notice that from these conditions it follows that $\mathbf{J}_{e,I} \in \mathcal{H}_{\varepsilon_I}(\Gamma, \partial\Omega; \Omega_I)^\perp$ and that the solution $\mathbf{E}_\delta^M$ of (2.9) belongs to $W_{\varepsilon_I}(\Omega_I; \Omega)$. In fact, by a direct computation from (2.10) we have $\operatorname{div}(\varepsilon_I \mathbf{E}_{\delta,I}^M) = 0$, and moreover $\varepsilon_I \mathbf{E}_{\delta,I}^M \cdot \mathbf{n} = 0$ on $\partial\Omega$, as from the boundary condition $\boldsymbol{\mu}^{-1} \operatorname{curl} \mathbf{E}_\delta^M \times \mathbf{n} = \mathbf{0}$ on $\partial\Omega$ it follows that $\operatorname{curl}(\boldsymbol{\mu}^{-1} \operatorname{curl} \mathbf{E}_\delta^M) \cdot \mathbf{n} = 0$ on $\partial\Omega$. Finally, it is clear that $\mathbf{E}_{\delta,I}^M \perp^{\varepsilon_I} \mathcal{H}_{\varepsilon_I}(\Gamma, \partial\Omega; \Omega_I)$, since for all $\mathbf{h}_I \in \mathcal{H}_{\varepsilon_I}(\Gamma, \partial\Omega; \Omega_I)$ we have

$$\begin{aligned} \int_{\Omega_I} \varepsilon_I \mathbf{E}_{\delta,I}^M \cdot \mathbf{h}_I &= \tfrac{1}{\omega^2\delta} \int_{\Omega_I} \operatorname{curl}(\boldsymbol{\mu}_I^{-1} \operatorname{curl} \mathbf{E}_{\delta,I}^M) \cdot \mathbf{h}_I + \tfrac{i}{\omega\delta} \int_{\Omega_I} \mathbf{J}_{e,I} \cdot \mathbf{h}_I \\ &= \tfrac{1}{\omega^2\delta} \int_{\Omega_I} \boldsymbol{\mu}_I^{-1} \operatorname{curl} \mathbf{E}_{\delta,I}^M \cdot \operatorname{curl} \mathbf{h}_I \\ &\quad + \tfrac{1}{\omega^2\delta}\omega \int_{\Gamma\cup\partial\Omega} \mathbf{n}_I \times \boldsymbol{\mu}_I^{-1} \operatorname{curl} \mathbf{E}_{\delta,I}^M \cdot \mathbf{h}_I = 0\,. \end{aligned}$$

We are thus in a position to prove the main result of this section.

Theorem 2.5. *There exists $\delta^* > 0$ such that if $0 < \delta \leq \delta^*$ one has*

$$\|\mathbf{E} - \mathbf{E}_\delta^M\|_{H(\text{curl};\Omega)} \leq C\delta\,,$$

for some constant $C > 0$ independent of δ.

Proof. Taking the difference between the first equations in (2.1) and (2.9), multiplying by a test function $\mathbf{z} \in H(\mathrm{curl};\Omega)$ and integrating by parts one obtains

$$\int_\Omega \boldsymbol{\mu}^{-1} \operatorname{curl}(\mathbf{E} - \mathbf{E}_\delta^M) \cdot \operatorname{curl} \overline{\mathbf{z}} + i\omega \int_{\Omega_C} \boldsymbol{\sigma}(\mathbf{E}_C - \mathbf{E}_{\delta,C}^M) \cdot \overline{\mathbf{z}_C} = -\delta\omega^2 \int_\Omega \boldsymbol{\varepsilon} \mathbf{E}_\delta^M \cdot \overline{\mathbf{z}} .$$

Since $\operatorname{div}(\boldsymbol{\varepsilon}_I(\mathbf{E}_I - \mathbf{E}_{\delta,I}^M)) = 0$ in Ω_I, from the coerciveness of the sesquilinear form $a_e^*(\cdot,\cdot)$ in $W_{\varepsilon_I}(\Omega_I;\Omega)$ it follows that there exists a constant $C_0 > 0$, independent of δ, such that

$$\begin{aligned} C_0 \|\mathbf{E} - \mathbf{E}_\delta^M\|_{H(\mathrm{curl};\Omega)}^2 &\le |a_e^*(\mathbf{E} - \mathbf{E}_\delta^M, \mathbf{E} - \mathbf{E}_\delta^M)| = \left|\delta\omega^2 \int_\Omega \boldsymbol{\varepsilon} \mathbf{E}_\delta^M \cdot (\overline{\mathbf{E}} - \overline{\mathbf{E}_\delta^M})\right| \\ &\le \delta\omega^2 \varepsilon_{\max} \|\mathbf{E}_\delta^M\|_{0,\Omega} \|\mathbf{E} - \mathbf{E}_\delta^M\|_{0,\Omega} , \end{aligned}$$

where $\varepsilon_{\max}$ is a uniform upper bound in Ω for the maximum eigenvalues of $\boldsymbol{\varepsilon}(\mathbf{x})$. Therefore

$$\|\mathbf{E} - \mathbf{E}_\delta^M\|_{H(\mathrm{curl};\Omega)} \le \delta \frac{\omega^2 \varepsilon_{\max}}{C_0} \|\mathbf{E}_\delta^M\|_{0,\Omega} . \tag{2.11}$$

Now we need to show that $\|\mathbf{E}_\delta^M\|_{0,\Omega}$ is bounded uniformly with respect to δ. To do that we proceed as follows: first of all $\mathbf{E}_\delta^M$ satisfies

$$\int_\Omega \boldsymbol{\mu}^{-1} \operatorname{curl} \mathbf{E}_\delta^M \cdot \operatorname{curl} \overline{\mathbf{z}} + i\omega \int_{\Omega_C} \boldsymbol{\sigma} \mathbf{E}_{\delta,C}^M \cdot \overline{\mathbf{z}_C} = \int_\Omega (-i\omega \mathbf{J}_e + \delta\omega^2 \boldsymbol{\varepsilon} \mathbf{E}_\delta^M) \cdot \overline{\mathbf{z}}$$

for all $\mathbf{z} \in H(\mathrm{curl};\Omega)$. Then, since $\operatorname{div}(\boldsymbol{\varepsilon}_I \mathbf{E}_{\delta,I}^M) = 0$ in Ω_I, using again the coerciveness of $a_e^*(\cdot,\cdot)$ in $W_{\varepsilon_I}(\Omega_I;\Omega)$ we have

$$C_0 \|\mathbf{E}_\delta^M\|_{H(\mathrm{curl};\Omega)} \le |\omega| \|\mathbf{J}_e\|_{0,\Omega} + \delta\omega^2 \varepsilon_{\max} \|\mathbf{E}_\delta^M\|_{0,\Omega} .$$

Now, taking for instance $\delta_* = \frac{C_0}{2\omega^2 \varepsilon_{\max}}$, for all $\delta \le \delta_*$ we find

$$\|\mathbf{E}_\delta^M\|_{H(\mathrm{curl};\Omega)} \le \frac{2|\omega|}{C_0} \|\mathbf{J}_e\|_{0,\Omega} ,$$

and by substituting in (2.11) we obtain the desired result. □

2.3 The eddy current model as the low-frequency limit

The eddy current model can also be considered as the low-frequency limit of the full Maxwell system. This statement must be properly understood, since the limit problem obtained by formally taking the frequency equal to 0 is in fact the magneto-electrostatic problem, where induced eddy currents are not present. The interpretation of the limit procedure we are interested in is that the difference between the solution of the full Maxwell system and the solution of the eddy current model is vanishing as the frequency is going to 0. A different asymptotic analysis is performed when focusing on the difference between the eddy current solution and the magneto-electrostatic solution: this problem is considered in Section 7.4.

In this section we assume that all the material parameters are fixed and we consider the asymptotic behaviour as the frequency goes to 0 of the difference between the solution of the full Maxwell system (2.8), denoted by $\mathbf{E}^M$, and the solution of the eddy current problem (2.1), denoted by $\mathbf{E}$. We focus on the magnetic boundary condition $\mathbf{H} \times \mathbf{n} = \mathbf{0}$ on $\partial\Omega$, but we recall that this result also holds true when considering the electric boundary condition $\mathbf{E} \times \mathbf{n} = \mathbf{E}^M \times \mathbf{n} = \mathbf{0}$ in $\partial\Omega$, as proved in Alonso [5].

We show that the norm in $L^2(\Omega)$ of the difference $\mathbf{E} - \mathbf{E}^M$ is of order $|\omega|$. We also give an estimate in terms of ω of the $L^2(\Omega)$-norm of the difference between the magnetic fields.

As a first step we obtain an estimate of the energy norm of $\mathbf{E} - \mathbf{E}^M$ in terms of a power of $|\omega|$ times the $L^2(\Omega)$-norm of $\mathbf{E}^M$. Since the solution $\mathbf{E}^M$ depends on ω, a second step is the proof that the $L^2(\Omega)$-norm of $\mathbf{E}^M$ is uniformly bounded in $|\omega|$.

Lemma 2.6. *There exists a constant $C > 0$, independent of ω, such that*

$$\| \operatorname{curl}(\mathbf{E} - \mathbf{E}^M)\|^2_{0,\Omega} + |\omega| \|\mathbf{E}_C - \mathbf{E}^M_C\|^2_{0,\Omega_C} \leq C|\omega|^3(|\omega| + 1)\|\mathbf{E}^M\|^2_{0,\Omega} \,.$$

Proof. As we have seen in the previous section, from (2.10) we know that $(\mathbf{E}-\mathbf{E}^M) \in W_\varepsilon(\Omega_I; \Omega)$ and that

$$\int_\Omega \boldsymbol{\mu}^{-1} \operatorname{curl}(\mathbf{E} - \mathbf{E}^M) \cdot \operatorname{curl}\overline{\mathbf{z}} + i\omega \int_{\Omega_C} \boldsymbol{\sigma}(\mathbf{E}_C - \mathbf{E}^M_C) \cdot \overline{\mathbf{z}_C} = -\omega^2 \int_\Omega \boldsymbol{\varepsilon}\mathbf{E}^M \cdot \overline{\mathbf{z}}$$

for all $\mathbf{z} \in H(\operatorname{curl}; \Omega)$.

Taking $\mathbf{z} = \mathbf{E} - \mathbf{E}^M$ we have

$$\begin{aligned} &\textstyle\int_\Omega \boldsymbol{\mu}^{-1} \operatorname{curl}(\mathbf{E} - \mathbf{E}^M) \cdot \operatorname{curl}(\overline{\mathbf{E}} - \overline{\mathbf{E}^M}) + i\omega \int_{\Omega_C} \boldsymbol{\sigma}(\mathbf{E}_C - \mathbf{E}^M_C) \cdot (\overline{\mathbf{E}_C} - \overline{\mathbf{E}^M_C}) \\ &\quad = -\omega^2 \textstyle\int_\Omega \boldsymbol{\varepsilon}\mathbf{E}^M \cdot (\overline{\mathbf{E}} - \overline{\mathbf{E}^M}) \,, \end{aligned}$$

hence

$$\begin{aligned} &\nu^2_{\min} \| \operatorname{curl}(\mathbf{E} - \mathbf{E}^M)\|^4_{0,\Omega} + \omega^2\sigma^2_{\min}\|\mathbf{E}_C - \mathbf{E}^M_C\|^4_{0,\Omega_C} \\ &\quad \leq \omega^4 \varepsilon^2_{\max}\|\mathbf{E}^M\|^2_{0,\Omega}\|\mathbf{E} - \mathbf{E}^M\|^2_{0,\Omega} \,. \end{aligned}$$

Since $\operatorname{div}(\boldsymbol{\varepsilon}_I(\mathbf{E}_I - \mathbf{E}^M_I)) = 0$ in Ω_I, from (2.7) we know that there exists a constant $C_1 > 0$, independent of ω, such that

$$\|\mathbf{E} - \mathbf{E}^M\|^2_{0,\Omega} \leq C_1 \left(\| \operatorname{curl}(\mathbf{E} - \mathbf{E}^M)\|^2_{0.\Omega} + \|\mathbf{E}_C - \mathbf{E}^M_C\|^2_{0,\Omega_C}\right) \,, \tag{2.12}$$

therefore we find, for a constant $C_3 > 0$ independent of ω,

$$\begin{aligned} &\| \operatorname{curl}(\mathbf{E} - \mathbf{E}^M)\|^4_{0,\Omega} + \omega^2\|\mathbf{E}_C - \mathbf{E}^M_C\|^4_{0,\Omega_C} \\ &\quad \leq C_3\omega^4\|\mathbf{E}^M\|^2_{0,\Omega}(\| \operatorname{curl}(\mathbf{E} - \mathbf{E}^M)\|^2_{0,\Omega} + \|\mathbf{E}_C - \mathbf{E}^M_C\|^2_{0,\Omega_C}) \,. \end{aligned}$$

Using that for each $\delta > 0$ it holds $AB \leq \frac{1}{2\delta}A^2 + \frac{\delta}{2}B^2$, we have

$$\begin{aligned} \| \operatorname{curl}(\mathbf{E} - \mathbf{E}^M)\|^4_{0,\Omega} + \omega^2\|\mathbf{E}_C - \mathbf{E}^M_C\|^4_{0,\Omega_C} &\leq \left(\tfrac{1}{2\delta_1} + \tfrac{1}{2\delta_2}\right)C_3^2\omega^8\|\mathbf{E}^M\|^4_{0,\Omega} \\ &\quad + \tfrac{\delta_1}{2}\| \operatorname{curl}(\mathbf{E} - \mathbf{E}^M)\|^4_{0,\Omega} + \tfrac{\delta_2}{2}\|\mathbf{E}_C - \mathbf{E}^M_C\|^4_{0,\Omega_C} \,, \end{aligned}$$

for $\delta_1 > 0$ and $\delta_2 > 0$. Taking in particular $\delta_1 = 1$ and $\delta_2 = \omega^2$ one finds

$$\|\operatorname{curl}(\mathbf{E} - \mathbf{E}^M)\|^4_{0,\Omega} + \omega^2 \|\mathbf{E}_C - \mathbf{E}^M_C\|^4_{0,\Omega_C} \leq C_3^2(\omega^2+1)\omega^6 \|\mathbf{E}^M\|^4_{0,\Omega} ,$$

hence the desired result. □

The following result provides a bound for $\|\mathbf{E}^M\|_{0,\Omega}$ that is uniform with respect to $|\omega|$.

Lemma 2.7. *There exists a constant* ω^*, $0 < \omega^* \leq 1$, *such that for* $|\omega| \leq \omega^*$ *one has*

$$\|\mathbf{E}^M\|_{0,\Omega} \leq C ,$$

for some constant $C > 0$ *independent of* ω.

Proof. Multiplying the first equation of (2.8) by $\mathbf{E}^M$ and integrating by parts we obtain

$$\begin{aligned} &\int_\Omega \boldsymbol{\mu}^{-1} \operatorname{curl} \mathbf{E}^M \cdot \operatorname{curl} \overline{\mathbf{E}^M} + i\omega \int_{\Omega_C} \boldsymbol{\sigma} \mathbf{E}^M_C \cdot \overline{\mathbf{E}^M_C} \\ &\qquad = -i\omega \int_\Omega \mathbf{J}_e \cdot \overline{\mathbf{E}^M} + \omega^2 \int_\Omega \boldsymbol{\varepsilon} \mathbf{E}^M \cdot \overline{\mathbf{E}^M} , \end{aligned}$$

hence

$$\begin{aligned} &\nu^2_{\min} \|\operatorname{curl} \mathbf{E}^M\|^4_{0,\Omega} + \omega^2 \sigma^2_{\min} \|\mathbf{E}^M_C\|^4_{0,\Omega_C} \\ &\qquad \leq 2\omega^2 \|\mathbf{J}_e\|^2_{0,\Omega} \|\mathbf{E}^M\|^2_{0,\Omega} + 2\omega^4 \varepsilon^2_{\max} \|\mathbf{E}^M\|^4_{0,\Omega} , \end{aligned}$$

or simply, for a suitable constant $\hat{C} > 0$, independent of ω,

$$\begin{aligned} &\|\operatorname{curl} \mathbf{E}^M\|^2_{0,\Omega} + |\omega| \|\mathbf{E}^M_C\|^2_{0,\Omega_C} \\ &\qquad \leq \hat{C}(|\omega| \|\mathbf{J}_e\|_{0,\Omega} \|\mathbf{E}^M\|_{0,\Omega} + \omega^2 \|\mathbf{E}^M\|^2_{0,\Omega}) . \end{aligned}$$

Since $\operatorname{div}(\varepsilon_I \mathbf{E}^M_I) = 0$ in Ω_I, as in (2.12) we have

$$\|\mathbf{E}^M\|^2_{0,\Omega} \leq C_1 \left(\|\operatorname{curl} \mathbf{E}^M\|^2_{0,\Omega} + \|\mathbf{E}^M_C\|^2_{0,\Omega_C} \right) .$$

Then, for $|\omega| \leq 1$,

$$\begin{aligned} \|\mathbf{E}^M\|^2_{0,\Omega} &\leq C_1 \tfrac{1}{|\omega|} (\|\operatorname{curl} \mathbf{E}^M\|^2_{0,\Omega} + |\omega| \|\mathbf{E}^M_C\|^2_{0,\Omega_C}) \\ &\leq \hat{C} C_1 \tfrac{1}{|\omega|} (|\omega| \|\mathbf{J}_e\|_{0,\Omega} \|\mathbf{E}^M\|_{0,\Omega} + \omega^2 \|\mathbf{E}^M\|^2_{0,\Omega}) \\ &\leq \tfrac{1}{2} \hat{C}^2 C_1^2 \|\mathbf{J}_e\|^2_{0,\Omega} + \tfrac{1}{2} \|\mathbf{E}^M\|^2_{0,\Omega} + \hat{C} C_1 |\omega| \|\mathbf{E}^M\|^2_{0,\Omega} . \end{aligned} \tag{2.13}$$

To finish the proof we have only to choose $|\omega| < \min\{1, \frac{1}{4\hat{C}C_1}\}$. □

In conclusion, we have obtained the following result.

Theorem 2.8. *There exists a constant* ω^*, $0 < \omega^* \leq 1$, *such that for* $|\omega| \leq \omega^*$ *one has*

$$\|\mathbf{E} - \mathbf{E}^M\|_{0,\Omega} \leq C|\omega|$$

$$\|\mathbf{H} - \mathbf{H}^M\|_{0,\Omega} \leq C|\omega|^{1/2} ,$$

for some constant $C > 0$ *independent of* ω.

Proof. From Lemma 2.6 and Lemma 2.7 for $|\omega| \leq \omega^*$ we have

$$\| \operatorname{curl}(\mathbf{E} - \mathbf{E}^M)\|^2_{0,\Omega} + |\omega| \|\mathbf{E}_C - \mathbf{E}^M_C\|^2_{0,\Omega_C} \leq C_* |\omega|^3 (|\omega| + 1) \tag{2.14}$$

for some constant $C_* > 0$ independent of ω. Hence proceeding as in (2.13), for $|\omega| \leq \omega^*$ we find

$$\|\mathbf{E} - \mathbf{E}^M\|^2_{0,\Omega} \leq C_1 C_* \omega^2 (|\omega| + 1) \leq 2 C_1 C_* \omega^2 .$$

From (2.14) it also follows that

$$\| \operatorname{curl}(\mathbf{E} - \mathbf{E}^M)\|_{0,\Omega} \leq C |\omega|^{3/2} .$$

Finally, from the Faraday law $\operatorname{curl}(\mathbf{E}-\mathbf{E}^M) = -i\omega\boldsymbol{\mu}(\mathbf{H}-\mathbf{H}^M)$ we have also obtained

$$\|\mathbf{H} - \mathbf{H}^M\|_{0,\Omega} \leq C |\omega|^{1/2} ,$$

which ends the proof. □

2.3.1 Higher order convergence

Under suitable additional assumptions the order of convergence can be improved. The following result can be found in Schmidt et al. [223], where the eddy current modeling error has been investigated under different points of view.

Lemma 2.9. *Suppose that* $\operatorname{div} \mathbf{J}_e = 0$ *in* Ω *and that* Ω_C *is simply-connected. There exists a constant* ω^*, $0 < \omega^* \leq 1$, *such that for* $|\omega| \leq \omega^*$ *one has*

$$\|\mathbf{E}^M\|_{0,\Omega} \leq C |\omega| , \tag{2.15}$$

for some constant $C > 0$, *independent of* ω.

Proof. For a while, let us proceed without making use of the assumptions that $\operatorname{div} \mathbf{J}_e = 0$ in Ω and Ω_C is simply-connected.

Since from Theorem 2.8 we have $\|\mathbf{E} - \mathbf{E}^M\|_{0,\Omega} \leq C|\omega|$, it is enough to show that $\|\mathbf{E}\|_{0,\Omega} \leq C|\omega|$. From the Ampère equation we have $\operatorname{div}(\boldsymbol{\sigma}\mathbf{E} + \mathbf{J}_e) = 0$ in Ω, hence

$$\operatorname{div}(\boldsymbol{\sigma}\mathbf{E}_C + \mathbf{J}_{e,C}) = 0 \qquad \text{in } \Omega_C$$

and

$$(\boldsymbol{\sigma}\mathbf{E}_C + \mathbf{J}_{e,C}) \cdot \mathbf{n}_C = -\mathbf{J}_{e,I} \cdot \mathbf{n}_I \qquad \text{on } \Gamma .$$

Proceeding as in Lemma 2.1 we obtain

$$\begin{aligned}
\|\mathbf{E}_C\|_{0,\Omega_C} &\leq C\Big(\| \operatorname{curl} \mathbf{E}_C\|_{0,\Omega_C} + \| \operatorname{div}(\boldsymbol{\sigma}\mathbf{E}_C)\|_{0,\Omega_C} \\
&\quad + \|\boldsymbol{\sigma}\mathbf{E}_C \cdot \mathbf{n}_C\|_{-1/2,\Gamma} + \textstyle\sum_{\beta=1}^{n_{\Omega_C}} \Big| \int_{\Omega_C} \boldsymbol{\sigma}\mathbf{E}_C \cdot \boldsymbol{\rho}^*_{\beta,C} \Big| \Big) \\
&\leq C\Big(\| \operatorname{curl} \mathbf{E}_C\|_{0,\Omega_C} + \| \operatorname{div} \mathbf{J}_{e,C}\|_{0,\Omega_C} \\
&\quad + \|\mathbf{J}_{e,C} \cdot \mathbf{n}_C + \mathbf{J}_{e,I} \cdot \mathbf{n}_I\|_{-1/2,\Gamma} + \textstyle\sum_{\beta=1}^{n_{\Omega_C}} \Big| \int_{\Omega_C} \boldsymbol{\sigma}\mathbf{E}_C \cdot \boldsymbol{\rho}^*_{\beta,C} \Big| \Big),
\end{aligned}$$

where $\boldsymbol{\rho}^*_{\beta,C}$, $\beta = 1, \ldots, n_{\Omega_C}$, are the basis functions of the space of harmonic fields $\mathcal{H}_\sigma(m; \Omega_C)$ defined as follows

$$\mathcal{H}_\sigma(m; \Omega_C) := \{\mathbf{z}_C \in (L^2(\Omega_C))^3 \mid \operatorname{curl} \mathbf{z}_C = \mathbf{0}, \operatorname{div}(\boldsymbol{\sigma} \mathbf{z}_C) = 0, \\ \boldsymbol{\sigma} \mathbf{z}_C \cdot \mathbf{n}_C = 0 \text{ on } \Gamma\} .$$

Moreover, from Lemma 2.2 we know that

$$\|\mathbf{E}_I\|_{0,\Omega_I} \leq C(\|\operatorname{curl} \mathbf{E}\|_{0,\Omega} + \|\mathbf{E}_C\|_{0,\Omega_C}) ,$$

so that we end up with

$$\begin{aligned} \|\mathbf{E}\|_{0,\Omega} \leq C\Big(\|\operatorname{curl} \mathbf{E}\|_{0,\Omega} &+ \|\operatorname{div} \mathbf{J}_{e,C}\|_{0,\Omega_C} \\ &+ \|\mathbf{J}_{e,C} \cdot \mathbf{n}_C + \mathbf{J}_{e,I} \cdot \mathbf{n}_I\|_{-1/2,\Gamma} \\ &+ \textstyle\sum_{\beta=1}^{n_{\Omega_C}} \left| \int_{\Omega_C} \boldsymbol{\sigma} \mathbf{E}_C \cdot \boldsymbol{\rho}^*_{\beta,C} \right| \Big) . \end{aligned} \tag{2.16}$$

From the Ampère equation we obtain by integration by parts

$$\int_\Omega \boldsymbol{\mu}^{-1} \operatorname{curl} \mathbf{E} \cdot \operatorname{curl} \overline{\mathbf{E}} + i\omega \int_{\Omega_C} \boldsymbol{\sigma} \mathbf{E}_C \cdot \overline{\mathbf{E}_C} = -i\omega \int_\Omega \mathbf{J}_e \cdot \overline{\mathbf{E}} ,$$

hence

$$\|\operatorname{curl} \mathbf{E}\|^2_{0,\Omega} \leq C|\omega| \|\mathbf{J}_e\|_{0,\Omega} \|\mathbf{E}\|_{0,\Omega} .$$

In conclusion, assuming that $\operatorname{div} \mathbf{J}_{e,C} = 0$ in Ω_C and $\mathbf{J}_{e,C} \cdot \mathbf{n}_C + \mathbf{J}_{e,I} \cdot \mathbf{n}_I = 0$ on Γ (which is equivalent to require that $\operatorname{div} \mathbf{J}_e = 0$ in Ω, as we have already assumed $\operatorname{div} \mathbf{J}_{e,I} = 0$ in Ω_I) and that Ω_C is simply-connected (so that the space $\mathcal{H}_\sigma(m; \Omega_C)$ is trivial), from (2.16) it follows

$$\|\mathbf{E}\|_{0,\Omega} \leq C\|\operatorname{curl} \mathbf{E}\|_{0,\Omega} \leq C|\omega|^{1/2} \|\mathbf{J}_e\|^{1/2}_{0,\Omega} \|\mathbf{E}\|^{1/2}_{0,\Omega} ,$$

hence

$$\|\mathbf{E}\|_{0,\Omega} \leq C|\omega| \|\mathbf{J}_e\|_{0,\Omega} ,$$

which ends the proof. □

Corollary 2.10. *Under the assumptions of Lemma 2.9, for $|\omega| \leq \omega^*$ one has*

$$\begin{aligned} \|\mathbf{E} - \mathbf{E}^M\|_{0,\Omega} &\leq C|\omega|^2 \\ \|\mathbf{H} - \mathbf{H}^M\|_{0,\Omega} &\leq C|\omega|^{3/2} . \end{aligned} \tag{2.17}$$

Proof. From (2.15) and Lemma 2.6 we find

$$\|\operatorname{curl}(\mathbf{E} - \mathbf{E}^M)\|^2_{0,\Omega} + |\omega| \|\mathbf{E}_C - \mathbf{E}^M_C\|^2_{0,\Omega_C} \leq C|\omega|^5 ,$$

and consequently, proceeding as in Theorem 2.8, the thesis follows. □

Remark 2.11. If we were able to prove that $\boldsymbol{\sigma}\mathbf{E}_C$ is $L^2(\Omega_C)$-orthogonal to $\mathcal{H}_\sigma(m;\Omega_C)$, in Lemma 2.9 we could avoid to require that Ω_C is simply-connected.

In Ammari et al. [23] an attempt is made to devise the necessary and sufficient conditions on $\mathbf{J}_e$ ensuring that $\boldsymbol{\sigma}\mathbf{E}_C$ is orthogonal to $\mathcal{H}_\sigma(m;\Omega_C)$; however, their argument is not conclusive, and to our knowledge a characterization of this orthogonality property in terms of $\mathbf{J}_e$ is not known. □

The estimate for the difference between the magnetic fields can be improved even if we do not impose additional conditions on $\mathbf{J}_e$ and Ω_C.

Theorem 2.12. *Suppose that the domain Ω is simply-connected. There exists a constant ω^*, $0 < \omega^* \leq 1$, such that for $|\omega| \leq \omega^*$ one has*

$$\|\mathbf{H} - \mathbf{H}^M\|_{0,\Omega} \leq C|\omega| \,. \tag{2.18}$$

Moreover, if estimate (2.15) *is satisfied one has*

$$\|\mathbf{H} - \mathbf{H}^M\|_{0,\Omega} \leq C\omega^2 \,. \tag{2.19}$$

In both cases the constant $C > 0$ is independent of ω.

Proof. To prove this result we use the formulation of the eddy current model in terms of a magnetic vector potential $\mathbf{A}$ and a electric scalar potential V_C (see Chapter 6). This means that we consider $\mathbf{A}$ and V_C such that

$$\operatorname{curl} \mathbf{A} = \boldsymbol{\mu}\mathbf{H} \quad \text{and} \quad \mathbf{E}_C = -i\omega\mathbf{A}_C - \operatorname{grad} V_C \,.$$

Since Ω is simply-connected, we can also do the same for the Maxwell equations, and introduce $\mathbf{A}^M$ and V_C^M such that

$$\operatorname{curl} \mathbf{A}^M = \boldsymbol{\mu}\mathbf{H}^M \quad \text{and} \quad \mathbf{E}_C^M = -i\omega\mathbf{A}_C^M - \operatorname{grad} V_C^M \,.$$

Setting now $(\mathbf{Z}, N_C) := (\mathbf{A} - \mathbf{A}^M, V_C - V_C^M)$, it is easily seen that it satisfies the problem

$$\begin{cases} \operatorname{curl}(\boldsymbol{\mu}^{-1} \operatorname{curl} \mathbf{Z}) - \mu_*^{-1} \operatorname{grad} \operatorname{div} \mathbf{Z} & \\ \qquad\qquad + i\omega\boldsymbol{\sigma}\mathbf{Z} + \boldsymbol{\sigma} \operatorname{grad} N_C = -i\omega\boldsymbol{\varepsilon}\mathbf{E}^M & \text{in } \Omega \\ \operatorname{div}(i\omega\boldsymbol{\sigma}\mathbf{Z}_C + \boldsymbol{\sigma} \operatorname{grad} N_C) = -i\omega \operatorname{div}(\boldsymbol{\varepsilon}\mathbf{E}^M) & \text{in } \Omega_C \\ (i\omega\boldsymbol{\sigma}\mathbf{Z}_C + \boldsymbol{\sigma} \operatorname{grad} N_C) \cdot \mathbf{n}_C & \\ \qquad\qquad = -i\omega(\boldsymbol{\varepsilon}_C\mathbf{E}_C^M \cdot \mathbf{n}_C + \boldsymbol{\varepsilon}_I\mathbf{E}_I^M \cdot \mathbf{n}_I) & \text{on } \Gamma \\ \mathbf{Z} \cdot \mathbf{n} = 0 & \text{on } \partial\Omega \\ (\boldsymbol{\mu}^{-1} \operatorname{curl} \mathbf{Z}) \times \mathbf{n} = \mathbf{0} & \text{on } \partial\Omega \,, \end{cases} \tag{2.20}$$

where N_C is determined up to an additive constant in each connected component of Ω_C. The corresponding weak formulation is the same than that presented in (6.12), with $\mathbf{J}_e$ replaced by $-i\omega\boldsymbol{\varepsilon}\mathbf{E}^M$.

Proceeding as in Section 6.1.2 (see in particular (6.36), (6.37), (6.38) and (6.39)), it can be proved that

$$\begin{aligned} \| \operatorname{curl} \mathbf{Z}\|_{0,\Omega}^2 + \| \operatorname{div} \mathbf{Z}\|_{0,\Omega}^2 + \|\mathbf{Z}\|_{0,\Omega}^2 + |\omega|^{-1}\tau\|N_C\|_{1,\Omega_C}^2 - C_4|\omega|\tau\|\mathbf{Z}\|_{0,\Omega}^2 \\ \leq C_4|\omega|\|\boldsymbol{\varepsilon}\mathbf{E}^M\|_{0,\Omega}\|\mathbf{Z}\|_{0,\Omega} + C_4\|\boldsymbol{\varepsilon}\mathbf{E}^M\|_{0,\Omega}\|N_C\|_{1,\Omega_C} \end{aligned}$$

for each $0 < \tau \leq 1/2$ and a suitable positive constant C_4, independent of ω. Then for each $\delta_1 > 0$ and $\delta_2 > 0$

$$\begin{aligned}(1 - C_4|\omega|\tau)\|\mathbf{Z}\|^2_{0,\Omega} + \|\operatorname{curl}\mathbf{Z}\|^2_{0,\Omega} + \|\operatorname{div}\mathbf{Z}\|^2_{0,\Omega} + |\omega|^{-1}\tau\|N_C\|^2_{1,\Omega_C}\\ \leq \tfrac{1}{2\delta_1}C_4^2\omega^2\|\boldsymbol{\varepsilon}\mathbf{E}^M\|^2_{0,\Omega} + \tfrac{\delta_1}{2}\|\mathbf{Z}\|^2_{0,\Omega} + \tfrac{1}{2\delta_2}C_4^2\|\boldsymbol{\varepsilon}\mathbf{E}^M\|^2_{0,\Omega} + \tfrac{\delta_2}{2}\|N_C\|^2_{1,\Omega_C}\,.\end{aligned}$$

Taking τ such that $1 - C_4|\omega|\tau > 0$ and choosing $\delta_1 = 1 - C_4|\omega|\tau$ and $\delta_2 = |\omega|^{-1}\tau$, we obtain that

$$|\omega|^{-1}\tau\|N_C\|^2_{1,\Omega_C} \leq C_4^2\omega^2\tfrac{1}{1-C_4|\omega|\tau}\|\boldsymbol{\varepsilon}\mathbf{E}^M\|^2_{0,\Omega} + C_4^2|\omega|\tfrac{1}{\tau}\|\boldsymbol{\varepsilon}\mathbf{E}^M\|^2_{0,\Omega}\,.$$

If we choose $\tau = \min\{\frac{1}{2}, \frac{1}{2C_4|\omega|}\}$, for $|\omega| \leq 1$ it is straightforward to verify that

$$\|N_C\|_{1,\Omega_C} \leq C_5|\omega|\|\boldsymbol{\varepsilon}\mathbf{E}^M\|_{0,\Omega}\,,$$

for some positive constant C_5, independent of ω.

Coming back to the weak formulation, we see that in particular we have

$$\begin{aligned}\int_\Omega \left(\boldsymbol{\mu}^{-1}\operatorname{curl}\mathbf{Z}\cdot\operatorname{curl}\overline{\mathbf{Z}} + \mu_*^{-1}|\operatorname{div}\mathbf{Z}|^2\right) + i\omega\int_{\Omega_C}\boldsymbol{\sigma}\mathbf{Z}_C\cdot\overline{\mathbf{Z}_C}\\ = -i\omega\int_\Omega \boldsymbol{\varepsilon}\mathbf{E}^M\cdot\overline{\mathbf{Z}} - \int_{\Omega_C}\boldsymbol{\sigma}\operatorname{grad}N_C\cdot\overline{\mathbf{Z}_C}\,.\end{aligned} \tag{2.21}$$

Hence, taking again into account (6.39), from (2.21) it is easy to see that

$$\begin{aligned}\|\boldsymbol{\mu}(\mathbf{H} - \mathbf{H}^M)\|_{0,\Omega} &= \|\operatorname{curl}\mathbf{Z}\|_{0,\Omega}\\ &\leq C(|\omega|\|\mathbf{E}^M\|_{0,\Omega} + \|N_C\|_{1,\Omega_C}) \leq C_6|\omega|\|\mathbf{E}^M\|_{0,\Omega}\end{aligned} \tag{2.22}$$

for some positive constant C_6, independent of ω. Thus, from Lemma 2.7,

$$\|\mathbf{H} - \mathbf{H}^M\|_{0,\Omega} = O(|\omega|)\,.$$

If we assume moreover that $\|\mathbf{E}^M\|_{0,\Omega} \leq C|\omega|$ is satisfied (see for instance Lemma 2.9), from (2.22) one readily obtains (2.19). □

Remark 2.13. The geometrical assumption in Theorem 2.12 can be relaxed.

First, the solution of the Maxwell equations can be written in terms of $\mathbf{A}^M$ and V_C^M in more general geometric situations; for instance, it is surely true if the domain Ω_C is contained in a simply-connected domain $\widehat{\Omega}$ which is contained in Ω (hence, for example, if Ω is simply-connected).

Moreover, the results in Section 6.1.2 hold under the quite general geometrical assumptions that are described there by requiring that (6.2) is satisfied and that $n_{\partial\Omega} = n_\Omega$ (in particular, these assumptions hold true if Ω is simply-connected). □

Remark 2.14. In Ammari et al. [23], the full Maxwell problem and eddy current problem are considered in $\mathbb{R}^3$ with the following asymptotic conditions at infinity: for the full Maxwell system $\left(\mathbf{H}^M \times \frac{\mathbf{x}}{|\mathbf{x}|} - \mathbf{E}^M\right)$ tends to $\mathbf{0}$ uniformly as $|\mathbf{x}|$ goes to infinity, and for the eddy current problem $\mathbf{H}(\mathbf{x}) = O(1/|\mathbf{x}|)$ and $\mathbf{E}(\mathbf{x}) = O(1/|\mathbf{x}|)$ uniformly as $|\mathbf{x}|$ tends to infinity (see Chapter 7 for a more detailed presentation of the eddy current problem in the whole space $\mathbb{R}^3$).

Formally expanding the solutions of both problems in power series with respect to ω, they show that the eddy current model is a first order approximation of the full Maxwell system

$$\begin{aligned} \|\mathbf{E} - \mathbf{E}^M\|_{0,\Omega_R} &\leq C|\omega| \\ \|\mathbf{H} - \mathbf{H}^M\|_{0,\Omega_R} &\leq C|\omega| \ , \end{aligned}$$

where $\Omega_R := (\mathbb{R}^3 \setminus \overline{\Omega_C}) \cap B_R$ and B_R is the open ball of radius R and center $\mathbf{0}$. In that paper the electric permittivity and the magnetic permeability are assumed to be constant outside B_R.

If additional conditions on the current source $\mathbf{J}_e$ and on Ω_C are fulfilled, they show that the eddy current model is in fact a second order approximation of the full Maxwell system

$$\begin{aligned} \|\mathbf{E} - \mathbf{E}^M\|_{0,\Omega_R} &\leq C\omega^2 \\ \|\mathbf{H} - \mathbf{H}^M\|_{0,\Omega_R} &\leq C\omega^2 \ . \end{aligned}$$

More precisely, having expanded $\mathbf{J}_e$ in the formal series

$$\mathbf{J}_e = \sum_{l \geq 0}^{\infty} \omega^l \mathbf{J}_{\rm e}^l \ ,$$

where as usual for all $l \geq 0$ one has required that $\operatorname{div} \mathbf{J}_{{\rm e},I}^l = 0$ and $\int_{\Gamma_j} \mathbf{J}_{{\rm e},I}^l \cdot \mathbf{n}_I = 0$ for all $j = 1, \ldots, p_\Gamma + 1$, the additional assumption on the leading term $\mathbf{J}_{\rm e}^0$ is the one we have already devised before for the complete field $\mathbf{J}_e$, namely, $\operatorname{div} \mathbf{J}_{\rm e}^0 = 0$ in $\mathbb{R}^3$. Moreover, to complete the proof of the second order approximation, the conductor Ω_C is assumed to be simply-connected.

Let us also note that in this case $\Omega = \mathbb{R}^3$ is simply-connected, therefore the asymptotic behaviours obtained by Ammari et al. [23] are in perfect agreement with those established by resorting to the vector potentials $\mathbf{A}$ and $\mathbf{A}^M$: namely, first order approximation under general geometrical assumptions, in particular when Ω is simply-connected, and second order approximation under the additional assumptions that $\operatorname{div} \mathbf{J}_e = 0$ in Ω and the conductor Ω_C is simply-connected. □

3

Existence and uniqueness of the solution

The proof of the existence and uniqueness of the solution to problems (1.22), (1.20) and (1.24) is quite similar. In this chapter, following Alonso Rodríguez et al. [11], we mainly focus on the magnetic boundary value problem (1.22), adding in Section 3.5 a few comments on the electric boundary value problem (1.20) and the no-flux boundary value problem (1.24).

The simplest way for obtaining the existence result is passing to a suitable weak formulation in terms of the sole magnetic field $\mathbf{H}$, and applying the Lax–Milgram lemma. Then the determination of the electric field is straightforward in Ω_C, while in Ω_I it requires the solution of the first order curl–div system. The solvability of this last problem is ensured if some compatibility conditions are satisfied, and these can be verified by writing in an explicit way the strong formulation of the eddy current model in terms of $\mathbf{H}$. Having determined $\mathbf{E}_I$, it is easy to prove the existence and uniqueness result for the complete eddy current model in its strong form (1.22).

It has to be noted that the existence and uniqueness of the solution of the eddy current problem can be proved in many different ways. In Chapter 2 we have essentially given the proof for the $\mathbf{E}$-based formulation (see Remark 2.4); in Chapters 4, 5, and 6 we derive a well-posedness analysis for some hybrid formulations, for the scalar potential formulation and for the vector potential formulation, respectively.

However, we think that the simplest proof is the one we present in this chapter, by focusing on the $\mathbf{H}$-based formulation. This permits also to clarify the problem to solve for the electric field $\mathbf{E}_I$ in the insulator, and at the same time to obtain the complete strong formulation in terms of the magnetic field only: a problem that was not completely understood in the literature (see Sections 3.3 and 3.3.1).

We also observe that the theoretical results we prove in Chapters 4, 5, and 6, though not specifically needed for showing well-posedness of the eddy current problem, are however useful for analyzing the convergence of the finite element schemes there proposed.

A. Alonso Rodríguez, A. Valli: Eddy Current Approximation of Maxwell Equations.

3.1 Weak formulation, existence and uniqueness for the magnetic field

In this chapter the geometrical situation is the one described in Section 1.3, and, as was the case there, we assume that the matrix $\boldsymbol{\mu}$ is symmetric and uniformly positive definite in Ω, with entries belonging to $L^\infty(\Omega)$, the matrix ε_I is symmetric and uniformly positive definite in Ω_I, with entries belonging to $L^\infty(\Omega_I)$, and the matrix $\boldsymbol{\sigma}$ is symmetric and uniformly positive definite in Ω_C, with entries belonging to $L^\infty(\Omega_C)$, whereas it vanishes in Ω_I.

Moreover, we suppose that the current density $\mathbf{J}_e \in (L^2(\Omega))^3$ and satisfies the (necessary) conditions

$$\operatorname{div} \mathbf{J}_{e,I} = 0 \text{ in } \Omega_I \ , \ \mathbf{J}_{e,I} \cdot \mathbf{n} = 0 \text{ on } \partial\Omega \ . \tag{3.1}$$

However, this is not enough: in fact, as shown in Section 1.5, two additional necessary conditions, related to the topology of Ω_I, have to be assumed, namely,

$$\begin{array}{ll} \int_{\Gamma_j} \mathbf{J}_{e,I} \cdot \mathbf{n}_I = 0 & \forall\, j = 1, \ldots, p_\Gamma \\ \int_{\Omega_I} \mathbf{J}_{e,I} \cdot \boldsymbol{\pi}_{k,I} = 0 & \forall\, k = 1, \ldots, n_{\partial\Omega} \ , \end{array} \tag{3.2}$$

where $\boldsymbol{\pi}_{k,I}$ are basis functions of the space of harmonic fields $\mathcal{H}_{\varepsilon_I}(\Gamma, \partial\Omega; \Omega_I)$ (to be precise, the basis functions of that space that are not expressed as the gradient of a potential: see Section 1.4).

As a consequence, Theorem 4.2 in Alonso and Valli [6] shows that there exists a vector field $\mathbf{H}_{e,I} \in H(\text{curl}; \Omega_I)$ satisfying

$$\begin{cases} \operatorname{curl} \mathbf{H}_{e,I} = \mathbf{J}_{e,I} \text{ in } \Omega_I \\ \mathbf{H}_{e,I} \times \mathbf{n} = \mathbf{0} \quad \text{on } \partial\Omega \ , \end{cases} \tag{3.3}$$

and we can also construct a vector field $\mathbf{H}_{e,C} \in H(\text{curl}; \Omega_C)$ such that

$$\mathbf{H}_{e,C} \times \mathbf{n}_C + \mathbf{H}_{e,I} \times \mathbf{n}_I = \mathbf{0} \quad \text{on } \Gamma \ . \tag{3.4}$$

We finally define $\mathbf{H}_e \in H_0(\text{curl}; \Omega)$ as

$$\mathbf{H}_e := \begin{cases} \mathbf{H}_{e,I} & \text{in } \Omega_I \\ \mathbf{H}_{e,C} & \text{in } \Omega_C \ ; \end{cases} \tag{3.5}$$

it can also be shown that $\mathbf{H}_e$ continuously depends on $\mathbf{J}_e$.

Let us introduce the (complex) vector space

$$V := \{\mathbf{v} \in H_0(\text{curl}; \Omega) \,|\, \operatorname{curl} \mathbf{v}_I = \mathbf{0} \text{ in } \Omega_I\} \ , \tag{3.6}$$

endowed with the norm

$$\|\mathbf{v}\|_V := \left(\int_{\Omega_C} |\operatorname{curl} \mathbf{v}_C|^2 + \int_\Omega |\mathbf{v}|^2 \right)^{1/2} \ .$$

Following the presentation in Bossavit [59], multiplying the Faraday equation by $\overline{\mathbf{v}}$, with $\mathbf{v} \in V$, integrating in Ω and integrating by parts we find

$$\int_{\Omega_C} \mathbf{E}_C \cdot \operatorname{curl} \overline{\mathbf{v}_C} + \int_{\Omega_I} \mathbf{E}_I \cdot \operatorname{curl} \overline{\mathbf{v}_I} + \int_{\partial\Omega} \mathbf{n} \times \mathbf{E} \cdot \overline{\mathbf{v}} + \int_{\Omega} i\omega\boldsymbol{\mu}\mathbf{H} \cdot \overline{\mathbf{v}} = 0 \ , \quad (3.7)$$

thus

$$\int_{\Omega_C} \mathbf{E}_C \cdot \operatorname{curl} \overline{\mathbf{v}_C} + \int_{\Omega} i\omega\boldsymbol{\mu}\mathbf{H} \cdot \overline{\mathbf{v}} = 0 \ , \quad (3.8)$$

as $\operatorname{curl} \mathbf{v}_I = \mathbf{0}$ in Ω_I and $\mathbf{v} \times \mathbf{n} = \mathbf{0}$ on $\partial\Omega$. Using the Ampère equation in Ω_C to express $\mathbf{E}_C$, we end up with the following problem

Find $(\mathbf{H} - \mathbf{H}_e) \in V$:

$$\int_{\Omega_C} \boldsymbol{\sigma}^{-1} \operatorname{curl} \mathbf{H}_C \cdot \operatorname{curl} \overline{\mathbf{v}_C} + \int_{\Omega} i\omega\boldsymbol{\mu}\mathbf{H} \cdot \overline{\mathbf{v}} = \int_{\Omega_C} \boldsymbol{\sigma}^{-1}\mathbf{J}_{e,C} \cdot \operatorname{curl} \overline{\mathbf{v}_C} \quad (3.9)$$

for each $\mathbf{v} \in V$.

This formulation is shown to be well-posed via the Lax–Milgram lemma (see, e.g., Bossavit [59], p. 313; Dautray and Lions [94], Chap. VI, Sect. 3, Theor. 7 and Rem. 8; Quarteroni and Valli [199], p. 133). In fact:

Theorem 3.1. *The sesquilinear form*

$$a(\mathbf{u}, \mathbf{v}) := \int_{\Omega_C} \boldsymbol{\sigma}^{-1} \operatorname{curl} \mathbf{u}_C \cdot \operatorname{curl} \overline{\mathbf{v}_C} + \int_{\Omega} i\omega\boldsymbol{\mu}\mathbf{u} \cdot \overline{\mathbf{v}} \quad (3.10)$$

is continuous and coercive in V.

Proof. The continuity follows at once from the boundedness of $\boldsymbol{\sigma}^{-1}$ and $\boldsymbol{\mu}$. The coerciveness reads

$$\begin{aligned} |a(\mathbf{v}, \mathbf{v})|^2 &= \left(\int_{\Omega_C} \boldsymbol{\sigma}^{-1} \operatorname{curl} \mathbf{v}_C \cdot \operatorname{curl} \overline{\mathbf{v}_C}\right)^2 + \omega^2 \left(\int_{\Omega} \boldsymbol{\mu}\mathbf{v} \cdot \overline{\mathbf{v}}\right)^2 \\ &\geq \sigma_{\max}^{-2} \left(\int_{\Omega_C} |\operatorname{curl} \mathbf{v}_C|^2\right)^2 + \omega^2 \mu_{\min}^2 \left(\int_{\Omega} |\mathbf{v}|^2\right)^2 , \end{aligned} \quad (3.11)$$

where $\sigma_{\max}$ is a uniform upper bound in Ω_C for the maximum eigenvalues of $\boldsymbol{\sigma}(\mathbf{x})$ and $\mu_{\min}$ is a uniform lower bound in Ω for the minimum eigenvalues of $\boldsymbol{\mu}(\mathbf{x})$. □

Since the anti-linear forms

$$\mathbf{v} \to \int_{\Omega_C} \boldsymbol{\sigma}^{-1}\mathbf{J}_{e,C} \cdot \operatorname{curl} \overline{\mathbf{v}_C}$$
$$\mathbf{v} \to \int_{\Omega_C} \boldsymbol{\sigma}^{-1} \operatorname{curl} \mathbf{H}_{e,C} \cdot \operatorname{curl} \overline{\mathbf{v}_C} + \int_{\Omega} i\omega\boldsymbol{\mu}\mathbf{H}_e \cdot \overline{\mathbf{v}}$$

are clearly continuous in V, from the Lax–Milgram lemma we have:

Theorem 3.2. *There exists a unique solution to problem* (3.9).

This result is the basis of the existence and uniqueness theory for the complete eddy current problem. However, let us note from the very beginning that it is not straightforward to devise a numerical algorithm based on this formulation, as the space V contains the differential constraint $\operatorname{curl} \mathbf{v}_I = \mathbf{0}$ in Ω_I.

3.2 Determination of the electric field

In the preceding section we have determined the magnetic field $\mathbf{H}$ solution to problem (3.9). The electric field in the conductor can be directly found by setting

$$\mathbf{E}_C = \boldsymbol{\sigma}^{-1}(\operatorname{curl}\mathbf{H}_C - \mathbf{J}_{e,C}) \quad \text{in } \Omega_C\,. \tag{3.12}$$

Therefore, we only need to find $\mathbf{E}_I$, that has to satisfy

$$\begin{cases} \operatorname{curl}\mathbf{E}_I = -i\omega\boldsymbol{\mu}_I\mathbf{H}_I & \text{in } \Omega_I \\ \operatorname{div}(\boldsymbol{\varepsilon}_I\mathbf{E}_I) = 0 & \text{in } \Omega_I \\ \boldsymbol{\varepsilon}_I\mathbf{E}_I\cdot\mathbf{n} = 0 & \text{on } \partial\Omega \\ \int_{\Gamma_j}\boldsymbol{\varepsilon}_I\mathbf{E}_I\cdot\mathbf{n}_I = 0 & \forall\, j = 1,\dots,p_\Gamma \\ \int_{\Omega_I}\boldsymbol{\varepsilon}_I\mathbf{E}_I\cdot\boldsymbol{\pi}_{k,I} = 0 & \forall\, k = 1,\dots,n_{\partial\Omega} \\ \mathbf{E}_I\times\mathbf{n}_I = -\mathbf{E}_C\times\mathbf{n}_C & \text{on } \Gamma\,. \end{cases} \tag{3.13}$$

We recall that the problem above becomes simpler if the boundary of the conductor Ω_C is connected, so that $p_\Gamma = 0$, and the domain Ω is simply-connected, so that $n_{\partial\Omega} = 0$: as an example, the reader can think to a connected conductor, possibly with some "handles", contained in a "box". This simplified geometrical situation is not very restrictive, as we require that the computational domain Ω (and not the conductor) has a simple shape, and indeed in many engineering applications Ω can be chosen freely.

It has to be noted that in solving equations $(3.13)_1$, $(3.13)_6$ some compatibility conditions on the data must be satisfied. In fact, one has

$$\operatorname{div}(i\omega\boldsymbol{\mu}_I\mathbf{H}_I) = -\operatorname{div}\operatorname{curl}\mathbf{E}_I = 0 \qquad \text{in } \Omega_I\,,$$

$$\begin{aligned} \operatorname{div}_\tau(\mathbf{E}_C\times\mathbf{n}_C) &= -\operatorname{div}_\tau(\mathbf{E}_I\times\mathbf{n}_I) \\ &= -\operatorname{curl}\mathbf{E}_I\cdot\mathbf{n}_I = i\omega\boldsymbol{\mu}_I\mathbf{H}_I\cdot\mathbf{n}_I \qquad \text{on } \Gamma\,, \end{aligned}$$

$$\int_{(\partial\Omega)_r} i\omega\boldsymbol{\mu}_I\mathbf{H}_I\cdot\mathbf{n} = -\int_{(\partial\Omega)_r}\operatorname{curl}\mathbf{E}_I\cdot\mathbf{n} = 0 \qquad \forall\, r = 1,\dots,p_{\partial\Omega}\,.$$

This is not enough: for any function $\mathbf{v}_I \in H(\operatorname{curl};\Omega_I)$ with $\operatorname{curl}\mathbf{v}_I = \mathbf{0}$ in Ω_I and $\mathbf{v}_I\times\mathbf{n} = \mathbf{0}$ on $\partial\Omega$ we have

$$\begin{aligned} -i\omega\int_{\Omega_I}\boldsymbol{\mu}_I\mathbf{H}_I\cdot\overline{\mathbf{v}_I} &= \int_{\Omega_I}\operatorname{curl}\mathbf{E}_I\cdot\overline{\mathbf{v}_I} \\ &= \int_\Gamma \mathbf{n}_I\times\mathbf{E}_I\cdot\overline{\mathbf{v}_I} = \int_\Gamma \mathbf{E}_C\times\mathbf{n}_C\cdot\overline{\mathbf{v}_I}\,. \end{aligned}$$

In particular, one can take $\mathbf{v}_I = \boldsymbol{\rho}_{l,I}$, $l = 1,\dots,n_\Gamma$, the basis functions of the space of harmonic fields $\mathcal{H}_{\mu_I}(\partial\Omega,\Gamma;\Omega_I)$ (precisely, the basis functions of that space that are not gradient of a potential: see Section 1.4). With that choice the latter compatibility conditions read

$$i\omega\int_{\Omega_I}\boldsymbol{\mu}_I\mathbf{H}_I\cdot\boldsymbol{\rho}_{l,I} = -\int_\Gamma\mathbf{E}_C\times\mathbf{n}_C\cdot\boldsymbol{\rho}_{l,I} \qquad \forall\, l = 1,\dots,n_\Gamma\,.$$

Therefore, for determining the electric field $\mathbf{E}$ satisfying the Faraday equation in the whole domain Ω we need to satisfy

$$\begin{array}{ll} \operatorname{div}(i\omega\boldsymbol{\mu}_I\mathbf{H}_I) = 0 & \text{in } \Omega_I \\ i\omega\boldsymbol{\mu}_I\mathbf{H}_I \cdot \mathbf{n}_I = \operatorname{div}_\tau(\mathbf{E}_C \times \mathbf{n}_C) & \text{on } \Gamma \\ \int_{(\partial\Omega)_r} i\omega\boldsymbol{\mu}_I\mathbf{H}_I \cdot \mathbf{n} = 0 & \forall\, r = 1, \ldots, p_{\partial\Omega} \\ \int_{\Omega_I} i\omega\boldsymbol{\mu}_I\mathbf{H}_I \cdot \boldsymbol{\rho}_{l,I} & \\ \quad = -\int_\Gamma (\mathbf{E}_C \times \mathbf{n}_C) \cdot \boldsymbol{\rho}_{l,I} & \forall\, l = 1, \ldots, n_\Gamma \,. \end{array} \tag{3.14}$$

What we are facing here is not a problem related to a point of secondary importance, namely, determining the electric field in the insulator Ω_I; it is instead a basic aspect in the solution of the eddy current problem in Ω. A formulation in terms of the magnetic field $\mathbf{H}$ alone (and with $\mathbf{E}_C = \boldsymbol{\sigma}^{-1}(\operatorname{curl}\mathbf{H}_C - \mathbf{J}_{e,C})$) is not correct if any of these compatibility conditions is missing.

In particular, we want to focus on the conditions

$$\int_{\Omega_I} i\omega\boldsymbol{\mu}_I\mathbf{H}_I \cdot \boldsymbol{\rho}_{l,I} = -\int_\Gamma (\mathbf{E}_C \times \mathbf{n}_C) \cdot \boldsymbol{\rho}_{l,I} \quad \forall\, l = 1, \ldots, n_\Gamma \,,$$

related to the topology of Ω_I. The fact that these are necessary conditions for finding the correct physical solution has been often overlooked in previous works on the subject. We will see in Section 3.3.1 that these conditions can be interpreted as the Faraday equation on the surfaces in Ω_I that are "cutting" the $\partial\Omega$-independent non-bounding cycles in Ω_I. We also analyze in Section 3.3.2 whether these conditions are satisfied when the eddy current problem is described through some other frequently used formulations.

It has been shown that conditions (3.14) are also sufficient for proving the existence and uniqueness of the solution of (3.13) (see, for example, Saranen [218], [219], Alonso and Valli [6]). More precisely, we have:

Theorem 3.3. *Assume that the electric field $\mathbf{E}_C$ and the magnetic field $\mathbf{H}_I$ satisfy the compatibility conditions (3.14). Then there exists a unique solution $\mathbf{E}_I$ to (3.13).*

Proof. As already shown in Section 1.5, for any basis function

$$\operatorname{grad} w_{j,I} \in \mathcal{H}_{\varepsilon_I}(\Gamma, \partial\Omega; \Omega_I), \quad j = 1, \ldots, p_\Gamma \,,$$

we have

$$\begin{aligned} &\int_{\Omega_I} \varepsilon_I\mathbf{E}_I \cdot \operatorname{grad} w_{j,I} \\ &\quad = -\int_{\Omega_I} \operatorname{div}(\varepsilon_I\mathbf{E}_I)\, w_{j,I} + \int_{\partial\Omega} \varepsilon_I\mathbf{E}_I \cdot \mathbf{n}\, w_{j,I} + \int_\Gamma \varepsilon_I\mathbf{E}_I \cdot \mathbf{n}_I\, w_{j,I} \\ &\quad = \int_{\Gamma_j} \varepsilon_I\mathbf{E}_I \cdot \mathbf{n}_I \,, \end{aligned}$$

as $\operatorname{div}(\varepsilon_I\mathbf{E}_I) = 0$ in Ω_I and $\varepsilon_I\mathbf{E}_I \cdot \mathbf{n} = 0$ on $\partial\Omega$. Therefore a solution $\mathbf{E}_I$ to (3.13) is orthogonal to the space $\mathcal{H}_{\varepsilon_I}(\Gamma, \partial\Omega; \Omega_I)$ with respect to the scalar product $(\mathbf{w}_I, \mathbf{z}_I)_{\varepsilon_I,\Omega_I} := \int_{\Omega_I} \varepsilon_I\mathbf{w}_I \cdot \mathbf{z}_I$. At the same time, when $\mathbf{H}_I = \mathbf{0} = \mathbf{E}_C$, we have $\mathbf{E}_I \in \mathcal{H}_{\varepsilon_I}(\Gamma, \partial\Omega; \Omega_I)$, hence uniqueness follows in a straightforward way.

We look for a solution of the form $\mathbf{E}_I = \varepsilon_I^{-1}\operatorname{curl}\mathbf{q}_I$. To construct it, let us introduce the space

$$Y_I := \left\{\mathbf{p}_I \in H_{0,\partial\Omega}(\operatorname{curl}; \Omega_I) \cap H_{0,\Gamma}(\operatorname{div}; \Omega_I) \,|\, \mathbf{p}_I \perp \mathcal{H}(\partial\Omega, \Gamma; \Omega_I)\right\}, \tag{3.15}$$

and consider the problem

$$\begin{aligned}
&\text{Find } \mathbf{q}_I \in Y_I \ : \\
&\quad \textstyle\int_{\Omega_I} (\boldsymbol{\varepsilon}_I^{-1} \operatorname{curl} \mathbf{q}_I \cdot \operatorname{curl} \overline{\mathbf{p}_I} + \operatorname{div} \mathbf{q}_I \operatorname{div} \overline{\mathbf{p}_I}) \\
&\qquad\qquad = - \textstyle\int_{\Omega_I} i\omega \boldsymbol{\mu}_I \mathbf{H}_I \cdot \overline{\mathbf{p}_I} - \int_\Gamma \mathbf{E}_C \times \mathbf{n}_C \cdot \overline{\mathbf{p}_I} \\
&\text{for each } \mathbf{p}_I \in Y_I \ .
\end{aligned} \tag{3.16}$$

As reported in (A.15), the following Poincaré-like inequality holds

$$\int_{\Omega_I} |\mathbf{p}_I|^2 \leq C \int_{\Omega_I} (|\operatorname{curl} \mathbf{p}_I|^2 + |\operatorname{div} \mathbf{p}_I|^2) \qquad \forall\, \mathbf{p}_I \in Y_I \ .$$

Therefore, the existence of a unique solution $\mathbf{q}_I$ to (3.16) is a consequence of the Lax–Milgram lemma.

We can prove that $\mathbf{q}_I$ satisfies (3.16) also for each $\mathbf{p}_I \in H_{0,\partial\Omega}(\operatorname{curl}; \Omega_I) \cap H_{0,\Gamma}(\operatorname{div}; \Omega_I)$, namely, without requiring $\mathbf{p}_I \perp \mathcal{H}(\partial\Omega, \Gamma; \Omega_I)$. To this aim, write $\mathbf{p}_I = (\mathbf{p}_I - \mathcal{P}\mathbf{p}_I) + \mathcal{P}\mathbf{p}_I$, where $\mathcal{P}\mathbf{p}_I$ is the $(L^2(\Omega))^3$-orthogonal projection of $\mathbf{p}_I$ over $\mathcal{H}(\partial\Omega, \Gamma; \Omega_I)$. Thus we have

$$\begin{aligned}
&\textstyle\int_{\Omega_I} (\boldsymbol{\varepsilon}_I^{-1} \operatorname{curl} \mathbf{q}_I \cdot \operatorname{curl} \overline{\mathbf{p}_I} + \operatorname{div} \mathbf{q}_I \operatorname{div} \overline{\mathbf{p}_I}) \\
&\quad = \textstyle\int_{\Omega_I} [\boldsymbol{\varepsilon}_I^{-1} \operatorname{curl} \mathbf{q}_I \cdot \operatorname{curl}(\overline{\mathbf{p}_I} - \mathcal{P}\overline{\mathbf{p}_I}) + \operatorname{div} \mathbf{q}_I \operatorname{div}(\overline{\mathbf{p}_I} - \mathcal{P}\overline{\mathbf{p}_I})] \\
&\quad = - \textstyle\int_{\Omega_I} i\omega \boldsymbol{\mu}_I \mathbf{H}_I \cdot (\overline{\mathbf{p}_I} - \mathcal{P}\overline{\mathbf{p}_I}) - \int_\Gamma \mathbf{E}_C \times \mathbf{n}_C \cdot (\overline{\mathbf{p}_I} - \mathcal{P}\overline{\mathbf{p}_I}) \ ,
\end{aligned}$$

and, to conclude, we need to prove that

$$\int_{\Omega_I} i\omega \boldsymbol{\mu}_I \mathbf{H}_I \cdot \mathcal{P}\overline{\mathbf{p}_I} + \int_\Gamma \mathbf{E}_C \times \mathbf{n}_C \cdot \mathcal{P}\overline{\mathbf{p}_I} = 0 \ .$$

It can be easily seen that each basis function of $\mathcal{H}(\partial\Omega, \Gamma; \Omega_I)$ differs from a basis function of $\mathcal{H}_{\mu_I}(\partial\Omega, \Gamma; \Omega_I)$ by a gradient of a function belonging to $H^1_{0,\partial\Omega}(\Omega_I)$. Therefore we can write

$$\mathcal{P}\overline{\mathbf{p}_I} = \sum_{r=1}^{p_{\partial\Omega}} a_{I,r} \operatorname{grad} z_{r,I} + \sum_{l=1}^{n_\Gamma} b_{I,l} \boldsymbol{\rho}_{l,I} + \operatorname{grad} \chi_I \ ,$$

where χ_I is a suitable function belonging to $H^1_{0,\partial\Omega}(\Omega_I)$, and we finally have

$$\begin{aligned}
&\textstyle\int_{\Omega_I} i\omega \boldsymbol{\mu}_I \mathbf{H}_I \cdot \operatorname{grad} \chi_I + \int_\Gamma \mathbf{E}_C \times \mathbf{n}_C \cdot \operatorname{grad} \chi_I \\
&\qquad = \textstyle\int_\Gamma i\omega \boldsymbol{\mu}_I \mathbf{H}_I \cdot \mathbf{n}_I\, \chi_I - \int_\Gamma \operatorname{div}_\tau (\mathbf{E}_C \times \mathbf{n}_C)\, \chi_I = 0
\end{aligned} \tag{3.17}$$

$$\begin{aligned}
&\textstyle\int_{\Omega_I} i\omega \boldsymbol{\mu}_I \mathbf{H}_I \cdot \operatorname{grad} z_{r,I} + \int_\Gamma \mathbf{E}_C \times \mathbf{n}_C \cdot \operatorname{grad} z_{r,I} \\
&\quad = \textstyle\int_{\partial\Omega} i\omega \boldsymbol{\mu}_I \mathbf{H}_I \cdot \mathbf{n}\, z_{r,I} + \int_\Gamma i\omega \boldsymbol{\mu}_I \mathbf{H}_I \cdot \mathbf{n}_I\, z_{r,I} - \int_\Gamma \operatorname{div}_\tau (\mathbf{E}_C \times \mathbf{n}_C)\, z_{r,I} \\
&\quad = \textstyle\int_{(\partial\Omega)_r} i\omega \boldsymbol{\mu}_I \mathbf{H}_I \cdot \mathbf{n} = 0 \qquad \forall\, r = 1, \ldots, p_{\partial\Omega}
\end{aligned}$$

and

$$\int_{\Omega_I} i\omega\boldsymbol{\mu}_I\mathbf{H}_I \cdot \boldsymbol{\rho}_{l,I} + \int_\Gamma \mathbf{E}_C \times \mathbf{n}_C \cdot \boldsymbol{\rho}_{l,I} = 0 \qquad \forall\, l = 1, \dots, n_\Gamma\,,$$

having used $(3.14)_3$ and $(3.14)_4$.

It can be also proved that $\operatorname{div} \mathbf{q}_I = 0$ in Ω_I. In fact, let us consider the function $v_I \in H^1(\Omega_I)$ solution to $\Delta v_I = \operatorname{div} \mathbf{q}_I$ in Ω_I, with $v_I = 0$ on $\partial\Omega$ and $\operatorname{grad} v_I \cdot \mathbf{n}_I = 0$ on Γ. It is readily seen that $\operatorname{grad} v_I$ belongs to $H_{0,\partial\Omega}(\mathrm{curl};\Omega_I) \cap H_{0,\Gamma}(\mathrm{div};\Omega_I)$, thus using $\mathbf{p}_I = \operatorname{grad} v_I$ as a test function in (3.16), taking into consideration $(3.14)_1$ and $(3.14)_2$ we find by integration by parts

$$\begin{aligned}\textstyle\int_{\Omega_I} |\operatorname{div} \mathbf{q}_I|^2 &= -\textstyle\int_{\Omega_I} i\omega\boldsymbol{\mu}_I\mathbf{H}_I \cdot \operatorname{grad} \overline{v_I} - \int_\Gamma \mathbf{E}_C \times \mathbf{n}_C \cdot \operatorname{grad} \overline{v_I} \\ &= -\textstyle\int_\Gamma i\omega\boldsymbol{\mu}_I\mathbf{H}_I \cdot \mathbf{n}_I\,\overline{v_I} + \int_\Gamma \operatorname{div}_\tau(\mathbf{E}_C \times \mathbf{n}_C)\,\overline{v_I} = 0\,,\end{aligned}$$

hence $\operatorname{div} \mathbf{q}_I = 0$ in Ω_I. Therefore $\mathbf{q}_I$ satisfies

$$\textstyle\int_{\Omega_I} \boldsymbol{\varepsilon}_I^{-1} \operatorname{curl} \mathbf{q}_I \cdot \operatorname{curl} \overline{\mathbf{p}_I} = -\int_{\Omega_I} i\omega\boldsymbol{\mu}_I\mathbf{H}_I \cdot \overline{\mathbf{p}_I} - \int_\Gamma \mathbf{E}_C \times \mathbf{n}_C \cdot \overline{\mathbf{p}_I} \tag{3.18}$$

for all $\mathbf{p}_I \in H_{0,\partial\Omega}(\mathrm{curl};\Omega_I) \cap H_{0,\Gamma}(\mathrm{div};\Omega_I)$.

As a further step, we want to prove that $\mathbf{q}_I$ indeed satisfies (3.18) as well for every $\mathbf{p}_I^* \in H_{0,\partial\Omega}(\mathrm{curl};\Omega_I)$. Define $v_I^* \in H^1(\Omega_I)$ to be the (weak) solution of $\Delta v_I^* = \operatorname{div} \mathbf{p}_I^*$ in Ω_I, with $v_I^* = 0$ on $\partial\Omega$ and $\operatorname{grad} v_I^* \cdot \mathbf{n}_I = \mathbf{p}_I^* \cdot \mathbf{n}_I$ on Γ. We easily check that $(\mathbf{p}_I^* - \operatorname{grad} v_I^*) \in H_{0,\partial\Omega}(\mathrm{curl};\Omega_I) \cap H_{0,\Gamma}(\mathrm{div};\Omega_I)$, and using it as a test function in (3.18) gives

$$\begin{aligned}&\textstyle\int_{\Omega_I} \boldsymbol{\varepsilon}_I^{-1} \operatorname{curl} \mathbf{q}_I \cdot \operatorname{curl} \overline{\mathbf{p}_I^*} \\ &\quad= -\textstyle\int_{\Omega_I} i\omega\boldsymbol{\mu}_I\mathbf{H}_I \cdot (\overline{\mathbf{p}_I^*} - \operatorname{grad} \overline{v_I^*}) - \int_\Gamma \mathbf{E}_C \times \mathbf{n}_C \cdot (\overline{\mathbf{p}_I^*} - \operatorname{grad} \overline{v_I^*}) \\ &\quad= -\textstyle\int_{\Omega_I} i\omega\boldsymbol{\mu}_I\mathbf{H}_I \cdot \overline{\mathbf{p}_I^*} - \int_\Gamma \mathbf{E}_C \times \mathbf{n}_C \cdot \overline{\mathbf{p}_I^*}\,,\end{aligned}$$

having used the fact that $v_I^* \in H^1_{0,\partial\Omega}(\Omega_I)$ and proceeding as in (3.17).

Taking now in (3.18) a test function $\mathbf{p}_I^* \in (C_0^\infty(\Omega_I))^3$, by integration by parts we find $\operatorname{curl}(\boldsymbol{\varepsilon}_I^{-1} \operatorname{curl} \mathbf{q}_I) = -i\omega\boldsymbol{\mu}_I\mathbf{H}_I$ in Ω_I; similarly, taking a test function $\mathbf{p}_I^* \in H_{0,\partial\Omega}(\mathrm{curl};\Omega_I)$ we have $(\boldsymbol{\varepsilon}_I^{-1} \operatorname{curl} \mathbf{q}_I) \times \mathbf{n}_I = -\mathbf{E}_C \times \mathbf{n}_C$ on Γ.

Setting $\mathbf{E}_I := \boldsymbol{\varepsilon}_I^{-1} \operatorname{curl} \mathbf{q}_I$, we have found the solution to $(3.13)_1$, $(3.13)_2$ and $(3.13)_6$. Moreover,

$$\boldsymbol{\varepsilon}_I\mathbf{E}_I \cdot \mathbf{n} = \operatorname{curl} \mathbf{q}_I \cdot \mathbf{n} = \operatorname{div}_\tau(\mathbf{q}_I \times \mathbf{n}) = 0 \qquad \text{on } \Gamma$$

$$\begin{aligned}\textstyle\int_{\Omega_I} \boldsymbol{\varepsilon}_I\mathbf{E}_I \cdot \boldsymbol{\pi}_{k,I} &= \textstyle\int_{\Omega_I} \operatorname{curl} \mathbf{q}_I \cdot \boldsymbol{\pi}_{k,I} = \int_{\Omega_I} \mathbf{q}_I \cdot \operatorname{curl} \boldsymbol{\pi}_{k,I} \\ &\quad+ \textstyle\int_{\partial\Omega} \mathbf{n} \times \mathbf{q}_I \cdot \boldsymbol{\pi}_{k,I} + \int_\Gamma \boldsymbol{\pi}_{k,I} \times \mathbf{n}_I \cdot \mathbf{q}_I = 0 \qquad \forall\, k = 1, \dots, n_{\partial\Omega}\end{aligned}$$

and

$$\int_{\Gamma_j} \boldsymbol{\varepsilon}_I\mathbf{E}_I \cdot \mathbf{n}_I = \int_{\Gamma_j} \operatorname{curl} \mathbf{q}_I \cdot \mathbf{n}_I = 0 \qquad \forall\, j = 1, \dots, p_\Gamma\,,$$

this last equation being the Stokes theorem for a closed surface. □

Remark 3.4. The weak problem (3.16) is not suitable for numerical approximation, due to the constraint of orthogonality to $\mathcal{H}(\partial\Omega, \Gamma; \Omega_I)$ that appears in the definition of the space Y_I; note also that this condition clearly depends on the topology of the conductor Ω_C. In Section 5.5 and Remark 6.12 we will propose a couple of weak formulations for the determination of $\mathbf{E}_I$ that are more convenient for numerical computations. □

Remark 3.5. In many real-life problems, the determination of $\mathbf{E}_I$ is not mandatory, as the knowledge of the magnetic field $\mathbf{H}$ in Ω and of the electric field $\mathbf{E}_C$ in Ω_C is often sufficient to simulate the physical process. In those cases, solving (3.16) is therefore not necessary. □

Remark 3.6. In particular, we have proved that the solution to (3.16) satisfies

$$\begin{cases} \operatorname{curl}(\varepsilon_I^{-1} \operatorname{curl} \mathbf{q}_I) = -i\omega\boldsymbol{\mu}_I \mathbf{H}_I & \text{in } \Omega_I \\ \operatorname{div} \mathbf{q}_I = 0 & \text{in } \Omega_I \\ \mathbf{q}_I \times \mathbf{n} = \mathbf{0} & \text{on } \partial\Omega \\ \mathbf{q}_I \cdot \mathbf{n}_I = 0 & \text{on } \Gamma \\ (\varepsilon_I^{-1} \operatorname{curl} \mathbf{q}_I) \times \mathbf{n}_I = -\mathbf{E}_C \times \mathbf{n}_C & \text{on } \Gamma \\ \mathbf{q}_I \perp \mathcal{H}(\partial\Omega, \Gamma; \Omega_I) \,. \end{cases} \tag{3.19}$$

Similar existence results are an important step for proving the orthogonal decompositions of $(L^2(\Omega_I))^3$ that are presented in Section A.3. □

Let us come now to show that the solution $\mathbf{H}$ to (3.9), with

$$\mathbf{E}_C = \boldsymbol{\sigma}^{-1}(\operatorname{curl} \mathbf{H}_C - \mathbf{J}_{e,C}),$$

indeed satisfies (3.14). To do that, we only need to choose suitable test functions $\mathbf{v}$ in the weak formulation (3.9). First, take $\mathbf{v} = \operatorname{grad} \chi$, where $\chi \in C_0^\infty(\Omega)$. Clearly $\operatorname{grad} \chi \in V$, and then

$$\int_\Omega \boldsymbol{\mu}\mathbf{H} \cdot \operatorname{grad} \overline{\chi} = 0 \,.$$

Integrating by parts we find

$$\operatorname{div}(\boldsymbol{\mu}\mathbf{H}) = 0 \qquad \text{in } \Omega \,,$$

in particular

$$\operatorname{div}(\boldsymbol{\mu}_I \mathbf{H}_I) = 0 \qquad \text{in } \Omega_I, \tag{3.20}$$

and

$$\boldsymbol{\mu}_I \mathbf{H}_I \cdot \mathbf{n}_I + \boldsymbol{\mu}_C \mathbf{H}_C \cdot \mathbf{n}_C = 0 \quad \text{on } \Gamma \,. \tag{3.21}$$

The second relation in (3.14) needs a preliminary result. Choosing in (3.9) a test function $\mathbf{v} \in V$ such that $\mathbf{v}_C \in (C_0^\infty(\Omega_C))^3$ and $\mathbf{v}_I = \mathbf{0}$, we find

$$\int_{\Omega_C} (\boldsymbol{\sigma}^{-1} \operatorname{curl} \mathbf{H}_C \cdot \operatorname{curl} \overline{\mathbf{v}_C} + i\omega\boldsymbol{\mu}_C \mathbf{H}_C \cdot \overline{\mathbf{v}_C}) = \int_{\Omega_C} \boldsymbol{\sigma}^{-1} \mathbf{J}_{e,C} \cdot \operatorname{curl} \overline{\mathbf{v}_C} \,.$$

Thus by integration by parts

$$\operatorname{curl}(\boldsymbol{\sigma}^{-1} \operatorname{curl} \mathbf{H}_C) + i\omega\boldsymbol{\mu}_C \mathbf{H}_C = \operatorname{curl}(\boldsymbol{\sigma}^{-1} \mathbf{J}_{e,C}) \qquad \text{in } \Omega_C \,. \tag{3.22}$$

In conclusion, from (3.22) and (3.21) it follows

$$\begin{aligned}\mathrm{div}_\tau(\mathbf{E}_C \times \mathbf{n}_C) &= \mathrm{div}_\tau([\boldsymbol{\sigma}^{-1}(\mathrm{curl}\,\mathbf{H}_C - \mathbf{J}_{e,C})] \times \mathbf{n}_C) \\ &= \mathrm{curl}[\boldsymbol{\sigma}^{-1}(\mathrm{curl}\,\mathbf{H}_C - \mathbf{J}_{e,C})] \cdot \mathbf{n}_C \\ &= -i\omega\boldsymbol{\mu}_C \mathbf{H}_C \cdot \mathbf{n}_C = i\omega\boldsymbol{\mu}_I \mathbf{H}_I \cdot \mathbf{n}_I \quad \text{on } \Gamma .\end{aligned}$$

Coming to $(3.14)_3$, for each basis function $\mathrm{grad}\, z_{r,I}$ of the space of harmonic fields $\mathcal{H}_{\mu_I}(\partial\Omega, \Gamma; \Omega_I)$, $r = 1, \ldots, p_{\partial\Omega}$, let us denote by $\mathbf{v}_{r,C}$ a function belonging to $H(\mathrm{curl}; \Omega_C)$ and satisfying $\mathbf{v}_{r,C} \times \mathbf{n}_C + \mathrm{grad}\, z_{r,I} \times \mathbf{n}_I = \mathbf{0}$ on Γ. Then, the function

$$\mathbf{v}_r := \begin{cases} \mathrm{grad}\, z_{r,I} & \text{in } \Omega_I \\ \mathbf{v}_{r,C} & \text{in } \Omega_C \end{cases}$$

belongs to V. By proceeding as before and using also (3.21), we easily find

$$\int_{(\partial\Omega)_r} \boldsymbol{\mu}_I \mathbf{H}_I \cdot \mathbf{n} = 0 \quad \forall\, r = 1, \ldots, p_{\partial\Omega} .$$

Finally, denote by $\mathbf{R}_{l,C} \in H(\mathrm{curl}; \Omega_C)$ a function satisfying $\mathbf{R}_{l,C} \times \mathbf{n}_C + \boldsymbol{\rho}_{l,I} \times \mathbf{n}_I = \mathbf{0}$ on Γ, where the function $\boldsymbol{\rho}_{l,I}$, $l = 1, \ldots, n_\Gamma$, is a basis function of the space of harmonic fields $\mathcal{H}_{\mu_I}(\partial\Omega, \Gamma; \Omega_I)$. The function

$$\mathbf{v}_l := \begin{cases} \boldsymbol{\rho}_{l,I} & \text{in } \Omega_I \\ \mathbf{R}_{l,C} & \text{in } \Omega_C \end{cases}$$

belongs to V; thus choosing it as a test function in (3.9), integrating by parts and taking into account (3.22) one obtains at once

$$\begin{aligned}&\int_{\Omega_I} i\omega\boldsymbol{\mu}_I \mathbf{H}_I \cdot \boldsymbol{\rho}_{l,I} \\ &\quad = -\int_{\Omega_C} [\boldsymbol{\sigma}^{-1}(\mathrm{curl}\,\mathbf{H}_C - \mathbf{J}_{e,C}) \cdot \mathrm{curl}\,\mathbf{R}_{l,C} + i\omega\boldsymbol{\mu}_C \mathbf{H}_C \cdot \mathbf{R}_{l,C}] \\ &\quad = -\int_\Gamma [\boldsymbol{\sigma}^{-1}(\mathrm{curl}\,\mathbf{H}_C - \mathbf{J}_{e,C})] \times \mathbf{n}_C \cdot \mathbf{R}_{l,C} \\ &\quad = -\int_\Gamma [\boldsymbol{\sigma}^{-1}(\mathrm{curl}\,\mathbf{H}_C - \mathbf{J}_{e,C})] \times \mathbf{n}_C \cdot \boldsymbol{\rho}_{l,I} \\ &\quad = -\int_\Gamma (\mathbf{E}_C \times \mathbf{n}_C) \cdot \boldsymbol{\rho}_{l,I} \quad \forall\, l = 1, \ldots, n_\Gamma ,\end{aligned}$$

namely, $(3.14)_4$.

3.3 Strong formulation for the magnetic field

The results in Section 3.2 also furnish the complete strong formulation for the magnetic field.

In fact, since $(\mathbf{H} - \mathbf{H}_e) \in V$, the solution to (3.9) satisfies $\mathrm{curl}\,\mathbf{H}_I = \mathrm{curl}\,\mathbf{H}_{e,I} = \mathbf{J}_{e,I}$ in Ω_I, $\mathbf{H}_I \times \mathbf{n} = \mathbf{0}$ on $\partial\Omega$, and $\mathbf{H}_C \times \mathbf{n}_C + \mathbf{H}_I \times \mathbf{n}_I = \mathbf{0}$ on Γ. Hence we have obtained

$$\begin{cases} \operatorname{curl}(\boldsymbol{\sigma}^{-1} \operatorname{curl} \mathbf{H}_C) + i\omega\boldsymbol{\mu}_C\mathbf{H}_C \\ \qquad\qquad = \operatorname{curl}(\boldsymbol{\sigma}^{-1}\mathbf{J}_{e,C}) & \text{in } \Omega_C \\ \operatorname{curl} \mathbf{H}_I = \mathbf{J}_{e,I} & \text{in } \Omega_I \\ \operatorname{div}(\boldsymbol{\mu}_I\mathbf{H}_I) = 0 & \text{in } \Omega_I \\ \int_{(\partial\Omega)_r} \boldsymbol{\mu}_I\mathbf{H}_I \cdot \mathbf{n} = 0 & \forall\, r = 1, \ldots, p_{\partial\Omega} \\ \int_{\Omega_I} i\omega\boldsymbol{\mu}_I\mathbf{H}_I \cdot \boldsymbol{\rho}_{l,I} \\ \quad = -\int_\Gamma [\boldsymbol{\sigma}^{-1}(\operatorname{curl} \mathbf{H}_C - \mathbf{J}_{e,C})] \times \mathbf{n}_C \cdot \boldsymbol{\rho}_{l,I} & \forall\, l = 1, \ldots, n_\Gamma \\ \mathbf{H}_I \times \mathbf{n} = \mathbf{0} & \text{on } \partial\Omega \\ \boldsymbol{\mu}_I\mathbf{H}_I \cdot \mathbf{n}_I + \boldsymbol{\mu}_C\mathbf{H}_C \cdot \mathbf{n}_C = 0 & \text{on } \Gamma \\ \mathbf{H}_I \times \mathbf{n}_I + \mathbf{H}_C \times \mathbf{n}_C = \mathbf{0} & \text{on } \Gamma\,. \end{cases} \tag{3.23}$$

A simpler situation occurs when the computational domain Ω has a connected boundary, so that $p_{\partial\Omega} = 0$, and when the conductor Ω_C is simply-connected, so that $n_\Gamma = 0$. However, the latter is an assumption on the shape of the conductor and it can be rather restrictive, as in many relevant engineering applications Ω_C has a complex topology.

Problem (3.23) has a unique solution: existence has been just proved, and uniqueness follows from the fact that, taking the solution $\mathbf{H}$ to (3.23) together with the electric field $\mathbf{E}$ constructed as in Section 3.2, we can repeat the arguments in Section 3.1 and hence are able to prove that $\mathbf{H}$ is a solution to the weak problem (3.9). Since this last problem has a unique solution, the solution to (3.23) is also unique, and the strong problem (3.23) is therefore equivalent to the weak problem (3.9).

Equations $(3.23)_4$ and $(3.23)_5$ take into account the topology of Ω_I. The physical interpretation of $(3.23)_4$ is simply that there is no "magnetic charge" hidden in the "holes" of Ω (namely, in the regions surrounded by $(\partial\Omega)_r$, $r = 1, \ldots, p_{\partial\Omega}$). The interpretation of $(3.23)_5$ will be given in Section 3.3.1.

It should be noted that the matching condition $(3.23)_7$ is weaker than the continuity of the tangential component of $\mathbf{E}$, namely,

$$\mathbf{E}_I \times \mathbf{n}_I + \mathbf{E}_C \times \mathbf{n}_C = \mathbf{0} \quad \text{on } \Gamma\,. \tag{3.24}$$

In fact, $(3.23)_7$ can be obtained from (3.24) taking the tangential divergence and using the Faraday equation separately in Ω_I and Ω_C, but the converse is not guaranteed in general topological situations.

Remark 3.7. We emphasize that, if we drop condition $(3.23)_5$ from problem (3.23), the remaining problem is not well-posed. In fact, uniqueness does not hold, as it can be seen from the following argument. Let us assume for simplicity that $n_\Gamma = 1$ (namely, in this case Ω_C is a torus, and there is only one basis cycle γ_1 in Ω_I: see Figure 3.1).

Consider the space

$$V_0 := \left\{ \mathbf{v} \in H_0(\operatorname{curl};\Omega) \,|\, \operatorname{curl} \mathbf{v}_I = \mathbf{0} \text{ in } \Omega_I\,,\ \int_{\gamma_1} \mathbf{v}_I \cdot d\boldsymbol{\tau} = 0 \right\}\,.$$

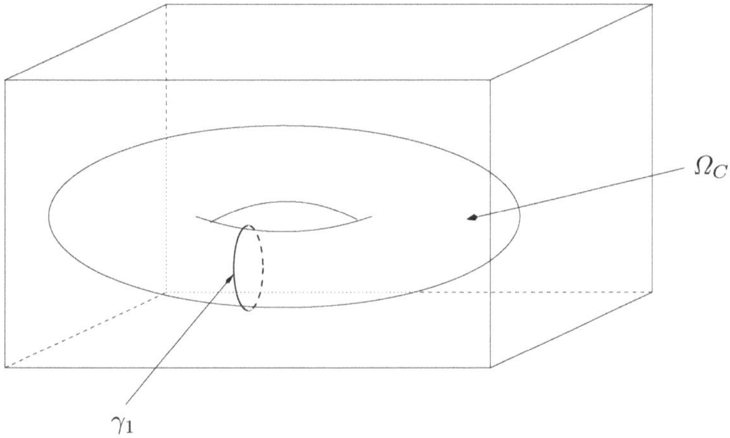

Fig. 3.1. The geometry of the problem

Using the Lax–Milgram lemma, for each complex number $q \neq 0$ one can find a unique solution of the problem

$$\mathbf{W} \in V_0 \ : \ a(\mathbf{W}, \mathbf{v}_0) = -qa(\boldsymbol{\rho}_1, \mathbf{v}_0) \quad \forall \, \mathbf{v}_0 \in V_0 \ ,$$

having defined $\boldsymbol{\rho}_1 \in H_0(\mathrm{curl}; \Omega)$ as

$$\boldsymbol{\rho}_1 := \begin{cases} \boldsymbol{\rho}_{1,I} & \text{in } \Omega_I \\ \mathbf{R}_{1,C} & \text{in } \Omega_C \ , \end{cases}$$

where $\boldsymbol{\rho}_{1,I}$ is the basis function of the space of harmonic fields $\mathcal{H}_{\mu_I}(\partial\Omega, \Gamma; \Omega_I)$ such that $\int_{\gamma_1} \boldsymbol{\rho}_{1,I} \cdot d\boldsymbol{\tau} = 1$, and the function $\mathbf{R}_{1,C} \in H(\mathrm{curl}; \Omega_C)$ satisfies $\mathbf{R}_{1,C} \times \mathbf{n}_C + \boldsymbol{\rho}_{1,I} \times \mathbf{n}_I = \mathbf{0}$ on Γ.

On the other hand, setting $\mathbf{H} := \mathbf{W} + q\boldsymbol{\rho}_1$, by proceeding as in Section 3.2 it is easily proved that, for each choice of the complex number q, $\mathbf{H}$ is a solution to $(3.23)_1$–$(3.23)_4$, $(3.23)_6$–$(3.23)_8$ for $\mathbf{J}_e = \mathbf{0}$ (it is enough to note that the test functions used in the proofs are always a gradient in Ω_I, therefore they satisfy the constraint $\int_{\gamma_1} \mathbf{v}_I \cdot d\boldsymbol{\tau} = 0$). Since $\int_{\gamma_1} \mathbf{H}_I \cdot d\boldsymbol{\tau} = q \neq 0$, it is apparent that $\mathbf{H} \neq \mathbf{0}$, thus uniqueness does not hold for $(3.23)_1$–$(3.23)_4$, $(3.23)_6$–$(3.23)_8$.

It is clear that by dropping $(3.23)_5$ we have lost the information determining the circulation of $\mathbf{H}_I$ along the basis cycle γ_1. By adding one equation for the circulation on γ_1 one could recover uniqueness (see Reissel [205]). But this would not yield a solution of the eddy current problem, as from the physical point of view what is really missing here is the Faraday equation for the surface which "cuts" γ_1 (see the following Section 3.3.1). □

Remark 3.8. In terms of the magnetic field $\mathbf{H}$ and of the electric field $\mathbf{E}_C$ introduced in (3.12), we can obviously rewrite (3.23) as

$$\begin{cases} \operatorname{curl} \mathbf{E}_C + i\omega\boldsymbol{\mu}_C \mathbf{H}_C = \mathbf{0} & \text{in } \Omega_C \\ \operatorname{curl} \mathbf{H}_C - \boldsymbol{\sigma}\mathbf{E}_C = \mathbf{J}_{e,C} & \text{in } \Omega_C \\ \operatorname{curl} \mathbf{H}_I = \mathbf{J}_{e,I} & \text{in } \Omega_I \\ \operatorname{div}(\boldsymbol{\mu}_I \mathbf{H}_I) = 0 & \text{in } \Omega_I \\ \int_{(\partial\Omega)_r} \boldsymbol{\mu}_I \mathbf{H}_I \cdot \mathbf{n} = 0 & \forall\, r = 1, \ldots, p_{\partial\Omega} \\ \int_{\Omega_I} i\omega\boldsymbol{\mu}_I \mathbf{H}_I \cdot \boldsymbol{\rho}_{l,I} = -\int_\Gamma \mathbf{E}_C \times \mathbf{n}_C \cdot \boldsymbol{\rho}_{l,I} & \forall\, l = 1, \ldots, n_\Gamma \\ \mathbf{H}_I \times \mathbf{n} = \mathbf{0} & \text{on } \partial\Omega \\ \boldsymbol{\mu}_I \mathbf{H}_I \cdot \mathbf{n}_I + \boldsymbol{\mu}_C \mathbf{H}_C \cdot \mathbf{n}_C = 0 & \text{on } \Gamma \\ \mathbf{H}_I \times \mathbf{n}_I + \mathbf{H}_C \times \mathbf{n}_C = \mathbf{0} & \text{on } \Gamma\,, \end{cases} \tag{3.25}$$

a formulation that will be useful in Chapters 6 and 7. □

3.3.1 The Faraday equation for the "cutting" surfaces

We want to clarify the physical meaning of the conditions

$$\int_{\Omega_I} i\omega\boldsymbol{\mu}_I \mathbf{H}_I \cdot \boldsymbol{\rho}_{l,I} = -\int_\Gamma [\boldsymbol{\sigma}^{-1}(\operatorname{curl} \mathbf{H}_C - \mathbf{J}_{e,C})] \times \mathbf{n}_C \cdot \boldsymbol{\rho}_{l,I} \quad \forall\, l = 1, \ldots, n_\Gamma\,.$$

The Faraday law, in its integral form, relates the flux of $i\omega\boldsymbol{\mu}\mathbf{H}$ through a surface with the line integral of $\mathbf{E}$ along the boundary of the surface. Since in (3.23) the Faraday equation is imposed in differential form in Ω_C, it is satisfied in integral form for any surface contained in Ω_C with boundary contained in $\overline{\Omega_C}$. Moreover, in formulation (3.23) the electric field in Ω_I is not yet available, and will be determined through (3.13). Thus we need not worry about the Faraday equation for surfaces in Ω_I with boundary in the interior of Ω_I.

But, since the electric field $\mathbf{E}_C = \boldsymbol{\sigma}^{-1}(\operatorname{curl} \mathbf{H}_C - \mathbf{J}_{e,C})$ has already been determined in $\overline{\Omega_C}$, and thus on Γ, we have indeed to take into consideration each surface Σ in Ω_I with boundary $\partial\Sigma$ on Γ.

If the boundary $\partial\Sigma$ is also the boundary of a surface $\hat{\Sigma}$ contained in Ω_C, we can argue in this way: due to the divergence-free condition for $i\omega\boldsymbol{\mu}\mathbf{H}$ in Ω, the flux of $i\omega\boldsymbol{\mu}_I\mathbf{H}_I$ through Σ is the same than the flux of $i\omega\boldsymbol{\mu}_C\mathbf{H}_C$ through $\hat{\Sigma}$, and, due to the Faraday equation in Ω_C, this flux equals the line integral of $\mathbf{E}_C$ on $\partial\hat{\Sigma} = \partial\Sigma$. Therefore, we have verified that the Faraday equation is satisfied for this type of surfaces (see Figure 3.2).

Having reached this point, the final question is: are there surfaces Σ in Ω_I with boundary on Γ such that $\partial\Sigma$ is not the boundary of any surface contained in Ω_C? If the answer is "yes", for this type of surfaces we have not yet imposed the Faraday equation, even if all the equations in $(3.23)_1$–$(3.23)_4$, $(3.23)_6$–$(3.23)_8$ are satisfied.

In general topological situations, these surfaces exist: they are the "cutting" surfaces Ξ_l, $l = 1, \ldots, n_\Gamma$, described in Section A.4. In fact, it is known that there is a kind of "duality" between "cutting" surfaces and non-bounding cycles: each "cutting"

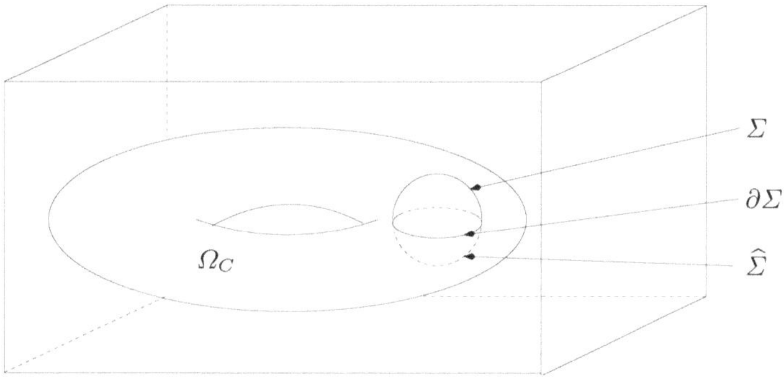

Fig. 3.2. A surface for which the Faraday equation is satisfied

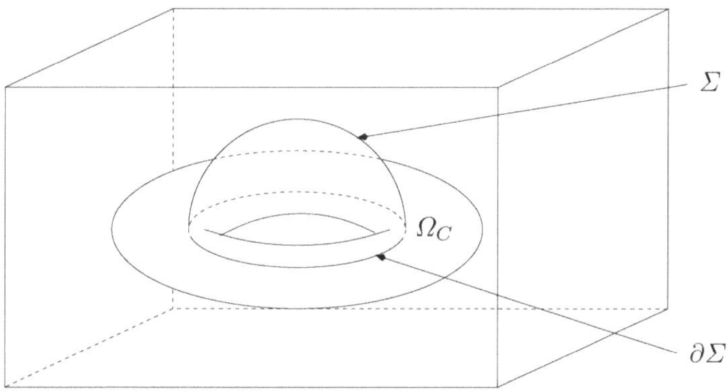

Fig. 3.3. A surface for which the Faraday equation is not satisfied

surface "cuts" a $\partial\Omega$ independent non bounding cycle in Ω_I, and, at the same time, its boundary is a non-bounding cycle in $\overline{\Omega_C}$ (see Figure 3.3).

We want thus to prove that conditions $(3.23)_5$ are equivalent, for a (regular enough) solution to $(3.23)_1$–$(3.23)_4$, $(3.23)_6$–$(3.23)_8$, to imposing the integral form of the Faraday equation for the surfaces Ξ_l and their boundaries $\partial\Xi_l$.

From Section A.4 (see in particular (A.34)), we know that the basis functions $\rho_{l,I}$ can be represented as the $(L^2(\Omega_I))^3$-extensions of $\operatorname{grad} p_{l,I}$, where $p_{l,I}$ are the solutions to

$$\begin{cases} \operatorname{div}(\boldsymbol{\mu}_I \operatorname{grad} p_{l,I}) = 0 & \text{in } \Omega_I \setminus \Xi_l \\ \boldsymbol{\mu}_I \operatorname{grad} p_{l,I} \cdot \mathbf{n}_I = 0 & \text{on } \Gamma \setminus \partial\Xi_l \\ p_{l,I} = 0 & \text{on } \partial\Omega \\ [\boldsymbol{\mu}_I \operatorname{grad} p_{l,I} \cdot \mathbf{n}_l]_{\Xi_l} = 0 & \\ [p_{l,I}]_{\Xi_l} = 1 \, , & \end{cases}$$

having denoted by $\mathbf{n}_l$ the unit normal vector on Ξ_l, and by $[\,\cdot\,]_{\Xi_l}$ the jump across the surface Ξ_l (say, $[p_{l,I}(\mathbf{x})]_{\Xi_l} = \lim_{s\to 0^+}[p_{l,I}(\mathbf{x} - s\mathbf{n}_l) - p_{l,I}(\mathbf{x} + s\mathbf{n}_l)]$).

For a solution to $(3.23)_1$–$(3.23)_4$, $(3.23)_6$–$(3.23)_8$, regular enough to give a meaning to the integrals we are planning to write, using integration by parts we have

$$\begin{aligned}\int_{\Omega_I} i\omega\boldsymbol{\mu}_I\mathbf{H}_I\cdot\boldsymbol{\rho}_{l,I} &= \int_{\Omega_I\setminus\Xi_l} i\omega\boldsymbol{\mu}_I\mathbf{H}_I\cdot\operatorname{grad} p_{l,I}\\ &= -\int_{\Omega_I\setminus\Xi_l} i\omega\operatorname{div}(\boldsymbol{\mu}_I\mathbf{H}_I)\,p_{l,I} + \int_{\partial\Omega} i\omega\boldsymbol{\mu}_I\mathbf{H}_I\cdot\mathbf{n}\,p_{l,I}\\ &\quad+\int_{\Gamma\setminus\partial\Xi_l} i\omega\boldsymbol{\mu}_I\mathbf{H}_I\cdot\mathbf{n}_I\,p_{l,I} + \int_{\Xi_l} i\omega\boldsymbol{\mu}_I\mathbf{H}_I\cdot\mathbf{n}_l\,[p_{l,I}]\\ &= \int_{\Gamma\setminus\partial\Xi_l} i\omega\boldsymbol{\mu}_I\mathbf{H}_I\cdot\mathbf{n}_I\,p_{l,I} + \int_{\Xi_l} i\omega\boldsymbol{\mu}_I\mathbf{H}_I\cdot\mathbf{n}_l,\end{aligned}$$

and

$$\begin{aligned}-\int_\Gamma(\mathbf{E}_C\times\mathbf{n}_C)\cdot\boldsymbol{\rho}_{l,I} &= -\int_{\Gamma\setminus\partial\Xi_l}(\mathbf{E}_C\times\mathbf{n}_C)\cdot\operatorname{grad} p_{l,I}\\ &= \int_{\Gamma\setminus\partial\Xi_l}\operatorname{div}_\tau(\mathbf{E}_C\times\mathbf{n}_C)\,p_{l,I} - \int_{\partial\Xi_l\cap\Gamma}(\mathbf{E}_C\times\mathbf{n}_C)\cdot\mathbf{n}_l\,[p_{l,I}]\\ &= \int_{\Gamma\setminus\partial\Xi_l}\operatorname{curl}\mathbf{E}_C\cdot\mathbf{n}_C\,p_{l,I} - \int_{\partial\Xi_l\cap\Gamma}(\mathbf{E}_C\times\mathbf{n}_C)\cdot\mathbf{n}_l\\ &= -\int_{\Gamma\setminus\partial\Xi_l} i\omega\boldsymbol{\mu}_C\mathbf{H}_C\cdot\mathbf{n}_C\,p_{l,I} - \int_{\partial\Xi_l\cap\Gamma}\mathbf{E}_C\cdot(\mathbf{n}_C\times\mathbf{n}_l)\\ &= \int_{\Gamma\setminus\partial\Xi_l} i\omega\boldsymbol{\mu}_I\mathbf{H}_I\cdot\mathbf{n}_I\,p_{l,I} - \int_{\partial\Xi_l\cap\Gamma}\mathbf{E}_C\cdot(\mathbf{n}_C\times\mathbf{n}_l)\,.\end{aligned}$$

Since $\partial\Xi_l\subset\Gamma$, we see that $(3.23)_5$ are equivalent to

$$\int_{\Xi_l} i\omega\boldsymbol{\mu}_I\mathbf{H}_I\cdot\mathbf{n}_l = -\int_{\partial\Xi_l}\mathbf{E}_C\cdot(\mathbf{n}_C\times\mathbf{n}_l) = -\int_{\partial\Xi_l}\mathbf{E}_C\cdot d\boldsymbol{\tau}\,,$$

namely, the Faraday equation for the surface Ξ_l.

Remark 3.9. We note that the additional conditions $(3.23)_5$ are not needed if the time-harmonic full Maxwell equations are considered. In fact, in that case the problem reads

$$\begin{cases}\operatorname{curl}\mathbf{H} - (i\omega\boldsymbol{\varepsilon}+\boldsymbol{\sigma})\mathbf{E} = \mathbf{J}_e & \text{in } \Omega\\ \operatorname{curl}\mathbf{E} + i\omega\boldsymbol{\mu}\mathbf{H} = \mathbf{0} & \text{in } \Omega\\ \mathbf{H}\times\mathbf{n} = \mathbf{0} & \text{on } \partial\Omega\,.\end{cases}$$

Since the matrix $\boldsymbol{\eta} := i\omega\boldsymbol{\varepsilon}+\boldsymbol{\sigma}$ is non-singular, we can rewrite the problem in terms of $\mathbf{H}$ only

$$\begin{cases}\operatorname{curl}(\boldsymbol{\eta}^{-1}\operatorname{curl}\mathbf{H}) + i\omega\boldsymbol{\mu}\mathbf{H} = \operatorname{curl}(\boldsymbol{\eta}^{-1}\mathbf{J}_e) & \text{in } \Omega\\ \mathbf{H}\times\mathbf{n} = \mathbf{0} & \text{on } \partial\Omega\,.\end{cases} \tag{3.26}$$

Problem (3.26) is uniquely solvable (see, for instance, Alonso and Valli [8], Alonso and Raffetto [15]), and setting $\mathbf{E} = \boldsymbol{\eta}^{-1}(\operatorname{curl}\mathbf{H} - \mathbf{J}_e)$ one verifies at once that the Faraday equation is satisfied in all of Ω: no additional condition related the geometry of Ω_I is coming into play. □

3.3.2 Suitability of other formulations

We want to investigate whether or not some frequently-used formulations for eddy current problems furnish a magnetic field $\mathbf{H}$ that satisfies $(3.23)_5$, namely,

$$\begin{aligned}&\int_{\Omega_I} i\omega\boldsymbol{\mu}_I\mathbf{H}_I\cdot\boldsymbol{\rho}_{l,I}\\ &\qquad= -\int_\Gamma[\boldsymbol{\sigma}^{-1}(\operatorname{curl}\mathbf{H}_C - \mathbf{J}_{e,C})]\times\mathbf{n}_C\cdot\boldsymbol{\rho}_{l,I}\quad\forall\, l=1,\ldots,n_\Gamma\,.\end{aligned}$$

(i) The $\mathbf{A}_C^* - \mathbf{A}_I$ formulation.

This formulation, reported in Bíró [48], is based on the unknowns $\mathbf{A}_C^*$ and $\mathbf{A}_I$ such that

$$i\omega\mathbf{A}_C^* = -\mathbf{E}_C \ , \quad \operatorname{curl}\mathbf{A}_I = \boldsymbol{\mu}_I\mathbf{H}_I \ ,$$

with the interface conditions on Γ

$$\begin{aligned}&\mathbf{A}_C^* \times \mathbf{n}_C + \mathbf{A}_I \times \mathbf{n}_I = \mathbf{0}\\ &(\boldsymbol{\mu}_C^{-1}\operatorname{curl}\mathbf{A}_C^*) \times \mathbf{n}_C + (\boldsymbol{\mu}_I^{-1}\operatorname{curl}\mathbf{A}_I) \times \mathbf{n}_I = \mathbf{0}\ .\end{aligned}$$

By integration by parts we have

$$\begin{aligned}\int_{\Omega_I} i\omega\boldsymbol{\mu}_I\mathbf{H}_I \cdot \boldsymbol{\rho}_{l,I} &= \int_{\Omega_I} i\omega\operatorname{curl}\mathbf{A}_I \cdot \boldsymbol{\rho}_{l,I} = i\omega\int_\Gamma(\mathbf{n}_I \times \mathbf{A}_I)\cdot\boldsymbol{\rho}_{l,I}\\ &= i\omega\int_\Gamma(\mathbf{A}_C^* \times \mathbf{n}_C)\cdot\boldsymbol{\rho}_{l,I} = -\int_\Gamma(\mathbf{E}_C \times \mathbf{n}_C)\cdot\boldsymbol{\rho}_{l,I}\ ,\end{aligned} \tag{3.27}$$

therefore condition $(3.23)_5$ is satisfied.

(ii) The $(\mathbf{A}_C, V_C) - \mathbf{A}_I$ formulation.

This formulation, reported in Bíró [48] and presented in detail in Chapter 6, is based on the unknowns $(\mathbf{A}_C, V_C)$ and $\mathbf{A}_I$ such that

$$i\omega\mathbf{A}_C + \operatorname{grad}V_C = -\mathbf{E}_C \ , \quad \operatorname{curl}\mathbf{A}_I = \boldsymbol{\mu}_I\mathbf{H}_I \ ,$$

with the interface conditions on Γ

$$\begin{aligned}&\mathbf{A}_C \times \mathbf{n}_C + \mathbf{A}_I \times \mathbf{n}_I = \mathbf{0}\\ &(\boldsymbol{\mu}_C^{-1}\operatorname{curl}\mathbf{A}_C) \times \mathbf{n}_C + (\boldsymbol{\mu}_I^{-1}\operatorname{curl}\mathbf{A}_I) \times \mathbf{n}_I = \mathbf{0}\ .\end{aligned}$$

With respect to the preceding case, on the right-hand side of (3.27) the only additional term is

$$-\int_\Gamma(\operatorname{grad}V_C \times \mathbf{n}_C)\cdot\boldsymbol{\rho}_{l,I}\ ;$$

however, $(3.23)_5$ is still satisfied, as the term above indeed vanishes. In fact, by integration by parts we have

$$\begin{aligned}&\int_\Gamma(\operatorname{grad}V_C \times \mathbf{n}_C)\cdot\boldsymbol{\rho}_{l,I}\\ &= \int_\Gamma(\boldsymbol{\rho}_{l,I} \times \mathbf{n}_I)\cdot\operatorname{grad}V_C\\ &= -\int_\Gamma \operatorname{div}_\tau(\boldsymbol{\rho}_{l,I} \times \mathbf{n}_I)\, V_{C|\Gamma}\\ &= -\int_\Gamma \operatorname{curl}\boldsymbol{\rho}_{l,I}\cdot\mathbf{n}_I\, V_{C|\Gamma} = 0\ ,\end{aligned}$$

as $\operatorname{curl}\boldsymbol{\rho}_{l,I} = \mathbf{0}$.

(iii) The $(\mathbf{T}_C^*, \Phi_C) - \mathbf{A}_I$ formulation.

This formulation, reported in Bíró [48] and described in Section 6.3, is based on the unknowns $(\mathbf{T}_C^*, \Phi_C)$ and $\mathbf{A}_I$ such that

$$\mathbf{T}_C^* + \operatorname{grad}\Phi_C = \mathbf{H}_C \ , \quad \operatorname{curl}\mathbf{A}_I = \boldsymbol{\mu}_I\mathbf{H}_I \ ,$$

with the interface conditions on Γ

$$\begin{aligned}
&[\boldsymbol{\sigma}^{-1}(\operatorname{curl} \mathbf{T}_C^* - \mathbf{J}_{e,C})] \times \mathbf{n}_C - i\omega \mathbf{A}_I \times \mathbf{n}_I = \mathbf{0} \\
&(\mathbf{T}_C^* + \operatorname{grad} \Phi_C) \times \mathbf{n}_C + (\boldsymbol{\mu}_I^{-1} \operatorname{curl} \mathbf{A}_I) \times \mathbf{n}_I = \mathbf{0} \,.
\end{aligned}$$

Therefore,

$$\begin{aligned}
\textstyle\int_{\Omega_I} i\omega \boldsymbol{\mu}_I \mathbf{H}_I \cdot \boldsymbol{\rho}_{l,I} &= \textstyle\int_{\Omega_I} i\omega \operatorname{curl} \mathbf{A}_I \cdot \boldsymbol{\rho}_{l,I} = i\omega \int_\Gamma (\mathbf{n}_I \times \mathbf{A}_I) \cdot \boldsymbol{\rho}_{l,I} \\
&= -\textstyle\int_\Gamma [\boldsymbol{\sigma}^{-1}(\operatorname{curl} \mathbf{T}_C^* - \mathbf{J}_{e,C})] \times \mathbf{n}_C \cdot \boldsymbol{\rho}_{l,I} \\
&= -\textstyle\int_\Gamma [\boldsymbol{\sigma}^{-1}(\operatorname{curl} \mathbf{H}_C - \mathbf{J}_{e,C})] \times \mathbf{n}_C \cdot \boldsymbol{\rho}_{l,I} \,,
\end{aligned}$$

that is $(3.23)_5$.

(iv) The $(\mathbf{H}_C, \mathbf{J}_C) - \mathbf{H}_I$ formulation.

Kanayama and Kikuchi [146] reported (and used in numerical computations) a formulation in which the unknowns are the magnetic field and the eddy current $\mathbf{J}_C = \boldsymbol{\sigma}\mathbf{E}_C + \mathbf{J}_{e,C}$, and the interface conditions on Γ are given by

$$\begin{aligned}
&\mathbf{H}_C \times \mathbf{n}_C + \mathbf{H}_I \times \mathbf{n}_I = \mathbf{0} \\
&\boldsymbol{\mu}_C \mathbf{H}_C \cdot \mathbf{n}_C + \boldsymbol{\mu}_I \mathbf{H}_I \cdot \mathbf{n}_I = 0 \,.
\end{aligned} \tag{3.28}$$

In this case condition $(3.23)_5$ is not imposed; therefore, this formulation cannot be employed for a general domain Ω without explicitly adding $(3.23)_5$, that, with respect to $\mathbf{J}_C$ and $\mathbf{H}_I$, reads

$$\int_{\Omega_I} i\omega \boldsymbol{\mu}_I \mathbf{H}_I \cdot \boldsymbol{\rho}_{l,I} + \int_\Gamma [\boldsymbol{\sigma}^{-1}(\mathbf{J}_C - \mathbf{J}_{e,C})] \times \mathbf{n}_C \cdot \boldsymbol{\rho}_{l,I} = 0$$

for $l = 1, \ldots, n_\Gamma$.

Clearly, the same considerations also apply to the $(\mathbf{H}_C, \mathbf{E}_C) - \mathbf{H}_I$ formulation (3.25), which is formally equivalent to the one here considered.

(v) The formulation considered by Reissel.

Reissel [205] considered the formulation given by $(3.23)_1$–$(3.23)_3$, $(3.23)_7$–$(3.23)_8$, namely, without $(3.23)_5$ (more precisely, the domain Ω_I is assumed to be an exterior domain, hence $\partial\Omega$ is empty, and conditions $(3.23)_4$ and $(3.23)_6$ are eliminated; on the other hand, the magnetic field $\mathbf{H}_I$ is assumed to vanish as $|\mathbf{x}|$ goes to ∞). The uniqueness of the magnetic field is obtained by imposing the additional conditions

$$\int_{\gamma_l} \mathbf{H}_I \cdot d\boldsymbol{\tau} = I_l^0 \ , \ \forall \, l = 1, \ldots, n_\Gamma \,, \tag{3.29}$$

where γ_l are the non-bounding cycles in $\overline{\Omega_I}$ (see Section 1.4).

Concerning this formulation, a remark is in order: as we will see in the following Section 3.4, the solution to the eddy current problem (1.22) is unique, therefore we are not free to impose another condition like (3.29). As a matter of fact, from a physical viewpoint $\int_{\gamma_l} \mathbf{H}_I \cdot d\boldsymbol{\tau}$ is the total intensity current crossing any surface having γ_l as a boundary (by the Ampère law): hence, in the present context, it is a quantity to be determined by solving the problem and not a datum that can be arbitrarily prescribed. (For problems in which the total intensity current is imposed, see Chapter 8: there the boundary conditions are different from those considered here, or else the given current density $\mathbf{J}_e$ is not a datum but an additional unknown that must be determined.)

Let us set $I_l^* := \int_{\gamma_l} \mathbf{H}_I \cdot d\boldsymbol{\tau}$, where $\mathbf{H}$ is the solution of the eddy current problem (1.22). Clearly, if the datum I_l^0 is different from I_l^* the solution of the problem proposed by Reissel is not the solution of the eddy current problem. Since all the other equations hold, what is wrong about the Reissel solution is that it does not satisfy the topological condition $(3.23)_5$: in other words, it does not satisfy the Faraday law on the surface Ξ_l which "cuts" the $\partial\Omega$-independent non-bounding cycle γ_l.

Other potential formulations for eddy current problems will be presented and analyzed in Section 6.3; in particular, there we will verify whether for those formulations condition $(3.23)_5$ is automatically satisfied or has to be explicitly imposed.

3.4 Existence and uniqueness for the complete eddy current model

This section is devoted to verify that the results of the preceding sections easily give the unique solution of the complete eddy current model.

First of all, let us show that putting together (3.13) and (3.25) is somehow redundant. In fact, $(3.25)_4$ and $(3.25)_5$ are a consequence of $(3.13)_1$ (the latter one by means of the Stokes theorem for closed surfaces). Also, from $(3.13)_6$ we have

$$\operatorname{div}_\tau(\mathbf{E}_I \times \mathbf{n}_I + \mathbf{E}_C \times \mathbf{n}_C) = 0 \quad \text{on } \Gamma \,,$$

thus

$$\operatorname{curl} \mathbf{E}_I \cdot \mathbf{n}_I + \operatorname{curl} \mathbf{E}_C \cdot \mathbf{n}_C = 0 \quad \text{on } \Gamma \,;$$

hence from $(3.13)_1$ and $(3.25)_1$ we obtain at once $(3.25)_8$. Finally, from $(3.13)_1$ and $(3.13)_6$ we obtain

$$\begin{aligned} \int_{\Omega_I} i\omega\boldsymbol{\mu}_I \mathbf{H}_I \cdot \boldsymbol{\rho}_{l,I} &= -\int_{\Omega_I} \operatorname{curl} \mathbf{E}_I \cdot \boldsymbol{\rho}_{l,I} = \int_{\partial\Omega_I} \mathbf{E}_I \times \mathbf{n}_I \cdot \boldsymbol{\rho}_{l,I} \\ &= -\int_{\Gamma} \mathbf{E}_C \times \mathbf{n}_C \cdot \boldsymbol{\rho}_{l,I} \,, \end{aligned}$$

namely, $(3.25)_6$.

Therefore, we can rewrite the global problem in the non-redundant form (1.22), namely,

$$\begin{cases} \operatorname{curl} \mathbf{H} - \boldsymbol{\sigma}\mathbf{E} = \mathbf{J}_e & \text{in } \Omega \\ \operatorname{curl} \mathbf{E} + i\omega\boldsymbol{\mu}\mathbf{H} = \mathbf{0} & \text{in } \Omega \\ \operatorname{div}(\boldsymbol{\varepsilon}_I \mathbf{E}_I) = 0 & \text{in } \Omega_I \\ \mathbf{H} \times \mathbf{n} = \mathbf{0} & \text{on } \partial\Omega \\ \boldsymbol{\varepsilon}_I \mathbf{E}_I \cdot \mathbf{n} = 0 & \text{on } \partial\Omega \\ \int_{\Gamma_j} \boldsymbol{\varepsilon}_I \mathbf{E}_I \cdot \mathbf{n}_I = 0 & \forall\, j = 1, \ldots, p_\Gamma \\ \int_{\Omega_I} \boldsymbol{\varepsilon}_I \mathbf{E}_I \cdot \boldsymbol{\pi}_{k,I} = 0 & \forall\, k = 1, \ldots, n_{\partial\Omega} \,. \end{cases} \tag{3.30}$$

Notice that all the equations of problem (3.13) are present, while those we have dropped are essentially the compatibility conditions of problem (3.13).

We conclude this section with the following theorem.

Theorem 3.10. *Problem* (3.30) *has a unique solution. Moreover,* $(\mathbf{E}, \mathbf{H})$ *is the solution to* (3.30) *if and only if* $\mathbf{H}$ *is the solution to* (3.23), $\mathbf{E}_C$ *is obtained by* (3.12) *and* $\mathbf{E}_I$ *is the solution to* (3.13).

Proof. We have already seen that a solution to (3.30) is given by $\mathbf{H}$, the solution to (3.9) (or, equivalently, to (3.23)), by $\mathbf{E}_C$ defined in (3.12) and by the solution $\mathbf{E}_I$ to (3.13). On the other hand, since any solution to (3.30) gives the solution to (3.9), the magnetic field $\mathbf{H}$ is uniquely determined. Consequently, using (3.12) also $\mathbf{E}_C$ is unique. Finally, uniqueness of $\mathbf{E}_I$ follows from that of problem (3.13).

The second statement follows straightforwordly noting that, if $(\mathbf{E}, \mathbf{H})$ is the solution to (3.30), then $\mathbf{H}$ is the unique solution to (3.23). □

Remark 3.11. The regularity of the solution of the eddy current problem (3.30) is not easy to be determined. In fact, due to the jump of the conductivity $\boldsymbol{\sigma}$ through Γ, (3.30) is essentially an interface problem, and in principle its solution is not very regular even if the coefficients $\boldsymbol{\mu}$, ε_I and $\boldsymbol{\sigma}$ are smooth scalar functions. Moreover, if Ω_C and Ω are polyhedral domains the solution can exhibit corner and edge singularities.

Since the general regularity result is rather technical, we do not state it here and refer to Costabel et al. [90] for a detailed presentation and a thorough analysis of this issue. □

3.5 Other boundary conditions

Analogous results to those presented in Sections 3.1, 3.2, 3.3 and 3.4 can be obtained for the eddy current problem in which the electric field $\mathbf{E}$ is subjected to the electric boundary condition

$$\mathbf{E} \times \mathbf{n} = \mathbf{0} \quad \text{on } \partial\Omega \,, \tag{3.31}$$

or to the no-flux boundary conditions

$$\begin{cases} \mu\mathbf{H} \cdot \mathbf{n} = 0 & \text{on } \partial\Omega \\ \varepsilon\mathbf{E} \cdot \mathbf{n} = 0 & \text{on } \partial\Omega \,. \end{cases} \tag{3.32}$$

Let us underline what we have to modify in the formulations and in the proofs, starting from the electric boundary condition (3.31). The first remark concerns the (necessary) assumptions on the current density. As before, we require that $\mathbf{J}_e \in (L^2(\Omega))^3$ and satisfies $\operatorname{div} \mathbf{J}_{e,I} = 0$ in Ω_I, but now, instead of $(3.1)_2$ and (3.2), we need

$$\begin{array}{ll} \int_{\Gamma_j} \mathbf{J}_{e,I} \cdot \mathbf{n}_I = 0 & \forall\, j = 1, \dots, p_\Gamma \\ \int_{(\partial\Omega)_r} \mathbf{J}_{e,I} \cdot \mathbf{n} = 0 & \forall\, r = 0, \dots, p_{\partial\Omega} \end{array} \tag{3.33}$$

(see Section 1.5).

The existence of a vector field $\mathbf{H}^*_{e,I} \in H(\mathrm{curl};\Omega_I)$ such that

$$\mathrm{curl}\,\mathbf{H}^*_{e,I} = \mathbf{J}_{e,I} \qquad \text{in } \Omega_I \tag{3.34}$$

is then ensured by well-known results (see, e.g., Alonso and Valli [6], Rem. 4.3), and we can also find a vector field $\mathbf{H}^*_{e,C} \in H(\mathrm{curl};\Omega_C)$ such that $\mathbf{H}^*_{e,C} \times \mathbf{n}_C + \mathbf{H}^*_{e,I} \times \mathbf{n}_I = \mathbf{0}$ on Γ. Summing up, we have constructed a vector field $\mathbf{H}^*_e \in H(\mathrm{curl};\Omega)$, continuously dependent on $\mathbf{J}_{e,I}$, given by

$$\mathbf{H}^*_e := \begin{cases} \mathbf{H}^*_{e,I} & \text{in } \Omega_I \\ \mathbf{H}^*_{e,C} & \text{in } \Omega_C \,. \end{cases} \tag{3.35}$$

Concerning the weak problem for the magnetic field $\mathbf{H}$, it becomes

$$\begin{aligned} &\text{Find } (\mathbf{H} - \mathbf{H}^*_e) \in V^* \text{ such that} \\ &a(\mathbf{H}, \mathbf{v}^*) = \textstyle\int_{\Omega_C} \boldsymbol{\sigma}^{-1}\mathbf{J}_{e,C} \cdot \mathrm{curl}\,\overline{\mathbf{v}^*_C} \\ &\text{for each } \mathbf{v}^* \in V^*, \end{aligned} \tag{3.36}$$

where

$$V^* := \{\mathbf{v}^* \in H(\mathrm{curl};\Omega) \,|\, \mathrm{curl}\,\mathbf{v}^*_I = \mathbf{0} \text{ in } \Omega_I\} \,. \tag{3.37}$$

As before, the unique solvability of this problem is ensured by the Lax–Milgram lemma.

Looking for the strong formulation, the electric field $\mathbf{E}_C$ in Ω_C is defined as usual by setting $\mathbf{E}_C = \boldsymbol{\sigma}^{-1}(\mathrm{curl}\,\mathbf{H}_C - \mathbf{J}_{e,C})$, while the electric field $\mathbf{E}_I$ now has to verify

$$\begin{cases} \mathrm{curl}\,\mathbf{E}_I = -i\omega\boldsymbol{\mu}_I\mathbf{H}_I & \text{in } \Omega_I \\ \mathrm{div}(\boldsymbol{\varepsilon}_I\mathbf{E}_I) = 0 & \text{in } \Omega_I \\ \mathbf{E}_I \times \mathbf{n} = \mathbf{0} & \text{on } \partial\Omega \\ \int_{\Gamma_j} \boldsymbol{\varepsilon}_I\mathbf{E}_I \cdot \mathbf{n}_I = 0 & \forall\, j = 1,\dots,p_\Gamma \\ \int_{(\partial\Omega)_r} \boldsymbol{\varepsilon}_I\mathbf{E}_I \cdot \mathbf{n} = 0 & \forall\, r = 0,1,\dots,p_{\partial\Omega} \\ \mathbf{E}_I \times \mathbf{n}_I = -\mathbf{E}_C \times \mathbf{n}_C & \text{on } \Gamma\,. \end{cases} \tag{3.38}$$

This problem is simpler if the boundaries of Ω and Ω_C are connected, so that $p_{\partial\Omega} = 0$ and $p_\Gamma = 0$: say, the computational domain and the conductor are connected and have no "holes" (but possibly have "handles").

It can be shown that there exists a unique solution $\mathbf{E}_I$ to (3.38), given by $\mathbf{E}_I := \boldsymbol{\varepsilon}_I^{-1}\,\mathrm{curl}\,\mathbf{q}^*_I$, where $\mathbf{q}^*_I \in Y^*_I$ is the solution to the problem

$$\begin{aligned} &\textstyle\int_{\Omega_I} (\boldsymbol{\varepsilon}_I^{-1}\,\mathrm{curl}\,\mathbf{q}^*_I \cdot \mathrm{curl}\,\overline{\mathbf{p}^*_I} + \mathrm{div}\,\mathbf{q}^*_I\;\mathrm{div}\,\overline{\mathbf{p}^*_I}) \\ &\quad = -\textstyle\int_{\Omega_I} i\omega\boldsymbol{\mu}_I\mathbf{H}_I \cdot \overline{\mathbf{p}^*_I} - \int_\Gamma \mathbf{E}_C \times \mathbf{n}_C \cdot \overline{\mathbf{p}^*_I} \qquad \text{for each } \mathbf{p}^*_I \in Y^*_I\,, \end{aligned} \tag{3.39}$$

where

$$Y^*_I := \left\{\mathbf{p}^*_I \in H(\mathrm{curl};\Omega_I) \cap H_0(\mathrm{div};\Omega_I) \,|\, \mathbf{p}^*_I \perp \mathcal{H}(m;\Omega_I)\right\}, \tag{3.40}$$

and the space of harmonic fields $\mathcal{H}(m;\Omega_I)$ has been introduced in Section 1.4.

The compatibility conditions ensuring the solvability of problem (3.39) are now

$$\begin{array}{ll} \operatorname{div}(i\omega\boldsymbol{\mu}_I\mathbf{H}_I) = 0 & \text{in } \Omega_I \\ i\omega\boldsymbol{\mu}_I\mathbf{H}_I\cdot\mathbf{n}_I = \operatorname{div}_\tau(\mathbf{E}_C\times\mathbf{n}_C) & \text{on } \Gamma \\ i\omega\boldsymbol{\mu}_I\mathbf{H}_I\cdot\mathbf{n} = 0 & \text{on } \partial\Omega \\ \int_{\Omega_I} i\omega\boldsymbol{\mu}_I\mathbf{H}_I\cdot\boldsymbol{\rho}^*_{\alpha,I} & \\ \quad = -\int_\Gamma(\mathbf{E}_C\times\mathbf{n}_C)\cdot\boldsymbol{\rho}^*_{\alpha,I} & \forall\,\alpha=1,\ldots,n_{\Omega_I}\,, \end{array} \tag{3.41}$$

where $\boldsymbol{\rho}^*_{\alpha,I}$ are the basis functions of the space of harmonic fields $\mathcal{H}_{\mu_I}(m;\Omega_I)$ (see Section 1.4). This space has dimension equal to n_{Ω_I}, the number of independent non-bounding cycles of Ω_I. Note also that each condition $(3.41)_4$ is equivalent to the Faraday equation on Ξ^*_α, the "cutting" surface of a non-bounding cycle.

To verify that conditions (3.41) are satisfied, one has to proceed in several steps. First of all, $(3.41)_1$ and $(3.41)_2$ can be proved exactly as for the magnetic boundary value problem. Moreover, if one takes in (3.36) the test function

$$\mathbf{v}^*_\eta := \begin{cases} \operatorname{grad}\chi^*_{\eta,I} & \text{in } \Omega_I \\ \mathbf{0} & \text{in } \Omega_C\,, \end{cases}$$

where $\chi^*_{\eta,I}\in H^1(\Omega_I)$, $\chi^*_{\eta,I}=0$ on Γ and $\chi^*_{\eta,I}=\eta$ on $\partial\Omega$, with η an arbitrary complex function defined on $\partial\Omega$, as an easy consequence one finds $\boldsymbol{\mu}_I\mathbf{H}_I\cdot\mathbf{n}=0$ on $\partial\Omega$. Finally, by taking as a test function in (3.36) a function whose restriction to Ω_I is equal to the basis function $\boldsymbol{\rho}^*_{\alpha,I}$ of $\mathcal{H}_{\mu_I}(m;\Omega_I)$, $\alpha=1,\ldots,n_{\Omega_I}$, it is easily seen that conditions $(3.41)_4$ are satisfied.

In conclusion, the strong problem for the magnetic field reads

$$\begin{cases} \operatorname{curl}(\boldsymbol{\sigma}^{-1}\operatorname{curl}\mathbf{H}_C)+i\omega\boldsymbol{\mu}_C\mathbf{H}_C & \\ \qquad = \operatorname{curl}(\boldsymbol{\sigma}^{-1}\mathbf{J}_{e,C}) & \text{in } \Omega_C \\ \operatorname{curl}\mathbf{H}_I = \mathbf{J}_{e,I} & \text{in } \Omega_I \\ \operatorname{div}(\boldsymbol{\mu}_I\mathbf{H}_I)=0 & \text{in } \Omega_I \\ \int_{\Omega_I} i\omega\boldsymbol{\mu}_I\mathbf{H}_I\cdot\boldsymbol{\rho}^*_{\alpha,I} & \\ \quad +\int_\Gamma[\boldsymbol{\sigma}^{-1}(\operatorname{curl}\mathbf{H}_C-\mathbf{J}_{e,C})]\times\mathbf{n}_C\cdot\boldsymbol{\rho}^*_{\alpha,I}=0 & \forall\,\alpha=1,\ldots,n_{\Omega_I} \\ \boldsymbol{\mu}_I\mathbf{H}_I\cdot\mathbf{n}=0 & \text{on } \partial\Omega \\ \boldsymbol{\mu}_I\mathbf{H}_I\cdot\mathbf{n}_I+\boldsymbol{\mu}_C\mathbf{H}_C\cdot\mathbf{n}_C=0 & \text{on } \Gamma \\ \mathbf{H}_I\times\mathbf{n}_I+\mathbf{H}_C\times\mathbf{n}_C=\mathbf{0} & \text{on } \Gamma\,. \end{cases} \tag{3.42}$$

This problem simplifies if the conductor Ω_C and the computational domain Ω are both simply-connected, so that $n_{\Omega_I}=0$. As already remarked, this assumption on Ω_C can be rather restrictive in many engineering problems.

Putting together (3.42) and (3.38), one finds the complete eddy current problem with electric boundary condition (1.20), namely,

$$\begin{cases} \operatorname{curl}\mathbf{H}-\boldsymbol{\sigma}\mathbf{E}=\mathbf{J}_e & \text{in } \Omega \\ \operatorname{curl}\mathbf{E}+i\omega\boldsymbol{\mu}\mathbf{H}=\mathbf{0} & \text{in } \Omega \\ \operatorname{div}(\boldsymbol{\varepsilon}_I\mathbf{E}_I)=0 & \text{in } \Omega_I \\ \mathbf{E}_I\times\mathbf{n}=\mathbf{0} & \text{on } \partial\Omega \\ \int_{\Gamma_j}\boldsymbol{\varepsilon}_I\mathbf{E}_I\cdot\mathbf{n}_I=0 & \forall\,j=1,\ldots,p_\Gamma \\ \int_{(\partial\Omega)_r}\boldsymbol{\varepsilon}_I\mathbf{E}_I\cdot\mathbf{n}=0 & \forall\,r=0,1,\ldots,p_{\partial\Omega}\,. \end{cases} \tag{3.43}$$

All the existence and uniqueness results proved in the preceding sections for the magnetic boundary value problem can be proved also for the electric boundary value problem (3.43).

Let us come now to the eddy current problem subject to the no-flux boundary conditions (3.32).

The (necessary) assumptions on the current density are the same as for the electric boundary condition, namely, $\mathbf{J}_e$ belongs to $(L^2(\Omega))^3$ and satisfies $\operatorname{div} \mathbf{J}_{e,I} = 0$ in Ω_I and (3.33); also the vector field $\mathbf{H}_e^* \in H(\operatorname{curl};\Omega)$ can be constructed as in that case.

To obtain the weak formulation for the magnetic field we need some preliminaries.

First of all, as explained in Remark 1.3, we recall that in Ω_I we have a collection of "cutting" surfaces Ξ_α^*, $\alpha = 1, \ldots, n_{\Omega_I}$, with $\partial\Xi_\alpha^* \subset \partial\Omega \cup \Gamma$. We have denoted by $n_{\Omega_I}^* \leq n_{\Omega_I}$ the number of the surfaces Ξ_α^* such that $\partial\Xi_\alpha^* \cap \partial\Omega \neq \emptyset$. Since the basis functions $\boldsymbol{\rho}_{\alpha,I}^*$ can be expressed as $\boldsymbol{\rho}_{\alpha,I}^* = \operatorname{grad} p_{\alpha,I}^*$ in $\Omega_I \setminus \Xi_\alpha^*$, for $\alpha \geq n_{\Omega_I}^* + 1$ one has $\boldsymbol{\rho}_{\alpha,I}^* = \operatorname{grad} p_{\alpha,I}^*$ in a neighborhood of $\partial\Omega$. Hence for $\alpha \geq n_{\Omega_I}^* + 1$

$$\begin{aligned} \textstyle\int_{\partial\Omega} \mathbf{E} \times \mathbf{n} \cdot \boldsymbol{\rho}_{\alpha,I}^* &= \textstyle\int_{\partial\Omega} \mathbf{E} \times \mathbf{n} \cdot \operatorname{grad} p_{\alpha,I}^* = -\int_{\partial\Omega} \operatorname{div}_\tau(\mathbf{E} \times \mathbf{n})\, p_{\alpha,I}^* \\ &= -\textstyle\int_{\partial\Omega} \operatorname{curl} \mathbf{E} \cdot \mathbf{n}\, p_{\alpha,I}^* = i\omega \int_{\partial\Omega} \boldsymbol{\mu}\mathbf{H} \cdot \mathbf{n}\, p_{\alpha,I}^* = 0 \,. \end{aligned} \tag{3.44}$$

Suppose now that $\mathbf{H}$ and $\mathbf{E}$ are solutions of the Faraday and Ampère equations. Looking for a weak formulation, multiply the Faraday equation by $\overline{\mathbf{v}^*}$, with $\mathbf{v}^* \in V^*$, the space introduced in (3.37), integrate in Ω and integrate by parts. Using also the Ampère equation in Ω_C it follows

$$\begin{aligned} &\textstyle\int_{\Omega_C} \boldsymbol{\sigma}^{-1} \operatorname{curl} \mathbf{H}_C \cdot \operatorname{curl} \overline{\mathbf{v}_C^*} + \int_\Omega i\omega\boldsymbol{\mu}\mathbf{H} \cdot \overline{\mathbf{v}^*} \\ &\quad = \textstyle\int_{\partial\Omega} \mathbf{E} \times \mathbf{n} \cdot \overline{\mathbf{v}^*} + \int_{\Omega_C} \boldsymbol{\sigma}^{-1} \mathbf{J}_{e,C} \cdot \operatorname{curl} \overline{\mathbf{v}_C^*} \,, \end{aligned}$$

as $\operatorname{curl} \mathbf{v}_I^* = \mathbf{0}$ in Ω_I. On the other hand, from the orthogonality result presented in Theorem A.8 we can write

$$\mathbf{v}_I^* = \operatorname{grad} \chi_I^* + \sum_{\alpha=1}^{n_{\Omega_I}} \theta_{I,\alpha}^* \boldsymbol{\rho}_{\alpha,I}^* \,.$$

Repeating the arguments leading to (3.44) we have

$$\int_{\partial\Omega} \mathbf{E}_I \times \mathbf{n} \cdot \operatorname{grad} \overline{\chi_I^*} = 0 \,.$$

Hence, using (3.44)

$$\begin{aligned} \textstyle\int_{\partial\Omega} \mathbf{E} \times \mathbf{n} \cdot \overline{\mathbf{v}^*} &= \textstyle\int_{\partial\Omega} \mathbf{E}_I \times \mathbf{n} \cdot \operatorname{grad} \overline{\chi_I^*} + \sum_{\alpha=1}^{n_{\Omega_I}} \overline{\theta_{I,\alpha}^*} \int_{\partial\Omega} \mathbf{E}_I \times \mathbf{n} \cdot \boldsymbol{\rho}_{\alpha,I}^* \\ &= \textstyle\sum_{\alpha=1}^{n_{\Omega_I}^*} \overline{\theta_{I,\alpha}^*} \int_{\partial\Omega} \mathbf{E}_I \times \mathbf{n} \cdot \boldsymbol{\rho}_{\alpha,I}^* \,. \end{aligned}$$

If we assume that the electric field also satisfies the conditions

$$\int_{\partial\Omega} \mathbf{E}_I \times \mathbf{n} \cdot \boldsymbol{\rho}_{\alpha,I}^* = 0 \qquad \forall\, \alpha = 1, \ldots, n_{\Omega_I}^* \,, \tag{3.45}$$

then we conclude that $\mathbf{H}$ is the solution of the weak problem (3.36).

We can now start from (3.36) in order to determine the complete set of equations constituting the strong problem. First of all, we have already seen that the unique solvability of (3.36) is ensured by the Lax–Milgram lemma. The magnetic field $\mathbf{H}$ clearly satisfies (3.22) and $\operatorname{curl} \mathbf{H}_I = \mathbf{J}_{e,I}$ in Ω_I. Then the electric field $\mathbf{E}_C$ in Ω_C is defined as usual by (3.12), while the electric field $\mathbf{E}_I$ is given by the solution to (3.13).

Conditions (3.14), that are necessary and sufficient for the solvability of problem (3.13), are verified. In fact, we have seen here above that the solution $\mathbf{H}$ of problem (3.36) satisfies conditions (3.41). Moreover, it is easily shown that $\mathbf{H}$ and $\mathbf{E}_C$ also satisfy

$$\int_{\Omega_I} i\omega \boldsymbol{\mu}_I \mathbf{H}_I \cdot \boldsymbol{\rho}_{l,I} = -\int_\Gamma (\mathbf{E}_C \times \mathbf{n}_C) \cdot \boldsymbol{\rho}_{l,I} \quad \forall\, l = 1, \dots, n_\Gamma \,.$$

In fact we can write

$$\boldsymbol{\rho}_{l,I} = \operatorname{grad} \chi_l^* + \sum_{\alpha=1}^{n_{\Omega_I}} \theta_{l,\alpha}^* \boldsymbol{\rho}_{\alpha,I}^* \,,$$

and

$$\begin{aligned}\int_{\Omega_I} i\omega \boldsymbol{\mu}_I \mathbf{H}_I \cdot \operatorname{grad} \chi_l^* &= -i\omega \int_{\Omega_I} \operatorname{div}(\boldsymbol{\mu}_I \mathbf{H}_I)\, \chi_l^* \\ &\quad + i\omega \int_{\partial\Omega} \boldsymbol{\mu}_I \mathbf{H}_I \cdot \mathbf{n}\, \chi_l^* + i\omega \int_\Gamma \boldsymbol{\mu}_I \mathbf{H}_I \cdot \mathbf{n}_I\, \chi_l^* \\ &= \int_\Gamma \operatorname{div}_\tau(\mathbf{E}_C \times \mathbf{n}_C)\, \chi_l^* = -\int_\Gamma \mathbf{E}_C \times \mathbf{n}_C \cdot \operatorname{grad} \chi_l^* \,.\end{aligned}$$

In conclusion, the strong formulation for the magnetic field $\mathbf{H}$ and the electric field $\mathbf{E}_C$ reads

$$\begin{cases} \operatorname{curl} \mathbf{E}_C + i\omega \boldsymbol{\mu}_C \mathbf{H}_C = \mathbf{0} & \text{in } \Omega_C \\ \operatorname{curl} \mathbf{H}_C - \boldsymbol{\sigma} \mathbf{E}_C = \mathbf{J}_{e,C} & \text{in } \Omega_C \\ \operatorname{curl} \mathbf{H}_I = \mathbf{J}_{e,I} & \text{in } \Omega_I \\ \operatorname{div}(\boldsymbol{\mu}_I \mathbf{H}_I) = 0 & \text{in } \Omega_I \\ \int_{\Omega_I} i\omega \boldsymbol{\mu}_I \mathbf{H}_I \cdot \boldsymbol{\rho}_{\alpha,I}^* + \int_\Gamma \mathbf{E}_C \times \mathbf{n}_C \cdot \boldsymbol{\rho}_{\alpha,I}^* = 0 & \forall\, \alpha = 1, \dots, n_{\Omega_I} \\ \boldsymbol{\mu}_I \mathbf{H}_I \cdot \mathbf{n} = 0 & \text{on } \partial\Omega \\ \boldsymbol{\mu}_I \mathbf{H}_I \cdot \mathbf{n}_I + \boldsymbol{\mu}_C \mathbf{H}_C \cdot \mathbf{n}_C = 0 & \text{on } \Gamma \\ \mathbf{H}_I \times \mathbf{n}_I + \mathbf{H}_C \times \mathbf{n}_C = \mathbf{0} & \text{on } \Gamma \,. \end{cases} \tag{3.46}$$

Putting together (3.46) and (3.13) one obtains the problem

$$\begin{cases} \operatorname{curl} \mathbf{H} - \boldsymbol{\sigma} \mathbf{E} = \mathbf{J}_e & \text{in } \Omega \\ \operatorname{curl} \mathbf{E} + i\omega \boldsymbol{\mu} \mathbf{H} = \mathbf{0} & \text{in } \Omega \\ \operatorname{div}(\varepsilon_I \mathbf{E}_I) = 0 & \text{in } \Omega_I \\ \boldsymbol{\mu} \mathbf{H} \cdot \mathbf{n} = 0 & \text{on } \partial\Omega \\ \varepsilon_I \mathbf{E}_I \cdot \mathbf{n} = 0 & \text{on } \partial\Omega \\ \int_{\Gamma_j} \varepsilon_I \mathbf{E}_I \cdot \mathbf{n}_I = 0 & \forall\, j = 1, \dots, p_\Gamma \\ \int_{\Omega_I} \varepsilon_I \mathbf{E}_I \cdot \boldsymbol{\pi}_{k,I} = 0 & \forall\, k = 1, \dots, n_{\partial\Omega} \\ \int_{\Omega_I} i\omega \boldsymbol{\mu}_I \mathbf{H}_I \cdot \boldsymbol{\rho}_{\alpha,I}^* + \int_\Gamma \mathbf{E}_C \times \mathbf{n}_C \cdot \boldsymbol{\rho}_{\alpha,I}^* = 0 & \forall\, \alpha = 1, \dots, n_{\Omega_I} \,. \end{cases}$$

Let us note that the last equations can be written in a different way: in fact, using the Faraday equation in Ω_I we have

$$\begin{aligned}\int_{\Omega_I} i\omega\boldsymbol{\mu}_I\mathbf{H}_I\cdot\boldsymbol{\rho}^*_{\alpha,I} &= -\int_{\Omega_I}\operatorname{curl}\mathbf{E}_I\cdot\boldsymbol{\rho}^*_{\alpha,I}\\ &= -\int_{\partial\Omega}\mathbf{n}\times\mathbf{E}_I\cdot\boldsymbol{\rho}^*_{\alpha,I} - \int_\Gamma \mathbf{n}_I\times\mathbf{E}_I\cdot\boldsymbol{\rho}^*_{\alpha,I}\\ &= -\int_{\partial\Omega}\mathbf{n}\times\mathbf{E}_I\cdot\boldsymbol{\rho}^*_{\alpha,I} - \int_\Gamma \mathbf{E}_C\times\mathbf{n}_C\cdot\boldsymbol{\rho}^*_{\alpha,I}\,,\end{aligned}$$

thus they can be expressed as

$$\int_{\partial\Omega}\mathbf{E}_I\times\mathbf{n}\cdot\boldsymbol{\rho}^*_{\alpha,I} = 0 \quad \forall\,\alpha = 1,\ldots,n_{\Omega_I}\,.$$

However, some of these relations are redundant: precisely, as proved in (3.44), those corresponding to $\alpha \geq n^*_{\Omega_I}+1$. We have thus obtained that the eddy current problem with no-flux boundary conditions reads

$$\begin{cases}\operatorname{curl}\mathbf{H} - \boldsymbol{\sigma}\mathbf{E} = \mathbf{J}_e & \text{in } \Omega\\ \operatorname{curl}\mathbf{E} + i\omega\boldsymbol{\mu}\mathbf{H} = \mathbf{0} & \text{in } \Omega\\ \operatorname{div}(\boldsymbol{\varepsilon}_I\mathbf{E}_I) = 0 & \text{in } \Omega_I\\ \boldsymbol{\mu}\mathbf{H}\cdot\mathbf{n} = 0 & \text{on } \partial\Omega\\ \boldsymbol{\varepsilon}_I\mathbf{E}_I\cdot\mathbf{n} = 0 & \text{on } \partial\Omega\\ \int_{\Gamma_j}\boldsymbol{\varepsilon}_I\mathbf{E}_I\cdot\mathbf{n}_I = 0 & \forall\,j = 1,\ldots,p_\Gamma\\ \int_{\Omega_I}\boldsymbol{\varepsilon}_I\mathbf{E}_I\cdot\boldsymbol{\pi}_{k,I} = 0 & \forall\,k = 1,\ldots,n_{\partial\Omega}\\ \int_{\partial\Omega}\mathbf{E}_I\times\mathbf{n}\cdot\boldsymbol{\rho}^*_{\alpha,I} = 0 & \forall\,\alpha = 1,\ldots,n^*_{\Omega_I}\,.\end{cases} \tag{3.47}$$

The problem becomes simpler if the boundary of the conductor Ω_C is connected, so that $p_\Gamma = 0$, and the computational domain Ω is simply-connected, so that $n_{\partial\Omega} = 0$ and $n^*_{\Omega_I} = 0$. An example of this situation is a connected conductor (possibly with "handles") inside a "box".

Let us show that problem (3.47) is well-posed. Taking the solution $\mathbf{H}$ of (3.36), $\mathbf{E}_C$ defined in (3.12) and the solution $\mathbf{E}_I$ of (3.13) we have just proved the existence of a solution to (3.47). Concerning uniqueness, it is enough to show that the magnetic field $\mathbf{H}$ is uniquely determined. The arguments before have shown that, starting from the Faraday and Ampère equations, we can construct a solution $\mathbf{H}$ of (3.36) provided that (3.45) are satisfied. Hence a solution $\mathbf{H}$ to (3.47) furnishes a solution to the weak problem (3.36), for which uniqueness holds.

Remark 3.12. By adapting the arguments reported in Remark 3.7, it is readily shown that conditions $(3.47)_8$ are necessary for obtaining the uniqueness of the solution to (3.47). □

4

Hybrid formulations for the electric and magnetic fields

The classical approaches to Maxwell and eddy current equations are often based on the introduction of a magnetic vector potential and an electric scalar potential, the latter being used only in the conducting region, or on the use of a magnetic scalar potential in the insulating region (see, e.g., Jackson [137], Silvester and Ferrari [227]). We present these formulations in Chapters 6 and 5 respectively.

Here we follow a less investigated path (at least for eddy currents), and we analyze the problem in terms of the original unknowns, namely, the electric and magnetic fields. This can yield some advantages at the numerical level, as we do not need to determine the fields by means of differentiation, a procedure that can lead to a loss of accuracy.

Moreover, we focus on hybrid formulations of the eddy current problem. As seen in the previous chapters, it is possible to reformulate the complete eddy current model by eliminating either the electric field or the magnetic field. We name "hybrid" those formulations where the eliminated field in the conducting region is different from the one eliminated in the insulating region, therefore the unknowns are the electric field in one subdomain and the magnetic field in the other one. These kind of formulations are particularly interesting in the context of a finite element approximation: for instance, since the two vector fields do not need to match on the interface Γ, it is possible to use independent meshes in Ω_C and Ω_I.

We have already noted that a particular feature of the eddy current model is the presence of differential constraints in the non-conducting region. In this chapter we propose to use a saddle-point approach to take these constraints into account.

In Sections 4.1 and 4.2 we start by considering the hybrid formulation that uses as main unknowns the electric field in the conductor and the magnetic field in the insulator. Hence we have to deal with the constraint $\operatorname{curl} \mathbf{H}_I = \mathbf{J}_{e,I}$ in Ω_I, that is enforced introducing a Lagrange multiplier. This multiplier turns out to be the electric field in the insulator, which is subjected to the differential constraint $\operatorname{div}(\varepsilon_I \mathbf{E}_I) = 0$ in Ω_I, so that we have to introduce a second Lagrange multiplier to enforce it.

The other hybrid formulation, the one based on the magnetic field in the conductor and the electric field in the insulator, is analyzed in Sections 4.4 and 4.5. Since its finite element discretization presents some difficulties, we also consider a modified problem, more suitable for numerical approximation, that instead of the electric field on Ω_I

A. Alonso Rodríguez, A. Valli: Eddy Current Approximation of Maxwell Equations.

provides a magnetic vector potential $\widetilde{\mathbf{E}}_I$ with the condition $\operatorname{div}\widetilde{\mathbf{E}}_I = 0$ in Ω_I. Also in this case we propose an augmented variational formulation introducing a (scalar) Lagrange multiplier to take into account this constraint.

Clearly, the saddle-point approach is not specific to hybrid formulations and it can be also used for the more classical formulations in terms of the magnetic field or the electric field. In Section 4.3 we use Lagrange multipliers to enforce the differential constraints in the $\mathbf{H}$-based formulation, adapting the arguments used for the $\mathbf{E}_C/\mathbf{H}_I$ problem, while in Section 4.6 we analyze a saddle-point approach for the $\mathbf{E}$-based formulation, analogous to the one adopted for the $\mathbf{H}_C/\widetilde{\mathbf{E}}_I$ case.

In this chapter we address the eddy current model with the magnetic boundary conditions (1.22). For the electric boundary condition (1.20) we only provide a series of remarks where the weak formulations of the corresponding problems are presented.

If not otherwise specified, the geometric situation is the same as that described in Section 1.3. However, let us note at once that for the $\mathbf{H}_C/\widetilde{\mathbf{E}}_I$ formulation we have not been able to avoid the assumption that the computational domain Ω is simply-connected. Moreover, but in this case for the sake of simplicity, the same assumption has been made for the finite element approximation of the $\mathbf{E}_C/\mathbf{H}_I$ formulation. With respect to this problem, the reader interested in the general geometrical case can adapt the arguments presented in Section 5.5, where a mixed finite element method is proposed for the approximation of the electric field in the non-conducting region, assuming that the electric field in the conductor is already known.

Concerning the material properties, as in the preceding chapters we suppose that the matrix $\boldsymbol{\mu}$ is symmetric and uniformly positive definite in Ω, with entries in $L^\infty(\Omega)$, the matrix $\boldsymbol{\varepsilon}_I$ is symmetric and uniformly positive definite in Ω_I, with entries in $L^\infty(\Omega_I)$, and the matrix $\boldsymbol{\sigma}$ is symmetric and uniformly positive definite in Ω_C, with entries in $L^\infty(\Omega_C)$, whereas it is vanishing in Ω_I. Finally, the current density is assumed to satisfy $\mathbf{J}_e \in (L^2(\Omega))^3$ and (1.23).

The reader mainly interested in numerical approximation and implementation can focus on problems (4.18), (4.22) and Remark 4.13 ($\mathbf{E}_C/\mathbf{H}_I$ formulation), on problems (4.35) and (4.36) ($\mathbf{H}$ formulation), on problems (4.49), (4.64) and Remark 4.26 ($\mathbf{H}_C/\widetilde{\mathbf{E}}_I$ formulation), on problems (4.78), (4.84) and Remark 4.38 ($\mathbf{E}$ formulation), and on Section 4.5.2.

4.1 Hybrid formulation using the magnetic field in the insulator

The first hybrid formulation we consider, following Alonso Rodríguez et al. [12], is obtained by eliminating the magnetic field in the conductor and the electric field in the insulator. It is the so-called $\mathbf{E}_C/\mathbf{H}_I$ formulation.

From the Faraday equation we know that the magnetic field can be written as

$$\mathbf{H} = -(i\omega)^{-1}\boldsymbol{\mu}^{-1}\operatorname{curl}\mathbf{E}\,,$$

and substituting it in the Ampère law we obtain an equation for the electric field that in the conductor Ω_C reads

$$\operatorname{curl}(\boldsymbol{\mu}_C^{-1}\operatorname{curl}\mathbf{E}_C) + i\omega\boldsymbol{\sigma}\mathbf{E}_C = -i\omega\mathbf{J}_{e,C}\,.$$

Multiplying by a test function $\mathbf{z}_C \in H(\mathrm{curl}; \Omega_C)$ one finds, by integration by parts,

$$\int_{\Omega_C} (\boldsymbol{\mu}_C^{-1} \,\mathrm{curl}\, \mathbf{E}_C \cdot \mathrm{curl}\, \overline{\mathbf{z}_C} + i\omega\boldsymbol{\sigma}\mathbf{E}_C \cdot \overline{\mathbf{z}_C}) - \int_\Gamma \boldsymbol{\mu}_C^{-1} \,\mathrm{curl}\, \mathbf{E}_C \times \mathbf{n}_C \cdot \overline{\mathbf{z}_C} \\ = -i\omega \int_{\Omega_C} \mathbf{J}_{e,C} \cdot \overline{\mathbf{z}_C} \,.$$

Using the Faraday law and the tangential continuity of the magnetic field across the boundary Γ of Ω_C we have

$$\boldsymbol{\mu}_C^{-1} \,\mathrm{curl}\, \mathbf{E}_C \times \mathbf{n}_C = -i\omega\mathbf{H}_C \times \mathbf{n}_C = -i\omega\mathbf{H}_I \times \mathbf{n}_C \,,$$

therefore

$$\int_{\Omega_C} (\boldsymbol{\mu}_C^{-1} \,\mathrm{curl}\, \mathbf{E}_C \cdot \mathrm{curl}\, \overline{\mathbf{z}_C} + i\omega\boldsymbol{\sigma}\mathbf{E}_C \cdot \overline{\mathbf{z}_C}) + i\omega \int_\Gamma \mathbf{H}_I \times \mathbf{n}_C \cdot \overline{\mathbf{z}_C} \\ = -i\omega \int_{\Omega_C} \mathbf{J}_{e,C} \cdot \overline{\mathbf{z}_C} \,.$$

On the other hand, multiplying the Faraday equation in Ω_I by a test function $\mathbf{v}_I$ such that $\mathrm{curl}\, \mathbf{v}_I = \mathbf{0}$ in Ω_I and $\mathbf{v}_I \times \mathbf{n} = \mathbf{0}$ on $\partial\Omega$, by integration by parts one has

$$i\omega \int_{\Omega_I} \boldsymbol{\mu}_I \mathbf{H}_I \cdot \overline{\mathbf{v}_I} = -\int_{\Omega_I} \mathrm{curl}\, \mathbf{E}_I \cdot \overline{\mathbf{v}_I} = -\int_\Gamma \mathbf{E}_C \times \mathbf{n}_C \cdot \overline{\mathbf{v}_I} \,,$$

where we have used the tangential continuity of the electric field across the interface Γ.

Let us now consider the spaces

$$V_I(\mathbf{J}_{e,I}) := \{\mathbf{v}_I \in H_{0,\partial\Omega}(\mathrm{curl}; \Omega_I) \mid \mathrm{curl}\, \mathbf{v}_I = \mathbf{J}_{e,I} \text{ in } \Omega_I\}, \tag{4.1}$$

and

$$V_I(\mathbf{0}) := \{\mathbf{v}_I \in H_{0,\partial\Omega}(\mathrm{curl}; \Omega_I) \mid \mathrm{curl}\, \mathbf{v}_I = \mathbf{0} \text{ in } \Omega_I\} \,.$$

We have thus obtained the following formulation

$$\begin{array}{l} \text{Find } (\mathbf{E}_C, \mathbf{H}_I) \in H(\mathrm{curl}; \Omega_C) \times V_I(\mathbf{J}_{e,I}) : \\ \quad \int_{\Omega_C} (\boldsymbol{\mu}_C^{-1} \,\mathrm{curl}\, \mathbf{E}_C \cdot \mathrm{curl}\, \overline{\mathbf{z}_C} + i\omega\boldsymbol{\sigma}\mathbf{E}_C \cdot \overline{\mathbf{z}_C}) \\ \qquad\quad -i\omega \int_\Gamma \overline{\mathbf{z}_C} \times \mathbf{n}_C \cdot \mathbf{H}_I = -i\omega \int_{\Omega_C} \mathbf{J}_{e,C} \cdot \overline{\mathbf{z}_C} \\ \quad -i\omega \int_\Gamma \mathbf{E}_C \times \mathbf{n}_C \cdot \overline{\mathbf{v}_I} + \omega^2 \int_{\Omega_I} \boldsymbol{\mu}_I \mathbf{H}_I \cdot \overline{\mathbf{v}_I} = 0 \\ \text{for all } (\mathbf{z}_C, \mathbf{v}_I) \in H(\mathrm{curl}; \Omega_C) \times V_I(\mathbf{0}) \,. \end{array} \tag{4.2}$$

Problem (4.2) is associated to the sesquilinear form

$$\begin{aligned} \mathcal{C}((\mathbf{w}_C, \mathbf{u}_I), (\mathbf{z}_C, \mathbf{v}_I)) := & \int_{\Omega_C} (\boldsymbol{\mu}_C^{-1} \,\mathrm{curl}\, \mathbf{w}_C \cdot \mathrm{curl}\, \overline{\mathbf{z}_C} + i\omega\boldsymbol{\sigma}\mathbf{w}_C \cdot \overline{\mathbf{z}_C}) \\ & -i\omega \int_\Gamma \overline{\mathbf{z}_C} \times \mathbf{n}_C \cdot \mathbf{u}_I - i\omega \int_\Gamma \mathbf{w}_C \times \mathbf{n}_C \cdot \overline{\mathbf{v}_I} \\ & +\omega^2 \int_{\Omega_I} \boldsymbol{\mu}_I \mathbf{u}_I \cdot \overline{\mathbf{v}_I} \,, \end{aligned} \tag{4.3}$$

that is coercive in $H(\mathrm{curl}; \Omega_C) \times V_I(\mathbf{0})$. In fact we have the following result.

Theorem 4.1. *There exists a positive constant* κ *such that*

$$|\mathcal{C}((\mathbf{w}_C, \mathbf{u}_I), (\mathbf{w}_C, \mathbf{u}_I))| \geq \kappa \left(\| \operatorname{curl} \mathbf{w}_C \|^2_{0,\Omega_C} + \| \mathbf{w}_C \|^2_{0,\Omega_C} + \| \mathbf{u}_I \|^2_{0,\Omega_I} \right)$$

for each $(\mathbf{w}_C, \mathbf{u}_I) \in H(\operatorname{curl}; \Omega_C) \times V_I(\mathbf{0})$.

Proof. We have

$$\begin{aligned} &|\operatorname{Re} \mathcal{C}((\mathbf{w}_C, \mathbf{u}_I), (\mathbf{w}_C, \mathbf{u}_I))| \\ &\quad = \int_{\Omega_C} \boldsymbol{\mu}_C^{-1} \operatorname{curl} \mathbf{w}_C \cdot \operatorname{curl} \overline{\mathbf{w}_C} + \omega^2 \int_{\Omega_I} \boldsymbol{\mu}_I \mathbf{u}_I \cdot \overline{\mathbf{u}_I} \\ &\quad \geq \mu_{\max}^{-1} \int_{\Omega_C} |\operatorname{curl} \mathbf{w}_C|^2 + \omega^2 \mu_{\min} \int_{\Omega_I} |\mathbf{u}_I|^2 , \end{aligned}$$

where $\mu_{\max}$ and $\mu_{\min}$ are a uniform upper bound in Ω for the maximum eigenvalues of $\boldsymbol{\mu}(\mathbf{x})$ and a uniform lower bound in Ω of the minimum eigenvalues for $\boldsymbol{\mu}(\mathbf{x})$, respectively. Moreover, for each $0 < \gamma \leq 1$ it holds

$$\begin{aligned} &|\operatorname{Im} \mathcal{C}((\mathbf{w}_C, \mathbf{u}_I), (\mathbf{w}_C, \mathbf{u}_I))| \\ &\quad = |\omega| \left| \int_{\Omega_C} \boldsymbol{\sigma} \mathbf{w}_C \cdot \overline{\mathbf{w}_C} - 2\operatorname{Re} \int_\Gamma \mathbf{w}_C \times \mathbf{n}_C \cdot \overline{\mathbf{u}_I} \right| \\ &\quad \geq \gamma |\omega| \int_{\Omega_C} \boldsymbol{\sigma} \mathbf{w}_C \cdot \overline{\mathbf{w}_C} - 2\gamma |\omega| \left| \int_\Gamma \mathbf{w}_C \times \mathbf{n}_C \cdot \overline{\mathbf{u}_I} \right| . \end{aligned}$$

From the duality estimate

$$\left| \int_\Gamma \mathbf{w}_C \times \mathbf{n}_C \cdot \overline{\mathbf{u}_I} \right| \leq C \| \mathbf{w}_C \times \mathbf{n}_C \|_{H^{-1/2}(\operatorname{div}_\tau; \Gamma)} \| \mathbf{n}_I \times \mathbf{u}_I \times \mathbf{n}_I \|_{H^{-1/2}(\operatorname{curl}_\tau; \Gamma)}$$

and the trace inequalities (A.10) and (A.11), taking into account that $\operatorname{curl} \mathbf{u}_I = \mathbf{0}$ we find

$$\left| \int_\Gamma \mathbf{w}_C \times \mathbf{n}_C \cdot \overline{\mathbf{u}_I} \right| \leq k_0 \left(\int_{\Omega_C} (|\mathbf{w}_C|^2 + |\operatorname{curl} \mathbf{w}_C|^2) \right)^{1/2} \left(\int_{\Omega_I} |\mathbf{u}_I|^2 \right)^{1/2} .$$

Moreover, since for each $\delta > 0$ the inequality $2|AB| \leq \delta A^2 + \delta^{-1} B^2$ holds, we finally have

$$\begin{aligned} &|\operatorname{Im} \mathcal{C}((\mathbf{w}_C, \mathbf{u}_I), (\mathbf{w}_C, \mathbf{u}_I))| \\ &\quad \geq \gamma |\omega| \sigma_{\min} \int_{\Omega_C} |\mathbf{w}_C|^2 - \gamma |\omega| k_0 \delta \int_{\Omega_C} |\mathbf{w}_C|^2 \\ &\quad\quad - \gamma |\omega| k_0 \delta \int_{\Omega_C} |\operatorname{curl} \mathbf{w}_C|^2 - \gamma |\omega| k_0 \delta^{-1} \int_{\Omega_I} |\mathbf{u}_I|^2 , \end{aligned}$$

where $\sigma_{\min}$ is a uniform lower bound in Ω_C for the minimum eigenvalues of $\boldsymbol{\sigma}(\mathbf{x})$. The proof follows by taking at first δ small enough and then γ small enough. □

4.2 A saddle-point approach for the $\mathrm{E}_C / \mathrm{H}_I$ formulation

A finite element method for approximating the solution to problem (4.2) has to deal with the constrained space $V_I(\mathbf{0})$. In this chapter we focus on approaches based on augmented variational equations; the more direct approach that incorporates the constraints into the variational space will be considered in Chapter 5. It is worth noting that adding extra equations can be a defect, because the finite element approximation of the resulting formulation presents more degrees of freedom. However this saddle-point approach avoids the construction of "cutting" surfaces in the insulating region, a procedure that on one side is indispensable for the approximation of the scalar magnetic potential used in the direct approach, on the other side can be cumbersome in complex geometry configurations.

To enforce the constraint $\operatorname{curl}\mathbf{H}_I = \mathbf{J}_{e,I}$ one can consider the following augmented problem

$$\begin{array}{l}\text{Find } (\mathbf{E}_C, \mathbf{H}_I, \mathbf{A}_I) \in H(\operatorname{curl}; \Omega_C) \times H_{0,\partial\Omega}(\operatorname{curl}; \Omega_I) \times (L^2(\Omega_I))^3 : \\ \quad \mathcal{C}((\mathbf{E}_C, \mathbf{H}_I), (\mathbf{z}_C, \mathbf{v}_I)) + \int_{\Omega_I} \operatorname{curl}\overline{\mathbf{v}_I} \cdot \mathbf{A}_I = L(\mathbf{z}_C) \\ \quad \int_{\Omega_I} \operatorname{curl}\mathbf{H}_I \cdot \overline{\mathbf{N}_I} = G(\mathbf{N}_I) \\ \text{for all } (\mathbf{z}_C, \mathbf{v}_I, \mathbf{N}_I) \in H(\operatorname{curl}; \Omega_C) \times H_{0,\partial\Omega}(\operatorname{curl}; \Omega_I) \times (L^2(\Omega_I))^3 ,\end{array} \tag{4.4}$$

where

$$L(\mathbf{z}_C) := -i\omega \int_{\Omega_C} \mathbf{J}_{e,C} \cdot \overline{\mathbf{z}_C} \ , \quad G(\mathbf{N}_I) := \int_{\Omega_I} \mathbf{J}_{e,I} \cdot \overline{\mathbf{N}_I} \ . \tag{4.5}$$

It is readily seen that for this problem uniqueness does not hold, as it is possible to add to $\mathbf{A}_I$ any function belonging to $H^0_{0,\Gamma}(\operatorname{curl}; \Omega_I)$ (for notation, see Section A.1).

Choosing as $\mathbf{v}_I$ a smooth vector function with compact support in Ω_I and $\mathbf{z}_C$ and $\mathbf{N}_I$ equal $\mathbf{0}$, we find that any solution of problem (4.4) satisfies $\operatorname{curl}\mathbf{A}_I = -\omega^2\boldsymbol{\mu}_I\mathbf{H}_I (= -i\omega \operatorname{curl}\mathbf{E}_I)$ in Ω_I. A similar choice of test functions with $\mathbf{v}_I$ vanishing only in a neighborhood of $\partial\Omega$ gives $\mathbf{A}_I \times \mathbf{n}_C = -i\omega\mathbf{E}_C \times \mathbf{n}_C (= -i\omega\mathbf{E}_I \times \mathbf{n}_C)$ on Γ. In order to deal with a well-posed saddle-point problem it is natural to look for $\mathbf{A}_I$ in the space W_I given by the functions $\mathbf{N}_I \in (L^2(\Omega_I))^3$ that satisfy

$$\begin{cases} \operatorname{div}(\varepsilon_I\mathbf{N}_I) = 0 & \text{in } \Omega_I \\ \varepsilon_I\mathbf{N}_I \cdot \mathbf{n} = 0 & \text{on } \partial\Omega \\ \int_{\Gamma_j} \varepsilon_I\mathbf{N}_I \cdot \mathbf{n}_I = 0 & \forall j = 1, \dots, p_\Gamma \\ \int_{\Omega_I} \varepsilon_I\mathbf{N}_I \cdot \boldsymbol{\pi}_{k,I} = 0 & \forall k = 1, \dots, n_{\partial\Omega} \ . \end{cases} \tag{4.6}$$

It is clear that in this way the Lagrange multiplier $\mathbf{A}_I$ turns out to be equal to $-i\omega\mathbf{E}_I$ (recall that the electric field in the insulator is the unique solution of the system of equations (3.13)).

Therefore, let us consider the space

$$W_I := \{\mathbf{N}_I \in (L^2(\Omega_I))^3 \,|\, \mathbf{N}_I \text{ satisfies } (4.6)\} \ .$$

Since the current density $\mathbf{J}_{e,I}$ satisfies the necessary conditions (1.23), the space $V_I(\mathbf{J}_{e,I})$ can also be defined in the following way

$$\begin{aligned} V_I(\mathbf{J}_{e,I}) = \{\mathbf{v}_I \in H_{0,\partial\Omega}(\operatorname{curl}; \Omega_I) \,|\, \\ \int_{\Omega_I} \operatorname{curl}\mathbf{v}_I \cdot \overline{\mathbf{N}_I} = \int_{\Omega_I} \mathbf{J}_{e,I} \cdot \overline{\mathbf{N}_I} \quad \forall\, \mathbf{N}_I \in W_I \,\} \,. \end{aligned}$$

This is easily seen by taking $\mathbf{N}_I = \varepsilon_I^{-1}(\operatorname{curl}\mathbf{v}_I - \mathbf{J}_{e,I})$.

Let us introduce the space

$$\Lambda := H(\mathrm{curl};\Omega_C) \times H_{0,\partial\Omega}(\mathrm{curl};\Omega_I)\,,$$

endowed with the norm $\|(\mathbf{z}_C,\mathbf{v}_I)\|_\Lambda := (\|\mathbf{z}_C\|^2_{H(\mathrm{curl};\Omega_C)} + \|\mathbf{v}_I\|^2_{H(\mathrm{curl};\Omega_I)})^{1/2}$. Instead of (4.4), we are thus led to consider the following problem

$$\begin{array}{l}\text{Find } (\mathbf{E}_C,\mathbf{H}_I) \in \Lambda\,,\ \mathbf{A}_I \in W_I :\\ \quad \mathcal{C}((\mathbf{E}_C,\mathbf{H}_I),(\mathbf{z}_C,\mathbf{v}_I)) + \int_{\Omega_I} \mathrm{curl}\,\overline{\mathbf{v}_I}\cdot\mathbf{A}_I = L(\mathbf{z}_C)\\ \quad \int_{\Omega_I} \mathrm{curl}\,\mathbf{H}_I\cdot\overline{\mathbf{N}_I} = G(\mathbf{N}_I)\\ \text{for all } (\mathbf{z}_C,\mathbf{v}_I)\in\Lambda\,,\ \mathbf{N}_I\in W_I\,.\end{array} \tag{4.7}$$

It can be proved that this problem has a unique solution $(\mathbf{E}_C,\mathbf{H}_I,\mathbf{A}_I)$. However, the space W_I is not easily approximated by finite element spaces. We obtain an alternative formulation by expressing in a different way conditions (4.6). First of all, $(4.6)_1$ and $(4.6)_2$ give that

$$\int_{\Omega_I} \varepsilon_I \mathbf{N}_I\cdot\mathrm{grad}\,\overline{\varphi_I} = 0 \quad \forall\,\varphi_I \in H^1_{0,\Gamma}(\Omega_I)\,.$$

Moreover, as shown at the end of Section 1.5, taking into account $(4.6)_1$ and $(4.6)_2$ we see that conditions $(4.6)_3$ and $(4.6)_4$ imply that

$$\int_{\Omega_I} \varepsilon_I \mathbf{N}_I\cdot\overline{\mathbf{h}_I} = 0 \quad \forall\,\mathbf{h}_I \in \mathcal{H}_{\varepsilon_I}(\Gamma,\partial\Omega;\Omega_I)\,.$$

Finally, using the orthogonality result presented in Theorem A.6, we have

$$\mathrm{grad}\,H^1_{0,\Gamma}(\Omega_I) \oplus \mathcal{H}_{\varepsilon_I}(\Gamma,\partial\Omega;\Omega_I) = H^0_{0,\Gamma}(\mathrm{curl};\Omega_I)\,. \tag{4.8}$$

We have thus seen that $\mathbf{N}_I \in W_I$ if and only if $\varepsilon_I\mathbf{N}_I$ is $(L^2(\Omega_I))^3$-orthogonal to $H^0_{0,\Gamma}(\mathrm{curl};\Omega_I)$.

We can therefore introduce another Lagrange multiplier and consider the problem

$$\begin{array}{l}\text{Find } (\mathbf{E}_C,\mathbf{H}_I) \in \Lambda\,,\ \mathbf{A}_I \in (L^2(\Omega_I))^3\,,\ \mathbf{r}_I \in H^0_{0,\Gamma}(\mathrm{curl};\Omega_I) :\\ \quad \mathcal{C}((\mathbf{E}_C,\mathbf{H}_I),(\mathbf{z}_C,\mathbf{v}_I)) + \int_{\Omega_I} \mathrm{curl}\,\overline{\mathbf{v}_I}\cdot\mathbf{A}_I = L(\mathbf{z}_C)\\ \quad \int_{\Omega_I} \mathrm{curl}\,\mathbf{H}_I\cdot\overline{\mathbf{N}_I} + \int_{\Omega_I}\varepsilon_I\overline{\mathbf{N}_I}\cdot\mathbf{r}_I = G(\mathbf{N}_I)\\ \quad \int_{\Omega_I}\varepsilon_I\mathbf{A}_I\cdot\overline{\mathbf{p}_I} = 0\\ \text{for all } (\mathbf{z}_C,\mathbf{v}_I)\in\Lambda\,,\ \mathbf{N}_I\in (L^2(\Omega_I))^3\,,\ \mathbf{p}_I \in H^0_{0,\Gamma}(\mathrm{curl};\Omega_I)\,.\end{array} \tag{4.9}$$

In order to analyze this formulation we can use the following result, which is Lemma 4.1 in Chen et al. [81] extended to complex Hilbert spaces.

Theorem 4.2. *Let X, Q, M be three complex Hilbert spaces and $a : X \times X \to \mathbb{C}$, $b : X \times Q \to \mathbb{C}$, $c : Q \times M \to \mathbb{C}$ be three sesquilinear forms. Given $f \in X'$, $g \in Q'$ and $l \in M'$, let us consider the saddle-point problem*

$$\begin{array}{l} \textit{Find } (\mathbf{Z}, \mathbf{A}, \mathbf{r}) \in X \times Q \times M : \\ \quad a(\mathbf{Z}, \mathbf{v}) + \overline{b(\mathbf{v}, \mathbf{A})} \qquad\qquad = \langle f, \mathbf{v} \rangle \\ \quad b(\mathbf{Z}, \mathbf{N}) \qquad\qquad + \overline{c(\mathbf{N}, \mathbf{r})} = \langle g, \mathbf{N} \rangle \\ \qquad\qquad c(\mathbf{A}, \mathbf{p}) \qquad\qquad = \langle l, \mathbf{p} \rangle \\ \textit{for all } (\mathbf{v}, \mathbf{N}, \mathbf{p}) \in X \times Q \times M . \end{array} \tag{4.10}$$

Assume that $a(\cdot, \cdot)$, $b(\cdot, \cdot)$ and $c(\cdot, \cdot)$ are continuous, i.e., there exists three positive constants c_1, c_2, c_3 such that

$$\begin{array}{ll} |a(\mathbf{v}, \mathbf{w})| \le c_1 \|\mathbf{v}\|_X \|\mathbf{w}\|_X & \forall\, \mathbf{v}, \mathbf{w} \in X \\ |b(\mathbf{v}, \mathbf{N})| \le c_2 \|\mathbf{v}\|_X \|\mathbf{N}\|_Q & \forall\, \mathbf{v} \in X, \mathbf{N} \in Q \\ |c(\mathbf{N}, \mathbf{p})| \le c_3 \|\mathbf{N}\|_Q \|\mathbf{p}\|_M & \forall\, \mathbf{N} \in Q, \mathbf{p} \in M . \end{array} \tag{4.11}$$

Moreover, setting

$$Q^0 := \{\mathbf{N} \in Q \,|\, c(\mathbf{N}, \mathbf{p}) = 0 \ \forall\, \mathbf{p} \in M\}$$

$$X^0 := \{\mathbf{v} \in X \,|\, b(\mathbf{v}, \mathbf{N}) = 0 \ \forall\, \mathbf{N} \in Q^0\} ,$$

assume that $a(\cdot, \cdot)$ is coercive in X^0, i.e., there exists a positive constant α such that

$$|a(\mathbf{v}, \mathbf{v})| \ge \alpha \|\mathbf{v}\|_X^2 \quad \forall\, \mathbf{v} \in X^0 , \tag{4.12}$$

and that the following inf–sup conditions hold

$$\inf_{\mathbf{N} \in Q^0} \sup_{\mathbf{v} \in X} \frac{|b(\mathbf{v}, \mathbf{N})|}{\|\mathbf{v}\|_X \|\mathbf{N}\|_Q} \ge \beta \tag{4.13}$$

$$\inf_{\mathbf{p} \in M} \sup_{\mathbf{N} \in Q} \frac{|c(\mathbf{N}, \mathbf{p})|}{\|\mathbf{N}\|_Q \|\mathbf{p}\|_M} \ge \gamma , \tag{4.14}$$

for some positive constants β and γ. Then problem (4.10) has a unique solution.

Now we are in position to prove the following result.

Theorem 4.3. *Problem (4.9) has a unique solution.*

Proof. We apply Theorem 4.2, with obvious notation. First of all, let us recall that the spaces W_I and $V_I(\mathbf{0})$ can be characterized as

$$W_I = \{\mathbf{N}_I \in (L^2(\Omega_I))^3 \,|\, \int_{\Omega_I} \varepsilon_I \mathbf{N}_I \cdot \overline{\mathbf{p}_I} = 0 \ \forall\, \mathbf{p}_I \in H^0_{0,\Gamma}(\mathrm{curl}; \Omega_I)\}$$

and

$$V_I(\mathbf{0}) = \{\mathbf{v}_I \in H_{0,\partial\Omega}(\mathrm{curl}; \Omega_I) \,|\, \int_{\Omega_I} \mathrm{curl}\, \mathbf{v}_I \cdot \overline{\mathbf{N}_I} = 0 \ \forall\, \mathbf{N}_I \in W_I\} .$$

Since the bilinear form $\mathcal{C}(\cdot,\cdot)$ is coercive in $H(\mathrm{curl};\Omega_C)\times V_I(\mathbf{0})$, we need only to show that the two inf–sup conditions are satisfied. More precisely, we have to prove that there exist two positive constants β and γ such that

$$\sup_{(\mathbf{z}_C,\mathbf{v}_I)\in\Lambda}\frac{|\int_{\Omega_I}\mathrm{curl}\,\mathbf{v}_I\cdot\overline{\mathbf{N}_I}|}{\|(\mathbf{z}_C,\mathbf{v}_I)\|_\Lambda}\geq\beta\|\mathbf{N}_I\|_{0,\Omega_I}\qquad\forall\mathbf{N}_I\in W_I$$

and

$$\sup_{\mathbf{N}_I\in(L^2(\Omega_I))^3}\frac{|\int_{\Omega_I}\varepsilon_I\mathbf{N}_I\cdot\overline{\mathbf{p}_I}|}{\|\mathbf{N}_I\|_{0,\Omega_I}}\geq\gamma\|\mathbf{p}_I\|_{0,\Omega_I}\qquad\forall\mathbf{p}_I\in H^0_{0,\Gamma}(\mathrm{curl};\Omega_I)\,.$$

Recalling that, from the orthogonal decomposition results (A.12) and (4.8), $\mathbf{N}_I\in W_I$ can be written as $\mathbf{N}_I=\varepsilon_I^{-1}\,\mathrm{curl}\,\mathbf{q}_I$, with $\mathbf{q}_I\in H_{0,\partial\Omega}(\mathrm{curl};\Omega_I)\cap H^0_{0,\Gamma}(\mathrm{div};\Omega_I)$, $\mathbf{q}_I\perp\mathcal{H}(\partial\Omega,\Gamma;\Omega_I)$, choosing $(\mathbf{z}_C,\mathbf{v}_I)=(\mathbf{0},\mathbf{q}_I)\in\Lambda$ we have

$$\frac{|\int_{\Omega_I}\mathrm{curl}\,\mathbf{q}_I\cdot\overline{\mathbf{N}_I}|}{\|\mathbf{q}_I\|_{H(\mathrm{curl};\Omega_I)}}=\frac{|\int_{\Omega_I}\mathrm{curl}\,\mathbf{q}_I\cdot\varepsilon_I^{-1}\,\mathrm{curl}\,\overline{\mathbf{q}_I}|}{\|\mathbf{q}_I\|_{H(\mathrm{curl};\Omega_I)}}\geq\frac{\varepsilon_{\max}^{-1}\int_{\Omega_I}|\,\mathrm{curl}\,\mathbf{q}_I|^2}{\|\mathbf{q}_I\|_{H(\mathrm{curl};\Omega_I)}}\,,$$

where $\varepsilon_{\max}$ is a uniform upper bound in Ω_I for the eigenvalues of $\varepsilon_I(\mathbf{x})$. We know that the Poincaré-like inequality

$$\int_{\Omega_I}|\,\mathrm{curl}\,\mathbf{q}_I|^2\geq C_0\int_{\Omega_I}(|\mathbf{q}_I|^2+|\,\mathrm{curl}\,\mathbf{q}_I|^2)\tag{4.15}$$

holds true for all $\mathbf{q}_I\in H_{0,\partial\Omega}(\mathrm{curl};\Omega_I)\cap H^0_{0,\Gamma}(\mathrm{div};\Omega_I)$, $\mathbf{q}_I\perp\mathcal{H}(\partial\Omega,\Gamma;\Omega_I)$ (see (A.15); for notation see Section A.1). Thus we can choose $\beta:=C_0\varepsilon_{\max}^{-1}\varepsilon_{\min}$, where $\varepsilon_{\min}$ is a uniform lower bound in Ω_I for the eigenvalues of $\varepsilon_I(\mathbf{x})$.

Concerning (4.14), it is enough to take $\mathbf{N}_I=\mathbf{p}_I$ and $\gamma=\varepsilon_{\min}$. □

Remark 4.4. It is worth noting that in (4.9) the Lagrange multiplier $\mathbf{r}_I$ is equal to $\mathbf{0}$. In fact, we have proved in Chapter 3 that the eddy current problem (3.30) has a unique solution $(\mathbf{E},\mathbf{H})\in\Lambda$. It is easy to see that $(\mathbf{E}_{|\Omega_C},\mathbf{H}_{|\Omega_I},-i\omega\mathbf{E}_{|\Omega_I},\mathbf{0})$ is a solution to (4.9), and, since this problem has a unique solution, it follows $\mathbf{r}_I=\mathbf{0}$. □

Remark 4.5. When considering the electric boundary condition (1.20), the problem to be solved reads

$$\begin{cases}\text{Find }(\mathbf{E}_C,\mathbf{H}_I)\in\Lambda^*\,,\ \mathbf{A}_I^*\in(L^2(\Omega_I))^3\,,\ \mathbf{r}_I^*\in H_0^0(\mathrm{curl};\Omega_I):\\ \mathcal{C}((\mathbf{E}_C,\mathbf{H}_I),(\mathbf{z}_C,\mathbf{v}_I))+\int_{\Omega_I}\mathrm{curl}\,\overline{\mathbf{v}_I}\cdot\mathbf{A}_I^*=L(\mathbf{z}_C)\\ \int_{\Omega_I}\mathrm{curl}\,\mathbf{H}_I\cdot\overline{\mathbf{N}_I^*}+\int_{\Omega_I}\varepsilon_I\overline{\mathbf{N}_I^*}\cdot\mathbf{r}_I^*=G(\mathbf{N}_I^*)\\ \int_{\Omega_I}\varepsilon_I\mathbf{A}_I^*\cdot\overline{\mathbf{p}_I^*}=0\\ \text{for all }(\mathbf{z}_C,\mathbf{v}_I)\in\Lambda^*\,,\ \mathbf{N}_I^*\in(L^2(\Omega_I))^3\,,\ \mathbf{p}_I^*\in H_0^0(\mathrm{curl};\Omega_I)\,,\end{cases}\tag{4.16}$$

where

$$\Lambda^*:=H(\mathrm{curl};\Omega_C)\times H(\mathrm{curl};\Omega_I)\,.$$

It is not difficult to adapt to this problem the analysis performed above. □

4.2.1 Finite element discretization

Our aim is to find a Galerkin finite element approximation of (4.9). In order to approximate the space $H^0_{0,\Gamma}(\text{curl}; \Omega_I) = \text{grad}\, H^1_{0,\Gamma}(\Omega_I) \oplus \mathcal{H}_{\varepsilon_I}(\Gamma, \partial\Omega; \Omega_I)$, we recall that the dimension of $\mathcal{H}_{\varepsilon_I}(\Gamma, \partial\Omega; \Omega_I)$ is equal to $p_\Gamma + n_{\partial\Omega}$, where $p_\Gamma + 1$ is the number of connected components of Γ and $n_{\partial\Omega}$ is the number of Γ-independent non-bounding cycles in Ω_I. A basis of $\mathcal{H}_{\varepsilon_I}(\Gamma, \partial\Omega; \Omega_I)$ is given by $\text{grad}\, w_{j,I}$ and $\boldsymbol{\pi}_{k,I}$, $j = 1, \ldots, p_\Gamma$, $k = 1, \ldots, n_{\partial\Omega}$. In particular $w_{j,I}$ belongs to $H^1(\Omega_I)$, $w_{j,I} = 0$ on $\Gamma \setminus \Gamma_j$ and $w_{j,I} = 1$ on Γ_j, $j = 1, \ldots, p_\Gamma$, where Γ_j are the connected components of Γ (see Sections 1.4 and A.1).

Setting

$$H^1_{*,\Gamma}(\Omega_I) := \{\xi_I \in H^1(\Omega_I) \mid \xi_{I|\Gamma_j} \text{ is constant } \forall\, j = 1, \ldots, p_\Gamma, \\ \xi_{I|\Gamma_{p_\Gamma+1}} = 0\}\,, \tag{4.17}$$

we have

$$H^0_{0,\Gamma}(\text{curl}; \Omega_I) = \text{grad}\, H^1_{*,\Gamma}(\Omega_I) \oplus \text{Span}\{\boldsymbol{\pi}_{k,I}\}_{k=1}^{n_{\partial\Omega}}\,.$$

Note that, if Ω_C is connected or, equivalently, Γ is connected, then $H^1_{*,\Gamma}(\Omega_I) = H^1_{0,\Gamma}(\Omega_I)$.

If we assume that the computational domain Ω is simply-connected, then we have $n_{\partial\Omega} = 0$ and any function belonging to $H^0_{0,\Gamma}(\text{curl}; \Omega_I)$ is the gradient of a function in $H^1_{*,\Gamma}(\Omega_I)$. On the other hand, if the computational domain Ω is not simply-connected, the number $n_{\partial\Omega}$ can be different from 0, and in this case one needs the construction of a suitable set of "cutting" surfaces in order to approximate the basis functions $\boldsymbol{\pi}_{k,I}$, $k = 1, \ldots, n_{\partial\Omega}$ (more details about this general geometrical situation can be found in Section 5.5).

In order to simplify the presentation, in the following we assume that *the computational domain Ω is simply-connected*, so that $n_{\partial\Omega} = 0$ and $H^0_{0,\Gamma}(\text{curl}; \Omega_I) = \text{grad}\, H^1_{*,\Gamma}(\Omega_I)$. Hence, problem (4.9) in fact reads

$$\begin{aligned} &\text{Find } (\mathbf{E}_C, \mathbf{H}_I) \in \Lambda\,,\ \mathbf{A}_I \in (L^2(\Omega_I))^3\,,\ \phi_I \in H^1_{*,\Gamma}(\Omega_I) : \\ &\quad \mathcal{C}((\mathbf{E}_C, \mathbf{H}_I), (\mathbf{z}_C, \mathbf{v}_I)) + \textstyle\int_{\Omega_I} \text{curl}\, \overline{\mathbf{v}_I} \cdot \mathbf{A}_I = L(\mathbf{z}_C) \\ &\quad \textstyle\int_{\Omega_I} \text{curl}\, \mathbf{H}_I \cdot \overline{\mathbf{N}_I} + \int_{\Omega_I} \varepsilon_I \overline{\mathbf{N}_I} \cdot \text{grad}\, \phi_I = G(\mathbf{N}_I) \\ &\quad \textstyle\int_{\Omega_I} \varepsilon_I \mathbf{A}_I \cdot \text{grad}\, \overline{\xi_I} = 0 \\ &\text{for all } (\mathbf{z}_C, \mathbf{v}_I) \in \Lambda\,,\ \mathbf{N}_I \in (L^2(\Omega_I))^3\,,\ \xi_I \in H^1_{*,\Gamma}(\Omega_I)\,. \end{aligned} \tag{4.18}$$

As noted in Remark 4.4, the Lagrange multiplier $\mathbf{r}_I$ is equal to $\mathbf{0}$: here, therefore, we have that the Lagrange multiplier ϕ_I is 0, too.

We assume that Ω, Ω_C, Ω_I are Lipschitz polyhedral domains and consider two families of regular tetrahedral meshes $\mathcal{T}_{C,h}$ and $\mathcal{T}_{I,h}$ of Ω_C and Ω_I, respectively. To approximate the functions in $H(\text{curl}; \Omega_L)$, $L = I, C$, we employ the complex-valued Nédélec curl-conforming edge elements of the lowest order

$$N^1_{L,h} := \{\mathbf{v}_h \in H(\text{curl}; \Omega_L) \mid \mathbf{v}_h(\mathbf{x})_{|K} = \mathbf{a}_K + \mathbf{b}_K \times \mathbf{x} \quad \forall\, K \in \mathcal{T}_{L,h}\},$$

where $\mathbf{a}_K$ and $\mathbf{b}_K$ are constant complex vectors (see Section A.2). The homogeneous boundary conditions on $\partial\Omega$ are incorporated by setting the degrees of freedom on $\partial\Omega$ equal to 0, leading to the space

$$X^1_{I,h} := N^1_{I,h} \cap H_{0,\partial\Omega}(\text{curl}; \Omega_I) .$$

We also set $\Lambda_h := N^1_{C,h} \times X^1_{I,h}$.

Let us denote by $\mathbb{P}_k$ the standard space of complex polynomials of total degree less than or equal to k with respect to the real variable $\mathbf{x}$. To discretize the Lagrange multiplier $\mathbf{A}_I \in (L^2(\Omega_I))^3$ we choose piecewise-constant vector functions in the space

$$Q_{I,h} := (C^0_{I,h})^3 ,$$

where

$$C^0_{I,h} := \{q_{I,h} \in L^2(\Omega_I) \mid q_{I,h|K} \in \mathbb{P}_0 \; \forall K \in \mathcal{T}_{I,h}\} .$$

In order to approximate the space $H^1_{*,\Gamma}(\Omega_I)$, we start from the non-conforming Crouzeix–Raviart elements defined as follows

$$U_{I,h} := \{\xi_{I,h} \in L^2(\Omega_I) \mid \xi_{I,h|K} \in \mathbb{P}_1 \; \forall K \in \mathcal{T}_{I,h} \text{ and } \xi_{I,h} \text{ is continuous at the centroid of any face } f \text{ common to two elements in } \mathcal{T}_{I,h}\} .$$

Then the discrete $\xi_{I,h}$ are chosen in the space $M_{I,h}$ defined as

$$M_{I,h} := \{\xi_{I,h} \in U_{I,h} \mid \xi_{I,h} \text{ takes the same value at all centroids of faces of } \Gamma_j,\ j = 1, \ldots, p_\Gamma, \text{ and } \xi_{I,h} = 0 \text{ at all centroids of faces of } \Gamma_{p_\Gamma+1}\} . \tag{4.19}$$

It is worth noting that we have limited ourselves to consider the lowest order finite element spaces because, to our knowledge, a stability and convergence analysis for higher order elements has not been performed.

Since functions in $U_{I,h}$ are no longer continuous, they are no longer in $H^1(\Omega_I)$. Therefore we must define a sesquilinear form $\mathcal{S}_h(\cdot,\cdot)$ acting also on $H^1_{*,\Gamma}(\Omega_I)+M_{I,h}$, and a norm on $H^1_{*,\Gamma}(\Omega_I) + M_{I,h}$. This can be done as follows: first, for each $\xi_I \in [H^1_{*,\Gamma}(\Omega_I) + M_{I,h}]$ we denote by $\widetilde{\text{grad}}\,\xi_I$ the function in $(L^2(\Omega_I))^3$ defined as

$$(\widetilde{\text{grad}}\,\xi_I)_{|K} := \text{grad}(\xi_{I|K}) \quad \forall K \in \mathcal{T}_{I,h}.$$

If $\xi_I \in H^1_{*,\Gamma}(\Omega_I)$. Clearly one has $\widetilde{\text{grad}}\,\xi_I = \text{grad}\,\xi_I$. Then, we define the norm in $H^1_{*,\Gamma}(\Omega_I) + M_{I,h}$ as

$$\|\xi_I\|^2_h := \sum_K \int_K |\,\text{grad}\,\xi_I|^2 = \|\widetilde{\text{grad}}\,\xi_I\|^2_{0,\Omega_I} .$$

For all $\mathbf{N}_I \in (L^2(\Omega_I))^3$ and $\xi_I \in [H^1_{*,\Gamma}(\Omega_I) + M_{I,h}]$ we set

$$\mathcal{S}_h(\mathbf{N}_I, \xi_I) := \int_{\Omega_I} \varepsilon_I \mathbf{N}_I \cdot \widetilde{\text{grad}}\,\overline{\xi_I} . \tag{4.20}$$

Introducing also the sesquilinear form

$$\mathcal{R}(\mathbf{v}_I, \mathbf{N}_I) := \int_{\Omega_I} \operatorname{curl} \mathbf{v}_I \cdot \overline{\mathbf{N}_I} \,, \tag{4.21}$$

the finite element approximation of (4.18) can be formulated as follows

$$\begin{aligned}
&\text{Find } (\mathbf{E}_{C,h}, \mathbf{H}_{I,h}) \in \Lambda_h, \mathbf{A}_{I,h} \in Q_{I,h}, \phi_{I,h} \in M_{I,h} : \\
&\mathcal{C}((\mathbf{E}_{C,h}, \mathbf{H}_{I,h}), (\mathbf{z}_{C,h}, \mathbf{v}_{I,h}) + \overline{\mathcal{R}(\mathbf{v}_{I,h}, \mathbf{A}_{I,h})} = L(\mathbf{z}_{C,h}) \\
&\mathcal{R}(\mathbf{H}_{I,h}, \mathbf{N}_{I,h}) + \overline{\mathcal{S}_h(\mathbf{N}_{I,h}, \phi_{I,h})} = G(\mathbf{N}_{I,h}) \\
&\mathcal{S}_h(\mathbf{A}_{I,h}, \xi_{I,h}) = 0 \\
&\text{for all } (\mathbf{z}_{C,h}, \mathbf{v}_{I,h}) \in \Lambda_h, \mathbf{N}_{I,h} \in Q_{I,h}, \xi_{I,h} \in M_{I,h} \,.
\end{aligned} \tag{4.22}$$

The following results will be crucial in the proof of existence and uniqueness of a solution to problem (4.22). The proof follows the one in Monk [178].

Lemma 4.6. *We have the $L^2(\Omega_I)$-orthogonal decomposition*

$$Q_{I,h} = \operatorname{curl} X^1_{I,h} \oplus \widetilde{\operatorname{grad}}\, M_{I,h} \,.$$

Proof. The proof has two parts. In the first part we show that for all $\mathbf{v}_{I,h} \in X^1_{I,h}$ and $\xi_{I,h} \in M_{I,h}$ it holds $\int_{\Omega_I} \operatorname{curl} \mathbf{v}_{I,h} \cdot \widetilde{\operatorname{grad}}\, \xi_{I,h} = 0$. In the second part we establish that $\dim(Q_{I,h}) = \dim(\operatorname{curl} X^1_{I,h}) + \dim(\widetilde{\operatorname{grad}}\, M_{I,h})$.

For any $\mathbf{v}_{I,h} \in X^1_{I,h}$ and $\xi_{I,h} \in M_{I,h}$ integration by parts yields

$$\begin{aligned}
\textstyle\int_{\Omega_I} \operatorname{curl} \mathbf{v}_{I,h} \cdot \widetilde{\operatorname{grad}}\, \xi_{I,h} &= \textstyle\sum_K \int_K \operatorname{curl} \mathbf{v}_{I,h} \cdot \operatorname{grad} \xi_{I,h} \\
&= \textstyle\sum_K \int_{\partial K} \operatorname{curl} \mathbf{v}_{I,h} \cdot \mathbf{n}_K\, \xi_{I,h} \\
&= \textstyle\sum_{f \in \mathcal{F}_{\text{int}}} \int_f \operatorname{curl} \mathbf{v}_{I,h} \cdot \mathbf{n}_f\, [\xi_{I,h}]_f \\
&\quad + \textstyle\sum_{f \in \mathcal{F}_{\partial\Omega}} \int_f \operatorname{curl} \mathbf{v}_{I,h} \cdot \mathbf{n}\, \xi_{I,h} \\
&\quad + \textstyle\sum_{j=1}^{p_\Gamma+1} \sum_{f \in \mathcal{F}_{\Gamma_j}} \int_f \operatorname{curl} \mathbf{v}_{I,h} \cdot \mathbf{n}_I\, \xi_{I,h} \,,
\end{aligned}$$

where $\mathcal{F}_{\text{int}}$ is the set of internal faces of the triangulation $\mathcal{T}_{I,h}$, $\mathcal{F}_{\partial\Omega}$ and $\mathcal{F}_{\Gamma_j}$ denote the set of faces of $\mathcal{T}_{I,h}$ on $\partial\Omega$ and Γ_j, respectively, and $[\xi_{I,h}]_f$ denotes the jump of $\xi_{I,h}$ across the face f. Note that, for all $f \in \mathcal{F}_{\text{int}}$, $(\operatorname{curl} \mathbf{v}_{I,h} \cdot \mathbf{n}_f)_{|f}$ is constant and $\int_f [\xi_{I,h}]_f = 0$, since $[\xi_{I,h}]_f$ is a linear function and it is equal to 0 at the centroid of f. Moreover $(\operatorname{curl} \mathbf{v}_{I,h} \cdot \mathbf{n})_{|f} = 0$ for all $f \in \mathcal{F}_{\partial\Omega}$, and, using that for all $j = 1, \ldots, p_\Gamma+1$ and for all faces $f \in \mathcal{F}_{\Gamma_j}$ one has $\int_f \xi_{I,h} = \xi_j \mathrm{meas}(f)$, where ξ_j is the value at the centroid of f, independent of $f \in \Gamma_j$, we finally find

$$\begin{aligned}
&\textstyle\sum_{f \in \mathcal{F}_{\Gamma_j}} \int_f \operatorname{curl} \mathbf{v}_{I,h} \cdot \mathbf{n}_I\, \xi_{I,h} = \sum_{f \in \mathcal{F}_{\Gamma_j}} (\operatorname{curl} \mathbf{v}_{I,h} \cdot \mathbf{n}_I)_{|f} \xi_j \mathrm{meas}(f) \\
&\qquad = \xi_j \textstyle\int_{\Gamma_j} \operatorname{curl} \mathbf{v}_{I,h} \cdot \mathbf{n}_I = 0 \,,
\end{aligned}$$

hence

$$\int_{\Omega_I} \operatorname{curl} \mathbf{v}_{I,h} \cdot \widetilde{\operatorname{grad}}\, \xi_{I,h} = 0 \,.$$

Let us introduce the Raviart–Thomas finite element space (see, e.g., Brezzi and Fortin [65], Chap. III)

$$RT_h := \{\mathbf{v}_h \in H(\mathrm{div}; \Omega_I) \mid \mathbf{v}_h(\mathbf{x})_{|K} = \mathbf{a}_K + b_K\,\mathbf{x} \quad \forall\, K \in \mathcal{T}_{I,h}\},$$

where $\mathbf{a}_K$ is a constant complex vector and b_K is a complex number, and the subspaces

$$RT_{0,\partial\Omega} := RT_h \cap H_{0,\partial\Omega}(\mathrm{div}, \Omega_I), \quad RT^0_{0,\partial\Omega} := RT_h \cap H^0_{0,\partial\Omega}(\mathrm{div}, \Omega_I).$$

By arguments from discrete cohomology, it can be proved (see Bossavit [59]) that, as vector spaces on $\mathbb{C}$,

$$\dim(\mathrm{curl}\, X^1_{I,h}) = \dim(RT^0_{0,\partial\Omega}) - p_\Gamma .$$

Let us denote by $\#K$ the number of tetrahedra of $\mathcal{T}_{I,h}$, by $\#\mathcal{F}$ the total number of faces of $\mathcal{T}_{I,h}$, and by $\#\mathcal{F}_{\partial\Omega}$ and $\#\mathcal{F}_\Gamma$ the number of faces of $\mathcal{T}_{I,h}$ on $\partial\Omega$ and on Γ, respectively. It is not difficult to prove that

$$\begin{aligned}
&\dim(RT^0_{0,\partial\Omega}) = \dim(RT_{0,\partial\Omega}) - \dim(\mathrm{div}(RT_{0,\partial\Omega}))\\
&\qquad = (\#\mathcal{F} - \#\mathcal{F}_{\partial\Omega}) - \#K\\
&\dim(M_{I,h}) = (\#\mathcal{F} - \#\mathcal{F}_\Gamma) + p_\Gamma = \dim(\widetilde{\mathrm{grad}}\,(M_{I,h}))\\
&\dim(Q_{I,h}) = 3\#K .
\end{aligned}$$

Since $4\#K = 2\#\mathcal{F} - (\#\mathcal{F}_{\partial\Omega} + \#\mathcal{F}_\Gamma)$, then

$$\begin{aligned}
&\dim(\mathrm{curl}\, X^1_{I,h}) + \dim(\widetilde{\mathrm{grad}}\, M_{I,h})\\
&\quad = [(\#\mathcal{F} - \#\mathcal{F}_{\partial\Omega}) - \#K - p_\Gamma] + [(\#\mathcal{F} - \#\mathcal{F}_\Gamma) + p_\Gamma]\\
&\quad = 2\#\mathcal{F} - (\#\mathcal{F}_{\partial\Omega} + \#\mathcal{F}_\Gamma) - \#K\\
&\quad = 4\#K - \#K\\
&\quad = \dim(Q_{I,h}).
\end{aligned}$$

Since, as it can be trivially checked, $\mathrm{curl}\, X^1_{I,h} \subset Q_{I,h}$ and $\widetilde{\mathrm{grad}}\, M_{I,h} \subset Q_{I,h}$, the proof is complete. □

The following lemma is the discrete counterpart of (4.15).

Lemma 4.7. *Let* $V^0_{I,h} := X^1_{I,h} \cap V_I(\mathbf{0})$ *and*

$$(V^0_{I,h})^\perp := \left\{\mathbf{p}_{I,h} \in X^1_{I,h} \mid \int_{\Omega_I} \mathbf{p}_{I,h} \cdot \overline{\mathbf{v}_{I,h}} = 0 \;\forall\, \mathbf{v}_{I,h} \in V^0_{I,h}\right\} .$$

There exists a constant $C_0 > 0$, *independent of* h, *such that for all* $\mathbf{p}_{I,h} \in (V^0_{I,h})^\perp$ *the following inequality holds*

$$\|\mathbf{p}_{I,h}\|_{0,\Omega_I} \leq C_0 \|\,\mathrm{curl}\, \mathbf{p}_{I,h}\|_{0,\Omega_I} .$$

Proof. Given $\mathbf{p}_{I,h} \in (V^0_{I,h})^\perp \subset (L^2(\Omega_I))^3$, taking $\boldsymbol{\mu}_I = \mathrm{Id}$ in the orthogonal decomposition result presented in Theorem A.7 we can write

$$\mathbf{p}_{I,h} = \operatorname{curl} \mathbf{Q}_I + \operatorname{grad} \chi_I + \mathbf{k}_I \,,$$

with

$$\mathbf{Q}_I \in H_{0,\Gamma}(\operatorname{curl}; \Omega_I) \cap H^0_{0,\partial\Omega}(\operatorname{div}; \Omega_I) \cap \mathcal{H}(\Gamma, \partial\Omega; \Omega_I)^\perp \,,$$
$$\chi_I \in H^1_{0,\partial\Omega}(\Omega_I)$$

and

$$\mathbf{k}_I \in \mathcal{H}(\partial\Omega, \Gamma; \Omega_I) \,.$$

Setting $\mathbf{U}_I = \operatorname{curl} \mathbf{Q}_I$, we have $\operatorname{curl} \mathbf{U}_I = \operatorname{curl} \mathbf{p}_{I,h}$, $\operatorname{div} \mathbf{U}_I = 0$, $\mathbf{U}_I \cdot \mathbf{n}_I = 0$ on Γ and $\mathbf{U}_I \times \mathbf{n} = \mathbf{0}$ on $\partial\Omega$. Hence $\mathbf{U}_I \in H_{0,\partial\Omega}(\operatorname{curl}; \Omega_I) \cap H_{0,\Gamma}(\operatorname{div}; \Omega_I)$, a space that is continuously embedded in $(H^s(\Omega_I))^3$ for some $s > 1/2$ (see Amrouche et al. [27]), so that there exists a positive constant C such that

$$\|\mathbf{U}_I\|_{s,\Omega_I} \le C(\|\mathbf{U}_I\|_{0,\Omega_I} + \|\operatorname{curl} \mathbf{U}_I\|_{0,\Omega_I}) \,.$$

Since it is easily verified that $\mathbf{U}_I \in \mathcal{H}(\partial\Omega, \Gamma; \Omega_I)^\perp$, the Poincaré-like inequality (A.15) gives

$$\|\mathbf{U}_I\|_{s,\Omega_I} \le C_1 \|\operatorname{curl} \mathbf{U}_I\|_{0,\Omega_I} \,. \tag{4.23}$$

Moreover, since $\operatorname{curl} \mathbf{U}_I = \operatorname{curl} \mathbf{p}_{I,h} \in (L^\infty(\Omega_I))^3$, the interpolant $\Pi_h \mathbf{U}_I$ in $N^1_{I,h}$ is well-defined (see Amrouche et al. [27]). Note that

$$\Pi_h \mathbf{p}_{I,h} = \mathbf{p}_{I,h} = \Pi_h \mathbf{U}_I + \Pi_h(\operatorname{grad} \chi_I + \mathbf{k}_I)$$

and $\Pi_h(\operatorname{grad} \chi_I + \mathbf{k}_I) \in V^0_{I,h}$, hence

$$\begin{aligned} \|\mathbf{p}_{I,h}\|^2_{0,\Omega_I} &= \int_{\Omega_I} \mathbf{p}_{I,h} \cdot \overline{\mathbf{p}_{I,h}} \\ &= \int_{\Omega_I} \mathbf{p}_{I,h} \cdot [\Pi_h \overline{\mathbf{U}_I} + \Pi_h(\operatorname{grad} \overline{\chi_I} + \overline{\mathbf{k}_I})] \\ &= \int_{\Omega_I} \mathbf{p}_{I,h} \cdot \Pi_h \overline{\mathbf{U}_I} \\ &\le \|\mathbf{p}_{I,h}\|_{0,\Omega_I} \|\Pi_h \mathbf{U}_I\|_{0,\Omega_I} . \end{aligned} \tag{4.24}$$

On the other hand, for all $K \in \mathcal{T}_{I,h}$

$$\begin{aligned} \|\Pi_h \mathbf{U}_I - \mathbf{U}_I\|_{0,K} &\le C_2 h^s (\|\mathbf{U}_I\|_{s,K} + \|\operatorname{curl} \mathbf{U}_I\|_{s,K}) \\ &\le C_2 h^s (\|\mathbf{U}_I\|_{s,K} + \|\operatorname{curl} \mathbf{p}_{I,h}\|_{s,K}) \\ &\le C_3 (h^s \|\mathbf{U}_I\|_{s,K} + \|\operatorname{curl} \mathbf{p}_{I,h}\|_{0,K}) \,, \end{aligned}$$

where we have used the local inverse estimate

$$\|\operatorname{curl} \mathbf{p}_{I,h}\|_{s,K} \le C h^{-s} \|\operatorname{curl} \mathbf{p}_{I,h}\|_{0,K} \,.$$

Using (4.23), this yields

$$\begin{aligned} \|\Pi_h \mathbf{U}_I\|_{0,\Omega_I} &\le \|\mathbf{U}_I\|_{0,\Omega_I} + \|\Pi_h \mathbf{U}_I - \mathbf{U}_I\|_{0,\Omega_I} \\ &\le C_4 (\|\mathbf{U}_I\|_{s,\Omega_I} + \|\operatorname{curl} \mathbf{p}_{I,h}\|_{0,\Omega_I}) \\ &\le C_5 \|\operatorname{curl} \mathbf{p}_{I,h}\|_{0,\Omega_I} \,, \end{aligned}$$

thus the thesis follows from (4.24). □

Now we are in a position to prove the main result of this section.

Theorem 4.8. *Assume that ε_I is a piecewise-constant matrix in Ω_I and that $\mathcal{T}_{I,h^0}$ is a triangulation such that $\varepsilon_{I|K}$ is a constant matrix for each $K \in \mathcal{T}_{I,h^0}$. Assuming that the triangulation $\mathcal{T}_{I,h}$ is a refinement of $\mathcal{T}_{I,h^0}$, problem* (4.22) *has a unique solution.*

Proof. As in the continuous case we use Theorem 4.2. Since for each $K \in \mathcal{T}_{I,h}$ we know that $\varepsilon_{I|K}$ is a constant matrix, $\varepsilon_I \mathbf{N}_{I,h} \in Q_{I,h}$ and $\varepsilon_I^{-1}\mathbf{N}_{I,h} \in Q_{I,h}$ for any $\mathbf{N}_{I,h} \in Q_{I,h}$. From Lemma 4.6 one has that $Q_{I,h} = \operatorname{curl} X^1_{I,h} \oplus \widetilde{\operatorname{grad}}\, M_{I,h}$, thus the space $Q^0_{I,h}$ defined by

$$\begin{aligned} Q^0_{I,h} &:= \{\mathbf{N}_{I,h} \in Q_{I,h} \,|\, \mathcal{S}_h(\mathbf{N}_{I,h}, \xi_{I,h}) = 0 \quad \forall\, \xi_{I,h} \in M_{I,h}\} \\ &= \{\mathbf{N}_{I,h} \in Q_{I,h} \,|\, \textstyle\int_{\Omega_I} \varepsilon_I \mathbf{N}_{I,h} \cdot \widetilde{\operatorname{grad}}\, \overline{\xi_{I,h}} = 0 \quad \forall\, \xi_{I,h} \in M_{I,h}\} \end{aligned}$$

is equal to $\varepsilon_I^{-1} \operatorname{curl} X^1_{I,h}$. Let us set

$$\Lambda^0_h := \{(\mathbf{z}_{C,h}, \mathbf{v}_{I,h}) \in \Lambda_h \,|\, \mathcal{R}(\mathbf{v}_{I,h}, \mathbf{N}_{I,h}) = 0 \quad \forall\, \mathbf{N}_{I,h} \in Q^0_{I,h}\}\,. \tag{4.25}$$

If $(\mathbf{z}_{C,h}, \mathbf{v}_{I,h}) \in \Lambda^0_h$, taking in particular in (4.25) $\mathbf{N}_{I,h} = \varepsilon_I^{-1} \operatorname{curl} \mathbf{v}_{I,h} \in Q^0_{I,h}$ it follows that $\operatorname{curl} \mathbf{v}_{I,h} = \mathbf{0}$ in Ω_I, hence $\Lambda^0_h = N^1_{C,h} \times V^0_{I,h} \subset H(\operatorname{curl}; \Omega_C) \times V_I(\mathbf{0})$, and (4.12) follows from Theorem 4.1.

Moreover, if $\mathbf{N}_{I,h} \in Q^0_{I,h}$ we have $\mathbf{N}_{I,h} = \varepsilon_I^{-1} \operatorname{curl} \mathbf{w}_{I,h}$ for some $\mathbf{w}_{I,h} \in X^1_{I,h}$. Taking the orthogonal projection $\mathbf{w}^*_{I,h}$ of $\mathbf{w}_{I,h}$ on $(V^0_{I,h})^\perp$, one clearly has $\mathbf{N}_{I,h} = \varepsilon_I^{-1} \operatorname{curl} \mathbf{w}^*_{I,h}$. Hence, choosing $(\mathbf{z}_{C,h}, \mathbf{v}_{I,h}) = (\mathbf{0}, \mathbf{w}^*_{I,h})$, by proceeding as in the continuous case and using Lemma 4.7 we obtain that the inf–sup condition (4.13) holds.

Concerning the inf–sup condition (4.14), let us note that for all $\xi_{I,h} \in M_{I,h}$ one has $\widetilde{\operatorname{grad}}\, \xi_{I,h} \in Q_{I,h}$, hence from the definition of the norm $\|\cdot\|_h$

$$\begin{aligned} \sup_{\mathbf{N}_{I,h} \in Q_{I,h}} \frac{|\mathcal{S}_h(\mathbf{N}_{I,h}, \xi_{I,h})|}{\|\mathbf{N}_{I,h}\|_{0,\Omega_I}} &\geq \frac{|\mathcal{S}_h(\widetilde{\operatorname{grad}}\, \xi_{I,h}, \xi_{I,h})|}{\|\widetilde{\operatorname{grad}}\, \xi_{I,h}\|_{0,\Omega_I}} \\ &= \frac{\int_{\Omega_I} \varepsilon_I \widetilde{\operatorname{grad}}\, \xi_{I,h} \cdot \widetilde{\operatorname{grad}}\, \overline{\xi_{I,h}}}{\|\widetilde{\operatorname{grad}} \xi_{I,h}\|_{0,\Omega_I}} \geq \varepsilon_{\min} \|\xi_{I,h}\|_h\,, \end{aligned}$$

and the proof is complete. □

We denote by c_1 and c_2 the continuity constants of the sesquilinear forms $\mathcal{C}(\cdot,\cdot)$ and $\mathcal{R}(\cdot,\cdot)$, respectively, by α the coerciveness constant in Λ^0_h of the sesquilinear form $\mathcal{C}(\cdot,\cdot)$, and by β and γ the two positive constants related to the discrete inf–sup conditions proved to hold in Theorem 4.8. It is easily shown that all these constants are independent of h. Then the convergence of the finite element approximation method can be proved.

Theorem 4.9. *Let the assumptions of Theorem 4.8 be satisfied. Suppose that* $(\mathbf{E}_C, \mathbf{H}_I) \in \Lambda$, $\mathbf{A}_I \in (L^2(\Omega_I))^3$ *and* $\phi_I = 0$ *are the solution of problem* (4.18) *and that*

$(\mathbf{E}_{C,h}, \mathbf{H}_{I,h}) \in \Lambda_h$, $\mathbf{A}_{I,h} \in Q_{I,h}$ and $\phi_{I,h} \in M_{I,h}$ are the solution of problem (4.22). Then the following error estimates hold

$$\begin{aligned} &\|(\mathbf{E}_C - \mathbf{E}_{C,h}, \mathbf{H}_I - \mathbf{H}_{I,h})\|_\Lambda \\ &\quad \leq \big(1 + \frac{c_1}{\alpha}\big)\big(1 + \frac{c_2}{\beta}\big) \inf_{(\mathbf{z}_{C,h}, \mathbf{v}_{I,h}) \in \Lambda_h} \|(\mathbf{E}_C - \mathbf{z}_{C,h}, \mathbf{H}_I - \mathbf{v}_{I,h})\|_\Lambda \end{aligned} \tag{4.26}$$

$$\begin{aligned} \|\mathbf{A}_I - \mathbf{A}_{I,h}\|_{0,\Omega_I} \leq \big(1 + \frac{c_2}{\beta}\big) &\inf_{\mathbf{N}_{I,h} \in Q^0_{I,h}} \|\mathbf{A}_I - \mathbf{N}_{I,h}\|_{0,\Omega_I} \\ &+ \frac{c_1}{\beta} \|(\mathbf{E}_C - \mathbf{E}_{C,h}, \mathbf{H}_I - \mathbf{H}_{I,h})\|_\Lambda \end{aligned} \tag{4.27}$$

$$\|\phi_I - \phi_{I,h}\|_h = \|\phi_{I,h}\|_h \leq \frac{c_2}{\gamma} \|(\mathbf{E}_C - \mathbf{E}_{C,h}, \mathbf{H}_I - \mathbf{H}_{I,h})\|_\Lambda \,. \tag{4.28}$$

In particular, the finite element approximation method is convergent.

Proof. The proof follows the lines of the results in Brezzi and Fortin [65], Chap. II. For all $(\mathbf{z}^*_{C,h}, \mathbf{v}^*_{I,h}), (\mathbf{z}_{C,h}, \mathbf{v}_{I,h}) \in \Lambda_h$ and $\mathbf{N}_{I,h} \in Q_{I,h}$ it holds

$$\begin{aligned} &\mathcal{C}((\mathbf{E}_{C,h} - \mathbf{z}^*_{C,h}, \mathbf{H}_{I,h} - \mathbf{v}^*_{I,h}), (\mathbf{z}_{C,h}, \mathbf{v}_{I,h})) + \overline{\mathcal{R}(\mathbf{v}_{I,h}, \mathbf{A}_{I,h} - \mathbf{N}_{I,h})} \\ &\quad = L(\mathbf{z}_{C,h}) - \mathcal{C}((\mathbf{z}^*_{C,h}, \mathbf{v}^*_{I,h}), (\mathbf{z}_{C,h}, \mathbf{v}_{I,h})) - \overline{\mathcal{R}(\mathbf{v}_{I,h}, \mathbf{N}_{I,h})} \\ &\quad = \mathcal{C}((\mathbf{E}_C - \mathbf{z}^*_{C,h}, \mathbf{H}_I - \mathbf{v}^*_{I,h}), (\mathbf{z}_{C,h}, \mathbf{v}_{I,h})) + \overline{\mathcal{R}(\mathbf{v}_{I,h}, \mathbf{A}_I - \mathbf{N}_{I,h})} \,. \end{aligned}$$

In particular, if $(\mathbf{z}_{C,h}, \mathbf{v}_{I,h}) \in \Lambda^0_h$ then we know that $\operatorname{curl} \mathbf{v}_{I,h} = \mathbf{0}$ in Ω_I, hence

$$\begin{aligned} &\mathcal{C}((\mathbf{E}_{C,h} - \mathbf{z}^*_{C,h}, \mathbf{H}_{I,h} - \mathbf{v}^*_{I,h}), (\mathbf{z}_{C,h}, \mathbf{v}_{I,h})) \\ &\quad = \mathcal{C}((\mathbf{E}_C - \mathbf{z}^*_{C,h}, \mathbf{H}_I - \mathbf{v}^*_{I,h}), (\mathbf{z}_{C,h}, \mathbf{v}_{I,h})) \,. \end{aligned}$$

Let us define

$$\Lambda^G_h := \{(\mathbf{z}_{C,h}, \mathbf{v}_{I,h}) \in \Lambda_h \mid \mathcal{R}(\mathbf{v}_{I,h}, \mathbf{N}_{I,h}) = G(\mathbf{N}_{I,h}) \ \forall \mathbf{N}_{I,h} \in Q^0_{I,h}\} \,.$$

Clearly we have $(\mathbf{E}_{C,h}, \mathbf{H}_{I,h}) \in \Lambda^G_h$, then for each $(\mathbf{z}^*_{C,h}, \mathbf{v}^*_{I,h}) \in \Lambda^G_h$ we obtain $(\mathbf{E}_{C,h} - \mathbf{z}^*_{C,h}, \mathbf{H}_{I,h} - \mathbf{v}^*_{I,h}) \in \Lambda^0_h$ and

$$\begin{aligned} &\mathcal{C}((\mathbf{E}_{C,h} - \mathbf{z}^*_{C,h}, \mathbf{H}_{I,h} - \mathbf{v}^*_{I,h}), (\mathbf{E}_{C,h} - \mathbf{z}^*_{C,h}, \mathbf{H}_{I,h} - \mathbf{v}^*_{I,h})) \\ &\quad = \mathcal{C}((\mathbf{E}_C - \mathbf{z}^*_{C,h}, \mathbf{H}_I - \mathbf{v}^*_{I,h}), (\mathbf{E}_{C,h} - \mathbf{z}^*_{C,h}, \mathbf{H}_{I,h} - \mathbf{v}^*_{I,h})) \,. \end{aligned}$$

From the continuity and the coerciveness of the sesquilinear form $\mathcal{C}(\cdot,\cdot)$ in Λ^0_h we conclude

$$\begin{aligned} &\|(\mathbf{E}_C - \mathbf{E}_{C,h}, \mathbf{H}_I - \mathbf{H}_{I,h})\|_\Lambda \\ &\quad \leq \|(\mathbf{E}_C - \mathbf{z}^*_{C,h}, \mathbf{H}_I - \mathbf{v}^*_{I,h})\|_\Lambda + \|(\mathbf{E}_{C,h} - \mathbf{z}^*_{C,h}, \mathbf{H}_{I,h} - \mathbf{v}^*_{I,h})\|_\Lambda \\ &\quad \leq \big(1 + \tfrac{c_1}{\alpha}\big) \|(\mathbf{E}_C - \mathbf{z}^*_{C,h}, \mathbf{H}_I - \mathbf{v}^*_{I,h})\|_\Lambda \end{aligned} \tag{4.29}$$

for all $(\mathbf{z}^*_{C,h}, \mathbf{v}^*_{I,h}) \in \Lambda^G_h$.

From the inf–sup condition (4.13), for all $\mathbf{v}_{I,h} \in X^1_{I,h}$ there exists a unique $(\mathbf{u}_{C,h}, \mathbf{w}_{I,h}) \in (\Lambda^0_h)^\perp = \{\mathbf{0}\} \times (V^0_{I,h})^\perp$ such that for all $\mathbf{N}_{I,h} \in Q^0_{I,h}$ one has $\mathcal{R}(\mathbf{w}_{I,h}, \mathbf{N}_{I,h}) = \mathcal{R}(\mathbf{H}_I - \mathbf{v}_{I,h}, \mathbf{N}_{I,h})$ and moreover

$$\|\mathbf{w}_{I,h}\|_{H(\mathrm{curl};\Omega_I)} \leq \frac{c_2}{\beta} \|\mathbf{H}_I - \mathbf{v}_{I,h}\|_{H(\mathrm{curl};\Omega_I)} \, .$$

Setting $\mathbf{v}^*_{I,h} := \mathbf{w}_{I,h} + \mathbf{v}_{I,h}$, for all $\mathbf{N}_{I,h} \in Q^0_{I,h}$ we have

$$\begin{aligned}\mathcal{R}(\mathbf{v}^*_{I,h}, \mathbf{N}_{I,h}) &= \mathcal{R}(\mathbf{H}_I, \mathbf{N}_{I,h}) \\ &= \mathcal{R}(\mathbf{H}_I, \mathbf{N}_{I,h}) + \textstyle\int_{\Omega_I} \varepsilon_I \overline{\mathbf{N}_{I,h}} \cdot \operatorname{grad} \phi_I = G(\mathbf{N}_{I,h}) \, ,\end{aligned}$$

as $\phi_I = 0$. Therefore, for each $(\mathbf{z}_{C,h}, \mathbf{v}_{I,h}) \in \Lambda_h$ we have $(\mathbf{z}_{C,h}, \mathbf{v}^*_{I,h}) \in \Lambda^G_h$ and

$$\begin{aligned}&\|(\mathbf{E}_C - \mathbf{z}_{C,h}, \mathbf{H}_I - \mathbf{v}^*_{I,h})\|_\Lambda \\ &\quad \leq \|(\mathbf{E}_C - \mathbf{z}_{C,h}, \mathbf{H}_I - \mathbf{v}_{I,h})\|_\Lambda + \|\mathbf{w}_{I,h}\|_{H(\mathrm{curl};\Omega_I)} \\ &\quad \leq \left(1 + \tfrac{c_2}{\beta}\right) \|(\mathbf{E}_C - \mathbf{z}_{C,h}, \mathbf{H}_I - \mathbf{v}_{I,h})\|_\Lambda \, .\end{aligned} \tag{4.30}$$

Hence (4.26) follows from (4.29) and (4.30).

To obtain (4.27) we use the inf–sup condition (4.13). For each $\mathbf{N}_{I,h} \in Q^0_{I,h}$ we find

$$\|\mathbf{A}_{I,h} - \mathbf{N}_{I,h}\|_{0,\Omega_I} \leq \frac{1}{\beta} \sup_{(\mathbf{z}_{C,h}, \mathbf{v}_{I,h}) \in \Lambda_h} \frac{|\mathcal{R}(\mathbf{v}_{I,h}, \mathbf{A}_{I,h} - \mathbf{N}_{I,h})|}{\|(\mathbf{z}_{C,h}, \mathbf{v}_{I,h})\|_\Lambda} \, .$$

On the other hand

$$\begin{aligned}&\mathcal{R}(\mathbf{v}_{I,h}, \mathbf{A}_{I,h} - \mathbf{N}_{I,h}) \\ &\quad = \overline{L(\mathbf{z}_{C,h}) - \mathcal{C}((\mathbf{E}_{C,h}, \mathbf{H}_{I,h}), (\mathbf{z}_{C,h}, \mathbf{v}_{I,h}))} - \mathcal{R}(\mathbf{v}_{I,h}, \mathbf{N}_{I,h}) \\ &\quad = \overline{\mathcal{C}((\mathbf{E}_C, \mathbf{H}_I), (\mathbf{z}_{C,h}, \mathbf{v}_{I,h}))} + \mathcal{R}(\mathbf{v}_{I,h}, \mathbf{A}_I) \\ &\qquad - \overline{\mathcal{C}((\mathbf{E}_{C,h}, \mathbf{H}_{I,h}), (\mathbf{z}_{C,h}, \mathbf{v}_{I,h}))} - \mathcal{R}(\mathbf{v}_{I,h}, \mathbf{N}_{I,h}) \, ,\end{aligned}$$

then

$$\begin{aligned}&\|\mathbf{A}_{I,h} - \mathbf{N}_{I,h}\|_{0,\Omega_I} \\ &\quad \leq \tfrac{c_1}{\beta} \|(\mathbf{E}_C - \mathbf{E}_{C,h}, \mathbf{H}_I - \mathbf{H}_{I,h})\|_\Lambda + \tfrac{c_2}{\beta} \|\mathbf{A}_I - \mathbf{N}_{I,h}\|_{0,\Omega_I} \, ,\end{aligned}$$

which yields (4.27).

To obtain (4.28) we use the inf–sup condition (4.14), that in particular gives

$$\|\phi_{I,h}\|_h \leq \frac{1}{\gamma} \sup_{\mathbf{N}_{I,h} \in Q_{I,h}} \frac{|\mathcal{S}_h(\mathbf{N}_{I,h}, \phi_{I,h})|}{\|\mathbf{N}_{I,h}\|_{0,\Omega_I}} \, .$$

On the other hand

$$\begin{aligned}\mathcal{S}_h(\mathbf{N}_{I,h}, \phi_{I,h}) &= \overline{G(\mathbf{N}_{I,h})} - \overline{\mathcal{R}(\mathbf{H}_{I,h}, \mathbf{N}_{I,h})} \\ &= \overline{\mathcal{R}(\mathbf{H}_I, \mathbf{N}_{I,h})} + \mathcal{S}(\mathbf{N}_{I,h}, \phi_I) - \overline{\mathcal{R}(\mathbf{H}_{I,h}, \mathbf{N}_{I,h})} \\ &= \overline{\mathcal{R}(\mathbf{H}_I - \mathbf{H}_{I,h}, \mathbf{N}_{I,h})} \, ,\end{aligned}$$

then

$$\|\phi_{I,h}\|_h \leq \frac{c_2}{\gamma} \|\mathbf{H}_I - \mathbf{H}_{I,h}\|_{H(\mathrm{curl};\Omega_I)} \, .$$

Since smooth functions are dense in $\Lambda \times (L^2(\Omega_I))^3$, the interpolation estimates in Section A.2 yield the convergence of the finite element approximation method. □

Assuming that the solution is more regular (for notation see Section A.1), the interpolation results in Section A.2 yield a precise error estimate.

Corollary 4.10. *Let the assumptions of Theorem 4.8 be satisfied. If the solution* $(\mathbf{E}_C, \mathbf{H}_I, \mathbf{A}_I, 0)$ *of problem* (4.18) *is smooth enough, namely,* $\mathbf{E}_C \in H^r(\text{curl}; \Omega_C)$, $\mathbf{H}_I \in H^r(\text{curl}; \Omega_I)$ *with* $r > 1/2$ *and* $\mathbf{A}_I \in (H^s(\Omega_I))^3$ *with* $s > 0$, *the following error estimates hold*

$$\begin{aligned}
&\|(\mathbf{E}_C - \mathbf{E}_{C,h}, \mathbf{H}_I - \mathbf{H}_{I,h})\|_\Lambda \leq C h^{\min(r,1)} \\
&\|\mathbf{A}_I - \mathbf{A}_{I,h}\|_{0,\Omega_I} \leq C h^{\min(r,s,1)} \\
&\|\phi_{I,h}\|_h \leq C h^{\min(r,1)} .
\end{aligned} \tag{4.31}$$

Remark 4.11. We note that problem (4.9) can be written as a more standard saddle-point problem. Setting

$$\Lambda_* := H(\text{curl}; \Omega_C) \times H_{0,\partial\Omega}(\text{curl}; \Omega_I) \times H^0_{0,\Gamma}(\text{curl}; \Omega_I) ,$$

and defining

$$\begin{aligned}
&\widehat{\mathcal{R}}(\cdot,\cdot) : \Lambda_* \times (L^2(\Omega_I))^3 \to \mathbb{C} \\
&\widehat{\mathcal{R}}((\mathbf{z}_C, \mathbf{v}_I, \mathbf{p}_I), \mathbf{N}_I) := \textstyle\int_{\Omega_I} (\text{curl}\, \mathbf{v}_I \cdot \overline{\mathbf{N}_I} + \varepsilon_I \overline{\mathbf{N}_I} \cdot \mathbf{p}_I) ,
\end{aligned}$$

and

$$\begin{aligned}
&\widehat{\mathcal{C}}(\cdot,\cdot) : \Lambda_* \times \Lambda_* \to \mathbb{C} \\
&\widehat{\mathcal{C}}((\mathbf{w}_C, \mathbf{u}_I, \mathbf{s}_I), (\mathbf{z}_C, \mathbf{v}_I, \mathbf{p}_I)) := \mathcal{C}((\mathbf{w}_C, \mathbf{u}_I), (\mathbf{z}_C, \mathbf{v}_I)) ,
\end{aligned}$$

problem (4.9) reads

$$\begin{aligned}
&\text{Find } [(\mathbf{E}_C, \mathbf{H}_I, \mathbf{r}_I), \mathbf{A}_I] \text{ in } \Lambda_* \times (L^2(\Omega_I))^3 : \\
&\ddot{\mathcal{C}}((\mathbf{E}_C, \mathbf{H}_I, \mathbf{r}_I), (\mathbf{z}_C, \mathbf{v}_I, \mathbf{p}_I)) + \overline{\widehat{\mathcal{R}}((\mathbf{z}_C, \mathbf{v}_I, \mathbf{p}_I), \mathbf{A}_I)} = L(\mathbf{z}_C) \\
&\widehat{\mathcal{R}}((\mathbf{E}_C, \mathbf{H}_I, \mathbf{r}_I), \mathbf{N}_I) = G(\mathbf{N}_I) \\
&\text{for all } [(\mathbf{z}_C, \mathbf{v}_I, \mathbf{p}_I), \mathbf{N}_I] \in \Lambda_* \times (L^2(\Omega_I))^3 .
\end{aligned} \tag{4.32}$$

This problem and its finite element approximation can be analyzed using the general theory of variational saddle-point problems (see Brezzi and Fortin [65]). However, with that approach we are able to prove the discrete inf–sup condition only if ε_I is a scalar constant. □

Remark 4.12. We do not address here the problem of determining a finite element approximation of the electric field $\mathbf{E}_I$ in Ω_I, referring instead to Section 5.5, where this problem is considered in a more general geometrical setting. □

Remark 4.13. For the electric boundary condition (1.20) the finite element problem is

$$\begin{array}{l}\text{Find } (\mathbf{E}_{C,h}, \mathbf{H}_{I,h}) \in \Lambda_h^*\,,\ \mathbf{A}_{I,h}^* \in Q_{I,h}\,,\ \mathbf{r}_{I,h}^* \in \widetilde{\text{grad}}\, M_{I,h}^* : \\ \quad \mathcal{C}((\mathbf{E}_{C,h}, \mathbf{H}_{I,h}), (\mathbf{z}_{C,h}, \mathbf{v}_{I,h})) + \int_{\Omega_I} \text{curl}\, \overline{\mathbf{v}_{I,h}} \cdot \mathbf{A}_{I,h}^* = L(\mathbf{z}_{C,h}) \\ \quad \int_{\Omega_I} \text{curl}\, \mathbf{H}_{I,h} \cdot \overline{\mathbf{N}_{I,h}^*} + \int_{\Omega_I} \varepsilon_I \overline{\mathbf{N}_{I,h}^*} \cdot \mathbf{r}_{I,h}^* = G(\mathbf{N}_{I,h}^*) \\ \quad \int_{\Omega_I} \varepsilon_I \mathbf{A}_{I,h}^* \cdot \overline{\mathbf{p}_{I,h}^*} = 0 \\ \text{for all } (\mathbf{z}_{C,h}, \mathbf{v}_{I,h}) \in \Lambda_h^*\,,\ \mathbf{N}_{I,h}^* \in Q_{I,h}\,,\ \mathbf{p}_{I,h}^* \in \widetilde{\text{grad}}\, M_{I,h}^*\,,\end{array} \tag{4.33}$$

where

$$\Lambda_h^* := N_{C,h}^1 \times N_{I,h}^1$$

and

$$\begin{array}{rl} M_{I,h}^* := \{\xi_{I,h}^* \in U_{I,h} \mid & \xi_{I,h}^* \text{ takes the same value at all centroids} \\ & \text{of faces of } \Gamma_j,\ j = 1, \dots, p_\Gamma, \text{ and of} \\ & (\partial\Omega)_r,\ r = 0, \dots, p_{\partial\Omega}, \text{ and } \xi_{I,h}^* = 0 \\ & \text{at all centroids of faces of } \Gamma_{p_\Gamma+1}\}\,. \end{array}$$

Note that, in this case, we have no need to assume that the computational domain Ω is simply-connected. □

4.3 A saddle-point approach for the H-based formulation

A saddle-point approach similar to the one analyzed in the previous section can also be used for the **H**-based formulation of the eddy current problem, as presented in Alonso Rodríguez et al. [13] (see also Guermond and Minev [116], who face the **H**-based formulation of the time-dependent eddy current problem by means of an analogous approach).

Let us consider the vector spaces

$$V(\mathbf{J}_{e,I}) := \{\mathbf{v} \in H_0(\text{curl};\Omega) \mid \text{curl}\, \mathbf{v}_I = \mathbf{J}_{e,I} \text{ in } \Omega_I\},$$

and, already introduced in (3.6),

$$V := \{\mathbf{v} \in H_0(\text{curl};\Omega) \mid \text{curl}\, \mathbf{v}_I = \mathbf{0} \text{ in } \Omega_I\}\,.$$

The weak form (3.9) of the **H**-based problem also reads

$$\begin{array}{l}\text{Find } \mathbf{H} \in V(\mathbf{J}_{e,I}) : \\ \quad \int_{\Omega_C} \boldsymbol{\sigma}^{-1} \text{curl}\, \mathbf{H}_C \cdot \text{curl}\, \overline{\mathbf{v}_C} + \int_\Omega i\omega\boldsymbol{\mu}\mathbf{H} \cdot \overline{\mathbf{v}} = \int_{\Omega_C} \boldsymbol{\sigma}^{-1} \mathbf{J}_{e,C} \cdot \text{curl}\, \overline{\mathbf{v}_C} \\ \text{for all } \mathbf{v} \in V\,.\end{array} \tag{4.34}$$

It can be reformulated in non-constrained vector spaces by introducing Lagrange multipliers. It is easy to see that (4.34) is equivalent to the following three fields formulation

$$
\begin{aligned}
&\text{Find } (\mathbf{H}, \mathbf{A}_I, \mathbf{r}_I) \text{ in } H_0(\mathrm{curl};\Omega) \times (L^2(\Omega_I))^3 \times H^0_{0,\Gamma}(\mathrm{curl};\Omega_I) : \\
&\quad a(\mathbf{H},\mathbf{v}) + \textstyle\int_{\Omega_I} \mathrm{curl}\,\overline{\mathbf{v}_I} \cdot \mathbf{A}_I = L^*(\mathbf{v}_C) \\
&\quad \textstyle\int_{\Omega_I} \mathrm{curl}\,\mathbf{H}_I \cdot \overline{\mathbf{N}_I} + \int_{\Omega_I} \varepsilon_I \overline{\mathbf{N}_I} \cdot \mathbf{r}_I = G(\mathbf{N}_I) \\
&\quad \textstyle\int_{\Omega_I} \varepsilon_I \mathbf{A}_I \cdot \overline{\mathbf{p}_I} = 0 \\
&\text{for all } (\mathbf{v}, \mathbf{N}_I, \mathbf{p}_I) \in H_0(\mathrm{curl};\Omega) \times (L^2(\Omega_I))^3 \times H^0_{0,\Gamma}(\mathrm{curl};\Omega_I) \,,
\end{aligned}
\tag{4.35}
$$

where

$$a(\mathbf{u},\mathbf{v}) := \int_{\Omega_C} \boldsymbol{\sigma}^{-1} \mathrm{curl}\,\mathbf{u}_C \cdot \mathrm{curl}\,\overline{\mathbf{v}_C} + \int_\Omega i\omega\boldsymbol{\mu}\mathbf{u}\cdot\overline{\mathbf{v}},$$

and

$$G(\mathbf{N}_I) := \int_{\Omega_I} \mathbf{J}_{e,I} \cdot \overline{\mathbf{N}_I}$$

have been introduced in (3.10) and (4.5), respectively, and

$$L^*(\mathbf{v}_C) := \int_{\Omega_C} \boldsymbol{\sigma}^{-1} \mathbf{J}_{e,C} \cdot \mathrm{curl}\,\overline{\mathbf{v}_C}\,.$$

Proceeding as in the case of the hybrid $\mathbf{E}_C/\mathbf{H}_I$ formulation of the previous section, it can be proved that problem (4.35) has a unique solution. It is in fact given by $(\mathbf{H}, \mathbf{E}_{|\Omega_I}, \mathbf{0})$, where $(\mathbf{H}, \mathbf{E})$ is the unique solution of (3.30), hence the Lagrange multiplier $\mathbf{A}_I$ is indeed the electric field $\mathbf{E}_I$, and the Lagrange multiplier $\mathbf{r}_I$ is equal to $\mathbf{0}$.

Concerning the finite element approximation of (4.35), as in the previous section we will assume for simplicity that *Ω is simply-connected*. In this way the space $H^0_{0,\Gamma}(\mathrm{curl};\Omega_I)$ is given by the gradients of functions belonging to $H^1_{*,\Gamma}(\Omega_I)$. To approximate the magnetic field we employ the (complex-valued) Nédélec curl-conforming edge elements of the lowest order N_h^1 (see Section A.2). In particular, we consider the space $X_h^1 := N_h^1 \cap H_0(\mathrm{curl};\Omega)$. The finite elements spaces for the Lagrange multipliers $\mathbf{A}_I$ and ϕ_I are the same spaces used in the hybrid case: the space $M_{I,h}$ of discontinuous piecewise-linear functions defined in (4.19), and the space $Q_{I,h}$ of piecewise-constant vector functions.

The finite element approximation of (4.35) can be formulated as follows

$$
\begin{aligned}
&\text{Find } (\mathbf{H}_h, \mathbf{A}_{I,h}, \phi_{I,h}) \text{ in } X_h^1 \times Q_{I,h} \times M_{I,h} : \\
&\quad a(\mathbf{H}_h,\mathbf{v}_h) + \overline{\mathcal{R}(\mathbf{v}_{I,h}, \mathbf{A}_{I,h})} = L^*(\mathbf{v}_{C,h}) \\
&\quad \mathcal{R}(\mathbf{H}_{I,h}, \mathbf{N}_{I,h}) + \overline{\mathcal{S}_h(\mathbf{N}_{I,h}, \phi_{I,h})} = G(\mathbf{N}_{I,h}) \\
&\quad \mathcal{S}_h(\mathbf{A}_{I,h}, \xi_{I,h}) = 0 \\
&\text{for all } (\mathbf{v}_h, \mathbf{N}_{I,h}, \xi_{I,h}) \in X_h^1 \times Q_{I,h} \times M_{I,h} \,,
\end{aligned}
\tag{4.36}
$$

where $\mathcal{R}(\cdot,\cdot)$ and $\mathcal{S}_h(\cdot,\cdot)$ have been defined in (4.21) and (4.20), respectively.

Using again Theorem 4.2 and proceeding as in the hybrid $\mathbf{E}_C/\mathbf{H}_I$ case we have the following result.

Theorem 4.14. *Under the assumptions of Theorem 4.8, problem* (4.36) *has a unique solution* $\mathbf{H}_h \in X_h^1$, $\mathbf{A}_{I,h} \in Q_{I,h}$ *and* $\phi_{I,h} \in M_{I,h}$. *Moreover, let* $\mathbf{H} \in H_0(\mathrm{curl};\Omega)$, $\mathbf{A}_I \in (L^2(\Omega_I))^3$ *be the solution of problem* (4.35). *Then the following error estimates hold*

$$\|\mathbf{H}-\mathbf{H}_h\|_{H(\mathrm{curl};\Omega)} \leq \left(1+\frac{c_1}{\alpha}\right)\left(1+\frac{c_2}{\beta}\right) \inf_{\mathbf{v}_h\in X_h^1} \|\mathbf{H}-\mathbf{v}_h\|_{H(\mathrm{curl};\Omega)}$$

$$\begin{aligned}\|\mathbf{A}_I-\mathbf{A}_{I,h}\|_{0,\Omega_I} \leq \left(1+\tfrac{c_2}{\beta}\right) \inf\nolimits_{\mathbf{N}_{I,h}\in Q_{I,h}^0} \|\mathbf{A}_I-\mathbf{N}_{I,h}\|_{0,\Omega_I} \\ +\tfrac{c_1}{\beta}\|\mathbf{H}-\mathbf{H}_h\|_{H(\mathrm{curl};\Omega)}\end{aligned}$$

$$\|\phi_{I,h}\|_h \leq \frac{c_2}{\gamma}\|\mathbf{H}-\mathbf{H}_h\|_{H(\mathrm{curl};\Omega)} ,$$

where all the constants are independent of h. In particular, the finite element method is convergent and, if $\mathbf{H} \in H^r(\mathrm{curl};\Omega)$ *with* $r > 1/2$ *and* $\mathbf{A}_I \in (H^s(\Omega_I))^3$ *with* $s > 0$, *it follows*

$$\begin{aligned}&\|\mathbf{H}-\mathbf{H}_h\|_{H(\mathrm{curl};\Omega)} \leq Ch^{\min(r,1)}\\ &\|\mathbf{A}_I-\mathbf{A}_{I,h}\|_{0,\Omega_I} \leq Ch^{\min(r,s,1)}\\ &\|\phi_{I,h}\|_h \leq Ch^{\min(r,1)} .\end{aligned}$$

4.4 Hybrid formulation using the electric field in the insulator

For the formulations using the magnetic field in the insulator as the main unknown, the saddle-point approach computes also the electric field in the insulator, because it is the Lagrange multiplier for $\mathbf{H}_I$. However, in many applications the electric field in the insulator is not an interesting physical quantity. In this case it would be convenient to avoid its approximation. This can be done using a second type of hybrid coupling, the $\mathbf{H}_C/\widetilde{\mathbf{E}}_I$ formulation, in which the main unknowns are the magnetic field in the conductor Ω_C and a vector magnetic potential $\widetilde{\mathbf{E}}_I$ in the insulating region Ω_I.

Here and in the next section we follow the presentation given in Alonso Rodríguez et al. [14]. The starting point is the hybrid formulation that has as main unknowns the magnetic field in the conductor and the electric field in the insulator.

For each $\mathbf{v}_C \in H(\mathrm{curl};\Omega_C)$, from the Faraday equation in Ω_C one finds by integration by parts

$$\int_{\Omega_C}(\mathbf{E}_C\cdot\mathrm{curl}\,\overline{\mathbf{v}_C} + i\omega\boldsymbol{\mu}_C\mathbf{H}_C\cdot\overline{\mathbf{v}_C}) - \int_\Gamma \mathbf{E}_C\times\mathbf{n}_C\cdot\overline{\mathbf{v}_C} = 0 .$$

Using the interface conditions for the electric field and the Ampère law in Ω_C leads to

$$\begin{aligned}&\int_{\Omega_C}(\boldsymbol{\sigma}^{-1}\,\mathrm{curl}\,\mathbf{H}_C\cdot\mathrm{curl}\,\overline{\mathbf{v}_C} + i\omega\boldsymbol{\mu}_C\mathbf{H}_C\cdot\overline{\mathbf{v}_C}) - \int_\Gamma \mathbf{E}_I\times\mathbf{n}_C\cdot\overline{\mathbf{v}_C}\\ &\quad = \int_{\Omega_C}\boldsymbol{\sigma}^{-1}\mathbf{J}_{e,C}\cdot\mathrm{curl}\,\overline{\mathbf{v}_C} .\end{aligned}$$

On the other hand, in the insulator one has $-i\omega \operatorname{curl} \mathbf{H}_I = -i\omega \mathbf{J}_{e,I}$, thus for each $\mathbf{z}_I \in H(\operatorname{curl}, \Omega_I)$ one finds by integration by parts

$$-i\omega \int_{\Omega_I} \mathbf{H}_I \cdot \operatorname{curl} \overline{\mathbf{z}_I} + i\omega \int_\Gamma \mathbf{H}_I \times \mathbf{n}_I \cdot \overline{\mathbf{z}_I} = -i\omega \int_{\Omega_I} \mathbf{J}_{e,I} \cdot \overline{\mathbf{z}_I} \, .$$

Using the Faraday law in Ω_I and the interface conditions one has

$$\int_{\Omega_I} \boldsymbol{\mu}_I^{-1} \operatorname{curl} \mathbf{E}_I \cdot \operatorname{curl} \overline{\mathbf{z}_I} - i\omega \int_\Gamma \mathbf{H}_C \times \mathbf{n}_C \cdot \overline{\mathbf{z}_I} = -i\omega \int_{\Omega_I} \mathbf{J}_{e,I} \cdot \overline{\mathbf{z}_I} \, .$$

Setting

$$Z_I := \{\mathbf{z}_I \in H(\operatorname{curl}; \Omega_I) \,|\, \mathbf{z}_I \text{ satisfies } (4.6)\} \, ,$$

the weak formulation of the hybrid $\mathbf{H}_C/\mathbf{E}_I$ formulation reads

$$\begin{aligned} &\text{Find } (\mathbf{H}_C, \mathbf{E}_I) \in H(\operatorname{curl}; \Omega_C) \times Z_I : \\ &\int_{\Omega_C} (\boldsymbol{\sigma}^{-1} \operatorname{curl} \mathbf{H}_C \cdot \operatorname{curl} \overline{\mathbf{v}_C} + i\omega \boldsymbol{\mu}_C \mathbf{H}_C \cdot \overline{\mathbf{v}_C}) \\ &\qquad\qquad + \int_\Gamma \overline{\mathbf{v}_C} \times \mathbf{n}_C \cdot \mathbf{E}_I = F(\mathbf{v}_C) \\ &\int_\Gamma \mathbf{H}_C \times \mathbf{n}_C \cdot \overline{\mathbf{z}_I} + i\omega^{-1} \int_{\Omega_I} \boldsymbol{\mu}_I^{-1} \operatorname{curl} \mathbf{E}_I \cdot \operatorname{curl} \overline{\mathbf{z}_I} = G(\mathbf{z}_I) \\ &\text{for all } (\mathbf{v}_C, \mathbf{z}_I) \in H(\operatorname{curl}; \Omega_C) \times Z_I \, , \end{aligned} \tag{4.37}$$

where

$$F(\mathbf{v}_C) := \int_{\Omega_C} \boldsymbol{\sigma}^{-1} \mathbf{J}_{e,C} \cdot \operatorname{curl} \overline{\mathbf{v}_C},$$

and, as in (4.5),

$$G(\mathbf{z}_I) := \int_{\Omega_I} \mathbf{J}_{e,I} \cdot \overline{\mathbf{z}_I} \, .$$

It can be shown, via the standard theory for saddle-point problems, that problem (4.37) has a unique solution.

Unfortunately, in the discrete setting it is not clear how to obtain an inf–sup condition for the pairing $(\mathbf{u}, \mathbf{w}) \to \int_\Gamma (\mathbf{u} \times \mathbf{n}_C) \cdot \overline{\mathbf{w}}$ on $H^{-1/2}(\operatorname{div}_\tau, \Gamma) \times H^{-1/2}(\operatorname{div}_\tau, \Gamma)$ which is uniform with respect to the mesh size h. For a detailed account of this problem see Christiansen and Nédélec [82], Sect. 3.

A remedy is offered by considering a different approach, in which one works on a smaller constrained space. The drawback is that in the alternative approach the solution obtained in Ω_I is not the physical electric field $\mathbf{E}_I$ but a suitable magnetic vector potential $\widetilde{\mathbf{E}}_I$.

In the following we assume for simplicity that $\mathbf{J}_{e,I} \cdot \mathbf{n}_I = 0$ on Γ (for less restrictive assumptions on $\mathbf{J}_{e,I}$ see Remark 4.18). By tangential continuity of $\mathbf{H}$ we can infer that

$$\operatorname{div}_\tau(\mathbf{H}_C \times \mathbf{n}_C) = \operatorname{curl} \mathbf{H}_I \cdot \mathbf{n}_C = 0 \quad \text{on } \Gamma \, .$$

Let us define the spaces

$$\widetilde{X}_C := \{\mathbf{v}_C \in H(\operatorname{curl}; \Omega_C) \,|\, \operatorname{div}_\tau(\mathbf{v}_C \times \mathbf{n}_C) = 0 \text{ on } \Gamma\},$$

and

$$\widetilde{Z}_I := \{\mathbf{z}_I \in H(\mathrm{curl};\Omega_I) \mid \int_{\Omega_I} \mathbf{z}_I \cdot \mathrm{grad}\,\overline{\xi_I} = 0 \text{ for all } \xi_I \in H^1(\Omega_I)\} \,.$$

Note that they are closed subspaces of $H(\mathrm{curl};\Omega_C)$ and $H(\mathrm{curl};\Omega_I)$, respectively. Moreover, $\widetilde{Z}_I$ is the space of $\mathbf{z}_I \in H(\mathrm{curl};\Omega_I)$ such that $\mathrm{div}\,\mathbf{z}_I = 0$ in Ω_I and $\mathbf{z}_I \cdot \mathbf{n} = 0$ on $\partial\Omega \cup \Gamma$.

We consider the following problem

$$\begin{aligned}
&\text{Find } (\mathbf{H}_C, \widetilde{\mathbf{E}}_I) \in \widetilde{X}_C \times \widetilde{Z}_I :\\
&\textstyle\int_{\Omega_C} (\boldsymbol{\sigma}^{-1} \mathrm{curl}\,\mathbf{H}_C \cdot \mathrm{curl}\,\overline{\mathbf{v}_C} + i\omega \boldsymbol{\mu}_C \mathbf{H}_C \cdot \overline{\mathbf{v}_C})\\
&\qquad\qquad\qquad\qquad \textstyle + \int_\Gamma \overline{\mathbf{v}_C} \times \mathbf{n}_C \cdot \widetilde{\mathbf{E}}_I = F(\mathbf{v}_C)\\
&\textstyle\int_\Gamma \mathbf{H}_C \times \mathbf{n}_C \cdot \overline{\mathbf{z}_I} + i\omega^{-1} \int_{\Omega_I} \boldsymbol{\mu}_I^{-1} \mathrm{curl}\,\widetilde{\mathbf{E}}_I \cdot \mathrm{curl}\,\overline{\mathbf{z}_I} = G(\mathbf{z}_I)\\
&\text{for all } (\mathbf{v}_C, \mathbf{z}_I) \in \widetilde{X}_C \times \widetilde{Z}_I \,.
\end{aligned} \tag{4.38}$$

In order to analyze this new problem we need some preliminary results. First of all we recall that, choosing $\boldsymbol{\mu}_I = \mathrm{Id}$ in Theorem A.8, any function $\mathbf{v}_I \in (L^2(\Omega_I))^3$ can be written as

$$\mathbf{v}_I = \mathrm{curl}\,\mathbf{Q}_I^* + \mathrm{grad}\,\chi_I^* + \mathbf{k}_I^* \,, \tag{4.39}$$

where $\mathbf{Q}_I^* \in H_0(\mathrm{curl};\Omega_I)$, $\chi_I^* \in H^1(\Omega_I)$ and $\mathbf{k}_I^* \in \mathcal{H}(m;\Omega_I)$, the finite dimensional space of harmonic vector fields in Ω_I introduced in Section 1.4. We also know that $\mathrm{grad}\,\chi_I^* = \mathbf{0}$ in Ω_I if and only if $\mathrm{div}\,\mathbf{v}_I = 0$ in Ω_I and $\mathbf{v}_I \cdot \mathbf{n} = 0$ on $\partial\Omega \cup \Gamma$.

It is readily seen that $\mathrm{curl}[H_0(\mathrm{curl};\Omega_I)] \subset H^0(\mathrm{curl};\Omega_I)^\perp$; moreover, from (4.39) one sees indeed that $\mathrm{curl}[H_0(\mathrm{curl};\Omega_I)] = H^0(\mathrm{curl};\Omega_I)^\perp$. Thus the following orthogonal decomposition holds

$$(L^2(\Omega_I))^3 = H^0(\mathrm{curl};\Omega_I)^\perp \oplus \mathrm{grad}\,H^1(\Omega_I) \oplus \mathcal{H}(m;\Omega_I) \,. \tag{4.40}$$

Moreover,

$$\begin{aligned}
&H(\mathrm{curl};\Omega_I)\\
&\quad = [H(\mathrm{curl};\Omega_I) \cap H^0(\mathrm{curl};\Omega_I)^\perp] \oplus \mathrm{grad}\,H^1(\Omega_I) \oplus \mathcal{H}(m;\Omega_I) \,,
\end{aligned} \tag{4.41}$$

and

$$\widetilde{Z}_I = [H(\mathrm{curl};\Omega_I) \cap H^0(\mathrm{curl};\Omega_I)^\perp] \oplus \mathcal{H}(m;\Omega_I) \,. \tag{4.42}$$

We also recall the following Poincaré-like inequality (see, e.g., Fernandes and Gilardi [104]): there exists a constant $C_1 > 0$ such that

$$\int_{\Omega_I} |\mathbf{v}_I|^2 \le C_1 \int_{\Omega_I} (|\,\mathrm{curl}\,\mathbf{v}_I|^2 + |\,\mathrm{div}\,\mathbf{v}_I|^2) \tag{4.43}$$

for all $\mathbf{v}_I \in H(\mathrm{curl};\Omega_I) \cap H_0(\mathrm{div};\Omega_I) \cap \mathcal{H}(m;\Omega_I)^\perp$.

If $\mathbf{v}_I \in H(\mathrm{curl};\Omega_I) \cap H^0(\mathrm{curl};\Omega_I)^\perp$ then, from the orthogonal decomposition (4.41), it follows that $\mathbf{v}_I \in H(\mathrm{curl};\Omega_I) \cap [\mathrm{grad}\, H^1(\Omega_I)]^\perp \cap \mathcal{H}(m;\Omega_I)^\perp$. As a consequence, $\mathrm{div}\, \mathbf{v}_I = 0$ in Ω_I and $\mathbf{v}_I \cdot \mathbf{n} = 0$ on $\partial\Omega \cup \Gamma$. In conclusion, for each $\mathbf{v}_I \in H(\mathrm{curl};\Omega_I) \cap H^0(\mathrm{curl};\Omega_I)^\perp$ we have

$$\int_{\Omega_I} |\mathbf{v}_I|^2 \le C_1 \int_{\Omega_I} |\,\mathrm{curl}\, \mathbf{v}_I|^2 . \tag{4.44}$$

We are now in a position to prove that the $\mathbf{H}_C/\widetilde{\mathbf{E}}_I$ formulation is well-posed.

Theorem 4.15. *Assume that Ω is simply-connected. The variational problem* (4.38) *has a unique solution.*

Proof. The proof is based on the classical saddle-point theory. Let us denote $\widetilde{Z}_I^\perp := H(\mathrm{curl};\Omega_I) \cap H^0(\mathrm{curl};\Omega_I)^\perp$. Since we have the direct sum decomposition (4.42), namely,

$$\widetilde{Z}_I = \widetilde{Z}_I^\perp \oplus \mathcal{H}(m;\Omega_I) , \tag{4.45}$$

we can rewrite (4.38) in equivalent form as

$$\begin{array}{l}
\text{Find } (\mathbf{H}_C, \widetilde{\mathbf{E}}_I^\perp) \in \widetilde{X}_C \times \widetilde{Z}_I^\perp \text{ and } \widetilde{\mathbf{E}}_I^{\mathcal{H}} \in \mathcal{H}(m;\Omega_I) : \\
\quad \mathcal{D}((\mathbf{H}_C, \widetilde{\mathbf{E}}_I^\perp), (\mathbf{v}_C, \mathbf{z}_I^\perp)) + \int_\Gamma \overline{\mathbf{v}_C} \times \mathbf{n}_C \cdot \widetilde{\mathbf{E}}_I^{\mathcal{H}} = F(\mathbf{v}_C) + G(\mathbf{z}_I^\perp) \\
\quad \int_\Gamma \mathbf{H}_C \times \mathbf{n}_C \cdot \overline{\mathbf{z}_I^{\mathcal{H}}} = G(\mathbf{z}_I^{\mathcal{H}}) \\
\text{for all } (\mathbf{v}_C, \mathbf{z}_I^\perp) \in \widetilde{X}_C \times \widetilde{Z}_I^\perp \text{ and } \mathbf{z}_I^{\mathcal{H}} \in \mathcal{H}(m;\Omega_I) .
\end{array}$$

The sesquilinear form $\mathcal{D}(\cdot,\cdot)$ is the sum of the two left-hand sides of (4.37), namely,

$$\begin{array}{l}
\mathcal{D}((\mathbf{u}_C, \mathbf{w}_I), (\mathbf{v}_C, \mathbf{z}_I)) \\
\qquad := \int_{\Omega_C} (\boldsymbol{\sigma}^{-1}\, \mathrm{curl}\, \mathbf{u}_C \cdot \mathrm{curl}\, \overline{\mathbf{v}_C} + i\omega \boldsymbol{\mu}_C \mathbf{u}_C \cdot \overline{\mathbf{v}_C}) \\
\qquad\quad + \int_\Gamma \mathbf{v}_C \times \mathbf{n}_C \cdot \mathbf{w}_I + \int_\Gamma \mathbf{u}_C \times \mathbf{n}_C \cdot \mathbf{z}_I \\
\qquad\quad + i\omega^{-1} \int_{\Omega_I} \boldsymbol{\mu}_I^{-1}\, \mathrm{curl}\, \mathbf{w}_I \cdot \mathrm{curl}\, \overline{\mathbf{z}_I} .
\end{array} \tag{4.46}$$

It can be proved that $\mathcal{D}(\cdot,\cdot)$ is coercive in $\widetilde{X}_C \times \widetilde{Z}_I^\perp$. In fact, based on (4.44), the proof is analogous to that presented for the sesquilinear form $\mathcal{C}(\cdot,\cdot)$ in the analysis of the $\mathbf{E}_C/\mathbf{H}_I$ formulation.

Now, we only need to check the inf–sup condition

$$\begin{array}{l}
\exists\, \beta > 0 \text{ such that} \\
\forall\, \mathbf{z}_I^{\mathcal{H}} \in \mathcal{H}(m;\Omega_I)\ \exists\, \mathbf{v}_C \in \widetilde{X}_C,\ \mathbf{v}_C \ne \mathbf{0} : \\
\left| \int_\Gamma \mathbf{v}_C \times \mathbf{n}_C \cdot \overline{\mathbf{z}_I^{\mathcal{H}}} \right| \ge \beta \|\mathbf{v}_C\|_{H(\mathrm{curl};\Omega_C)} \|\mathbf{z}_I^{\mathcal{H}}\|_{0,\Omega_I} .
\end{array} \tag{4.47}$$

For simplicity we start assuming that Ω_C is a torus, a set which has only one independent non-bounding cycle, or, equivalently, whose first Betti number is equal to 1. Since Ω is simply-connected, the non-bounding cycle of Ω_I are on Γ. We know that for $L = I, C$ we can find an orientable two-dimensional surface Σ_L such that both

$\Omega_L \setminus \Sigma_L$ have trivial first homology group. Setting $\partial \Sigma_C = \gamma_C$ and $\partial \Sigma_I = \gamma_I$, we choose the orientation of these two cycles in such a way that, if the normal vector on Σ_C has the same direction of γ_I, then γ_C is oriented counterclockwise.

Denote by p_L^* a harmonic function in $H^1(\Omega_L \setminus \Sigma_L)$ that has a jump of height 1 across Σ_L (namely, the line integral of $\widetilde{\operatorname{grad} p_C^*}$ along γ_I has value 1, and similarly for $\widetilde{\operatorname{grad} p_I^*}$), so that $\widetilde{\operatorname{grad} p_L^*}$ is the basis function of $\mathcal{H}(m; \Omega_L)$. Then $\mathbf{z}_I^{\mathcal{H}} = c \, \widetilde{\operatorname{grad} p_I^*}$, with $c \in \mathbb{C}$. Then we can use as special candidate for $\mathbf{v}_C \in \widetilde{X}_C$ the harmonic vector field $c \, \widetilde{\operatorname{grad} p_C^*}$: in fact we have

$$
\begin{aligned}
\int_\Gamma c \, \widetilde{\operatorname{grad} p_C^*} \times \mathbf{n}_C \cdot \overline{\mathbf{z}_I^{\mathcal{H}}} &= c \int_{\Gamma \setminus \gamma_C} \operatorname{grad} p_C^* \times \mathbf{n}_C \cdot \overline{\mathbf{z}_I^{\mathcal{H}}} \\
&= c \int_{\Gamma \setminus \gamma_C} p_C^* \operatorname{div}_\tau (\overline{\mathbf{z}_I^{\mathcal{H}}} \times \mathbf{n}_C) + c \int_{\gamma_C} \overline{\mathbf{z}_I^{\mathcal{H}}} \cdot d\boldsymbol{\tau} \\
&= c \int_{\gamma_C} \overline{\mathbf{z}_I^{\mathcal{H}}} \cdot d\boldsymbol{\tau} = |c|^2 \int_{\gamma_C} \widetilde{\operatorname{grad} p_I^*} \cdot d\boldsymbol{\tau} = |c|^2 \, .
\end{aligned}
$$

Since $\|\mathbf{z}_I^{\mathcal{H}}\|_{0,\Omega_I} = |c| \|\widetilde{\operatorname{grad} p_I^*}\|_{0,\Omega_I}$, we obtain the inf–sup condition with $\beta = (\|\widetilde{\operatorname{grad} p_C^*}\|_{0,\Omega_C} \|\widetilde{\operatorname{grad} p_I^*}\|_{0,\Omega_I})^{-1}$.

For the sake of simplicity, here below the first Betti number n_{Ω_C} of Ω_C will be simply denoted by n_C. If $n_C > 1$, since Ω is simply-connected the non-bounding cycles of Ω_I are on Γ and we can find $2n_C$ independent non-bounding cycles $\gamma_1, \ldots, \gamma_{2n_C}$ that represent generators of the first homology group on Γ. They can be chosen such that $\gamma_k = \partial \Sigma_k$, $k = 1, \ldots, n_C$ (see, e.g., Hiptmair and Ostrowski [128]). Moreover, $\gamma_{n_C+1}, \ldots, \gamma_{2n_C}$ can be chosen dual to $\gamma_1, \ldots, \gamma_{n_C}$, which implies

$$
\int_{\gamma_{n_C+k}} \widetilde{\operatorname{grad} p_{j,I}^*} \cdot d\boldsymbol{\tau} = \delta_{kj}, \quad k, j \in \{1, \ldots, n_C\} \, . \tag{4.48}
$$

By proceeding in a similar way, one can easily see that the constant β in the inf–sup condition is given by

$$
\beta = \min_{\mathbf{c} \in \mathbb{C}^{n_C}, \mathbf{c} \neq \mathbf{0}} \frac{|\mathbf{c}|^2}{(M^C \mathbf{c} \cdot \overline{\mathbf{c}})^{1/2} (M^I \mathbf{c} \cdot \mathbf{c})^{1/2}} \, ,
$$

where M^L, $L = I, C$, is the matrix given by $M^L_{kj} = \int_{\Omega_L} \widetilde{\operatorname{grad} p_{k,L}^*} \cdot \widetilde{\operatorname{grad} p_{j,L}^*}$, $k, j = 1, \ldots, n_C$. □

Remark 4.16. It is worth noting that the solution $\widetilde{\mathbf{E}}_I$ to (4.38) is not the physical electric field we are looking for. In fact, what we have determined satisfies the interface condition $\widetilde{\mathbf{E}}_I \cdot \mathbf{n}_I = 0$ on Γ, which is not the case for the correct electric field. Therefore, it has to be interpreted as a vector potential for the magnetic field $\mathbf{H}_I = i\omega^{-1} \boldsymbol{\mu}_I^{-1} \operatorname{curl} \widetilde{\mathbf{E}}_I$.

In order to be sure that we have really solved the eddy current problem, we have thus to check that the magnetic field $(\mathbf{H}_C, \mathbf{H}_I)$ satisfies (3.23).

Choosing as $\mathbf{v}_C$ a smooth vector function with compact support we find $(3.23)_1$, the Faraday equation in Ω_C. Moreover, $(3.23)_3$ and $(3.23)_4$ are trivial from the definition of $\mathbf{H}_I$. In order to verify $(3.23)_2$, the Ampère equation in Ω_I, we notice that the second equation of (4.38) holds true also for $\mathbf{z}_I = \operatorname{grad} \xi_I$ with

$\xi_I \in H^1(\Omega_I)$. In fact $\int_\Gamma \mathbf{H}_C \times \mathbf{n}_C \cdot \operatorname{grad} \overline{\xi_I} = 0$ because $\operatorname{div}_\tau(\mathbf{H}_C \times \mathbf{n}_C) = 0$ on Γ, and $\int_{\Omega_I} \mathbf{J}_{e,I} \cdot \operatorname{grad} \overline{\xi_I} = 0$ because $\operatorname{div} \mathbf{J}_{e,I} = 0$ and $\mathbf{J}_{e,I} \cdot \mathbf{n}_I = 0$ on $\partial\Omega_I$. Therefore, using the orthogonal decomposition (4.41), we conclude that the second equation in (4.38) holds for each $\mathbf{z}_I \in H(\operatorname{curl}; \Omega_I)$. Taking now as $\mathbf{z}_I$ a smooth vector function with compact support we obtain $(3.23)_2$, and then a similar choice with $\mathbf{z}_I$ vanishing only in the neighborhood of Γ gives $\mathbf{H}_I \times \mathbf{n} = \mathbf{0}$ on $\partial\Omega$, hence $(3.23)_6$.

Concerning the interface conditions, the choice $\mathbf{v}_C = \operatorname{grad} \eta_C$, where η_C is an arbitrary function in $H^1(\Omega_C)$ (so that $\operatorname{div}_\tau(\operatorname{grad} \eta_C \times \mathbf{n}_C) = \operatorname{curl} \operatorname{grad} \eta_C \cdot \mathbf{n}_C = 0$ on Γ), gives easily that $i\omega\boldsymbol{\mu}_C \mathbf{H}_C \cdot \mathbf{n}_C + \operatorname{curl} \widetilde{\mathbf{E}}_I \cdot \mathbf{n}_C = 0$ on Γ, hence $(3.23)_7$. On the other hand, choosing $\mathbf{z}_I \in H(\operatorname{curl}; \Omega_I)$ and using the Ampère equation in Ω_I and the boundary condition $\mathbf{H}_I \times \mathbf{n} = \mathbf{0}$ on $\partial\Omega$ one finds at once $\mathbf{H}_C \times \mathbf{n}_C - i\omega^{-1}\boldsymbol{\mu}_I^{-1} \operatorname{curl} \widetilde{\mathbf{E}}_I \times \mathbf{n}_C = \mathbf{0}$ on Γ, namely, $(3.23)_8$.

The last condition $(3.23)_5$ follows by taking $\mathbf{v}_C = \mathbf{R}_{l,C}$, an extension of the trace $\boldsymbol{\rho}_{l,I} \times \mathbf{n}_C$ in Ω_C. Clearly $\mathbf{R}_{l,C} \in \widetilde{X}_C$, as $\operatorname{div}_\tau(\mathbf{R}_{l,C} \times \mathbf{n}_C) = \operatorname{div}_\tau(\boldsymbol{\rho}_{l,I} \times \mathbf{n}_C) = \operatorname{curl} \boldsymbol{\rho}_{l,I} \cdot \mathbf{n}_C = 0$ on Γ. This choice and the Faraday equation in Ω_C yield

$$
\begin{aligned}
0 &= \textstyle\int_{\Omega_C} [\boldsymbol{\sigma}^{-1}(\operatorname{curl} \mathbf{H}_C - \mathbf{J}_{e,C}) \cdot \operatorname{curl} \mathbf{R}_{l,C} + i\omega\boldsymbol{\mu}_C \mathbf{H}_C \cdot \mathbf{R}_{l,C}] \\
&\quad + \textstyle\int_\Gamma \mathbf{R}_{l,C} \times \mathbf{n}_C \cdot \widetilde{\mathbf{E}}_I \\
&= \textstyle\int_\Gamma [\boldsymbol{\sigma}^{-1}(\operatorname{curl} \mathbf{H}_C - \mathbf{J}_{e,C}) - \widetilde{\mathbf{E}}_I] \times \mathbf{n}_C \cdot \mathbf{R}_{l,C} \,,
\end{aligned}
$$

and therefore $(3.23)_5$, recalling that $\mathbf{R}_{l,C} \times \mathbf{n}_C = \boldsymbol{\rho}_{l,I} \times \mathbf{n}_C$ on Γ and $\operatorname{curl} \widetilde{\mathbf{E}}_I = -i\omega\boldsymbol{\mu}_I \mathbf{H}_I$ in Ω_I. □

4.5 A saddle-point approach for the $\mathbf{H}_C/\widetilde{\mathbf{E}}_I$ formulation

In order to get rid of the constrained space $\widetilde{X}_C \times \widetilde{Z}_I$ appearing in (4.38), we can make use of the fact that both the constraints can be included in an augmented variational problem as extra linear conditions. Let us define the space

$$
X_C^* := \{\mathbf{v}_C \in H(\operatorname{curl}; \Omega_C) \mid \operatorname{div}_\tau(\mathbf{v}_C \times \mathbf{n}_C) \in L^2(\Gamma)\} \,,
$$

endowed with the graph norm

$$
\|\mathbf{v}_C\|_{X_C^*} := \|\mathbf{v}_C\|_{H(\operatorname{curl};\Omega_C)} + \|\operatorname{div}_\tau(\mathbf{v}_C \times \mathbf{n}_C)\|_{0,\Gamma}
$$

(which coincides with $\|\mathbf{v}_C\|_{H(\operatorname{curl};\Omega_C)}$ when $\mathbf{v}_C \in \widetilde{X}_C$). Let us also set

$$
L^2_\sharp(\Gamma) := \prod_{j=1}^{p_\Gamma+1} (L^2(\Gamma_j)/\mathbb{C}) \,.
$$

The unconstrained variational problem that we consider is

$$
\begin{aligned}
&\text{Find } \mathbf{H}_C \in X_C^*, \widetilde{\mathbf{E}}_I \in H(\text{curl}; \Omega_I), Q \in L^2_\sharp(\Gamma), \phi_I \in H^1(\Omega_I)/\mathbb{C} : \\
&\quad \mathcal{D}((\mathbf{H}_C, \widetilde{\mathbf{E}}_I), (\mathbf{v}_C, \mathbf{z}_I)) \\
&\qquad - \int_\Gamma \text{div}_\tau(\overline{\mathbf{v}_C} \times \mathbf{n}_C)\, Q - \int_{\Omega_I} \overline{\mathbf{z}_I} \cdot \text{grad}\, \phi_I = F(\mathbf{v}_C) + G(\mathbf{z}_I) \\
&\quad \int_\Gamma \text{div}_\tau(\mathbf{H}_C \times \mathbf{n}_C)\, \overline{P} = 0 \\
&\quad \int_{\Omega_I} \widetilde{\mathbf{E}}_I \cdot \text{grad}\, \overline{\xi_I} = 0 \\
&\text{for all } \mathbf{v}_C \in X_C^*, \mathbf{z}_I \in H(\text{curl}; \Omega_I), P \in L^2_\sharp(\Gamma), \xi_I \in H^1(\Omega_I)/\mathbb{C} .
\end{aligned}
\tag{4.49}
$$

Note that, since $\int_{\Gamma_j} \text{div}_\tau(\mathbf{v}_C \times \mathbf{n}_C) = 0$ for each $\mathbf{v}_C \in X_C^*$ and $j = 1, \ldots, p_\Gamma + 1$, this problem is indeed well-defined for $Q, P \in L^2_\sharp(\Gamma)$, namely, adding on Γ_j a constant to Q or P does not change the problem. In particular, any function belonging to $L^2(\Gamma)$ can be chosen as test function P.

Theorem 4.17. *Assume that Ω is simply-connected. Problem* (4.49) *has a unique solution* $(\mathbf{H}_C, \widetilde{\mathbf{E}}_I, Q, \phi_I)$ *in* $X_C^* \times H(\text{curl}; \Omega_I) \times L^2_\sharp(\Gamma) \times H^1(\Omega_I)/\mathbb{C}$*, and* $(\mathbf{H}_C, \widetilde{\mathbf{E}}_I)$ *is the solution to problem* (4.38)*. Moreover, the Lagrange multiplier ϕ_I is* 0.

Proof. Choosing $P = \text{div}_\tau(\mathbf{H}_C \times \mathbf{n}_C)$, it is clear that the two last equations in (4.49) imply that $(\mathbf{H}_C, \widetilde{\mathbf{E}}_I) \in \widetilde{X}_C \times \widetilde{Z}_I$, thus $(\mathbf{H}_C, \widetilde{\mathbf{E}}_I)$ is the solution to problem (4.38). From Theorem 4.15 and the classical theory of saddle-point problems (see Brezzi and Fortin [65]) we also see that, to prove existence and uniqueness of a solution to problem (4.49), it is sufficient to verify the inf–sup condition

$$
\begin{aligned}
&\exists\, \beta^* > 0 \text{ such that} \\
&\forall\, (P, \xi_I) \in L^2_\sharp(\Gamma) \times H^1(\Omega_I)/\mathbb{C} \quad \exists\, (\mathbf{v}_C, \mathbf{z}_I) \in X_C^* \times H(\text{curl}; \Omega_I), \\
&(\mathbf{v}_C, \mathbf{z}_I) \neq (\mathbf{0}, \mathbf{0}) : \\
&\left| \int_\Gamma \text{div}_\tau(\mathbf{v}_C \times \mathbf{n}_C)\, \overline{P} + \int_{\Omega_I} \mathbf{z}_I \cdot \text{grad}\, \overline{\xi_I} \right| \\
&\quad \geq \beta^* (\|\mathbf{v}_C\|_{X_C^*} + \|\mathbf{z}_I\|_{H(\text{curl};\Omega_I)}) \\
&\qquad \times (\textstyle\sum_{j=1}^{p_\Gamma+1} \|P_{|\Gamma_j}\|_{L^2(\Gamma_j)/\mathbb{C}} + \|\xi_I\|_{H^1(\Omega_I)/\mathbb{C}}) .
\end{aligned}
\tag{4.50}
$$

Since we can assume that $\int_{\Gamma_j} P_{|\Gamma_j} = 0$ for each $j = 1, \ldots, p_\Gamma + 1$, we can take $\mathbf{v}_C$ such that $\text{div}_\tau(\mathbf{v}_C \times \mathbf{n}_C) = P$. More precisely, this can be done by using the Laplace–Beltrami operator Δ_τ (see Section A.1) and considering the solution $\lambda_j \in H^1(\Gamma_j)/\mathbb{C}$ of $\Delta_\tau \lambda_j = P_{|\Gamma_j}$ on Γ, $j = 1, \ldots, p_\Gamma + 1$, a solution that satisfies $\|\,\text{grad}_\tau \lambda_j\|_{0,\Gamma_j} \leq C_0 \|P_{|\Gamma_j}\|_{0,\Gamma_j}$. Then we take for $\mathbf{v}_C$ a continuous extension in $H(\text{curl}; \Omega_C)$ of $\text{grad}_\tau \lambda_j \in H^{-1/2}(\text{div}_\tau; \Gamma_j)$, namely, a function $\mathbf{v}_C$ satisfying $(\mathbf{v}_C \times \mathbf{n}_C)_{|\Gamma_j} = \text{grad}_\tau \lambda_j$. In particular $\mathbf{v}_C \in X_C^*$ and

$$
\|\mathbf{v}_C\|_{X_C^*} \leq C_1 \sum_{j=1}^{p_\Gamma+1} (\|\,\text{grad}_\tau \lambda_j\|_{0,\Gamma_j} + \|P_{|\Gamma_j}\|_{0,\Gamma_j}) \leq C_2 \|P\|_{0,\Gamma} .
$$

Choosing $\mathbf{z}_I = \operatorname{grad} \xi_I$ (where, without restriction, we can also assume that $\int_{\Omega_I} \xi_I = 0$), we have

$$\begin{aligned}\left|\int_\Gamma \operatorname{div}_\tau(\mathbf{v}_C \times \mathbf{n}_C)\overline{P} + \int_{\Omega_I} \mathbf{z}_I \cdot \operatorname{grad}\overline{\xi_I}\right| &= \int_\Gamma |P|^2 + \int_{\Omega_I} |\operatorname{grad} \xi_I|^2 \\ &\geq \tfrac{1}{C_2 C_3}\|\mathbf{v}_C\|_{X_C^*} \textstyle\sum_j \|P_{|\Gamma_j}\|_{0,\Gamma_j} + \|\mathbf{z}_I\|_{H(\mathrm{curl};\Omega_I)}\|\operatorname{grad}\xi_I\|_{0,\Omega_I} \,,\end{aligned}$$

since $\sum_j \|P_{|\Gamma_j}\|_{0,\Gamma_j} \leq C_3\|P\|_{0,\Gamma}$. Recalling that $\|P_{|\Gamma_j}\|_{L^2(\Gamma_j)/\mathbb{C}} = \|P_{|\Gamma_j}\|_{0,\Gamma_j}$, as $\int_{\Gamma_j} P_{|\Gamma_j} = 0$, and moreover that $\|\xi_I\|_{H^1(\Omega_I)/\mathbb{C}} = \|\xi_I\|_{1,\Omega_I}$ if $\int_{\Omega_I} \xi_I = 0$ (see Section A.1), the inf–sup condition (4.50) follows from the Poincaré inequality

$$\int_{\Omega_I} |\operatorname{grad} \xi_I|^2 \geq c_0 \int_{\Omega_I} (|\xi_I|^2 + |\operatorname{grad}\xi_I|^2)\,, \tag{4.51}$$

which is valid for $\xi_I \in H^1(\Omega_I)$ with $\int_{\Omega_I} \xi_I = 0$ (see, e.g., Dautray and Lions [94], Chap. IV, Sect. 7, Prop. 2).

To show that the Lagrange multiplier ϕ_I is 0, we take as test functions in (4.49) $\mathbf{v}_C = \mathbf{0}$, $\mathbf{z}_I = \operatorname{grad}\phi_I$, $P = 0$ and $\xi_I = 0$. Then we have

$$\begin{aligned}\int_\Gamma \mathbf{H}_C \times \mathbf{n}_C \cdot \operatorname{grad}\overline{\phi_I} - \int_{\Omega_I} |\operatorname{grad}\phi_I|^2 &= \int_{\Omega_I} \mathbf{J}_{e,I}\cdot\operatorname{grad}\overline{\phi_I} \\ &= -\int_{\Omega_I} \operatorname{div}\mathbf{J}_{e,I}\,\overline{\phi_I} + \int_{\partial\Omega\cup\Gamma} \mathbf{J}_{e,I}\cdot\mathbf{n}\,\overline{\phi_I} = 0\,,\end{aligned}$$

and also, from $(4.49)_2$,

$$\int_\Gamma \mathbf{H}_C \times \mathbf{n}_C \cdot \operatorname{grad}\overline{\phi_I} = -\int_\Gamma \operatorname{div}_\tau(\mathbf{H}_C \times \mathbf{n}_C)\,\overline{\phi_I} = 0\,,$$

so that $\operatorname{grad}\phi_I = \mathbf{0}$. □

Remark 4.18. If $\mathbf{J}_{e,I}\cdot\mathbf{n}_I \neq 0$ on Γ, we can consider the function $\mathbf{H}_e \in H_0(\mathrm{curl};\Omega)$ defined in (3.5) and, setting $\mathbf{W}_C = \mathbf{H}_C - \mathbf{H}_{e,C}$, reformulate problem (4.37) as

$$\begin{aligned}&\text{Find } (\mathbf{W}_C, \mathbf{E}_I) \in H(\mathrm{curl};\Omega_C)\times Z_I : \\ &\int_{\Omega_C}(\boldsymbol{\sigma}^{-1}\operatorname{curl}\mathbf{W}_C\cdot\operatorname{curl}\overline{\mathbf{v}_C} + i\omega\boldsymbol{\mu}_C\mathbf{W}_C\cdot\overline{\mathbf{v}_C}) \\ &\qquad\qquad + \int_\Gamma \overline{\mathbf{v}_C}\times\mathbf{n}_C\cdot\mathbf{E}_I = \widetilde{F}(\mathbf{v}_C) \\ &\int_\Gamma \mathbf{W}_C\times\mathbf{n}_C\cdot\overline{\mathbf{z}_I} + i\omega^{-1}\int_{\Omega_I}\boldsymbol{\mu}_I^{-1}\operatorname{curl}\mathbf{E}_I\cdot\operatorname{curl}\overline{\mathbf{z}_I} = \widetilde{G}(\mathbf{z}_I) \\ &\text{for all } (\mathbf{v}_C,\mathbf{z}_I)\in H(\mathrm{curl};\Omega_C)\times Z_I\,,\end{aligned} \tag{4.52}$$

where

$$\widetilde{F}(\mathbf{v}_C) := \int_{\Omega_C}\left[\boldsymbol{\sigma}^{-1}(\mathbf{J}_{e,C} - \operatorname{curl}\mathbf{H}_{e,C})\cdot\operatorname{curl}\overline{\mathbf{v}_C} - i\omega\boldsymbol{\mu}_C\mathbf{H}_{e,C}\cdot\overline{\mathbf{v}_C})\right]\,,$$

and

$$\begin{aligned}\widetilde{G}(\mathbf{z}_I) &= \int_{\Omega_I}\mathbf{J}_{e,I}\cdot\overline{\mathbf{z}_I} - \int_\Gamma \mathbf{H}_{e,C}\times\mathbf{n}_C\cdot\overline{\mathbf{z}_I} \\ &= \int_{\Omega_I}\operatorname{curl}\mathbf{H}_{e,I}\cdot\overline{\mathbf{z}_I} + \int_\Gamma \mathbf{H}_{e,I}\times\mathbf{n}_I\cdot\overline{\mathbf{z}_I} \\ &= \int_{\Omega_I}\mathbf{H}_{e,I}\cdot\operatorname{curl}\overline{\mathbf{z}_I}\,,\end{aligned}$$

since $\mathbf{H}_{e,I}\times\mathbf{n} = \mathbf{0}$ on $\partial\Omega$.

The new unknown $\mathbf{W}_C$ belongs to $\widetilde{X}_C$ and we can proceed as in the case in which we have assumed $\mathbf{J}_{e_I} \cdot \mathbf{n}_I = 0$ on Γ. In particular, the unconstrained variational problem reads

$$
\begin{aligned}
&\text{Find } \mathbf{W}_C \in X_C^*, \widetilde{\mathbf{E}}_I \in H(\text{curl}; \Omega_I), Q \in L^2_\sharp(\Gamma), \phi_I \in H^1(\Omega_I)/\mathbb{C} : \\
&\quad \mathcal{D}((\mathbf{W}_C, \widetilde{\mathbf{E}}_I), (\mathbf{v}_C, \mathbf{z}_I)) - \textstyle\int_\Gamma \operatorname{div}_\tau(\overline{\mathbf{v}_C} \times \mathbf{n}_C)\, Q \\
&\qquad\qquad - \textstyle\int_{\Omega_I} \overline{\mathbf{z}_I} \cdot \operatorname{grad} \phi_I = \widetilde{F}(\mathbf{v}_C) + \widetilde{G}(\mathbf{z}_I) \\
&\quad \textstyle\int_\Gamma \operatorname{div}_\tau(\mathbf{W}_C \times \mathbf{n}_C)\, \overline{P} = 0 \\
&\quad \textstyle\int_{\Omega_I} \widetilde{\mathbf{E}}_I \cdot \operatorname{grad} \overline{\xi_I} = 0 \\
&\text{for all } \mathbf{v}_C \in X_C^*, \mathbf{z}_I \in H(\text{curl}; \Omega_I), P \in L^2_\sharp(\Gamma), \xi_I \in H^1(\Omega_I)/\mathbb{C} .
\end{aligned}
\tag{4.53}
$$

If $\mathbf{J}_{e,I} \cdot \mathbf{n}_I \in L^2(\Gamma)$, then in fact $\mathbf{H}_C \in X_C^*$ and it is possible to avoid the use of $\mathbf{H}_e$, considering the problem

$$
\begin{aligned}
&\text{Find } \mathbf{H}_C \in X_C^*, \widetilde{\mathbf{E}}_I \in H(\text{curl}; \Omega_I), Q \in L^2_\sharp(\Gamma), \phi_I \in H^1(\Omega_I)/\mathbb{C} : \\
&\quad \mathcal{D}((\mathbf{H}_C, \widetilde{\mathbf{E}}_I), (\mathbf{v}_C, \mathbf{z}_I)) - \textstyle\int_\Gamma \operatorname{div}_\tau(\overline{\mathbf{v}_C} \times \mathbf{n}_C)\, Q \\
&\qquad\qquad - \textstyle\int_{\Omega_I} \overline{\mathbf{z}_I} \cdot \operatorname{grad} \phi_I = F(\mathbf{v}_C) + G(\mathbf{z}_I) \\
&\quad \textstyle\int_\Gamma \operatorname{div}_\tau(\mathbf{H}_C \times \mathbf{n}_C)\, \overline{P} = - \textstyle\int_\Gamma \mathbf{J}_{e,I} \cdot \mathbf{n}_I\, \overline{P} \\
&\quad \textstyle\int_{\Omega_I} \widetilde{\mathbf{E}}_I \cdot \operatorname{grad} \overline{\xi_I} = 0 \\
&\text{for all } \mathbf{v}_C \in X_C^*, \mathbf{z}_I \in H(\text{curl}; \Omega_I), P \in L^2_\sharp(\Gamma), \xi_I \in H^1(\Omega_I)/\mathbb{C} .
\end{aligned}
\tag{4.54}
$$

Both problems (4.53) and (4.54) have the same structure of (4.49). □

Remark 4.19. For the electric boundary condition (1.20) the variational problem is

$$
\begin{aligned}
&\text{Find } \mathbf{H}_C \in X_C^*, \widetilde{\mathbf{E}}_I \in H_{0,\partial\Omega}(\text{curl}; \Omega_I), Q^* \in L^2_\sharp(\Gamma) \\
&\qquad \phi_I^* \in H^1_{0,\partial\Omega}(\Omega_I) : \\
&\quad \mathcal{D}((\mathbf{H}_C, \widetilde{\mathbf{E}}_I), (\mathbf{v}_C, \mathbf{z}_I)) \\
&\qquad - \textstyle\int_\Gamma \operatorname{div}_\tau(\overline{\mathbf{v}_C} \times \mathbf{n}_C)\, Q^* - \textstyle\int_{\Omega_I} \overline{\mathbf{z}_I} \cdot \operatorname{grad} \phi_I^* = F(\mathbf{v}_C) + G(\mathbf{z}_I) \\
&\quad \textstyle\int_\Gamma \operatorname{div}_\tau(\mathbf{H}_C \times \mathbf{n}_C)\, \overline{P^*} = 0 \\
&\quad \textstyle\int_{\Omega_I} \widetilde{\mathbf{E}}_I \cdot \operatorname{grad} \overline{\xi_I^*} = 0 \\
&\text{for all } \mathbf{v}_C \in X_C^*, \mathbf{z}_I \in H_{0,\partial\Omega}(\text{curl}; \Omega_I), P^* \in L^2_\sharp(\Gamma), \\
&\qquad \xi_I^* \in H^1_{0,\partial\Omega}(\Omega_I) ,
\end{aligned}
\tag{4.55}
$$

where, as for the magnetic boundary conditions, $\widetilde{\mathbf{E}}_I$ is a vector potential of $-i\omega\boldsymbol{\mu}_I\mathbf{H}_I$. It is easy to verify, by proceeding as in Remark 4.16, that $\mathbf{H}_C$ and $\mathbf{H}_I = -(i\omega)^{-1}\boldsymbol{\mu}_I^{-1} \operatorname{curl} \mathbf{E}_I$ are a solution of the strong problem (3.42).

It can be shown that in the present case it is not necessary to assume that Ω is simply-connected. □

4.5.1 Finite element discretization

Let $\mathbb{P}_k$ be the set of complex polynomials of degree less than or equal to k in x_1, x_2, x_3, and $\widetilde{\mathbb{P}}_k$ the set of homogeneous complex polynomials of degree k. Then for $k \geq 1$ we consider the (complex-valued) Nédélec curl-conforming edge elements

$$N_h^k := \{\mathbf{z}_h \in H(\text{curl}; \Omega) \mid \mathbf{z}_{h|K} \in R_k \; \forall \; K \in \mathcal{T}_h\} \,, \tag{4.56}$$

where $R_k := (\mathbb{P}_{k-1})^3 \oplus S_k$ and $S_k := \{\mathbf{p} \in (\widetilde{\mathbb{P}}_k)^3 \mid \mathbf{p}(\mathbf{x}) \cdot \mathbf{x} = 0\}$ (see Section A.2).

In order to obtain a finite element approximation of the variational problem (4.38), we introduce the finite element spaces

$$\widetilde{X}_{C,h} := \{\mathbf{v}_{C,h} \in N_{C,h}^k \mid \text{div}_\tau(\mathbf{v}_{C,h} \times \mathbf{n}_C) = 0 \text{ on } \Gamma\}, \tag{4.57}$$

and

$$\widetilde{Z}_{I,h} := \{\mathbf{z}_{I,h} \in N_{I,h}^k \mid \int_{\Omega_I} \mathbf{z}_{I,h} \cdot \text{grad}\, \xi_{I,h} = 0 \; \forall \; \xi_{I,h} \in L_{I,h}^k\} \,, \tag{4.58}$$

where $N_{C,h}^k$ and $N_{I,h}^k$ are the Nédélec curl-conforming edge elements related to the domains Ω_C and Ω_I, respectively. We also denote by

$$L_{I,h}^k := \{\xi_{I,h} \in C^0(\Omega_I) \mid \xi_{I,h|K} \in \mathbb{P}_k \; \forall \; K \in \mathcal{T}_{I,h}\}$$

the standard piecewise-polynomial Lagrange nodal elements (see Section A.2).

We consider the following finite element approximation of the hybrid $\mathbf{H}_C/\mathbf{E}_I$ formulation

$$\begin{aligned}
&\text{Find } (\mathbf{H}_{C,h}, \widetilde{\mathbf{E}}_{I,h}) \in \widetilde{X}_{C,h} \times \widetilde{Z}_{I,h} : \\
&\int_{\Omega_C} (\boldsymbol{\sigma}^{-1} \text{curl}\, \mathbf{H}_{C,h} \cdot \text{curl}\, \overline{\mathbf{v}_{C,h}} + i\omega \boldsymbol{\mu}_C \mathbf{H}_{C,h} \cdot \overline{\mathbf{v}_{C,h}}) \\
&\qquad\qquad + \int_\Gamma \overline{\mathbf{v}_{C,h}} \times \mathbf{n}_C \cdot \widetilde{\mathbf{E}}_{I,h} = F(\mathbf{v}_{C,h}) \\
&\int_\Gamma \mathbf{H}_{C,h} \times \mathbf{n}_C \cdot \overline{\mathbf{z}_{I,h}} + i\omega^{-1} \int_{\Omega_I} \boldsymbol{\mu}_I^{-1} \text{curl}\, \widetilde{\mathbf{E}}_{I,h} \cdot \text{curl}\, \overline{\mathbf{z}_{I,h}} = G(\mathbf{z}_{I,h}) \\
&\text{for all } (\mathbf{v}_{C,h}, \mathbf{z}_{I,h}) \in \widetilde{X}_{C,h} \times \widetilde{Z}_{I,h} \,.
\end{aligned} \tag{4.59}$$

Theorem 4.20. *Assume that Ω is simply-connected and that the families of triangulations $\mathcal{T}_{C,h}$ and $\mathcal{T}_{I,h}$ are obtained as a refinement of coarse triangulations $\mathcal{T}_{C,h^0}$ and $\mathcal{T}_{I,h^0}$ for a fixed h^0. Then the variational problem* (4.59) *has a unique solution in $\widetilde{X}_{C,h} \times \widetilde{Z}_{I,h}$.*

Proof. First we note that the space $\widetilde{Z}_{I,h}$ can be decomposed similarly to $\widetilde{Z}_I$ in (4.45). In fact, we have $\text{grad}\, L_{I,h}^k \subset H^0(\text{curl}; \Omega_I) \cap N_{I,h}^k$ (note that in general geometrical configurations these two spaces do not coincide). Consequently, the functions in $N_{I,h}^k$ that are $(L^2(\Omega_I))^3$-orthogonal to $H^0(\text{curl}; \Omega_I) \cap N_{I,h}^k$ are included in $\widetilde{Z}_{I,h}$, namely, $[H^0(\text{curl}; \Omega_I) \cap N_{I,h}^k]^\perp \subset \widetilde{Z}_{I,h}$. Denoting by

$$\mathcal{H}_{I,h} := H^0(\text{curl}; \Omega_I) \cap \widetilde{Z}_{I,h}$$

(in the discrete case this corresponds to the space of harmonic vector fields $\mathcal{H}(m; \Omega_I)$, though it is not a subspace of it), we finally have

$$\widetilde{Z}_{I,h} = [H^0(\text{curl}; \Omega_I) \cap N^k_{I,h}]^\perp \oplus \mathcal{H}_{I,h} \, . \tag{4.60}$$

It is also checked easily that $\dim \mathcal{H}(m; \Omega_I) = \dim \mathcal{H}_{I,h}$.

Based on (4.60) we can rewrite (4.59) in equivalent form as

$$\begin{aligned}
&\text{Find } \mathbf{H}_{C,h} \in \widetilde{X}_{C,h},\ \widetilde{\mathbf{E}}^\perp_{I,h} \in [H^0(\text{curl}; \Omega_I) \cap N^k_{I,h}]^\perp,\ \widetilde{\mathbf{E}}^{\mathcal{H}}_{I,h} \in \mathcal{H}_{I,h} : \\
&\quad \mathcal{D}((\mathbf{H}_{C,h}, \widetilde{\mathbf{E}}^\perp_{I,h}), (\mathbf{v}_{C,h}, \mathbf{z}^\perp_{I,h})) + \textstyle\int_\Gamma \overline{\mathbf{v}_{C,h}} \times \mathbf{n}_C \cdot \widetilde{\mathbf{E}}^{\mathcal{H}}_{I,h} = F(\mathbf{v}_{C,h}) + G(\mathbf{z}^\perp_{I,h}) \\
&\quad \textstyle\int_\Gamma \mathbf{H}_{C,h} \times \mathbf{n}_C \cdot \overline{\mathbf{z}^{\mathcal{H}}_{I,h}} = G(\mathbf{z}^{\mathcal{H}}_{I,h}) \\
&\text{for all } \mathbf{v}_{C,h} \in \widetilde{X}_{C,h},\ \mathbf{z}^\perp_{I,h} \in [H^0(\text{curl}; \Omega_I) \cap N^k_{I,h}]^\perp,\ \mathbf{z}^{\mathcal{H}}_{I,h} \in \mathcal{H}_{I,h} \, .
\end{aligned}$$

To prove that $\mathcal{D}(\cdot, \cdot)$ is uniformly coercive in $\widetilde{X}_{C,h} \times [H^0(\text{curl}; \Omega_I) \cap N^k_{I,h}]^\perp$ we can proceed as in the continuous case, since there exists a constant C_2, independent of h, such that

$$\|\mathbf{v}_{I,h}\|_{0,\Omega_I} \leq C_2 \| \operatorname{curl} \mathbf{v}_{I,h}\|_{0,\Omega_I} \ \forall\, \mathbf{v}_{I,h} \in [H^0(\text{curl}; \Omega_I) \cap N^k_{I,h}]^\perp$$

(the proof is similar to that of Lemma 4.7).

Now we need to check the discrete inf–sup condition

$$\begin{aligned}
&\exists\, \widetilde{\beta} > 0 \text{ such that} \\
&\forall\, \mathbf{z}^{\mathcal{H}}_{I,h} \in \mathcal{H}_{I,h}\ \exists\, \mathbf{v}_{C,h} \in \widetilde{X}_{C,h}\, , \mathbf{v}_{C,h} \neq \mathbf{0} : \\
&\left| \textstyle\int_\Gamma \mathbf{v}_{C,h} \times \mathbf{n}_C \cdot \overline{\mathbf{z}^{\mathcal{H}}_{I,h}} \right| \geq \widetilde{\beta} \|\mathbf{v}_{C,h}\|_{H(\text{curl};\Omega_C)} \, \|\mathbf{z}^{\mathcal{H}}_{I,h}\|_{0,\Omega_I} \, .
\end{aligned} \tag{4.61}$$

For the sake of simplicity, we start assuming that Ω_C is a torus, so its first Betti number n_{Ω_C} is equal to 1. Since Ω is simply-connected, the non-bounding cycles of Ω_I are on Γ and the first Betti number n_{Ω_I} of Ω_I is also equal to 1. As in the proof of Theorem 4.15, we can consider the "cutting" surfaces Σ_L, $L = I, C$. Let us denote by Π_L the piecewise-linear function, defined on a coarse mesh $\mathcal{T}_{L,h^0}$, which takes value 1 at the nodes on one side of Σ_L and 0 at all the other nodes. Then, if the family of triangulations $\mathcal{T}_{L,h}$ are obtained by refining the coarse mesh $\mathcal{T}_{L,h^0}$, $\widetilde{\text{grad}}\, \Pi_I$ belongs to $H^0(\text{curl}; \Omega_I) \cap N^1_{I,h}$ (thus to $H^0(\text{curl}; \Omega_I) \cap N^k_{I,h}$), but clearly, due to the jump of Π_I on Σ_I, $\widetilde{\text{grad}}\, \Pi_I \notin \text{grad}\, L^1_{I,h}$. Any function $\mathbf{z}^{\mathcal{H}}_{I,h}$ can be written as $\mathbf{z}^{\mathcal{H}}_{I,h} = c\ \widetilde{\text{grad}}\, \Pi_I + \text{grad}\, \xi_{I,h}$ for some $c \in \mathbb{C}$ and $\xi_{I,h} \in L^k_{I,h}$. Therefore we choose $\mathbf{v}_{C,h} = c\ \widetilde{\text{grad}}\, \Pi_C \in \widetilde{X}_{C,h}$, and we can proceed as in the continuous case. At first, we have to note that for $\text{grad}\, \xi_{I,h}$ the line integral on a closed cycle is always vanishing. Moreover, we have

$$\begin{aligned}
\|\mathbf{z}^{\mathcal{H}}_{I,h}\|^2_{0,\Omega_I} &= \textstyle\int_{\Omega_I} \mathbf{z}^{\mathcal{H}}_{I,h} \cdot \overline{\mathbf{z}^{\mathcal{H}}_{I,h}} = \overline{c} \int_{\Omega_I} \mathbf{z}^{\mathcal{H}}_{I,h} \cdot \widetilde{\text{grad}}\, \Pi_I \\
&\leq |c| \|\widetilde{\text{grad}}\, \Pi_I\|_{0,\Omega_I} \|\mathbf{z}^{\mathcal{H}}_{I,h}\|_{0,\Omega_I} \, ,
\end{aligned}$$

so that $\|\mathbf{z}^{\mathcal{H}}_{I,h}\|_{0,\Omega_I} \leq |c| \|\widetilde{\text{grad}}\, \Pi_I\|_{0,\Omega_I}$. Therefore, the proof ends as in the continuous case.

For the Betti number $n_{\Omega_C} = n_C > 1$, one arrives at the inf–sup constant

$$\widetilde{\beta} = \min_{\mathbf{c}\in\mathbb{C}^{n_C},\mathbf{c}\neq\mathbf{0}} \frac{|\mathbf{c}|^2}{(\widetilde{M}^C\mathbf{c}\cdot\overline{\mathbf{c}})^{1/2}(\widetilde{M}^I\mathbf{c}\cdot\overline{\mathbf{c}})^{1/2}} ,$$

where $\widetilde{M}^L$, $L = I, C$, is the matrix given by $\widetilde{M}^L_{lj} = \int_{\Omega^L} \widetilde{\operatorname{grad} \Pi_{l,L}} \cdot \widetilde{\operatorname{grad} \Pi_{j,L}}$, $l, j = 1, \dots, n_C$. □

Remark 4.21. From the arguments of Theorem 4.20 we readily derive that for all $(\mathbf{F}_h, \mathbf{G}_h) \in (\widetilde{X}_{C,h})' \times (\widetilde{Z}_{I,h})'$ there exists a unique solution of the problem

$$\begin{aligned}
&\text{Find } (\mathbf{u}_{C,h}, \mathbf{w}_{I,h}) \in \widetilde{X}_{C,h} \times \widetilde{Z}_{I,h} : \\
&\quad \mathcal{D}((\mathbf{u}_{C,h}, \mathbf{w}_{I,h}), (\mathbf{v}_{C,h}, \mathbf{z}_{I,h})) = \langle \mathbf{F}_h, \mathbf{v}_{C,h}\rangle + \langle \mathbf{G}_h, \mathbf{z}_{I,h}\rangle \\
&\text{for all } (\mathbf{v}_{C,h}, \mathbf{z}_{I,h}) \in \widetilde{X}_{C,h} \times \widetilde{Z}_{I,h} ,
\end{aligned} \tag{4.62}$$

where $\langle\cdot,\cdot\rangle$ denotes the duality pairing. Moreover, the solution is bounded as follows

$$\|\mathbf{u}_{C,h}\|_{H(\text{curl};\Omega_C)} + \|\mathbf{w}_{I,h}\|_{H(\text{curl};\Omega_I)} \leq C_*(\|\mathbf{F}_h\|_{(\widetilde{X}_{C,h})'} + \|\mathbf{G}_h\|_{(\widetilde{Z}_{I,h})'}) .$$

The constant C_* depends on the continuity constant of the bilinear form $\mathcal{D}(\cdot,\cdot)$ in $\widetilde{X}_C \times (H^0(\text{curl}; \Omega_I))^\perp$, on its coerciveness constant in $\widetilde{X}_{C,h} \times [H^0(\text{curl}; \Omega_I) \cap \mathbf{X}_{I,h}]^\perp$, and on the constant $\widetilde{\beta}$ in (4.61), hence it is independent of h. As a consequence, it is easily shown that

$$\begin{aligned}
&\exists\, \alpha > 0, \text{ independent of } h, \text{ such that} \\
&\forall\, (\mathbf{u}_{C,h}, \mathbf{w}_{I,h}) \in \widetilde{X}_{C,h} \times \widetilde{Z}_{I,h}\ \exists\, (\mathbf{v}_{C,h}, \mathbf{z}_{I,h}) \in \widetilde{X}_{C,h} \times \widetilde{Z}_{I,h} , \\
&(\mathbf{v}_{C,h}, \mathbf{z}_{I,h}) \neq (\mathbf{0}, \mathbf{0}) : \\
&\mathcal{D}((\mathbf{u}_{C,h}, \mathbf{w}_{I,h}), (\mathbf{v}_{C,h}, \mathbf{z}_{I,h})) \\
&\quad \geq \alpha(\|\mathbf{v}_{C,h}\|_{H(\text{curl};\Omega_C)} + \|\mathbf{z}_{I,h}\|_{H(\text{curl};\Omega_I)}) \\
&\qquad\qquad \times(\|\mathbf{u}_{C,h}\|_{H(\text{curl};\Omega_C)} + \|\mathbf{w}_{I,h}\|_{H(\text{curl};\Omega_I)}) ,
\end{aligned} \tag{4.63}$$

that is a uniform inf–sup condition. □

For devising a suitable conforming finite element approximation of the unconstrained problem (4.49) we need another discrete space. We start from

$$C^{k-1}_{\Gamma_j,h} := \{P_{h,j} \in L^2(\Gamma_j) \mid P_{h,j|T} \in \mathbb{P}_{k-1}\ \forall T \in \mathcal{T}_{\Gamma_j,h}\} ,$$

where $\mathcal{T}_{\Gamma_j,h}$ is the restriction to Γ_j of the mesh $\mathcal{T}_{C,h}$, and then we define

$$Y^{k-1}_{\Gamma,h} := \prod_{j=1}^{p_\Gamma+1} (C^{k-1}_{\Gamma_j,h}/\mathbb{C}) .$$

We consider the following problem

$$
\begin{aligned}
&\text{Find } \mathbf{H}_{C,h} \in N^k_{C,h}, \widetilde{\mathbf{E}}_{I,h} \in N^k_{I,h}, Q_h \in Y^{k-1}_{\Gamma,h}, \phi_{I,h} \in L^k_{I,h}/\mathbb{C} : \\
&\mathcal{D}((\mathbf{H}_{C,h}, \widetilde{\mathbf{E}}_{I,h}), \mathbf{v}_{C,h}, \mathbf{z}_{I,h})) - \textstyle\int_\Gamma \operatorname{div}_\tau(\overline{\mathbf{v}_{C,h}} \times \mathbf{n}_C)\, Q_h \\
&\qquad\qquad - \textstyle\int_{\Omega_I} \overline{\mathbf{z}_{I,h}} \cdot \operatorname{grad} \phi_{I,h} = F(\mathbf{v}_{C,h}) + G(\mathbf{z}_{I,h}) \\
&\textstyle\int_\Gamma \operatorname{div}_\tau(\mathbf{H}_{C,h} \times \mathbf{n}_C)\, \overline{P_h} = 0 \\
&\textstyle\int_{\Omega_I} \widetilde{\mathbf{E}}_{I,h} \cdot \operatorname{grad} \overline{\xi_{I,h}} = 0 \\
&\text{for all } \mathbf{v}_{C,h} \in N^k_{C,h}, \mathbf{z}_{I,h} \in N^k_{I,h}, P_h \in Y^{k-1}_{\Gamma,h}, \xi_{I,h} \in L^k_{I,h}/\mathbb{C} .
\end{aligned}
\tag{4.64}
$$

Theorem 4.22. *Under the assumptions of Theorem 4.20, problem* (4.64) *has a unique solution* $(\mathbf{H}_{C,h}, \widetilde{\mathbf{E}}_{I,h}, Q_h, \phi_{I,h})$ *in* $N^k_{C,h} \times N^k_{I,h} \times Y^{k-1}_{\Gamma,h} \times L^k_{I,h}/\mathbb{C}$*, and* $(\mathbf{H}_{C,h}, \widetilde{\mathbf{E}}_{I,h})$ *is the solution to problem* (4.59)*. Moreover, the Lagrange multiplier* $\phi_{I,h}$ *is* 0*.*

Proof. Since $\operatorname{div}_\tau(\mathbf{H}_{C,h} \times \mathbf{n}_C)_{|\Gamma_j} \in C^{k-1}_{\Gamma_j,h}$, choosing $P_h = \operatorname{div}_\tau(\mathbf{H}_{C,h} \times \mathbf{n}_C)$ we have that $(\mathbf{H}_{C,h}, \widetilde{\mathbf{E}}_{I,h}) \in \widetilde{X}_{C,h} \times \widetilde{Z}_{I,h}$, and we conclude at once that $(\mathbf{H}_{C,h}, \widetilde{\mathbf{E}}_{I,h})$ is the solution to (4.59). Moreover, estimate (4.63) holds, and, as in the proof of Theorem 4.17, we only need to verify the uniform discrete inf–sup condition

$$
\begin{aligned}
&\exists\, \beta_* > 0, \text{ independent of } h, \text{ such that} \\
&\forall\, P_h \in Y^{k-1}_{\Gamma,h}, \xi_{I,h} \in L^k_{I,h}/\mathbb{C}\ \exists\, (\mathbf{v}_{C,h}, \mathbf{z}_{I,h}) \in N^k_{C,h} \times N^k_{I,h}, \\
&(\mathbf{v}_{C,h}, \mathbf{z}_{I,h}) \neq (\mathbf{0}, \mathbf{0}) : \\
&\left| \textstyle\int_\Gamma \operatorname{div}_\tau(\mathbf{v}_{C,h} \times \mathbf{n}_C)\, \overline{P_h} + \int_{\Omega_I} \mathbf{z}_{I,h} \cdot \operatorname{grad} \overline{\xi_{I,h}} \right| \\
&\quad \geq \beta_* (\|\mathbf{v}_{C,h}\|_{X^*_C} + \|\mathbf{z}_{I,h}\|_{H(\mathrm{curl};\Omega_I)}) \\
&\qquad\quad \times (\textstyle\sum_{j=1}^{p_\Gamma+1} \|P_{h|\Gamma_j}\|_{L^2(\Gamma_j)/\mathbb{C}} + \|\xi_{I,h}\|_{H^1(\Omega_I)/\mathbb{C}}) .
\end{aligned}
\tag{4.65}
$$

Without restriction we can assume that $\int_{\Gamma_j} P_{h|\Gamma_j} = 0$ and $\int_{\Omega_I} \xi_{I,h} = 0$. Then we choose $\mathbf{z}_{I,h} = \operatorname{grad} \xi_{I,h}$ and $\mathbf{v}_{C,h}$ such that $\operatorname{div}_\tau(\mathbf{v}_{C,h} \times \mathbf{n}_C) = P_h$. More precisely, let us denote by $\mathcal{X}_{\Gamma_j,h}$ the space of tangential traces on Γ_j of $N^k_{C,h}$ (namely, the Raviart–Thomas finite elements on Γ_j) and by $\mathcal{X}^0_{\Gamma_j,h}$ the kernel of the div_τ operator in $\mathcal{X}_{\Gamma_j,h}$, $j = 1, \ldots, p_\Gamma + 1$. Since

$$
\operatorname{div}_\tau \mathcal{X}_{\Gamma_j,h} = \left\{ P_{h,j} \in C^{k-1}_{\Gamma_j,h} \mid \int_{\Gamma_j} P_{h,j} = 0 \right\},
$$

there exists a function $\mathbf{r}_{h,j} \in (\mathcal{X}^0_{\Gamma_j,h})^\perp$ such that $\operatorname{div}_\tau \mathbf{r}_{h,j} = P_{h|\Gamma_j}$. We can take for $\mathbf{v}_{C,h}$ a uniformly continuous extension in $N^k_{C,h}$ of $\mathbf{r}_{h,j}$ (see Alonso and Valli [9]), so that $(\mathbf{v}_{C,h} \times \mathbf{n}_C)_{|\Gamma_j} = \mathbf{r}_{h,j}$.

It can be shown that there exists a constant $C_0 > 0$, independent of h, such that

$$
\|\mathbf{s}_{h|\Gamma_j}\|_{0,\Gamma_j} \leq C_0 \|\operatorname{div}_\tau \mathbf{s}_{h|\Gamma_j}\|_{0,\Gamma_j} \quad \forall\, \mathbf{s}_{h|\Gamma_j} \in (\mathcal{X}^0_{\Gamma_j,h})^\perp .
$$

In fact, let us consider the Laplace–Beltrami operator Δ_τ (see Section A.1) and denote by $p_j \in H^1(\Gamma_j)/\mathbb{C}$ the solution of $\Delta_\tau p_j = \operatorname{div}_\tau \mathbf{s}_{h|\Gamma_j}$ on Γ_j, $j = 1, \ldots, p_\Gamma + 1$. This solution satisfies $\| \operatorname{grad}_\tau p_j\|_{\delta,\Gamma_j} \leq C_0 \| \operatorname{div}_\tau \mathbf{s}_{h|\Gamma_j}\|_{0,\Gamma_j}$ for a suitable $\delta > 0$. We have

$$\|\mathbf{s}_{h|\Gamma_j}\|_{0,\Gamma_j} \leq \|\mathbf{s}_{h|\Gamma_j} - \operatorname{grad}_\tau p_j\|_{0,\Gamma_j} + \| \operatorname{grad}_\tau p_j\|_{0,\Gamma_j} ;$$

moreover, denoting by $\mathbf{I}_{h,\Gamma_j}$ the interpolation operator in $\mathcal{X}_{\Gamma_j,h}$, from the fact that $(\mathbf{s}_{h|\Gamma_j} - \mathbf{I}_{h,\Gamma_j} \operatorname{grad}_\tau p_j) \in \mathcal{X}^0_{\Gamma_j,h}$ we have

$$\begin{aligned}
&\|\mathbf{s}_{h|\Gamma_j} - \operatorname{grad}_\tau p_j\|^2_{0,\Gamma_j} \\
&\quad = \textstyle\int_{\Gamma_j} (\mathbf{s}_{h|\Gamma_j} - \operatorname{grad}_\tau p_j) \cdot (\mathbf{s}_{h|\Gamma_j} - \mathbf{I}_{h,\Gamma_j} \operatorname{grad}_\tau p_j) \\
&\qquad + \textstyle\int_{\Gamma_j} (\mathbf{s}_{h|\Gamma_j} - \operatorname{grad}_\tau p_j) \cdot (\mathbf{I}_{h,\Gamma_j} \operatorname{grad}_\tau p_j - \operatorname{grad}_\tau p_j) \\
&\quad = \textstyle\int_{\Gamma_j} (\mathbf{s}_{h|\Gamma_j} - \operatorname{grad}_\tau p_j) \cdot (\mathbf{I}_{h,\Gamma_j} \operatorname{grad}_\tau p_j - \operatorname{grad}_\tau p_j) \\
&\quad \leq \|\mathbf{s}_{h|\Gamma_j} - \operatorname{grad}_\tau p_j\|_{0,\Gamma_j} \|\mathbf{I}_{h,\Gamma_j} \operatorname{grad}_\tau p_j - \operatorname{grad}_\tau p_j\|_{0,\Gamma_j} .
\end{aligned}$$

Finally, following Buffa et al. [72], Theor. 4.2, we find that

$$\|\mathbf{I}_{h,\Gamma_j} \operatorname{grad}_\tau p_j - \operatorname{grad}_\tau p_j\|_{0,\Gamma_j}$$

can be estimated by $\| \operatorname{div}_\tau \mathbf{s}_{h|\Gamma_j}\|_{0,\Gamma_j}$.

In conclusion,

$$\begin{aligned}
\|\mathbf{v}_{C,h}\|_{X^*_C} &= \|\mathbf{v}_{C,h}\|_{H(\mathrm{curl};\Omega_C)} + \| \operatorname{div}_\tau(\mathbf{v}_{C,h} \times \mathbf{n}_C)\|_{0,\Gamma} \\
&\leq C_1 \textstyle\sum_j (\|\mathbf{r}_{h,j}\|_{0,\Gamma_j} + \|P_{h|\Gamma_j}\|_{0,\Gamma_j}) \\
&\leq C_2 \|P_h\|_{0,\Gamma} ,
\end{aligned}$$

and, by proceeding as in the continuous case, we have (4.65).

The proof that the Lagrange multiplier $\phi_{I,h} \in L^k_{I,h} \setminus \mathbb{C}$ is equal to 0 is easily done. In fact, we have already noted that from the second equation in (4.64) it follows that $\operatorname{div}_\tau(\mathbf{H}_{C,h} \times \mathbf{n}_C) = 0$ on Γ. Now taking $(\mathbf{0}, \operatorname{grad} \phi_{I,h}) \in N^k_{C,h} \times N^k_{I,h}$ as test function in the first equation of (4.64) we have that

$$\int_\Gamma \mathbf{H}_{C,h} \times \mathbf{n}_C \cdot \operatorname{grad} \overline{\phi_{I,h}} - \int_{\Omega_I} | \operatorname{grad} \phi_{I,h}|^2 = \int_{\Omega_I} \mathbf{J}_{e,I} \cdot \operatorname{grad} \overline{\phi_{I,h}} .$$

Since $\operatorname{div} \mathbf{J}_{e,I} = 0$ in Ω_I and $\mathbf{J}_{e,I} \cdot \mathbf{n} = 0$ on $\partial\Omega \cup \Gamma$, integrating by parts the first and third term we find $\int_{\Omega_I} | \operatorname{grad} \phi_{I,h}|^2 = 0$, hence $\phi_{I,h} = 0$ in $L^k_{I,h} \setminus \mathbb{C}$. □

The convergence of the solution of problem (4.64) to the solution of problem (4.49) is a consequence of the standard theory of saddle-point problems (see Brezzi and Fortin [65]). An usual density argument and the interpolation estimates in Section A.2 yields the following convergence theorem (see Section A.1 for notation).

Theorem 4.23. *Let the assumptions of Theorem 4.20 be satisfied. Let $\mathbf{H}_C$, $\widetilde{\mathbf{E}}_I$, Q and $\phi_I = 0$ be the solution of problem* (4.49) *and $\mathbf{H}_{C,h}$, $\widetilde{\mathbf{E}}_{I,h}$, Q_h and $\phi_{I,h} = 0$ be the solution of problem* (4.64). *The finite element approximation method is convergent and, if $\mathbf{H}_C \in H^r(\mathrm{curl};\Omega_C)$, $\widetilde{\mathbf{E}}_I \in H^r(\mathrm{curl};\Omega_I)$ and $Q \in H^s(\Gamma)$ with $r > 1/2$ and $s > 0$, the following error estimate holds*

$$\begin{aligned}
&\|\mathbf{H}_C - \mathbf{H}_{C,h}\|_{H(\mathrm{curl};\Omega_C)} \leq C h^{\min(r,k)} \\
&\|\widetilde{\mathbf{E}}_I - \widetilde{\mathbf{E}}_{I,h}\|_{H(\mathrm{curl};\Omega_I)} \leq C h^{\min(r,k)} \\
&\|Q - Q_h\|_{0,\Gamma} \leq C h^{\min(r,s,k)} .
\end{aligned} \tag{4.66}$$

Remark 4.24. A suitable finite element basis of the constrained space $\widetilde{X}_{C,h}$ for $k = 1$ is presented in Section 7.6.2, and could be used for avoiding the introduction of the Lagrange multiplier Q_h. □

Remark 4.25. Since in (4.49) the vector field $\widetilde{\mathbf{E}}_I$ is not the physical electric field, the numerical solution $\widetilde{\mathbf{E}}_{I,h}$ is not a correct approximation of $\mathbf{E}_I$. If interested in that, see Section 5.5. □

Remark 4.26. For the electric boundary condition (1.20) the discrete variational problem is

$$
\begin{aligned}
&\text{Find } \mathbf{H}_{C,h} \in N^k_{C,h}, \widetilde{\mathbf{E}}_{I,h} \in X^k_{I,h}, Q^*_h \in Y^{k-1}_{\Gamma,h}, \\
&\qquad \phi^*_{I,h} \in L^k_{I,h} \cap H^1_{0,\partial\Omega}(\Omega_I) : \\
&\mathcal{D}((\mathbf{H}_{C,h}, \widetilde{\mathbf{E}}_{I,h}), (\mathbf{v}_{C,h}, \mathbf{z}_{I,h})) - \textstyle\int_\Gamma \operatorname{div}_\tau(\overline{\mathbf{v}_{C,h}} \times \mathbf{n}_C)\, Q^*_h \\
&\qquad - \textstyle\int_{\Omega_I} \overline{\mathbf{z}_{I,h}} \cdot \operatorname{grad} \phi^*_{I,h} = F(\mathbf{v}_{C,h}) + G(\mathbf{z}_{I,h}) \\
&\textstyle\int_\Gamma \operatorname{div}_\tau(\mathbf{H}_{C,h} \times \mathbf{n}_C)\, \overline{P^*_h} = 0 \\
&\textstyle\int_{\Omega_I} \widetilde{\mathbf{E}}_{I,h} \cdot \operatorname{grad} \overline{\xi^*_{I,h}} = 0 \\
&\text{for all } \mathbf{v}_{C,h} \in N^k_{C,h}, \mathbf{z}_{I,h} \in X^k_{I,h}, P^*_h \in Y^{k-1}_{\Gamma,h}, \\
&\qquad \xi^*_{I,h} \in L^k_{I,h} \cap H^1_{0,\partial\Omega}(\Omega_I),
\end{aligned}
\tag{4.67}
$$

where $X^k_{I,h} := N^k_{I,h} \cap H_{0,\partial\Omega}(\text{curl}; \Omega_I)$. Note that it is not necessary to assume that Ω is simply-connected. □

4.5.2 Some remarks on implementation

In this section, following Alonso Rodríguez and Vázquez Hernández [21], we focus on the resolution of the linear system arising from (4.64) in the case $k = 1$. Let us consider the following bases $\{\mathbf{v}^l_{C,h}\}_{l=1}^{N_C}$, basis of $N^1_{C,h}$, $\{\mathbf{z}^l_{I,h}\}_{l=1}^{N_I}$, basis of $N^1_{I,h}$, $\{P^l_h\}_{l=1}^{K}$, basis of $Y^0_{\Gamma,h}$ and $\{\xi^l_{I,h}\}_{l=1}^{M_I}$, basis of $L^1_{I,h}/\mathbb{C}$. We also consider the following matrices

$$M_C = \{m^C_{i,j}\} \ , \ m^C_{i,j} := \omega \int_{\Omega_C} \boldsymbol{\mu} \mathbf{v}^j_{C,h} \cdot \overline{\mathbf{v}^i_{C,h}}$$

$$S_C = \{s^C_{i,j}\} \ , \ s^C_{i,j} := \int_{\Omega_C} \boldsymbol{\sigma}^{-1} \operatorname{curl} \mathbf{v}^j_{C,h} \cdot \operatorname{curl} \overline{\mathbf{v}^i_{C,h}}$$

$$S_I = \{s^I_{i,j}\} \ , \ s^I_{i,j} := \omega^{-1} \int_{\Omega_I} \boldsymbol{\mu}^{-1} \operatorname{curl} \mathbf{z}^j_{I,h} \cdot \operatorname{curl} \overline{\mathbf{z}^i_{I,h}}$$

$$D = \{d_{i,j}\} \ , \ d_{i,j} := \int_\Gamma \mathbf{v}^j_{C,h} \times \mathbf{n}_C \cdot \overline{\mathbf{z}^i_{I,h}}$$

$$B_C = \{b^C_{i,j}\} \ , \ b^C_{i,j} := \int_\Gamma \operatorname{div}_\tau(\mathbf{v}^j_{C,h} \times \mathbf{n}_C) \overline{P^i_h}$$

$$B_I = \{b^I_{i,j}\} \ , \ b^I_{i,j} := \int_{\Omega_I} \mathbf{z}^j_{I,h} \cdot \operatorname{grad} \overline{\xi^i_{I,h}} \, ,$$

and finally we set $A_C := S_C + iM_C$ and $A_I = iS_I$.

System (4.64) can be written as

$$\begin{bmatrix} A_C & D^T & B_C^T & \\ D & A_I & & B_I^T \\ B_C & & & \\ & B_I & & \end{bmatrix} \begin{bmatrix} H_C \\ \widetilde{E}_I \\ Q \\ \Phi_I \end{bmatrix} = \begin{bmatrix} F_C \\ G_I \\ 0 \\ 0 \end{bmatrix} . \tag{4.68}$$

The complex vectors H_C, $\widetilde{E}_I$, Q and Φ_I are the coefficients of $\mathbf{H}_{C,h}$, $\widetilde{\mathbf{E}}_{I,h}$, Q_h and $\phi_{I,h}$ in the chosen bases of $N^1_{C,h}$, $N^1_{I,h}$, $Y^0_{\Gamma,h}$ and $L^1_{I,h} \backslash \mathbb{C}$, respectively. The complex vectors F_C and G_I are obtained by applying the functionals F and G to the elements of the basis of $N^1_{C,h}$ and $N^1_{I,h}$, respectively. All the matrices S_C, S_I, M_C, D, B_C and B_I are real. S_C and S_I are symmetric and positive semi-definite, while M_C is symmetric and positive definite.

Problem (4.68) is an indefinite system that arises from a saddle-point problem. It has the form

$$\begin{bmatrix} \mathsf{A} & \mathsf{B}^T \\ \mathsf{B} & \end{bmatrix} \begin{bmatrix} x \\ y \end{bmatrix} = \begin{bmatrix} f \\ 0 \end{bmatrix} ,$$

with A and B block-structured matrices. A is a complex matrix with symmetric and positive semi-definite real and imaginary parts. It can be solved using, for instance, the method presented in Hu and Zou [135] (see also Benzi et al. [37] for a review of numerical methods for the solution of saddle-point problems). However, to take advantage of the fact that the system arises from an eddy current problem with two different subdomains, we rearrange system (4.68) in the following way

$$\begin{bmatrix} A_C & B_C^T & D^T & \\ B_C & & & \\ D & & A_I & B_I^T \\ & & B_I & \end{bmatrix} \begin{bmatrix} H_C \\ Q \\ \widetilde{E}_I \\ \Phi_I \end{bmatrix} = \begin{bmatrix} F_C \\ 0 \\ G_I \\ 0 \end{bmatrix} . \tag{4.69}$$

Proceeding as in Theorem 4.22, it is easy to see that the block $\begin{bmatrix} A_C & B_C^T \\ B_C & \end{bmatrix}$ is non-singular.

Since we know that $\Phi_I = 0$, it is possible to eliminate this unknown considering the reduced system

$$\begin{bmatrix} A_C & B_C^T & D^T \\ B_C & & \\ D & & A_I + i\gamma B_I^T B_I \end{bmatrix} \begin{bmatrix} H_C \\ Q \\ \widetilde{E}_I \end{bmatrix} = \begin{bmatrix} F_C \\ 0 \\ G_I \end{bmatrix} , \tag{4.70}$$

where the parameter γ is any positive real number, and we have taken into account the equation $B_I \widetilde{E}_I = 0$.

Lemma 4.27. *System* (4.69) *and system* (4.70) *are equivalent.*

Proof. Since system (4.69) has a unique solution $(H_C, Q, \widetilde{E}_I, \Phi_I)$ with $\Phi_I = 0$, and in particular $B_I \widetilde{E}_I = 0$, it is clear that $(H_C, Q, \widetilde{E}_I)$ is solution of (4.70). Hence it is enough to show that (4.70) has a unique solution. Let (V_C, P, Z_I) be a solution to

$$\begin{bmatrix} A_C & B_C^T & D^T \\ B_C & & \\ D & & A_I + i\gamma B_I^T B_I \end{bmatrix} \begin{bmatrix} V_C \\ P \\ Z_I \end{bmatrix} = \begin{bmatrix} 0 \\ 0 \\ 0 \end{bmatrix} ;$$

recalling that $A_C = S_C + iM_C$ and $A_I = iS_I$, we have

$$\begin{aligned} &(S_C + iM_C)V_C \cdot \overline{V_C} + B_C^T P \cdot \overline{V_C} + D^T Z_I \cdot \overline{V_C} = 0 \\ &B_C V_C \cdot \overline{P} = 0 \\ &D V_C \cdot \overline{Z_I} = -i(S_I + \gamma B_I^T B_I) Z_I \cdot \overline{Z_I} . \end{aligned} \tag{4.71}$$

Replacing $B_C^T P \cdot \overline{V_C} = B_C \overline{V_C} \cdot P$ and $D^T Z_I \cdot \overline{V_C} = D\overline{V_C} \cdot Z_I$ in the first equation of (4.71) by the values given by the second and third equation we get

$$S_C V_C \cdot \overline{V_C} + i(M_C V_C \cdot \overline{V_C} + S_I Z_I \cdot \overline{Z_I} + \gamma B_I^T B_I Z_I \cdot \overline{Z_I}) = 0 .$$

In particular, since M_C is symmetric and positive definite and S_I is symmetric and positive semi-definite, it follows that $B_I Z_I = 0$. This means that

$$\begin{bmatrix} A_C & B_C^T & D^T & \\ B_C & & & \\ D & & A_I & B_I^T \\ & & B_I & \end{bmatrix} \begin{bmatrix} V_C \\ P \\ Z_I \\ 0 \end{bmatrix} = \begin{bmatrix} 0 \\ 0 \\ 0 \\ 0 \end{bmatrix} ,$$

hence we have found a solution to the homogeneous system (4.69). □

Even if for any value of $\gamma > 0$ systems (4.69) and (4.70) are equivalent, the computed solutions for small values of γ could be different, because in the limit case $\gamma = 0$ the reduced system is singular. On the other hand, for big values of γ the matrix of the reduced system is ill-conditioned. The convergence rate of the resolution algorithms depends on the choice of this parameter, more precisely γ should be chosen such that the matrices S_I and $\gamma B_I^T B_I$ are balanced in norm.

Remark 4.28. It is worth noting that the matrix $A_I + i\gamma B_I^T B_I = i(S_I + \gamma B_I^T B_I)$ is invertible if and only if Ω_I is simply-connected. In fact, let us consider the space $\widetilde{Z}_{I,h}$ introduced in (4.58), that, as shown in Theorem 4.20, can be decomposed as the following direct sum

$$\widetilde{Z}_{I,h} = [H^0(\text{curl}; \Omega_I) \cap N^1_{I,h}]^\perp \oplus \mathcal{H}_{I,h} ,$$

where

$$\mathcal{H}_{I,h} := H^0(\text{curl}; \Omega_I) \cap \widetilde{Z}_{I,h} .$$

Given $Z_I \in \mathbb{C}^n$, let us denote by $\mathbf{z}_{I,h}$ the function in $N^1_{I,h}$ with coefficients Z_I. From the definitions of S_I and B_I we see that $(S_I + \gamma B_I^T B_I)\, Z_I = 0$ if and only if curl $\mathbf{z}_{I,h} = \mathbf{0}$ and $\mathbf{z}_{I,h} \in \widetilde{Z}_{I,h}$, which means that $\mathbf{z}_{I,h} \in \mathcal{H}_{I,h}$. Moreover it can be proved that $\dim \mathcal{H}_{I,h} = \dim \mathcal{H}(m; \Omega_I)$, which is 0 if and only if Ω_I is simply-connected.

On the other hand, let us consider the perturbed matrix $S_I + \gamma B_I^T B_I + \epsilon D D^T$. It is possible to prove that this matrix is non-singular for each $\epsilon > 0$, and this result is true in any geometrical configuration. In fact, if $(S_I + \gamma B_I^T B_I + \epsilon D D^T) Z_I = 0$ we have in particular $(S_I + \gamma B_I^T B_I) Z_I = 0$ and $D^T Z_I = 0$, hence $\mathbf{z}_{I,h} \in \mathcal{H}_{I,h}$ and

$$\int_\Gamma \mathbf{v}_{C,h} \times \mathbf{n}_C \cdot \overline{\mathbf{z}_{I,h}} = 0$$

for all $\mathbf{v}_{C,h} \in N^1_{C,h}$. From the discrete inf–sup condition (4.61) it follows that $\mathbf{z}_{I,h} = \mathbf{0}$, hence $Z_I = 0$. □

We present now two different algorithms for solving system (4.70), both of them taking advantage from the fact that the problem is formulated in two subdomains.

Modified SOR method

It is a block-version of the SOR method. If the domain Ω_I is not simply-connected, the subproblem in the insulating region is modified by adding the term $i\epsilon D D^T$, so that we have non-singular matrices on the diagonal of the block decomposition (see Remark 4.28). In Kanayama et al. [147] a similar idea has been used to solve the problem formulated in terms of a magnetic vector potential in the whole domain Ω. In that paper the problem is perturbed by adding a term of the form ϵM, where M is the mass matrix in Ω, namely, the matrix that corresponds to the scalar product $\int_\Omega \mathbf{w} \cdot \overline{\mathbf{z}}$. In our experience, using in the insulator the matrix $D D^T$ instead of the mass matrix M_I improves the convergence of the method. The algorithm reads

Algorithm 4.29. Given H_C^0, Q^0 and $\widetilde{E}_I^0$, for $m \geq 0$ solve

$$\begin{bmatrix} A_C & B_C^T \\ B_C & \end{bmatrix} \begin{bmatrix} H_C^{m+1/2} \\ Q^{m+1/2} \end{bmatrix} = \begin{bmatrix} F_C - D^T \widetilde{E}_I^m \\ 0 \end{bmatrix},$$

set

$$\begin{bmatrix} H_C^{m+1} \\ Q^{m+1} \end{bmatrix} = (1-\theta) \begin{bmatrix} H_C^m \\ Q^m \end{bmatrix} + \theta \begin{bmatrix} H_C^{m+1/2} \\ Q^{m+1/2} \end{bmatrix},$$

then solve

$$i(S_I + \gamma B_I^T B_I + \epsilon D D^T) \widetilde{E}_I^{m+1/2} = G_I - D H_C^{m+1} + i\epsilon D D^T \widetilde{E}_I^m,$$

and finally set

$$\widetilde{E}_I^{m+1} = (1-\theta) \widetilde{E}_I^m + \theta \widetilde{E}_I^{m+1/2}.$$

The real number θ is the relaxation parameter of the SOR method and has to satisfy $0 < \theta < 2$. The parameter ϵ is taken equal to 0 if the subdomain Ω_I is simply-connected; otherwise we choose a suitable $\epsilon > 0$. The performance of the algorithm depends on the appropriate selection of both parameters (see Table 4.2).

At each iteration of the algorithm one needs to solve a linear system in each subdomain. To solve the subproblem in the insulator we use the preconditioned conjugate gradient method, taking as preconditioner an incomplete Cholesky factorization of $S_I + \gamma B_I^T B_I + \epsilon D D^T$.

For the subproblem in the conductor we take into account its saddle-point structure, and solve it with an inexact Uzawa algorithm with variable relaxation parameters (see Hu and Zou [134] for the real case). For a system of the general form

$$\begin{bmatrix} A_C & B_C^T \\ B_C & \end{bmatrix} \begin{bmatrix} h \\ q \end{bmatrix} = \begin{bmatrix} f \\ 0 \end{bmatrix}$$

the algorithm reads: given h^0 and q^0, for $j \geq 0$ set

$$h^{j+1} = h^j + \omega_j \hat{A}_C^{-1} \left[f - (A_C h^j + B_C^T q^j) \right] ,$$

and

$$q^{j+1} = q^j + \tau_j \hat{P}_C^{-1} B_C h^{j+1},$$

where $\hat{A}_C$ is a preconditioner for A_C and $\hat{P}_C$ is a preconditioner for $B_C \hat{A}_C^{-1} B_C^T$. In particular, we take as $\hat{A}_C$ an incomplete LU factorization of A_C, and as $\hat{P}_C$ an incomplete LU factorization of $B_C \Lambda_C^{-1} B_C^T$, where Λ_C is the diagonal matrix with the elements of the diagonal of A_C. The parameters ω_j and τ_j are computed dynamically at each iteration, as it is done in Hu and Zou [134].

Uzawa-like method

An alternative approach for the solution of (4.70) is to consider an Uzawa-like method. Since this kind of method does not require the (2,2)-block of the matrix (in our case the block $i(S_I + \gamma B_I^T B_I)$) to be invertible, it can be used without penalization even in the case of a conductor with general topology. In particular, we consider a preconditioned Uzawa method with variable relaxation parameter. We formally adapt to system (4.70) the algorithm analyzed in Hu and Zou [134] for a real system with null (2,2)-block, i.e., a real system of the form $\begin{bmatrix} K & D^T \\ D & \end{bmatrix}$, with K symmetric and positive definite. The algorithm that we propose reads as follows

Algorithm 4.30. Given $\widetilde{E}_I^0$, for $m \geq 0$ solve

$$\begin{bmatrix} A_C & B_C^T \\ B_C & \end{bmatrix} \begin{bmatrix} H_C^{m+1} \\ Q^{m+1} \end{bmatrix} = \begin{bmatrix} F_C - D^T \widetilde{E}_I^m \\ 0 \end{bmatrix},$$

then compute

$$r_m = D H_C^{m+1} + i(S_I + \gamma B_I^T B_I) \widetilde{E}_I^m - G_I ,$$

$$d_m = \hat{N}^{-1} r_m ,$$

and

$$\tau_m = \frac{r_m \cdot \overline{d_m}}{\hat{A}_C^{-1} D^T d_m \cdot \overline{D^T d_m} - i(S_I + \gamma B_I^T B_I) d_m \cdot \overline{d_m}},$$

and set

$$\widetilde{E}_I^{m+1} = \widetilde{E}_I^m + \eta_m \tau_m d_m \,.$$

Here η_m is a parameter that, for an exact Uzawa algorithm, has to satisfy $0 < \eta_m \leq \frac{1}{2}$; the matrix $\hat{N}$ is a preconditioner for

$$N = \begin{bmatrix} D & 0 \end{bmatrix} \begin{bmatrix} A_C & B_C^T \\ B_C & 0 \end{bmatrix}^{-1} \begin{bmatrix} D^T \\ 0 \end{bmatrix} - i(S_I + \gamma B_I^T B_I) \,, \tag{4.72}$$

and $\hat{A}_C$ is a preconditioner for A_C.

In all the numerical tests presented in Section 4.5.3 we have set $\eta_m = 1/2$, and we have taken as preconditioner $\hat{A}_C$ an incomplete LU factorization of the matrix A_C, and as preconditioner $\hat{N}$ an incomplete LU factorization of the matrix $S_I + \gamma B_I^T B_I$ (the first term in (4.72) has a structure that is not easy to treat, therefore in the choice of the preconditioner we have taken into account only the second term).

4.5.3 Numerical results

The finite element methods and the algorithms introduced in the previous section have been implemented in MATLAB by Alonso Rodríguez and Vázquez Hernández [21]. In the following we present some numerical tests illustrating how the algorithms perform. In the first set of numerical experiments we solve a problem with a known analytical solution to validate the computer code and test the convergence properties of the methods. In the second and third numerical test we consider a torus-shaped coil inducing eddy currents in a conductor which is a torus in the second test problem and a trefoil knot in the third one. The last case concerns the benchmark problem number 7 in the TEAM Workshop, which deals with an asymmetrical conductor with a hole (see Fujiwara and Nakata [107], Kanayama et al. [147]).

All the simulations have been run on a single processor Intel Xeon QuadCore 5430 2.66GHz. The stopping test for the modified SOR and the Uzawa-like solvers of the linear system (4.70) (written here as $\mathsf{A}\mathbf{x} = \mathbf{b}$) is

$$\frac{\max_i |b_i - (\mathsf{A}\mathbf{x})_i|}{\max_i |b_i|} < 10^{-6} \,.$$

A problem with a known analytical solution

In this set of tests the conductor Ω_C and the domain Ω are two cubes centered at the origin and with edge length equal to 2 and 10, respectively. We shall construct an analytical solution $(\mathbf{H}_C, \widetilde{\mathbf{E}}_I)$ which will consist of two C^2-functions with compact supports in Ω_C and Ω_I, respectively.

Let us suppose that ω, μ and $\sigma_{|\Omega_C}$ are positive constants equal to 1, and that $\sigma_{|\Omega_I} = 0$. Given a closed ball centered at $\mathbf{x}_0 \in \Omega$ and with radius r_0, we define the function $p(\mathbf{x})$ with support in this ball as follows

$$p(\mathbf{x}) = \begin{cases} q\left(\frac{|\mathbf{x}-\mathbf{x}_0|}{r_0}\right), & \text{if } |\mathbf{x}-\mathbf{x}_0| \leq r_0, \\ 0, & \text{if } |\mathbf{x}-\mathbf{x}_0| > r_0, \end{cases}$$

q being the unique eighth degree polynomial such that $q(0) = 1$, $q(1) = q(-1) = 0$ and with its first three derivatives vanishing at the points 1 and -1. It is easily seen that q is given by the expression

$$q(t) = t^8 - 4t^6 + 6t^4 - 4t^2 + 1\,.$$

Now, let Θ_C be the closed ball centered at the origin with radius $r_0 = 0.9$, and Θ_I the ball with center at $\mathbf{x}_0 = (0, 3, 0)$ and radius $r_0 = 1.9$. Obviously, the two balls are disjoint and they are strictly contained in Ω_C and Ω_I, respectively. Let us denote by p_C and p_I the functions corresponding to the balls Θ_C and Θ_I, and define the "electric" field in the insulator as

$$\widetilde{\mathbf{E}}_I := \operatorname{curl}(0, 0, p_I) = \left(\frac{\partial p_I}{\partial y}, -\frac{\partial p_I}{\partial x}, 0 \right),$$

and the electric field and the magnetic field in the conductor as

$$\begin{aligned}
\mathbf{E}_C &:= \operatorname{curl}(0, 0, p_C) = \left(\tfrac{\partial p_C}{\partial y}, -\tfrac{\partial p_C}{\partial x}, 0 \right) \\
\mathbf{H}_C &:= i \operatorname{curl}(\operatorname{curl}(0, 0, p_C)) = i \left(\tfrac{\partial^2 p_C}{\partial x \partial z}, \tfrac{\partial^2 p_C}{\partial y \partial z}, -\tfrac{\partial^2 p_C}{\partial x^2} - \tfrac{\partial^2 p_C}{\partial y^2} \right).
\end{aligned}$$

Now one can easily compute $\mathbf{J}_{e,C}$ and $\mathbf{J}_{e,I}$, and check that the excitation current density $\mathbf{J}_e$ satisfies the three compatibility conditions.

The program has been tested by solving this academic problem with four successively refined meshes, with grid sizes corresponding to h, $h/2$, $h/3$ and $h/4$ and setting the parameter $\gamma = 1$ in the four cases. In Table 4.1 we present the relative error between the computed and the exact solutions. More precisely, we set

$$e_{\mathbf{H}} = \frac{\|\mathbf{H}_C - \mathbf{H}_{C,h}\|_{H(\mathrm{curl};\Omega_C)}}{\|\mathbf{H}_C\|_{H(\mathrm{curl};\Omega_C)}}, \qquad e_{\mathbf{E}} = \frac{\|\widetilde{\mathbf{E}}_I - \widetilde{\mathbf{E}}_{I,h}\|_{H(\mathrm{curl};\Omega_I)}}{\|\widetilde{\mathbf{E}}_I\|_{H(\mathrm{curl};\Omega_I)}}.$$

Figure 4.1 shows the plots in a log–log scale of the relative errors of $e_{\mathbf{H}}$ and $e_{\mathbf{E}}$ versus the mesh size h. As it can be seen, the error is reduced when the mesh is refined, and linear convergence can be observed for the computations on the last meshes. However, the relative error for the finest mesh is still quite large. One of the reasons for these large errors is that the solution of our problem is a polynomial of seventh degree, and the support is concentrated in a small part of the domain.

In Table 4.1 we also present the number of iterations and the computational time needed for solving system (4.70) using the SOR method with relaxation parameter $\theta = 0.6$ and using the Uzawa method. For this test case, since Ω_I is simply-connected, in the SOR method we take the perturbation parameter $\epsilon = 0$. The computational time includes the calculation of the preconditioners.

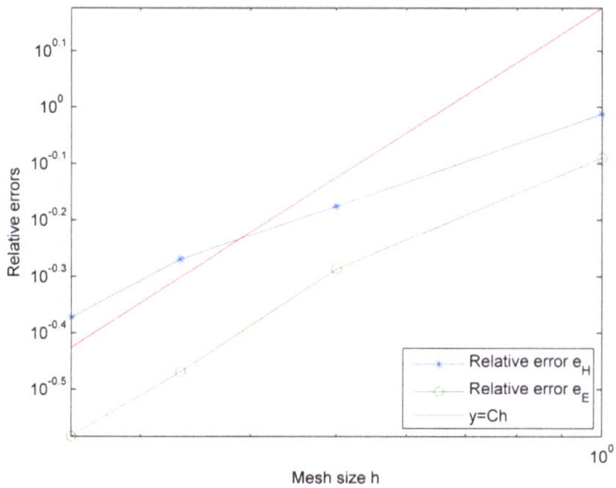

Fig. 4.1. Relative errors versus mesh size h

Table 4.1. Results for problem with known analytical solution

				SOR ($\theta = 0.6$)		*Uzawa*	
Elements	*DoF*	$e_{\mathbf{H}}$	$e_{\mathbf{E}}$	*iterations*	*time [s]*	*iterations*	*time [s]*
1155	1721	0.9710	0.8128	16	2.0	18	1.2
9240	12245	0.6668	0.5163	16	38.4	24	24.2
31185	39656	0.5377	0.3399	16	317.1	73	201.0
73920	92039	0.4253	0.2607	16	2118.6	239	1089.9

Two tori

In the second test the computational domain is a cube with edge length 27 cm. We consider two coaxial tori of square section with edge length 1 cm and radius 6.5 cm: the upper torus is a coil, included in the insulating region Ω_I, where we impose a clockwise current density $\mathbf{J}_{e,I}$ of magnitude 10^6 A/m^2, whereas the second torus is the conductor. We are taking $\mu = \mu_0 = 4\pi \times 10^{-7}$ H/m, i.e., the magnetic permeability of the air, the electric conductivity $\sigma = 10^7$ S/m and the angular frequency $\omega = 2\pi \times 50$ rad/s. The parameter γ is set equal to $10^6/(\omega\mu)$. In this case, since the insulating region Ω_I is not simply-connected (precisely, there is one non-bounding cycle in Ω_I), the parameter ϵ in the SOR method must be positive.

We present in Table 4.2 the convergence results for different choices of ϵ and θ for a non-uniform mesh with 27152 elements (the mesh is finer in the coil and the conductor). Then we take the best values found for both parameters ($\theta = 0.5, \epsilon = 10^5$) and use them in three different meshes to compare the behavior of the modified SOR and the Uzawa methods. The results are summarized in Table 4.3. We also include in Table 4.4 the results obtained with a different choice of γ, specifically $\gamma = \|S_I\|_2/\|B_I\|_2^2$. Notice that the cost of one iteration of the SOR method depends on γ, because the condition number of $S_I + \gamma B_I^T B_I + \epsilon DD^T$ is quite sensitive to it.

Table 4.2. Two tori. Results for the SOR method with several values of θ and ϵ (NC: not convergent)

	$\epsilon = 10^5$		$\epsilon = 10^6$		$\epsilon = 10^7$	
	iterations	*time [s]*	*iterations*	*time [s]*	*iterations*	*time [s]*
$\theta = 0.25$	51	552.1	195	975.7	1749	4693.5
$\theta = 0.5$	22	340.7	96	574.4	873	2435.9
$\theta = 0.75$	NC	NC	63	436.9	581	1701.3
$\theta = 1$	NC	NC	47	367.9	435	1321.3
$\theta = 1.25$	NC	NC	NC	NC	358	1127.4

Table 4.3. Two tori. Comparison of SOR and Uzawa methods ($\gamma = 10^6/(\omega\mu) = 2.5 \times 10^9$)

		SOR ($\theta = 0.5$, $\epsilon = 10^5$)		*Uzawa*	
Elements	*DoF*	*iterations*	*time [s]*	*iterations*	*time [s]*
3394	4763	24	20.9	137	18.6
27152	34795	22	340.7	59	250.8
91638	113853	25	2230.6	136	1928.0

Table 4.4. Two tori. Comparison of SOR and Uzawa methods ($\gamma = \|S_I\|_2/\|B_I\|_2^2$)

			SOR ($\theta = 0.5$, $\epsilon = 10^5$)		*Uzawa*	
Elements	*DoF*	γ	*iterations*	*time [s]*	*iterations*	*time [s]*
3394	4763	$2.3\,10^8$	34	17.4	95	12.8
27152	34795	$9.6\,10^8$	26	352.8	92	274.3
91638	113853	$1.8\,10^9$	29	2275.0	221	2262.7

The trefoil knot

We want to show how the $(\mathbf{H}_C, \widetilde{\mathbf{E}}_I)$ formulation performs for problems with complicated geometries. In particular, we consider a problem in which the conductor is a trefoil knot. It is well known (see, e.g., Bossavit [59], Hiptmair [126], Gross and Kotiuga [115]) that there exists one surface which "cuts" the basic non-bounding cycle in Ω_I, but its construction, on a given mesh, can be a non-trivial task. We show in Figure 4.2 a visualization of this surface (known as the Seifert surface associated to the trefoil knot).

We suppose that Ω_C is a trefoil knot formed joining cubes of edge length 1 cm. A torus-shaped coil is placed above the conductor. The sizes of the coil and the computational domain Ω are the same that in the previous test case. The physical magnitudes and the source current are also taken as in the two tori test case.

In Table 4.5, having chosen $\gamma = 10^6/(\omega\mu)$, we present the number of iterations and the CPU time for the modified SOR method, with relaxation parameter $\theta = 1$ and $\epsilon = 10^6$ (that we have verified to be a reasonable choice for this example), and for the Uzawa-like scheme.

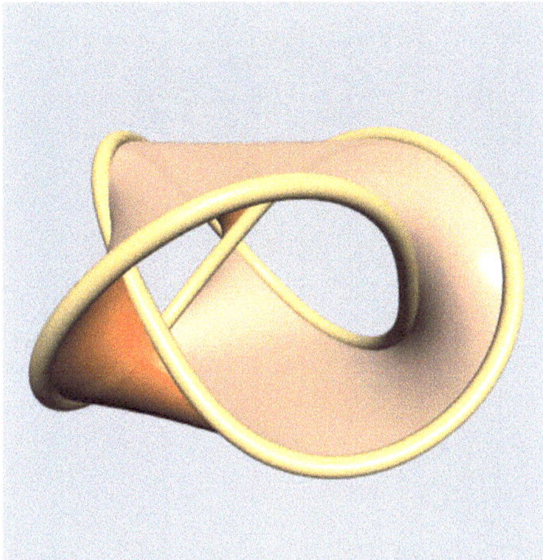

Fig. 4.2. Seifert surface for the trefoil knot (courtesy of J.J. van Wijk, Eindhoven University of Technology)

Table 4.5. Results for the trefoil knot

		SOR ($\theta = 1$, $\epsilon = 10^6$)		*Uzawa*	
Elements	*DoF*	*iterations*	*CPU time [s]*	*iterations*	*CPU time [s]*
37057	51665	402	3456.7	392	2114.7

In Figure 4.3 and Figure 4.4 we show the real part and the imaginary part of the current density $\mathbf{J}_C = \operatorname{curl} \mathbf{H}_C$ on the surface of the knot, respectively.

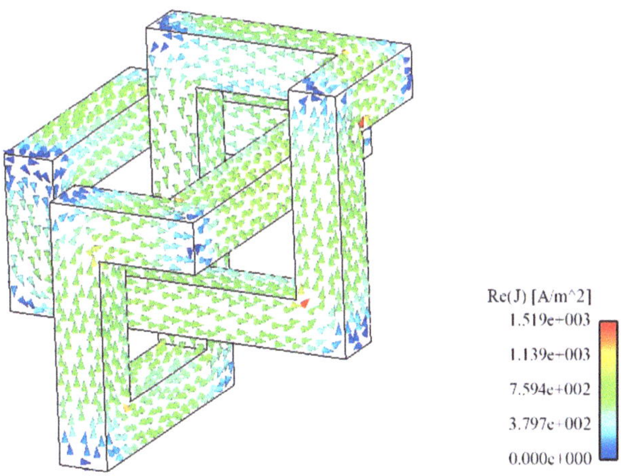

Fig. 4.3. The current density in the trefoil knot: real part

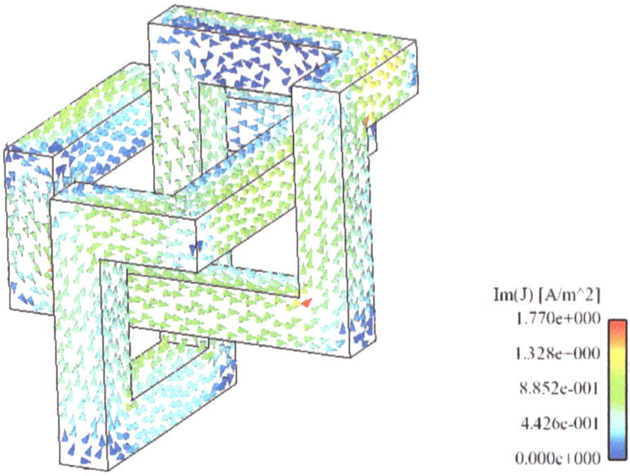

Fig. 4.4. The current density in the trefoil knot: imaginary part

Benchmark problem 7 in the TEAM Workshop

Our last test corresponds to benchmark problem number 7 in the TEAM workshop (see Fujiwara and Nakata [107]). It consists of a conductor Ω_C given by a thick aluminum plate with an eccentrically placed hole, subjected to an asymmetric magnetic field. The field is produced by an exciting current traversing a coil above the plate (see Figures 4.5, 4.6).

The plate and the coil are strictly inside the computational domain, which is a hexahedron with edges length $460 \times 460 \times 309$ mm. The conductor is centered in the horizontal plane and it is 80 mm far from the bottom of the computational domain, while the coil is 80 mm far from the top of the computational domain. The magnetic

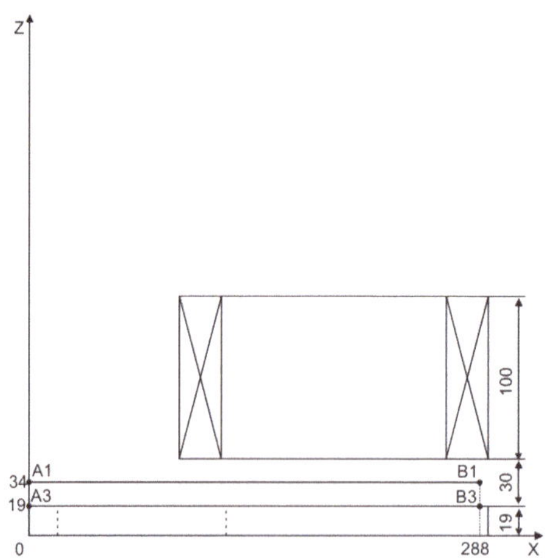

Fig. 4.5. Geometry of the TEAM model: elevation

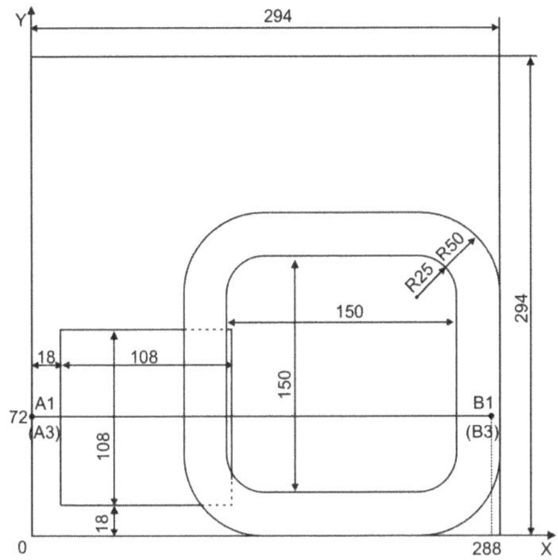

Fig. 4.6. Geometry of the TEAM model: plan

permeability is $\mu = \mu_0 = 4\pi \times 10^{-7}$ H/m, the electric conductivity is $\sigma = 3.526 \times 10^7$ S/m, the angular frequency is $\omega = 2\pi \times 50$ rad/s, and the absolute value of the real part (respectively, imaginary part) of the excitation current density $\mathbf{J}_{e,I}$ is 1.0968×10^6 (respectively, 0) A/m^2. The parameter γ is set equal to $5 \times 10^6/(\omega\mu)$.

In Table 4.6 we present the number of iterations and CPU time for the modified SOR method with relaxation parameter $\theta = 1$ and $\epsilon = 10^6$, and for the Uzawa-like scheme.

Table 4.6. Results for problem 7 in the TEAM Workshop

		SOR ($\theta = 1$, $\epsilon = 10^6$)		*Uzawa*	
Elements	*DoF*	*iterations*	*CPU time [s]*	*iterations*	*CPU time [s]*
98857	131434	134	10950.7	167	11222.3

We also present a comparison of our numerical results with the experimental data reported in Fujiwara and Nakata [107]. The z component of the magnetic induction $\mathbf{B}_I$ $(= \frac{i}{\omega} \operatorname{curl} \widetilde{\mathbf{E}}_I)$ along a straight line in the insulating region, with $y = 72$ mm and $z = 34$ mm, is represented in Figure 4.7. The y component of the current density $\mathbf{J}_C$ $(= \operatorname{curl} \mathbf{H}_C)$ along a straight line on the surface of the conductor, with $y = 72$ mm and $z = 19$ mm, is represented in Figure 4.8.

We notice that the results for the magnetic induction in the insulator are not completely satisfactory. This is due to the fact that $\mathbf{B}_I$ is calculated from the curl of the computed $\widetilde{\mathbf{E}}_I$ and, since we are using first order edge elements, this curl is constant on each element. The oscillations are due to the relative position of the elements of the mesh where we compute the numerical values, with respect to the line where the field is measured.

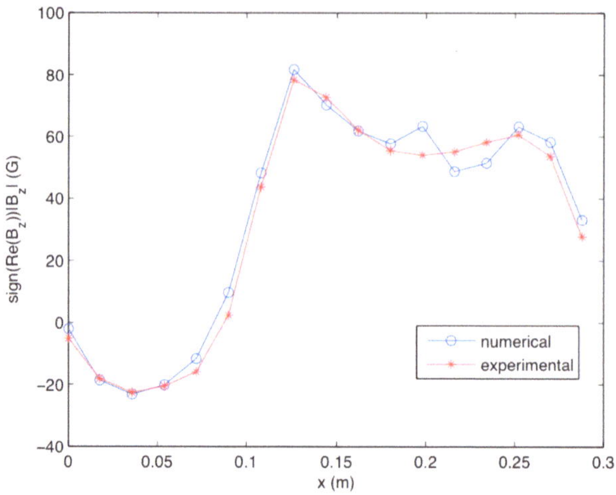

Fig. 4.7. z component of $\mathbf{B}_I$ along line A1–B1

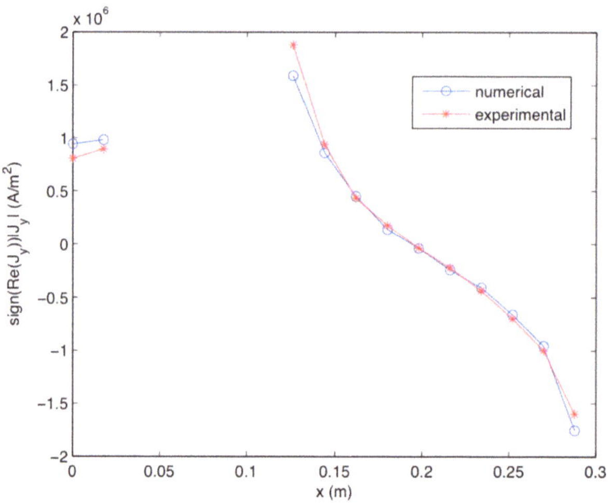

Fig. 4.8. y component of $\mathbf{J}_C$ along line A3–B3

4.6 A saddle-point approach for the E-based formulation

To conclude this chapter we consider, as in Alonso Rodríguez and Valli [17], the **E**-based formulation of the magnetic eddy current problem

$$\begin{cases} \operatorname{curl}(\boldsymbol{\mu}^{-1} \operatorname{curl} \mathbf{E}) + i\omega\boldsymbol{\sigma}\mathbf{E} = -i\omega\mathbf{J}_e & \text{in } \Omega \\ \operatorname{div}(\varepsilon_I \mathbf{E}_I) = 0 & \text{in } \Omega_I \\ \boldsymbol{\mu}^{-1} \operatorname{curl} \mathbf{E} \times \mathbf{n} = \mathbf{0} & \text{on } \partial\Omega \\ \varepsilon_I \mathbf{E}_I \cdot \mathbf{n} = 0 & \text{on } \partial\Omega \\ \int_{\Gamma_j} \varepsilon_I \mathbf{E}_I \cdot \mathbf{n}_I = 0 & \forall j = 1, \dots, p_\Gamma \\ \int_{\Omega_I} \varepsilon_I \mathbf{E}_I \cdot \boldsymbol{\pi}_{k,I} = 0 & \forall k = 1, \dots, n_{\partial\Omega} \, . \end{cases} \tag{4.73}$$

As already noted, the problem is simpler if the boundary of the conductor Ω_C is connected, so that $p_\Gamma = 0$, and the domain Ω is simply-connected, so that $n_{\partial\Omega} = 0$. Since in many cases the computational domain Ω can be chosen freely, this simplified geometrical situation often occurs in applications.

In Remark 2.4 (see also Section 6.1.5) we have considered a weak formulation of this problem, where the condition $\operatorname{div}(\varepsilon_I \mathbf{E}_I) = 0$ is imposed by penalization. An alternative approach, similar to the one that we have used for the hybrid formulations, is to impose the set of conditions $(4.73)_2$, $(4.73)_4$, $(4.73)_5$ and $(4.73)_6$ by means of Lagrange multipliers. We consider the space

$$Z := \{\mathbf{z} \in H(\operatorname{curl}; \Omega) \mid \mathbf{z}_I \text{ satisfies } (4.6)\}, \tag{4.74}$$

and the following weak formulation

$$\begin{aligned} &\text{Find } \mathbf{E} \in Z : \\ &\int_\Omega \boldsymbol{\mu}^{-1} \operatorname{curl} \mathbf{E} \cdot \operatorname{curl} \overline{\mathbf{z}} + i\omega \int_{\Omega_C} \boldsymbol{\sigma} \mathbf{E}_C \cdot \overline{\mathbf{z}}_C = -i\omega \int_\Omega \mathbf{J}_e \cdot \overline{\mathbf{z}} \\ &\text{for all } \mathbf{z} \in Z \,. \end{aligned} \tag{4.75}$$

Let us introduce the sesquilinear form

$$a_e(\mathbf{w}, \mathbf{z}) := \int_\Omega \boldsymbol{\mu}^{-1} \operatorname{curl} \mathbf{w} \cdot \operatorname{curl} \overline{\mathbf{z}} + i\omega \int_{\Omega_C} \boldsymbol{\sigma} \mathbf{w}_C \cdot \overline{\mathbf{z}}_C \,. \tag{4.76}$$

Using that $\mathbf{z} \in Z$ if and only if $\mathbf{z} \in H(\operatorname{curl}; \Omega)$ and $\int_{\Omega_I} \varepsilon_I \mathbf{z}_I \cdot \overline{\mathbf{p}_I} = 0$ for all $\mathbf{p}_I \in H^0_{0,\Gamma}(\operatorname{curl}; \Omega_I)$, we have the equivalent formulation

$$\begin{aligned} &\text{Find } (\mathbf{E}, \mathbf{r}_I) \in H(\operatorname{curl}; \Omega) \times H^0_{0,\Gamma}(\operatorname{curl}; \Omega_I) : \\ &\quad a_e(\mathbf{E}, \mathbf{z}) + \int_{\Omega_I} \varepsilon_I \overline{\mathbf{z}_I} \cdot \mathbf{r}_I = -i\omega \int_\Omega \mathbf{J}_e \cdot \overline{\mathbf{z}} \\ &\quad \int_{\Omega_I} \varepsilon_I \mathbf{E}_I \cdot \overline{\mathbf{p}_I} = 0 \\ &\text{for all } (\mathbf{z}, \mathbf{p}_I) \in H(\operatorname{curl}; \Omega) \times H^0_{0,\Gamma}(\operatorname{curl}; \Omega_I) \,. \end{aligned} \tag{4.77}$$

If *the computational domain Ω is simply-connected*, then

$$H^0_{0,\Gamma}(\operatorname{curl}; \Omega_I) = \operatorname{grad} H^1_{*,\Gamma}(\Omega_I),$$

where the space $H^1_{*,\Gamma}(\Omega_I)$ has been introduced in (4.17); hence the space Z can be expressed as

$$Z = \{\mathbf{z} \in H(\operatorname{curl}; \Omega) \mid \int_{\Omega_I} \varepsilon_I \mathbf{z}_I \cdot \operatorname{grad} \overline{\xi_I} = 0 \text{ for all } \xi_I \in H^1_{*,\Gamma}(\Omega_I)\} \,.$$

Therefore, problem (4.77) can be rewritten in the following form, that is suitable for finite element approximation

$$\begin{aligned} &\text{Find } (\mathbf{E}, \phi_I) \in H(\operatorname{curl}; \Omega) \times H^1_{*,\Gamma}(\Omega_I) : \\ &\quad a_e(\mathbf{E}, \mathbf{z}) + \int_{\Omega_I} \varepsilon_I \overline{\mathbf{z}_I} \cdot \operatorname{grad} \phi_I = -i\omega \int_\Omega \mathbf{J}_e \cdot \overline{\mathbf{z}} \\ &\quad \int_{\Omega_I} \varepsilon_I \mathbf{E}_I \cdot \operatorname{grad} \overline{\xi_I} = 0 \\ &\text{for all } (\mathbf{z}, \xi_I) \in H(\operatorname{curl}; \Omega) \times H^1_{*,\Gamma}(\Omega_I) \,. \end{aligned} \tag{4.78}$$

Proceeding as in Lemma 2.2 it is easy to prove that the sesquilinear form $a_e(\cdot,\cdot)$ is coercive in the space Z. On the other hand, the inf–sup condition is satisfied. In fact, given $\xi_I \in H^1_{*,\Gamma}(\Omega_I)$, the function $\mathbf{z}^*$ defined as

$$\mathbf{z}^* := \begin{cases} \operatorname{grad} \xi_I & \text{in } \Omega_I \\ \mathbf{0} & \text{in } \Omega_C \end{cases}$$

belongs to $H(\operatorname{curl};\Omega)$ and satisfies

$$\begin{aligned} \textstyle\int_{\Omega_I} \varepsilon_I \mathbf{z}_I^* \cdot \operatorname{grad} \overline{\xi_I} &= \textstyle\int_{\Omega_I} \varepsilon_I \mathbf{z}_I^* \cdot \overline{\mathbf{z}_I^*} \geq C_1 \|\mathbf{z}_I^*\|^2_{0,\Omega_I} \\ &= C_1 \|\mathbf{z}^*\|_{H(\operatorname{curl};\Omega)} \|\operatorname{grad} \xi_I\|_{0,\Omega_I} \\ &\geq C_2 \|\mathbf{z}^*\|_{H(\operatorname{curl};\Omega)} \|\xi_I\|_{1,\Omega_I} \ , \end{aligned}$$

having used the Poincaré inequality

$$\int_{\Omega_I} (|\xi_I|^2 + |\operatorname{grad} \xi_I|^2) \leq C_0 \int_{\Omega_I} |\operatorname{grad} \xi_I|^2 \tag{4.79}$$

which is valid in $H^1_{*,\Gamma}(\Omega_I)$ (see, e.g., Dautray and Lions [94], Chap. IV, Sect. 7, Rem. 4).

As a consequence, the well known theory of saddle-point problems (see, e.g., Brezzi and Fortin [65]) tells us that for each $\mathbf{J}_e \in (L^2(\Omega))^3$ there exists a unique solution to (4.78).

We can also prove that the Lagrange multiplier ϕ_I is equal to 0. In fact, let us take as test function in (4.78) $\mathbf{z} \in H(\operatorname{curl};\Omega)$ given by

$$\mathbf{z} := \begin{cases} \operatorname{grad} \phi_I & \text{in } \Omega_I \\ \mathbf{0} & \text{in } \Omega_C \ . \end{cases}$$

Since the current density $\mathbf{J}_e$ satisfies the necessary assumptions (3.1) and (3.2), one finds

$$\int_{\Omega_I} \varepsilon \overline{\mathbf{z}_I} \cdot \operatorname{grad} \phi_I = 0 \ ,$$

hence $\phi_I = 0$.

Remark 4.31. For the electric boundary condition $\mathbf{E} \times \mathbf{n} = \mathbf{0}$ on $\partial\Omega$ the problem can be written in strong form as

$$\begin{cases} \operatorname{curl}(\boldsymbol{\mu}^{-1} \operatorname{curl} \mathbf{E}) + i\omega\boldsymbol{\sigma}\mathbf{E} = -i\omega\mathbf{J}_e & \text{in } \Omega \\ \operatorname{div}(\varepsilon_I \mathbf{E}_I) = 0 & \text{in } \Omega_I \\ \mathbf{E} \times \mathbf{n} = \mathbf{0} & \text{on } \partial\Omega \\ \int_{\Gamma_j} \varepsilon_I \mathbf{E}_I \cdot \mathbf{n}_I = 0 & \forall j = 1, \ldots, p_\Gamma \\ \int_{(\partial\Omega)_r} \varepsilon_I \mathbf{E}_I \cdot \mathbf{n} = 0 & \forall r = 0, 1, \ldots, p_{\partial\Omega}, \end{cases} \tag{4.80}$$

and in weak form as

$$\begin{aligned} &\text{Find } (\mathbf{E}, \phi_I) \in H_0(\operatorname{curl};\Omega) \times H^1_*(\Omega_I) : \\ &\quad a_e(\mathbf{E}, \mathbf{z}) + \textstyle\int_{\Omega_I} \varepsilon_I \overline{\mathbf{z}_I} \cdot \operatorname{grad} \phi_I = -i\omega \int_\Omega \mathbf{J}_e \cdot \overline{\mathbf{z}} \\ &\quad \textstyle\int_{\Omega_I} \varepsilon_I \mathbf{E}_I \cdot \operatorname{grad} \overline{\xi_I} = 0 \\ &\text{for all } (\mathbf{z}, \xi_I) \in H_0(\operatorname{curl};\Omega) \times H^1_*(\Omega_I) \ , \end{aligned} \tag{4.81}$$

where

$$H^1_*(\Omega_I) := \{\xi_I \in H^1(\Omega_I) \mid \xi_{I|\Gamma_j} \text{ is constant } \forall\, j = 1, \dots, p_\Gamma, \\ \xi_{I|(\partial\Omega)_r} \text{ is constant } \forall\, r = 0, 1, \dots, p_{\partial\Omega}, \\ \xi_{I|\Gamma_{p_\Gamma+1}} = 0\}\,. \tag{4.82}$$

Note that no assumption is needed on the geometry of Ω, as conditions $(4.80)_2$–$(4.80)_5$ can always be expressed as orthogonality conditions to grad $H^1_*(\Omega_I)$.

Moreover, if the boundaries of the conductor Ω_C and of the computational domain Ω are both connected, the last two equations in (4.80) reduce to $\int_{\partial\Omega} \varepsilon_I \mathbf{E}_I \cdot \mathbf{n} = 0$, and the space $H^1_*(\Omega_I)$ is simply constituted by the functions vanishing on Γ and constant on $\partial\Omega$. □

To approximate the electric field we employ the Nédélec curl-conforming edge elements and to approximate the Lagrange multiplier ϕ_I we employ piecewise-polynomial continuous functions.

Let Ω, Ω_C, Ω_I be Lipschitz polyhedra and consider a family of regular tetrahedral meshes $\{\mathcal{T}_h\}_h$ of Ω such that each element $K \in \mathcal{T}_h$ is contained either in $\overline{\Omega_C}$ or in $\overline{\Omega_I}$. We denote $\mathcal{T}_{C,h}$, $\mathcal{T}_{I,h}$ the restriction of $\mathcal{T}_h$ to Ω_C and Ω_I, respectively.

We consider the (complex-valued) Nédélec curl-conforming edge elements defined in Section A.2, the standard Lagrange nodal elements

$$L^k_{I,h} := \{\xi_{I,h} \in C^0(\Omega_I) \mid \xi_{I,h|K} \in \mathbb{P}_k \;\forall\, K \in \mathcal{T}_{I,h}\}\,,$$

and the finite element space

$$H^k_{I,h} := L^k_{I,h} \cap H^1_{*,\Gamma}(\Omega_I)\,. \tag{4.83}$$

The finite element discretization of problem (4.78) reads

$$\begin{aligned} &\text{Find } (\mathbf{E}_h, \phi_{I,h}) \in N^k_h \times H^k_{I,h} : \\ &\quad a_e(\mathbf{E}_h, \mathbf{z}_h) + \textstyle\int_{\Omega_I} \varepsilon_I \overline{\mathbf{z}_{I,h}} \cdot \operatorname{grad} \phi_{I,h} = -i\omega \int_\Omega \mathbf{J}_e \cdot \overline{\mathbf{z}_h} \\ &\quad \textstyle\int_{\Omega_I} \varepsilon_I \mathbf{E}_{I,h} \cdot \operatorname{grad} \overline{\xi_{I,h}} = 0 \\ &\text{for all } (\mathbf{z}_h, \xi_{I,h}) \in N^k_h \times H^k_{I,h}. \end{aligned} \tag{4.84}$$

The analysis of this problem is based on the standard theory of mixed finite element methods.

We start recalling the following interpolation error estimate for the curl-conforming edge elements (see Alonso and Valli [9], Monk [179], Lemma 5.38 and Theor. 5.41).

Lemma 4.32. *If* $\mathbf{z}_{|K} \in (H^{1/2+r}(K))^3$, $0 < r \leq 1/2$, *and* $\operatorname{curl} \mathbf{z}_{|K} \in (\mathbb{P}_{k-1})^3 \oplus \widetilde{\mathbb{P}}_{k-1}\mathbf{x}$, *then the interpolation operator* $\Pi_h \mathbf{z}$ *is well-defined and*

$$\|\mathbf{z} - \Pi_h \mathbf{z}\|_{0,K} \leq C(h_K^{1/2+r} \|\mathbf{z}\|_{1/2+r,K} + h_K \|\operatorname{curl} \mathbf{z}\|_{0,K}).$$

Since $\Gamma \cap \partial\Omega = \emptyset$, a straightforward consequence of Theorem 4.3 and Theorem 4.4 in Alonso and Valli [9] is that the following regularity result holds true.

Lemma 4.33. *Let Ω_I be a Lipschitz polyhedron. Then for each $\delta > 0$ and small enough the space*

$$\mathcal{X}^\delta(\Omega_I) := \{\mathbf{z}_I \in H(\text{curl}; \Omega_I) \cap H_{0,\partial\Omega}(\text{div}; \Omega_I) \mid (\mathbf{z}_I \times \mathbf{n}_I)_{|\Gamma} \in (H^\delta(\Gamma))^3\},$$

is continuously imbedded in $(H^{1/2+\delta}(\Omega_I))^3$.

Now we are in position to verify that problem (4.84) satisfies the assumptions of the theory of discrete saddle-point problems. Defining

$$Z_h := \{\mathbf{z}_h \in N_h^k \mid \int_{\Omega_I} \varepsilon_I \mathbf{z}_{I,h} \cdot \text{grad}\, \overline{\xi_{I,h}} = 0 \;\forall\, \xi_{I,h} \in H_{I,h}^k\},$$

the following theorem holds.

Theorem 4.34. *Assuming that ε_I is a scalar constant and that $\{\mathcal{T}_h\}_h$ induces on Γ a quasi-uniform family of triangulations, there exist positive constants C_1 and C_2, independent of h, such that*

$$|a_e(\mathbf{z}_h, \mathbf{z}_h)| \geq C_1 \|\mathbf{z}_h\|^2_{H(\text{curl};\Omega)} \quad \forall\, \mathbf{z}_h \in Z_h \tag{4.85}$$

$$\sup_{\mathbf{z}_h \in N_h^k} \frac{\left|\int_{\Omega_I} \varepsilon_I \mathbf{z}_h \cdot \text{grad}\, \overline{\xi_{I,h}}\right|}{\|\mathbf{z}_h\|_{H(\text{curl};\Omega)}} \geq C_2 \|\xi_{I,h}\|_{1,\Omega_I} \quad \forall\, \xi_{I,h} \in H_{I,h}^k\,. \tag{4.86}$$

Proof. From (A.12), taking into account that Ω is simply-connected we know that $\mathbf{z}_{I,h} \in Z_{I,0}^h$ can be written as

$$\mathbf{z}_{I,h} = \varepsilon_I^{-1} \,\text{curl}\, \mathbf{q}_I + \text{grad}\, \varphi_I + \sum_{j=1}^{p_\Gamma} c_{I,j}\, \text{grad}\, w_{j,I}\,,$$

where $\mathbf{q}_I \in H_{0,\partial\Omega}(\text{curl}; \Omega_I)$ and $\varphi_I \in H^1_{0,\Gamma}(\Omega_I)$.

Let us set $\nu_I := \varphi_I + \sum_{j=1}^{p_\Gamma} c_{I,j} w_{j,I}$ and $\mathbf{U}_I := \mathbf{z}_{I,h} - \text{grad}\, \nu_I = \varepsilon_I^{-1}\, \text{curl}\, \mathbf{q}_I$. It is readily seen that $\nu_I \in H^1_{*,\Gamma}(\Omega_I)$ and that $\mathbf{U}_I \times \mathbf{n}_I = \mathbf{z}_{I,h} \times \mathbf{n}_I$ on Γ. Therefore the function given by

$$\mathbf{U} := \begin{cases} \mathbf{U}_I & \text{in } \Omega_I \\ \mathbf{z}_{C,h} & \text{in } \Omega_C \end{cases}$$

belongs to Z. Moreover, since $\text{curl}\, \mathbf{U}_I = \text{curl}\, \mathbf{z}_{I,h}$, from the coerciveness of $a_e(\cdot,\cdot)$ in Z we have

$$\begin{aligned} |a_e(\mathbf{z}_h, \mathbf{z}_h)| &= |a_e(\mathbf{U}, \mathbf{U})| \geq c_0 \|\mathbf{U}\|^2_{H(\text{curl};\Omega)} \\ &= c_0 \left(\|\mathbf{z}_{C,h}\|^2_{H(\text{curl};\Omega_C)} + \|\mathbf{U}_I\|^2_{H(\text{curl};\Omega_I)} \right). \end{aligned}$$

Thus it remains to show that there exists a positive constant C, independent of h, such that

$$\|\mathbf{z}_{I,h}\|_{0,\Omega_I} \leq C(\|\mathbf{U}_I\|_{H(\text{curl};\Omega_I)} + \|\mathbf{z}_{C,h}\|_{H(\text{curl};\Omega_C)})\,. \tag{4.87}$$

Having assumed that ε_I is a scalar constant in Ω_I, it follows that $\operatorname{div} \mathbf{U}_I = 0$ in Ω_I, and thus $\mathbf{U}_I \in \mathcal{X}^\delta(\Omega_I) \subset (H^{1/2+\delta}(\Omega_I))^3$ for some $\delta > 0$ and small enough. Moreover $\operatorname{curl} \mathbf{U}_I = \operatorname{curl} \mathbf{z}_{I,h}$ and $\operatorname{curl} \mathbf{z}_{I,h|K} \in (\mathbb{P}_{k-1})^3$, hence by Lemma 4.32 the interpolant $\Pi_h \mathbf{U}_I$ is well-defined and then also $\Pi_h \operatorname{grad} \nu_I$ is well-defined. Moreover, since $\operatorname{curl}(\Pi_h \operatorname{grad} \nu_I) = \mathbf{0}$ and $\Pi_h \operatorname{grad} \nu_I \times \mathbf{n}_I = \mathbf{0}$ on Γ, there exists $\nu_{I,h} \in H^k_{I,h}$ such that $\Pi_h \operatorname{grad} \nu_I = \operatorname{grad} \nu_{I,h}$ (see, e.g., Monk [179], Lemma 5.28). Then

$$\mathbf{z}_{I,h} = \Pi_h \mathbf{z}_{I,h} = \Pi_h \mathbf{U}_I + \Pi_h \operatorname{grad} \nu_I = \Pi_h \mathbf{U}_I + \operatorname{grad} \nu_{I,h},$$

and

$$\|\mathbf{z}_{I,h}\|^2_{0,\Omega_I} \leq C \int_{\Omega_I} \varepsilon_I \mathbf{z}_{I,h} \cdot \overline{\mathbf{z}_{I,h}} = C \int_{\Omega_I} \varepsilon_I \mathbf{z}_{I,h} \cdot \Pi_h \overline{\mathbf{U}_I}\,,$$

therefore

$$\begin{aligned}\|\mathbf{z}_{I,h}\|_{0,\Omega_I} &\leq C\varepsilon_I \|\Pi_h \mathbf{U}_I\|_{0,\Omega_I} \\ &\leq C\varepsilon_I (\|\Pi_h \mathbf{U}_I - \mathbf{U}_I\|_{0,\Omega_I} + \|\mathbf{U}_I\|_{0,\Omega_I}).\end{aligned}$$

By combining Lemma 4.32 and Lemma 4.33 we have

$$\begin{aligned}&\|\Pi_h \mathbf{U}_I - \mathbf{U}_I\|_{0,\Omega_I} \\ &\quad \leq C h^{1/2+\delta} (\|\mathbf{U}_I\|_{0,\Omega_I} + \|\operatorname{curl} \mathbf{U}_I\|_{0,\Omega_I} + \|\mathbf{U}_I \times \mathbf{n}_I\|_{\delta,\Gamma})\,.\end{aligned}$$

Since $\mathcal{T}_h$ induces a quasi–uniform family of triangulations on Γ, the following inverse inequality holds true (see Alonso and Valli [9])

$$\begin{aligned}&h^{1/2+\delta} \|\mathbf{U}_I \times \mathbf{n}_I\|_{\delta,\Gamma} = h^{1/2+\delta} \|\mathbf{z}_{I,h} \times \mathbf{n}_I\|_{\delta,\Gamma} \\ &\quad \leq C \|\mathbf{z}_{I,h} \times \mathbf{n}_I\|_{H^{-1/2}(\operatorname{div}_\tau,\Gamma)} = C \|\mathbf{z}_{C,h} \times \mathbf{n}_C\|_{H^{-1/2}(\operatorname{div}_\tau,\Gamma)}\,,\end{aligned}$$

$H^{-1/2}(\operatorname{div}_\tau, \Gamma)$ being the space of tangential traces on Γ of $H(\operatorname{curl}; \Omega_C)$ and $H(\operatorname{curl}; \Omega_I)$ (see section A.1).

Then

$$\begin{aligned}&\|\Pi_h \mathbf{U}_I - \mathbf{U}_I\|_{0,\Omega_I} \\ &\quad \leq C \left[h^{1/2+\delta} (\|\mathbf{U}_I\|_{0,\Omega_I} + \|\operatorname{curl} \mathbf{U}_I\|_{0,\Omega_I}) + \|\mathbf{z}_{C,h} \times \mathbf{n}_C\|_{H^{-1/2}(\operatorname{div}_\tau,\Gamma)} \right]\,,\end{aligned}$$

and from the trace inequality (A.10) it follows that (4.87) holds.

For what is concerned with the inf–sup condition, given $\xi_{I,h} \in H^k_{I,h}$ the function $\mathbf{z}^*_h$ defined by

$$\mathbf{z}^*_h := \begin{cases} \operatorname{grad} \xi_{I,h} & \text{in } \Omega_I \\ \mathbf{0} & \text{in } \Omega_C \end{cases}$$

belongs to N^k_h, and we conclude as in the continuous case. □

Well known results provide the existence and uniqueness of the solution (see, e.g., Brezzi and Fortin [65]). Moreover, by adapting the proof used for problem (4.78) it is easy to show that $\phi_{I,h} = 0$.

A standard density argument and the interpolation estimates in Section A.2 permit to obtain the following convergence result (see Section A.1 for notation).

Corollary 4.35. *Let the assumptions of Theorem 4.34 be satisfied. The finite element approximation method is convergent and, if the solution* $\mathbf{E}$ *of problem* (4.78) *is smooth enough, namely,* $\mathbf{E} \in H^r(\mathrm{curl}; \Omega)$ *with* $r > 1/2$, *the following error estimate holds*

$$\|\mathbf{E} - \mathbf{E}_h\|_{H(\mathrm{curl};\Omega)} \leq C h^{\min(r,k)} .$$

Remark 4.36. Golias et al. [112] has proposed an $\mathbf{E}$-based formulation which has the same form of (4.75), but with the space Z replaced by $H(\mathrm{curl}; \Omega)$.

The problem turns out to be singular, as the electric field is not uniquely determined in Ω_I. However, some numerical experiments show that its finite element approximation via edge elements is furnishing reasonable results (in this respect, see also Remark 6.10). □

Remark 4.37. The numerical approximation of (4.78) by a discontinuous Galerkin finite element scheme has been proposed and analyzed by Houston et al. [133]. □

Remark 4.38. The discrete problem for the electric boundary condition (1.20) reads

$$\begin{array}{l} \text{Find } (\mathbf{E}_h, \phi_{I,h}) \in X_h^k \times \widehat{H}_{I,h}^k : \\ \quad a_e(\mathbf{E}_h, \mathbf{z}_h) + \int_{\Omega_I} \varepsilon_I \overline{\mathbf{z}_{I,h}} \cdot \mathrm{grad}\, \phi_{I,h} = -i\omega \int_\Omega \mathbf{J}_e \cdot \overline{\mathbf{z}_h} \\ \quad \int_{\Omega_I} \varepsilon_I \mathbf{E}_{I,h} \cdot \mathrm{grad}\, \overline{\xi_{I,h}} = 0 \\ \text{for all } (\mathbf{z}_h, \xi_{I,h}) \in X_h^k \times \widehat{H}_{I,h}^k , \end{array} \tag{4.88}$$

where $X_h^k := N_h^k \cap H_0(\mathrm{curl}; \Omega)$ and $\widehat{H}_{I,h}^k := L_{I,h}^k \cap H_*^1(\Omega_I)$ (for the definition of $H_*^1(\Omega_I)$ see (4.82)). □

5

Formulations via scalar potentials

As we have already remarked in the preceding chapters, a specific feature of eddy current problems is the presence of differential constraints acting in the non-conducting part of the domain: namely, $\operatorname{curl} \mathbf{H}_I = \mathbf{J}_{e,I}$ in Ω_I and $\operatorname{div}(\varepsilon_I \mathbf{E}_I) = 0$ in Ω_I.

The use of Lagrangian multipliers to take into account these constraints has been illustrated in Chapter 4. Here we want to describe a different approach, which on one hand is very natural and on the other hand leads the introduction of a scalar unknown instead of a vector-valud one, thus yielding a cheaper algorithm for numerical approximation.

The starting point is to rewrite the constraint on $\operatorname{curl} \mathbf{H}_I$ as an irrotationality constraint: for instance, in the case of the magnetic boundary value problem we construct the vector function $\mathbf{H}_{e,I}$ defined in (3.3), and consider $\mathbf{H}_I - \mathbf{H}_{e,I}$, which is curl-free. The other boundary value problems are treated in a similar way.

If the topology of the domain Ω_I is simple, say, Ω_I is simply-connected, we know that a curl-free vector function can be expressed as the gradient of a scalar function, called a scalar potential. Therefore, we reduce the problem to a new unknown ψ_I such that $\operatorname{grad} \psi_I = \mathbf{H}_I - \mathbf{H}_{e,I}$ in Ω_I (when the vector function $\mathbf{H}_{e,I}$ is non-vanishing, in the engineering literature the scalar function ψ_I is often called a reduced scalar magnetic potential; instead, even if the relation between $\mathbf{H}_I$ and ψ_I is not exactly $\mathbf{H}_I = \operatorname{grad} \psi_I$, we will simply call ψ_I a scalar magnetic potential).

However, some problems arise when the insulator Ω_I has a more complicated topology (and this is very often the case in real-life problems). In fact, in this case a curl-free vector function is no longer guaranteed to be a gradient of a potential, and some additional terms have to be taken into account: more precisely, harmonic fields are coming into play.

In this chapter, mainly following Alonso Rodríguez et al. [10], we address this problem, showing how to determine these harmonic fields and devising some weak formulations that are related to this approach. We also propose domain decomposition algorithms for the effective solution of the problem, in which one alternates between a solution step in the conductor and another solution step in the insulator.

We start by describing a first method, in which the chosen unknowns are the magnetic field $\mathbf{H}_C$ in the conductor and the scalar magnetic potential ψ_I, plus a harmonic

A. Alonso Rodríguez, A. Valli: Eddy Current Approximation of Maxwell Equations.

field in the insulator. We also present an alternative approach in terms of the electric field $\mathbf{E}_C$ in the conductor: this has the advantage that the matching between the unknowns on the interface Γ is of weak type, thus in numerical computations independent meshes could be used in Ω_C and Ω_I.

The basic points in the proof are the $(L^2(\Omega_I))^3$-orthogonal decomposition results in Section A.3, which, in addition to the weak formulations presented in Chapter 3, lead to the weak formulations in terms of the scalar magnetic potentials. In Sections 5.1 and 5.3 we also describe some alternative weak formulations, proposed by Leonard and Rodger [167] and Alonso Rodríguez et al. [10], that do not require the knowledge of the harmonic fields but only use some easily computable interpolants; the latter formulations are the most suited for numerical approximation.

Based on the weak formulations thus obtained, the numerical approximation of the weak problems is performed by means of edge finite elements in the conductor and (scalar) nodal finite elements in the insulator. The positiveness of the associated sesquilinear forms leads in a straightforward way to the proof of an optimal error estimate. However, as already recalled, a preliminary step is the explicit construction of the vector function $\mathbf{H}_{e,I}$; this is not always easily achieved, hence in some case one must slightly modify the approach in order to devise a viable numerical approximation scheme.

Motivated by the fact that, when using a scalar magnetic potential, the electric field in the insulator is computed in a second step, we end the chapter with the presentation of a finite element approximation scheme for the electric field $\mathbf{E}_I$ in Ω_I. The starting point is a saddle-point formulation similar to that used in Section 4.6 for describing the eddy current problem in terms of the electric field $\mathbf{E}$.

In the whole chapter the geometrical assumptions on Ω, Ω_C and Ω_I are the same as those of Section 1.3. Moreover, we again assume that the matrix $\boldsymbol{\mu}$ is symmetric and uniformly positive definite in Ω, with entries belonging to $L^\infty(\Omega)$, the matrix ε_I is symmetric and uniformly positive definite in Ω_I, with entries belonging to $L^\infty(\Omega_I)$, and the matrix $\boldsymbol{\sigma}$ is symmetric and uniformly positive definite in Ω_C, with entries belonging to $L^\infty(\Omega_C)$, whereas it is vanishing in Ω_I.

To a reader interested in numerical approximation and implementation we suggest to focus on problems (5.21) and (5.71) ($\mathbf{H}_C/\widehat{\psi}_I$ formulation), on problems (5.67) and (5.75) ($\mathbf{H}_C/\widehat{\psi}_I^*$ formulation), on problems (5.58) and (5.80) ($\mathbf{E}_C/\widehat{\psi}_I^*$ formulation), and on the domain decomposition procedures described in Sections 5.2.1 and 5.3.1.

5.1 The weak formulation in terms of $\mathbf{H}_C$ and ψ_I

From Section 3.1 let us recall that, if we are considering the magnetic boundary value problem, the following necessary assumptions have to be imposed on the current density $\mathbf{J}_e \in (L^2(\Omega))^3$

$$\operatorname{div} \mathbf{J}_{e,I} = 0 \quad \text{in } \Omega_I \ , \ \mathbf{J}_{e,I} \cdot \mathbf{n} = \mathbf{0} \quad \text{on } \partial\Omega \tag{5.1}$$

$$\begin{aligned} &\textstyle\int_{\Gamma_j} \mathbf{J}_{e,I} \cdot \mathbf{n}_I = 0 \quad \forall\, j = 1, \ldots, p_\Gamma \\ &\textstyle\int_{\Omega_I} \mathbf{J}_{e,I} \cdot \boldsymbol{\pi}_{k,I} = 0 \quad \forall\, k = 1, \ldots, n_{\partial\Omega} \ . \end{aligned} \tag{5.2}$$

As noted there, these assumptions ensure that there exist two vector fields $\mathbf{H}_{e,I} \in H(\text{curl}; \Omega_I)$ and $\mathbf{H}_{e,C} \in H(\text{curl}; \Omega_C)$ satisfying the properties described in (3.3) and (3.4), namely,

$$\begin{cases} \text{curl}\, \mathbf{H}_{e,I} = \mathbf{J}_{e,I} & \text{in } \Omega_I \\ \mathbf{H}_{e,I} \times \mathbf{n} = \mathbf{0} & \text{on } \partial\Omega, \end{cases} \tag{5.3}$$

and

$$\mathbf{H}_{e,C} \times \mathbf{n}_C + \mathbf{H}_{e,I} \times \mathbf{n}_I = \mathbf{0} \quad \text{on } \Gamma \,. \tag{5.4}$$

As in (3.5) we will denote by $\mathbf{H}_e \in H_0(\text{curl}; \Omega)$ the vector field defined by

$$\mathbf{H}_e := \begin{cases} \mathbf{H}_{e,I} & \text{in } \Omega_I \\ \mathbf{H}_{e,C} & \text{in } \Omega_C \,. \end{cases} \tag{5.5}$$

In Chapter 3 we have obtained and proved the well-posedness of the following weak problem

Find $(\mathbf{H} - \mathbf{H}_e) \in V$ such that

$$\int_{\Omega_C} \boldsymbol{\sigma}^{-1} \text{curl}\, \mathbf{H}_C \cdot \text{curl}\, \overline{\mathbf{v}_C} + \int_\Omega i\omega \boldsymbol{\mu} \mathbf{H} \cdot \overline{\mathbf{v}} = \int_{\Omega_C} \boldsymbol{\sigma}^{-1} \mathbf{J}_{e,C} \cdot \text{curl}\, \overline{\mathbf{v}_C} \tag{5.6}$$

for each $\mathbf{v} \in V$,

where

$$V := \{\mathbf{v} \in H_0(\text{curl}; \Omega) \,|\, \text{curl}\, \mathbf{v}_I = \mathbf{0} \text{ in } \Omega_I\} \,. \tag{5.7}$$

We want to rewrite this problem using the orthogonal decomposition presented in Theorem A.7. First of all, define in $H(\text{curl}; \Omega) \times H(\text{curl}; \Omega)$ the sesquilinear form $a(\cdot, \cdot)$ as

$$a(\mathbf{u}, \mathbf{v}) := a_C(\mathbf{u}_C, \mathbf{v}_C) + a_I(\mathbf{u}_I, \mathbf{v}_I) \,, \tag{5.8}$$

where

$$a_C(\mathbf{u}_C, \mathbf{v}_C) := \int_{\Omega_C} (\boldsymbol{\sigma}^{-1} \text{curl}\, \mathbf{u}_C \cdot \text{curl}\, \overline{\mathbf{v}_C} + i\omega \boldsymbol{\mu}_C \mathbf{u}_C \cdot \overline{\mathbf{v}_C}), \tag{5.9}$$

and

$$a_I(\mathbf{u}_I, \mathbf{v}_I) := i\omega \int_{\Omega_I} \boldsymbol{\mu}_I \mathbf{u}_I \cdot \overline{\mathbf{v}_I} \,. \tag{5.10}$$

Up to the factor $i\omega$, the sesquilinear form $a_I(\cdot, \cdot)$ is the scalar product $(\cdot, \cdot)_{\mu_I, \Omega_I}$ for complex valued vector functions, see (A.20).

For the sake of simplicity, let us also introduce the notation $\mathbf{Z} := \mathbf{H} - \mathbf{H}_e$. The restriction $\mathbf{Z}_I$ can be written as

$$\mathbf{H}_I - \mathbf{H}_{e,I} = \mathbf{Z}_I = \text{grad}\, \psi_I + \sum_{r=1}^{p_{\partial\Omega}} \alpha_{I,r} \,\text{grad}\, z_{r,I} + \sum_{l=1}^{n_\Gamma} \beta_{I,l} \boldsymbol{\rho}_{l,I} \,, \tag{5.11}$$

where the complex valued function ψ_I and the complex constants $\alpha_{I,r}$, $\beta_{I,l}$ are defined as in (A.23) and (A.24), respectively, replacing $\mathbf{v}_I$ with $\mathbf{Z}_I$. Note here that this formula simplifies if $p_{\partial\Omega} = 0$ and $n_\Gamma = 0$. The first relation means that the boundary of Ω is connected, the second is satisfied if Ω_C is simply-connected.

In a similar way, for any $\mathbf{v} \in V$ we have

$$\mathbf{v}_I = \operatorname{grad} \chi_I + \sum_{r=1}^{p_{\partial\Omega}} a_{I,r} \operatorname{grad} z_{r,I} + \sum_{l=1}^{n_\Gamma} b_{I,l} \boldsymbol{\rho}_{l,I} \,. \tag{5.12}$$

To simplify the notation, we will denote by $\boldsymbol{\eta}_I \in \mathbb{C}^N$, $N = p_{\partial\Omega} + n_\Gamma$, the complex vector with components $\eta_{I,r} = \alpha_{I,r}$, $r = 1, \dots, p_{\partial\Omega}$, $\eta_{I,p_{\partial\Omega}+l} = \beta_{I,l}$, $l = 1, \dots, n_\Gamma$, and similarly denote by $\boldsymbol{\theta}_I \in \mathbb{C}^N$ the complex vector with components $\theta_{I,r} = a_{I,r}$, $r = 1, \dots, p_{\partial\Omega}$, $\theta_{I,p_{\partial\Omega}+l} = b_{I,l}$, $l = 1, \dots, n_\Gamma$. Moreover, the basis functions of the space $\mathcal{H}_{\mu_I}(\partial\Omega, \Gamma; \Omega_I)$ will be indicated by $\boldsymbol{\omega}_{q,I}$, $q = 1, \dots, N$, where $\boldsymbol{\omega}_{r,I} = \operatorname{grad} z_{r,I}$, $r = 1, \dots, p_{\partial\Omega}$, and $\boldsymbol{\omega}_{p_{\partial\Omega}+l,I} = \boldsymbol{\rho}_{l,I}$, $l = 1, \dots, n_\Gamma$. Finally, we will denote by $[\cdot, \cdot]$ the scalar product in $\mathbb{C}^N$.

We have:

Theorem 5.1. *Let* $\mathbf{Z}$ *be the solution to* (5.6), *and decompose* $\mathbf{Z}_I$ *as in* (5.11), *namely,* $\mathbf{Z}_I = \operatorname{grad} \psi_I + \sum_{q=1}^N \eta_{I,q} \boldsymbol{\omega}_{q,I}$. *Then* $(\mathbf{Z}_C, \psi_I, \boldsymbol{\eta}_I)$ *is the unique solution to*

$$\begin{aligned}
&\textit{Find } (\mathbf{Z}_C, \psi_I, \boldsymbol{\eta}_I) \in W \textit{ such that}\\
&a_C(\mathbf{Z}_C, \mathbf{v}_C) + a_I(\operatorname{grad} \psi_I, \operatorname{grad} \chi_I) + i\omega[A\,\boldsymbol{\eta}_I, \boldsymbol{\theta}_I]\\
&\quad = -a_C(\mathbf{H}_{e,C}, \mathbf{v}_C) - a_I(\mathbf{H}_{e,I}, \operatorname{grad} \chi_I)\\
&\qquad -a_I\Big(\mathbf{H}_{e,I}, \textstyle\sum_{q=1}^N \theta_{I,q}\, \boldsymbol{\omega}_{q,I}\Big) + \int_{\Omega_C} \boldsymbol{\sigma}^{-1} \mathbf{J}_{e,C} \cdot \operatorname{curl} \overline{\mathbf{v}_C}\\
&\textit{for each } (\mathbf{v}_C, \chi_I, \boldsymbol{\theta}_I) \in W \,,
\end{aligned} \tag{5.13}$$

where the matrix A *is defined in* (A.25) *and*

$$\begin{aligned}
W := \{&(\mathbf{v}_C, \chi_I, \boldsymbol{\theta}_I) \in H(\operatorname{curl}; \Omega_C) \times H^1_{0,\partial\Omega}(\Omega_I) \times \mathbb{C}^N \mid\\
&\mathbf{v}_C \times \mathbf{n}_C + \operatorname{grad} \chi_I \times \mathbf{n}_I + \textstyle\sum_{q=1}^N \theta_{I,q}\, \boldsymbol{\omega}_{q,I} \times \mathbf{n}_I = \mathbf{0} \textit{ on } \Gamma\} \,.
\end{aligned} \tag{5.14}$$

Proof. From (5.12) we have $\mathbf{v}_I = \operatorname{grad} \chi_I + \sum_{q=1}^N \theta_{I,q} \boldsymbol{\omega}_{q,I}$; therefore, recalling the orthogonality result given in Theorem A.7, it is easily seen that the solution to problem (5.6) gives a solution $(\mathbf{Z}_C, \psi_I, \boldsymbol{\eta}_I)$ to (5.13). On the other hand, the sesquilinear form at the left hand side of (5.13) is clearly continuous and coercive in W, endowed with the natural norm. Therefore, from Lax–Milgram lemma, there exists a unique solution to (5.13). □

A conforming finite element approximation based directly on (5.13) is not a viable option, as it would require that the functions $\boldsymbol{\omega}_{q,I}$, $q = 1, \dots, N$, be explicitly known. An alternative approach, that overcomes this difficulty and has been proposed by Leonard and Rodger [167] and Alonso Rodríguez et al. [10], is based on a different decomposition of $\mathbf{H}_I$. Let us suppose that Ω is a polyhedral domain, and that there is a triangulation $\mathcal{T}_{h^0}$ such that $\overline{\Omega} = \cup_{K \in \mathcal{T}_{h^0}} K$, where K is a tetrahedron or a parallelepiped and $h^0 > 0$ is the (fixed) mesh size. From Bossavit [59], Hiptmair [126], Gross and Kotiuga [115] we know that in Ω_I there exists a system of "cutting" surfaces Ξ_l, $l = 1, \dots, n_\Gamma$, with $\partial \Xi_l \subset \Gamma$, such that every function $\mathbf{z}_I \in H^0_{0,\partial\Omega}(\operatorname{curl}; \Omega_I)$ restricted to $\Omega_I \setminus \cup_{l=1}^{n_\Gamma} \Xi_l$ is the gradient of a function belonging to $H^1(\Omega_I \setminus \cup_{l=1}^{n_\Gamma} \Xi_l)$.

It is not restrictive to assume that the triangulation $\mathcal{T}_{h^0}$ induces a triangulation on each surface Ξ_l. Let us denote by Π_q the piecewise-polynomial function taking value 1 at the nodes on $(\partial\Omega)_q$ and 0 at all the other nodes for $q = 1, \ldots, p_{\partial\Omega}$, and taking value 1 at the nodes on one side of $\Xi_{q-p_{\partial\Omega}}$, say $\Xi^+_{q-p_{\partial\Omega}}$, and 0 at all the other nodes (including those on $\Xi^-_{q-p_{\partial\Omega}}$, the other side of $\Xi_{q-p_{\partial\Omega}}$) for $q = p_{\partial\Omega} + 1, \ldots, p_{\partial\Omega} + n_\Gamma$. Set also

$$\boldsymbol{\lambda}_q := \begin{cases} \operatorname{grad} \Pi_q & \text{for } q = 1, \ldots, p_{\partial\Omega} \\ \widetilde{\operatorname{grad} \Pi_q} & \text{for } q = p_{\partial\Omega} + 1, \ldots, p_{\partial\Omega} + n_\Gamma \,, \end{cases} \tag{5.15}$$

where $\widetilde{\operatorname{grad} \Pi_q}$ denotes the $(L^2(\Omega_I))^3$-extension of $\operatorname{grad} \Pi_q$ computed in $\Omega_I \setminus \Xi_q$. It is verified at once that for each $q = 1, \ldots, N$ the vector function $\boldsymbol{\lambda}_q$ belongs to $(L^2(\Omega_I))^3$ and satisfies

$$\begin{cases} \operatorname{curl} \boldsymbol{\lambda}_q = \mathbf{0} & \text{in } \Omega_I \\ \boldsymbol{\lambda}_q \times \mathbf{n} = \mathbf{0} & \text{on } \partial\Omega \,. \end{cases}$$

Denoting by $g_q \in H^1(\Omega_I)$ the solution to

$$\begin{cases} \operatorname{div}(\boldsymbol{\mu}_I \operatorname{grad} g_q) = \operatorname{div}(\boldsymbol{\mu}_I \boldsymbol{\lambda}_q) & \text{in } \Omega_I \\ \boldsymbol{\mu}_I \operatorname{grad} g_q \cdot \mathbf{n}_I = \boldsymbol{\mu}_I \boldsymbol{\lambda}_q \cdot \mathbf{n}_I & \text{on } \Gamma \\ g_q = 0 & \text{on } \partial\Omega \,, \end{cases} \tag{5.16}$$

one can easily check that the basis functions $\boldsymbol{\omega}_{q,I}$ of the space $\mathcal{H}_{\mu_I}(\partial\Omega, \Gamma; \Omega_I)$ can be written as

$$\boldsymbol{\omega}_{q,I} = \boldsymbol{\lambda}_q - \operatorname{grad} g_q \,. \tag{5.17}$$

Using this result in the representation formula (5.11) one finds

$$\begin{aligned} \mathbf{Z}_I &= \operatorname{grad} \psi_I + \textstyle\sum_{q=1}^N \eta_{I,q} \boldsymbol{\omega}_{q,I} = \operatorname{grad} \psi_I + \textstyle\sum_{q=1}^N \eta_{I,q} (\boldsymbol{\lambda}_q - \operatorname{grad} g_q) \\ &= \operatorname{grad} \widehat{\psi}_I + \textstyle\sum_{q=1}^N \eta_{I,q} \boldsymbol{\lambda}_q \,, \end{aligned} \tag{5.18}$$

having defined

$$\widehat{\psi}_I := \psi_I - \sum_{q=1}^N \eta_{I,q} g_q \,. \tag{5.19}$$

If we set

$$\widehat{a}_I(\widehat{\zeta}_I, \boldsymbol{\gamma}_I; \widehat{\chi}_I, \boldsymbol{\theta}_I) := a_I\Big(\operatorname{grad} \widehat{\zeta}_I + \sum_{q=1}^N \gamma_{I,q} \boldsymbol{\lambda}_q, \operatorname{grad} \widehat{\chi}_I + \sum_{q=1}^N \theta_{I,q} \boldsymbol{\lambda}_q\Big) \,, \tag{5.20}$$

the weak problem (5.13) can thus be rewritten as follows

Find $(\mathbf{Z}_C, \widehat{\psi}_I, \boldsymbol{\eta}_I) \in \widehat{W}$ such that

$$\begin{aligned} & a_C(\mathbf{Z}_C, \mathbf{v}_C) + \widehat{a}_I(\widehat{\psi}_I, \boldsymbol{\eta}_I; \widehat{\chi}_I, \boldsymbol{\theta}_I) \\ & \quad = -a_C(\mathbf{H}_{e,C}, \mathbf{v}_C) - a_I(\mathbf{H}_{e,I}, \operatorname{grad} \widehat{\chi}_I) \\ & \qquad - a_I\Big(\mathbf{H}_{e,I}, \textstyle\sum_{q=1}^N \theta_{I,q} \boldsymbol{\lambda}_q\Big) + \int_{\Omega_C} \boldsymbol{\sigma}^{-1} \mathbf{J}_{e,C} \cdot \operatorname{curl} \overline{\mathbf{v}_C} \end{aligned} \tag{5.21}$$

for each $(\mathbf{v}_C, \widehat{\chi}_I, \boldsymbol{\theta}_I) \in \widehat{W}$, where

$$\widehat{W} := \{(\mathbf{v}_C, \widehat{\chi}_I, \boldsymbol{\theta}_I) \in H(\mathrm{curl}; \Omega_C) \times H^1_{0,\partial\Omega}(\Omega_I) \times \mathbb{C}^N \mid \\ \mathbf{v}_C \times \mathbf{n}_C + \mathrm{grad}\, \widehat{\chi}_I \times \mathbf{n}_I + \sum_{q=1}^N \theta_{I,q} \boldsymbol{\lambda}_q \times \mathbf{n}_I = \mathbf{0} \text{ on } \Gamma\} . \tag{5.22}$$

It is clear that the two problems (5.13) and (5.21) are equivalent, therefore (5.21) is well-posed. Moreover, it is worth noting that the sesquilinear form at the left hand side in (5.21) is continuous in $\widehat{W}$, and it can also be seen that it is coercive in $\widehat{W}$. In fact,

$$a_C(\mathbf{v}_C, \mathbf{v}_C) + \widehat{a}_I(\widehat{\chi}_I, \boldsymbol{\theta}_I; \widehat{\chi}_I, \boldsymbol{\theta}_I) = a(\mathbf{v}, \mathbf{v}) ,$$

where $\mathbf{v}_I = \mathrm{grad}\, \widehat{\chi}_I + \sum_{q=1}^N \theta_{I,q} \boldsymbol{\lambda}_q$, and $a(\cdot, \cdot)$ is coercive in V, namely,

$$|a(\mathbf{v}, \mathbf{v})| \geq \kappa\Big(\int_{\Omega_C} (|\mathbf{v}_C|^2 + |\,\mathrm{curl}\, \mathbf{v}_C|^2) + \int_{\Omega_I} \boldsymbol{\mu}_I \mathbf{v}_I \cdot \overline{\mathbf{v}}_I \Big)$$

(see Theorem 3.1). Thus the following lemma, that will be also useful in the sequel, permits to complete the proof.

Lemma 5.2. *The function* $\mathbf{v}_I = \mathrm{grad}\, \widehat{\chi}_I + \sum_{q=1}^N \theta_{I,q} \boldsymbol{\lambda}_q$ *satisfies*

$$\int_{\Omega_I} \boldsymbol{\mu}_I \mathbf{v}_I \cdot \overline{\mathbf{v}}_I \geq C_0 \Big(\int_{\Omega_I} |\,\mathrm{grad}\, \widehat{\chi}_I|^2 + |\boldsymbol{\theta}_I|^2 \Big) . \tag{5.23}$$

Proof. Since $\boldsymbol{\lambda}_q = \mathrm{grad}\, g_q + \boldsymbol{\omega}_{q,I}$, we have $\mathbf{v}_I = \mathrm{grad}\, \widehat{\chi}_I + \sum_{q=1}^N \theta_{I,q}\, \mathrm{grad}\, g_q + \sum_{q=1}^N \theta_{I,q} \boldsymbol{\omega}_{q,I}$. Due to the orthogonality properties presented in Theorem A.7 and the fact that the matrix A defined in (A.25) is positive definite, we have

$$\begin{aligned} \int_{\Omega_I} \boldsymbol{\mu}_I \mathbf{v}_I \cdot \overline{\mathbf{v}}_I &\geq C_1 \int_{\Omega_I} |\,\mathrm{grad}\, \widehat{\chi}_I + \textstyle\sum_{q=1}^N \theta_{I,q}\, \mathrm{grad}\, g_q|^2 \\ &\quad + \textstyle\sum_{p,q=1}^N \theta_{I,q} \overline{\theta_{I,p}} \int_{\Omega_I} \boldsymbol{\mu}_I \boldsymbol{\omega}_{q,I} \cdot \boldsymbol{\omega}_{p,I} \\ &\geq C_2 \big(\int_{\Omega_I} |\,\mathrm{grad}\, \widehat{\chi}_I + \textstyle\sum_{q=1}^N \theta_{I,q}\, \mathrm{grad}\, g_q|^2 + |\boldsymbol{\theta}_I|^2\big) . \end{aligned}$$

Recalling that for each $0 < \delta < 1$ one has $|2ab| \leq \delta |a|^2 + \delta^{-1} |b|^2$, we find

$$\begin{aligned} &\int_{\Omega_I} |\,\mathrm{grad}\, \widehat{\chi}_I + \textstyle\sum_{q=1}^N \theta_{I,q}\, \mathrm{grad}\, g_q|^2 \\ &\quad \geq (1-\delta) \int_{\Omega_I} |\,\mathrm{grad}\, \widehat{\chi}_I|^2 - (1-\delta)\delta^{-1} \int_{\Omega_I} |\textstyle\sum_{q=1}^N \theta_{I,q}\, \mathrm{grad}\, g_q|^2 \\ &\quad \geq (1-\delta) \int_{\Omega_I} |\,\mathrm{grad}\, \widehat{\chi}_I|^2 - C_3 (1-\delta)\delta^{-1} |\boldsymbol{\theta}_I|^2 . \end{aligned}$$

Thus the proof is complete by choosing δ such that $C_3/(1+C_3) < \delta < 1$. □

From Lemma 5.2 the proof of the coerciveness of the sesquilinear form at the left hand side in (5.21) follows at once from the Poincaré inequality

$$\int_{\Omega_I} |\widehat{\chi}_I|^2 \leq C \int_{\Omega_I} |\,\mathrm{grad}\, \widehat{\chi}_I|^2 , \tag{5.24}$$

which holds in $H^1_{0,\partial\Omega}(\Omega_I)$ (see, e.g., Dautray and Lions [94], Chap. IV, Sect. 7, Rem. 4). In particular, we have also proved that the sesquilinear form $\widehat{a}_I(\cdot;\cdot)$ introduced in (5.20) is coercive in $H^1_{0,\partial\Omega}(\Omega_I) \times \mathbb{C}^N$.

Remark 5.3. The unknowns $\eta_{I,p_{\partial\Omega}+l} = \beta_{I,l}$, $l = 1, \ldots, n_\Gamma$, in (5.13) or (5.21) are related to the current intensity through suitable sections of Ω_C. Solving (5.13) or (5.21) permits to determine these in such a way that the complete eddy current problem is satisfied.

Instead, imposing them as data for the problem leads to the violation of the Faraday equation on the "cutting" surfaces Ξ_l, $l = 1, \ldots, n_\Gamma$ (see Section 3.3.2, (v)). In this respect, see also the considerations presented in Section 6.3, where the $(\mathbf{T}_C, \psi_C) - \psi_I$ formulation is described.

Addressing the eddy current problem with assigned current intensity or voltage needs a particular geometrical setting and specific boundary conditions: for this, see Chapter 8. □

5.2 The strong formulation in terms of $\mathbf{H}_C$ and ψ_I

As proved in Chapter 3, the strong formulation of the eddy current problem in terms of the magnetic field reads

$$\begin{cases} \operatorname{curl}(\boldsymbol{\sigma}^{-1}\operatorname{curl}\mathbf{H}_C) + i\omega\boldsymbol{\mu}_C\mathbf{H}_C & \\ \qquad\qquad = \operatorname{curl}(\boldsymbol{\sigma}^{-1}\mathbf{J}_{e,C}) & \text{in } \Omega_C \\ \operatorname{curl}\mathbf{H}_I = \mathbf{J}_{e,I} & \text{in } \Omega_I \\ \operatorname{div}(\boldsymbol{\mu}_I\mathbf{H}_I) = 0 & \text{in } \Omega_I \\ \int_{(\partial\Omega)_r} \boldsymbol{\mu}_I\mathbf{H}_I\cdot\mathbf{n} = 0 & \forall\, r = 1, \ldots, p_{\partial\Omega} \\ \int_{\Omega_I} i\omega\boldsymbol{\mu}_I\mathbf{H}_I\cdot\boldsymbol{\rho}_{l,I} & \\ \quad = -\int_\Gamma[\boldsymbol{\sigma}^{-1}(\operatorname{curl}\mathbf{H}_C - \mathbf{J}_{e,C})]\times\mathbf{n}_C\cdot\boldsymbol{\rho}_{l,I} & \forall\, l = 1, \ldots, n_\Gamma \\ \mathbf{H}_I\times\mathbf{n} = \mathbf{0} & \text{on } \partial\Omega \\ \boldsymbol{\mu}_I\mathbf{H}_I\cdot\mathbf{n}_I + \boldsymbol{\mu}_C\mathbf{H}_C\cdot\mathbf{n}_C = 0 & \text{on } \Gamma \\ \mathbf{H}_I\times\mathbf{n}_I + \mathbf{H}_C\times\mathbf{n}_C = \mathbf{0} & \text{on } \Gamma\,. \end{cases} \tag{5.25}$$

It must be noted that the conditions $(5.25)_4$ and $(5.25)_5$ are necessary for determining the correct projection of the solution $\mathbf{Z}_I = \mathbf{H}_I - \mathbf{H}_{e,I}$ over the space of harmonic fields $\mathcal{H}_{\mu_I}(\partial\Omega, \Gamma; \Omega_I)$ (see the orthogonal decomposition formula (5.11)).

We are now interested in rewriting this problem in an equivalent form, in term of $\mathbf{H}_C$, ψ_I and $\boldsymbol{\eta}_I = (\boldsymbol{\alpha}_I, \boldsymbol{\beta}_I)$ (see (5.11)).

Theorem 5.4. *Problem* (5.25) *is equivalent to the following one*

$$\begin{cases} \operatorname{curl}(\boldsymbol{\sigma}^{-1}\operatorname{curl}\mathbf{H}_C) + i\omega\boldsymbol{\mu}_C\mathbf{H}_C = \operatorname{curl}(\boldsymbol{\sigma}^{-1}\mathbf{J}_{e,C}) & \textit{in } \Omega_C \\ \mathbf{H}_C\times\mathbf{n}_C = -\operatorname{grad}\psi_I\times\mathbf{n}_I & \\ \qquad -\sum_{q=1}^N \eta_{I,q}\,\boldsymbol{\omega}_{q,I}\times\mathbf{n}_I - \mathbf{H}_{e,I}\times\mathbf{n}_I & \textit{on } \Gamma \end{cases} \tag{5.26}$$

$$\begin{cases} \operatorname{div}(\boldsymbol{\mu}_I\operatorname{grad}\psi_I) = -\operatorname{div}(\boldsymbol{\mu}_I\mathbf{H}_{e,I}) & \textit{in } \Omega_I \\ \boldsymbol{\mu}_I\operatorname{grad}\psi_I\cdot\mathbf{n}_I = -\boldsymbol{\mu}_C\mathbf{H}_C\cdot\mathbf{n}_C - \boldsymbol{\mu}_I\mathbf{H}_{e,I}\cdot\mathbf{n}_I & \textit{on } \Gamma \\ \psi_I = 0 & \textit{on } \partial\Omega \end{cases} \tag{5.27}$$

$$\begin{aligned} i\omega(A\boldsymbol{\eta}_I)_q = &-\int_\Gamma[\boldsymbol{\sigma}^{-1}(\operatorname{curl}\mathbf{H}_C - \mathbf{J}_{e,C})]\times\mathbf{n}_C\cdot\boldsymbol{\omega}_{q,I} \\ &-a_I(\mathbf{H}_{e,I}, \boldsymbol{\omega}_{q,I}) \qquad \forall\, q = 1, \ldots, N\,. \end{aligned} \tag{5.28}$$

Proof. Equations (5.26) and (5.27) are easily obtained from $(5.25)_1$ and $(5.25)_8$, $(5.25)_3$ and $(5.25)_7$, respectively, and the representation formula (5.11). We are therefore left with the proof of (5.28).

Recalling that $z_{r,I} = 1$ on $(\partial\Omega)_r$ and $z_{r,I} = 0$ on $\partial\Omega \setminus (\partial\Omega)_r$, for $r = 1, \ldots, p_{\partial\Omega}$, and taking into account $(5.25)_3$, $(5.25)_4$, $(5.25)_7$ and $(5.25)_1$, by integrating by parts we find that

$$\begin{aligned} i\omega \textstyle\int_{\Omega_I} \boldsymbol{\mu}_I \mathbf{H}_I \cdot \operatorname{grad} z_{r,I} &= i\omega \textstyle\int_\Gamma \boldsymbol{\mu}_I \mathbf{H}_I \cdot \mathbf{n}_I \, z_{r,I} \\ &= -i\omega \textstyle\int_\Gamma \boldsymbol{\mu}_C \mathbf{H}_C \cdot \mathbf{n}_C \, z_{r,I} \\ &= \textstyle\int_\Gamma \operatorname{curl}[\boldsymbol{\sigma}^{-1}(\operatorname{curl} \mathbf{H}_C - \mathbf{J}_{e,C})] \cdot \mathbf{n}_C \, z_{r,I} \\ &= \textstyle\int_\Gamma \operatorname{div}_\tau([\boldsymbol{\sigma}^{-1}(\operatorname{curl} \mathbf{H}_C - \mathbf{J}_{e,C})] \times \mathbf{n}_C) \, z_{r,I} \\ &= -\textstyle\int_\Gamma [\boldsymbol{\sigma}^{-1}(\operatorname{curl} \mathbf{H}_C - \mathbf{J}_{e,C})] \times \mathbf{n}_C \cdot \operatorname{grad} z_{r,I} \, . \end{aligned} \tag{5.29}$$

In other words, putting together $(5.25)_5$ and (5.29), we have

$$i\omega \int_{\Omega_I} \boldsymbol{\mu}_I \mathbf{H}_I \cdot \boldsymbol{\omega}_{q,I} = -\int_\Gamma [\boldsymbol{\sigma}^{-1}(\operatorname{curl} \mathbf{H}_C - \mathbf{J}_{e,C})] \times \mathbf{n}_C \cdot \boldsymbol{\omega}_{q,I} \tag{5.30}$$

for each $q = 1, \ldots, N$.

Recalling the definition of the matrix A in (A.25), the representation formula (5.11) and the orthogonality of $\operatorname{grad} \psi_I$ and $\boldsymbol{\omega}_{q,I}$ with respect to the scalar product $(\mathbf{u}_I, \mathbf{v}_I)_{\mu_I, \Omega_I} := \int_{\Omega_I} \boldsymbol{\mu}_I \mathbf{u}_I \cdot \mathbf{v}_I$, we finally find

$$\begin{aligned} i\omega(A\boldsymbol{\eta}_I)_q &= i\omega \textstyle\int_{\Omega_I} \boldsymbol{\mu}_I \sum_{p=1}^N \eta_{I,p} \, \boldsymbol{\omega}_{p,I} \cdot \boldsymbol{\omega}_{q,I} \\ &= i\omega \textstyle\int_{\Omega_I} \boldsymbol{\mu}_I (\mathbf{H}_I - \mathbf{H}_{e,I}) \cdot \boldsymbol{\omega}_{q,I} \\ &= -\textstyle\int_\Gamma [\boldsymbol{\sigma}^{-1}(\operatorname{curl} \mathbf{H}_C - \mathbf{J}_{e,C})] \times \mathbf{n}_C \cdot \boldsymbol{\omega}_{q,I} \\ &\quad -i\omega \textstyle\int_{\Omega_I} \boldsymbol{\mu}_I \mathbf{H}_{e,I} \cdot \boldsymbol{\omega}_{q,I} \, , \end{aligned} \tag{5.31}$$

which ends the proof. □

An alternative strong formulation derives from the representation formula (5.18).

Theorem 5.5. *Problem* (5.25) *is equivalent to the following one*

$$\begin{cases} \operatorname{curl}(\boldsymbol{\sigma}^{-1} \operatorname{curl} \mathbf{H}_C) + i\omega \boldsymbol{\mu}_C \mathbf{H}_C = \operatorname{curl}(\boldsymbol{\sigma}^{-1} \mathbf{J}_{e,C}) & \textit{in } \Omega_C \\ \mathbf{H}_C \times \mathbf{n}_C = -\operatorname{grad} \widehat{\psi}_I \times \mathbf{n}_I & \\ \qquad - \sum_{q=1}^N \eta_{I,q} \boldsymbol{\lambda}_q \times \mathbf{n}_I - \mathbf{H}_{e,I} \times \mathbf{n}_I & \textit{on } \Gamma \end{cases} \tag{5.32}$$

$$\begin{cases} \operatorname{div}(\boldsymbol{\mu}_I \operatorname{grad} \widehat{\psi}_I) = -\operatorname{div}(\boldsymbol{\mu}_I \mathbf{H}_{e,I}) - \sum_{q=1}^N \eta_{I,q} \operatorname{div}(\boldsymbol{\mu}_I \boldsymbol{\lambda}_q) & \textit{in } \Omega_I \\ \boldsymbol{\mu}_I \operatorname{grad} \widehat{\psi}_I \cdot \mathbf{n}_I = -\boldsymbol{\mu}_C \mathbf{H}_C \cdot \mathbf{n}_C - \boldsymbol{\mu}_I \mathbf{H}_{e,I} \cdot \mathbf{n}_I & \\ \qquad - \sum_{q=1}^N \eta_{I,q} \boldsymbol{\mu}_I \boldsymbol{\lambda}_q \cdot \mathbf{n}_I & \textit{on } \Gamma \\ \widehat{\psi}_I = 0 & \textit{on } \partial\Omega \end{cases} \tag{5.33}$$

$$\begin{aligned} i\omega(\widehat{A}\boldsymbol{\eta}_I)_q = &-\textstyle\int_\Gamma [\boldsymbol{\sigma}^{-1}(\operatorname{curl} \mathbf{H}_C - \mathbf{J}_{e,C})] \times \mathbf{n}_C \cdot \boldsymbol{\lambda}_q \\ &-a_I(\mathbf{H}_{e,I}, \boldsymbol{\lambda}_q) - a_I(\operatorname{grad} \widehat{\psi}_I, \boldsymbol{\lambda}_q) \qquad \forall\, q = 1, \ldots, N \, , \end{aligned} \tag{5.34}$$

where the matrix $\widehat{A}$ *is defined as*

$$\widehat{A}_{pq} := \int_{\Omega_I} \boldsymbol{\mu}_I \boldsymbol{\lambda}_p \cdot \boldsymbol{\lambda}_q \ , \quad p, q = 1, \ldots, N \, . \tag{5.35}$$

Proof. Equations (5.32) and (5.33) follow by inserting the relations (5.17) and (5.19) in (5.26) and (5.27), respectively.

In order to obtain the last equation, we start by noting that

$$\begin{aligned} i\omega \textstyle\int_{\Omega_I} \boldsymbol{\mu}_I \mathbf{H}_I \cdot \operatorname{grad} g_q &= -i\omega \textstyle\int_{\Omega_I} \operatorname{div}(\boldsymbol{\mu}_I \mathbf{H}_I)\, g_q + i\omega \textstyle\int_\Gamma \boldsymbol{\mu}_I \mathbf{H}_I \cdot \mathbf{n}_I\, g_q \\ &= -i\omega \textstyle\int_\Gamma \boldsymbol{\mu}_C \mathbf{H}_C \cdot \mathbf{n}_C\, g_q \,, \end{aligned}$$

having used $(5.25)_3$ and $(5.25)_7$. Moreover, by integration by parts

$$\begin{aligned} &\textstyle\int_\Gamma [\boldsymbol{\sigma}^{-1}(\operatorname{curl} \mathbf{H}_C - \mathbf{J}_{e,C})] \times \mathbf{n}_C \cdot \operatorname{grad} g_q \\ &\quad = -\textstyle\int_\Gamma \operatorname{div}_\tau([\boldsymbol{\sigma}^{-1}(\operatorname{curl} \mathbf{H}_C - \mathbf{J}_{e,C})] \times \mathbf{n}_C)\, g_q \\ &\quad = -\textstyle\int_\Gamma \operatorname{curl}[\boldsymbol{\sigma}^{-1}(\operatorname{curl} \mathbf{H}_C - \mathbf{J}_{e,C})] \cdot \mathbf{n}_C\, g_q \\ &\quad = i\omega \textstyle\int_\Gamma \boldsymbol{\mu}_C \mathbf{H}_C \cdot \mathbf{n}_C\, g_q \,, \end{aligned}$$

having used $(5.25)_1$. Therefore, putting (5.17) in (5.30) we have found

$$i\omega \int_{\Omega_I} \boldsymbol{\mu}_I \mathbf{H}_I \cdot \boldsymbol{\lambda}_q = -\int_\Gamma [\boldsymbol{\sigma}^{-1}(\operatorname{curl} \mathbf{H}_C - \mathbf{J}_{e,C})] \times \mathbf{n}_C \cdot \boldsymbol{\lambda}_q \tag{5.36}$$

for each $q = 1, \ldots, N$.

From this relation and the representation formula (5.18) we obtain

$$\begin{aligned} i\omega(\widehat{A}\boldsymbol{\eta}_I)_q &= i\omega \textstyle\int_{\Omega_I} \boldsymbol{\mu}_I \sum_{p=1}^N \eta_{I,p} \boldsymbol{\lambda}_p \cdot \boldsymbol{\lambda}_q \\ &= i\omega \textstyle\int_{\Omega_I} \boldsymbol{\mu}_I (\mathbf{H}_I - \mathbf{H}_{e,I} - \operatorname{grad} \widehat{\psi}_I) \cdot \boldsymbol{\lambda}_q \\ &= -\textstyle\int_\Gamma [\boldsymbol{\sigma}^{-1}(\operatorname{curl} \mathbf{H}_C - \mathbf{J}_{e,C})] \times \mathbf{n}_C \cdot \boldsymbol{\lambda}_q \\ &\quad -i\omega \textstyle\int_{\Omega_I} \boldsymbol{\mu}_I \mathbf{H}_{e,I} \cdot \boldsymbol{\lambda}_q - i\omega \textstyle\int_{\Omega_I} \boldsymbol{\mu}_I \operatorname{grad} \widehat{\psi}_I \cdot \boldsymbol{\lambda}_q \,, \end{aligned} \tag{5.37}$$

the result we have to prove. □

It is easily seen that the matrix $\widehat{A}$ is symmetric and positive definite, as the vectors $\boldsymbol{\lambda}_q$ are linearly independent.

Remark 5.6. The scalar magnetic potential can be introduced also when considering the electric boundary condition $\mathbf{E} \times \mathbf{n} = \mathbf{0}$ on $\partial\Omega$ (see Alonso Rodríguez et al. [10]). We are not considering this formulation here; instead, for this boundary value problem we present in the next section a formulation in terms of a scalar magnetic potential in Ω_I and of the electric field $\mathbf{E}_C$ in Ω_C. □

5.2.1 A domain decomposition procedure

The strong formulations presented in Theorems 5.4 and 5.5 can suggest some iterative procedures for solving (5.26)–(5.28) and (5.32)–(5.34). To give an idea of the procedure, let us focus on the second case. The iteration-by-subdomain algorithm reads as follows: given $\mathbf{e}_\Gamma^{\rm old}$ on Γ, first solve

$$\begin{cases} \operatorname{div}(\boldsymbol{\mu}_I \operatorname{grad} \widehat{\psi}_I) + \sum_{q=1}^N \eta_{I,q} \operatorname{div}(\boldsymbol{\mu}_I \boldsymbol{\lambda}_q) = -\operatorname{div}(\boldsymbol{\mu}_I \mathbf{H}_{e,I}) & \text{in } \Omega_I \\ i\omega(\widehat{A}\boldsymbol{\eta}_I)_q + a_I(\operatorname{grad} \psi_I, \boldsymbol{\lambda}_q) & \\ \quad = -\int_\Gamma \mathbf{e}_\Gamma^{\rm old} \cdot \boldsymbol{\lambda}_q - a_I(\mathbf{H}_{e,I}, \boldsymbol{\lambda}_q) \qquad \forall\, q = 1, \ldots, N & \\ \boldsymbol{\mu}_I \operatorname{grad} \widehat{\psi}_I \cdot \mathbf{n}_I + \sum_{q=1}^N \eta_{I,q} \boldsymbol{\mu}_I \boldsymbol{\lambda}_q \cdot \mathbf{n}_I & \\ \quad = -i\omega^{-1} \operatorname{div}_\tau \mathbf{e}_\Gamma^{\rm old} - \boldsymbol{\mu}_I \mathbf{H}_{e,I} \cdot \mathbf{n}_I & \text{on } \Gamma \\ \widehat{\psi}_I = 0 & \text{on } \partial\Omega\,, \end{cases} \tag{5.38}$$

which is well-posed since its weak formulation is expressed in terms of the sesquilinear form $\widehat{a}_I(\cdot;\cdot)$, that is coercive in $H^1_{0,\partial\Omega}(\Omega_I) \times \mathbb{C}^N$. Then solve (5.32), namely

$$\begin{cases} \operatorname{curl}(\boldsymbol{\sigma}^{-1} \operatorname{curl} \mathbf{H}_C) + i\omega\boldsymbol{\mu}_C \mathbf{H}_C = \operatorname{curl}(\boldsymbol{\sigma}^{-1} \mathbf{J}_{e,C}) & \text{in } \Omega_C \\ \mathbf{H}_C \times \mathbf{n}_C = -\operatorname{grad} \widehat{\psi}_I \times \mathbf{n}_I & \\ \qquad - \sum_{q=1}^N \eta_{I,q} \boldsymbol{\lambda}_q \times \mathbf{n}_I - \mathbf{H}_{e,I} \times \mathbf{n}_I & \text{on } \Gamma\,. \end{cases} \tag{5.39}$$

Finally set

$$\mathbf{e}_\Gamma^{\rm new} = (1-\delta)\mathbf{e}_\Gamma^{\rm old} + \delta\, [\boldsymbol{\sigma}^{-1}(\operatorname{curl} \mathbf{H}_C - \mathbf{J}_{e,C})] \times \mathbf{n}_C \quad \text{on } \Gamma\,, \tag{5.40}$$

and iterate until convergence (here $\delta > 0$ is an acceleration parameter). Clearly, in the limit one has $\mathbf{e}_\Gamma^\infty = [\boldsymbol{\sigma}^{-1}(\operatorname{curl} \mathbf{H}_C - \mathbf{J}_{e,C})] \times \mathbf{n}_C = \mathbf{E}_C \times \mathbf{n}_C$ on Γ.

The possible advantage of this approach is that, when passing to the numerical approximation, the single reduced problems (5.38) and (5.39) can be easier to solve than the global problem (5.32)–(5.34). In fact, first of all they have a smaller size, and moreover they have a more classical nature (for a simple topological situation, namely, when $N = 0$, (5.38) is a mixed Dirichlet–Neumann problem for a uniformly elliptic operator, whereas (5.39) is a standard Dirichlet boundary value problem for a curl curl-like operator).

An analogous iterative scheme can be proposed for formulation (5.26)–(5.28). In this case, a further interesting point is that the problems (5.27) and (5.28) can be solved in parallel, as they are independent.

We will not dwell on these iterative schemes here, referring the interested reader to Alonso Rodríguez et al. [10]. However, in next section we will focus on an iteration-by-subdomain procedure for a problem where, instead of the magnetic field $\mathbf{H}_C$, the unknown in the conductor is the electric field $\mathbf{E}_C$.

5.3 The formulation in terms of $\mathbf{E}_C$ and ψ_I^*

In this section we want to present a formulation in terms of the electric field $\mathbf{E}_C$ in Ω_C and of a magnetic potential in Ω_I. For the sake of variety, we focus here on the electric boundary value problem, namely, we impose $\mathbf{E} \times \mathbf{n} = \mathbf{0}$ on $\partial\Omega$.

Let us start recalling the necessary conditions that have to be imposed to the current density $\mathbf{J}_e \in (L^2(\Omega))^3$ in this case

$$\operatorname{div} \mathbf{J}_{e,I} = 0 \quad \text{in } \Omega_I \tag{5.41}$$

$$\begin{aligned} &\textstyle\int_{\Gamma_j} \mathbf{J}_{e,I} \cdot \mathbf{n}_I = 0 \qquad \forall\, j = 1, \ldots, p_\Gamma \\ &\textstyle\int_{(\partial\Omega)_r} \mathbf{J}_{e,I} \cdot \mathbf{n} = 0 \qquad \forall\, r = 0, \ldots, p_{\partial\Omega} \,. \end{aligned} \tag{5.42}$$

The existence of a vector field $\mathbf{H}_{e,I}^* \in H(\mathrm{curl};\Omega_I)$ such that

$$\mathrm{curl}\, \mathbf{H}_{e,I}^* = \mathbf{J}_{e,I} \qquad \text{in } \Omega_I \tag{5.43}$$

is then ensured by well-known results (see, e.g., Alonso and Valli [6], Rem. 4.3).

In order to devise the formulation in terms of $\mathbf{E}_C$ and a scalar magnetic potential, the weak form of the eddy current problem in terms of $\mathbf{E}_C$ and $\mathbf{H}_I$ is the right starting point: adapting to the electric boundary condition what we have derived in Chapter 4, it reads

$$\begin{aligned} &\text{Find } (\mathbf{E}_C, \mathbf{H}_I) \in H(\mathrm{curl};\Omega_C) \times V_I^*(\mathbf{J}_{e,I}) : \\ &\textstyle\int_{\Omega_C} (\boldsymbol{\mu}_C^{-1}\, \mathrm{curl}\, \mathbf{E}_C \cdot \mathrm{curl}\, \overline{\mathbf{z}_C} + i\omega\boldsymbol{\sigma}\mathbf{E}_C \cdot \overline{\mathbf{z}_C}) \\ &\qquad\qquad -i\omega \textstyle\int_\Gamma \mathbf{H}_I \cdot \overline{\mathbf{z}_C} \times \mathbf{n}_C = -i\omega \int_{\Omega_C} \mathbf{J}_{e,C} \cdot \overline{\mathbf{z}_C} \\ &-i\omega \textstyle\int_\Gamma \mathbf{E}_C \times \mathbf{n}_C \cdot \overline{\mathbf{v}_I^*} + \omega^2 \int_{\Omega_I} \boldsymbol{\mu}_I \mathbf{H}_I \cdot \overline{\mathbf{v}_I^*} = 0 \\ &\text{for all } (\mathbf{z}_C, \mathbf{v}_I^*) \in H(\mathrm{curl};\Omega_C) \times V_I^*(\mathbf{0}) \,, \end{aligned} \tag{5.44}$$

where

$$V_I^*(\mathbf{J}_{e,I}) := \{\mathbf{v}_I^* \in H(\mathrm{curl};\Omega_I) \mid \mathrm{curl}\, \mathbf{v}_I^* = \mathbf{J}_{e,I} \text{ in } \Omega_I\} \,,$$

and similarly for $V_I^*(\mathbf{0})$.

Now we could use the decomposition result (A.26), but, as we have already explained, for numerical approximation it is better to consider a decomposition for which the harmonic fields $\boldsymbol{\rho}_{\alpha,I}^*$ are not employed. Therefore, as in the preceding section, let us suppose that Ω is a polyhedral domain, and that there is a triangulation $\mathcal{T}_{h^0}$ such that $\overline{\Omega} = \cup_{K \in \mathcal{T}_{h^0}} K$, where K is a tetrahedron or a parallelepiped and $h^0 > 0$ is the (fixed) mesh size. From Bossavit [59], Hiptmair [126], Gross and Kotiuga [115] we know that in Ω_I there exists a system of "cutting" surfaces Ξ_α^*, $\alpha = 1, \ldots, n_{\Omega_I}$, such that every function $\mathbf{z}_I \in H^0(\mathrm{curl};\Omega_I)$ restricted to $\Omega_I \setminus \cup_{\alpha=1}^{n_{\Omega_I}} \Xi_\alpha^*$ is the gradient of a function belonging to $H^1(\Omega_I \setminus \cup_{\alpha=1}^{n_{\Omega_I}} \Xi_\alpha^*)$. It is not restrictive to suppose that the triangulation $\mathcal{T}_{h^0}$ induces a triangulation on these surfaces. For $\alpha = 1, \ldots, n_{\Omega_I}$ let us denote by Π_α^* the piecewise-polynomial function taking value 1 at the nodes on one side of Ξ_α^*, say $(\Xi_\alpha^*)^+$, and 0 at all the other nodes (including those on $(\Xi_\alpha^*)^-$, the other side of Ξ_α^*). Set also

$$\boldsymbol{\lambda}_\alpha^* := \widetilde{\mathrm{grad}}\, \Pi_\alpha^* \quad \text{for } \alpha = 1, \ldots, n_{\Omega_I} \,, \tag{5.45}$$

where $\widetilde{\mathrm{grad}}\, \Pi_\alpha^*$ denotes the $(L^2(\Omega_I))^3$-extension of $\mathrm{grad}\, \Pi_\alpha^*$ computed in $\Omega_I \setminus \Xi_\alpha^*$. It is verified at once that for each $\alpha = 1, \ldots, n_{\Omega_I}$ the vector function $\boldsymbol{\lambda}_\alpha^*$ belongs to $(L^2(\Omega_I))^3$ and satisfies

$$\mathrm{curl}\, \boldsymbol{\lambda}_\alpha^* = \mathbf{0} \quad \text{in } \Omega_I \,.$$

Denoting by $g_\alpha^* \in H^1(\Omega_I)/\mathbb{C}$ the solution to

$$\begin{cases} \mathrm{div}(\boldsymbol{\mu}_I\, \mathrm{grad}\, g_\alpha^*) = \mathrm{div}(\boldsymbol{\mu}_I \boldsymbol{\lambda}_\alpha^*) & \text{in } \Omega_I \\ \boldsymbol{\mu}_I\, \mathrm{grad}\, g_\alpha^* \cdot \mathbf{n} = \boldsymbol{\mu}_I \boldsymbol{\lambda}_\alpha^* \cdot \mathbf{n} & \text{on } \Gamma \cup \partial\Omega \,, \end{cases} \tag{5.46}$$

one can easily check that the basis functions $\boldsymbol{\rho}^*_{\alpha,I}$ of the space $\mathcal{H}_{\mu_I}(m;\Omega_I)$ can be written as

$$\boldsymbol{\rho}^*_{\alpha,I} = \boldsymbol{\lambda}^*_\alpha - \operatorname{grad} g^*_\alpha \,. \tag{5.47}$$

In Theorem A.8 we have proved the following representation formula for the vector field $\mathbf{Z}^*_I = \mathbf{H}_I - \mathbf{H}^*_{e,I}$ (where $\mathbf{H}^*_{e,I}$ is defined in (5.43))

$$\mathbf{H}_I - \mathbf{H}^*_{e,I} = \mathbf{Z}^*_I = \operatorname{grad} \psi^*_I + \sum_{\alpha=1}^{n_{\Omega_I}} \eta^*_{I,\alpha} \boldsymbol{\rho}^*_{\alpha,I} \,. \tag{5.48}$$

Note that this formula simplifies if $n_{\Omega_I} = 0$. This means that Ω_I is simply-connected (in particular, that Ω_C is simply-connected).

Using (5.47) one finds

$$\begin{aligned} \mathbf{Z}^*_I &= \operatorname{grad} \psi^*_I + \textstyle\sum_{\alpha=1}^{n_{\Omega_I}} \eta^*_{I,\alpha}(\boldsymbol{\lambda}^*_\alpha - \operatorname{grad} g^*_\alpha) \\ &= \operatorname{grad} \widehat{\psi}^*_I + \textstyle\sum_{\alpha=1}^{n_{\Omega_I}} \eta^*_{I,\alpha} \boldsymbol{\lambda}^*_\alpha \,, \end{aligned} \tag{5.49}$$

having defined

$$\widehat{\psi}^*_I := \psi^*_I - \sum_{\alpha=1}^{n_{\Omega_I}} \eta^*_{I,\alpha} g^*_\alpha \,. \tag{5.50}$$

Similarly, each test function $\mathbf{v}^*_I \in V^*_I(\mathbf{0})$ can be written as $\mathbf{v}^*_I = \operatorname{grad} \widehat{\chi}^*_I + \sum_{\alpha=1}^{n_{\Omega_I}} \theta^*_{I,\alpha} \boldsymbol{\lambda}^*_\alpha$. Inserting these expressions for $\mathbf{H}_I = \mathbf{Z}^*_I + \mathbf{H}^*_{e,I}$ and $\mathbf{v}^*_I$ in (5.44), we find

$$\begin{aligned} &\textstyle\int_{\Omega_C} (\boldsymbol{\mu}_C^{-1} \operatorname{curl} \mathbf{E}_C \cdot \operatorname{curl} \overline{\mathbf{z}_C} + i\omega \boldsymbol{\sigma} \mathbf{E}_C \cdot \overline{\mathbf{z}_C}) \\ &\qquad -i\omega \textstyle\int_\Gamma \operatorname{grad} \widehat{\psi}^*_I \cdot \overline{\mathbf{z}_C} \times \mathbf{n}_C - i\omega \sum_{\alpha=1}^{n_{\Omega_I}} \eta^*_{I,\alpha} \int_\Gamma \boldsymbol{\lambda}^*_\alpha \cdot \overline{\mathbf{z}_C} \times \mathbf{n}_C \\ &\quad = -i\omega \textstyle\int_{\Omega_C} \mathbf{J}_{e,C} \cdot \overline{\mathbf{z}_C} + i\omega \int_\Gamma \mathbf{H}^*_{e,I} \cdot \overline{\mathbf{z}_C} \times \mathbf{n}_C \,, \end{aligned} \tag{5.51}$$

$$\begin{aligned} &-i\omega \textstyle\int_\Gamma \mathbf{E}_C \times \mathbf{n}_C \cdot \operatorname{grad} \overline{\widehat{\chi}^*_I} + \omega^2 \int_{\Omega_I} \boldsymbol{\mu}_I \operatorname{grad} \widehat{\psi}^*_I \cdot \operatorname{grad} \overline{\widehat{\chi}^*_I} \\ &\qquad + \omega^2 \textstyle\sum_{\alpha=1}^{n_{\Omega_I}} \eta^*_{I,\alpha} \int_{\Omega_I} \boldsymbol{\mu}_I \boldsymbol{\lambda}^*_\alpha \cdot \operatorname{grad} \overline{\widehat{\chi}^*_I} \\ &\quad = -\omega^2 \textstyle\int_{\Omega_I} \boldsymbol{\mu}_I \mathbf{H}^*_{e,I} \cdot \operatorname{grad} \overline{\widehat{\chi}^*_I}, \end{aligned} \tag{5.52}$$

and

$$\begin{aligned} &-i\omega \textstyle\sum_{\beta=1}^{n_{\Omega_I}} \overline{\theta^*_{I,\beta}} \int_\Gamma \mathbf{E}_C \times \mathbf{n}_C \cdot \boldsymbol{\lambda}^*_\beta + \omega^2 \sum_{\beta=1}^{n_{\Omega_I}} \overline{\theta^*_{I,\beta}} \int_{\Omega_I} \boldsymbol{\mu}_I \operatorname{grad} \widehat{\psi}^*_I \cdot \boldsymbol{\lambda}^*_\beta \\ &\qquad + \omega^2 \textstyle\sum_{\alpha,\beta=1}^{n_{\Omega_I}} \Big(\int_{\Omega_I} \boldsymbol{\mu}_I \boldsymbol{\lambda}^*_\alpha \cdot \boldsymbol{\lambda}^*_\beta \Big) \eta^*_{I,\alpha} \overline{\theta^*_{I,\beta}} \\ &\quad = -\omega^2 \textstyle\sum_{\beta=1}^{n_{\Omega_I}} \overline{\theta^*_{I,\beta}} \int_{\Omega_I} \boldsymbol{\mu}_I \mathbf{H}^*_{e,I} \cdot \boldsymbol{\lambda}^*_\beta \,. \end{aligned} \tag{5.53}$$

The variational space where we are looking for the solution is clearly

$$U^* := H(\operatorname{curl};\Omega_C) \times H^1(\Omega_I)/\mathbb{C} \times \mathbb{C}^{n_{\Omega_I}} \,. \tag{5.54}$$

This shows that we have no longer strong matching on the interface Γ, hence the choice of finite elements for numerical approximation will be easier than that we will propose for problems (5.13) and (5.21).

Setting

$$a_{e,C}(\mathbf{w}_C, \mathbf{z}_C) := \int_{\Omega_C} (\boldsymbol{\mu}_C^{-1} \operatorname{curl} \mathbf{w}_C \cdot \operatorname{curl} \overline{\mathbf{z}_C} + i\omega\boldsymbol{\sigma}\mathbf{w}_C \cdot \overline{\mathbf{z}_C}) , \tag{5.55}$$

the sesquilinear form at the left hand side of (5.51), (5.52) and (5.53) is given by

$$\begin{aligned}
&\mathcal{K}((\mathbf{w}_C, \widehat{\zeta}_I^*, \boldsymbol{\gamma}_I^*), (\mathbf{z}_C, \widehat{\chi}_I^*, \boldsymbol{\theta}_I^*)) \\
&\quad := a_{e,C}(\mathbf{w}_C, \mathbf{z}_C) - i\omega \textstyle\int_\Gamma \operatorname{grad} \widehat{\zeta}_I^* \cdot \overline{\mathbf{z}_C} \times \mathbf{n}_C \\
&\quad\quad - i\omega \textstyle\sum_{\alpha=1}^{n_{\Omega_I}} \gamma_{I,\alpha}^* \int_\Gamma \boldsymbol{\lambda}_\alpha^* \cdot \overline{\mathbf{z}_C} \times \mathbf{n}_C - i\omega \int_\Gamma \operatorname{grad} \overline{\widehat{\chi}_I^*} \cdot \mathbf{w}_C \times \mathbf{n}_C \\
&\quad\quad + \omega^2 \textstyle\int_{\Omega_I} \boldsymbol{\mu}_I \operatorname{grad} \widehat{\zeta}_I^* \cdot \operatorname{grad} \overline{\widehat{\chi}_I^*} + \omega^2 \sum_{\alpha=1}^{n_{\Omega_I}} \gamma_{I,\alpha}^* \int_{\Omega_I} \boldsymbol{\mu}_I \boldsymbol{\lambda}_\alpha^* \cdot \operatorname{grad} \overline{\widehat{\chi}_I^*} \\
&\quad\quad - i\omega \textstyle\sum_{\alpha=1}^{n_{\Omega_I}} \overline{\theta_{I,\alpha}^*} \int_\Gamma \boldsymbol{\lambda}_\alpha^* \cdot \mathbf{w}_C \times \mathbf{n}_C \\
&\quad\quad + \omega^2 \textstyle\sum_{\alpha=1}^{n_{\Omega_I}} \overline{\theta_{I,\alpha}^*} \int_{\Omega_I} \boldsymbol{\mu}_I \boldsymbol{\lambda}_\alpha^* \cdot \operatorname{grad} \widehat{\zeta}_I^* + \omega^2 [\widehat{A}^* \boldsymbol{\gamma}_I^*, \boldsymbol{\theta}_I^*] ,
\end{aligned} \tag{5.56}$$

where the matrix $\widehat{A}^*$ is defined as

$$\widehat{A}_{\alpha\beta}^* := \int_{\Omega_I} \boldsymbol{\mu}_I \boldsymbol{\lambda}_\alpha^* \cdot \boldsymbol{\lambda}_\beta^* \ , \ \ \alpha, \beta = 1, \ldots, n_{\Omega_I} \ , \tag{5.57}$$

and $[\cdot, \cdot]$ denotes the scalar product in $\mathbb{C}^{n_{\Omega_I}}$.

The weak formulation of the eddy current problem reads

$$\begin{aligned}
&\text{Find } (\mathbf{E}_C, \widehat{\psi}_I^*, \boldsymbol{\eta}_I^*) \in U^* \text{ such that} \\
&\mathcal{K}((\mathbf{E}_C, \widehat{\psi}_I^*, \boldsymbol{\eta}_I^*), (\mathbf{z}_C, \widehat{\chi}_I^*, \boldsymbol{\theta}_I^*)) \\
&\quad = -i\omega \textstyle\int_{\Omega_C} \mathbf{J}_{e,C} \cdot \overline{\mathbf{z}_C} + i\omega \int_\Gamma \mathbf{H}_{e,I}^* \cdot \overline{\mathbf{z}_C} \times \mathbf{n}_C \\
&\quad\quad - \omega^2 \textstyle\int_{\Omega_I} \boldsymbol{\mu}_I \mathbf{H}_{e,I}^* \cdot \operatorname{grad} \overline{\widehat{\chi}_I^*} \\
&\quad\quad - \omega^2 \textstyle\sum_{\alpha=1}^{n_{\Omega_I}} \overline{\theta_{I,\alpha}^*} \int_{\Omega_I} \boldsymbol{\mu}_I \mathbf{H}_{e,I}^* \cdot \boldsymbol{\lambda}_\alpha^* \\
&\text{for each } (\mathbf{z}_C, \widehat{\chi}_I^*, \boldsymbol{\theta}_I^*) \in U^* ,
\end{aligned} \tag{5.58}$$

The well-posedness of this weak problem is a consequence of Lax–Milgram lemma, as the following coerciveness result holds:

Proposition 5.7. *The sesquilinear form $\mathcal{K}(\cdot, \cdot)$ is coercive in U^*, namely, there exists a constant $\kappa_0 > 0$ such that*

$$\begin{aligned}
&|\mathcal{K}((\mathbf{z}_C, \widehat{\chi}_I^*, \boldsymbol{\theta}_I^*), (\mathbf{z}_C, \widehat{\chi}_I^*, \boldsymbol{\theta}_I^*))| \\
&\quad \geq \kappa_0 \big(\textstyle\int_{\Omega_C} (|\mathbf{z}_C|^2 + |\operatorname{curl} \mathbf{z}_C|^2) + \int_{\Omega_I} (|\widehat{\chi}_I^*|^2 + |\operatorname{grad} \widehat{\chi}_I^*|^2) + |\boldsymbol{\theta}_I^*|^2 \big)
\end{aligned}$$

for each $(\mathbf{z}_C, \widehat{\chi}_I^*, \boldsymbol{\theta}_I^*) \in U^*$ satisfying $\int_{\Omega_I} \widehat{\chi}_I^* = 0$.

Proof. We have already proved in Theorem 4.1 that the sesquilinear form $\mathcal{C}(\cdot,\cdot)$ at the right hand side of (5.44) is coercive in $H(\text{curl}; \Omega_C) \times V_I(\mathbf{0})$, where

$$V_I(\mathbf{0}) := \{\mathbf{v}_I \in H_{0,\partial\Omega}(\text{curl}; \Omega_I) \mid \text{curl}\, \mathbf{v}_I = \mathbf{0} \text{ in } \Omega_I\} \, .$$

However, the fact that $\mathbf{v}_I \in V_I(\mathbf{0})$ satisfies $\mathbf{v}_I \times \mathbf{n} = \mathbf{0}$ on $\partial\Omega$ does not play any role in the proof, hence the same result holds for $H(\text{curl}; \Omega_C) \times V_I^*(\mathbf{0})$. On the other hand, writing $\mathbf{v}_I^* = \text{grad}\, \widehat{\chi}_I^* + \sum_{\alpha=1}^{n_{\Omega_I}} \theta_{I,\alpha}^* \boldsymbol{\lambda}_\alpha^*$, we can repeat the proof of Lemma 5.2 and we obtain

$$\int_{\Omega_I} \boldsymbol{\mu}_I \mathbf{v}_I^* \cdot \overline{\mathbf{v}_I^*} \geq C_0 \Big(\int_{\Omega_I} |\, \text{grad}\, \widehat{\chi}_I^*|^2 + |\boldsymbol{\theta}_I^*|^2 \Big) \, .$$

From the Poincaré inequality (4.51) the proof is thus complete. □

Concerning the strong formulation, from the decomposition result (5.49) for the magnetic field $\mathbf{H}_I$ we easily find that $\mathbf{E}_C$, $\widehat{\psi}_I^*$ (determined up to an additive constant) and $\boldsymbol{\eta}_I^*$ satisfy the strong problem

$$\begin{cases} \text{curl}(\boldsymbol{\mu}_C^{-1} \text{curl}\, \mathbf{E}_C) + i\omega\boldsymbol{\sigma}\mathbf{E}_C = -i\omega \mathbf{J}_{e,C} & \text{in } \Omega_C \\ (\boldsymbol{\mu}_C^{-1} \text{curl}\, \mathbf{E}_C) \times \mathbf{n}_C = i\omega\, \text{grad}\, \widehat{\psi}_I^* \times \mathbf{n}_I & \\ \qquad\qquad + i\omega \sum_{\alpha=1}^{n_{\Omega_I}} \eta_{I,\alpha}^* \boldsymbol{\lambda}_\alpha^* \times \mathbf{n}_I + i\omega \mathbf{H}_{e,I}^* \times \mathbf{n}_I & \text{on } \Gamma \end{cases} \tag{5.59}$$

$$\begin{cases} \text{div}(\boldsymbol{\mu}_I\, \text{grad}\, \widehat{\psi}_I^*) = -\,\text{div}(\boldsymbol{\mu}_I \mathbf{H}_{e,I}^*) - \sum_{\alpha=1}^{n_{\Omega_I}} \eta_{I,\alpha}^*\, \text{div}(\boldsymbol{\mu}_I \boldsymbol{\lambda}_\alpha^*) & \text{in } \Omega_I \\ \boldsymbol{\mu}_I\, \text{grad}\, \widehat{\psi}_I^* \cdot \mathbf{n}_I = -i\omega^{-1}\, \text{curl}\, \mathbf{E}_C \cdot \mathbf{n}_C - \boldsymbol{\mu}_I \mathbf{H}_{e,I}^* \cdot \mathbf{n}_I & \\ \qquad\qquad - \sum_{\alpha=1}^{n_{\Omega_I}} \eta_{I,\alpha}^* \boldsymbol{\mu}_I \boldsymbol{\lambda}_\alpha^* \cdot \mathbf{n}_I & \text{on } \Gamma \\ \boldsymbol{\mu}_I\, \text{grad}\, \widehat{\psi}_I^* \cdot \mathbf{n} = -\boldsymbol{\mu}_I \mathbf{H}_{e,I}^* \cdot \mathbf{n} - \sum_{\alpha=1}^{n_{\Omega_I}} \eta_{I,\alpha}^* \boldsymbol{\mu}_I \boldsymbol{\lambda}_\alpha^* \cdot \mathbf{n} & \text{on } \partial\Omega \end{cases} \tag{5.60}$$

$$\begin{aligned} i\omega(\widehat{A}^* \boldsymbol{\eta}_I^*)_\beta = &-\textstyle\int_\Gamma \mathbf{E}_C \times \mathbf{n}_C \cdot \boldsymbol{\lambda}_\beta^* \\ &- a_I(\mathbf{H}_{e,I}^*, \boldsymbol{\lambda}_\beta^*) - a_I(\text{grad}\, \widehat{\psi}_I^*, \boldsymbol{\lambda}_\beta^*) \quad \forall\, \beta = 1, \ldots, n_{\Omega_I} \, . \end{aligned} \tag{5.61}$$

5.3.1 A domain decomposition procedure

Starting from (5.59)–(5.61) it is easy to devise an iteration-by-subdomain procedure for solving the eddy current problem (for this type of domain decomposition approach, in different contexts, see, e.g., Quarteroni and Valli [200]). Here we focus on the strong forms of the problems: the corresponding weak formulations can be easily determined.

Given $\mathbf{e}_\Gamma^{\text{old}}$ on Γ, find the solution $(\widehat{\psi}_I^*, \boldsymbol{\eta}_I^*)$ ($\widehat{\psi}_I^*$ determined up to an additive constant) of the coupled problem

$$\begin{cases} \text{div}(\boldsymbol{\mu}_I\, \text{grad}\, \widehat{\psi}_I^*) + \sum_{\alpha=1}^{n_{\Omega_I}} \eta_{I,\alpha}^*\, \text{div}(\boldsymbol{\mu}_I \boldsymbol{\lambda}_\alpha^*) = -\,\text{div}(\boldsymbol{\mu}_I \mathbf{H}_{e,I}^*) & \text{in } \Omega_I \\ a_I(\text{grad}\, \widehat{\psi}_I^*, \boldsymbol{\lambda}_\beta^*) + i\omega(\widehat{A}^* \boldsymbol{\eta}_I^*)_\beta & \\ \qquad = -\int_\Gamma \mathbf{e}_\Gamma^{\text{old}} \cdot \boldsymbol{\lambda}_\beta^* - a_I(\mathbf{H}_{e,I}^*, \boldsymbol{\lambda}_\beta^*) \quad \forall\, \beta = 1, \ldots, n_{\Omega_I} & \\ \boldsymbol{\mu}_I\, \text{grad}\, \widehat{\psi}_I^* \cdot \mathbf{n}_I + \sum_{\alpha=1}^{n_{\Omega_I}} \eta_{I,\alpha}^* \boldsymbol{\mu}_I \boldsymbol{\lambda}_\alpha^* \cdot \mathbf{n}_I & \\ \qquad = -i\omega^{-1}\, \text{div}_\tau\, \mathbf{e}_\Gamma^{\text{old}} - \boldsymbol{\mu}_I \mathbf{H}_{e,I}^* \cdot \mathbf{n}_I & \text{on } \Gamma \\ \boldsymbol{\mu}_I\, \text{grad}\, \widehat{\psi}_I^* \cdot \mathbf{n} + \sum_{\alpha=1}^{n_{\Omega_I}} \eta_{I,\alpha}^* \boldsymbol{\mu}_I \boldsymbol{\lambda}_\alpha^* \cdot \mathbf{n} = -\boldsymbol{\mu}_I \mathbf{H}_{e,I}^* \cdot \mathbf{n} & \text{on } \partial\Omega \, , \end{cases} \tag{5.62}$$

then solve

$$\begin{cases} \operatorname{curl}(\boldsymbol{\mu}_C^{-1} \operatorname{curl} \mathbf{E}_C) + i\omega\boldsymbol{\sigma}\mathbf{E}_C = -i\omega\mathbf{J}_{e,C} & \text{in } \Omega_C \\ (\boldsymbol{\mu}_C^{-1} \operatorname{curl} \mathbf{E}_C) \times \mathbf{n}_C = i\omega \operatorname{grad} \widehat{\psi}_I^* \times \mathbf{n}_I & \\ \qquad + i\omega \sum_{\alpha=1}^{n_{\Omega_I}} \eta_{I,\alpha}^* \boldsymbol{\lambda}_\alpha^* \times \mathbf{n}_I + i\omega\mathbf{H}_{e,I}^* \times \mathbf{n}_I & \text{on } \Gamma , \end{cases} \tag{5.63}$$

finally set

$$\mathbf{e}_\Gamma^{\text{new}} = (1-\delta)\mathbf{e}_\Gamma^{\text{old}} + \delta\, \mathbf{E}_C \times \mathbf{n}_C \quad \text{on } \Gamma , \tag{5.64}$$

and iterate until convergence (again, $\delta > 0$ is an acceleration parameter). In the limit one has $\mathbf{e}_\Gamma^\infty = \mathbf{E}_C \times \mathbf{n}_C$ on Γ. Note that the solvability of problem (5.62) is ensured from the fact that the sesquilinear form associated to it is coercive, as a consequence of Lemma 5.2 (for this, see the last part of the proof of Proposition 5.7). Moreover, problem (5.63) is a standard Neumann boundary value problem for a curl curl-like operator.

This algorithm has been analyzed by Alonso and Valli [7], in the case in which Ω and Ω_C are topologically simple (namely, $n_{\Omega_I} = 0$). There it is proved that the iterative procedure converges, provided that the parameter δ is chosen in a suitable interval $(0, \delta_0)$. Moreover, when considering a finite element approximation, the convergence is proved to be independent of the mesh size h. In the next section we present some numerical results concerned with this domain decomposition procedure.

As we have already noted when we have considered the formulation in terms of $\mathbf{H}_C$ and $\widehat{\psi}_I$, the advantage of this approach is that, at the numerical level, the reduced problems (5.62) and (5.63) can be easier to solve than the global coupled problem (5.59)–(5.61), as they have a smaller size and, moreover, they have a standard structure (at least, for a simple topological situation, that is, when $n_{\Omega_I} = 0$; when instead $n_{\Omega_I} \neq 0$, we can anyway see that (5.62) is a simple variant of a classical elliptic problem). Therefore, at each step we can employ our favourite solvers for elliptic problems, and, if the number of iterations is not too large, this procedure could be computationally more efficient than the one based on a direct discretization of (5.59)–(5.61).

5.4 Numerical approximation

The weak formulations of the problems presented in the preceding sections are based on sesquilinear forms that are coercive in a suitable Hilbert space, say, X. Therefore, the numerical approximation is somehow standard: choose an internal approximation of the space X, and estimate the error between the exact solution and the discrete solution by means of the Céa lemma.

However, a preliminary question to be faced is the determination of the vector fields $\mathbf{H}_{e,I}$ and $\mathbf{H}_{e,I}^*$ (see (5.3) and (5.43), respectively), that appear at the right hand side of the weak formulations, and that are also needed to find the right magnetic field.

5.4.1 The determination of a vector potential for the density current $\mathbf{J}_{e,I}$

A first obvious remark is that we have nothing to do if $\mathbf{J}_{e,I} = \mathbf{0}$ in Ω_I: for some real-life problems this is indeed the case (see, e.g., Sections 9.2 and 9.4). On the contrary, when $\mathbf{J}_{e,I} \neq \mathbf{0}$, we can use the Biot–Savart law and define for each $\mathbf{x} \in \Omega_I$

$$\begin{aligned} \mathbf{H}^*_{e,I}(\mathbf{x}) &:= \operatorname{curl}\Big(\int_{\Omega_I} \tfrac{1}{4\pi|\mathbf{x}-\mathbf{y}|} \mathbf{J}_{e,I}(\mathbf{y})\, d\mathbf{y} \Big) \\ &= \int_{\Omega_I} \tfrac{\mathbf{y}-\mathbf{x}}{4\pi|\mathbf{x}-\mathbf{y}|^3} \times \mathbf{J}_{e,I}(\mathbf{y})\, d\mathbf{y} \,. \end{aligned} \tag{5.65}$$

However, this has some flaws. In fact, one has

$$\begin{aligned} \operatorname{curl} \mathbf{H}^*_{e,I} &= \operatorname{curl}\operatorname{curl}\Big(\int_{\Omega_I} \tfrac{1}{4\pi|\mathbf{x}-\mathbf{y}|} \mathbf{J}_{e,I}(\mathbf{y})\, d\mathbf{y} \Big) \\ &= -\Delta\Big(\int_{\Omega_I} \tfrac{1}{4\pi|\mathbf{x}-\mathbf{y}|} \mathbf{J}_{e,I}(\mathbf{y})\, d\mathbf{y} \Big) \\ &\quad + \operatorname{grad}\operatorname{div}\Big(\int_{\Omega_I} \tfrac{1}{4\pi|\mathbf{x}-\mathbf{y}|} \mathbf{J}_{e,I}(\mathbf{y})\, d\mathbf{y} \Big) \,, \end{aligned}$$

hence $\operatorname{curl} \mathbf{H}^*_{e,I} = \mathbf{J}_{e,I}$ in Ω_I provided that

$$\operatorname{div}\Big(\int_{\Omega_I} \frac{1}{4\pi|\mathbf{x}-\mathbf{y}|} \mathbf{J}_{e,I}(\mathbf{y})\, d\mathbf{y} \Big) = 0 \qquad \text{in } \Omega_I \,.$$

Since $\operatorname{div} \mathbf{J}_{e,I} = 0$ in Ω_I, this is ensured when the conditions $\mathbf{J}_{e,I} \cdot \mathbf{n}_I = 0$ on Γ and $\mathbf{J}_{e,I} \cdot \mathbf{n} = 0$ on $\partial\Omega$ are satisfied, so that one can extend $\mathbf{J}_{e,I}$ by $\mathbf{0}$ outside Ω_I still keeping a divergence-free vector field. In this respect, note that $\mathbf{J}_{e,I} \cdot \mathbf{n}_I = 0$ on Γ is not a necessary condition for solvability of the eddy current problem, while $\mathbf{J}_{e,I} \cdot \mathbf{n} = 0$ on $\partial\Omega$ is a necessary condition only for the magnetic boundary value problem: therefore, for the sake of completeness, we have to take into account also the case in which these conditions are not satisfied. In this latter situation, one has to extend the vector field $\mathbf{J}_{e,I}$ to a domain $\Omega_I^\dagger$, as small as possible but containing $\overline{\Omega_I}$, in such a way that it is still divergence-free, but now with vanishing normal component on the boundary of $\Omega_I^\dagger$, and then use the Biot–Savart formula in $\Omega_I^\dagger$ for this extended current density field.

The extension of $\mathbf{J}_{e,I}$ can be achieved by taking the gradient of the solutions $\Phi_{\partial\Omega}$ and Φ_Γ of the Neumann problems

$$\begin{cases} \Delta\Phi_{\partial\Omega} = 0 & \text{in } \Omega_I^\dagger \setminus \overline{\Omega} \\ \operatorname{grad}\Phi_{\partial\Omega} \cdot \mathbf{n} = \mathbf{J}_{e,I} \cdot \mathbf{n} & \text{on } \partial\Omega \\ \operatorname{grad}\Phi_{\partial\Omega} \cdot \mathbf{n} = \mathbf{0} & \text{on } \partial\Omega_I^\dagger \cap (\mathbb{R}^3 \setminus \Omega) \end{cases}$$

$$\begin{cases} \Delta\Phi_\Gamma = 0 & \text{in } \Omega_I^\dagger \cap \Omega_C \\ \operatorname{grad}\Phi_\Gamma \cdot \mathbf{n}_I = \mathbf{J}_{e,I} \cdot \mathbf{n}_I & \text{on } \Gamma \\ \operatorname{grad}\Phi_\Gamma \cdot \mathbf{n} = \mathbf{0} & \text{on } \partial\Omega_I^\dagger \cap \Omega_C \,, \end{cases}$$

(solutions determined up to an additive constant in each connected component of $\Omega_I^\dagger \setminus \overline{\Omega}$ and $\Omega_I^\dagger \cap \Omega_C$, respectively), and at the discrete level this procedure has a small computational cost. Note that the solvability conditions for the Neumann problems here above are satisfied for both the magnetic and the electric boundary value problem, as follows from (5.1), $(5.2)_1$ and (5.41), (5.42), respectively.

For the electric boundary value problem we have thus obtained the vector field $\mathbf{H}^*_{e,I}$. Instead, for the magnetic boundary value problem we have not yet completed the construction, as the vector $\mathbf{H}_{e,I}$ has to satisfy not only $\operatorname{curl}\mathbf{H}_{e,I} = \mathbf{J}_{e,I}$ in Ω_I, but also $\mathbf{H}_{e,I} \times \mathbf{n} = \mathbf{0}$ on $\partial\Omega$, and this last condition does not hold for $\mathbf{H}^*_{e,I}$ defined as in (5.65).

Here we have some different alternatives: the first one could be interesting for practitioners, and consists in dropping the boundary condition $\mathbf{H}_{e,I} \times \mathbf{n} = \mathbf{0}$ on $\partial\Omega$, just keeping $\mathbf{H}^*_{e,I}$. Clearly, this does not introduce a too large error provided that $\mathbf{H}^*_{e,I} \times \mathbf{n}$ is small enough on $\partial\Omega$. Denoting by $S_{e,I}$ the support of $\mathbf{J}_{e,I}$, namely, the region where $\mathbf{J}_{e,I}$ is different from 0, and setting $l = \operatorname{dist}(S_{e,I}, \partial\Omega)$, from (5.65) we readily have

$$\left(\int_{\partial\Omega} |\mathbf{H}^*_{e,I} \times \mathbf{n}|^2\right)^{1/2} \leq \frac{1}{4\pi l^2} (\operatorname{meas}(\partial\Omega))^{1/2} \int_{S_{e,I}} |\mathbf{J}_{e,I}| \,,$$

hence this approach can be followed if the quantity at the right hand side is small enough.

Another possibility is to forget the Biot–Savart law (5.65) and to look for $\mathbf{H}_{e,I} = \operatorname{curl}\mathbf{q}_{e,I}$, where $\mathbf{q}_{e,I} \in H(\operatorname{curl};\Omega_I) \cap H(\operatorname{div};\Omega_I)$ is the solution to

$$\begin{cases} \operatorname{curl}\operatorname{curl}\mathbf{q}_{e,I} = \mathbf{J}_{e,I} & \text{in } \Omega_I \\ \operatorname{div}\mathbf{q}_{e,I} = 0 & \text{in } \Omega_I \\ \mathbf{q}_{e,I} \times \mathbf{n}_I = \mathbf{0} & \text{on } \Gamma \\ \mathbf{q}_{e,I} \cdot \mathbf{n} = 0 & \text{on } \partial\Omega \\ \operatorname{curl}\mathbf{q}_{e,I} \times \mathbf{n} = \mathbf{0} & \text{on } \partial\Omega \\ \mathbf{q}_{e,I} \perp \mathcal{H}(\Gamma, \partial\Omega; \Omega_I)\,. \end{cases} \tag{5.66}$$

This problem, which is similar to (A.14) or to (A.22), can be formulated as a penalization problem, by adding the term $-\operatorname{grad}\operatorname{div}\mathbf{q}_{e,I}$ at the left hand side of $(5.66)_1$, or else as a saddle point problem, where the constraint on the divergence is imposed by means of a Lagrange multiplier. Focusing on this latter choice, the numerical approximation of (5.66) is performed using edge elements for $\mathbf{q}_{e,I}$ and nodal elements for the Lagrange multiplier, and, though the whole procedure is standard, it has however a computational cost that is not negligible.

A third possibility consists in considering $\mathbf{H}^*_{e,I}$ defined as in (5.65), and writing $\mathbf{H}_I - \mathbf{H}^*_{e,I}$ in terms of the orthogonal decomposition (5.49), namely,

$$\mathbf{H}_I - \mathbf{H}^*_{e,I} = \operatorname{grad}\widehat{\psi}^*_I + \sum_{\alpha=1}^{n_{\Omega_I}} \eta^*_{I,\alpha}\boldsymbol{\lambda}^*_\alpha \,.$$

Note that, though $(\mathbf{H}_I - \mathbf{H}^*_{e,I}) \times \mathbf{n} \neq \mathbf{0}$ on $\partial\Omega$, the terms $\mathbf{Q}^*_I$ in (A.26) is vanishing, as $\operatorname{curl}(\mathbf{H}_I - \mathbf{H}^*_{e,I}) = \mathbf{0}$ in Ω_I.

Denoting by $\mathbf{H}^*_{e,C} \in H(\mathrm{curl}; \Omega_C)$ a vector field such that $\mathbf{H}^*_{e,C} \times \mathbf{n}_C + \mathbf{H}^*_{e,I} \times \mathbf{n}_I = \mathbf{0}$ on Γ, we are now looking for the solution to the following problem

$$
\begin{aligned}
&\text{Find } (\mathbf{Z}^*_C, \widehat{\psi}^*_I, \boldsymbol{\eta}^*_I) \in \widehat{W}^* \text{ such that} \\
&\mathrm{grad}\, \widehat{\psi}^*_I \times \mathbf{n} + \textstyle\sum_{\alpha=1}^{n_{\Omega_I}} \eta^*_{I,\alpha} \boldsymbol{\lambda}^*_\alpha \times \mathbf{n} = -\mathbf{H}^*_{e,I} \times \mathbf{n} \text{ on } \partial\Omega \\
&a_C(\mathbf{Z}^*_C, \mathbf{v}^*_C) + \widehat{a}^*_I(\widehat{\psi}^*_I, \boldsymbol{\eta}^*_I; \widehat{\chi}^*_I, \boldsymbol{\theta}^*_I) \\
&\qquad = -a_C(\mathbf{H}^*_{e,C}, \mathbf{v}^*_C) - a_I(\mathbf{H}^*_{e,I}, \mathrm{grad}\, \widehat{\chi}^*_I) \\
&\qquad\quad - a_I\Big(\mathbf{H}^*_{e,I}, \textstyle\sum_{\alpha=1}^{n_{\Omega_I}} \theta^*_{I,\alpha} \boldsymbol{\lambda}^*_\alpha\Big) + \int_{\Omega_C} \boldsymbol{\sigma}^{-1} \mathbf{J}_{e,C} \cdot \mathrm{curl}\, \overline{\mathbf{v}^*_C} \\
&\text{for each } (\mathbf{v}^*_C, \widehat{\chi}^*_I, \boldsymbol{\theta}^*_I) \in \widehat{W}^*_0 \,,
\end{aligned}
\tag{5.67}
$$

where we have introduced the spaces

$$
\begin{aligned}
\widehat{W}^* := \{&(\mathbf{v}^*_C, \widehat{\chi}^*_I, \boldsymbol{\theta}^*_I) \in H(\mathrm{curl}; \Omega_C) \times H^1(\Omega_I)/\mathbb{C} \times \mathbb{C}^{n_{\Omega_I}} \mid \\
&\mathbf{v}^*_C \times \mathbf{n}_C + \mathrm{grad}\, \widehat{\chi}^*_I \times \mathbf{n}_I + \textstyle\sum_{\alpha=1}^{n_{\Omega_I}} \theta^*_{I,\alpha} \boldsymbol{\lambda}^*_\alpha \times \mathbf{n}_I = \mathbf{0} \text{ on } \Gamma\}
\end{aligned}
\tag{5.68}
$$

$$
\begin{aligned}
\widehat{W}^*_0 := \{&(\mathbf{v}^*_C, \widehat{\chi}^*_I, \boldsymbol{\theta}^*_I) \in \widehat{W}^* \mid \\
&\mathrm{grad}\, \widehat{\chi}^*_I \times \mathbf{n} + \textstyle\sum_{\alpha=1}^{n_{\Omega_I}} \theta^*_{I,\alpha} \boldsymbol{\lambda}^*_\alpha \times \mathbf{n} = \mathbf{0} \text{ on } \partial\Omega\},
\end{aligned}
\tag{5.69}
$$

and the sesquilinear form

$$
\widehat{a}^*_I(\widehat{\zeta}^*_I, \boldsymbol{\gamma}^*_I; \widehat{\chi}^*_I, \boldsymbol{\theta}^*_I) := a_I\Big(\mathrm{grad}\, \widehat{\zeta}^*_I + \sum_{\alpha=1}^{n_{\Omega_I}} \gamma^*_{I,\alpha} \boldsymbol{\lambda}^*_\alpha, \mathrm{grad}\, \widehat{\chi}^*_I + \sum_{\alpha=1}^{n_{\Omega_I}} \theta^*_{I,\alpha} \boldsymbol{\lambda}^*_\alpha \Big) \,. \tag{5.70}
$$

With respect to (5.21), what is new here is that we have to satisfy the additional constraint $\mathrm{grad}\, \widehat{\psi}^*_I \times \mathbf{n} + \sum_{\alpha=1}^{n_{\Omega_I}} \eta^*_{I,\alpha} \boldsymbol{\lambda}^*_\alpha \times \mathbf{n} = -\mathbf{H}^*_{e,I} \times \mathbf{n}$ on $\partial\Omega$. In the next section, following Bermúdez et al. [42], we will describe an efficient way to devise a finite element approximation of this problem, based on the introduction of a Lagrange multiplier for taking into account the boundary condition on $\partial\Omega$.

Remark 5.8. A method for determining a suitable approximation of the vector fields $\mathbf{H}_{e,I}$ and $\mathbf{H}^*_{e,I}$ has been proposed by Webb and Forghani [241]. To be precise, an edge element vector field $\widehat{\mathbf{H}}_{I,h}$ is constructed, such that $\mathrm{curl}\, \widehat{\mathbf{H}}_{I,h}$ is equal to a suitable interpolant of the current density $\mathbf{J}_{e,I}$. In a simple topological configuration the proposed algorithm leads to a uniquely determined edge element. Instead, in the general case some degrees of freedom remain undefined: one for each non-bounding cycle contained in Ω_I. □

5.4.2 Finite element approximation

A finite element numerical approximation of (5.13), (5.21), (5.58) or (5.67) must be clearly based on (scalar) nodal elements in Ω_I and edge elements in Ω_C.

However, any approximation procedure starting from the weak formulation (5.13) has some drawbacks, as it is also necessary either compute explicitly $\omega_{q,I}$ (and in general this is not feasible), or to approximate them through the problems described in Section A.4, see (A.33) and (A.34), and this is rather expensive, especially for large N. Moreover, the matching condition on Γ, that is included in the definition of the space W, cannot be satisfied at the discrete level, as $\boldsymbol{\omega}_{q,I} \times \mathbf{n}_I$ is not a discrete function on Γ. Therefore one should consider the matching condition for a suitable interpolant (or projection) of $\boldsymbol{\omega}_{q,I}$ on Γ, thus resulting in a non-conforming finite element approximation.

Instead, the numerical approximation based on (5.21), (5.58) or (5.67) is much easier to implement, as the construction of each $\boldsymbol{\lambda}_q$ and $\boldsymbol{\lambda}^*_\alpha$ is straightforward, once the "cutting" surfaces are available; moreover, if the finite element triangulation $\mathcal{T}_h$ is a refinement of the triangulation $\mathcal{T}_{h^0}$ used in the construction of $\boldsymbol{\lambda}_q$, the matching condition on Γ can be imposed directly, as in this case $\boldsymbol{\lambda}_q \times \mathbf{n}$ is a discrete function also in any finer mesh.

Remark 5.9. It is worth noting that the determination of the vector fields $\boldsymbol{\lambda}_q$ and $\boldsymbol{\lambda}^*_\alpha$ requires the explicit construction of the "cutting" surfaces Ξ_l or Ξ^*_α, which can be troublesome in general topological configurations. Some algorithms have been proposed by Kotiuga [152], [153], [154], Leonard et al. [165], Gross and Kotiuga [114], Ren [207]. A detailed presentation of one of these algorithm can be found in Gross and Kotiuga [115], Chap. 6. An up-to-date review of the methods used for constructing "cutting" surfaces, together with the proposal of a new algorithm, are presented in Dłotko et al. [97]. □

In the sequel we focus on the numerical approximation of (5.21), (5.67) and (5.58), and we assume that Ω_I and Ω_C are polyhedrical domains, and that $\mathcal{T}_{I,h}$ and $\mathcal{T}_{C,h}$ are two regular families of triangulations of Ω_I and Ω_C, respectively; for the sake of simplicity, we suppose that each element K of $\mathcal{T}_{I,h}$ and $\mathcal{T}_{C,h}$ is a tetrahedron. When considering the numerical approximation of (5.21) and (5.67), we also assume that these triangulations match on Γ, and that they are a refinement of the triangulation $\mathcal{T}_{h^0}$ used in the construction of $\boldsymbol{\lambda}_q$.

In Ω_C we employ the Nédélec curl-conforming edge element space $N^k_{C,h}$, that is defined as

$$N^k_{C,h} := \{\mathbf{z}_{C,h} \in H(\mathrm{curl};\Omega_C) \mid \mathbf{z}_{C,h|K} \in R_k \quad \forall K \in \mathcal{T}_{C,h}\} ,$$

where $R_k := (\mathbb{P}_{k-1})^3 \oplus S_k$ and $S_k := \{\mathbf{p} \in (\widetilde{\mathbb{P}}_k)^3 \mid \mathbf{p}(\mathbf{x}) \cdot \mathbf{x} = 0\}$. Other choices would be possible, for instance the second family of curl-conforming finite element spaces introduced and analyzed by Nédélec [186] (for notation and a detailed description of these spaces see Section A.2). We also consider $N^k_{I,h}$, the correspondent space in Ω_I, and the trace space

$$\mathcal{X}_{\Gamma,h} := \{(\mathbf{v}_{I,h} \times \mathbf{n}_I)_{|\Gamma} \mid \mathbf{v}_{I,h} \in N^k_{I,h}\} = \{(\mathbf{v}_{C,h} \times \mathbf{n}_C)_{|\Gamma} \mid \mathbf{v}_{C,h} \in N^k_{C,h}\} .$$

Finally, we consider the space of scalar Lagrange nodal elements

$$L^k_{I,h} := \{\chi_{I,h} \in C^0(\Omega_I) \mid \chi_{I,h|K} \in \mathbb{P}_k \, \forall \, K \in \mathcal{T}_{I,h}\} .$$

Remark 5.10. Note that one has $\operatorname{grad}\chi_{I,h} \in N^k_{I,h}$ for all $\chi_{I,h} \in L^k_{I,h}$ and that $\boldsymbol{\lambda}_q \in N^k_{I,h}$ for $q = 1, \ldots, N$. Moreover, if $\mathbf{v}_{I,h} \in N^k_{I,h}$ is such that $\operatorname{curl}\mathbf{v}_{I,h} = \mathbf{0}$ in Ω_I and $\mathbf{v}_{I,h}\times\mathbf{n} = \mathbf{0}$ on $\partial\Omega$, from Theorem A.7 and (5.17) there exist $\widehat{\chi}_I \in H^1_{0,\partial\Omega}(\Omega_I)$ and $\boldsymbol{\theta}_I \in \mathbb{C}^N$ such that $\mathbf{v}_{I,h} = \operatorname{grad}\widehat{\chi}_I + \sum_{q=1}^N \theta_{I,q}\boldsymbol{\lambda}_q$. Since $\boldsymbol{\lambda}_q \in N^k_{I,h}$, then also $\operatorname{grad}\widehat{\chi}_I \in N^k_{I,h}$ and thus $\operatorname{grad}\widehat{\chi}_{I|K} \in R_k$ for each $K \in \mathcal{T}_{I,h}$. From Girault and Raviart [111], Chap. III, Lemma 5.5, it follows that $\widehat{\chi}_{I|K} \in \mathbb{P}_k$, hence $\widehat{\chi}_I \in L^k_{I,h} \cap H^1_{0,\partial\Omega}(\Omega_I)$. □

Let us now consider the space

$$\begin{aligned}\widehat{W}_h := \{(\mathbf{v}_{C,h}, \widehat{\chi}_{I,h}, \boldsymbol{\theta}_I) \in N^k_{C,h} \times L^k_{I,h} \times \mathbb{C}^N \,|\, \widehat{\chi}_{I,h|\partial\Omega} = 0,\\ \mathbf{v}_{C,h} \times \mathbf{n}_C + \operatorname{grad}\widehat{\chi}_{I,h} \times \mathbf{n}_I + \textstyle\sum_{q=1}^N \theta_{I,q}\,\boldsymbol{\lambda}_q \times \mathbf{n}_I = \mathbf{0} \text{ on } \Gamma\}\,.\end{aligned}$$

It is well-defined, since $\mathbf{v}_{C,h} \times \mathbf{n}_C$, $\operatorname{grad}\widehat{\chi}_{I,h} \times \mathbf{n}_I$ and $\sum_{q=1}^N \theta_{I,q}\,\boldsymbol{\lambda}_q \times \mathbf{n}_I$ are in the same space $\mathcal{X}_{\Gamma,h}$. The matching condition on the interface can be imposed by eliminating the degrees of freedom of $\mathbf{v}_{C,h}$ associated to the edges and faces on Γ in terms of those of $\operatorname{grad}\widehat{\chi}_{I,h} + \sum_{q=1}^N \theta_{I,q}\,\boldsymbol{\lambda}_q$. Moreover, $\widehat{W}_h$ is clearly contained in $\widehat{W}$.

Assuming that we have explicitly constructed $\mathbf{H}_{e,I}$ and $\mathbf{H}_{e,C}$, the numerical approximation of (5.21) reads

$$\begin{aligned}&\text{Find } (\mathbf{Z}_{C,h}, \widehat{\psi}_{I,h}, \boldsymbol{\eta}_{I,h}) \in \widehat{W}_h \text{ such that}\\ &\quad a_C(\mathbf{Z}_{C,h}, \mathbf{v}_{C,h}) + \widehat{a}_I(\widehat{\psi}_{I,h}, \boldsymbol{\eta}_{I,h}; \widehat{\chi}_{I,h}, \boldsymbol{\theta}_I)\\ &\qquad = -a_C(\mathbf{H}_{e,C}, \mathbf{v}_{C,h}) - a_I(\mathbf{H}_{e,I}, \operatorname{grad}\widehat{\chi}_{I,h})\\ &\qquad\quad -a_I\Big(\mathbf{H}_{e,I}, \textstyle\sum_{q=1}^N \theta_{I,q}\,\boldsymbol{\lambda}_q\Big) + \int_{\Omega_C} \boldsymbol{\sigma}^{-1}\mathbf{J}_{e,C} \cdot \operatorname{curl}\overline{\mathbf{v}_{C,h}}\\ &\text{for each } (\mathbf{v}_{C,h}, \widehat{\chi}_{I,h}, \boldsymbol{\theta}_I) \in \widehat{W}_h\,.\end{aligned} \tag{5.71}$$

The following theorem can be easily proved.

Theorem 5.11. *Assume that the families of triangulations $\mathcal{T}_{C,h}$ and $\mathcal{T}_{I,h}$ are obtained as a refinement of the coarse triangulation $\mathcal{T}_{h^0}$. The solution $(\mathbf{Z}_{C,h}, \widehat{\psi}_{I,h}, \boldsymbol{\eta}_{I,h})$ of problem (5.71) exists and is unique. If the solution of problem (5.21) satisfies $\mathbf{Z}_C \in H^r(\mathrm{curl}; \Omega_C)$ and $\widehat{\psi}_I \in H^{1+r}(\Omega_I)$ with $r > 1/2$, the following error estimate holds*

$$\begin{aligned}&\|\mathbf{Z}_C - \mathbf{Z}_{C,h}\|_{H(\mathrm{curl};\Omega_C)} + \|\widehat{\psi}_I - \widehat{\psi}_{I,h}\|_{1,\Omega_I} + |\boldsymbol{\eta}_I - \boldsymbol{\eta}_{I.h}|\\ &\quad \leq c_* h^{\min(r,k)}(\|\mathbf{Z}_C\|_{H^r(\mathrm{curl};\Omega_C)} + \|\widehat{\psi}_I\|_{1+r,\Omega_I})\,.\end{aligned} \tag{5.72}$$

Proof. Existence and uniqueness of the solution follow from the fact that the sesquilinear form $a_C(\cdot,\cdot) + \widehat{a}_I(\cdot;\cdot)$ is continuous and coercive in $\widehat{W}$.

Using the Céa lemma (see, e.g., Ciarlet [83], p. 104; Quarteroni and Valli [199], p. 137) one finds at once

$$\begin{aligned}&\|\mathbf{Z}_C - \mathbf{Z}_{C,h}\|_{H(\mathrm{curl};\Omega_C)} + \|\widehat{\psi}_I - \widehat{\psi}_{I,h}\|_{1,\Omega_I} + |\boldsymbol{\eta}_I - \boldsymbol{\eta}_{I.h}|\\ &\quad \leq c^*(\|\mathbf{Z}_C - \mathbf{v}_{C,h}\|_{H(\mathrm{curl};\Omega_C)} + \|\widehat{\psi}_I - \widehat{\chi}_{I,h}\|_{1,\Omega_I} + |\boldsymbol{\eta}_I - \boldsymbol{\theta}_I|)\end{aligned} \tag{5.73}$$

for any choice of $(\mathbf{v}_{C,h}, \widehat{\chi}_{I,h}, \boldsymbol{\theta}_I)$ in $\widehat{W}_h$.

We note that, since $\mathbf{Z}_I = \operatorname{grad}\widehat{\psi}_I + \sum_{q=1}^N \eta_{I,q}\boldsymbol{\lambda}_q$, the interpolant $\mathbf{r}_{I,h}(\mathbf{Z}_I) = \mathbf{r}_{I,h}(\operatorname{grad}\widehat{\psi}_I) + \sum_{q=1}^N \eta_{I,q}\boldsymbol{\lambda}_q$ is well-defined, and moreover

$$\mathbf{r}_{I,h}(\operatorname{grad}\widehat{\psi}_I) = \operatorname{grad}\pi_{I,h}(\widehat{\psi}_I)$$

holds, where $\pi_{I,h}$ is the interpolation operator in $L^k_{I,h}$ (see Monk [179], Theor. 5.49). Hence in (5.73) one can choose $\mathbf{v}_{C,h}$ the interpolant of $\mathbf{Z}_C$ in $N^k_{C,h}$, $\widehat{\chi}_{I,h}$ the interpolant of $\widehat{\psi}_I$ in $L^k_{I,h}$ and $\boldsymbol{\theta}_I = \boldsymbol{\eta}_I$, since with this choice the matching condition on Γ is satisfied. Thus, as a consequence of the corresponding interpolation estimates (see Alonso and Valli [9] and Section A.2) one readily finds (5.72). □

The difficulties that can arise in the construction of the vector field $\mathbf{H}_{e,I}$ have lead us to propose the weak formulation (5.67) for the magnetic boundary value problem. Now we are interested in its finite element approximation. As before, we employ the edge finite element spaces $N^k_{C,h}$ in Ω_C and $N^k_{I,h}$ in Ω_I, and the space of scalar nodal elements $L^k_{I,h}$ in Ω_I.

Since at the discrete level the boundary condition cannot be satisfied exactly, we need to consider an approximation of the boundary data $-\mathbf{H}^*_{e,I} \times \mathbf{n}$ in the finite dimensional space

$$\mathcal{Y}^0_{\partial\Omega,h} := \{(\mathbf{v}_{I,h} \times \mathbf{n})_{|\partial\Omega} \mid \mathbf{v}_{I,h} \in N^k_{I,h} \cap H^0(\operatorname{curl};\Omega_I)\}\,.$$

Proceeding as in Remark 5.10 it is easy to see that each function $\mathbf{v}_{I,h} \in N^k_{I,h}$ such that $\operatorname{curl}\mathbf{v}_{I,h} = \mathbf{0}$ in Ω_I can be decomposed as

$$\mathbf{v}_{I,h} = \operatorname{grad}\widehat{\chi}^*_{I,h} + \sum_{\alpha=1}^{n_{\Omega_I}} \theta^*_{I,\alpha}\boldsymbol{\lambda}^*_\alpha\,, \tag{5.74}$$

for some $\widehat{\chi}^*_{I,h} \in L^k_{I,h}$ and $\boldsymbol{\theta}^*_I \in \mathbb{C}^{n_{\Omega_I}}$. Let us set

$$L^k_{\partial\Omega,h} := \{\nu_h \in C^0(\partial\Omega) \mid \nu_{h|T} \in \mathbb{P}_k \;\forall\, K \in \mathcal{T}_{\partial\Omega,h}\}\,,$$

where $\mathcal{T}_{\partial\Omega,h}$ is the triangulation induced by $\mathcal{T}_{I,h}$ on $\partial\Omega$. Then by (5.74) $\boldsymbol{\xi}_h \in \mathcal{Y}^0_{\partial\Omega,h}$ if and only if $\mathbf{n} \times \boldsymbol{\xi}_h = \operatorname{grad}_\tau \nu_h + \sum_{\alpha=1}^{n_{\Omega_I}} \theta^*_{I,\alpha}\mathbf{n} \times \boldsymbol{\lambda}^*_\alpha \times \mathbf{n}$ for some $\nu_h \in L^k_{\partial\Omega,h}$ and $\boldsymbol{\theta}^*_I \in \mathbb{C}^{n_{\Omega_I}}$ (see Section A.1 for the definition of the operator grad_τ in $H^{1/2}(\partial\Omega)$).

From well-known results of potential theory (see, e.g., Dautray and Lions [93], Chap. II, Sect. 3), since $\mathbf{J}_{e,I} \in (L^2(\Omega_I))^3$ (and also its extension belongs to $(L^2(\Omega^\dagger_I))^3$), then $\mathbf{H}^*_{e,I} \in (H^1(\Omega_I))^3$. Moreover $\operatorname{curl}\mathbf{H}^*_{e,I} = \mathbf{J}_{e,I}$, hence assuming that $\mathbf{J}_{e,I} \in (L^p(\Omega))^3$ for some $p > 2$ the interpolant $\mathbf{r}_{I,h}(\mathbf{H}^*_{e,I})$ is well-defined (see Monk [179], Lemma 5.38). Our aim is to choose the value $-\left(\mathbf{r}_{I,h}(\mathbf{H}^*_{e,I}) \times \mathbf{n}\right)_{|\partial\Omega}$ as discrete boundary datum belonging to $\mathcal{Y}^0_{\partial\Omega,h}$.

In this respect, let us note that, if there exists a function $\mathbf{G}_I \in (H^{1/2+\delta}(\Omega_I))^3$, $\delta > 0$, such that $\operatorname{curl}\mathbf{G}_I = \mathbf{J}_{e,I}$ in Ω_I and $\mathbf{G}_I \times \mathbf{n} = \mathbf{0}$ on $\partial\Omega$, then it is easy to see that $-\left(\mathbf{r}_{I,h}(\mathbf{H}^*_{e,I}) \times \mathbf{n}\right)_{|\partial\Omega}$ is an element of $\mathcal{Y}^0_{\partial\Omega,h}$. In fact, one readily verifies that

$\mathbf{r}_{I,h}(\mathbf{G}_I - \mathbf{H}^*_{e,I}) \in N^k_{I,h}$, $\mathrm{curl}(\mathbf{r}_{I,h}(\mathbf{G}_I - \mathbf{H}^*_{e,I})) = \mathbf{0}$ in Ω_I and $\mathbf{r}_{I,h}(\mathbf{G}_I) \times \mathbf{n} = \mathbf{0}$ on $\partial\Omega$. In the following we assume that such a function $\mathbf{G}_I$ exists (for instance, it is enough to know that the solution $\mathbf{H}_I$ belongs to $(H^{1/2+\delta}(\Omega_I))^3$).

We thus consider the following discrete problem

$$
\begin{aligned}
&\text{Find } (\mathbf{Z}^*_{C,h}, \widehat{\psi}^*_{I,h}, \boldsymbol{\eta}^*_{I,h}) \in \widehat{W}^*_h \text{ such that}\\
&\quad \mathrm{grad}\, \widehat{\psi}^*_{I,h} \times \mathbf{n} + \textstyle\sum_{\alpha=1}^{n_{\Omega_I}} (\eta^*_{I,h})_\alpha \boldsymbol{\lambda}^*_\alpha \times \mathbf{n} = -\mathbf{r}_{I,h}(\mathbf{H}^*_{e,I}) \times \mathbf{n} \text{ on } \partial\Omega\\
&\quad a_C(\mathbf{Z}^*_{C,h}, \mathbf{v}^*_{C,h}) + \widehat{a}^*_I(\widehat{\psi}^*_{I,h}, \boldsymbol{\eta}^*_{I,h}; \widehat{\chi}^*_{I,h}, \boldsymbol{\theta}^*_I)\\
&\qquad = -a_C(\mathbf{H}^*_{e,C}, \mathbf{v}^*_{C,h}) - a_I(\mathbf{H}^*_{e,I}, \mathrm{grad}\, \widehat{\chi}^*_{I,h})\\
&\qquad\quad - a_I\left(\mathbf{H}^*_{e,I}, \textstyle\sum_{\alpha=1}^{n_{\Omega_I}} \theta^*_{I,\alpha} \boldsymbol{\lambda}^*_\alpha\right) + \int_{\Omega_C} \boldsymbol{\sigma}^{-1} \mathbf{J}_{e,C} \cdot \mathrm{curl}\, \overline{\mathbf{v}^*_{C,h}}\\
&\text{for each } (\mathbf{v}^*_{C,h}, \widehat{\chi}^*_{I,h}, \boldsymbol{\theta}^*_I) \in \widehat{W}^*_{0,h}\,,
\end{aligned}
\tag{5.75}
$$

where

$$
\begin{aligned}
\widehat{W}^*_h := \{(\mathbf{v}^*_{C,h}, \widehat{\chi}^*_{I,h}, \boldsymbol{\theta}^*_I) \in N^k_{C,h} \times L^k_{I,h}/\mathbb{C} \times \mathbb{C}^{n_{\Omega_I}} \mid\\
\mathbf{v}^*_{C,h} \times \mathbf{n}_C + \mathrm{grad}\, \widehat{\chi}^*_{I,h} \times \mathbf{n}_I + \textstyle\sum_{\alpha=1}^{n_{\Omega_I}} \theta^*_{I,\alpha} \boldsymbol{\lambda}^*_\alpha \times \mathbf{n}_I = \mathbf{0} \text{ on } \Gamma\},
\end{aligned}
$$

and

$$
\widehat{W}^*_{0,h} := \{(\mathbf{v}^*_{C,h}, \widehat{\chi}^*_{I,h}, \boldsymbol{\theta}^*_I) \in \widehat{W}^*_h \mid \mathrm{grad}\, \widehat{\chi}^*_{I,h} \times \mathbf{n} + \sum_{\alpha=1}^{n_{\Omega_I}} \theta^*_{I,\alpha} \boldsymbol{\lambda}^*_\alpha \times \mathbf{n} = \mathbf{0} \text{ on } \partial\Omega\}\,.
$$

It is straightforward to obtain the following result.

Theorem 5.12. *Assume that the families of triangulations $\mathcal{T}_{C,h}$ and $\mathcal{T}_{I,h}$ are obtained as a refinement of the coarse triangulation $\mathcal{T}_{h^0}$ and that $\mathbf{r}_{I,h}(\mathbf{H}^*_{e,I}) \times \mathbf{n} \in \mathcal{Y}^0_{\partial\Omega,h}$. Then there exists a unique solution $(\mathbf{Z}^*_{C,h}, \widehat{\psi}^*_{I,h}, \boldsymbol{\eta}^*_{I,h})$ to problem* (5.75). *Moreover, if the solution to problem* (5.67) *satisfies $\mathbf{Z}^*_C \in H^r(\mathrm{curl}; \Omega_C)$ and $\widehat{\psi}^*_I \in H^{1+r}(\Omega_I)$ with $r > 1/2$, the following error estimate holds*

$$
\begin{aligned}
&\|\mathbf{Z}^*_C - \mathbf{Z}^*_{C,h}\|_{H(\mathrm{curl};\Omega_C)} + \|\widehat{\psi}^*_I - \widehat{\psi}^*_{I,h}\|_{H^1(\Omega_I)/\mathbb{C}} + |\boldsymbol{\eta}^*_I - \boldsymbol{\eta}^*_{I.h}|\\
&\quad \le c_* h^{\min(r,k)} (\|\mathbf{Z}^*_C\|_{H^r(\mathrm{curl};\Omega_C)} + \|\,\mathrm{grad}\, \widehat{\psi}^*_I\|_{r,\Omega_I})\,.
\end{aligned}
\tag{5.76}
$$

Proof. The sesquilinear form $a_C(\cdot,\cdot) + \widehat{a}^*_I(\cdot,\cdot)$ is clearly continuous in $\widehat{W}^*$, and proceeding as in Proposition 5.7 it is easily seen that it is also coercive in $\widehat{W}^*$. Since $\widehat{W}^*_{0,h} \subset \widehat{W}^*_0 \subset \widehat{W}^*$ and $\mathbf{r}_{I,h}(\mathbf{H}^*_{e,I}) \times \mathbf{n} \in \mathcal{Y}^0_{\partial\Omega,h}$, well-posedness of the discrete problem follows at once. Moreover, by using Céa lemma we find that

$$
\begin{aligned}
&\|\mathbf{Z}^*_C - \mathbf{Z}^*_{C,h}\|_{H(\mathrm{curl};\Omega_C)} + \|\widehat{\psi}^*_I - \widehat{\psi}^*_{I,h}\|_{H^1(\Omega_I)/\mathbb{C}} + |\boldsymbol{\eta}^*_I - \boldsymbol{\eta}^*_{I,h}|\\
&\quad \le c^*(\|\mathbf{Z}^*_C - \mathbf{v}^*_{C,h}\|_{H(\mathrm{curl};\Omega_C)} + \|\widehat{\psi}^*_I - \widehat{\chi}^*_{I,h}\|_{H^1(\Omega_I)/\mathbb{C}} + |\boldsymbol{\eta}^*_I - \boldsymbol{\theta}^*_I|)
\end{aligned}
$$

for any choice of test functions $(\mathbf{v}^*_{C,h}, \widehat{\chi}^*_{I,h}, \boldsymbol{\theta}^*_I) \in \widehat{W}^*_0$ such that $\mathrm{grad}\, \widehat{\chi}^*_{I,h} \times \mathbf{n} + \sum_{\alpha=1}^{n_{\Omega_I}} \theta^*_{I,\alpha} \boldsymbol{\lambda}^*_\alpha \times \mathbf{n} = -\mathbf{r}_{I,h}(\mathbf{H}^*_{e,I}) \times \mathbf{n}$ on $\partial\Omega$. Since $\mathbf{H}_I = \mathbf{H}^*_{e,I} + \mathrm{grad}\, \widehat{\psi}^*_I +$

$\sum_{\alpha=1}^{n_{\Omega_I}} \eta^*_{I,\alpha} \boldsymbol{\lambda}^*_\alpha$, we can choose $\mathbf{v}^*_{C,h} = \mathbf{r}_{C,h}(\mathbf{Z}^*_C)$, grad $\widehat{\chi}^*_{I,h} = \mathbf{r}_{I,h}(\text{grad}\,\widehat{\psi}^*_I)$ and $\boldsymbol{\theta}^*_I = \boldsymbol{\eta}^*_I$, where $\mathbf{r}_{C,h}$ and $\mathbf{r}_{I,h}$ are the interpolation operators in $N^k_{C,h}$ and $N^k_{I,h}$, respectively; therefore, employing the standard interpolation results, we obtain at once the error estimate (5.76). □

To complete our analysis, let us specify how to impose the non-homogeneous magnetic boundary condition. First, we note now that $n_{\Omega_I} \geq n_\Gamma$, n_Γ being the number of $\partial\Omega$-independent non-bounding cycles in Ω_I. Therefore, we can arrange the independent non-bounding cycles in Ω_I in such a way that the $\partial\Omega$-independent non-bounding cycles are numbered with α from 1 to n_Γ.

Let us also *assume* that the system of "cutting" surfaces Ξ^*_α is such that $\partial\Xi^*_\alpha \subset \Gamma$ for $\alpha = 1, \ldots, n_\Gamma$ and $\partial\Xi^*_\alpha \subset \partial\Omega$ for $\alpha = n_\Gamma + 1, \ldots, n_{\Omega_I}$; note that this is not true, for instance, in the case in which Ω and Ω_C are two co-axial tori. Then, for $\alpha = n_\Gamma + 1, \ldots, n_{\Omega_I}$, the coefficient $(\eta^*_{I,h})_\alpha$ can be computed in advance from the data of the problem. In fact, denoting by $\mathbf{Z}^*_{I,h} := \text{grad}\,\widehat{\psi}^*_{I,h} + \sum_{\alpha=1}^{n_{\Omega_I}} (\eta^*_{I,h})_\alpha \, \boldsymbol{\lambda}^*_\alpha$, it is easy to see that

$$(\eta^*_{I,h})_\alpha = \int_{\varphi_\alpha} \mathbf{Z}^*_{I,h} \cdot d\boldsymbol{\tau} \,,$$

hence for $\alpha = n_\Gamma + 1, \ldots, n_{\Omega_I}$ one has

$$(\eta^*_{I,h})_\alpha = -\int_{\varphi_\alpha} (\mathbf{n} \times \mathbf{r}_{I,h}(\mathbf{H}^*_{e,I}) \times \mathbf{n}) \cdot d\boldsymbol{\tau} \,. \tag{5.77}$$

Moreover, for $\alpha = 1, \ldots, n_\Gamma$, it holds $\boldsymbol{\lambda}^*_\alpha \times \mathbf{n} = \mathbf{0}$ on $\partial\Omega$, hence $(5.75)_1$ is in fact a boundary condition for grad $\widehat{\psi}^*_{I,h}$

$$\text{grad}\,\widehat{\psi}^*_{I,h} \times \mathbf{n} = -\left(\mathbf{r}_{I,h}(\mathbf{H}^*_{e,I}) + \sum_{\alpha=n_\Gamma+1}^{n_{\Omega_I}} (\eta^*_{I,h})_\alpha \boldsymbol{\lambda}^*_\alpha \right) \times \mathbf{n} \,, \tag{5.78}$$

with $(\eta^*_{I,h})_\alpha$ obtained in (5.77).

An efficient procedure for imposing this boundary condition has been proposed by Bermúdez et al. [42] and it is based on the use of Lagrange multipliers. Since $\mathbf{r}_{I,h}(\mathbf{H}^*_{e,I}) \times \mathbf{n} \in \mathcal{Y}^0_{\partial\Omega,h}$, $(\eta^*_{I,h})_\alpha = -\int_{\varphi_\alpha} (\mathbf{n} \times \mathbf{r}_{I,h}(\mathbf{H}^*_{e,I}) \times \mathbf{n}) \cdot d\boldsymbol{\tau}$ for $\alpha = n_\Gamma + 1, \ldots, n_{\Omega_I}$ and $\boldsymbol{\lambda}^*_\alpha \times \mathbf{n} = \mathbf{0}$ on $\partial\Omega$ for $\alpha = 1, \ldots, n_\Gamma$, it follows that $\mathbf{n} \times \left(\mathbf{r}_{I,h}(\mathbf{H}^*_{e,I}) + \sum_{\alpha=n_\Gamma+1}^{n_{\Omega_I}} (\eta^*_{I,h})_\alpha \boldsymbol{\lambda}^*_\alpha \right) \times \mathbf{n} = \text{grad}_\tau \eta_h$ for some $\eta_h \in L^k_{\partial\Omega,h}$. Hence it is clear that (5.78) holds if and only if

$$\int_{\partial\Omega} \mathbf{n} \times \text{grad}\,\widehat{\psi}^*_{I,h} \times \mathbf{n} \cdot \text{grad}_\tau \overline{\nu_h} = -\int_{\partial\Omega} \text{grad}_\tau \eta_h \cdot \text{grad}_\tau \overline{\nu_h} \,,$$

for all $\nu_h \in L^k_{\partial\Omega,h}/\mathbb{C}$.

Therefore we consider the following problem

$$
\begin{aligned}
&\text{Find } (\mathbf{Z}^*_{C,h}, \widehat{\psi}^*_{I,h}, \boldsymbol{\eta}^*_{I,h}) \in \widehat{W}^*_h, \kappa_h \in L^k_{\partial\Omega,h}/\mathbb{C} \text{ such that}\\
&\quad (\eta^*_{I,h})_\alpha = -\textstyle\int_{\varphi_\alpha} \big(\mathbf{n} \times \mathbf{r}_{I,h}(\mathbf{H}^*_{e,I}) \times \mathbf{n}\big) \cdot d\boldsymbol{\tau}\\
&\qquad\qquad\qquad\qquad\qquad \text{for } \alpha = n_\Gamma + 1, \dots, n_{\Omega_I}\\
&\quad a_C(\mathbf{Z}^*_{C,h}, \mathbf{v}^*_{C,h}) + \widehat{a}^*_I(\widehat{\psi}^*_{I,h}, \boldsymbol{\eta}^*_{I,h}; \widehat{\chi}^*_{I,h}, \boldsymbol{\theta}^*_{I,h})\\
&\qquad + \textstyle\int_{\partial\Omega} \operatorname{grad}_\tau \kappa_h \cdot \operatorname{grad}_\tau \overline{\widehat{\chi}^*_{I,h}}\\
&\qquad = -a_C(\mathbf{H}^*_{e,C}, \mathbf{v}^*_{C,h}) - a_I(\mathbf{H}^*_{e,I}, \operatorname{grad} \widehat{\chi}^*_{I,h})\\
&\qquad\quad - a_I\left(\mathbf{H}^*_{e,I}, \textstyle\sum_{\alpha=1}^{n_{\Omega_I}} \theta^*_{I,\alpha} \boldsymbol{\lambda}^*_\alpha\right) + \textstyle\int_{\Omega_C} \boldsymbol{\sigma}^{-1} \mathbf{J}_{e,C} \cdot \operatorname{curl} \overline{\mathbf{v}^*_{C,h}}\\
&\quad \textstyle\int_{\partial\Omega} \operatorname{grad}_\tau \widehat{\psi}^*_{I,h} \cdot \operatorname{grad}_\tau \overline{\nu_h}\\
&\qquad = -\textstyle\int_{\partial\Omega} \mathbf{n} \times \mathbf{r}_{I,h}(\mathbf{H}^*_{e,I}) \times \mathbf{n} \cdot \operatorname{grad}_\tau \overline{\nu_h}\\
&\qquad\quad - \textstyle\int_{\partial\Omega} \mathbf{n} \times \sum_{\alpha=n_\Gamma+1}^{n_{\Omega_I}} (\eta^*_{I,h})_\alpha \boldsymbol{\lambda}^*_\alpha \times \mathbf{n} \cdot \operatorname{grad}_\tau \overline{\nu_h}\\
&\text{for each } (\mathbf{v}^*_{C,h}, \widehat{\chi}^*_{I,h}, \boldsymbol{\theta}^*_I) \in \widehat{W}^*_{0,h}, \nu_h \in L^k_{\partial\Omega,h}/\mathbb{C}\,.
\end{aligned}
\tag{5.79}
$$

The following theorem shows that this problem is well-posed and that it is equivalent to problem (5.75).

Theorem 5.13. *Let the assumptions of Theorem 5.12 be satisfied and let the system of "cutting" surfaces Ξ^*_α be such that $\partial\Xi^*_\alpha \subset \Gamma$ for $\alpha = 1, \dots, n_\Gamma$ and $\partial\Xi^*_\alpha \subset \partial\Omega$ for $\alpha = n_\Gamma + 1, \dots, n_{\Omega_I}$. Under these assumptions problem (5.79) has a unique solution. Moreover if we consider the unique solution $(\mathbf{Z}^*_{C,h}, \widehat{\psi}^*_{I,h}, \boldsymbol{\eta}^*_{I,h}) \in \widehat{W}^*_h$ of problem (5.75), then $(\mathbf{Z}^*_{C,h}, \widehat{\psi}^*_{I,h}, \boldsymbol{\eta}^*_{I,h}, 0) \in \widehat{W}^*_h \times L^k_{\partial\Omega,h}/\mathbb{C}$ is the solution of (5.79).*

Proof. To prove the existence and uniqueness of the solution it is enough to show that the homogeneous problem has only the trivial solution. Assuming that $\mathbf{H}^*_{e,I} = \mathbf{0}$, $\mathbf{H}^*_{e,C} = \mathbf{0}$ and $\mathbf{J}_{e,C} = \mathbf{0}$, from $(5.79)_1$ it follows that $(\eta^*_{I,h})_\alpha = 0$ for $\alpha = n_\Gamma + 1, \dots, n_{\Omega_I}$ and from $(5.79)_3$ we see that $\operatorname{grad}_\tau \widehat{\psi}^*_{I,h} = \mathbf{0}$ on $\partial\Omega$. Taking in $(5.79)_2$ $(\mathbf{Z}^*_{C,h}, \widehat{\psi}^*_{I,h}, \boldsymbol{\eta}^*_{I,h})$ as test function, from the coerciveness of the sesquilinear form $a_C(\cdot,\cdot) + \widehat{a}^*_I(\cdot,\cdot)$ we also have $\mathbf{Z}^*_{C,h} = \mathbf{0}$ in Ω_C and $\operatorname{grad} \widehat{\psi}^*_{I,h} + \sum_{\alpha=1}^{n_{\Omega_I}} (\eta^*_{I,h})_\alpha \boldsymbol{\lambda}^*_\alpha = \operatorname{grad} \widehat{\psi}^*_{I,h} + \sum_{\alpha=1}^{n_\Gamma} (\eta^*_{I,h})_\alpha \boldsymbol{\lambda}^*_\alpha = \mathbf{0}$ in Ω_I. Integrating this last equation along the cycle φ_α, $\alpha = 1, \dots, n_\Gamma$, we easily find $(\eta^*_{I,h})_\alpha = 0$, and therefore also $\operatorname{grad} \widehat{\psi}^*_{I,h} = \mathbf{0}$ in Ω_I.

The second part of the theorem follows from (5.77) and (5.78). □

Let us finally focus on the numerical approximation of (5.58). It reads

$$
\begin{aligned}
&\text{Find } (\mathbf{E}_{C,h}, \widehat{\psi}^*_{I,h}, \boldsymbol{\eta}^*_{I,h}) \in U^*_h \text{ such that} \\
&\mathcal{K}((\mathbf{E}_{C,h}, \widehat{\psi}^*_{I,h}, \boldsymbol{\eta}^*_{I,h}), (\mathbf{z}_{C,h}, \widehat{\chi}^*_{I,h}, \boldsymbol{\theta}^*_I)) \\
&\quad = -i\omega \int_{\Omega_C} \mathbf{J}_{e,C} \cdot \overline{\mathbf{z}_{C,h}} + i\omega \int_\Gamma \mathbf{H}^*_{e,I} \cdot \overline{\mathbf{z}_{C,h}} \times \mathbf{n}_C \\
&\quad\quad -\omega^2 \int_{\Omega_I} \boldsymbol{\mu}_I \mathbf{H}^*_{e,I} \cdot \overline{\operatorname{grad} \widehat{\chi}^*_{I,h}} \\
&\quad\quad -\omega^2 \sum_{\alpha=1}^{n_{\Omega_I}} \overline{\theta^*_{I,\alpha}} \int_{\Omega_I} \boldsymbol{\mu}_I \mathbf{H}^*_{e,I} \cdot \boldsymbol{\lambda}^*_\alpha \\
&\text{for each } (\mathbf{z}_{C,h}, \widehat{\chi}^*_{I,h}, \boldsymbol{\theta}^*_I) \in U^*_h \, ,
\end{aligned}
\tag{5.80}
$$

where the finite element space $U^*_h \subset U^*$ is given by

$$
U^*_h := N^k_{C,h} \times L^k_{I,h}/\mathbb{C} \times \mathbb{C}^{n_{\Omega_I}} \, .
$$

Since no matching condition is required on Γ, the meshes induced on Γ by $\mathcal{T}_{I,h}$ and $\mathcal{T}_{C,h}$ can be independent.

Since the sesquilinear form $\mathcal{K}(\cdot,\cdot)$ is coercive in U^* and smooth functions are dense in U^*, we can repeat the same arguments presented before and we end up with the following result.

Theorem 5.14. *Assume that the families of triangulations $\mathcal{T}_{C,h}$ and $\mathcal{T}_{I,h}$ are obtained as a refinement of coarse triangulations $\mathcal{T}_{C,h^0}$ and $\mathcal{T}_{I,h^0}$. Then the solution $(\mathbf{E}_{C,h}, \widehat{\psi}^*_{I,h}, \boldsymbol{\eta}^*_{I,h})$ of (5.80) exists, is unique and converges to the solution $(\mathbf{E}_C, \widehat{\psi}^*_I, \boldsymbol{\eta}^*_I)$ of problem (5.58). Moreover, if $\mathbf{E}_C \in H^r(\mathrm{curl}; \Omega_C)$ and $\widehat{\psi}^*_I \in H^{1+r}(\Omega_I)$ for $r > 1/2$, it follows*

$$
\begin{aligned}
&\|\mathbf{E}_C - \mathbf{E}_{C,h}\|_{H(\mathrm{curl};\Omega_C)} + \|\widehat{\psi}^*_I - \widehat{\psi}^*_{I,h}\|_{H^1(\Omega_I)/\mathbb{C}} + |\boldsymbol{\eta}^*_I - \boldsymbol{\eta}^*_{I.h}| \\
&\quad \leq c_* h^{\min(r,k)} (\|\mathbf{E}_C\|_{H^r(\mathrm{curl};\Omega_C)} + \|\operatorname{grad} \widehat{\psi}^*_I\|_{r,\Omega_I}) \, .
\end{aligned}
\tag{5.81}
$$

Remark 5.15. The weak formulations (5.21) and (5.67) are somehow different from those presented in Bermúdez et al. [42], even in the case of homogeneous boundary conditions, as we explicitly construct the part of the magnetic field that is not a gradient (namely, $\sum_{q=1}^N \eta_{I,q} \boldsymbol{\lambda}_q$ in (5.18)), while Bermúdez et al. [42] work with a multivalued potential. For other ways in which the magnetic scalar potential in multiply-connected regions is dealt with, see also Leonard and Rodger [167], [168] and the references therein. □

Remark 5.16. Following the approach described in Remark 5.8, one could decompose the discrete magnetic field as $\mathbf{H}_{I,h} = \widehat{\mathbf{H}}_{I,h} + \operatorname{grad} \Psi_{I,h}$, where $\widehat{\mathbf{H}}_{I,h}$ is an edge element with the property that $\operatorname{curl} \widehat{\mathbf{H}}_{I,h}$ is an interpolant of $\mathbf{J}_{e,I}$. Using this representation, instead of $\mathbf{H}_{I,h} = \mathbf{H}_{e,I} + \operatorname{grad} \widehat{\psi}_{I,h} + \sum_{q=1}^N (\eta_{I,h})_q \boldsymbol{\lambda}_q$ as in (5.71) or $\mathbf{H}_{I,h} = \mathbf{H}^*_{e,I} + \operatorname{grad} \widehat{\psi}^*_{I,h} + \sum_{\alpha=1}^{n_{\Omega_I}} (\eta^*_{I,h})_\alpha \boldsymbol{\lambda}^*_\alpha$ as in (5.79) and (5.80), can lead to devise suitable numerical approximation schemes. This procedure is adopted by Webb and Forghani [242], though a way for determining the degrees of freedom in $\widehat{\mathbf{H}}_{I,h}$ that are associated to the non-bounding cycles contained in Ω_I is not clearly described. □

We end this section by presenting some numerical results obtained by using the domain decomposition procedure described in Section 5.3.1 (a more detailed presentation can be found in Alonso Rodríguez and Valli [16]).

For testing the iterative algorithm (5.62), (5.63) and (5.64) we consider a model problem with scalar constant parameters μ and σ. The computational domain is the parallelepiped $\Omega = (0,2) \times (0,1) \times (0,1)$ and we take $\Omega_I := (0, x_\Gamma) \times (0,1) \times (0,1)$ and $\Omega_C := (x_\Gamma, 2) \times (0,1) \times (0,1)$ (for the sake of geometric simplicity, in this example we have chosen a conductor Ω_C not strictly contained in Ω; moreover, we have no harmonic fields into play, namely, $n_{\Omega_I} = 0$). The numerical mesh is uniform, and each element of the grid is a cube of side h. The employed finite elements are the first order curl-conforming hexahedral edge elements of Nédélec (see Nédélec [185] and Section A.2).

Aim of these numerical experiments is to verify the effectiveness of the iteration-by-subdomain procedure. Therefore it is not restrictive to study a model problem with current density $\mathbf{J}_e = \mathbf{0}$ and, starting with a non-zero initial datum, analyze the convergence of the solution to 0.

For each index m the iterations can be rewritten in the following way: given $\mathbf{e}^m_{\Gamma,h}$, the first step furnishes $\xi^m_{I,h} := i\omega\mu\widehat{\psi}^{*,m}_{I,h}$, the finite element approximate solution of the problem

$$\begin{cases} \Delta\xi^m_I = 0 & \text{in } \Omega_I \\ \frac{\partial \xi^m_I}{\partial n} = \operatorname{div}_\tau \mathbf{e}^m_{\Gamma,h} & \text{on } \Gamma \\ \frac{\partial \xi^m_I}{\partial n} = 0 & \text{on } \partial\Omega \cap \partial\Omega_I\,. \end{cases}$$

Then the second step gives $\mathbf{E}^m_{C,h}$, the approximate solution of

$$\begin{cases} \operatorname{curl}\operatorname{curl}\mathbf{E}^m_C + i\omega\mu\sigma\mathbf{E}^m_C = \mathbf{0} & \text{in } \Omega_C \\ \operatorname{curl}\mathbf{E}^m_C \times \mathbf{n}_C = \operatorname{grad}\xi^m_{I,h} \times \mathbf{n}_I & \text{on } \Gamma \\ \operatorname{curl}\mathbf{E}^m_C \times \mathbf{n} = \mathbf{0} & \text{on } \partial\Omega \cap \partial\Omega_C\,. \end{cases}$$

Finally, the new datum on Γ is defined as

$$\mathbf{e}^{m+1}_{\Gamma,h} = (1-\delta)\mathbf{e}^m_{\Gamma,h} + \delta\,\mathbf{E}^m_{C,h} \times \mathbf{n}_C \qquad \text{on } \Gamma\,.$$

The iterations are interrupted when the stopping test

$$\|\operatorname{grad}\xi^{m+1}_{I,h} - \operatorname{grad}\xi^m_{I,h}\|^2_{0,\Omega_I} + \|\mathbf{E}^{m+1}_{C,h} - \mathbf{E}^m_{C,h}\|^2_{H(\mathrm{curl};\Omega_C)} \leq 10^{-8}$$

is satisfied.

The initial datum $\mathbf{e}^0_{\Gamma,h}$ is a finite element approximation of $(\mathbf{E}^0_C \times \mathbf{n}_C)_{|\Gamma}$, for suitable vector fields $\mathbf{E}^0_C$. In the test case 1 we take a real function $\mathbf{E}^0_C$ given by

$$\mathbf{E}^0_{C,test\,1} = (e^z \sin(xy),\ e^x(y+z),\ \cos(xz))\,,$$

whereas in the test case 2 the initial value is complex:

$$\mathbf{E}^0_{C,test\,2} = (e^z + i\sin(xy), yz + ie^x, i\cos(xz)).$$

In Table 5.1 and Table 5.2 we show the number of iterations for the test case 1 and for the test case 2, respectively, with different values of h and x_Γ. The coefficient $\kappa := \omega\sigma\mu$, the only active parameter in this test problem, is set equal to 50 and the relaxation parameter δ is equal to 1. It can be seen that the number of iterations required to achieve convergence is almost constant with respect to h (in fact, it decreases slightly as the mesh size decreases).

Table 5.1. Number of iterations. Test case 1, $\delta = 1$

x_Γ \ h	*1/4*	*1/6*	*1/8*	*1/10*	*1/12*
1/2	5	4	4	4	3
1	4	4	4	4	3
3/2	4	4	4	4	3

Table 5.2. Number of iterations. Test case 2, $\delta = 1$

x_Γ \ h	*1/4*	*1/6*	*1/8*	*1/10*	*1/12*
1/2	5	4	4	4	4
1	4	4	3	3	3
3/2	4	4	4	4	4

From Table 5.3 we can see that the choice of the relaxation parameter $\delta = 1$ seems to be optimal.

Table 5.3. Number of iterations. Test case 1, $h = 1/8$

x_Γ \ δ	*0.9*	*0.95*	1	*1.05*	*1.1*
1/2	5	4	4	5	5
1	5	5	4	5	6
3/2	5	5	4	5	6

Finally, we consider different values of the coefficient κ and we observe (see Table 5.4) that the number of iterations required to achieve convergence does not change with κ.

Table 5.4. Number of iterations. Test case 2, $\delta = 1$, $h = 1/8$

x_Γ \ κ	10^{-1}	1	10	10^2	10^3
1/2	4	4	4	4	4
1	3	3	3	3	3
3/2	4	4	4	4	4

In the second set of numerical experiments we construct the data in the following way: we take a function $\mathbf{E}_C \in H(\text{curl};\Omega_C)$, we compute its interpolant $\mathbf{E}_{C,h}$ in the finite element space, and using this function we determine $\xi_{I,h}$ by solving the discrete Neumann problem for the Laplace operator in Ω_I, where the only non-vanishing datum is $\text{div}_\tau(\mathbf{E}_{C,h} \times \mathbf{n}_C)$ on Γ. Then we calculate the data of the problem (5.63) (namely, $-i\omega\mu\mathbf{J}_{e,C}$ in Ω_C, $i\omega\mu\mathbf{H}^*_{e,I} \times \mathbf{n}_I$ on Γ and, in the particular geometry we are considering for the numerical tests, $\mathbf{E}_{C,h} \times \mathbf{n}$ on $\partial\Omega_C \cap \partial\Omega$) in such a way that $\mathbf{E}_{C,h}$ is the discrete solution of the electric field problem. Precisely, this means that $-i\omega\mu\mathbf{J}_{e,C} = \text{curl}\,\text{curl}\,\mathbf{E}_C + i\omega\mu\sigma\mathbf{E}_C$, $i\omega\mu\mathbf{H}^*_{e,I} \times \mathbf{n}_I = \text{curl}\,\mathbf{E}_{C,h} \times \mathbf{n}_C - \text{grad}\,\xi_{I,h} \times \mathbf{n}_I$.

In particular, we consider

$$\mathbf{E}_C(x, y, z) = (i\,y(1-y)z(1-z)\sin(xy), z(1-z)e^x, y(1-y)(1+i\cos z))$$

(so that $\mathbf{E}_C \times \mathbf{n} = \mathbf{0}$ on $y = 0$, $y = 1$, $z = 0$ and $z = 1$), and constant coefficients $\omega = 50$, $\mu = 10^{-6}$, $\sigma = 10^6$. In this set of experiments we initialize the iterations with $\mathbf{e}^0_\Gamma = \mathbf{0}$ and we study the convergence to 0 of the relative difference between two subsequent iterates

$$\frac{\|\,\text{grad}\,\xi^{m+1}_{I,h} - \text{grad}\,\xi^m_{I,h}\|^2_{0,\Omega_I}}{\|\,\text{grad}\,\xi^{m+1}_{I,h}\|^2_{0,\Omega_I}} + \frac{\|\mathbf{E}^{m+1}_{C,h} - \mathbf{E}^m_{C,h}\|^2_{H(\text{curl};\Omega_C)}}{\|\mathbf{E}^{m+1}_{C,h}\|^2_{H(\text{curl};\Omega_C)}}\,.$$

In Figure 5.1 we see the convergence histories for $x_\Gamma = 1$ and different values of h, whereas in Figure 5.2 and in Figure 5.3 we set $h = 1/10$ and consider different values of x_Γ.

It can be thus concluded that this algorithm for solving the eddy current problem is efficient: its convergence is fast and it seems to converge even better for smaller

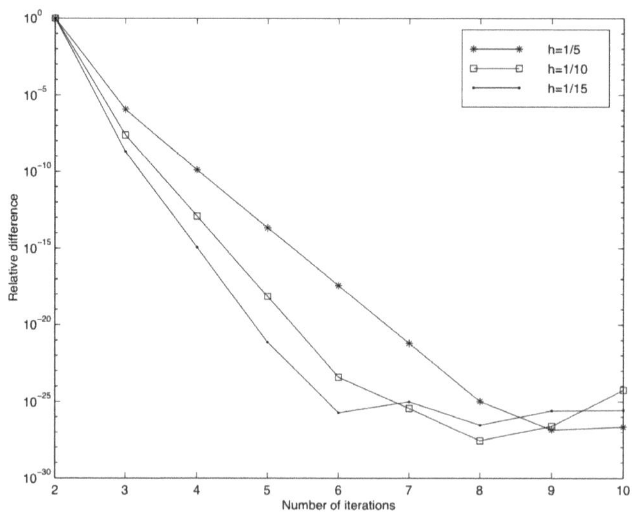

Fig. 5.1. Convergence history for $x_\Gamma = 1$ and different values of h

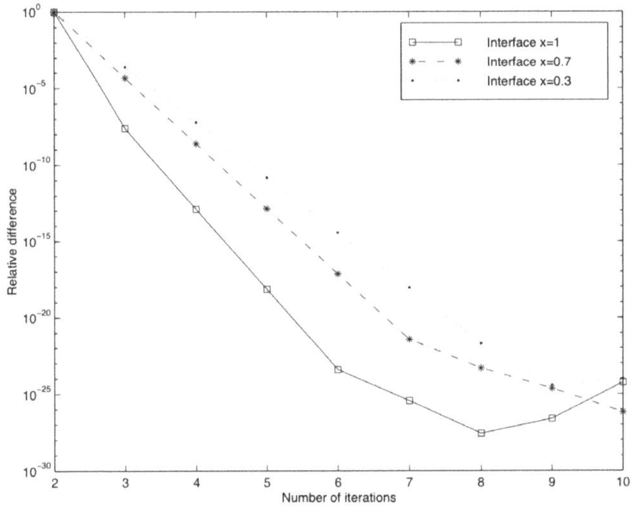

Fig. 5.2. Convergence history for $h = 1/10$ and different values of x_Γ

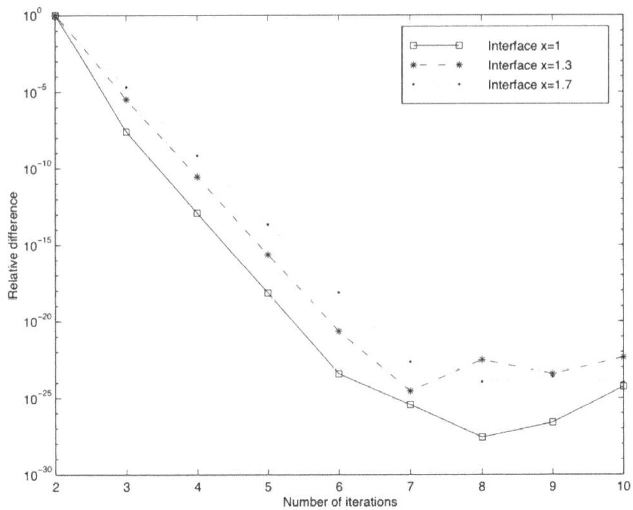

Fig. 5.3. Convergence history for $h = 1/10$ and different values of x_Γ (continued)

values of the mesh size. The iteration-by-subdomain procedure is quite insensitive to the position of the interface; it converges faster when the two subdomains are of the same size but the performance of the method is good in other cases as well. Finally, at least in the case of constant coefficients, the algorithm is rather insensitive to the coefficients of the problem.

5.5 The finite element approximation of $\mathbf{E}_I$

If the magnetic field is known, the electric field in the conductor Ω_C can be directly found by setting

$$\mathbf{E}_C = \boldsymbol{\sigma}^{-1}(\operatorname{curl} \mathbf{H}_C - \mathbf{J}_{e,C}) \, .$$

Instead, the electric field in the insulator is determined by solving (3.38) (for the electric boundary condition) or (3.13) (for the magnetic boundary conditions). However, for numerical computations it is better to consider a different problem.

Focusing, for the sake of definiteness, on the magnetic boundary value problem, we easily see that the electric field $\mathbf{E}_I$ is a solution to the system of equations

$$\begin{cases} \operatorname{curl}(\boldsymbol{\mu}_I^{-1} \operatorname{curl} \mathbf{E}_I) = -i\omega \mathbf{J}_I & \text{in } \Omega_I \\ \operatorname{div}(\boldsymbol{\varepsilon}_I \mathbf{E}_I) = 0 & \text{in } \Omega_I \\ \boldsymbol{\mu}^{-1} \operatorname{curl} \mathbf{E}_I \times \mathbf{n} = \mathbf{0} & \text{on } \partial\Omega \\ \boldsymbol{\varepsilon}_I \mathbf{E}_I \cdot \mathbf{n} = 0 & \text{on } \partial\Omega \\ \int_{\Gamma_j} \boldsymbol{\varepsilon}_I \mathbf{E}_I \cdot \mathbf{n}_I = 0 & \forall\, j = 1, \ldots, p_\Gamma \\ \int_{\Omega_I} \boldsymbol{\varepsilon}_I \mathbf{E}_I \cdot \boldsymbol{\pi}_{k,I} = 0 & \forall\, k = 1, \ldots, n_{\partial\Omega} \\ \mathbf{E}_I \times \mathbf{n}_I = -\mathbf{E}_C \times \mathbf{n}_C & \text{on } \Gamma \, . \end{cases} \tag{5.82}$$

We already noted that this problem simplifies if the boundary of the conductor Ω_C is connected and the domain Ω is simply-connected , as $p_\Gamma = 0$ and $n_{\partial\Omega} = 0$. This happens, for instance, if one considers a connected conductor, possibly with some "handles", contained in a "box": a situation which is often verified in applications.

Problem (5.82) can be written in a weak form, following a saddle-point approach similar to the one used for the **E**-based formulation in Section 4.6. Let us consider the spaces

$$Z_I := \{\mathbf{z}_I \in H(\operatorname{curl}; \Omega_I) \,|\, \mathbf{z}_I \text{ satisfies } (4.6)\},$$

and $Z_{I,0} := Z_I \cap H_{0,\Gamma}(\operatorname{curl}; \Omega_I)$. The weak formulation of (5.82) reads

$$\begin{aligned} &\text{Find } \mathbf{E}_I \in Z_I : \\ &\quad \mathbf{E}_I \times \mathbf{n}_I = -\mathbf{E}_C \times \mathbf{n}_C \quad \text{on } \Gamma \\ &\quad \int_{\Omega_I} \boldsymbol{\mu}_I^{-1} \operatorname{curl} \mathbf{E}_I \cdot \operatorname{curl} \overline{\mathbf{z}_I} = -i\omega \int_{\Omega_I} \mathbf{J}_{e,I} \cdot \overline{\mathbf{z}_I} \\ &\text{for all } \mathbf{z}_I \in Z_{I,0} \, . \end{aligned} \tag{5.83}$$

From Lemma 2.1 it follows at once that the sesquilinear form

$$a_{e,I}(\mathbf{w}_I, \mathbf{z}_I) := \int_{\Omega_I} \boldsymbol{\mu}_I^{-1} \operatorname{curl} \mathbf{w}_I \cdot \operatorname{curl} \overline{\mathbf{z}_I} \tag{5.84}$$

is coercive in $Z_{I,0}$, therefore (5.83) has a unique solution.

Recalling that $\mathbf{z}_I \in Z_I$ if and only if $\mathbf{z}_I \in H(\text{curl}; \Omega_I)$ and $\int_{\Omega_I} \varepsilon_I \mathbf{z}_I \cdot \overline{\mathbf{p}_I} = 0$ for all $\mathbf{p}_I \in H^0_{0,\Gamma}(\text{curl}; \Omega_I)$ (see Section 4.2), we have the equivalent formulation

$$\begin{array}{l} \text{Find } (\mathbf{E}_I, \mathbf{r}_I) \in H(\text{curl}; \Omega_I) \times H^0_{0,\Gamma}(\text{curl}; \Omega_I) : \\ \quad \mathbf{E}_I \times \mathbf{n}_I = -\mathbf{E}_C \times \mathbf{n}_C \quad \text{on } \Gamma \\ \quad \int_{\Omega_I} \boldsymbol{\mu}_I^{-1} \,\text{curl}\, \mathbf{E}_I \cdot \text{curl}\, \overline{\mathbf{z}_I} + \int_{\Omega_I} \varepsilon_I \overline{\mathbf{z}_I} \cdot \mathbf{r}_I = -i\omega \int_{\Omega_I} \mathbf{J}_{e,I} \cdot \overline{\mathbf{z}_I} \\ \quad \int_{\Omega_I} \varepsilon_I \mathbf{E}_I \cdot \overline{\mathbf{p}_I} = 0 \\ \text{for all } (\mathbf{z}_I, \mathbf{p}_I) \in H_{0,\Gamma}(\text{curl}; \Omega_I) \times H^0_{0,\Gamma}(\text{curl}; \Omega_I) . \end{array} \tag{5.85}$$

The inf–sup condition for this saddle-point problem is trivially verified: given $\mathbf{r}_I \in H^0_{0,\Gamma}(\text{curl}; \Omega_I)$, one chooses $\mathbf{z}_I = \mathbf{r}_I$ and obtains

$$\begin{aligned} \int_{\Omega_I} \varepsilon_I \overline{\mathbf{z}_I} \cdot \mathbf{r}_I &= \int_{\Omega_I} \varepsilon_I \overline{\mathbf{r}_I} \cdot \mathbf{r}_I \\ &\geq \varepsilon_{I,\min} \|\mathbf{r}_I\|^2_{0,\Omega_I} = \varepsilon_{I,\min} \|\mathbf{r}_I\|_{H(\text{curl};\Omega_I)} \|\mathbf{z}_I\|_{H(\text{curl};\Omega_I)} , \end{aligned}$$

where $\varepsilon_{I,\min}$ is a uniform lower bound for the eigenvalues of $\varepsilon_I(\mathbf{x})$ in Ω_I.

Therefore, problem (5.85) has a unique solution. Moreover, if $\mathbf{E}_I$ is the solution of problem (5.83), it is seen at once that $(\mathbf{E}_I, \mathbf{0})$ is a solution to (5.85); hence in (5.85) one has $\mathbf{r}_I = \mathbf{0}$.

In order to find a finite element approximation of the space $H^0_{0,\Gamma}(\text{curl}; \Omega_I)$ we recall that

$$H^0_{0,\Gamma}(\text{curl}; \Omega_I) = \text{grad}\, H^1_{0,\Gamma}(\Omega_I) \oplus \mathcal{H}_{\varepsilon_I}(\Gamma, \partial\Omega; \Omega_I) ,$$

(see (4.8)) and that a basis of $\mathcal{H}_{\varepsilon_I}(\Gamma, \partial\Omega; \Omega_I)$ is given by

$$\{\text{grad}\, w_{j,I}\}_{j=1}^{p_\Gamma} \cup \{\boldsymbol{\pi}_{k,I}\}_{k=1}^{n_{\partial\Omega}}$$

(see Sections 1.4 and A.4). Hence any function $\mathbf{p}_I \in H^0_{0,\Gamma}(\text{curl}; \Omega_I)$ can be decomposed as

$$\mathbf{p}_I = \text{grad}\, \varphi_I + \sum_{j=1}^{p_\Gamma} c_{I,j}\, \text{grad}\, w_{j,I} + \sum_{k=1}^{n_{\partial\Omega}} d_{I,k} \boldsymbol{\pi}_{k,I} ,$$

where $\varphi_I \in H^1_{0,\Gamma}(\Omega_I)$ and $c_{I,j}$, $d_{I,k}$, $j = 1, \ldots, p_\Gamma$, $k = 1, \ldots, n_{\partial\Omega}$, are complex numbers.

As proved in Bossavit [59], Hiptmair [126], Gross and Kotiuga [115], in Ω_I we can consider a system of "cutting" surfaces Σ_k, $k = 1, \ldots, n_{\partial\Omega}$, with $\Sigma_k \subset \Omega_I$ and $\partial\Sigma_k \subset \partial\Omega$, such that every curl-free vector field in Ω_I with vanishing tangential component on Γ has a global potential in $\Omega_I \setminus \cup_k \Sigma_k$ (for their explicit construction, which is needed for numerical approximation, see Remark 5.9). Let us suppose that Ω_I is a polyhedral domain and that there is a triangulation $\mathcal{T}_{I,h^0}$ of Ω_I, where $h^0 > 0$ is the (fixed) mesh size, that induces a triangulation on the surfaces Σ_k. For $k = 1, \ldots, n_{\partial\Omega}$ let us denote by Π_k the piecewise-polynomial function taking value 1 at the nodes on one side of Σ_k and 0 at all the other nodes (including those on the other side of Σ_k). Notice that $\Pi_{k|\Gamma} = 0$ since $\partial\Sigma_k \cap \Gamma = \emptyset$. Let us set $\boldsymbol{\pi}^0_k = \widetilde{\text{grad}}\, \Pi_k$, where

$\widetilde{\text{grad}\,\Pi_k}$ denotes the $(L^2(\Omega_I))^3$-extension of $\text{grad}\,\Pi_k$ computed in $\Omega_I \setminus \Sigma_k$. It follows that $\boldsymbol{\pi}_k^0 \in H_{0,\Gamma}(\text{curl}, \Omega_I)$ and that $\boldsymbol{\pi}_{k,I} = \boldsymbol{\pi}_k^0 + \text{grad}\, g_k$, with $g_k \in H^1_{0,\Gamma}(\Omega_I)$.

Any function $\mathbf{p}_I \in H^0_{0,\Gamma}(\text{curl}; \Omega_I)$ can be thus written as

$$\mathbf{p}_I = \text{grad}\, \xi_I + \sum_{k=1}^{n_{\partial\Omega}} d_{I,k} \boldsymbol{\pi}_k^0 \,,$$

with $\xi_I = \varphi_I + \sum_{j=1}^{p_\Gamma} c_{I,j} w_{j,I} + \sum_{k=1}^{n_{\partial\Omega}} d_{I,k} g_k$, and $\xi_I \in H^1_{*,\Gamma}(\Omega_I)$ (see (4.17)).

For the finite element approximation we write (5.85) in the following equivalent way

$$
\begin{aligned}
&\text{Find } (\mathbf{E}_I, \phi_I, \mathbf{m}_I) \in H(\text{curl}; \Omega_I) \times H^1_{*,\Gamma}(\Omega_I) \times \mathbb{C}^{n_{\partial\Omega}} : \\
&\quad \mathbf{E}_I \times \mathbf{n}_I = -\mathbf{E}_C \times \mathbf{n}_C \quad \text{on } \Gamma \\
&\quad \textstyle\int_{\Omega_I} \boldsymbol{\mu}_I^{-1} \,\text{curl}\, \mathbf{E}_I \cdot \text{curl}\, \overline{\mathbf{z}_I} \\
&\qquad\quad + \textstyle\int_{\Omega_I} \varepsilon_I \overline{\mathbf{z}_I} \cdot \text{grad}\, \phi_I + \sum_{k=1}^{n_{\partial\Omega}} m_{I,k} \int_{\Omega_I} \varepsilon_I \overline{\mathbf{z}_I} \cdot \boldsymbol{\pi}_k^0 = L_I(\mathbf{z}_I) \\
&\quad \textstyle\int_{\Omega_I} \varepsilon_I \mathbf{E}_I \cdot \text{grad}\, \overline{\xi_I} = 0 \\
&\quad \textstyle\sum_{k=1}^{n_{\partial\Omega}} \overline{d_{I,k}} \int_{\Omega_I} \varepsilon_I \mathbf{E}_I \cdot \boldsymbol{\pi}_k^0 = 0 \\
&\text{for all } (\mathbf{z}_I, \xi_I, \mathbf{d}_I) \in H_{0,\Gamma}(\text{curl}; \Omega_I) \times H^1_{*,\Gamma}(\Omega_I) \times \mathbb{C}^{n_{\partial\Omega}} \,,
\end{aligned}
\tag{5.86}
$$

where

$$L_I(\mathbf{z}_I) := -i\omega \int_{\Omega_I} \mathbf{J}_{e,I} \cdot \overline{\mathbf{z}_I} \,.$$

It is worth noting that $\text{grad}\, \phi_I + \sum_{k=1}^{n_{\partial\Omega}} m_{I,k} \boldsymbol{\pi}_k^0 = \mathbf{0}$. This is easily verified by choosing in (5.86) the test function

$$\mathbf{z}_I = \text{grad}\, \phi_I + \sum_{k=1}^{n_{\partial\Omega}} m_{I,k} \boldsymbol{\pi}_k^0 \,,$$

and recalling that $\mathbf{J}_{e,I}$ satisfies (5.1) and (5.2).

We employ the Nédélec curl-conforming edge elements $N^k_{I,h}$ (see Section A.2) to approximate the electric field and the piecewise-polynomial continuous functions $H^k_{I,h}$ (see (4.83)) to approximate the Lagrange multiplier ϕ_I. Let us also set

$$Y^k_{I,h} := N^k_{I,h} \cap H_{0,\Gamma}(\text{curl}; \Omega_I) \,.$$

We assume that a suitable finite element approximation of the tangential trace of the electric field on the interface Γ is known. Denoting it by $\mathbf{E}_{C,h} \times \mathbf{n}_C$, we suppose that $\mathbf{E}_{C,h} \times \mathbf{n}_C \in \mathcal{X}_{\Gamma,h} := \{(\mathbf{z}_{I,h} \times \mathbf{n}_I)_{|\Gamma} \mid \mathbf{z}_{I,h} \in N^k_{I,h}\}$.

The discrete problem reads

$$
\begin{aligned}
&\text{Find } (\mathbf{E}_{I,h}, \phi_{I,h}, \mathbf{m}_I^h) \in N_{I,h}^k \times H_{I,h}^k \times \mathbb{C}^{n_{\partial\Omega}} : \\
&\quad \mathbf{E}_{I,h} \times \mathbf{n}_I = -\mathbf{E}_{C,h} \times \mathbf{n}_C \quad \text{on } \Gamma \\
&\quad \int_{\Omega_I} \boldsymbol{\mu}_I^{-1} \operatorname{curl} \mathbf{E}_{I,h} \cdot \operatorname{curl} \overline{\mathbf{z}_{I,h}} \\
&\qquad + \int_{\Omega_I} \varepsilon_I \overline{\mathbf{z}_{I,h}} \cdot \operatorname{grad} \phi_{I,h} + \sum_{k=1}^{n_{\partial\Omega}} m_{I,k}^h \int_{\Omega_I} \varepsilon_I \overline{\mathbf{z}_{I,h}} \cdot \boldsymbol{\pi}_k^0 = L_I(\mathbf{z}_{I,h}) \qquad (5.87) \\
&\quad \int_{\Omega_I} \varepsilon_I \mathbf{E}_{I,h} \cdot \operatorname{grad} \overline{\xi_{I,h}} = 0 \\
&\quad \sum_{k=1}^{n_{\partial\Omega}} \overline{d_{I,k}} \int_{\Omega_I} \varepsilon_I \mathbf{E}_{I,h} \cdot \boldsymbol{\pi}_k^0 = 0 \\
&\text{for all } (\mathbf{z}_{I,h}, \xi_{I,h}, \mathbf{d}_I) \in Y_{I,h}^k \times H_{I,h}^k \times \mathbb{C}^{n_{\partial\Omega}} .
\end{aligned}
$$

Similarly to the result proved for problem (5.86), we can observe that $\operatorname{grad} \phi_{I,h} + \sum_{k=1}^{n_{\partial\Omega}} m_{I,k}^h \boldsymbol{\pi}_k^0 = \mathbf{0}$.

Our aim now is to apply the standard theory of mixed finite elements. For that, we need to introduce the space $Z_{I,0}^h$ of functions $\mathbf{z}_{I,h} \in Y_{I,h}^k$ such that

$$
\int_{\Omega_I} \varepsilon_I \mathbf{z}_{I,h} \cdot \left[\operatorname{grad} \overline{\xi_{I,h}} + \sum_{k=1}^{n_{\partial\Omega}} \overline{d_{I,k}} \boldsymbol{\pi}_k^0 \right] = 0
$$

for all $(\xi_{I,h}, \mathbf{d}_I) \in H_{I,h}^k \times \mathbb{C}^{n_{\partial\Omega}}$.

The following result is the key point for the proof of well-posedness and convergence of the finite element approximation scheme.

Lemma 5.17. *Assuming that ε_I is a scalar constant and that the triangulation $\mathcal{T}_{I,h}$ is a refinement of the coarse triangulation $\mathcal{T}_{I,h^0}$, there exist positive constants C_1, C_2, independent of h, such that*

$$
\int_{\Omega_I} \boldsymbol{\mu}_I^{-1} \operatorname{curl} \mathbf{z}_{I,h} \cdot \operatorname{curl} \overline{\mathbf{z}_{I,h}} \geq C_1 \|\mathbf{z}_{I,h}\|^2_{H(\operatorname{curl};\Omega_I)} \qquad (5.88)
$$

for all $\mathbf{z}_{I,h} \in Z_{I,0}^h$, and

$$
\begin{aligned}
\sup_{\mathbf{z}_{I,h} \in Y_{I,h}^k} & \frac{\int_{\Omega_I} \varepsilon_I \mathbf{z}_{I,h} \cdot \left[\operatorname{grad} \overline{\xi_{I,h}} + \sum_{k=1}^{n_{\partial\Omega}} \overline{d_{I,k}} \boldsymbol{\pi}_k^0 \right]}{\|\mathbf{z}_{I,h}\|_{H(\operatorname{curl};\Omega_I)}} \qquad (5.89) \\
& \geq C_2 (\|\xi_{I,h}\|_{1,\Omega_I} + |\mathbf{d}_I|)
\end{aligned}
$$

for all $(\xi_{I,h}, \mathbf{d}_I) \in H_{I,h}^k \times \mathbb{C}^{n_{\partial\Omega}}$.

Proof. The discrete inf–sup condition (5.89) is easily obtained by taking $\mathbf{z}_{I,h} = \operatorname{grad} \xi_{I,h} + \sum_{k=1}^{n_{\partial\Omega}} d_{I,k} \boldsymbol{\pi}_k^0$ and using Lemma 5.2 (suitably adapted to the present situation).

In order to prove that (5.88) holds, we recall that from (A.12) any function in $(L^2(\Omega_I))^3$, and in particular any function $\mathbf{z}_{I,h} \in Z_{I,0}^h$, can be written as

$$
\mathbf{z}_{I,h} = \varepsilon_I^{-1} \operatorname{curl} \mathbf{q}_I + \operatorname{grad} \xi_I + \sum_{k=1}^{n_{\partial\Omega}} d_{I,k} \boldsymbol{\pi}_k^0 ,
$$

where $\mathbf{q}_I \in H_{0,\partial\Omega}(\operatorname{curl}; \Omega_I)$, $\xi_I \in H_{*,\Gamma}^1(\Omega_I)$ and $\boldsymbol{\pi}_k^0 = \widetilde{\operatorname{grad}} \Pi_k$.

Let us set $\mathbf{U}_I := \varepsilon_I^{-1} \operatorname{curl} \mathbf{q}_I$. Since $\mathbf{z}_{I,h} \in H_{0,\Gamma}(\operatorname{curl}; \Omega_I)$, we have $\mathbf{U}_I \in Z_{I,0}$. As already recalled, we know from Lemma 2.1 that the sesquilinear form $a_{e,I}(\cdot,\cdot)$ introduced in (5.84) is continuous and coercive in $Z_{I,0}$, i.e., there exists a positive constant c_1 such that

$$\int_{\Omega_I} \boldsymbol{\mu}_I^{-1} \operatorname{curl} \mathbf{U}_I \cdot \operatorname{curl} \overline{\mathbf{U}_I} \geq c_1 \|\mathbf{U}_I\|_{H(\operatorname{curl};\Omega_I)} .$$

Taking into account that $\operatorname{curl} \mathbf{z}_{I,h} = \operatorname{curl} \mathbf{U}_I$, to obtain (5.88) we have only to show that there exists a constant $c_2 > 0$, independent of h, such that

$$\|\mathbf{z}_{I,h}\|_{0,\Omega_I} \leq c_2 \|\mathbf{U}_I\|_{H(\operatorname{curl};\Omega_I)} .$$

The procedure is similar to the one presented in the saddle-point approach for the **E**-based formulation (see Section 4.6). Since ε_I is a scalar constant, from Lemma 4.33 we know that $Z_{I,0} \subset (H^{1/2+\delta}(\Omega_I))^3$ for some $\delta > 0$ small enough. Moreover $\operatorname{curl} \mathbf{U}_{I|K} = \operatorname{curl} \mathbf{z}_{I,h|K} \in (\mathbb{P}_{k-1})^3$, hence, by Lemma 4.32, the interpolant $\Pi_{I,h}\mathbf{U}_I$ is well-defined.

Since $\boldsymbol{\pi}_k^0 \in N_{I,h}^k$, this means that also $\Pi_{I,h}(\operatorname{grad} \xi_I)$ is well-defined. We have

$$\operatorname{curl}[\Pi_{I,h}(\operatorname{grad} \xi_I)] = 0 \text{ in } \Omega_I \ , \ \Pi_{I,h}(\operatorname{grad} \xi_I) \times \mathbf{n}_I = \mathbf{0} \text{ on } \Gamma ,$$

and $\int_{\gamma_k} \Pi_{I,h}(\operatorname{grad} \xi_I) \cdot d\tau = 0$ for all Γ-independent non-bounding cycles γ_k, $k = 1, \ldots, n_{\partial\Omega}$. Hence $\Pi_{I,h}(\operatorname{grad} \xi_I) = \operatorname{grad} \xi_{I,h}$ for some $\xi_{I,h} \in H_{I,h}^k$ (see, e.g., Monk [179], Lemma 5.28).

Since $\mathbf{z}_{I,h} \in Z_{I,h}$

$$\begin{aligned} \textstyle\int_{\Omega_I} \varepsilon_I \mathbf{z}_{I,h} \cdot \overline{\mathbf{z}_{I,h}} &= \textstyle\int_{\Omega_I} \varepsilon_I \mathbf{z}_{I,h} \cdot [\Pi_{I,h}\overline{\mathbf{U}_I} + \operatorname{grad} \overline{\xi_{I,h}} + \sum_{k=1}^{n_{\partial\Omega}} \overline{d_{I,k}\boldsymbol{\pi}_k^0}] \\ &= \textstyle\int_{\Omega_I} \varepsilon_I \mathbf{z}_{I,h} \cdot \Pi_{I,h}\overline{\mathbf{U}_I} , \end{aligned}$$

hence

$$\|\mathbf{z}_{I,h}\|_{0,\Omega_I} \leq C \, \|\Pi_{I,h}\mathbf{U}_I\|_{0,\Omega_I} \leq C \left(\|\Pi_{I,h}\mathbf{U}_I - \mathbf{U}_I\|_{0,\Omega_I} + \|\mathbf{U}_I\|_{0,\Omega_I}\right) .$$

Since $\operatorname{div} \mathbf{U}_I = 0$ in Ω_I, by combining Lemma 4.32 and Lemma 4.33 we find

$$\begin{aligned} \|\Pi_{I,h}\mathbf{U}_I - \mathbf{U}_I\|_{0,\Omega_I} &\leq Ch^{1/2+\delta}(\|\mathbf{U}_I\|_{1/2+\delta,\Omega_I} + \|\operatorname{curl} \mathbf{U}_I\|_{0,\Omega_I}) \\ &\leq \hat{C}(\|\mathbf{U}_I\|_{0,\Omega_I} + \|\operatorname{curl} \mathbf{U}_I\|_{0,\Omega_I}) , \end{aligned}$$

that concludes the proof. □

Remark 5.18. For the electric boundary condition the strong problem reads

$$\begin{cases} \operatorname{curl}(\boldsymbol{\mu}_I^{-1} \operatorname{curl} \mathbf{E}_I) = -i\omega \mathbf{J}_I & \text{in } \Omega_I \\ \operatorname{div}(\varepsilon_I \mathbf{E}_I) = 0 & \text{in } \Omega_I \\ \mathbf{E}_I \times \mathbf{n} = \mathbf{0} & \text{on } \partial\Omega \\ \int_{\Gamma_j} \varepsilon_I \mathbf{E}_I \cdot \mathbf{n}_I = 0 & \forall j = 1, \ldots, p_\Gamma \\ \int_{(\partial\Omega)_r} \varepsilon_I \mathbf{E}_I \cdot \mathbf{n} = 0 & \forall r = 0, 1, \ldots, p_{\partial\Omega} \\ \mathbf{E}_I \times \mathbf{n}_I = -\mathbf{E}_C \times \mathbf{n}_C & \text{on } \Gamma , \end{cases} \tag{5.90}$$

(which simplifies if the boundaries of Ω_C and Ω are connected, so that $p_\Gamma = p_{\partial\Omega} = 0$), while the weak problem is

$$
\begin{array}{l}
\text{Find } (\mathbf{E}_I, \phi_I^*) \in H_{0,\partial\Omega}(\mathrm{curl}; \Omega_I) \times H_*^1(\Omega_I) : \\
\quad \mathbf{E}_I \times \mathbf{n}_I = -\mathbf{E}_C \times \mathbf{n}_C \quad \text{on } \Gamma \\
\quad \int_{\Omega_I} \mu_I^{-1} \,\mathrm{curl}\, \mathbf{E}_I \cdot \mathrm{curl}\, \overline{\mathbf{z}_I} + \int_{\Omega_I} \varepsilon_I \overline{\mathbf{z}_I} \cdot \mathrm{grad}\, \phi_I^* = L_I(\mathbf{z}_I) \\
\quad \int_{\Omega_I} \varepsilon_I \mathbf{E}_I \cdot \mathrm{grad}\, \overline{\xi_I^*} = 0 \\
\text{for all } (\mathbf{z}_I, \xi_I^*) \in H_0(\mathrm{curl}; \Omega_I) \times H_*^1(\Omega_I) ,
\end{array}
\tag{5.91}
$$

where $H_*^1(\Omega_I)$ has been defined in (4.82).

The finite element approximation is given by

$$
\begin{array}{l}
\text{Find } (\mathbf{E}_{I,h}, \phi_{I,h}^*) \in X_{I,h}^k \times \widehat{H}_{I,h}^k : \\
\quad \mathbf{E}_{I,h} \times \mathbf{n}_I = -\mathbf{E}_{C,h} \times \mathbf{n}_C \quad \text{on } \Gamma \\
\quad \int_{\Omega_I} \mu_I^{-1} \,\mathrm{curl}\, \mathbf{E}_{I,h} \cdot \mathrm{curl}\, \overline{\mathbf{z}_{I,h}} + \int_{\Omega_I} \varepsilon_I \overline{\mathbf{z}_{I,h}} \cdot \mathrm{grad}\, \phi_{I,h}^* = L_I(\mathbf{z}_{I,h}) \\
\quad \int_{\Omega_I} \varepsilon_I \mathbf{E}_{I,h} \cdot \mathrm{grad}\, \overline{\xi_{I,h}^*} = 0 \\
\text{for all } (\mathbf{z}_{I,h}, \xi_{I,h}^*) \in \widehat{X}_{I,h}^k \times \widehat{H}_{I,h}^k ,
\end{array}
\tag{5.92}
$$

where $X_{I,h}^k := N_{I,h}^k \cap H_{0,\partial\Omega}(\mathrm{curl}; \Omega_I)$, $\widehat{X}_{I,h}^k := N_{I,h}^k \cap H_0(\mathrm{curl}; \Omega_I)$ and $\widehat{H}_{I,h}^k := L_{I,h}^k \cap H_*^1(\Omega_I)$.

As for the magnetic boundary value problem, it is readily shown that one has $\phi_I^* = 0$ in (5.91) and $\phi_{I,h}^* = 0$ in (5.92). □

Remark 5.19. Another algorithm for computing the finite element approximation of the electric field $\mathbf{E}_I$ is presented in Remark 6.12; it is based on a vector magnetic potential formulation of the eddy current problem. □

6

Formulations via vector potentials

Motivated by the fact that the magnetic induction $\mathbf{B} = \boldsymbol{\mu}\mathbf{H}$ is divergence-free in Ω, a classical approach to the Maxwell equations and to eddy current problems is based on the introduction of a vector magnetic potential $\mathbf{A}$ such that $\operatorname{curl} \mathbf{A} = \boldsymbol{\mu}\mathbf{H}$. Often, this is also accompanied by the use of a scalar electric potential V_C in the conductor Ω_C, satisfying $i\omega\mathbf{A}_C + \operatorname{grad} V_C = -\mathbf{E}_C$ (see Silvester and Ferrari [227]; for the engineering literature, see, e.g., Chari et al. [79], Biddlecombe et al. [47], Morisue [180], Bíró and Preis [49]).

This approach opens up the problem of determining correct *gauge* conditions, namely, some conditions ensuring the uniqueness of $\mathbf{A}$ and V_C. Although in principle all that needs to be required is only that $\operatorname{curl} \mathbf{A}$ and $\mathbf{E}_C$ be uniquely determined, the choice of suitable gauge conditions may also be necessary when considering numerical approximation, in order to avoid that the discrete problem becomes singular.

Writing $\mathbf{H} = \boldsymbol{\mu}^{-1} \operatorname{curl} \mathbf{A}$ and $\mathbf{E}_C = -i\omega\mathbf{A}_C - \operatorname{grad} V_C$ one sees that the Faraday equation in Ω_C is straightforwardly satisfied, whereas the Ampère law in Ω becomes a second order partial differential equation for the vector potential $\mathbf{A}$, thus leading to a variational problem with a nice structure.

Since $\operatorname{curl} \mathbf{A}$ is assigned, a natural gauge condition for $\mathbf{A}$ is to impose its divergence. In early papers on this subject this choice was accompanied by the remark that $-\Delta\mathbf{A} = \operatorname{curl}\operatorname{curl} \mathbf{A} - \operatorname{grad}\operatorname{div} \mathbf{A}$, so that for a constant magnetic permeability μ the Ampère equation reduces to a vector Poisson problem for $\mathbf{A}$. However, at the variational level this is not a true advantage, as the boundary conditions for $\mathbf{A}$ remain those associated to the first order curl–div system, that are different from the Dirichlet condition, for which the entire vector $\mathbf{A}$ is assigned on $\partial\Omega$. Therefore the variational problem cannot be written in terms of a sesquilinear form like $\int_\Omega \sum_{i,j} D_i z_j D_i \overline{w}_j$, but it has to keep the structure $\int_\Omega (\operatorname{curl} \mathbf{z} \cdot \operatorname{curl} \overline{\mathbf{w}} + \operatorname{div} \mathbf{z} \operatorname{div} \overline{\mathbf{w}})$.

In Sections 6.1 and 6.2 we present two different gauge conditions for the approach based on the vector magnetic potential: the Coulomb gauge and the Lorenz gauge. In both cases, the divergence of $\mathbf{A}$ is required to satisfy a suitable equation. In Section 6.3 we conclude the chapter by describing some other vector potential formulations for eddy current problems.

A. Alonso Rodríguez, A. Valli: Eddy Current Approximation of Maxwell Equations.

Although all these methods have been used in many engineering applications, up to now their analysis has not been fully performed. In general, only uniqueness results were presented (see, e.g., Bíró and Preis [49]). Our aim here is to furnish a complete theory concerning well-posedness of these formulations and convergence of their finite element numerical approximations.

Concerning the material coefficients, in this chapter we will assume that the matrix $\boldsymbol{\mu}$ is symmetric and uniformly positive definite in Ω, with entries belonging to $L^\infty(\Omega)$, the matrix ε_I is symmetric and uniformly positive definite in Ω_I, with entries belonging to $L^\infty(\Omega_I)$, and the matrix $\boldsymbol{\sigma}$ is symmetric and uniformly positive definite in Ω_C, with entries belonging to $L^\infty(\Omega_C)$, whereas it is vanishing in Ω_I.

The reader mainly interested in numerical approximation and implementation can focus on problems (6.12) and (6.45) (magnetic boundary conditions for the $(\mathbf{A}, V_C)$ formulation), on problems (6.32) and (6.47) (electric boundary condition for the $(\mathbf{A}, V_C)$ formulation), on problem (6.50) (penalized $\mathbf{E}$ formulation), on problem (6.92) ($(\mathbf{A}, V_C) - \psi_I$ formulation), on problem (6.102) ($(\mathbf{T}_C, \psi_C) - \psi_I$ formulation), on problem (6.108) ($(\mathbf{T}_C^*, \Phi_C) - \mathbf{A}_I$ formulation) and on Section 6.1.4.

6.1 Formulation for the Coulomb gauge and its numerical approximation

Let us suppose that Ω, Ω_C and Ω_I satisfy the assumptions of Section 1.3, and, for the sake of definiteness, let us consider the magnetic boundary value problem (1.22). As usual the current density $\mathbf{J}_e \in (L^2(\Omega))^3$ is assumed to satisfy the necessary conditions (1.23).

In this section, following Bíró and Valli [54], we present and analyze the formulation in which one looks for a magnetic vector potential $\mathbf{A}$ and a scalar electric potential V_C such that

$$\mathbf{E}_C = -i\omega \mathbf{A}_C - \operatorname{grad} V_C \ , \ \ \boldsymbol{\mu}\mathbf{H} = \operatorname{curl} \mathbf{A} \ . \tag{6.1}$$

Let us verify which equations in (3.25) are satisfied, and which instead have to be imposed. We see at once that $\operatorname{curl} \mathbf{E}_C = -i\omega \operatorname{curl} \mathbf{A}_C = -i\omega \boldsymbol{\mu}_C \mathbf{H}_C$, and therefore $(3.25)_1$, namely, the Faraday equation in Ω_C, is satisfied. Moreover, $\boldsymbol{\mu}\mathbf{H}$ is equal to $\operatorname{curl} \mathbf{A}$ in Ω, therefore it is a solenoidal vector field in Ω and has a vanishing flux through any closed surface in $\overline{\Omega}$: hence equations $(3.25)_4$, $(3.25)_5$ and $(3.25)_8$ are satisfied. We also know that from $(6.1)_2$ the topological conditions $(3.25)_6$ are satisfied (see Section 3.3.2).

Hence we have only to require that the Ampère equation is satisfied in Ω, so that $(3.25)_2$, $(3.25)_3$ and $(3.25)_9$ are satisfied, and to impose the magnetic boundary condition $(3.25)_7$.

However, if we want to devise a well-posed problem, this is not enough: in fact, we also have to consider the fact that the introduction of the new additional unknown V_C and the necessity of obtaining a unique vector potential $\mathbf{A}$ lead us to impose some additional conditions, usually called gauge conditions. The most frequently used is the

Coulomb gauge

$$\operatorname{div} \mathbf{A} = 0 \qquad \text{in } \Omega\,,$$

with the boundary condition

$$\mathbf{A} \cdot \mathbf{n} = 0 \qquad \text{on } \partial\Omega\,.$$

In a general geometrical situation, even these additional conditions can be not enough for determining a unique vector potential $\mathbf{A}$ in Ω. In fact, there exist non-trivial irrotational, solenoidal and tangential vector fields, namely, the elements of the finite dimensional space of harmonic fields

$$\mathcal{H}(m;\Omega) := \{\mathbf{w} \in (L^2(\Omega))^3 \mid \operatorname{curl} \mathbf{w} = \mathbf{0}, \operatorname{div} \mathbf{w} = 0, \mathbf{w} \cdot \mathbf{n} = 0 \text{ on } \partial\Omega\}\,.$$

In Section 1.4 we have denoted by n_Ω the dimension of this vector space: it is a topological invariant, the so-called first Betti number of the domain Ω, namely, the number of independent non-bounding cycles in Ω (see, e.g., Bossavit [59], Hiptmair [126], Gross and Kotiuga [115]). In particular, it is proved that in Ω there exist n_Ω connected orientable Lipschitz surfaces $\hat{\Sigma}_t$ with $\partial\hat{\Sigma}_t \subset \partial\Omega$ and such that every curl-free vector in Ω has a global potential in $\widetilde{\Omega} := \Omega \setminus \cup_t \hat{\Sigma}_t$.

In other words, each surface $\hat{\Sigma}_t$, $t = 1, \ldots, n_\Omega$, "cuts" a cycle on $\partial\Omega$ that is not bounding a surface contained in Ω. Let us note that the explicit construction of these "cutting" surfaces in general topological situation is not a trivial task: some algorithms have been proposed by Kotiuga [152], [153], [154], Leonard et al. [165], Gross and Kotiuga [114], Ren [207].

In this context, we are in a position to make precise the additional conditions that we have to impose in order to determine a unique vector potential $\mathbf{A}$ in Ω. To achieve this, first of all we need to introduce the family of $n_{\partial\Omega}$ "cuts" Σ_k that are more precisely defined in Section A.4. Then we require that the following geometrical condition is satisfied.

> It is assumed that $n_{\partial\Omega} \le n_\Omega$. Moreover, the family of "cuts" $\hat{\Sigma}_t$ coincides with the family of "cuts" Σ_k for $t, k = 1, \ldots, n_{\partial\Omega}$ (in particular, $\hat{\Sigma}_t \subset \Omega_I$ for each $t = 1, \ldots, n_{\partial\Omega}$), whereas $\hat{\Sigma}_t \cap \Omega_C \neq \emptyset$ for each $t = n_{\partial\Omega} + 1, \ldots, n_\Omega$. Finally, the "cuts" $\hat{\Sigma}_t$ are assumed to be disjoint for $t = 1, \ldots, n_{\partial\Omega}$. (6.2)

Let us note that we do not know any example in which this assumption fails to hold; on the other hand, we have not found a proof that it always holds.

When $n_{\partial\Omega} = n_\Omega$, this is telling us that we can choose the "cuts" in Ω associated to the vector space $\mathcal{H}(m;\Omega)$ without intersecting Ω_C (see Figure 6.1). Conversely, when $n_{\partial\Omega} < n_\Omega$ some of the "cuts" have to intersect the conductor Ω_C: for example, this happens in Figure 6.2, where Ω and Ω_C are two coaxial tori ($n_{\partial\Omega} = 0$, $n_\Omega = 1$), and in Figure 6.3, where Ω is a double torus and Ω_C is a torus, co-axial to one of the two handles of Ω ($n_{\partial\Omega} = 1$, $n_\Omega = 2$).

We are going to prove that the number of the needed additional conditions to be imposed to the vector field $\mathbf{A}$ is $n_{\partial\Omega}$. Let us temporarily formulate these conditions in the abstract form

$$\mathcal{G}_k(\mathbf{A}) = 0 \qquad \forall\, k = 1, \ldots, n_{\partial\Omega}\,,$$

Fig. 6.1. The computational domain: the dark rectangle is the "cutting" surface $\hat{\Sigma}_1 = \Sigma_1$

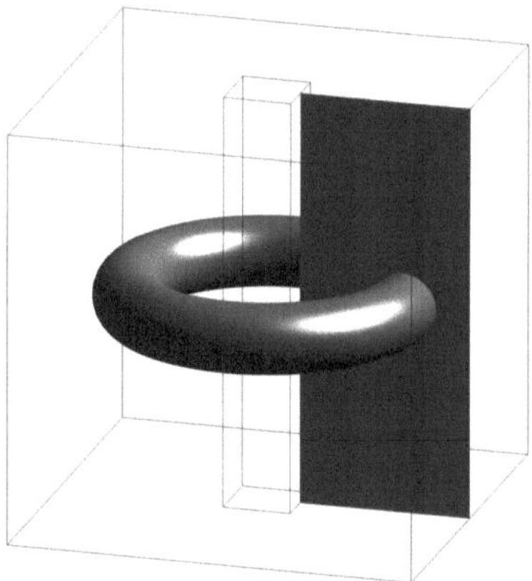

Fig. 6.2. The computational domain: the dark rectangle is the "cutting" surface $\hat{\Sigma}_1$

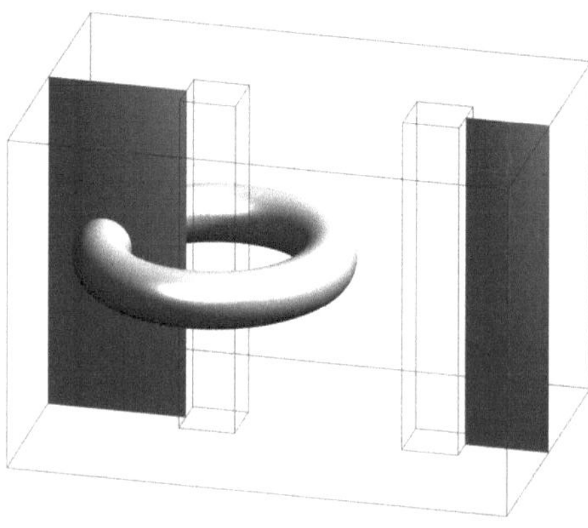

Fig. 6.3. The computational domain: the dark rectangles are the "cutting" surfaces ($\hat{\Sigma}_1 = \Sigma_1$ on the right, $\hat{\Sigma}_2$ on the left)

where $\mathcal{G}_t(\cdot)$, $t = 1, \ldots, n_\Omega$, are a suitable set of linear functionals that we will make precise in the sequel (see (6.29) and (6.30)).

In conclusion, we are left with the problem

$$\begin{cases} \operatorname{curl}(\boldsymbol{\mu}^{-1} \operatorname{curl} \mathbf{A}) + i\omega\boldsymbol{\sigma}\mathbf{A} + \boldsymbol{\sigma} \operatorname{grad} V_C = \mathbf{J}_e & \text{in } \Omega \\ \operatorname{div} \mathbf{A} = 0 & \text{in } \Omega \\ \mathcal{G}_k(\mathbf{A}) = 0 & \forall\, k = 1, \ldots, n_{\partial\Omega} \\ \mathbf{A} \cdot \mathbf{n} = 0 & \text{on } \partial\Omega \\ (\boldsymbol{\mu}^{-1} \operatorname{curl} \mathbf{A}) \times \mathbf{n} = \mathbf{0} & \text{on } \partial\Omega \,, \end{cases} \tag{6.3}$$

where the notation $\boldsymbol{\sigma} \operatorname{grad} V_C$ means

$$\boldsymbol{\sigma} \operatorname{grad} V_C := \begin{cases} \boldsymbol{\sigma}_{|\Omega_C} \operatorname{grad} V_C & \text{in } \Omega_C \\ \mathbf{0} & \text{in } \Omega_I \,, \end{cases}$$

and clearly V_C is determined up to an additive constant in each connected component $\Omega_{C,j}$ of Ω_C, $j = 1, \ldots, p_\Gamma + 1$.

Problem (6.3) simplifies if $n_{\partial\Omega} = 0$: this is true, for instance, if the computational domain Ω is simply-connected, an assumption that is not very restrictive in engineering computations.

The necessity of introducing a gauge on the vector magnetic potential $\mathbf{A}$ leads to the presence in (6.3) of the differential constraint $\operatorname{div} \mathbf{A} = 0$ in Ω. Like the constraint on the magnetic vector field $\operatorname{curl} \mathbf{H}_I = \mathbf{J}_{e,I}$ in Ω_I, that appears in the eddy current problem (1.22), this equation is not easy to treat at the discrete level, as it is not simple to construct a suitable space of finite elements which are divergence-free. In the preceding Chapters 4 and 5 we have seen two different approaches to help us treat differential constraints: the introduction of Lagrange multipliers in Chapter 4, and the use of a scalar magnetic potential in Chapter 5.

Here we use a different idea: the addition of a *penalization* term. This can be done as follows (see, e.g., Coulomb [91], Morisue [180], Bíró and Preis [49]): introducing the constant $\mu_* > 0$, representing a suitable average in Ω of the entries of the matrix $\boldsymbol{\mu}$, the Coulomb gauge condition $\operatorname{div} \mathbf{A} = 0$ in Ω can be incorporated in the Ampère equation, which becomes

$$\operatorname{curl}(\boldsymbol{\mu}^{-1} \operatorname{curl} \mathbf{A}) - \mu_*^{-1} \operatorname{grad} \operatorname{div} \mathbf{A} + i\omega\boldsymbol{\sigma}\mathbf{A} + \boldsymbol{\sigma} \operatorname{grad} V_C = \mathbf{J}_e \qquad \text{in } \Omega \,;$$

moreover one adds the two equations

$$\begin{cases} \operatorname{div}(i\omega\boldsymbol{\sigma}\mathbf{A}_C + \boldsymbol{\sigma} \operatorname{grad} V_C) = \operatorname{div} \mathbf{J}_{e,C} & \text{in } \Omega_C \\ (i\omega\boldsymbol{\sigma}\mathbf{A}_C + \boldsymbol{\sigma} \operatorname{grad} V_C) \cdot \mathbf{n}_C = \mathbf{J}_{e,C} \cdot \mathbf{n}_C + \mathbf{J}_{e,I} \cdot \mathbf{n}_I & \text{on } \Gamma \,, \end{cases}$$

being necessary as, due to the modification in the Ampère equation, it is no longer guaranteed now that the electric field $\mathbf{E}_C = -i\omega\mathbf{A}_C - \operatorname{grad} V_C$ satisfies the necessary conditions $\operatorname{div}(\boldsymbol{\sigma}\mathbf{E}_C) = -\operatorname{div} \mathbf{J}_{e,C}$ in Ω_C and $\boldsymbol{\sigma}\mathbf{E}_C \cdot \mathbf{n}_C = -\mathbf{J}_{e,C} \cdot \mathbf{n}_C - \mathbf{J}_{e,I} \cdot \mathbf{n}_I$ on Γ.

The complete $(\mathbf{A}, V_C)$ formulation is therefore

$$\begin{cases} \operatorname{curl}(\boldsymbol{\mu}^{-1} \operatorname{curl} \mathbf{A}) - \mu_*^{-1} \operatorname{grad} \operatorname{div} \mathbf{A} \\ \qquad\qquad + i\omega\boldsymbol{\sigma}\mathbf{A} + \boldsymbol{\sigma} \operatorname{grad} V_C = \mathbf{J}_e & \text{in } \Omega \\ \operatorname{div}(i\omega\boldsymbol{\sigma}\mathbf{A}_C + \boldsymbol{\sigma} \operatorname{grad} V_C) = \operatorname{div} \mathbf{J}_{e,C} & \text{in } \Omega_C \\ (i\omega\boldsymbol{\sigma}\mathbf{A}_C + \boldsymbol{\sigma} \operatorname{grad} V_C) \cdot \mathbf{n}_C \\ \qquad\qquad = \mathbf{J}_{e,C} \cdot \mathbf{n}_C + \mathbf{J}_{e,I} \cdot \mathbf{n}_I & \text{on } \Gamma \\ \mathcal{G}_k(\mathbf{A}) = 0 & \forall\, k = 1, \dots, n_{\partial\Omega} \\ \mathbf{A} \cdot \mathbf{n} = 0 & \text{on } \partial\Omega \\ (\boldsymbol{\mu}^{-1} \operatorname{curl} \mathbf{A}) \times \mathbf{n} = \mathbf{0} & \text{on } \partial\Omega \,, \end{cases} \tag{6.4}$$

where, as before, V_C is determined up to an additive constant in each connected component $\Omega_{C,j}$ of Ω_C, $j = 1, \dots, p_\Gamma + 1$.

In the next section, starting from (6.4), we obtain a suitable weak problem, which is shown to be equivalent to (6.3). Hence it is not strictly necessary to prove that from (6.4) we can recover (6.3). However, for the sake of completeness, here is the result:

Lemma 6.1. *For any solution* $(\mathbf{A}, V_C)$ *to* (6.4) *one has* $\operatorname{div} \mathbf{A} = 0$ *in* Ω, *therefore* $(\mathbf{A}, V_C)$ *is indeed a solution to* (6.3).

Proof. Taking the divergence of $(6.4)_1$ and using $(6.4)_2$ we have $-\Delta \operatorname{div} \mathbf{A}_C = 0$ in Ω_C. Moreover, recalling that the current density $\mathbf{J}_e$ by assumption satisfies $\operatorname{div} \mathbf{J}_{e,I} = 0$ in Ω_I, one also obtains $-\Delta \operatorname{div} \mathbf{A}_I = 0$ in Ω_I. On the other hand, using $(6.4)_3$, on the interface Γ we have

$$\begin{aligned} -\mu_*^{-1} \operatorname{grad} \operatorname{div} \mathbf{A}_C \cdot \mathbf{n}_C &= -\mathbf{J}_{e,I} \cdot \mathbf{n}_I - \operatorname{curl}(\boldsymbol{\mu}_C^{-1} \operatorname{curl} \mathbf{A}_C) \cdot \mathbf{n}_C \\ &= -\mathbf{J}_{e,I} \cdot \mathbf{n}_I - \operatorname{div}_\tau[(\boldsymbol{\mu}_C^{-1} \operatorname{curl} \mathbf{A}_C) \times \mathbf{n}_C] \,, \end{aligned} \tag{6.5}$$

and also

$$\begin{aligned} -\mu_*^{-1} \operatorname{grad} \operatorname{div} \mathbf{A}_I \cdot \mathbf{n}_I &= \mathbf{J}_{e,I} \cdot \mathbf{n}_I - \operatorname{curl}(\boldsymbol{\mu}_I^{-1} \operatorname{curl} \mathbf{A}_I) \cdot \mathbf{n}_I \\ &= \mathbf{J}_{e,I} \cdot \mathbf{n}_I - \operatorname{div}_\tau[(\boldsymbol{\mu}_I^{-1} \operatorname{curl} \mathbf{A}_I) \times \mathbf{n}_I] \,. \end{aligned} \tag{6.6}$$

Moreover, a solution to $(6.4)_1$ satisfies on the interface Γ

$$\begin{aligned} &(\boldsymbol{\mu}_C^{-1} \operatorname{curl} \mathbf{A}_C) \times \mathbf{n}_C + \mu_*^{-1} \operatorname{div} \mathbf{A}_C \, \mathbf{n}_C \\ &\qquad + (\boldsymbol{\mu}_I^{-1} \operatorname{curl} \mathbf{A}_I) \times \mathbf{n}_I + \mu_*^{-1} \operatorname{div} \mathbf{A}_I \, \mathbf{n}_I = \mathbf{0} \,, \end{aligned}$$

therefore, due to orthogonality,

$$(\boldsymbol{\mu}_C^{-1} \operatorname{curl} \mathbf{A}_C) \times \mathbf{n}_C + (\boldsymbol{\mu}_I^{-1} \operatorname{curl} \mathbf{A}_I) \times \mathbf{n}_I = \mathbf{0} \;,\; \operatorname{div} \mathbf{A}_C = \operatorname{div} \mathbf{A}_I \,.$$

Hence from (6.5) and (6.6) we also have

$$\operatorname{grad} \operatorname{div} \mathbf{A}_C \cdot \mathbf{n}_C + \operatorname{grad} \operatorname{div} \mathbf{A}_I \cdot \mathbf{n}_I = 0 \quad \text{on } \Gamma \,.$$

This last condition and the matching of $\operatorname{div} \mathbf{A}$ on Γ furnish that $\operatorname{div} \mathbf{A}$ is a harmonic function in all of Ω. Moreover, using $(6.4)_6$, its Neumann value on the boundary $\partial\Omega$ satisfies

$$\begin{aligned} -\mu_*^{-1} \operatorname{grad} \operatorname{div} \mathbf{A} \cdot \mathbf{n} &= \mathbf{J}_{e,I} \cdot \mathbf{n} - \operatorname{curl}(\boldsymbol{\mu}^{-1} \operatorname{curl} \mathbf{A}) \cdot \mathbf{n} \\ &= \mathbf{J}_{e,I} \cdot \mathbf{n} - \operatorname{div}_\tau[(\boldsymbol{\mu}^{-1} \operatorname{curl} \mathbf{A}) \times \mathbf{n}] = 0 \,, \end{aligned}$$

as by assumption the current density satisfies $\mathbf{J}_{e,I} \cdot \mathbf{n} = 0$ on $\partial\Omega$. As a consequence, we find that $\operatorname{div} \mathbf{A}$ is a constant c_0 in Ω, and finally

$$c_0 \operatorname{meas}(\Omega) = \int_\Omega \operatorname{div} \mathbf{A} = \int_{\partial\Omega} \mathbf{A} \cdot \mathbf{n} = 0 \,,$$

having used $(6.4)_5$. □

In conclusion, we have proved that any solution to (6.4) yields a solution to (6.3), and consequently, by virtue of (6.1), it is the solution to the eddy current problem (3.25).

Remark 6.2. It is worth noting that, after having solved the Coulomb gauged problem (6.3), hence having determined $\mathbf{A}$ and V_C from the data of the problem, we are also in a condition to find the electric field $\mathbf{E}_I$ in Ω_I.

In fact first we solve the mixed boundary value problem

$$\begin{cases} -\operatorname{div}(\varepsilon_I \operatorname{grad} V_I^\dagger) = i\omega \operatorname{div}(\varepsilon_I \mathbf{A}_I) & \text{in } \Omega_I \\ V_I^\dagger = V_C & \text{on } \Gamma \\ \varepsilon_I \operatorname{grad} V_I^\dagger \cdot \mathbf{n} = -i\omega\varepsilon_I \mathbf{A}_I \cdot \mathbf{n} & \text{on } \partial\Omega \ . \end{cases} \tag{6.7}$$

Then we determine the vector $(c^\dagger_{I,j}, d^\dagger_{I,k})$, $j = 1, \ldots, p_\Gamma$, $k = 1, \ldots, n_{\partial\Omega}$, the solution of the linear system

$$A^\dagger \begin{pmatrix} c^\dagger_{I,j} \\ d^\dagger_{I,k} \end{pmatrix} = \begin{pmatrix} \int_{\Omega_I} \varepsilon_I (i\omega \mathbf{A}_I + \operatorname{grad} V_I^\dagger) \cdot \operatorname{grad} w_{g,I} \\ \int_{\Omega_I} \varepsilon_I (i\omega \mathbf{A}_I + \operatorname{grad} V_I^\dagger) \cdot \boldsymbol{\pi}_{i,I} \end{pmatrix} , \tag{6.8}$$

$g = 1, \ldots, p_\Gamma$, $i = 1, \ldots, n_{\partial\Omega}$, where, as in (A.19), $A^\dagger := \begin{pmatrix} D^\dagger & B^\dagger \\ (B^\dagger)^T & C^\dagger \end{pmatrix}$ with

$$\begin{aligned} D^\dagger_{gj} &:= \textstyle\int_{\Omega_I} \varepsilon_I \operatorname{grad} w_{j,I} \cdot \operatorname{grad} w_{g,I} \\ B^\dagger_{gk} &:= \textstyle\int_{\Omega_I} \varepsilon_I \boldsymbol{\pi}_{k,I} \cdot \operatorname{grad} w_{g,I} \\ C^\dagger_{ik} &:= \textstyle\int_{\Omega_I} \varepsilon_I \boldsymbol{\pi}_{k,I} \cdot \boldsymbol{\pi}_{i,I} \,, \end{aligned} \tag{6.9}$$

and the harmonic vector fields $\operatorname{grad} w_{j,I}$ and $\boldsymbol{\pi}_{k,I}$ are the basis functions of the space $\mathcal{H}_{\varepsilon_I}(\Gamma, \partial\Omega; \Omega_I)$ introduced in Section 1.4.

It is easily proved that the matrix $A^\dagger$ is symmetric and positive definite, as the matrix $\varepsilon_I(\mathbf{x})$ is symmetric and positive definite, uniformly with respect to $\mathbf{x}$, and the functions $\boldsymbol{\pi}_{k,I}$ and $\operatorname{grad} w_{j,I}$ are linearly independent.

Then, defining $\mathbf{h}_I^\dagger := \sum_{j=1}^{p_\Gamma} c^\dagger_{I,j} \operatorname{grad} w_{j,I} + \sum_{k=1}^{n_{\partial\Omega}} d^\dagger_{I,k} \boldsymbol{\pi}_{k,I}$ and

$$\mathbf{E}_I := -i\omega \mathbf{A}_I - \operatorname{grad} V_I^\dagger + \mathbf{h}_I^\dagger \,,$$

taking into account (6.1) it is easily checked that $\operatorname{curl} \mathbf{E}_I = -i\omega \boldsymbol{\mu}_I \mathbf{H}_I$ in Ω_I, $\operatorname{div}(\varepsilon_I \mathbf{E}_I) = 0$ in Ω_I,

$$\begin{aligned} \mathbf{E}_I \times \mathbf{n}_I &= -i\omega \mathbf{A}_I \times \mathbf{n}_I - \operatorname{grad} V_I^\dagger \times \mathbf{n}_I \\ &= i\omega \mathbf{A}_C \times \mathbf{n}_C + \operatorname{grad} V_C \times \mathbf{n}_C = -\mathbf{E}_C \times \mathbf{n}_C \quad \text{on } \Gamma \,, \end{aligned}$$

$\varepsilon_I \mathbf{E}_I \cdot \mathbf{n} = 0$ on $\partial\Omega$, $\int_{\Gamma_g} \varepsilon_I \mathbf{E}_I \cdot \mathbf{n}_I = \int_{\Omega_I} \varepsilon_I \mathbf{E}_I \cdot \operatorname{grad} w_{g,I} = 0$ for all $g = 1, \dots, p_\Gamma$ and $\int_{\Omega_I} \varepsilon_I \mathbf{E}_I \cdot \boldsymbol{\pi}_{i,I} = 0$ for all $i = 1, \dots, n_{\partial\Omega}$, therefore $\mathbf{E}_I$ is the electric field in Ω_I (see (3.13)). □

6.1.1 The weak formulation

We are now interested in finding a suitable weak formulation of (6.4).

First of all, taking a test function $\mathbf{w} \in H(\mathrm{curl}; \Omega) \cap H_0(\mathrm{div}; \Omega)$, multiplying $(6.4)_1$ by $\overline{\mathbf{w}}$ and integrating in Ω, we obtain by integration by parts

$$\begin{aligned}\int_\Omega (\boldsymbol{\mu}^{-1} \operatorname{curl} \mathbf{A} \cdot \operatorname{curl} \overline{\mathbf{w}} + \mu_*^{-1} \operatorname{div} \mathbf{A} \operatorname{div} \overline{\mathbf{w}}) \\ + \int_{\Omega_C} (i\omega \boldsymbol{\sigma} \mathbf{A}_C \cdot \overline{\mathbf{w}_C} + \boldsymbol{\sigma} \operatorname{grad} V_C \cdot \overline{\mathbf{w}_C}) \\ = \int_\Omega \mathbf{J}_e \cdot \overline{\mathbf{w}} \,,\end{aligned}$$

where we have used $(6.4)_6$.

Let us now multiply $(6.4)_2$ by $i\omega^{-1}\overline{Q_C}$, where $Q_C \in H^1(\Omega_C)$, and integrate in Ω_C: by integration by parts we find

$$\begin{aligned}\int_{\Omega_C} (-\boldsymbol{\sigma} \mathbf{A}_C \cdot \operatorname{grad} \overline{Q_C} + i\omega^{-1} \boldsymbol{\sigma} \operatorname{grad} V_C \cdot \operatorname{grad} \overline{Q_C}) \\ = i\omega^{-1} \int_{\Omega_C} \mathbf{J}_{e,C} \cdot \operatorname{grad} \overline{Q_C} + i\omega^{-1} \int_\Gamma \mathbf{J}_{e,I} \cdot \mathbf{n}_I \overline{Q_C} \,,\end{aligned} \tag{6.10}$$

having used $(6.4)_3$.

Introducing the sesquilinear form

$$\begin{aligned}&\mathcal{A}[(\mathbf{z}, U_C), (\mathbf{w}, Q_C)] \\ &\quad := \int_\Omega (\boldsymbol{\mu}^{-1} \operatorname{curl} \mathbf{z} \cdot \operatorname{curl} \overline{\mathbf{w}} + \mu_*^{-1} \operatorname{div} \mathbf{z} \operatorname{div} \overline{\mathbf{w}}) \\ &\quad\quad + \int_{\Omega_C} (i\omega \boldsymbol{\sigma} \mathbf{z}_C \cdot \overline{\mathbf{w}_C} + \boldsymbol{\sigma} \operatorname{grad} U_C \cdot \overline{\mathbf{w}_C}) \\ &\quad\quad + \int_{\Omega_C} (-\boldsymbol{\sigma} \mathbf{z}_C \cdot \operatorname{grad} \overline{Q_C} + i\omega^{-1} \boldsymbol{\sigma} \operatorname{grad} U_C \cdot \operatorname{grad} \overline{Q_C}) \\ &\quad = \int_\Omega (\boldsymbol{\mu}^{-1} \operatorname{curl} \mathbf{z} \cdot \operatorname{curl} \overline{\mathbf{w}} + \mu_*^{-1} \operatorname{div} \mathbf{z} \operatorname{div} \overline{\mathbf{w}}) \\ &\quad\quad + i\omega^{-1} \int_{\Omega_C} \boldsymbol{\sigma} (i\omega \mathbf{z}_C + \operatorname{grad} U_C) \cdot (-i\omega \overline{\mathbf{w}_C} + \operatorname{grad} \overline{Q_C}) \,,\end{aligned} \tag{6.11}$$

we have finally rewritten (6.4) as

$$\begin{aligned}&\text{Find } (\mathbf{A}, V_C) \in W_\sharp \times H^1_\sharp(\Omega_C) \text{ such that} \\ &\quad \mathcal{A}[(\mathbf{A}, V_C), (\mathbf{w}, Q_C)] = \int_\Omega \mathbf{J}_e \cdot \overline{\mathbf{w}} \\ &\quad\quad + i\omega^{-1} \int_{\Omega_C} \mathbf{J}_{e,C} \cdot \operatorname{grad} \overline{Q_C} + i\omega^{-1} \int_\Gamma \mathbf{J}_{e,I} \cdot \mathbf{n}_I \overline{Q_C} \\ &\text{for all } (\mathbf{w}, Q_C) \in W_\sharp \times H^1_\sharp(\Omega_C) \,,\end{aligned} \tag{6.12}$$

where

$$\begin{aligned}W_\sharp := \{\mathbf{w} \in H(\mathrm{curl}; \Omega) \cap H_0(\mathrm{div}; \Omega) \mid \\ \mathcal{G}_k(\mathbf{w}) = 0 \ \forall\, k = 1, \dots, n_{\partial\Omega}\} \,,\end{aligned} \tag{6.13}$$

and

$$H^1_\sharp(\Omega_C) := \prod_{j=1}^{p_\Gamma + 1} H^1(\Omega_{C,j}) / \mathbb{C} \,. \tag{6.14}$$

Before starting the proof that the weak problem (6.12) has a unique solution (for that, see Section 6.1.2), it is useful to show that a solution to (6.12) is indeed a solution to (6.3).

Let us first introduce a couple of assumptions on the linear functional $\mathcal{G}_t$. The first one reads

$$\text{For } \mathbf{w}_m \in \mathcal{H}(m;\Omega), \text{ the conditions } \mathcal{G}_t(\mathbf{w}_m) = 0 \text{ for each } t = 1,\ldots,n_\Omega \text{ give } \mathbf{w}_m = \mathbf{0} \text{ in } \Omega. \tag{6.15}$$

The second one requires some preliminaries: we recall that a set of basis functions of $\mathcal{H}(m;\Omega)$ is given by $\widetilde{\boldsymbol{\pi}}_t$, $t = 1,\ldots,n_\Omega$, the $(L^2(\Omega))^3$-extension of $\operatorname{grad}\hat{q}_t$, where $\hat{q}_t$, defined in $\Omega\setminus\hat{\Sigma}_t$, is the real-valued function, determined up to an additive constant, solution of

$$\begin{cases} \Delta\hat{q}_t = 0 & \text{in } \Omega\setminus\hat{\Sigma}_t \\ \operatorname{grad}\hat{q}_t\cdot\mathbf{n} = 0 & \text{on } \partial\Omega\setminus\partial\hat{\Sigma}_t \\ [\operatorname{grad}\hat{q}_t\cdot\mathbf{n}_\Sigma]_{\hat{\Sigma}_t} = 0 & \\ [\hat{q}_t]_{\hat{\Sigma}_t} = 1\,, & \end{cases} \tag{6.16}$$

having denoted by $[\,\cdot\,]_{\hat{\Sigma}_t}$ the jump across the surface $\hat{\Sigma}_t$ (see, e.g., Foias and Temam [106]; see also Section A.4). Moreover, when (6.15) is satisfied we can introduce the real-valued vector functions $\boldsymbol{\eta}_t$, the basis functions of the space $\mathcal{H}(m;\Omega)$ such that $\mathcal{G}_q(\boldsymbol{\eta}_t) = \delta_{tq}$ (the Kronecker symbol).

The second assumption on the functionals $\mathcal{G}_t$ is given by

The basis $\boldsymbol{\eta}_t$ associated to the linear functionals $\mathcal{G}_t(\cdot)$ can be expressed in terms of the basis $\widetilde{\boldsymbol{\pi}}_q$ by means of a matrix $\{\beta_{tq}\}$, $t,q = 1,\ldots,n_\Omega$, such that its principal minor for $t,q = n_{\partial\Omega}+1,\ldots,n_\Omega$ is non-singular. (6.17)

At the end of this section we present some possible choices of the functionals $\mathcal{G}_t$ satisfying these conditions.

Theorem 6.3. *Let* $\mathbf{J}_e \subset (L^2(\Omega))^3$ *satisfy the necessary conditions* (1.23). *If* $n_{\partial\Omega} > 0$, *assume that the geometrical condition* (6.2) *is verified and that the linear functionals* $\mathcal{G}_t$ *satisfy the conditions* (6.15) *and* (6.17) *(this last one when* $0 < n_{\partial\Omega} < n_\Omega$*). Then any solution to the weak problem* (6.12) *is a solution to the strong problem* (6.3).

Proof. Let us first prove that a solution to (6.12) satisfies

$$\begin{aligned}\mathcal{A}[(\mathbf{A},V_C),(\mathbf{w},Q_C)] = {}& \textstyle\int_\Omega \mathbf{J}_e\cdot\overline{\mathbf{w}} \\ &+ i\omega^{-1}\textstyle\int_{\Omega_C}\mathbf{J}_{e,C}\cdot\operatorname{grad}\overline{Q_C} + i\omega^{-1}\int_\Gamma \mathbf{J}_{e,I}\cdot\mathbf{n}_I\,\overline{Q_C}\end{aligned} \tag{6.18}$$

also for any test function $(\mathbf{w},Q_C)$ such that $\mathbf{w}\in H(\operatorname{curl};\Omega)\cap H_0(\operatorname{div};\Omega)$ and $Q_C \in H^1(\Omega_C)$, namely, without any additional constraint.

First of all, equation (6.12) does not change if we add to Q_C a different constant in each connected component $\Omega_{C,j}$ of Ω_C. In fact, we know that $\int_{\Gamma_j}\mathbf{J}_{e,I}\cdot\mathbf{n}_I = 0$ for each $j = 1,\ldots,p_\Gamma$, and moreover $\operatorname{div}\mathbf{J}_{e,I} = 0$ in Ω_I and $\mathbf{J}_{e,I}\cdot\mathbf{n} = 0$ on $\partial\Omega$, so that

$\int_{\Gamma_j} \mathbf{J}_{e,I} \cdot \mathbf{n}_I = 0$ for $j = p_\Gamma + 1$, too. Hence a solution $(\mathbf{A}, V_C)$ of (6.12) satisfies the integral equation not only for every $Q_C \in H^1_\sharp(\Omega_C)$ but also for each $Q_C \in H^1(\Omega_C)$.

As a consequence, taking $\mathbf{w} = \mathbf{0}$ a first general result is that any solution to (6.12) satisfies

$$\begin{cases} \operatorname{div}(i\omega\boldsymbol{\sigma}\mathbf{A}_C + \boldsymbol{\sigma}\operatorname{grad} V_C) = \operatorname{div}\mathbf{J}_{e,C} & \text{in } \Omega_C \\ (i\omega\boldsymbol{\sigma}\mathbf{A}_C + \boldsymbol{\sigma}\operatorname{grad} V_C)\cdot\mathbf{n}_C = \mathbf{J}_{e,C}\cdot\mathbf{n}_C + \mathbf{J}_{e,I}\cdot\mathbf{n}_I & \text{on } \Gamma\,. \end{cases} \tag{6.19}$$

Setting

$$\mathbf{J} := \begin{cases} -i\omega\boldsymbol{\sigma}\mathbf{A}_C - \boldsymbol{\sigma}\operatorname{grad} V_C + \mathbf{J}_{e,C} & \text{in } \Omega_C \\ \mathbf{J}_{e,I} & \text{in } \Omega_I \end{cases},$$

the assumptions $\operatorname{div}\mathbf{J}_{e,I} = 0$ in Ω_I and $\mathbf{J}_{e,I}\cdot\mathbf{n} = 0$ on $\partial\Omega$ are telling us that $\operatorname{div}\mathbf{J} = 0$ in Ω and $\mathbf{J}\cdot\mathbf{n} = 0$ on $\partial\Omega$.

In the case $n_{\partial\Omega} = 0$ we have finished the proof that (6.18) is satisfied for any $\mathbf{w} \in H(\operatorname{curl};\Omega) \cap H_0(\operatorname{div};\Omega)$ and $Q_C \in H^1(\Omega_C)$. Instead, if $n_{\partial\Omega} > 0$ we need additional informations.

Due to assumption (6.15), for any $\mathbf{w} \in H(\operatorname{curl};\Omega) \cap H_0(\operatorname{div};\Omega)$ we can define by $\mathbf{w}_m$ the harmonic field in $\mathcal{H}(m;\Omega)$ satisfying $\mathcal{G}_t(\mathbf{w}_m) = \mathcal{G}_t(\mathbf{w})$ for each $t = 1,\ldots,n_\Omega$. Clearly, the difference $\mathbf{w} - \mathbf{w}_m$ belongs to $W_\sharp$. Hence

$$\begin{aligned} &\mathcal{A}[(\mathbf{A}, V_C), (\mathbf{w}, Q_C)] \\ &\quad = \mathcal{A}[(\mathbf{A}, V_C), (\mathbf{w} - \mathbf{w}_m, Q_C)] + \mathcal{A}[(\mathbf{A}, V_C), (\mathbf{w}_m, 0)] \\ &\quad = \textstyle\int_\Omega \mathbf{J}_e \cdot (\overline{\mathbf{w}} - \overline{\mathbf{w}_m}) + i\omega^{-1}\int_{\Omega_C} \mathbf{J}_{e,C}\cdot\operatorname{grad}\overline{Q_C} \\ &\qquad + i\omega^{-1}\textstyle\int_\Gamma \mathbf{J}_{e,I}\cdot\mathbf{n}_I\,\overline{Q_C} + \int_{\Omega_C}(i\omega\boldsymbol{\sigma}\mathbf{A}_C + \boldsymbol{\sigma}\operatorname{grad} V_C)\cdot\overline{\mathbf{w}_{m,C}} \\ &\quad = \textstyle\int_\Omega \mathbf{J}_e\cdot\overline{\mathbf{w}} + i\omega^{-1}\int_{\Omega_C}\mathbf{J}_{e,C}\cdot\operatorname{grad}\overline{Q_C} \\ &\qquad + i\omega^{-1}\textstyle\int_\Gamma \mathbf{J}_{e,I}\cdot\mathbf{n}_I\,\overline{Q_C} - \int_\Omega \mathbf{J}\cdot\overline{\mathbf{w}_m}\,. \end{aligned} \tag{6.20}$$

Therefore, the only result that remains to be proved is

$$\int_\Omega \mathbf{J}\cdot\overline{\mathbf{w}_m} = 0\,, \tag{6.21}$$

or, equivalently, that

$$\int_\Omega \mathbf{J}\cdot\hat{\boldsymbol{\pi}}_t = 0 \qquad \forall\, t = 1,\ldots,n_\Omega\,, \tag{6.22}$$

where $\hat{\boldsymbol{\pi}}_t$ are the basis function of $\mathcal{H}(m;\Omega)$ introduced in (6.16).

We have not yet used the assumptions

$$\int_{\Omega_I} \mathbf{J}_{e,I}\cdot\boldsymbol{\pi}_{k,I} = 0 \quad \forall\, k = 1,\ldots,n_{\partial\Omega}\,,$$

where $\boldsymbol{\pi}_{k,I}$ are basis functions of the space $\mathcal{H}_{\varepsilon_I}(\Gamma,\partial\Omega;\Omega_I)$ that are not gradients. Similarly to what was done for the basis functions $\hat{\boldsymbol{\pi}}_t$ in (6.16) (see also Section A.4), it is possible to express $\boldsymbol{\pi}_{k,I}$ as the $(L^2(\Omega_I))^3$-extension of grad $q_{k,I}$, where $q_{k,I}$, defined in $\Omega_I \setminus \Sigma_k$, is the real-valued function solution of (A.32), namely,

$$\begin{cases} \operatorname{div}(\varepsilon_I \operatorname{grad} q_{k,I}) = 0 & \text{in } \Omega_I \setminus \Sigma_k \\ \varepsilon_I \operatorname{grad} q_{k,I} \cdot \mathbf{n}_I = 0 & \text{on } \partial\Omega \setminus \partial\Sigma_k \\ q_{k,I} = 0 & \text{on } \Gamma \\ [\varepsilon_I \operatorname{grad} q_{k,I} \cdot \mathbf{n}_\Sigma]_{\Sigma_k} = 0 \\ [q_{k,I}]_{\Sigma_k} = 1 \, , \end{cases}$$

having denoted by $[\cdot]_{\Sigma_k}$ the jump across the surface Σ_k. Therefore, since $\operatorname{div} \mathbf{J}_{e,I} = 0$ in Ω_I and $\mathbf{J}_{e,I} \cdot \mathbf{n} = 0$ on $\partial\Omega$, we have

$$\begin{aligned} 0 = \int_{\Omega_I} \mathbf{J}_{e,I} \cdot \boldsymbol{\pi}_{k,I} &= \int_{\Omega_I \setminus \Sigma_k} \mathbf{J}_{e,I} \cdot \operatorname{grad} q_{k,I} \\ &= -\int_{\Omega_I \setminus \Sigma_k} (\operatorname{div} \mathbf{J}_{e,I}) q_{k,I} + \int_{\partial\Omega_I \setminus \partial\Sigma_k} \mathbf{J}_{e,I} \cdot \mathbf{n}_I \, q_{k,I} \\ &\qquad + \int_{\Sigma_k} \mathbf{J}_{e,I} \cdot \mathbf{n}_\Sigma \, [q_{k,I}]_{\Sigma_k} \\ &= \int_{\Sigma_k} \mathbf{J}_{e,I} \cdot \mathbf{n}_\Sigma \, . \end{aligned}$$

On the other hand, proceeding in the same way and recalling that $\operatorname{div} \mathbf{J} = 0$ in Ω, $\mathbf{J} \cdot \mathbf{n} = 0$ on $\partial\Omega$ and that from assumption (6.2) for $k = 1, \ldots, n_{\partial\Omega}$ the "cuts" $\hat{\Sigma}_k$ are coincident with Σ_k, we find that

$$\begin{aligned} 0 &= \int_{\Sigma_k} \mathbf{J}_{e,I} \cdot \mathbf{n}_\Sigma = \int_{\Sigma_k} \mathbf{J} \cdot \mathbf{n}_\Sigma \\ &= \int_{\hat{\Sigma}_k} \mathbf{J} \cdot \mathbf{n}_{\hat{\Sigma}} = \int_\Omega \mathbf{J} \cdot \hat{\boldsymbol{\pi}}_k \qquad \forall\, k = 1, \ldots, n_{\partial\Omega} \, . \end{aligned} \tag{6.23}$$

This result ends the proof in the case $0 < n_{\partial\Omega} = n_\Omega$. On the other hand, if $0 < n_{\partial\Omega} < n_\Omega$ it is easily seen that the basis functions $\boldsymbol{\eta}_t$ introduced in (6.17) satisfy $\boldsymbol{\eta}_t \in W_\sharp$ for $t = n_{\partial\Omega} + 1, \ldots, n_\Omega$. Using these test functions (and $Q_C = 0$) in (6.12) we find

$$\int_\Omega \mathbf{J} \cdot \boldsymbol{\eta}_t = 0 \qquad \forall\, t = n_{\partial\Omega} + 1, \ldots, n_\Omega \, . \tag{6.24}$$

We can write the basis $\boldsymbol{\eta}_t$ in terms of the basis $\hat{\boldsymbol{\pi}}_t$

$$\boldsymbol{\eta}_t = \sum_{q=1}^{n_\Omega} \beta_{tq} \hat{\boldsymbol{\pi}}_q \, .$$

Hence, using (6.23) and (6.24), for $t = n_{\partial\Omega} + 1, \ldots, n_\Omega$ one has

$$0 = \int_\Omega \mathbf{J} \cdot \boldsymbol{\eta}_t = \sum_{q=1}^{n_\Omega} \beta_{tq} \int_\Omega \mathbf{J} \cdot \hat{\boldsymbol{\pi}}_q = \sum_{q=n_{\partial\Omega}+1}^{n_\Omega} \beta_{tq} \int_\Omega \mathbf{J} \cdot \hat{\boldsymbol{\pi}}_q \, . \tag{6.25}$$

As a consequence, from (6.23), (6.25) and condition (6.17) it follows that (6.22) holds, and we have finally proved that any solution to (6.12) indeed satisfies equation (6.18) for any test function $(\mathbf{w}, Q_C)$ such that $\mathbf{w} \in H(\mathrm{curl}; \Omega) \cap H_0(\mathrm{div}; \Omega)$ and $Q_C \in H^1(\Omega_C)$.

Now an additional step is to prove that any solution to (6.12) satisfies $\operatorname{div} \mathbf{A} = 0$ in Ω. In fact, let us denote by $\eta \in H^1(\Omega)/\mathbb{C}$ the solution of the Neumann problem $\Delta\eta =$

$\operatorname{div} \mathbf{A}$ in Ω and $\operatorname{grad} \eta \cdot \mathbf{n} = 0$ on $\partial\Omega$. Clearly $\mathbf{w} = \operatorname{grad} \eta \in H(\operatorname{curl}; \Omega) \cap H_0(\operatorname{div}; \Omega)$, and $\eta_C \in H^1(\Omega_C)$. Using $\mathbf{w}$ and $-i\omega\eta_C$ as test functions in (6.18) gives

$$\begin{aligned}\mu_*^{-1} \int_\Omega |\operatorname{div} \mathbf{A}|^2 &= \int_\Omega \mathbf{J}_e \cdot \operatorname{grad} \overline{\eta} - \int_{\Omega_C} \mathbf{J}_{e,C} \cdot \operatorname{grad} \overline{\eta_C} - \int_\Gamma \mathbf{J}_{e,I} \cdot \mathbf{n}_I \overline{\eta_C} \\ &= \int_{\Omega_I} \mathbf{J}_{e,I} \cdot \operatorname{grad} \overline{\eta_I} - \int_\Gamma \mathbf{J}_{e,I} \cdot \mathbf{n}_I \overline{\eta_C} = 0 \,,\end{aligned}$$

as

$$\begin{aligned}\int_{\Omega_I} \mathbf{J}_{e,I} \cdot \operatorname{grad} \overline{\eta_I} &= -\int_{\Omega_I} (\operatorname{div} \mathbf{J}_{e,I}) \overline{\eta_I} + \int_{\partial\Omega\cup\Gamma} \mathbf{J}_{e,I} \cdot \mathbf{n}_I \overline{\eta_I} \\ &= \int_\Gamma \mathbf{J}_{e,I} \cdot \mathbf{n}_I \overline{\eta_C} \,,\end{aligned}$$

hence $\operatorname{div} \mathbf{A} = 0$ in Ω.

We have thus proved that a solution $(\mathbf{A}, V_C)$ to problem (6.12) satisfies $\operatorname{div} \mathbf{A} = 0$ in Ω,

$$\begin{aligned}\int_\Omega \boldsymbol{\mu}^{-1} \operatorname{curl} \mathbf{A} \cdot \operatorname{curl} \overline{\mathbf{w}} + \int_{\Omega_C} (i\omega\boldsymbol{\sigma} \mathbf{A}_C \cdot \overline{\mathbf{w}_C} + \boldsymbol{\sigma} \operatorname{grad} V_C \cdot \overline{\mathbf{w}_C}) \\ = \int_\Omega \mathbf{J}_e \cdot \overline{\mathbf{w}}\end{aligned} \tag{6.26}$$

for all $\mathbf{w} \in H(\operatorname{curl}; \Omega) \cap H_0(\operatorname{div}; \Omega)$, and

$$\begin{aligned}\int_{\Omega_C} (i\omega\boldsymbol{\sigma} \mathbf{A}_C \cdot \operatorname{grad} \overline{Q_C} + \boldsymbol{\sigma} \operatorname{grad} V_C \cdot \operatorname{grad} \overline{Q_C}) \\ = \int_{\Omega_C} \mathbf{J}_{e,C} \cdot \operatorname{grad} \overline{Q_C} + \int_\Gamma \mathbf{J}_{e,I} \cdot \mathbf{n}_I \overline{Q_C}\end{aligned} \tag{6.27}$$

for each $Q_C \in H^1(\Omega_C)$. Indeed, we can prove something more: namely, equation (6.26) is satisfied for any function $\mathbf{w}^* \in H(\operatorname{curl}; \Omega)$. In fact, take the solution $\eta^* \in H^1(\Omega)/\mathbb{C}$ of $\Delta\eta^* = \operatorname{div} \mathbf{w}^*$ in Ω and $\operatorname{grad} \eta^* \cdot \mathbf{n} = \mathbf{w}^* \cdot \mathbf{n}$ on $\partial\Omega$. Using in (6.26) the test function $\mathbf{w} = (\mathbf{w}^* - \operatorname{grad} \eta^*) \in H(\operatorname{curl}; \Omega) \cap H_0(\operatorname{div}; \Omega)$, we have

$$\begin{aligned}\int_\Omega \boldsymbol{\mu}^{-1} \operatorname{curl} \mathbf{A} \cdot \operatorname{curl} \overline{\mathbf{w}^*} &+ \int_{\Omega_C} (i\omega\boldsymbol{\sigma} \mathbf{A}_C \cdot \overline{\mathbf{w}_C^*} + \boldsymbol{\sigma} \operatorname{grad} V_C \cdot \overline{\mathbf{w}_C^*}) \\ &= \int_\Omega \boldsymbol{\mu}^{-1} \operatorname{curl} \mathbf{A} \cdot \operatorname{curl} \overline{\mathbf{w}} + \int_{\Omega_C} (i\omega\boldsymbol{\sigma} \mathbf{A}_C \cdot \overline{\mathbf{w}_C} + \boldsymbol{\sigma} \operatorname{grad} V_C \cdot \overline{\mathbf{w}_C}) \\ &\quad + \int_{\Omega_C} (i\omega\boldsymbol{\sigma} \mathbf{A}_C \cdot \operatorname{grad} \overline{\eta_C^*} + \boldsymbol{\sigma} \operatorname{grad} V_C \cdot \operatorname{grad} \overline{\eta_C^*}) \\ &= \int_\Omega \mathbf{J}_e \cdot \overline{\mathbf{w}} + \int_{\Omega_C} \mathbf{J}_{e,C} \cdot \operatorname{grad} \overline{\eta_C^*} + \int_\Gamma \mathbf{J}_{e,I} \cdot \mathbf{n}_I \overline{\eta_C^*} \,,\end{aligned} \tag{6.28}$$

having used (6.27). Since

$$\begin{aligned}\int_\Omega \mathbf{J}_e \cdot \operatorname{grad} \overline{\eta^*} &= \int_{\Omega_C} \mathbf{J}_{e,C} \cdot \operatorname{grad} \overline{\eta_C^*} + \int_{\Omega_I} \mathbf{J}_{e,I} \cdot \operatorname{grad} \overline{\eta_I^*} \\ &= \int_{\Omega_C} \mathbf{J}_{e,C} \cdot \operatorname{grad} \overline{\eta_C^*} - \int_{\Omega_I} (\operatorname{div} \mathbf{J}_{e,I}) \overline{\eta_I^*} + \int_{\partial\Omega\cup\Gamma} \mathbf{J}_{e,I} \cdot \mathbf{n}_I \overline{\eta_I^*} \\ &= \int_{\Omega_C} \mathbf{J}_{e,C} \cdot \operatorname{grad} \overline{\eta_C^*} + \int_\Gamma \mathbf{J}_{e,I} \cdot \mathbf{n}_I \overline{\eta_C^*} \,,\end{aligned}$$

the right hand side in (6.28) is $\int_\Omega \mathbf{J}_e \cdot \overline{\mathbf{w}^*}$.

Taking now in (6.26) a test function $\mathbf{w}^* \in (C_0^\infty(\Omega))^3$, by integration by parts we find at once that

$$\operatorname{curl}(\boldsymbol{\mu}^{-1} \operatorname{curl} \mathbf{A}) + i\omega\boldsymbol{\sigma}\mathbf{A} + \boldsymbol{\sigma} \operatorname{grad} V_C = \mathbf{J}_e \quad \text{in } \Omega \,.$$

Repeating the same argument for $\mathbf{w}^* \in H(\operatorname{curl}; \Omega)$ gives $(\boldsymbol{\mu}^{-1} \operatorname{curl} \mathbf{A}) \times \mathbf{n} = \mathbf{0}$ on $\partial\Omega$.

Therefore we have concluded our proof: any solution $(\mathbf{A}, V_C)$ of the weak problem (6.12) is a solution of the strong problem (6.3). □

Let us finish this section by showing some examples of choices of the linear functionals $\mathcal{G}_t(\cdot)$ for which conditions (6.15) and (6.17) are satisfied.

- *Case 1:* $n_{\partial\Omega} = 0$

This is the simplest case: no constraint has to be imposed in (6.4) (or, equivalently, we are working in the space $H(\mathrm{curl};\Omega) \cap H_0(\mathrm{div};\Omega)$), and no condition has to be verified.

A geometrical example of this situation is the case in which Ω is a ball-like set ($n_{\partial\Omega} = 0, n_\Omega = 0$), or Ω and Ω_C two co-axial tori ($n_{\partial\Omega} = 0, n_\Omega = 1$: see Figure 6.2).

- *Case 2:* $0 < n_{\partial\Omega} = n_\Omega$

Also this case is simple: in fact, we claim that condition (6.15) is satisfied with anyone of the following choices

$$\mathcal{G}_k(\mathbf{w}) = \int_{\Sigma_k} \mathbf{w} \cdot \mathbf{n}_\Sigma \ ; \ \mathcal{G}_k(\mathbf{w}) = \int_\Omega \mathbf{w} \cdot \hat{\boldsymbol{\pi}}_k \ ; \ \mathcal{G}_k(\mathbf{w}) = \int_\Omega \mathbf{w} \cdot \hat{\boldsymbol{\Psi}}_k \ . \tag{6.29}$$

Here $\hat{\boldsymbol{\Psi}}_k$ is the $(L^2(\Omega))^3$-extension of $\mathrm{grad}\,\hat{\psi}_k$, where $\hat{\psi}_k$ is any (continuous and piecewise-regular) real-valued function, multivalued on $\hat{\Sigma}_k$, that satisfies $[\hat{\psi}_k]_{\hat{\Sigma}_k} = 1$ (for instance, one can take the finite element Lagrange interpolant having 0 value everywhere, except at the nodes on $\hat{\Sigma}_k$, where it has the double value 0 and 1).

Using the fact that both $\hat{\boldsymbol{\pi}}_k$ and $\hat{\boldsymbol{\Psi}}_k$ are the extensions of the gradient of a function having jump equal to 1 on $\hat{\Sigma}_k$, it is easily seen that, for a divergence-free and tangential vector field $\mathbf{w}$, the three functionals above are the same, and for all of them the set of constraints $\mathcal{G}_k(\mathbf{w}) = 0$ express the orthogonality of $\mathbf{w}$ to $\mathcal{H}(m;\Omega)$. Condition (6.15) is thus satisfied.

Let us note that the solution $\mathbf{A}$ to (6.12) is the same for any choice of $\mathcal{G}_k(\cdot)$ in (6.29) (as it is a divergence-free and tangential vector field). The only difference will be, when approximating the solution by means of finite elements, the algebraic structure of the stiffness matrix associated to (6.12). In this respect, it can be also noted that the choice $\mathcal{G}_k(\mathbf{w}) = \int_\Omega \mathbf{w} \cdot \hat{\boldsymbol{\pi}}_k$ is not suitable for numerical approximation, as the basis functions $\hat{\boldsymbol{\pi}}_k$ are not explicitly available: in the discrete case it is thus better to focus on one of the two other alternatives.

A geometrical example of this situation is the case in which Ω is a torus and Ω_C is either a ball-like set or a torus, but with a different axis ($n_{\partial\Omega} = 1$, $n_\Omega = 1$: see Figure 6.1).

- *Case 3:* $0 < n_{\partial\Omega} < n_\Omega$

Also in this case we propose the following three alternative choices of the linear functionals $\mathcal{G}_t(\cdot)$, $t = 1, \ldots, n_\Omega$

$$\mathcal{G}_t(\mathbf{w}) = \int_{\hat{\Sigma}_t} \mathbf{w} \cdot \mathbf{n}_\Sigma \ ; \ \mathcal{G}_t(\mathbf{w}) = \int_\Omega \mathbf{w} \cdot \hat{\boldsymbol{\pi}}_t \ ; \ \mathcal{G}_t(\mathbf{w}) = \int_\Omega \mathbf{w} \cdot \hat{\boldsymbol{\Psi}}_t \ . \tag{6.30}$$

Proceeding as in the case before, we easily verify that (6.15) holds. To show that (6.17) is satisfied, let us introduce the matrix $\{\gamma_{qt}\}$ given by $\gamma_{qt} := \int_\Omega \hat{\boldsymbol{\pi}}_q \cdot \hat{\boldsymbol{\pi}}_t$, and denote by $\{\theta_{qt}\}$ its inverse matrix. Since, as we already noted, the functionals above are the same for a divergence-free and tangential vector field, we can always write $\gamma_{qt} = \mathcal{G}_t(\hat{\boldsymbol{\pi}}_q)$.

We can now easily check that

$$\boldsymbol{\eta}_t = \sum_{p=1}^{n_\Omega} \theta_{tp} \hat{\boldsymbol{\pi}}_p \,.$$

In fact,

$$\mathcal{G}_q\Big(\sum_{p=1}^{n_\Omega} \theta_{tp} \hat{\boldsymbol{\pi}}_p\Big) = \sum_{p=1}^{n_\Omega} \theta_{tp} \mathcal{G}_q(\hat{\boldsymbol{\pi}}_p) = \sum_{p=1}^{n_\Omega} \theta_{tp} \gamma_{pq} = \delta_{tq} \,.$$

Therefore, the matrix $\{\beta_{tq}\}$ appearing in condition (6.17) is given by $\{\theta_{tq}\}$. Since $\{\gamma_{tq}\}$ is symmetric and positive definite, the same holds for the inverse matrix $\{\theta_{tq}\}$, and also for all the principal minors of it. Condition (6.17) is thus satisfied.

A geometrical example of this situation is the case in which Ω is a double torus and Ω_C is a torus, co-axial to one of the two handles of Ω ($n_{\partial\Omega} = 1$, $n_\Omega = 2$: see Figure 6.3).

Remark 6.4. Let us also present, without giving further details, the strong and weak vector potential formulations for the eddy current problem subject to the electric boundary condition (1.5).

The strong formulation reads

$$\begin{cases} \operatorname{curl}(\boldsymbol{\mu}^{-1} \operatorname{curl} \mathbf{A}) - \mu_*^{-1} \operatorname{grad} \operatorname{div} \mathbf{A} & \\ \qquad\qquad + i\omega\boldsymbol{\sigma}\mathbf{A} + \boldsymbol{\sigma} \operatorname{grad} V_C = \mathbf{J}_e & \text{in } \Omega \\ \operatorname{div}(i\omega\boldsymbol{\sigma}\mathbf{A}_C + \boldsymbol{\sigma} \operatorname{grad} V_C) = \operatorname{div} \mathbf{J}_{e,C} & \text{in } \Omega_C \\ (i\omega\boldsymbol{\sigma}\mathbf{A}_C + \boldsymbol{\sigma} \operatorname{grad} V_C) \cdot \mathbf{n}_C & \\ \qquad\qquad = \mathbf{J}_{e,C} \cdot \mathbf{n}_C + \mathbf{J}_{e,I} \cdot \mathbf{n}_I & \text{on } \Gamma \\ \int_{(\partial\Omega)_r} \mathbf{A} \cdot \mathbf{n} = 0 & \forall\, r = 1, \ldots, p_{\partial\Omega} \\ \operatorname{div} \mathbf{A} = 0 & \text{on } \partial\Omega \\ \mathbf{A} \times \mathbf{n} = \mathbf{0} & \text{on } \partial\Omega \,, \end{cases} \tag{6.31}$$

having already inserted in the first equation the penalization term assuring that $\operatorname{div} \mathbf{A} = 0$ in Ω. It can be noted that, if the vector field $\mathbf{A}$ satisfies $\operatorname{div} \mathbf{A} = 0$ in Ω, the conditions $\int_{(\partial\Omega)_r} \mathbf{A} \cdot \mathbf{n} = 0$, $r = 1, \ldots, p_{\partial\Omega}$, are equivalent to the orthogonality to the space of harmonic fields

$$\mathcal{H}(e;\Omega) := \{\mathbf{w} \in (L^2(\Omega))^3 \mid \operatorname{curl} \mathbf{w} = \mathbf{0}, \operatorname{div} \mathbf{w} = 0, \mathbf{w} \times \mathbf{n} = 0 \text{ on } \partial\Omega\} \,;$$

in fact, a basis for this space is given by $\operatorname{grad} \hat{z}_r$, $r = 1, \ldots, p_{\partial\Omega}$, where $\hat{z}_r$ is the solution to

$$\begin{cases} \Delta \hat{z}_r = 0 & \text{in } \Omega \\ \hat{z}_r = 0 & \text{on } \partial\Omega \setminus (\partial\Omega)_r \\ \hat{z}_r = 1 & \text{on } (\partial\Omega)_r \,. \end{cases}$$

Note also that problem (6.31) simplifies if $p_{\partial\Omega} = 0$: namely, if the computational domain Ω has a connected boundary, which is very often the case in engineering problems.

The weak formulation is given by

$$
\begin{aligned}
&\text{Find } (\mathbf{A}, V_C) \in W_\sharp^* \times H_\sharp^1(\Omega_C) \text{ such that} \\
&\qquad \mathcal{A}[(\mathbf{A}, V_C), (\mathbf{w}, Q_C)] = \int_\Omega \mathbf{J}_e \cdot \overline{\mathbf{w}} \\
&\qquad\qquad + i\omega^{-1} \int_{\Omega_C} \mathbf{J}_{e,C} \cdot \operatorname{grad} \overline{Q_C} + i\omega^{-1} \int_\Gamma \mathbf{J}_{e,I} \cdot \mathbf{n}_I \overline{Q_C} \\
&\text{for all } (\mathbf{w}, Q_C) \in W_\sharp^* \times H_\sharp^1(\Omega_C) ,
\end{aligned}
\tag{6.32}
$$

where

$$
\begin{aligned}
W_\sharp^* := \{ \mathbf{w} \in H_0(\operatorname{curl}; \Omega) \cap H(\operatorname{div}; \Omega) \mid \\
\int_{(\partial\Omega)_r} \mathbf{A} \cdot \mathbf{n} = 0 \; \forall \, r = 1, \dots, p_{\partial\Omega} \} .
\end{aligned}
\tag{6.33}
$$

In comparison with the magnetic boundary value problem this case is simpler to treat, as we do not need to distinguish between different geometrical configurations. □

6.1.2 Existence and uniqueness of the solution to the weak formulation

The proof of existence and uniqueness is different in the three cases considered at the end of Section 6.1.1. Let us recall that here and in the sequel we assume that (6.2) is satisfied and the functionals $\mathcal{G}_t(\cdot)$ are as in (6.29) or (6.30).

- *Case $n_{\partial\Omega} = n_\Omega$ (Case 2 and Case 1 with $n_\Omega = 0$)*

  In these cases, the existence and uniqueness result derives from the Lax–Milgram lemma. Since the sesquilinear form $\mathcal{A}[\cdot,\cdot]$ is clearly continuous in $W_\sharp \times H_\sharp^1(\Omega_C)$, we have only to check that the right hand side in (6.12) is a continuous (antilinear) functional in $W_\sharp \times H_\sharp^1(\Omega_C)$ and that $\mathcal{A}[\cdot,\cdot]$ is coercive in $W_\sharp \times H_\sharp^1(\Omega_C)$, namely, that there exists a constant $\kappa_0 > 0$ such that for each $(\mathbf{w}, Q_C) \in W_\sharp \times H^1(\Omega_C)$ with $\int_{\Omega_{C,j}} Q_{C|\Omega_{C,j}} = 0$, $j = 1, \dots, p_\Gamma + 1$, it holds

$$
\begin{aligned}
|\mathcal{A}[(\mathbf{w}, Q_C), (\mathbf{w}, Q_C)]| \geq \kappa_0 \Big(\int_\Omega (|\mathbf{w}|^2 + |\operatorname{curl} \mathbf{w}|^2 + |\operatorname{div} \mathbf{w}|^2) \\
+ \int_{\Omega_C} (|Q_C|^2 + |\operatorname{grad} Q_C|^2) \Big) .
\end{aligned}
\tag{6.34}
$$

On the right hand side of (6.12) the only term to check is the third one. We have

$$
\begin{aligned}
\left| \int_\Gamma \mathbf{J}_{e,I} \cdot \mathbf{n}_I \overline{Q_C} \right| &\leq C_1 \|\mathbf{J}_{e,I} \cdot \mathbf{n}_I\|_{-1/2,\Gamma} \|Q_C\|_{1/2,\Gamma} \\
&\leq C_2 \|\mathbf{J}_{e,I}\|_{0,\Omega_I} \|Q_C\|_{1,\Omega_C} ,
\end{aligned}
\tag{6.35}
$$

having used the trace theorems from $H(\operatorname{div}; \Omega_I)$ onto $H^{-1/2}(\Gamma)$ and from $H^1(\Omega_C)$ onto $H^{1/2}(\Gamma)$ (see (A.9) and (A.8), respectively).

Concerning (6.34), we have

$$\begin{aligned} &\mathcal{A}[(\mathbf{w}, Q_C), (\mathbf{w}, Q_C)] \\ &\quad = \int_\Omega (\boldsymbol{\mu}^{-1} \operatorname{curl} \mathbf{w} \cdot \operatorname{curl} \overline{\mathbf{w}} + \mu_*^{-1} |\operatorname{div} \mathbf{w}|^2) \\ &\qquad + i\omega^{-1} \int_{\Omega_C} \boldsymbol{\sigma}(i\omega \mathbf{w}_C + \operatorname{grad} Q_C) \cdot (-i\omega \overline{\mathbf{w}_C} + \operatorname{grad} \overline{Q_C}) \,. \end{aligned} \tag{6.36}$$

Given a pair of real numbers a and b, for each $0 < \delta < 1$ there holds

$$|2ab| \leq \delta a^2 + \delta^{-1} b^2 \,;$$

hence one has

$$\begin{aligned} &|\omega|^{-1} \int_{\Omega_C} \boldsymbol{\sigma}(i\omega \mathbf{w}_C + \operatorname{grad} Q_C) \cdot (-i\omega \overline{\mathbf{w}_C} + \operatorname{grad} \overline{Q_C}) \\ &\quad \geq |\omega|^{-1} \sigma_{\min} \int_{\Omega_C} [|\operatorname{grad} Q_C|^2 + \omega^2 |\mathbf{w}_C|^2 \\ &\qquad\qquad + 2 \operatorname{Re}(i\omega \mathbf{w}_C \cdot \operatorname{grad} \overline{Q_C})] \\ &\quad \geq |\omega|^{-1} \sigma_{\min} (1 - \delta) \int_{\Omega_C} |\operatorname{grad} Q_C|^2 \\ &\qquad - |\omega| \sigma_{\min} (1 - \delta) \delta^{-1} \int_{\Omega_C} |\mathbf{w}_C|^2 \,, \end{aligned} \tag{6.37}$$

where $\sigma_{\min}$ is a uniform lower bound in Ω_C for the minimum eigenvalues of $\boldsymbol{\sigma}(\mathbf{x})$.

Using the Poincaré inequality (see, for instance, Dautray and Lions [94], Chap. IV, Sect. 7, Prop. 2), one can conclude that there exists a constant $K_1 > 0$ such that for each $Q_C \in H^1(\Omega_C)$ with $\int_{\Omega_{C,j}} Q_{C|\Omega_{C,j}} = 0$ we have

$$\begin{aligned} \int_{\Omega_C} |\operatorname{grad} Q_C|^2 &= \sum_{j=1}^{p_\Gamma+1} \int_{\Omega_{C,j}} |\operatorname{grad} Q_{C|\Omega_{C,j}}|^2 \\ &\geq K_1 \sum_{j=1}^{p_\Gamma+1} \int_{\Omega_{C,j}} (|\operatorname{grad} Q_{C|\Omega_{C,j}}|^2 + |Q_{C|\Omega_{C,j}}|^2) \\ &= K_1 \int_{\Omega_C} (|\operatorname{grad} Q_C|^2 + |Q_C|^2) \,. \end{aligned} \tag{6.38}$$

Moreover, and this is the point in which Case 2 (or Case 1 with $0 = n_{\partial\Omega} = n_\Omega$) differs from the other cases, there exists a constant $K_2 > 0$ such that for any function $\mathbf{w} \in W_\sharp$ one has the Poincaré-like inequality

$$\begin{aligned} &\int_\Omega (\boldsymbol{\mu}^{-1} \operatorname{curl} \mathbf{w} \cdot \operatorname{curl} \overline{\mathbf{w}} + \mu_*^{-1} |\operatorname{div} \mathbf{w}|^2) \\ &\quad \geq \int_\Omega (\mu_{\max}^{-1} |\operatorname{curl} \mathbf{w}|^2 + \mu_*^{-1} |\operatorname{div} \mathbf{w}|^2) \\ &\quad \geq K_2 \int_\Omega (|\operatorname{curl} \mathbf{w}|^2 + |\operatorname{div} \mathbf{w}|^2 + |\mathbf{w}|^2) \,, \end{aligned} \tag{6.39}$$

where $\mu_{\max}$ is a uniform upper bound in Ω for the maximum eigenvalues of $\boldsymbol{\mu}(\mathbf{x})$ (see, for instance, Girault and Raviart [111], Chap. I, Lemma 3.6; the proof can be easily extended to the present geometrical situation, by proceeding as in Alonso and Valli [6], Lemma 3.3, and noting that, for a divergence-free and tangential vector field and for $\mathcal{G}_k(\cdot)$ defined as in (6.29), the conditions $\mathcal{G}_k(\mathbf{w}) = 0$ for $k = 1, \ldots, n_{\partial\Omega} = n_\Omega$ are equivalent to the orthogonality to $\mathcal{H}(m; \Omega)$).

Choosing $(1 - \delta)$ so small that $|\omega| \sigma_{\min}(1 - \delta) < K_2 \delta$, from (6.36)–(6.39) we find at once (6.34).

- *Case $n_{\partial\Omega} < n_\Omega$ (Case 3 and Case 1 with $n_\Omega > 0$)*

In this case coerciveness of $\mathcal{A}[\cdot,\cdot]$ in $W_\sharp \times H^1_\sharp(\Omega_C)$ is questionable, as we do not know if estimate (6.39) holds (the conditions $\mathcal{G}_k(\mathbf{w}) = 0$ for $k = 1,\ldots,n_{\partial\Omega}$ are not equivalent to the orthogonality to $\mathcal{H}(m;\Omega)$). However, we find that $\mathcal{A}[(\mathbf{w},Q_C),(\mathbf{w},Q_C)] = 0$ for $(\mathbf{w},Q_C) \in W_\sharp \times H^1_\sharp(\Omega_C)$ implies $\mathbf{w} = \mathbf{0}$ and $Q_C = 0$, or, in other words, that

$$|\mathcal{A}[(\mathbf{w},Q_C),(\mathbf{w},Q_C)]| > 0$$

for $(\mathbf{w},Q_C) \in W_\sharp \times H^1_\sharp(\Omega_C)$, $(\mathbf{w},Q_C) \neq (\mathbf{0},0)$.

This can be achieved as follows: assuming that $\mathcal{A}[(\mathbf{w},Q_C),(\mathbf{w},Q_C)] = 0$, from (6.11) we have that $\operatorname{curl}\mathbf{w} = \mathbf{0}$ and $\operatorname{div}\mathbf{w} = 0$ in Ω, and $i\omega\mathbf{w}_C + \operatorname{grad} Q_C = \mathbf{0}$ in Ω_C. Therefore $\mathbf{w} \in \mathcal{H}(m;\Omega)$ and we can express it in terms of the basis functions of $\mathcal{H}(m;\Omega)$, say, $\mathbf{w} = \sum_{t=1}^{n_\Omega} \alpha_t \hat{\boldsymbol{\pi}}_t$. In particular, in Ω_C we have $\mathbf{w}_C = \sum_{t=1}^{n_\Omega} \alpha_t \hat{\boldsymbol{\pi}}_{t,C} = i\omega^{-1} \operatorname{grad} Q_C$, and taking the line integral along the non-bounding cycle contained in Ω_C and associated to the "cutting" surface $\hat{\Sigma}_q$, $q = n_{\partial\Omega}+1,\ldots,n_\Omega$, we find that the coefficient α_q is vanishing for $q = n_{\partial\Omega}+1,\ldots,n_\Omega$.

Moreover, taking into account the definition of $\mathcal{G}_t(\cdot)$ in (6.30) and the relation $\mathcal{G}_k(\mathbf{w}) = 0$ for $k = 1,\ldots,n_{\partial\Omega}$, we also have

$$0 = \mathcal{G}_k(\mathbf{w}) = \sum_{t=1}^{n_\Omega} \alpha_t \mathcal{G}_k(\hat{\boldsymbol{\pi}}_t) = \sum_{i=1}^{n_{\partial\Omega}} \alpha_i \gamma_{ki} \ , \ k = 1,\ldots,n_{\partial\Omega} \ .$$

Since the matrix $\{\mathcal{G}_q(\hat{\boldsymbol{\pi}}_t)\} = \{\gamma_{tq}\}$ is symmetric and positive definite, the same holds for its principal minor $\{\gamma_{ki}\}$, $k,i = 1,\ldots,n_{\partial\Omega}$, therefore it follows $\alpha_i = 0$ for each $i = 1,\ldots,n_{\partial\Omega}$. In conclusion, $\mathbf{w} = \mathbf{0}$ in Ω and consequently $\operatorname{grad} Q_C = \mathbf{0}$ in Ω_C.

This concludes the proof of the uniqueness of the solution. Its existence can be demonstrated as follows. For the sake of definiteness, the reader can refer to the geometric situation illustrated in Figure 6.4.

We employ a procedure that will be used in Section 6.2.2 for showing the well-posedness of the formulation based on the Lorenz gauge. We start from the electric field $\mathbf{E}$ and the magnetic field $\mathbf{H}$, the solutions of the eddy current problem we know

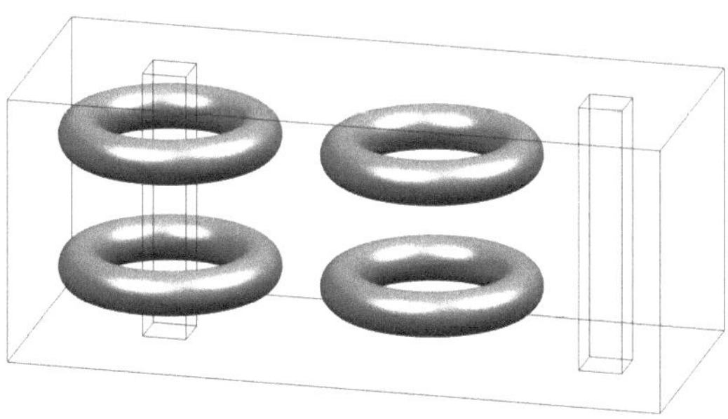

Fig. 6.4. A computational domain for which $n_{\partial\Omega} = 1$, $n_\Omega = 2$, $n_{\Omega_C} = 4$

to exist, and we first solve

$$\begin{cases} \operatorname{curl} \mathbf{A} = \boldsymbol{\mu}\mathbf{H} & \text{in } \Omega \\ \operatorname{div} \mathbf{A} = 0 & \text{in } \Omega \\ \mathbf{A} \cdot \mathbf{n} = 0 & \text{on } \partial\Omega \,. \end{cases} \tag{6.40}$$

Since Ω is not simply-connected, this solution is not unique. However, for each choice of $\mathbf{A}$ one has that $\mathbf{E} + i\omega\mathbf{A}$ is curl-free in Ω. Denote by $\boldsymbol{\rho}^*_{\beta,C}$, $\beta = 1, \ldots, n_{\Omega_C}$, the basis functions of the space of harmonic fields

$$\mathcal{H}(m; \Omega_C) := \{\mathbf{z}_C \in (L^2(\Omega_C))^3 \mid \operatorname{curl} \mathbf{z}_C = \mathbf{0}, \operatorname{div} \mathbf{z}_C = 0, \\ \mathbf{z}_C \cdot \mathbf{n}_C = 0 \text{ on } \Gamma\} \,,$$

We know that n_{Ω_C} is the first Betti number of Ω_C, namely, the number of independent non-bounding cycles in Ω_C.

Proceeding as in Theorem A.8 we can write in Ω_C

$$\mathbf{E}_C + i\omega\mathbf{A}_C = \operatorname{grad} \chi^*_C + \sum_{\beta=1}^{n_{\Omega_C}} \theta^*_{C,\beta} \boldsymbol{\rho}^*_{\beta,C} \,,$$

where $\theta^*_{C,\beta}$ represent the line integrals of $\mathbf{E}_C + i\omega\mathbf{A}_C$ along the non-bounding cycles γ^*_β of Ω_C.

Among the independent non-bounding cycles of Ω_C there are those that are non-bounding cycles also in Ω and those that are the boundary of a surface contained in Ω (in Figure 6.4 the number of the cycles in the first group is 2, the number of the cycles in the second group is 2). For the latter set we clearly have $\theta^*_{C,\beta} = 0$, as $\mathbf{E} + i\omega\mathbf{A}$ is curl-free in Ω. The former set can be subdivided between those cycles that form an independent set of non-bounding cycles in Ω, and the remaining ones; the number of the non-bounding cycles of Ω_C that are independent non-bounding cycles in Ω is $n_\Omega - n_{\partial\Omega}$ (in Figure 6.4 this number is 1).

We impose that $\theta^*_{C,\beta} = 0$ for all the indices β corresponding to these non-bounding cycles of Ω_C that form an independent set of non-bounding cycles in Ω. We are thus imposing $n_\Omega - n_{\partial\Omega}$ conditions, and these conditions ensure that $\theta^*_{C,\beta} = 0$ for all $\beta = 1, \ldots, n_{\Omega_C}$: in fact, if we take an index β not belonging to the set of indices related to the non-bounding cycles independent in Ω, it is associated either to a bounding cycle of Ω or to a non-bounding cycle of Ω that is dependent on the cycles for which the circulation vanishes. As an example, in Figure 6.4 the non-bounding cycles in Ω_C are 4, those that are non-bounding also in Ω are 2, but in Ω they are dependent: hence we set $\theta^*_{C,\beta} = 0$ only for one of them.

In conclusion, having shown that $\theta^*_{C,\beta} = 0$ for all $\beta = 1, \ldots, n_{\Omega_C}$, we see that in Ω_C the field $\mathbf{E}_C + i\omega\mathbf{A}_C$ is the gradient of some potential, say $\chi^*_C = -V_C$.

We have not yet indicated how to select a unique solution $\mathbf{A}$ to (6.40). Applying Theorem A.8 in the domain Ω, we know that a solution of problem (6.40) has the form

$$i\omega\mathbf{A} = -\mathbf{E} + \operatorname{grad} \Phi + \sum_{k=1}^{n_{\partial\Omega}} \alpha_k \hat{\boldsymbol{\pi}}_k + \sum_{t=n_{\partial\Omega}+1}^{n_\Omega} \alpha_t \hat{\boldsymbol{\pi}}_t \,,$$

and Φ is uniquely determined by $\mathbf{E}$. Moreover, α_t, $t = n_{\partial\Omega}+1,\ldots,n_\Omega$, are the line integrals of $i\omega\mathbf{A}+\mathbf{E}$ along the non-bounding cycles of Ω that are also non-bounding cycles in Ω_C, hence they are the quantities $\theta^*_{C,\beta}$ we have imposed to vanish. We thus finish requiring that

$$\mathcal{G}_i(\mathbf{A}) = 0 \qquad \forall\, i = 1,\ldots,n_{\partial\Omega}\,,$$

and from these conditions the coefficients α_k are uniquely determined for all $k = 1,\ldots,n_{\partial\Omega}$.

The potentials $\mathbf{A}$ and V_C we have constructed are clearly the solutions to problem (6.3), hence to the weak problem (6.12).

Remark 6.5. It should be noted that, even if the constraints $\mathcal{G}_k(\mathbf{A}) = 0$ for $k = 1,\ldots,n_{\partial\Omega}$ are not imposed, from $\mathcal{A}[(\mathbf{A},V_C),(\mathbf{A},V_C)] = 0$ and (6.11) we always obtain $\mathbf{A}\in\mathcal{H}(m;\Omega)$ and $i\omega\mathbf{A}_C+\operatorname{grad}V_C = \mathbf{0}$ in Ω_C. Therefore, $\mathbf{H} = \boldsymbol{\mu}^{-1}\operatorname{curl}\mathbf{A} = \mathbf{0}$ in Ω and $\mathbf{E}_C = -i\omega\mathbf{A}_C - \operatorname{grad}V_C = \mathbf{0}$ in Ω_C, and the uniqueness of the magnetic and electric fields is in any case verified.

In other words, the constraints $\mathcal{G}_k(\mathbf{A}) = 0$ seem not to play any role in determining the right physical solution. This is true, but, since they are needed for well-posedness, they can have a role in the efficiency of the numerical algorithm used for approximation.

Indeed, as reported in Remark 6.8, it will be clear that for the finite element approximation well-posedness is satisfied even without imposing these constraints. However, the numerical computations presented in Section 6.1.4 are showing that in fact the efficiency of the numerical algorithm is better when the constraints are satisfied. □

6.1.3 Numerical approximation

In this section we present the finite element numerical approximation of problem (6.12). It is naturally based on nodal finite elements, as imposing matching conditions on the interelements for both $\mathbf{A}\times\mathbf{n}$ and $\mathbf{A}\cdot\mathbf{n}$ is equivalent to requiring the continuity of the whole vector $\mathbf{A}$.

In the sequel we assume that Ω, Ω_C and Ω_I are Lipschitz polyhedra, and that $\mathcal{T}_{I,h}$ and $\mathcal{T}_{C,h}$ are two regular families of triangulations of Ω_I and Ω_C, respectively. For the sake of simplicity, we suppose that each element K of $\mathcal{T}_{I,h}$ and $\mathcal{T}_{C,h}$ is a tetrahedron; however, the results below also hold for hexahedral elements (and for second order hexahedral "serendipity" elements). We also assume that these triangulations match on Γ, so that they furnish a family of triangulations $\mathcal{T}_h$ of Ω.

Let $\mathbb{P}_k$, $k\geq 1$, be the space of polynomials of degree less than or equal to k. For $r\geq 1$ and $s\geq 1$ we introduce the discrete spaces of Lagrange nodal elements defined as

$$\begin{aligned} W_h^r := \{\mathbf{w}_h \in (C^0(\Omega))^3 \mid \mathbf{w}_{h|K} \in (\mathbb{P}_r)^3\ \forall\, K\in\mathcal{T}_h,\\ \mathbf{w}_h\cdot\mathbf{n} = 0 \text{ on } \partial\Omega\}\,, \end{aligned} \tag{6.41}$$

and

$$L^s_{C,h} := \{Q_{C,h}\in C^0(\Omega_C) \mid Q_{C,h|K}\in\mathbb{P}_s\ \forall\, K\in\mathcal{T}_{C,h}\}\,, \tag{6.42}$$

and we employ

$$(W_h^r)_\sharp := W_h^r \cap W_\sharp \tag{6.43}$$

$$(L^s_{C,h})_\sharp := L^s_{C,h} \cap H^1_\sharp(\Omega_C)\,, \tag{6.44}$$

hence we are considering a conforming finite element approximation.

The discrete problem is given by

$$\begin{array}{l}
\text{Find } (\mathbf{A}_h, V_{C,h}) \in (W^r_h)_\sharp \times (L^s_{C,h})_\sharp \text{ such that} \\
\qquad \mathcal{A}[(\mathbf{A}_h, V_{C,h}), (\mathbf{w}_h, Q_{C,h})] = \int_\Omega \mathbf{J}_e \cdot \overline{\mathbf{w}_h} \\
\qquad\qquad +i\omega^{-1} \int_{\Omega_C} \mathbf{J}_{e,C} \cdot \operatorname{grad} \overline{Q_{C,h}} + i\omega^{-1} \int_\Gamma \mathbf{J}_{e,I} \cdot \mathbf{n}_I \overline{Q_{C,h}} \\
\text{for all } (\mathbf{w}_h, Q_{C,h}) \in (W^r_h)_\sharp \times (L^s_{C,h})_\sharp \,.
\end{array} \tag{6.45}$$

Here below we establish the well-posedness of this discrete problem and investigate the convergence of the discrete solution to the exact solution.

- *Case $n_{\partial\Omega} = n_\Omega$ (Case 2 and Case 1 with $n_\Omega = 0$)*

Since the sesquilinear form $\mathcal{A}[\cdot,\cdot]$ is continuous and coercive, we have that the discrete solution exists and is unique; moreover, via Céa lemma for each $\mathbf{w}_h \in (W^r_h)_\sharp$ and $Q_{C,h} \in (L^s_{C,h})_\sharp$ we have

$$\begin{aligned}
\kappa_0 \Big(\int_\Omega (|\mathbf{A} - \mathbf{A}_h|^2 + |\operatorname{curl}(\mathbf{A} - \mathbf{A}_h)|^2 + |\operatorname{div}(\mathbf{A} - \mathbf{A}_h)|^2) \\
+ \int_{\Omega_C} |\operatorname{grad}(V_C - V_{C,h})|^2 \Big)^{1/2} \\
\leq C_0 \Big(\int_\Omega (|\mathbf{A} - \mathbf{w}_h|^2 + |\operatorname{curl}(\mathbf{A} - \mathbf{w}_h)|^2 + |\operatorname{div}(\mathbf{A} - \mathbf{w}_h)|^2) \\
+ \int_{\Omega_C} |\operatorname{grad}(V_C - Q_{C,h})|^2 \Big)^{1/2} ,
\end{aligned}$$

where $C_0 > 0$ is the continuity constant of $\mathcal{A}[\cdot,\cdot]$.

Therefore, at least when the constraints are expressed by the relations $\mathcal{G}_k(\mathbf{w}) = \int_{\Sigma_k} \mathbf{w} \cdot \mathbf{n}_\Sigma = 0$, and provided that the solutions $\mathbf{A}$ and V_C are regular enough, by means of well-known interpolation results it is possible to find the error estimate

$$\begin{aligned}
\Big(\int_\Omega (|\mathbf{A} - \mathbf{A}_h|^2 + |\operatorname{curl}(\mathbf{A} - \mathbf{A}_h)|^2 + |\operatorname{div}(\mathbf{A} - \mathbf{A}_h)|^2) \\
+ \int_{\Omega_C} |\operatorname{grad}(V_C - V_{C,h})|^2 \Big)^{1/2} \\
\leq C h^{\min(r,s)} \,.
\end{aligned} \tag{6.46}$$

In fact, in (6.12) and in (6.45) we can take $Q_{C,h} \in L^s_{C,h}$, thus we can choose as $Q_{C,h}$ the interpolant of V_C. Instead, the interpolant $\mathbf{I}_h\mathbf{A}$ of $\mathbf{A}$ cannot be selected as $\mathbf{w}_h$, as this test function has to satisfy the constraints $\mathcal{G}_k(\mathbf{w}_h) = 0$ for each $k = 1, \ldots, n_{\partial\Omega}$. Then one proceeds in the following way: since in (6.2) we have assumed that for $k = 1, \ldots, n_{\partial\Omega}$ the surfaces Σ_k are disjoint, one can easily find a set of discrete functions $\mathbf{v}_k$ (defined on a fixed coarse mesh $\mathcal{T}_{h^0}$) such that $\mathcal{G}_i(\mathbf{v}_k) = \delta_{ik}$ (the Kronecker symbol). Choosing the triangulation $\mathcal{T}_h$ as a refinement of $\mathcal{T}_{h^0}$ and $\mathbf{w}_h = \mathbf{I}_h\mathbf{A} - \sum_k \mathcal{G}_k(\mathbf{I}_h\mathbf{A})\mathbf{v}_k$, one clearly finds $\mathcal{G}_i(\mathbf{w}_h) = 0$ for each $i = 1, \ldots, n_{\partial\Omega}$, and moreover

$$\begin{aligned}\|\mathbf{A}-\mathbf{w}_h\|_W &= \|\mathbf{A}-\mathbf{I}_h\mathbf{A}+\textstyle\sum_k \mathcal{G}_k(\mathbf{I}_h\mathbf{A})\mathbf{v}_k\|_W\\ &\le \|\mathbf{A}-\mathbf{I}_h\mathbf{A}\|_W+\textstyle\sum_k \|\mathcal{G}_k(\mathbf{I}_h\mathbf{A})\mathbf{v}_k-\mathcal{G}_k(\mathbf{A})\mathbf{v}_k\|_W\\ &\le \|\mathbf{A}-\mathbf{I}_h\mathbf{A}\|_W+\textstyle\sum_k |\mathcal{G}_k(\mathbf{A}-\mathbf{I}_h\mathbf{A})|\|\mathbf{v}_k\|_W\\ &\le \|\mathbf{A}-\mathbf{I}_h\mathbf{A}\|_W+(\textstyle\sum_k \|\mathcal{G}_k\|_{\mathcal{L}(W;\mathbb{C})}\|\mathbf{v}_k\|_W)\|\mathbf{A}-\mathbf{I}_h\mathbf{A}\|_W\,,\end{aligned}$$

having denoted by $\|\cdot\|_W$ the norm in $H(\text{curl};\Omega)\cap H(\text{div};\Omega)$ and by $\|\cdot\|_{\mathcal{L}(W;\mathbb{C})}$ the norm of a linear functional from W to $\mathbb{C}$.

Remark 6.6. Concerning this convergence result, it has to be noted that the regularity of $\mathbf{A}$ is not ensured if Ω has reentrant corners or edges, namely, if it is a non-convex polyhedron (see Costabel and Dauge [87], Costabel et al. [90]). More important, in that case the space $H^1_\tau(\Omega) := (H^1(\Omega))^3 \cap H_0(\text{div};\Omega)$ turns out to be a proper *closed* subspace of $H(\text{curl};\Omega)\cap H_0(\text{div};\Omega)$ ($H^1_\tau(\Omega)$ and $H(\text{curl};\Omega)\cap H_0(\text{div};\Omega)$ coincide if and only if Ω is convex). Hence the nodal finite element approximate solution $\mathbf{A}_h \in W^r_h \subset H^1_\tau(\Omega)$ cannot approach an exact solution $\mathbf{A}\in H(\text{curl};\Omega)\cap H_0(\text{div};\Omega)$ with $\mathbf{A}\notin H^1_\tau(\Omega)$, and convergence in $H(\text{curl};\Omega)\cap H(\text{div};\Omega)$ is lost: this is a general problem for the nodal finite element approximation of Maxwell equations.

However, the result we have proved here above ensures that for the Coulomb gauged vector potential formulation of the eddy current problem the nodal finite element approximation is convergent either if the solution is regular (and this information could be available even for a non-convex polyhedron Ω) or if the domain Ω is a convex polyhedron, as in this case the space of smooth tangential vector fields is dense in $H^1_\tau(\Omega) = H(\text{curl};\Omega)\cap H_0(\text{div};\Omega)$, and one can apply Céa lemma in the standard way.

Let us also note that the assumption that Ω is convex is not a severe restriction, as in most real-life applications $\partial\Omega$ arises from a somehow arbitrary truncation of the whole space. Hence, reentrant corners and edges of Ω can be easily avoided. □

Remark 6.7. It is worth noting that a cure for the lack of convergence of nodal finite element approximations in the presence of re-entrant corners and edges has been proposed by Costabel and Dauge [88]. They introduce a special weight in the grad div penalization term, thus making it possible to use standard nodal finite elements in a numerically efficient way. □

- *Case $n_{\partial\Omega} < n_\Omega$ (Case 3 and Case 1 with $n_\Omega > 0$)*

In this case, we limit ourselves to the proof of the existence and uniqueness of the solution, without providing an error estimate. Since the problem is finite dimensional, the proof of uniqueness is enough.

Thus, let us consider a solution $(\mathbf{A}_h, V_{C,h})$ to (6.45) with a vanishing right hand side. As in the infinite dimensional case, from (6.11) we find that $\mathbf{A}_h \in \mathcal{H}(m;\Omega)$ and $i\omega\mathbf{A}_{h|\Omega_C} + \text{grad}\, V_{C,h} = \mathbf{0}$ in Ω_C.

Since the harmonic fields in $\mathcal{H}(m;\Omega)$ are C^∞ vector functions in Ω, we deduce that the piecewise-polynomial $\mathbf{A}_h$ is indeed a global polynomial $(\mathbb{P}_r)^3$ in Ω. Consequently, curl $\mathbf{A}_h$ is a global polynomial $(\mathbb{P}_{r-1})^3$ in Ω, and there it is vanishing. Thus we have $\operatorname{curl} \mathbf{A}_h = \mathbf{0}$ in $\mathbb{R}^3$, and $\mathbf{A}_h = \operatorname{grad} U$ in $\mathbb{R}^3$. In particular, $\mathbf{A}_h = \operatorname{grad} U_{|\Omega}$ in Ω, and the conditions $\operatorname{div} \mathbf{A}_h = 0$ in Ω and $\mathbf{A}_h \cdot \mathbf{n} = 0$ on $\partial\Omega$ tell us that $U_{|\Omega}$ is a harmonic function with vanishing normal derivative on the boundary, therefore is a constant. In conclusion, $\mathbf{A}_h = \mathbf{0}$ in Ω and therefore $\operatorname{grad} V_{C,h} = \mathbf{0}$ in Ω_C.

Remark 6.8. It can be noted that at the discrete level, in all the geometrical Cases 1, 2 and 3, one could formulate the problem by replacing $(W_h^r)_\sharp$ with W_h^r, namely, working in the unconstrained space $H(\operatorname{curl}; \Omega) \cap H_0(\operatorname{div}; \Omega)$, still obtaining existence and uniqueness.

In fact, if the right hand side of the discrete equation is vanishing, from (6.11) one always finds $\mathbf{A}_h \in \mathcal{H}(m; \Omega)$ and $i\omega \mathbf{A}_{h|\Omega_C} + \operatorname{grad} V_{C,h} = \mathbf{0}$ in Ω_C. Therefore, proceeding as before, one shows that $\mathbf{A}_h = \operatorname{grad} U_{|\Omega}$ in Ω, and the uniqueness of the discrete solution again follows.

A natural question therefore arises: from the computational point of view, the constrained discrete approximation in the space $W_\sharp$ is more efficient than the unconstrained one in the space $H(\operatorname{curl}; \Omega) \cap H_0(\operatorname{div}; \Omega)$? One argument in favour of the constrained formulation is that, at least in the case $n_{\partial\Omega} = n_\Omega$ and for a regular exact solution, we are able to prove an error estimate, therefore convergence is ensured.

In the next section we are going to present some numerical results that confirm this assertion. □

Remark 6.9. In the numerical implementation, imposing the boundary condition $\mathbf{A}_h \cdot \mathbf{n} = 0$ on $\partial\Omega$ (or else $\mathbf{A}_h \times \mathbf{n} = \mathbf{0}$ on $\partial\Omega$, as in the approximation of problem (6.32)) is clearly straightforward if the boundary of the computational domain Ω is formed by planar surfaces, parallel to the reference planes.

If this is not the case, for each node $\mathbf{p}$ on $\partial\Omega$ it is possible to introduce a local system of coordinates with one axis aligned with $\mathbf{n}_a$, a suitable average of the normals to the surface elements containing $\mathbf{p}$, and to express, through a rotation, the vector $\mathbf{A}_h$ with respect to that system: the condition $\mathbf{A}_h \cdot \mathbf{n}_a = 0$ is then trivially imposed. The details of this procedure can be found in Rodger and Eastham [212].

Another possible approach, which avoids the arbitrariness inherent in the averaging process of the normals at corner points, is described (and at the same time criticized) by Bossavit [60]. It is based on imposing $\mathbf{A}_h \cdot \mathbf{n} = 0$ at the center of the element faces on $\partial\Omega$: the drawback is that it results in a constrained problem, requiring the introduction of as many Lagrange multipliers as the number of surface elements on $\partial\Omega$. □

Remark 6.10. The need for imposing the gauge condition $\operatorname{div} \mathbf{A} = 0$ in Ω in order to obtain a unique vector potential $\mathbf{A}$ has led to the use of nodal finite elements for numerical approximation. A possible alternative has been described by Bíró [48] (see also Fujiwara et al. [108], Ren [206], Kameari and Koganezawa [145], Ren and Ida [209], Ren and Razek [210], Hollaus and Bíró [132]): edge elements are employed for the approximation of the potential $\mathbf{A}$, without requiring that the gauge condition is satisfied.

Clearly, in this way the resulting linear system is singular: however, in many cases the right-hand sides turn out to be compatible, so that suitable iterative algebraic solvers can still be convergent, though, to the best of our knowledge, a complete theory assuring the effectiveness of this procedure is not available. □

Remark 6.11. In Jin [140], Chap. 5, Sect. 5.7.4, it is underlined that a finite element approximation based on a weak form in which the term $\int_\Omega \mu^{-1} \operatorname{div} \mathbf{z} \operatorname{div} \overline{\mathbf{w}}$ is present can be inefficient if the coefficient μ has jumps. In this respect, it should be noted that in (6.12) the sesquilinear form $\mathcal{A}[\cdot,\cdot]$ contains $\int_\Omega \mu_*^{-1} \operatorname{div} \mathbf{z} \operatorname{div} \overline{\mathbf{w}}$, but μ_* is an auxiliary constant which is not required to be equal to the physical magnetic permeability $\boldsymbol{\mu}$, so that jumps are avoided. □

Remark 6.12. The procedure presented in Remark 6.2 can be used for computing the numerical approximation of the electric field $\mathbf{E}_I$ in a way different from that described in Section 5.5.

In fact, it is sufficient to insert $\mathbf{A}_{I,h}$ and $V_{C,h}$ at the right hand side of (6.7) and compute the finite element approximation of $V^\dagger_{I,h}$, using standard scalar nodal elements. Then, putting $\mathbf{A}_{I,h}$ and $V^\dagger_{I,h}$ at the right hand side of (6.8), one finds the complex numbers $(c^\dagger_{I,j,h}, d^\dagger_{I,k,h})$, $j = 1, \ldots, p_\Gamma$, $k = 1, \ldots, n_{\partial\Omega}$, from which one is led to define

$$\mathbf{E}_{I,h} := -i\omega \mathbf{A}_{I,h} - \operatorname{grad} V^\dagger_{I,h} + \sum_{j=1}^{p_\Gamma} c^\dagger_{I,j,h} \operatorname{grad} w_{j,I} + \sum_{k=1}^{n_{\partial\Omega}} d^\dagger_{I,k,h} \boldsymbol{\pi}_{k,I} \,.$$

As described in Section 5.1, it is also possible to avoid the use of the basis functions $\operatorname{grad} w_{j,I}$ and $\boldsymbol{\pi}_{k,I}$, by replacing them with the gradients of suitable interpolants. In that case, problems (6.7) and (6.8) are no longer decoupled, but, using Lemma 5.2, it can be seen that the coupled problem generated from these problems is still associated to a coercive sesquilinear form. Thus it is solvable. □

Remark 6.13. Houston et al. [133] have proposed and analyzed an approximation algorithm for (6.3) based on the discontinuous Galerkin finite element method. □

Remark 6.14. For the electric boundary condition, the numerical approximation of the vector potential formulation reads

$$\begin{array}{l}
\text{Find } (\mathbf{A}_h, V_{C,h}) \in (W^r_h)^*_\sharp \times (L^s_{C,h})_\sharp \text{ such that} \\
\quad \mathcal{A}[(\mathbf{A}_h, V_{C,h}), (\mathbf{w}_h, Q_{C,h})] = \int_\Omega \mathbf{J}_e \cdot \overline{\mathbf{w}_h} \\
\qquad + i\omega^{-1} \int_{\Omega_C} \mathbf{J}_{e,C} \cdot \operatorname{grad} \overline{Q_{C,h}} + i\omega^{-1} \int_\Gamma \mathbf{J}_{e,I} \cdot \mathbf{n}_I \overline{Q_{C,h}} \\
\text{for all } (\mathbf{w}_h, Q_{C,h}) \in (W^r_h)^*_\sharp \times (L^s_{C,h})_\sharp \,,
\end{array} \tag{6.47}$$

where

$$(W^r_h)^*_\sharp := W^r_h \cap W^*_\sharp \,,$$

and $W^*_\sharp$, W^r_h and $(L^s_{C,h})_\sharp$ have been introduced in (6.33), (6.41) and (6.44), respectively. □

6.1.4 Numerical results

The numerical results we present here have been obtained in Bíró and Valli [54], where the numerical approximation of problem (6.12) in Case 2, with $n_{\partial\Omega} = n_\Omega = 1$ (Ω is a torus and Ω_C is a ball-like set) has been considered.

Aim of these numerical tests is to analyze the influence of the constraints $\mathcal{G}_k(\mathbf{w}) = 0$ on the efficiency of the computational algorithm. The finite elements employed are second order hexahedral "serendipity" elements, with 20 nodes (8 at the vertices and 12 at the midpoints of each edge), for all the components of $\mathbf{A}_h$ and for V_h (see, e.g., Ciarlet [83]).

The values of the physical coefficients are as follows: $\mu = \mu_* = 4\pi \times 10^{-7}$ H/m, $\sigma = 5.7 \times 10^7$ S/m, $\omega = 2\pi \times f = 100\pi$ rad/s, i.e., $f = 50$ Hz.

The CG iterations are stopped when the norm of the residual (normalized by the norm of the right hand side) is under a given tolerance. For the first two examples below, this tolerance is 10^{-10}, while for the third example is 10^{-6}.

In the first example, half of the domain is described in Figure 6.5. The coils (the support of $\mathbf{J}_{e,I}$) are the yellow sets on the left, while the conductor Ω_C is the red half-cylinder on the right; the "cutting" surface Σ_1 is green. We remark that all the results presented in this chapter still hold true even if the basis of the conducting domain Ω_C is touching the boundary $\partial\Omega$ as in Figure 6.5.

The difference between the constrained and the unconstrained finite element spaces resides only in one degree of freedom, the one associated to the "cut" Σ_1, "cutting" the equator of the torus Ω. More precisely, in the constrained case we are assuming that trial and test functions satisfy $\int_{\Sigma_1} \mathbf{w}_h \cdot \mathbf{n}_\Sigma = 0$. This can be achieved very easily; in fact, let us denote by $\boldsymbol{\phi}_i$ the basis function of the unconstrained finite element space and set $c_i := \int_{\Sigma_1} \boldsymbol{\phi}_i \cdot \mathbf{n}_\Sigma$. If $c_i = 0$ for each index i, there is nothing to do, as the unconstrained and the constrained space coincide. Conversely, if for some index, say $i = 1$, one has $c_1 \neq 0$, for $i \geq 2$ define $\widehat{\boldsymbol{\phi}}_i := \boldsymbol{\phi}_i - \frac{c_i}{c_1}\boldsymbol{\phi}_1$. These functions are easily proved to be the basis functions of the constrained space.

The current density is given by $\mathbf{J}_{e,C} = \mathbf{0}$ and $\mathbf{J}_{e,I} = J_{e,I}\mathbf{e}_\phi$, where $\mathbf{e}_\phi$ is the azymuthal unit vector in the cylindrical system centered at the point (100,0,0), oriented

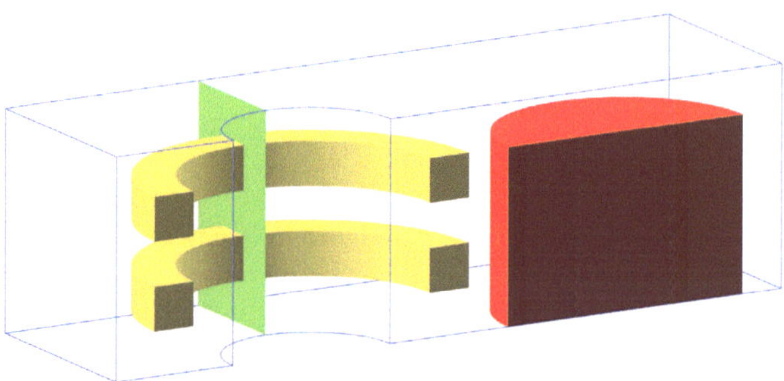

Fig. 6.5. The computational domain for the first example

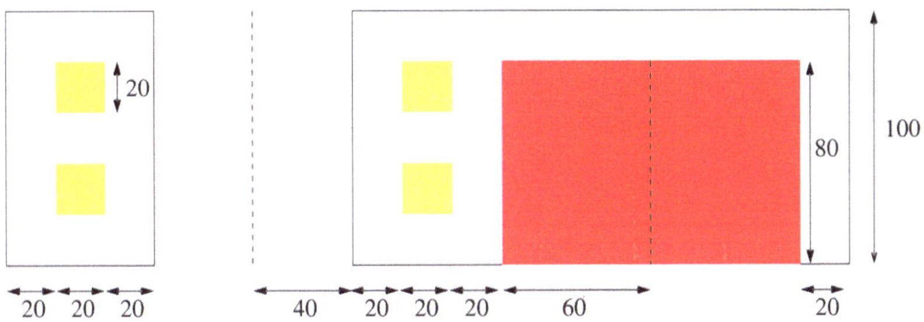

Fig. 6.6. The dimensions of the computational domain for the first example (projection on the (x, z)-plane)

counterclockwise, and

$$J_{e,I} = \begin{cases} 10^6 \text{ A/m}^2 & \text{if } 60 < r < 80 \, , \; 60 < z < 80 \\ -10^6 \text{ A/m}^2 & \text{if } 60 < r < 80 \, , \; 20 < z < 40 \\ 0 & \text{otherwise} \end{cases}$$

(see Figure 6.6 for a precise description of the geometry of Ω).

The solution has been computed for seven meshes, the coarsest one with 290 elements, the finest one with 99470 elements. With respect to the grid size, the seven meshes correspond to the choices h, $h/2$, ..., $h/7$.

For finding a reference solution, the problem has been also solved by means of edge elements on the finest grid: this solution is called $\mathbf{A}_{\text{edge}}$, $V_{C,\text{edge}}$. For this computation, the so-called quadratic 36-edge elements proposed by Kameari [144] have been used, writing the problem in terms of a ungauged magnetic vector potential and an electric scalar potential, namely, using the functional in (6.11) but dropping the term containing the divergence (see also Remark 6.10).

In Table 6.1, for each of the meshes described above, the error between the computed solution $\mathbf{A}_h$, $V_{C,h}$ and the reference solution is presented. More precisely, we have set

$$e_{\mathbf{J}} := \frac{\|\mathbf{J}_{C,h} - \mathbf{J}_{C,\text{edge}}\|_{0,\Omega_C}}{\|\mathbf{J}_{C,\text{edge}}\|_{0,\Omega_C}} \, , \; e_{\mathbf{B}} := \frac{\|\mathbf{B}_h - \mathbf{B}_{\text{edge}}\|_{0,\Omega}}{\|\mathbf{B}_{\text{edge}}\|_{0,\Omega}} \, ,$$

where $\mathbf{J}_{C,\text{edge}} := -i\omega\sigma\mathbf{A}_{\text{edge}|\Omega_C} - \sigma \operatorname{grad} V_{C,\text{edge}}$, $\mathbf{B}_{\text{edge}} := \operatorname{curl} \mathbf{A}_{\text{edge}}$, and similarly for $\mathbf{J}_{C,h}$ and $\mathbf{B}_h$. The number of conjugate gradient iterations needed to compute the approximate solution is also indicated. The computations are repeated twice, at first for the *unconstrained* approximate solution (namely, we have not imposed that the flux of the vector potential is vanishing on the "cutting" surface), and then for the *constrained* approximate solution. Clearly, in the latter case there is one degree of freedom less.

It can be seen that the CG iterations are always approximately 10% fewer when computing the constrained solution, while the accuracy is quite similar in both cases.

Table 6.1. Relatives errors $e_{\mathbf{J}}$ and $e_{\mathbf{B}}$ for the first example (see the text for further explanations)

Elements	*DoF*	*Iterations*	$e_{\mathbf{J}}$	$e_{\mathbf{B}}$
290	3,939	108	$1.602\ 10^{-1}$	$7.812\ 10^{-2}$
	3,938	97	$1.602\ 10^{-1}$	$7.812\ 10^{-2}$
2,320	31,337	206	$5.634\ 10^{-2}$	$2.125\ 10^{-2}$
	31,336	185	$5.659\ 10^{-2}$	$2.125\ 10^{-2}$
7,830	105,571	325	$2.786\ 10^{-2}$	$1.015\ 10^{-2}$
	105,570	294	$2.783\ 10^{-2}$	$1.015\ 10^{-2}$
18,560	250,017	448	$1.605\ 10^{-2}$	$7.228\ 10^{-3}$
	250,016	419	$1.602\ 10^{-2}$	$7.225\ 10^{-3}$
36,250	488,051	597	$1.054\ 10^{-2}$	$5.286\ 10^{-3}$
	488,050	540	$1.052\ 10^{-2}$	$5.284\ 10^{-3}$
62,640	843,049	739	$7.603\ 10^{-3}$	$4.729\ 10^{-3}$
	843,048	666	$7.588\ 10^{-3}$	$4.727\ 10^{-3}$
99,470	1,338,387	885	$5.959\ 10^{-3}$	$4.221\ 10^{-3}$
	1,338,386	793	$5.948\ 10^{-3}$	$4.219\ 10^{-3}$

In Figures 6.7 and 6.8 some details of the computed solution for the finest mesh are presented, and these pictures show a good agreement with the expected physical behaviour of the solution.

The second example is an academic one, useful for illustrating the rate of convergence of the method. It is based on the applied current density $\mathbf{J}_e$ computed starting from the smooth exact solution

$$\mathbf{A}_{\text{ex}} = \begin{cases} \text{curl}(0, 0, \exp(r^2/Q)) & \text{where } Q < 0 \\ 0 & \text{otherwise} \end{cases}$$

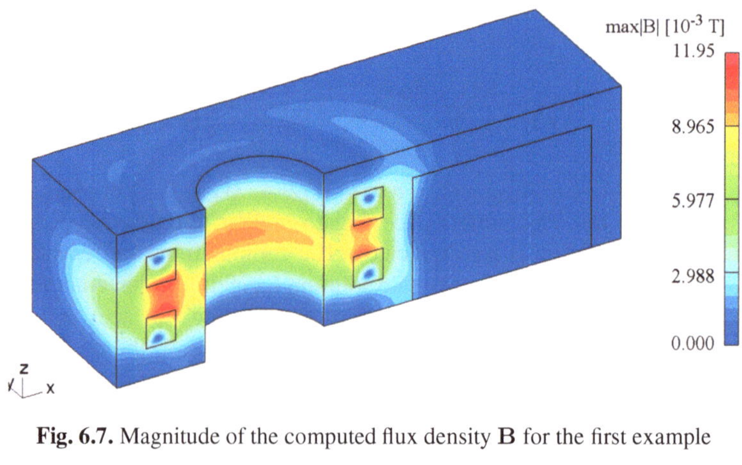

Fig. 6.7. Magnitude of the computed flux density **B** for the first example

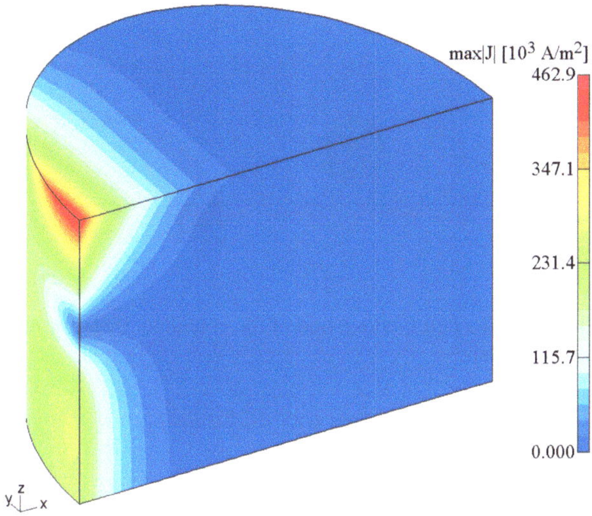

Fig. 6.8. Magnitude of the computed current density $\mathbf{J}_C := -i\omega\sigma\mathbf{A}_C - \sigma \operatorname{grad} V_C$ for the first example

$$\operatorname{grad} V_{C,\text{ex}} = \begin{cases} i\omega \operatorname{grad} \exp(r^2/Q) & \text{where } Q < 0 \\ 0 & \text{otherwise} \, , \end{cases}$$

where

$$Q := (x - x_0)^2 + (y - y_0)^2 + (z - z_0)^2 - r^2 \, ,$$

and $(x_0, y_0, z_0) \in \Omega$, $r > 0$ can be chosen freely. Clearly, if the ball $\{Q < 0\}$ is contained in Ω_I, as it will be the case here below, we have $\operatorname{grad} V_{C,\text{ex}} = \mathbf{0}$ and $\mathbf{E}_{C,\text{ex}} := -i\omega\mathbf{A}_{\text{ex}|\Omega_C} - \operatorname{grad} V_{C,\text{ex}} = \mathbf{0}$. In particular, in this case the coil is the ball $\{Q < 0\}$.

The same domains Ω, Ω_C and Ω_I of the first example (illustrated in Figure 6.5) are considered, but now the coil is the ball $\{Q < 0\}$, where we choose $(x_0, y_0, z_0) = (60/\sqrt{2} + 100, 60/\sqrt{2}, 60)$ and $r = 19$ (see Figure 6.9).

In Table 6.2 the error between the computed solution $\mathbf{A}_h$, $V_{C,h}$ and the exact solution is presented, setting $\mathbf{E}_{C,h} := -i\omega\mathbf{A}_{h|\Omega_C} - \operatorname{grad} V_{C,h}$, $\mathbf{B}_h := \operatorname{curl} \mathbf{A}_h$, $\mathbf{B}_{\text{ex}} := \operatorname{curl} \mathbf{A}_{\text{ex}}$, and

$$\widehat{e}_{\mathbf{E}} := \|\mathbf{E}_{C,h}\|_{0,\Omega_C} \; , \; e_{\mathbf{B}} := \frac{\|\mathbf{B}_h - \mathbf{B}_{\text{ex}}\|_{0,\Omega}}{\|\mathbf{B}_{\text{ex}}\|_{0,\Omega}} \, .$$

This time the coarsest mesh is of size h and is constituted by 150 elements, and then the mesh size is taken equal to $h/3$, $h/5$, $h/7$ and $h/9$. The computations are repeated two times: for the unconstrained algorithm and for the constrained algorithm.

This second example shows again that the CG iterations for the constrained algorithm are less than in the other case; moreover, the accuracy of the constrained algorithm is much better than that of the unconstrained algorithm. In particular, when using the unconstrained approximation the absolute error for the electric field is not

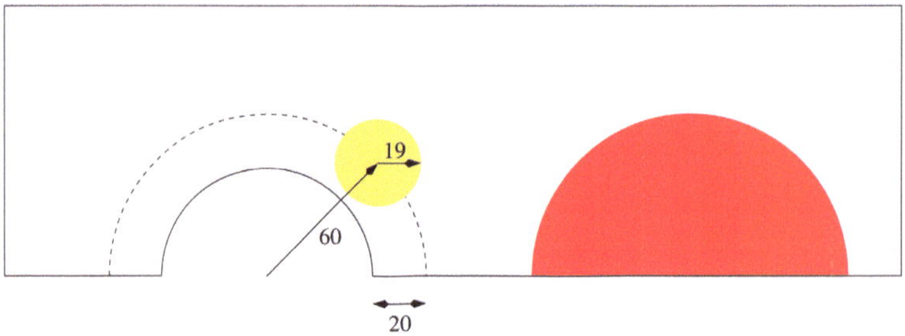

Fig. 6.9. The position of the coil (yellow) and of the conductor (red) for the second example: projection on the (x, y)-plane

Table 6.2. Absolute error $\widehat{e}_{\mathbf{E}}$ and relative error $e_{\mathbf{B}}$ for the second example (see the text for further explanations)

Elements	*DoF*	*Iterations*	$\widehat{e}_{\mathbf{E}}$	*Rate*	$e_{\mathbf{B}}$	*Rate*
150	2,080	67	$6.061\ 10^{1}$	-	$1.955\ 10^{0}$	-
	2,079	59	$7.485\ 10^{0}$	-	$1.680\ 10^{0}$	-
4,050	55,192	199	$1.434\ 10^{2}$	-0.78	$3.105\ 10^{0}$	-0.42
	55,191	181	$1.727\ 10^{0}$	1.33	$9.784\ 10^{-1}$	0.49
18,750	254,536	357	$1.022\ 10^{1}$	5.16	$7.898\ 10^{-1}$	2.67
	254,535	319	$3.893\ 10^{-1}$	2.91	$6.343\ 10^{-1}$	0.84
51,450	697,264	553	$3.488\ 10^{0}$	3.19	$5.743\ 10^{-1}$	0.94
	697,263	459	$1.840\ 10^{-1}$	2.22	$4.870\ 10^{-1}$	0.78
109,350	1,480,528	674	$2.022\ 10^{0}$	2.16	$4.815\ 10^{-1}$	0.70
	1,480,527	591	$5.053\ 10^{-2}$	5.14	$3.856\ 10^{-1}$	0.93

at all satisfactory even on the finest grid. Therefore, the advantage of the constrained algorithm is evident from this example.

It must be noted that for the magnetic field both algorithms are still far from being satisfactory; in our opinion, this is due to the fact that the coil is quite small, and even on the finest mesh it is not represented in a good way.

The estimated convergence rate, when passing from a mesh to the subsequent one, does not seem to be constant (for some other computations we have verified even larger oscillations in the errors). However, it appears to be asymptotically quadratic for the electric field, though not even linear for the magnetic field (whereas the theoretical estimate is quadratic, see (6.46)): in fact, passing from the coarsest to the finest mesh the global rate of convergence for the electric field is 1.54 for the unconstrained algorithm and 2.27 for the constrained one, while for the magnetic field it is 0.63 and 0.67, respectively. In this respect, we note that a better order of convergence is achieved in the next example.

The third example is related to the exact solution of the form we described before, but for a different domain, described in Figures 6.10, 6.11.

As indicated in these figures, this time we choose $(x_0, y_0, z_0) = (0, 0, 0), r = 0.29$. The main difference with respect to the preceding situation is that now the coil is larger (and the eddy current region is smaller), so that the numerical approximation does not need very fine meshes for being satisfactory. The coarsest mesh is of size h and is constituted by 66 elements, and then mesh size is taken equal to $h/2$, $h/4$, $h/8$ and $h/16$. As before, the computations are repeated two times: for the unconstrained algorithm and for the constrained algorithm. The results are presented in Table 6.3.

Table 6.3. Absolute error $\widehat{e}_{\mathbf{E}}$ and relative error $e_{\mathbf{B}}$ for the third example (see the text for further explanations)

Elements	*DoF*	*Iterations*	$\widehat{e}_{\mathbf{E}}$	*Rate*	$e_{\mathbf{B}}$	*Rate*
66	927	46	$6.524\ 10^{1}$	-	$7.595\ 10^{-1}$	-
	926	41	$6.424\ 10^{1}$	-	$7.676\ 10^{-1}$	-
528	6,922	74	$1.654\ 10^{1}$	1.98	$1.064\ 10^{0}$	−0.49
	6,921	63	$1.071\ 10^{1}$	2.58	$9.372\ 10^{-1}$	−0.29
4,224	53,322	135	$7.186\ 10^{0}$	1.20	$6.489\ 10^{-1}$	0.71
	53,321	117	$6.649\ 10^{0}$	0.69	$6.207\ 10^{-1}$	0.59
33,792	418,162	256	$1.415\ 10^{-1}$	5.67	$2.222\ 10^{-1}$	1.54
	418,161	227	$1.415\ 10^{-1}$	5.55	$2.222\ 10^{-1}$	1.48
270,336	3,311,202	401	$2.244\ 10^{-2}$	2.66	$6.885\ 10^{-2}$	1.69
	3,311,201	237	$2.225\ 10^{-2}$	2.67	$6.884\ 10^{-2}$	1.69

In this last case, the accuracy of the unconstrained and constrained approximations is similar, and is good enough. The rate of convergence now appears to be asymptotically quadratic also for the magnetic field approximation. The number of CG iterations is lower for the constrained algorithm than for the unconstrained one, and the difference is particularly significative for the finest mesh.

In order to compare the performance of (constrained) nodal element and edge element approximations, in the second and third examples we have also computed the solution by means of edge elements, employing a ungauged formulation as described in Remark 6.10.

The results are presented in Table 6.4. For each choice of the number of elements, the first row represents the nodal case, the second row the edge case. As a consequence of the fact that for the edge element approximation the linear system to solve is singular, it can be observed that the CG iterations for the nodal approximation are almost always less than for the edge approximation. In the second example the number of CG iterations is about 7% lower, in the third example the difference becomes much more significative as the mesh is finer.

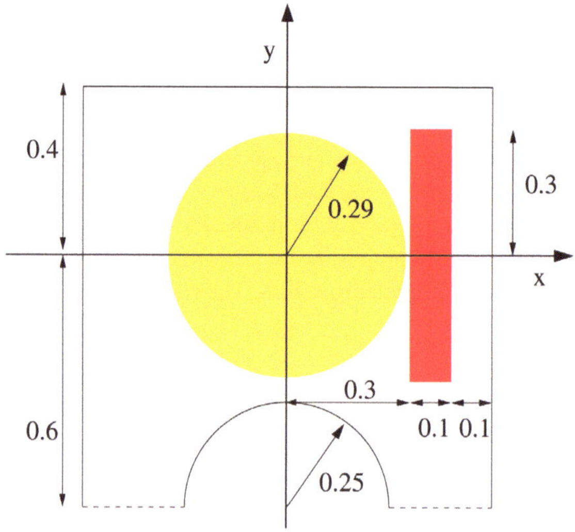

Fig. 6.10. The computational domain for the third example (the conductor is red and the coil is yellow): projection on the (x, y)-plane

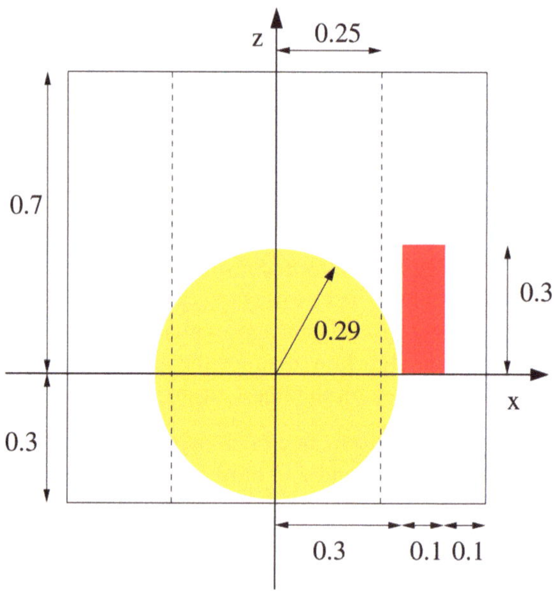

Fig. 6.11. The computational domain for the third example (the conductor is red and the coil is yellow): projection on the (x, z)-plane

Table 6.4. Number of iterations for the second example (left) and the third example (right). First row: nodal elements; second row: edge elements

Elements	*DoF*	*Iterations*
150	2,079	59
	1,640	57
4,050	55,191	181
	43,164	195
18,750	254,535	319
	198,640	346
51,450	697,263	459
	543,620	490
109,350	1,480,527	591
	1,153,656	635

Elements	*DoF*	*Iterations*
66	926	41
	789	25
528	6,921	63
	5,539	62
4,224	53,321	117
	41,373	137
33,792	418,161	227
	319,513	283
270,336	3,311,201	237
	2,510,769	583

6.1.5 A penalized formulation for the electric field

The procedure based on the introduction of a magnetic vector potential $\mathbf{A}$ can be adapted in order to devise a formulation of the eddy current problem (2.1), in which the constraint $\operatorname{div}(\varepsilon_I \mathbf{E}_I) = 0$ in Ω_I is imposed via a penalization term acting only in Ω_I. The problem reads

$$\begin{cases} \operatorname{curl}(\boldsymbol{\mu}_C^{-1} \operatorname{curl} \mathbf{E}_C) + i\omega\boldsymbol{\sigma}\mathbf{E}_C = -i\omega\mathbf{J}_{e,C} & \text{in } \Omega_C \\ \operatorname{curl}(\boldsymbol{\mu}_I^{-1} \operatorname{curl} \mathbf{E}_I) & \\ \qquad -c_0^* \varepsilon_I \operatorname{grad} \operatorname{div}(\varepsilon_I \mathbf{E}_I) = -i\omega\mathbf{J}_{e,I} & \text{in } \Omega_I \\ (\boldsymbol{\mu}_I^{-1} \operatorname{curl} \mathbf{E}_I) \times \mathbf{n} = \mathbf{0} & \text{on } \partial\Omega \\ \varepsilon_I \mathbf{E}_I \cdot \mathbf{n} = 0 & \text{on } \partial\Omega \\ \int_{\Gamma_j} \varepsilon_I \mathbf{E}_I \cdot \mathbf{n}_I = 0 & \forall\, j = 1, \ldots, p_\Gamma \\ \int_{\Omega_I} \varepsilon_I \mathbf{E}_I \cdot \boldsymbol{\pi}_{k,I} = 0 & \forall\, k = 1, \ldots, n_{\partial\Omega} \\ \operatorname{div}(\varepsilon_I \mathbf{E}_I) = 0 & \text{on } \Gamma \\ \mathbf{E}_C \times \mathbf{n}_C + \mathbf{E}_I \times \mathbf{n}_I = \mathbf{0} & \text{on } \Gamma \\ (\boldsymbol{\mu}_C^{-1} \operatorname{curl} \mathbf{E}_C) \times \mathbf{n}_C + (\boldsymbol{\mu}_I^{-1} \operatorname{curl} \mathbf{E}_I) \times \mathbf{n}_I = \mathbf{0} & \text{on } \Gamma\,, \end{cases} \tag{6.48}$$

where $c_0^* > 0$ is a dimensional constant. Note that, due to the presence of the penalization term in $(6.48)_2$, the constraint $\operatorname{div}(\varepsilon_I \mathbf{E}_I) = 0$ has to be kept only on the interface Γ, namely, it is necessary to impose equation $(6.48)_7$.

We have already noted that $p_\Gamma = 0$ and $n_{\partial\Omega} = 0$ if the boundary of the conductor Ω_C is connected and the computational domain Ω is simply-connected: for example, a connected conductor (possibly with "handles") contained in a "box".

In order to arrive at a weak formulation that leads to a more efficient numerical approximation, it is better to reformulate the conditions $\int_{\Omega_I} \varepsilon_I \mathbf{E}_I \cdot \boldsymbol{\pi}_{k,I} = 0$. From the Appendix we know that in $\Omega_I \setminus \Sigma_k$ one can write $\boldsymbol{\pi}_{k,I} = \operatorname{grad} q_{k,I}$ (see problem

(A.32)). Therefore, taking into account the properties of $\mathbf{E}_I$ and $q_{k,I}$, we have

$$\begin{aligned}
\int_{\Omega_I} \varepsilon_I \mathbf{E}_I \cdot \boldsymbol{\pi}_{k,I} &= \int_{\Omega_I \setminus \Sigma_k} \varepsilon_I \mathbf{E}_I \cdot \operatorname{grad} q_{k,I} \\
&= -\int_{\Omega_I \setminus \Sigma_k} \operatorname{div}(\varepsilon_I \mathbf{E}_I)\, q_{k,I} + \int_{\partial\Omega \setminus \partial\Sigma_k} \varepsilon_I \mathbf{E}_I \cdot \mathbf{n}\, q_{k,I} \\
&\quad + \int_\Gamma \varepsilon_I \mathbf{E}_I \cdot \mathbf{n}\, q_{k,I} + \int_{\Sigma_k} \varepsilon_I \mathbf{E}_I \cdot \mathbf{n}_\Sigma \\
&= \int_{\Sigma_k} \varepsilon_I \mathbf{E}_I \cdot \mathbf{n}_\Sigma \, .
\end{aligned}$$

Define the space

$$\begin{aligned}
\widehat{W}_{\varepsilon_I}(\Omega_I;\Omega) := \{\mathbf{z} \in H(\mathrm{curl};\Omega) \,|\, & \mathbf{z}_I \in H_{0,\partial\Omega}(\varepsilon_I, \mathrm{div};\Omega_I), \\
& \textstyle\int_{\Gamma_j} \varepsilon_I \mathbf{z}_I \cdot \mathbf{n}_I = 0 \ \forall\, j = 1, \ldots, p_\Gamma, \\
& \textstyle\int_{\Sigma_k} \varepsilon_I \mathbf{z}_I \cdot \mathbf{n}_\Sigma = 0 \ \forall\, k = 1, \ldots, n_{\partial\Omega}\} \, .
\end{aligned} \tag{6.49}$$

Then it is readily verified that the solution $\mathbf{E}$ to (6.48) satisfies the weak formulation

$$\begin{aligned}
&\text{Find } \mathbf{E} \in \widehat{W}_{\varepsilon_I}(\Omega_I;\Omega) \text{ such that} \\
&\quad \textstyle\int_\Omega \boldsymbol{\mu}^{-1} \operatorname{curl} \mathbf{E} \cdot \operatorname{curl} \overline{\mathbf{z}} + i\omega \int_{\Omega_C} \boldsymbol{\sigma} \mathbf{E}_C \cdot \overline{\mathbf{z}_C} \\
&\qquad\quad \textstyle + c_0^* \int_{\Omega_I} \operatorname{div}(\varepsilon_I \mathbf{E}_I)\, \operatorname{div}(\varepsilon_I \overline{\mathbf{z}_I}) = -i\omega \int_\Omega \mathbf{J}_e \cdot \overline{\mathbf{z}} \\
&\text{for each } \mathbf{z} \in \widehat{W}_{\varepsilon_I}(\Omega_I;\Omega) \, .
\end{aligned} \tag{6.50}$$

In Section 2.1 it has been proved that the sesquilinear form

$$\begin{aligned}
a_e^*(\mathbf{w},\mathbf{z}) := \textstyle\int_\Omega \boldsymbol{\mu}^{-1} \operatorname{curl} \mathbf{w} \cdot \operatorname{curl} \overline{\mathbf{z}} + i\omega \int_{\Omega_C} \boldsymbol{\sigma} \mathbf{w}_C \cdot \overline{\mathbf{z}_C} \\
\textstyle + c_0^* \int_{\Omega_I} \operatorname{div}(\varepsilon_I \mathbf{w}_I)\, \operatorname{div}(\varepsilon_I \overline{\mathbf{z}_I})
\end{aligned}$$

is coercive in $W_{\varepsilon_I}(\Omega_I;\Omega)$, where $W_{\varepsilon_I}(\Omega_I;\Omega)$ has been introduced in (2.3), and reads

$$\begin{aligned}
W_{\varepsilon_I}(\Omega_I;\Omega) := \{\mathbf{z} \in H(\mathrm{curl};\Omega) \,|\, & \mathbf{z}_I \in H_{0,\partial\Omega}(\varepsilon_I, \mathrm{div};\Omega_I), \\
& \mathbf{z}_I \perp^{\varepsilon_I} \mathcal{H}_{\varepsilon_I}(\Gamma, \partial\Omega;\Omega_I)\} \, .
\end{aligned} \tag{6.51}$$

We now show that the same result is true in the space $\widehat{W}_{\varepsilon_I}(\Omega_I;\Omega)$. The crucial point is proving something similar to Lemma 2.1. We have:

Lemma 6.15. *There exists a constant $C > 0$ such that*

$$\begin{aligned}
\|\mathbf{z}_I\|_{0,\Omega_I} \leq C\Big(& \|\operatorname{curl} \mathbf{z}_I\|_{0,\Omega_I} + \|\operatorname{div}(\varepsilon_I \mathbf{z}_I)\|_{0,\Omega_I} + \|\mathbf{z}_I \times \mathbf{n}_I\|_{H^{-1/2}(\mathrm{div}_\tau;\Gamma)} \\
& \textstyle + \sum_{j=1}^{p_\Gamma} \left| \int_{\Gamma_j} \varepsilon_I \mathbf{z}_I \cdot \mathbf{n}_I \right| + \sum_{k=1}^{n_{\partial\Omega}} \left| \int_{\Sigma_k} \varepsilon_I \mathbf{z}_I \cdot \mathbf{n}_\Sigma \right| \Big)
\end{aligned}$$

for all $\mathbf{z}_I \in H(\mathrm{curl};\Omega_I) \cap H_{0,\partial\Omega}(\varepsilon_I, \mathrm{div};\Omega_I)$.

Proof. From the representation formula (A.12), namely,

$$\mathbf{z}_I = \varepsilon_I^{-1} \operatorname{curl} \mathbf{q}_I + \operatorname{grad} \varphi_I + \mathbf{h}_I \, ,$$

and from Lemma 2.1 we realize that the only term to be estimated is $\mathbf{h}_I$. Since it belongs to the finite dimensional space $\mathcal{H}_{\varepsilon_I}(\Gamma, \partial\Omega; \Omega_I)$ and depends linearly on $\int_{\Omega_I} \varepsilon_I \mathbf{z}_I \cdot \operatorname{grad} w_{j,I}$ and $\int_{\Omega_I} \varepsilon_I \mathbf{z}_I \cdot \boldsymbol{\pi}_{k,I}$, we have at once

$$\|\mathbf{h}_I\|_{0,\Omega_I} \leq C \left(\sum_{j=1}^{p_\Gamma} \left| \int_{\Omega_I} \varepsilon_I \mathbf{z}_I \cdot \operatorname{grad} w_{j,I} \right| + \sum_{k=1}^{n_{\partial\Omega}} \left| \int_{\Omega_I} \varepsilon_I \mathbf{z}_I \cdot \boldsymbol{\pi}_{k,I} \right| \right) .$$

On the other hand we have $\varepsilon_I \mathbf{z}_I \cdot \mathbf{n} = 0$ on $\partial\Omega$ and $w_{j,I} = 0$ on $\Gamma \setminus \Gamma_j$, $w_{j,I} = 1$ on Γ_j, thus integration by parts gives

$$\int_{\Omega_I} \varepsilon_I \mathbf{z}_I \cdot \operatorname{grad} w_{j,I} = - \int_{\Omega_I} \operatorname{div}(\varepsilon_I \mathbf{z}_I)\, w_{j,I} + \int_{\Gamma_j} \varepsilon_I \mathbf{z}_I \cdot \mathbf{n}_I .$$

Moreover, proceeding as in (6.1.5), since $\varepsilon_I \mathbf{z}_I \cdot \mathbf{n} = 0$ on $\partial\Omega$ we find

$$\textstyle\int_{\Omega_I} \varepsilon_I \mathbf{z}_I \cdot \boldsymbol{\pi}_{k,I} = - \int_{\Omega_I \setminus \Sigma_k} \operatorname{div}(\varepsilon_I \mathbf{z}_I)\, q_{k,I} + \int_{\Sigma_k} \varepsilon_I \mathbf{z}_I \cdot \mathbf{n}_\Sigma .$$

Since the functions $w_{j,I}$ and $q_{k,I}$ only depend on Ω_I and ε_I, the Hölder inequality thus gives

$$\begin{aligned} \|\mathbf{h}_I\|_{0,\Omega_I} \leq C \Big(& \| \operatorname{div}(\varepsilon_I \mathbf{z}_I)\|_{0,\Omega_I} \\ & + \textstyle\sum_{j=1}^{p_\Gamma} \left| \int_{\Gamma_j} \varepsilon_I \mathbf{z}_I \cdot \mathbf{n}_I \right| + \sum_{k=1}^{n_{\partial\Omega}} \left| \int_{\Sigma_k} \varepsilon_I \mathbf{z}_I \cdot \mathbf{n}_\Sigma \right| \Big) , \end{aligned}$$

and the proof is complete. □

Showing that the sesquilinear form $a_e^*(\cdot,\cdot)$ is coercive in $\widehat{W}_{\varepsilon_I}(\Omega_I; \Omega)$ is now an easy task, by repeating the arguments presented in Lemma 2.2 and Theorem 2.3: hence there exists a unique solution $\mathbf{E}$ to (6.50).

We also know from Section 3.4 that there exists a unique solution to the eddy current problem (1.22), and moreover it is straightforward to verify that the electric field determined there satisfies (6.48), and consequently also (6.50). Since for all these problems we have proved that uniqueness holds, we can thus conclude that the electric field in (1.22), (6.48) and (6.50) is always the same. Thus we deduce that for the solution to (6.48) and (6.50) the constraint $\operatorname{div}(\varepsilon_I \mathbf{E}_I) = 0$ in Ω_I is indeed satisfied. (A direct proof of this fact could also be obtained by adapting the arguments in the proof of Lemma 6.1.)

Concerning numerical approximation, when ε_I is a matrix with smooth entries a finite element scheme based on the formulation (6.50) can be easily devised, using nodal elements in Ω_I and edge elements in Ω_C. The constraints on the fluxes on Γ_j and Σ_k of the test functions can be satisfied by proceeding as in Section 6.1.4, by explicitly constructing basis functions with vanishing fluxes.

On the other hand, if ε_I is not smooth, even a smooth $\mathbf{z}_I$ could violate the condition $\operatorname{div}(\varepsilon_I \mathbf{z}_I) \in L^2(\Omega_I)$, and also for a piecewise-smooth ε_I it is not easy to construct finite elements satisfying the necessary matching condition $[\varepsilon_I \mathbf{z}_{I,h} \cdot \mathbf{n}] = 0$ on the interelements.

In any case, the approximation scheme here described could have a major drawback, even for a scalar constant ε_I: in fact, as noted in Remark 6.6, it can happen that the space

$$H^1_{\tau,\partial\Omega}(\Omega_I) = \{\mathbf{z}_I \in (H^1(\Omega_I))^3 \,|\, \mathbf{z}_I \cdot \mathbf{n} = 0 \text{ on } \partial\Omega\}$$

is a proper subspace of $H(\text{curl}; \Omega_I) \cap H_{0,\partial\Omega}(\text{div}; \Omega_I)$, and that the solution $\mathbf{E}_I$ lies in $H(\text{curl}; \Omega_I) \cap H_{0,\partial\Omega}(\text{div}; \Omega_I)$ but not in $H^1_{\tau,\partial\Omega}(\Omega_I)$. If $H^1_{\tau,\partial\Omega}(\Omega_I)$ is a closed proper subspace of $H(\text{curl}; \Omega_I) \cap H_{0,\partial\Omega}(\text{div}; \Omega_I)$, the nodal finite element approximate solution $\mathbf{E}_{I,h}$ (which clearly is always contained in $H^1_{\tau,\partial\Omega}(\Omega_I)$) cannot converge to the exact solution.

This problem occurs if the domain Ω has re-entrant corners (see Costabel and Dauge [87], Costabel et al. [90]). Instead, if Ω is convex, the convergence could be achieved, though there are always re-entrant corners for the insulator Ω_I on the interface Γ. In fact, since no boundary condition is imposed on Γ, it is likely that the space of smooth tangential vector fields is dense in the space

$$\widehat{W}_\tau(\Omega) := \{\mathbf{z} \in H(\text{curl}; \Omega) \,|\, \mathbf{z}_I \in H_{0,\partial\Omega}(\text{div}; \Omega_I),\ \mathbf{z} \times \mathbf{n} \in (L^2(\Gamma))^3\}$$

(see Nicaise [188]). Therefore, provided that for the solution $\mathbf{E}$ it holds $\mathbf{E} \times \mathbf{n} \in (L^2(\Gamma))^3$, which does not seem to be a very restrictive assumption, Céa lemma and a density argument would ensure that the Galerkin method is convergent.

Remark 6.16. A hp discontinuous Galerkin approximation of the eddy current problem (6.48) has been proposed by Perugia and Schötzau [191]. With that approach the discrete solution does not belong to $H(\text{curl}; \Omega) \cap H(\text{div}; \Omega)$; however, in that paper it is proved that the finite element scheme is convergent, no matter if the exact solution does not belong to $(H^1(\Omega))^3$. The price to pay is a larger number of degrees of freedom. □

6.2 Formulation for the Lorenz gauge and its numerical approximation

In this section we suppose that Ω, Ω_C and Ω_I satisfy the assumptions of Section 1.3, and moreover we assume that Ω is *simply-connected* and that Ω_C is *connected* (some informations regarding more general geometrical configurations are presented in Remark 6.24). For the sake of definiteness, we also limit ourselves to considering the magnetic boundary value problem (1.22).

Following the analysis presented in Fernandes and Valli [105], we deal with a different gauge condition for the magnetic vector potential $\mathbf{A}$: the so-called Lorenz gauge

$$\text{div}\,\mathbf{A}_C + \mu_*\sigma^* V_C = 0 \ \text{ in } \Omega_C \ , \quad \text{div}\,\mathbf{A}_I = 0 \ \text{ in } \Omega_I \ , \qquad (6.52)$$

where σ^* is assumed to be a scalar function, satisfying $0 < \sigma_1^* \leq \sigma^*(\mathbf{x}) \leq \sigma_2^*$ in Ω_C and $\sigma^*(\mathbf{x}) = 0$ in Ω_I, and $\mu_* > 0$ is a dimensional constant. (For instance, one can think that μ_* is a suitable average in Ω of the entries of the matrix $\boldsymbol{\mu}$, and that

$\sigma^* = \frac{\kappa}{3}\,\mathrm{trace}(\boldsymbol{\sigma})$ for a suitable non-dimensional constant $\kappa > 0$, so that $\sigma_1^* = \kappa\sigma_{\min}$, $\sigma_2^* = \kappa\sigma_{\max}$, where $\sigma_{\min}$ and $\sigma_{\max}$ are a uniform lower bound for the minimum eigenvalues of $\boldsymbol{\sigma}(\mathbf{x})$ in Ω_C and a uniform upper bound for the maximum eigenvalues of $\boldsymbol{\sigma}(\mathbf{x})$ in Ω_C, respectively.)

The Lorenz gauge has been originally proposed with the aim of decoupling the equation for $\mathbf{A}$ from the equation for V_C, substituting $\boldsymbol{\sigma}$ grad V_C with

$$-\boldsymbol{\sigma}\,\mathrm{grad}[(\mu_*\sigma^*)^{-1}\,\mathrm{div}\,\mathbf{A}_C]\ .$$

In particular, an additional feature of this approach is that, for a scalar constant $\sigma_{|\Omega_C} = \sigma^*$ and a scalar constant $\mu_C = \mu_*$, the latter term simplifies to $-\mu_*^{-1}\,\mathrm{grad}\,\mathrm{div}\,\mathbf{A}_C$, which, added to $\mu_*^{-1}\,\mathrm{curl}\,\mathrm{curl}\,\mathbf{A}_C$, gives at last $-\mu_*^{-1}\Delta\mathbf{A}_C$.

However, as also noted by Bíró and Preis [51], in Section 6.2.1 we will see that this decoupling is difficult to handle for a non-constant conductivity $\boldsymbol{\sigma}_{|\Omega_C}$, as one ends up facing a problem which looks hard to solve. Moreover, in the opposite situation, namely, when $\sigma_{|\Omega_C}$ is a scalar constant, we will see that, when the current density satisfies the standard assumption $\mathrm{div}\,\mathbf{J}_e = 0$ in Ω, decoupling is always leading to a problem for which $\mathrm{div}\,\mathbf{A}_C = 0$ and $V_C = 0$ in Ω_C, hence to a formulation in terms of the sole electric field $\mathbf{E}_C = -i\omega\mathbf{A}_C$.

In other words, when considering an approach based on "genuine" magnetic vector potentials and scalar electric potentials, the Lorenz gauge is never furnishing a formulation which is at the same time well-posed and decoupled, and it has simply to be thought as a variant of the Coulomb gauge. However, it is also worth noting that sometimes the Lorenz gauged formulation has shown better performances in numerical computations (see, e.g., Bryant et al. [67], Morisue [181]).

Let us now go on in completing the setting. Besides the gauging condition (6.52), we have to add a boundary condition, which, when considering the magnetic boundary condition $\mathbf{H}\times\mathbf{n} = \mathbf{0}$, is given by

$$\mathbf{A}\cdot\mathbf{n} = 0 \qquad \text{on } \partial\Omega\ . \tag{6.53}$$

Moreover, we have also to impose an additional condition on the interface Γ. In fact, differently from the Coulomb gauge, the constraints on the divergence have been given separately in Ω_C and Ω_I. We consider three possible alternatives, that have been frequently proposed in the literature devoted to this subject (see, e.g., Fernandes [103]): the first one is the "slip" condition

$$\mathbf{A}_C\cdot\mathbf{n}_C = 0 \qquad \text{on } \Gamma\ ; \tag{6.54}$$

the second one is the Dirichlet condition

$$V_C = 0 \qquad \text{on } \Gamma\ ; \tag{6.55}$$

the last one is the matching condition

$$\mathbf{A}_I\cdot\mathbf{n}_I + \mathbf{A}_C\cdot\mathbf{n}_C = 0 \qquad \text{on } \Gamma\ . \tag{6.56}$$

The interface conditions (6.54) and (6.55) were indeed proposed with the aim of decoupling the equations for $\mathbf{A}$ from that for V_C: however, as we have just noted, for a non-constant $\boldsymbol{\sigma}_{|\Omega_C}$ the decoupled problem turns out to be hard to handle, while for a scalar constant $\sigma_{|\Omega_C}$ and for $\operatorname{div} \mathbf{J}_e = 0$ in Ω it reduces to an electric field formulation. Hence, the use of these interface conditions has to be looked at as a possible alternative to the more natural condition (6.56), most often leading to a coupled problem (similar to the one we will present in Section 6.2.3 for the interface condition (6.56)).

The matching condition $\mathbf{A}_C \cdot \mathbf{n}_C = k^2 V_C$ on Γ has been also proposed (see Bryant et al. [68]). We could adapt the following presentation also to this case, but in our opinion, for a non-constant $\boldsymbol{\sigma}_{|\Omega_C}$, its use leads to a less simple algorithm for numerical approximation, due to the necessity of constructing finite elements satisfying that matching on Γ. Besides, for a scalar constant $\sigma_{|\Omega_C}$ and for $\operatorname{div} \mathbf{J}_e = 0$ in Ω it also reduces to an electric field formulation with $\mathbf{A}_C \cdot \mathbf{n}_C = 0$ on Γ.

We are now in a position to formulate the problem. Let us start specifying in detail the formulation associated to the matching condition (6.56). First, note that $(6.52)_2$, (6.53) and (6.56) imply that $\int_\Gamma \mathbf{A}_I \cdot \mathbf{n}_I = 0 = \int_\Gamma \mathbf{A}_C \cdot \mathbf{n}_C$, hence $\int_{\Omega_C} \operatorname{div} \mathbf{A}_C = 0$. As a consequence, we can also impose

$$\int_{\Omega_C} \sigma^* V_C = 0 \tag{6.57}$$

without actually introducing any further constraint.

In conclusion, writing the eddy current equations in terms of these unknowns we are left with the problem

$$\begin{cases} \operatorname{curl}(\boldsymbol{\mu}^{-1} \operatorname{curl} \mathbf{A}) + i\omega\boldsymbol{\sigma}\mathbf{A} + \boldsymbol{\sigma} \operatorname{grad} V_C = \mathbf{J}_e & \text{in } \Omega \\ \operatorname{div} \mathbf{A}_C + \mu_* \sigma^* V_C = 0 & \text{in } \Omega_C \\ \operatorname{div} \mathbf{A}_I = 0 & \text{in } \Omega_I \\ \mathbf{A}_I \cdot \mathbf{n}_I + \mathbf{A}_C \cdot \mathbf{n}_C = 0 & \text{on } \Gamma \\ \int_{\Omega_C} \sigma^* V_C = 0 & \\ \mathbf{A}_I \cdot \mathbf{n} = 0 & \text{on } \partial\Omega \\ (\boldsymbol{\mu}_I^{-1} \operatorname{curl} \mathbf{A}_I) \times \mathbf{n} = \mathbf{0} & \text{on } \partial\Omega \ , \end{cases} \tag{6.58}$$

where, as in (6.3), the notation $\boldsymbol{\sigma} \operatorname{grad} V_C$ means

$$\boldsymbol{\sigma} \operatorname{grad} V_C := \begin{cases} \boldsymbol{\sigma}_{|\Omega_C} \operatorname{grad} V_C & \text{in } \Omega_C \\ \mathbf{0} & \text{in } \Omega_I \ . \end{cases}$$

If we replace the interface condition $(6.58)_4$ (i.e., (6.56)) with (6.54), we obtain another problem, that will be denoted by (6.58)*. Moreover, if we replace $(6.58)_4$ with (6.55) and we drop the average condition $(6.58)_5$, we obtain a third problem, that will be denoted by (6.58)**.

Defining

$$\mathbf{J} := \begin{cases} \mathbf{J}_{e,C} - i\omega\boldsymbol{\sigma}\mathbf{A}_C - \boldsymbol{\sigma} \operatorname{grad} V_C & \text{in } \Omega_C \\ \mathbf{J}_{e,I} & \text{in } \Omega_I \ , \end{cases} \tag{6.59}$$

as a consequence of (6.58) (or (6.58)*, or (6.58)**) we also have

$$\operatorname{curl}(\boldsymbol{\mu}^{-1} \operatorname{curl} \mathbf{A}) = \mathbf{J} \qquad \text{in } \Omega \ ,$$

therefore

$$\begin{cases} \operatorname{div}(i\omega\boldsymbol{\sigma}\mathbf{A}_C + \boldsymbol{\sigma} \operatorname{grad} V_C - \mathbf{J}_{e,C}) = 0 & \text{in } \Omega_C \\ (i\omega\boldsymbol{\sigma}\mathbf{A}_C + \boldsymbol{\sigma} \operatorname{grad} V_C - \mathbf{J}_{e,C}) \cdot \mathbf{n}_C = \mathbf{J}_{e,I} \cdot \mathbf{n}_I & \text{on } \Gamma \ . \end{cases} \tag{6.60}$$

Remark 6.17. As we have already noted, the condition $\int_{\Omega_C} \sigma^* V_C = 0$ follows from the gauge conditions $(6.58)_2$, $(6.58)_3$, $(6.58)_4$ and $(6.58)_6$. Therefore, we could omit it in (6.58). However, this vanishing average condition is useful in Section 6.2.3 when we analyze the weak formulation of the Lorenz gauged eddy current problem; hence we prefer to keep it in formulation (6.58). The same remark applies to the formulation (6.58)*. □

6.2.1 Decoupled weak formulations and alternative gauge conditions

As a starting point, with the aim of making clear the superiority of our choice in Section 6.2.3, in this section we discuss in detail some of the weak formulations that have been previously proposed for problem (6.58) (or (6.58)*, or (6.58)**). Let us point out that we are not assuming that $\boldsymbol{\sigma}$ is smooth, but only that each of its entries belongs to $L^\infty(\Omega_C)$.

In the following, in order to give a meaning to the integrals we are going to consider, we assume, as will be proved in Section 6.2.2, that there exists a solution to (6.58) (or (6.58)*, or (6.58)**), satisfying $(\mathbf{A}, V_C) \in U_0 \times H^1(\Omega_C)$, where

$$U_0 := \{\mathbf{w} \in H(\mathrm{curl};\Omega) \mid \mathrm{div}\,\mathbf{w}_C \in L^2(\Omega_C),\, \mathrm{div}\,\mathbf{w}_I \in L^2(\Omega_I),\\ \mathbf{w}_I \cdot \mathbf{n} = 0 \text{ on } \partial\Omega\}\ ,$$

and moreover we consider the space of test functions

$$W_0 := \{\mathbf{w} \in H(\mathrm{curl};\Omega) \mid \mathrm{div}(\boldsymbol{\sigma}\mathbf{w}_C) \in L^2(\Omega_C),\, \mathrm{div}\,\mathbf{w}_I \in L^2(\Omega_I),\\ \mathbf{w}_I \cdot \mathbf{n} = 0 \text{ on } \partial\Omega\}$$

(for a smooth scalar conductivity $\sigma_{|\Omega_C}$ we have $U_0 = W_0$).

Multiply $(6.58)_1$ by a test function $\mathbf{w} \in W_0$ and integrate in Ω. Integration by parts yields

$$\begin{aligned} &\int_\Omega [\boldsymbol{\mu}^{-1}\,\mathrm{curl}\,\mathbf{A}\cdot\mathrm{curl}\,\overline{\mathbf{w}} + \int_{\Omega_C}[i\omega\boldsymbol{\sigma}\mathbf{A}_C\cdot\overline{\mathbf{w}_C} - V_C\,\mathrm{div}(\boldsymbol{\sigma}\overline{\mathbf{w}_C})] \\ &\quad + \int_\Gamma V_C\boldsymbol{\sigma}\overline{\mathbf{w}_C}\cdot\mathbf{n}_C \\ &= \int_\Omega \mathbf{J}_e\cdot\overline{\mathbf{w}}\,. \end{aligned} \tag{6.61}$$

Using the Lorenz gauge in Ω_C permits to replace the unknown V_C and, taking also into account that $\mathrm{div}\,\mathbf{A}_I = 0$ in Ω_I, we end up with

$$\begin{aligned} &\int_\Omega \boldsymbol{\mu}^{-1}\,\mathrm{curl}\,\mathbf{A}\cdot\mathrm{curl}\,\overline{\mathbf{w}} + \int_{\Omega_C}(\mu_*\sigma^*)^{-1}\,\mathrm{div}\,\mathbf{A}_C\,\mathrm{div}(\boldsymbol{\sigma}\overline{\mathbf{w}_C}) \\ &\quad + \int_{\Omega_C} i\omega\boldsymbol{\sigma}\mathbf{A}_C\cdot\overline{\mathbf{w}_C} + \int_{\Omega_I}\mu_*^{-1}\,\mathrm{div}\,\mathbf{A}_I\,\mathrm{div}\,\overline{\mathbf{w}_I} \\ &\quad + \int_\Gamma V_C\boldsymbol{\sigma}\overline{\mathbf{w}_C}\cdot\mathbf{n}_C \\ &= \int_\Omega \mathbf{J}_e\cdot\overline{\mathbf{w}} \qquad \forall\,\mathbf{w}\in W_0\,. \end{aligned} \tag{6.62}$$

To conclude, let us obtain the weak formulation for the scalar electric potential. From (6.60), we see that V_C satisfies

$$\begin{aligned} 0 &= -\int_{\Omega_C}\mathrm{div}(i\omega\boldsymbol{\sigma}\mathbf{A}_C + \boldsymbol{\sigma}\,\mathrm{grad}\,V_C - \mathbf{J}_{e,C})\,\overline{Q_C} \\ &= \int_{\Omega_C}(i\omega\boldsymbol{\sigma}\mathbf{A}_C + \boldsymbol{\sigma}\,\mathrm{grad}\,V_C - \mathbf{J}_{e,C})\cdot\mathrm{grad}\,\overline{Q_C} \\ &\qquad - \int_\Gamma(i\omega\boldsymbol{\sigma}\mathbf{A}_C + \boldsymbol{\sigma}\,\mathrm{grad}\,V_C - \mathbf{J}_{e,C})\cdot\mathbf{n}_C\,\overline{Q_C} \\ &= \int_{\Omega_C}(i\omega\boldsymbol{\sigma}\mathbf{A}_C + \boldsymbol{\sigma}\,\mathrm{grad}\,V_C - \mathbf{J}_{e,C})\cdot\mathrm{grad}\,\overline{Q_C} - \int_\Gamma \mathbf{J}_{e,I}\cdot\mathbf{n}_I\,\overline{Q_C} \end{aligned}$$

for each $Q_C \in H^1(\Omega_C)$, namely,

$$\begin{aligned}\int_{\Omega_C} \boldsymbol{\sigma} \operatorname{grad} V_C \cdot \operatorname{grad} \overline{Q_C} &= - \int_{\Omega_C} i\omega\boldsymbol{\sigma} \mathbf{A}_C \cdot \operatorname{grad} \overline{Q_C} \\ &+ \int_{\Omega_C} \mathbf{J}_{e,C} \cdot \operatorname{grad} \overline{Q_C} + \int_\Gamma \mathbf{J}_{e,I} \cdot \mathbf{n}_I \overline{Q_C} \qquad \forall\, Q_C \in H^1(\Omega_C) \,.\end{aligned} \tag{6.63}$$

Now, in order to obtain a formulation which looks feasible and for which the unknowns $\mathbf{A}$ and V_C are decoupled, we have to eliminate the term containing V_C in (6.62). This can be done either assuming that the test function $\mathbf{w}$ belongs to W_{00}, where

$$W_{00} := \{\mathbf{w} \in W_0 \mid \boldsymbol{\sigma}\mathbf{w}_C \cdot \mathbf{n}_C = 0 \text{ on } \Gamma\} \,,$$

or else using the interface condition (6.55), i.e., $V_C = 0$ on Γ.

In the first case the final problem, associated to the interface conditions (6.54) or (6.56), is

$$\begin{aligned}\mathbf{A} \in U_0 \ : \ &\int_\Omega \boldsymbol{\mu}^{-1} \operatorname{curl} \mathbf{A} \cdot \operatorname{curl} \overline{\mathbf{w}} + \int_{\Omega_C} (\mu_* \sigma^*)^{-1} \operatorname{div} \mathbf{A}_C \operatorname{div}(\boldsymbol{\sigma}\overline{\mathbf{w}_C}) \\ &+ \int_{\Omega_C} i\omega\boldsymbol{\sigma} \mathbf{A}_C \cdot \overline{\mathbf{w}_C} + \int_{\Omega_I} \mu_*^{-1} \operatorname{div} \mathbf{A}_I \operatorname{div} \overline{\mathbf{w}_I} \\ &= \int_\Omega \mathbf{J}_e \cdot \overline{\mathbf{w}} \qquad \forall\, \mathbf{w} \in W_{00} \,,\end{aligned} \tag{6.64}$$

followed by

$$\begin{aligned}V_C \in H^1_\dagger(\Omega_C) \ : \ &\int_{\Omega_C} \boldsymbol{\sigma} \operatorname{grad} V_C \cdot \operatorname{grad} \overline{Q_C} \\ &= - \int_{\Omega_C} i\omega\boldsymbol{\sigma} \mathbf{A}_C \cdot \operatorname{grad} \overline{Q_C} \\ &\quad + \int_{\Omega_C} \mathbf{J}_{e,C} \cdot \operatorname{grad} \overline{Q_C} \\ &\quad + \int_\Gamma \mathbf{J}_{e,I} \cdot \mathbf{n}_I \overline{Q_C} \quad \forall\, Q_C \in H^1_\dagger(\Omega_C) \,,\end{aligned} \tag{6.65}$$

where

$$H^1_\dagger(\Omega_C) := \left\{ Q_C \in H^1(\Omega_C) \mid \int_{\Omega_C} \sigma^* Q_C = 0 \right\} . \tag{6.66}$$

In the latter case the problem, associated to the interface condition (6.55), is

$$\begin{aligned}\mathbf{A} \in U_0 \ : \ &\int_\Omega \boldsymbol{\mu}^{-1} \operatorname{curl} \mathbf{A} \cdot \operatorname{curl} \overline{\mathbf{w}} + \int_{\Omega_C} (\mu_* \sigma^*)^{-1} \operatorname{div} \mathbf{A}_C \operatorname{div}(\boldsymbol{\sigma}\overline{\mathbf{w}_C}) \\ &+ \int_{\Omega_C} i\omega\boldsymbol{\sigma} \mathbf{A}_C \cdot \overline{\mathbf{w}_C} + \int_{\Omega_I} \mu_*^{-1} \operatorname{div} \mathbf{A}_I \operatorname{div} \overline{\mathbf{w}_I} \\ &= \int_\Omega \mathbf{J}_e \cdot \overline{\mathbf{w}} \qquad \forall\, \mathbf{w} \in W_0 \,,\end{aligned} \tag{6.67}$$

followed by

$$\begin{aligned}V_C \in H^1_0(\Omega_C) \ : \ &\int_{\Omega_C} \boldsymbol{\sigma} \operatorname{grad} V_C \cdot \operatorname{grad} \overline{Q_C} \\ &= - \int_{\Omega_C} i\omega\boldsymbol{\sigma} \mathbf{A}_C \cdot \operatorname{grad} \overline{Q_C} \\ &\quad + \int_{\Omega_C} \mathbf{J}_{e,C} \cdot \operatorname{grad} \overline{Q_C} \qquad \forall\, Q_C \in H^1_0(\Omega_C) \,.\end{aligned} \tag{6.68}$$

While problems (6.65) and (6.68) are classical elliptic boundary value problems, without additional assumptions the formulations (6.64) or (6.67) are not easy to handle. A favourable situation appears when σ is a scalar constant in Ω_C, as, first of all, in this case one has $U_0 = W_0$ and, moreover, for the interface condition (6.54) we know that $\mathbf{A} \in W_{00}$. Therefore, in problems (6.64) (for the interface condition

(6.54)) and (6.67) (for the interface condition (6.55)), the space of trial functions and the space of test functions are the same (on the contrary, even for $\sigma_{|\Omega_C} = \text{const}$ this is not the case for the interface condition (6.56)). Furthermore, if in addition one chooses $\sigma^* = \sigma_{|\Omega_C} = \text{const}$ in Ω_C, one also has $\int_{\Omega_C} (\mu_* \sigma^*)^{-1} \operatorname{div} \mathbf{A}_C \operatorname{div}(\sigma \overline{\mathbf{w}_C}) = \int_{\Omega_C} \mu_*^{-1} \operatorname{div} \mathbf{A}_C \operatorname{div} \overline{\mathbf{w}_C}$, so that the first order terms in the sesquilinear forms at the left hand side of (6.64) and (6.67) are hermitian and positive definite.

An analysis of these two formulations for $\sigma_{|\Omega_C} = \text{const}$ and $\sigma^* = \sigma_{|\Omega_C}$ is presented here below, for a slightly generalized form of the Lorenz gauge proposed by Bossavit [60], that indeed for $\sigma_{|\Omega_C} = \text{const}$ coincides with the usual one. However, in the general case of a non-constant $\boldsymbol{\sigma}_{|\Omega_C}$, the formulations (6.64) and (6.67) are not suitable: for instance, it is not clear that a uniqueness result holds for them, even if in (6.64) we use the additional information that the solution satisfies (6.54) or (6.56).

For arriving to a decoupled and well-suited formulation even in the case of a non-constant $\boldsymbol{\sigma}$ a change of the point of view is thus in order. Bossavit [60], assuming that σ is a *scalar function*, proposed to modify the Lorenz gauge in Ω_C in the following way

$$\operatorname{div}(\sigma \mathbf{A}_C) + \mu_* \sigma^2 V_C = 0 \qquad \text{in } \Omega_C \ , \tag{6.69}$$

which, as we already noted, for a scalar constant value of $\sigma_{|\Omega_C}$ reduces to the usual Lorenz gauge (6.52) (for the choice $\sigma^* = \sigma_{|\Omega_C}$). Accordingly, instead of the interface condition (6.54) one has to consider $\sigma \mathbf{A}_C \cdot \mathbf{n}_C = 0$ on Γ, while the condition (6.55) is kept unchanged (in the following, the interface condition (6.56) or its variant $\sigma \mathbf{A}_C \cdot \mathbf{n}_C + \mathbf{A}_I \cdot \mathbf{n}_I = 0$ on Γ will not be considered, as they do not lead to a decoupled problem).

Let us suppose that there exists a solution $(\mathbf{A}, V_C) \in W_{00} \times H^1(\Omega_C)$ (for the interface condition $\sigma \mathbf{A}_C \cdot \mathbf{n}_C = 0$ on Γ) or $(\mathbf{A}, V_C) \in W_0 \times H_0^1(\Omega_C)$ (for the interface condition (6.55)) to these Bossavit–Lorenz gauged problems; without giving further details, we note that we could adapt the proofs reported in Section 6.2.2 to show that these existence results are in fact true.

Proceeding as before, for the interface condition $\sigma \mathbf{A}_C \cdot \mathbf{n}_C = 0$ on Γ the corresponding weak formulation now reads

$$\begin{aligned} \mathbf{A} \in W_{00} \ : \ & \textstyle\int_\Omega \boldsymbol{\mu}^{-1} \operatorname{curl} \mathbf{A} \cdot \operatorname{curl} \overline{\mathbf{w}} \\ & + \textstyle\int_{\Omega_C} \mu_*^{-1} \sigma^{-2} \operatorname{div}(\sigma \mathbf{A}_C) \operatorname{div}(\sigma \overline{\mathbf{w}_C}) \\ & + \textstyle\int_{\Omega_C} i\omega\sigma \mathbf{A}_C \cdot \overline{\mathbf{w}_C} + \int_{\Omega_I} \mu_*^{-1} \operatorname{div} \mathbf{A}_I \operatorname{div} \overline{\mathbf{w}_I} \\ & = \textstyle\int_\Omega \mathbf{J}_e \cdot \overline{\mathbf{w}} \qquad \forall\, \mathbf{w} \in W_{00} \ . \end{aligned} \tag{6.70}$$

Similarly, for the interface condition (6.55) one can write

$$\begin{aligned} \mathbf{A} \in W_0 \ : \ & \textstyle\int_\Omega \boldsymbol{\mu}^{-1} \operatorname{curl} \mathbf{A} \cdot \operatorname{curl} \overline{\mathbf{w}} \\ & + \textstyle\int_{\Omega_C} \mu_*^{-1} \sigma^{-2} \operatorname{div}(\sigma \mathbf{A}_C) \operatorname{div}(\sigma \overline{\mathbf{w}_C}) \\ & + \textstyle\int_{\Omega_C} i\omega\sigma \mathbf{A}_C \cdot \overline{\mathbf{w}_C} + \int_{\Omega_I} \mu_*^{-1} \operatorname{div} \mathbf{A}_I \operatorname{div} \overline{\mathbf{w}_I} \\ & = \textstyle\int_\Omega \mathbf{J}_e \cdot \overline{\mathbf{w}} \qquad \forall\, \mathbf{w} \in W_0 \ . \end{aligned} \tag{6.71}$$

Indeed these weak problems look easier to handle. First of all, it is easy to prove that they are well-posed, namely, that uniqueness holds. In fact, for $\mathbf{J}_e = \mathbf{0}$ it follows at once that $\mathbf{A}_C = \mathbf{0}$ in Ω_C; consequently, $\mathbf{A}_I$ satisfies $\operatorname{curl} \mathbf{A}_I = \mathbf{0}$ in Ω_I, $\operatorname{div} \mathbf{A}_I = 0$ in Ω_I, $\mathbf{A}_I \cdot \mathbf{n} = 0$ on $\partial\Omega$ and finally $\mathbf{A}_I \times \mathbf{n}_I = -\mathbf{A}_C \times \mathbf{n}_C = \mathbf{0}$ on Γ, hence $\mathbf{A}_I = \mathbf{0}$ in Ω_I.

Another relevant result is the following: the solution $\mathbf{A}$ to (6.70) or (6.71) enjoys the property that $\mu_*^{-1}\sigma^{-2} \operatorname{div}(\sigma \mathbf{A}_C)$ has a distributional gradient belonging to $(L^2(\Omega_C))^3$ and, moreover, that $\operatorname{div} \mathbf{A}_I = 0$ in Ω_I. (Since, as we have already noted, it is possible to show that there exists a solution to the Bossavit–Lorenz gauged vector potential problems, these results are indeed trivial, as $\mu_*^{-1}\sigma^{-2} \operatorname{div}(\sigma \mathbf{A}_C) = -V_C \in H^1(\Omega_C)$, and $\operatorname{div} \mathbf{A}_I = 0$ is the gauge condition in Ω_I; however, it is useful to show that they follow directly from the intrinsic structure of the weak problems (6.70) or (6.71).)

In fact, take $\mathbf{q}_C \in (C_0^\infty(\Omega_C))^3$ and let $u_C \in H^1(\Omega_C)$ be the solution of the Neumann problem

$$\begin{cases} \operatorname{div}(\sigma \operatorname{grad} u_C) + i\omega\mu_*\sigma^2 u_C = \operatorname{div} \mathbf{q}_C & \text{in } \Omega_C \\ \sigma \operatorname{grad} u_C \cdot \mathbf{n}_C = 0 & \text{on } \Gamma \end{cases} \tag{6.72}$$

(for the interface condition $\sigma \mathbf{A}_C \cdot \mathbf{n}_C = 0$ on Γ) or of the Dirichlet problem

$$\begin{cases} \operatorname{div}(\sigma \operatorname{grad} u_C) + i\omega\mu_*\sigma^2 u_C = \operatorname{div} \mathbf{q}_C & \text{in } \Omega_C \\ u_C = 0 & \text{on } \Gamma \end{cases} \tag{6.73}$$

(for the interface condition (6.55)). Then, for $g_I \in L^2(\Omega_I)$ let $u_I \in H^1(\Omega_I)$ be the solution of the mixed problem

$$\begin{cases} \Delta u_I = \mu_* g_I & \text{in } \Omega_I \\ u_I = u_C & \text{on } \Gamma \\ \operatorname{grad} u_I \cdot \mathbf{n} = 0 & \text{on } \partial\Omega\,. \end{cases} \tag{6.74}$$

Setting

$$u := \begin{cases} u_C & \text{in } \Omega_C \\ u_I & \text{in } \Omega_I\,, \end{cases} \tag{6.75}$$

we have $\operatorname{grad} u \in W_{00}$ (respectively, $\operatorname{grad} u \in W_0$) for the interface condition $\sigma \mathbf{A}_C \cdot \mathbf{n}_C = 0$ on Γ (respectively, for the interface condition (6.55)). Choosing the test function $\mathbf{w} = \operatorname{grad} u$ in (6.70) or in (6.71) gives

$$\begin{aligned} \int_{\Omega_C} i\omega\sigma \mathbf{A}_C \cdot \operatorname{grad} \overline{u_C} &= -\int_{\Omega_C} i\omega \operatorname{div}(\sigma \mathbf{A}_C)\, \overline{u_C} + \int_\Gamma i\omega\sigma \mathbf{A}_C \cdot \mathbf{n}_C\, \overline{u_C} \\ &= -\int_{\Omega_C} i\omega \operatorname{div}(\sigma \mathbf{A}_C)\, \overline{u_C}, \end{aligned}$$

and

$$\begin{aligned} \int_\Omega \mathbf{J}_e \cdot \operatorname{grad} \overline{u} &= \int_{\Omega_C} \mathbf{J}_{e,C} \cdot \operatorname{grad} \overline{u_C} + \int_{\Omega_I} \mathbf{J}_{e,I} \cdot \operatorname{grad} \overline{u_I} \\ &= \int_{\Omega_C} \mathbf{J}_{e,C} \cdot \operatorname{grad} \overline{u_C} + \int_\Gamma \mathbf{J}_{e,I} \cdot \mathbf{n}_I\, \overline{u_I}\,, \end{aligned}$$

therefore

$$\begin{aligned} &\int_{\Omega_C} \mu_*^{-1}\sigma^{-2} \operatorname{div}(\sigma \mathbf{A}_C) \operatorname{div} \overline{\mathbf{q}_C} + \int_{\Omega_I} \operatorname{div} \mathbf{A}_I\, \overline{g_I} \\ &\quad = \int_{\Omega_C} \mathbf{J}_{e,C} \cdot \operatorname{grad} \overline{u_C} + \int_\Gamma \mathbf{J}_{e,I} \cdot \mathbf{n}_I\, \overline{u_C}\,. \end{aligned} \tag{6.76}$$

Taking $\mathbf{q}_C = \mathbf{0}$ we have $u_C = 0$ in Ω_C, hence the right hand side in (6.76) is vanishing, and we conclude that $\operatorname{div} \mathbf{A}_I = 0$ in Ω_I.

The map $\mathbf{q}_C \to \int_{\Omega_C} \mathbf{J}_{e,C} \cdot \operatorname{grad} \overline{u_C} + \int_\Gamma \mathbf{J}_{e,I} \cdot \mathbf{n}_I \overline{u_C}$ is anti-linear and continuous with respect to the norm in $(L^2(\Omega_C))^3$. Therefore, it can be extended by density to $\mathbf{q}_C \in (L^2(\Omega_C))^3$, and then, by the Riesz theorem, represented as $\int_{\Omega_C} \mathbf{G}_C \cdot \overline{\mathbf{q}_C}$ for a suitable $\mathbf{G}_C \in (L^2(\Omega_C))^3$. In conclusion, we have $\operatorname{grad}[\mu_*^{-1}\sigma^{-2} \operatorname{div}(\sigma \mathbf{A}_C)] = -\mathbf{G}_C \in (L^2(\Omega_C))^3$ in Ω_C.

However, the most interesting property of the solution $\mathbf{A}$ to the weak problems (6.70) or (6.71) arises when the current density satisfies the assumption $\operatorname{div} \mathbf{J}_e = 0$ in Ω (namely, in addition to the necessary assumption $\operatorname{div} \mathbf{J}_{e,I} = 0$ in Ω_I, also $\operatorname{div} \mathbf{J}_{e,C} = 0$ in Ω_C and $\mathbf{J}_{e,C} \cdot \mathbf{n}_C + \mathbf{J}_{e,I} \cdot \mathbf{n}_I = 0$ on Γ). In this case the formulations (6.70) and (6.71) are not related to a "genuine" Lorenz gauged problem, as we find $\operatorname{div}(\sigma \mathbf{A}_C) = 0$ and $V_C = 0$ in Ω_C.

In fact, by integrating by parts one easily sees that the right hand side of (6.76) is vanishing, and, repeating the arguments above by replacing $\operatorname{div} \mathbf{q}_C$ in (6.72) and (6.73) with $\mu_* \sigma^2 g_C$, where $g_C \in L^2(\Omega_C)$, we end up with

$$\int_{\Omega_C} \operatorname{div}(\sigma \mathbf{A}_C) \overline{g_C} = 0 \ ,$$

hence $\operatorname{div}(\sigma \mathbf{A}_C) = 0$ in Ω_C. Furthermore, from the weak problem (6.65) (with $H^1_\dagger(\Omega_C)$ replaced by $H^1_\ddagger(\Omega_C) := \{Q_C \in H^1(\Omega_C) \,|\, \int_{\Omega_C} \sigma^2 Q_C = 0\}$, in order to be consistent with the gauge condition (6.69)) or from the weak problem (6.68) it follows that $V_C = 0$ in Ω_C.

In conclusion, under the very common assumption $\operatorname{div} \mathbf{J}_e = 0$ in Ω the formulations (6.70) and (6.71) are not "genuine" Lorenz gauged formulations, since they both essentially reduce to a formulation in terms of the modified magnetic vector potential $\mathbf{A}_C^* = i\omega^{-1} \mathbf{E}_C$ in Ω_C, keeping the vector potential $\mathbf{A}_I$ in Ω_I.

For numerical approximation, which has to be performed by means of nodal elements, these formulations could be interesting only for smooth $\sigma_{|\Omega_C}$, as in the opposite case it is not ensured that $\operatorname{div}(\sigma \mathbf{w}_C)$ belongs to $L^2(\Omega_C)$ even for a smooth $\mathbf{w}_C$. A better situation occurs when $\sigma_{|\Omega_C}$ is piecewise-smooth, but also in that case it is not an easy task to construct finite elements $\mathbf{w}_{C,h}$ such that the matching condition $[\sigma \mathbf{w}_{C,h} \cdot \mathbf{n}] = 0$ is satisfied on the interelements where σ is jumping.

Moreover, the usual difficulties coming from the presence of re-entrant corners occur. In this respect, convergence for formulations (6.70) and (6.71) is ensured provided that Ω is convex and, for formulation (6.70) only, that Ω_C is convex, too.

In conclusion, an approach in terms of the electric field $\mathbf{E}$ like (6.50) looks similar but generally better suited than (6.70) or (6.71): first of all, solving (6.50) one also finds $\mathbf{H} = -(i\omega\boldsymbol{\mu})^{-1} \operatorname{curl} \mathbf{E}$, while the solution of (6.70) or (6.71) does not furnish the electric field $\mathbf{E}_I$ (and for $\operatorname{div} \mathbf{J}_e \neq 0$ it also needs the computation of the electric scalar potential V_C); secondly, in (6.50) it is not assumed that $\boldsymbol{\sigma}$ is a scalar function; finally, in real-life problems it is more likely that $\boldsymbol{\varepsilon}_I$ is smooth than that $\sigma_{|\Omega_C}$ is smooth.

Remark 6.18. The assumption $\operatorname{div} \mathbf{J}_e = 0$ in Ω is not needed to solve the eddy current problem, as the necessary and sufficient conditions for solving it are $\operatorname{div} \mathbf{J}_{e,I} = 0$ in Ω_I and $\mathbf{J}_e \cdot \mathbf{n} = 0$ on $\partial\Omega$. However, this divergence-free condition is not particularly restrictive, and is very often satisfied from current densities used in real-life applications. As a matter of fact, it is automatically satisfied whenever the support of $\mathbf{J}_e$ is contained in Ω_I (and, clearly, $\operatorname{div} \mathbf{J}_{e,I} = 0$ in Ω_I). □

6.2.2 Well-posed formulations based on the Lorenz gauge

In this section, following a general approach proposed in Fernandes [103] and developed in Fernandes and Valli [105], we present a "genuine" Lorenz gauged formulation for which we are able to prove well-posedness. However, it must be noted that we do not obtain a decoupled problem for $\mathbf{A}$ and V_C. Let us also underline that here we are only assuming that the conductivity $\boldsymbol{\sigma}$ is a symmetric matrix, with entries belonging to $L^\infty(\Omega)$, uniformly positive definite in Ω_C and vanishing in Ω_I.

For the sake of simplicity, in the following we focus on the interface condition (6.56), namely, $\mathbf{A}_C \cdot \mathbf{n}_C + \mathbf{A}_I \cdot \mathbf{n}_I = 0$ on Γ. The reader interested in the alternative interface conditions (6.54) and (6.55) is referred to Fernandes and Valli [105] (where however the problems are considered for Ω and Ω_C having the simplest geometrical shape). There one can also find some motivations, related to numerical approximation, that lead to the consideration of this choice of the interface condition and find it to be the most suitable Lorenz gauged magnetic vector potential formulation to use.

With the aim of obtaining a more general result, the existence theory we are going to develop in this section does not follow the same guidelines of that we have presented in Section 6.1.2 for the Coulomb gauge: in this way we can prove the existence and uniqueness of the solution to the Lorenz gauged problem for any choice of the function σ^* (see Theorem 6.21). On the other hand, the proof of the coerciveness of the sesquilinear form associated to the weak problem and the convergence of a finite element approximation will require a suitable choice of σ^* (see Proposition 6.23 and Section 6.2.4).

We start from the unique solution $(\mathbf{H}, \mathbf{E}) \in H(\mathrm{curl}; \Omega) \times H(\mathrm{curl}; \Omega)$ of the eddy current problem (1.22) (the existence of such a solution is proved in Section 3.4).

The first step of the procedure consists in solving

$$\begin{cases} -\Delta V + i\omega\mu_*\sigma^* V = \operatorname{div} \mathbf{E} & \text{in } \Omega \\ \dfrac{\partial V}{\partial n} = -\mathbf{E} \cdot \mathbf{n} & \text{on } \partial\Omega \\ \int_{\Omega_C} \sigma^* V_C = 0 \,, \end{cases} \tag{6.77}$$

to be intended in the weak sense made precise in the following Proposition 6.19. The second step is finding the solution to

$$\begin{cases} \operatorname{curl} \mathbf{A} = \boldsymbol{\mu}\mathbf{H} & \text{in } \Omega \\ \operatorname{div} \mathbf{A} = -\mu_*\sigma^* V & \text{in } \Omega \\ \mathbf{A} \cdot \mathbf{n} = 0 & \text{on } \partial\Omega \end{cases} \tag{6.78}$$

(note that this implicitly says that $\mathbf{A}_C \cdot \mathbf{n}_C + \mathbf{A}_I \cdot \mathbf{n}_I = 0$ on Γ).

The necessary solvability conditions for (6.78) are $\operatorname{div}(\boldsymbol{\mu}\mathbf{H}) = 0$ in Ω and $\int_{(\partial\Omega)_r} \boldsymbol{\mu}\mathbf{H} \cdot \mathbf{n} = 0$ for each $r = 1, \ldots, p_{\partial\Omega}$, as usual following from the Faraday equation and the Stokes theorem for closed surfaces, and $\int_\Omega \sigma^* V = 0$, namely, $\int_{\Omega_C} \sigma^* V_C = 0$, that is satisfied due to $(6.77)_3$.

Hence, it remains to show that (6.77) is well-posed.

Proposition 6.19. *There exists a unique solution of the Neumann problem* (6.77).

Proof. We start showing that the following weak problem has a unique solution: find $V \in H^1(\Omega)$ with $\int_{\Omega_C} \sigma^* V_C = 0$ such that

$$\int_\Omega \operatorname{grad} V \cdot \operatorname{grad} \overline{\eta_0} + i\omega \int_\Omega \mu_* \sigma^* V \, \overline{\eta_0} = -\int_\Omega \mathbf{E} \cdot \operatorname{grad} \overline{\eta_0} \tag{6.79}$$

for all $\eta_0 \in H^1(\Omega)$ with $\int_{\Omega_C} \sigma^* \eta_{0,C} = 0$.

The existence and uniqueness of the solution to (6.79) is a consequence of the Lax–Milgram lemma, as it is easy to prove that the Poincaré inequality

$$\int_\Omega |\operatorname{grad} \eta_0|^2 \geq K_3 \int_\Omega (|\eta_0|^2 + |\operatorname{grad} \eta_0|^2) \tag{6.80}$$

holds for functions $\eta_0 \in H^1(\Omega)$ with $\int_{\Omega_C} \sigma^* \eta_{0,C} = 0$ (one can adapt, for instance, the proof reported in Dautray and Lions [94], Chap. IV, Sect. 7, Prop. 2, where the function η_0 is assumed to satisfy $\int_\Omega \eta_0 = 0$ instead of $\int_{\Omega_C} \sigma^* \eta_{0,C} = 0$). Taking now $\eta \in H^1(\Omega)$, we set

$$\eta_0 := \eta - \left(\int_{\Omega_C} \sigma^* \right)^{-1} \left(\int_{\Omega_C} \sigma^* \eta_C \right) ;$$

clearly, η_0 can be used as a test function in (6.79). Therefore we have

$$\int_\Omega \operatorname{grad} V \cdot \operatorname{grad} \overline{\eta} + i\omega \int_\Omega \mu_* \sigma^* V \, \overline{\eta} = -\int_\Omega \mathbf{E} \cdot \operatorname{grad} \overline{\eta} ,$$

as $\operatorname{grad} \eta_0 = \operatorname{grad} \eta$ and $\int_\Omega \sigma^* V = \int_{\Omega_C} \sigma^* V_C = 0$.

Integrating by parts, one gets easily that $\operatorname{div}(\operatorname{grad} V + \mathbf{E}) = i\omega\mu_*\sigma^* V$ in Ω and $(\operatorname{grad} V + \mathbf{E}) \cdot \mathbf{n} = 0$ on $\partial\Omega$, namely, V is the solution to (6.77).

The uniqueness of the solution to (6.77) follows from the fact that each solution V to (6.77) is clearly a solution to (6.79). □

We can thus obtain:

Proposition 6.20. *There exists a unique solution* $(\mathbf{A}, V) \in H(\text{curl}; \Omega) \times H^1(\Omega)$ *to the problem*

$$\begin{cases} \operatorname{curl} \mathbf{A} = \boldsymbol{\mu} \mathbf{H} & \text{in } \Omega \\ i\omega \mathbf{A} + \operatorname{grad} V = -\mathbf{E} & \text{in } \Omega \\ \operatorname{div} \mathbf{A} + \mu_* \sigma^* V = 0 & \text{in } \Omega \\ \mathbf{A} \cdot \mathbf{n} = 0 & \text{on } \partial\Omega , \end{cases} \tag{6.81}$$

and it is given by the solution to (6.77), (6.78).

Proof. For the existence of the solution we only need to show that $(6.81)_2$ is satisfied. Setting $\mathbf{N} := i\omega\mathbf{A} + \operatorname{grad} V + \mathbf{E}$, from the Faraday equation, (6.77) and (6.78) we have $\operatorname{curl}\mathbf{N} = \mathbf{0}$ in Ω, $\operatorname{div}\mathbf{N} = 0$ in Ω and $\mathbf{N}\cdot\mathbf{n} = 0$ on $\partial\Omega$, therefore $\mathbf{N} = \mathbf{0}$ in Ω (remember that we have assumed that Ω is simply-connected).

To prove the uniqueness it is enough to observe that, putting $\mathbf{H} = \mathbf{0}$ and $\mathbf{E} = \mathbf{0}$ in (6.81), V satisfies

$$\begin{cases} -\Delta V + i\omega\mu_*\sigma^* V = 0 & \text{in } \Omega \\ \dfrac{\partial V}{\partial n} = 0 & \text{on } \partial\Omega \\ \int_{\Omega_C} \sigma^* V_C = 0\,, \end{cases}$$

hence $V = 0$ in Ω and $\mathbf{A} = \mathbf{0}$ in Ω. □

We finally have an existence result in terms of the current density $\mathbf{J}_e$ alone.

Theorem 6.21. *There exists a solution $(\mathbf{A}, V) \in H(\operatorname{curl};\Omega)\times H^1(\Omega)$ to the Lorenz gauged problem*

$$\begin{cases} \operatorname{curl}(\boldsymbol{\mu}^{-1}\operatorname{curl}\mathbf{A}) + i\omega\boldsymbol{\sigma}\mathbf{A} + \boldsymbol{\sigma}\operatorname{grad} V = \mathbf{J}_e & \text{in } \Omega \\ \operatorname{div}\mathbf{A} + \mu_*\sigma^* V = 0 & \text{in } \Omega \\ \int_{\Omega_C} \sigma^* V_C = 0 & \\ \mathbf{A}\cdot\mathbf{n} = 0 & \text{on } \partial\Omega \\ (\boldsymbol{\mu}^{-1}\operatorname{curl}\mathbf{A})\times\mathbf{n} = \mathbf{0} & \text{on } \partial\Omega\,, \end{cases} \tag{6.82}$$

and it is given by the solution to (6.81). *In particular,* $\mathbf{A}$ *and* $V_C := V_{|\Omega_C}$ *are a solution to the Lorenz gauged problem* (6.58). *Moreover, the solution* $(\mathbf{A}, V_C)$ *to* (6.58) *is uniquely determined.*

Proof. The proof of the existence is trivial, as one has only to write the eddy current problem in terms of the solutions to (6.81). The uniqueness of the solution $(\mathbf{A}, V_C)$ of problem (6.58) is proved as follows. Assume that $\mathbf{J}_e = \mathbf{0}$ in Ω, multiply $(6.58)_1$ by $\overline{\mathbf{A}}$ and integrate in Ω: by integration by parts one obtains

$$\int_\Omega \boldsymbol{\mu}^{-1}\operatorname{curl}\mathbf{A}\cdot\operatorname{curl}\overline{\mathbf{A}} + i\omega\int_{\Omega_C}\boldsymbol{\sigma}\mathbf{A}_C\cdot\overline{\mathbf{A}_C} + \int_{\Omega_C}\boldsymbol{\sigma}\operatorname{grad} V_C\cdot\overline{\mathbf{A}_C} = 0\,.$$

Since $\operatorname{div}\operatorname{curl}(\boldsymbol{\mu}^{-1}\operatorname{curl}\mathbf{A}) = 0$ in Ω, one also has

$$\operatorname{curl}(\boldsymbol{\mu}_C^{-1}\operatorname{curl}\mathbf{A}_C)\cdot\mathbf{n}_C = -\operatorname{curl}(\boldsymbol{\mu}_I^{-1}\operatorname{curl}\mathbf{A}_I)\cdot\mathbf{n}_I = 0 \qquad \text{on } \Gamma\,;$$

thus, multiplying by $(i\omega)^{-1}\operatorname{grad}\overline{V_C}$ the equation $(6.58)_1$ (restricted to Ω_C) and integrating by parts in Ω_C, one finds

$$\int_{\Omega_C}\boldsymbol{\sigma}\mathbf{A}_C\cdot\operatorname{grad}\overline{V_C} + (i\omega)^{-1}\int_{\Omega_C}\boldsymbol{\sigma}\operatorname{grad} V_C\cdot\operatorname{grad}\overline{V_C} = 0\,.$$

Therefore, $\operatorname{Re}(\int_{\Omega_C}\boldsymbol{\sigma}\mathbf{A}_C\cdot\operatorname{grad}\overline{V_C}) = 0$ and $\operatorname{Re}(\int_{\Omega_C}\boldsymbol{\sigma}\operatorname{grad} V_C\cdot\overline{\mathbf{A}}_C) = 0$, hence $\int_\Omega\boldsymbol{\mu}^{-1}\operatorname{curl}\mathbf{A}\cdot\operatorname{curl}\overline{\mathbf{A}} = 0$ and consequently $\operatorname{curl}\mathbf{A} = \mathbf{0}$ in Ω. In addition, inserting this result in $(6.58)_1$, we obtain $i\omega\boldsymbol{\sigma}\mathbf{A}_C + \boldsymbol{\sigma}\operatorname{grad} V_C = \mathbf{0}$ in Ω_C.

Since we have assumed that Ω is simply-connected, the curl-free condition guarantees the existence of a function $\mathcal{U} \in H^1(\Omega)$ such that $i\omega\mathbf{A} = -\operatorname{grad}\mathcal{U}$ in Ω; moreover, since Ω_C is connected, it is not restrictive to suppose that $\mathcal{U}_C = V_C$ in Ω_C. Hence we have

$$-\Delta\mathcal{U} = i\omega \operatorname{div}\mathbf{A} = -i\omega\mu_*\sigma^* V_C = -i\omega\mu_*\sigma^*\mathcal{U} \quad \text{in } \Omega$$

and $\operatorname{grad}\mathcal{U} \cdot \mathbf{n} = -i\omega\mathbf{A} \cdot \mathbf{n} = 0$ on $\partial\Omega$, therefore $\mathcal{U}$ is a solution to the homogeneous Neumann problem (6.77). We thus have $\mathcal{U} = 0$ in Ω, and consequently $\mathbf{A} = \mathbf{0}$ in Ω and $V_C = 0$ in Ω_C. □

6.2.3 Weak formulations and positiveness

In order to devise a finite element approximation scheme, we are now interested in deriving the weak formulation of the Lorenz gauged magnetic vector potential problem (6.58).

Using the boundary and interface conditions in (6.58), the usual integration by parts gives

$$\begin{aligned}\textstyle\int_\Omega \boldsymbol{\mu}^{-1} \operatorname{curl}\mathbf{A} \cdot \operatorname{curl}\overline{\mathbf{w}} + \int_{\Omega_C}(i\omega\boldsymbol{\sigma}\mathbf{A}_C \cdot \overline{\mathbf{w}_C} + \boldsymbol{\sigma} \operatorname{grad} V_C \cdot \overline{\mathbf{w}_C})\\ = \textstyle\int_\Omega \mathbf{J}_e \cdot \overline{\mathbf{w}} \qquad \forall\, \mathbf{w} \in H(\operatorname{curl}, \Omega)\,.\end{aligned}$$

Due to the Lorenz gauge, one can add three other terms, finding

$$\begin{aligned}&\textstyle\int_\Omega \boldsymbol{\mu}^{-1} \operatorname{curl}\mathbf{A} \cdot \operatorname{curl}\overline{\mathbf{w}}\\ &\quad \textstyle+ \int_{\Omega_C} \mu_*^{-1} \operatorname{div}\mathbf{A}_C \operatorname{div}\overline{\mathbf{w}_C} + \int_{\Omega_I} \mu_*^{-1} \operatorname{div}\mathbf{A}_I \operatorname{div}\overline{\mathbf{w}_I}\\ &\quad \textstyle+ \int_{\Omega_C}(i\omega\boldsymbol{\sigma}\mathbf{A}_C \cdot \overline{\mathbf{w}_C} + \boldsymbol{\sigma} \operatorname{grad} V_C \cdot \overline{\mathbf{w}_C} + \sigma^* V_C \operatorname{div}\overline{\mathbf{w}_C})\\ &\textstyle= \int_\Omega \mathbf{J}_e \cdot \overline{\mathbf{w}} \qquad \forall\, \mathbf{w} \in H(\operatorname{curl}; \Omega) \cap H(\operatorname{div}; \Omega)\,.\end{aligned} \tag{6.83}$$

For a solution satisfying the Lorenz gauge, this equation is formally equivalent to (6.62); however, its structure is unlike, and moreover the functional framework is now different, as we are requiring that $\mathbf{w} \in H(\operatorname{curl}; \Omega) \cap H(\operatorname{div}; \Omega)$ and not that $\mathbf{w} \in W_0$.

On the other hand, using the Lorenz gauge equation in (6.63) and multiplying by $i\omega^{-1}$ yields

$$\begin{aligned}&\textstyle\int_{\Omega_C}(i\omega^{-1}\boldsymbol{\sigma} \operatorname{grad} V_C - \boldsymbol{\sigma}\mathbf{A}_C) \cdot \operatorname{grad}\overline{Q_C}\\ &\quad \textstyle+ \int_{\Omega_C} \sigma^*(\operatorname{div}\mathbf{A}_C + \mu_*\sigma^* V_C)\,\overline{Q_C}\\ &\textstyle= i\omega^{-1} \int_{\Omega_C} \mathbf{J}_{e,C} \cdot \operatorname{grad}\overline{Q_C}\\ &\quad \textstyle+ i\omega^{-1} \int_\Gamma \mathbf{J}_{e,I} \cdot \mathbf{n}_I\, \overline{Q_C} \quad \forall\, Q_C \in H^1(\Omega_C)\,.\end{aligned} \tag{6.84}$$

The two additional terms that are present here and are not contained in (6.63) are useful for obtaining a symmetric problem.

Let us consider the variational spaces $H^1_\dagger(\Omega_C)$, introduced in (6.66), and

$$W := H(\operatorname{curl}; \Omega) \cap H_0(\operatorname{div}; \Omega)\,, \tag{6.85}$$

the sesquilinear form

$$\begin{aligned}
\mathcal{B}[(\mathbf{z},U_C),(\mathbf{w},Q_C)] &:= \textstyle\int_\Omega \boldsymbol{\mu}^{-1}\operatorname{curl}\mathbf{z}\cdot\operatorname{curl}\overline{\mathbf{w}} \\
&\quad + \textstyle\int_{\Omega_I} \mu_*^{-1}\operatorname{div}\mathbf{z}_I\operatorname{div}\overline{\mathbf{w}_I} \\
&\quad + \textstyle\int_{\Omega_C} \mu_*^{-1}(\operatorname{div}\mathbf{z}_C + \mu_*\sigma^* U_C)(\operatorname{div}\overline{\mathbf{w}_C} + \mu_*\sigma^*\overline{Q_C}) \\
&\quad + i\omega^{-1}\textstyle\int_{\Omega_C} \boldsymbol{\sigma}(i\omega\mathbf{z}_C + \operatorname{grad}U_C)\cdot(-i\omega\overline{\mathbf{w}_C} + \operatorname{grad}\overline{Q_C}) \\
&= \textstyle\int_\Omega \boldsymbol{\mu}^{-1}\operatorname{curl}\mathbf{z}\cdot\operatorname{curl}\overline{\mathbf{w}} + \int_\Omega \mu_*^{-1}\operatorname{div}\mathbf{z}\operatorname{div}\overline{\mathbf{w}} \\
&\quad + \textstyle\int_{\Omega_C}(i\omega\boldsymbol{\sigma}\mathbf{z}_C\cdot\overline{\mathbf{w}_C} + \boldsymbol{\sigma}\operatorname{grad}U_C\cdot\overline{\mathbf{w}_C} + \sigma^* U_C\operatorname{div}\overline{\mathbf{w}_C}) \\
&\quad + \textstyle\int_{\Omega_C}(i\omega^{-1}\boldsymbol{\sigma}\operatorname{grad}U_C\cdot\operatorname{grad}\overline{Q_C} + \mu_*(\sigma^*)^2 U_C\overline{Q_C}) \\
&\quad + \textstyle\int_{\Omega_C}(\sigma^*\operatorname{div}\mathbf{z}_C\overline{Q_C} - \boldsymbol{\sigma}\mathbf{z}_C\cdot\operatorname{grad}\overline{Q_C})
\end{aligned} \tag{6.86}$$

defined (and continuous) in $[H(\mathrm{curl};\Omega)\cap H(\mathrm{div};\Omega)]\times H^1(\Omega_C)$, and the anti-linear functional

$$\mathcal{F}(\mathbf{w},Q_C) := \int_\Omega \mathbf{J}_e\cdot\overline{\mathbf{w}} + i\omega^{-1}\left(\int_{\Omega_C}\mathbf{J}_{e,C}\cdot\operatorname{grad}\overline{Q_C} + \int_\Gamma \mathbf{J}_{e,I}\cdot\mathbf{n}_I\overline{Q_C}\right)$$

defined (and continuous) in $L^2(\Omega)\times H^1(\Omega_C)$.

Theorem 6.22. *There exists a unique solution to the weak problem*

$$\begin{aligned}
(\mathbf{A},V_C)\in W\times H^1_\dagger(\Omega_C)\ :\ &\mathcal{B}[(\mathbf{A},V_C),(\mathbf{w},Q_C)] = \mathcal{F}(\mathbf{w},Q_C) \\
&\forall\,(\mathbf{w},Q_C)\in W\times H^1_\dagger(\Omega_C)\,.
\end{aligned} \tag{6.87}$$

Proof. Following the procedure just described, the existence is an easy consequence of the fact that, as proved in Theorem 6.21, problem (6.58) is well-posed.

Uniqueness follows from the fact that, if $\mathcal{B}[(\mathbf{A},V_C),(\mathbf{w},Q_C)] = 0$ for each $(\mathbf{w},Q_C)\in W\times H^1_\dagger(\Omega_C)$, by choosing $\mathbf{w}=\mathbf{A}$, $Q_C=V_C$ one finds

$$\begin{aligned}
&\textstyle\int_\Omega \boldsymbol{\mu}^{-1}\operatorname{curl}\mathbf{A}\cdot\operatorname{curl}\overline{\mathbf{A}} + \int_{\Omega_I}\mu_*^{-1}|\operatorname{div}\mathbf{A}_I|^2 \\
&\quad + \textstyle\int_{\Omega_C}\mu_*^{-1}|\operatorname{div}\mathbf{A}_C + \mu_*\sigma^* V_C|^2 \\
&\quad + i\omega^{-1}\textstyle\int_{\Omega_C}\boldsymbol{\sigma}(i\omega\mathbf{A}_C + \operatorname{grad}V_C)\cdot(-i\omega\overline{\mathbf{A}_C} + \operatorname{grad}\overline{V_C}) = 0\,.
\end{aligned}$$

Therefore one has obtained a solution to the homogeneous problem (6.58). Since this problems has a unique solution (see Theorem 6.21), the thesis follows. □

For numerical approximation, it is useful to check that the sesquilinear form $\mathcal{B}[\cdot,\cdot]$ is coercive in $W\times H^1_\dagger(\Omega_C)$. Before starting, let us recall that the Poincaré-like inequality (6.39) is valid in W, as for $n_\Omega = n_{\partial\Omega} = 0$ we have $W = W_\sharp$. Moreover, by proceeding as done for inequality (6.80), taking into account that Ω_C is connected we also have that

$$\int_{\Omega_C}|\operatorname{grad}Q_C|^2 \ge K_4\int_{\Omega_C}(|Q_C|^2 + |\operatorname{grad}Q_C|^2) \qquad \forall\,Q_C\in H^1_\dagger(\Omega_C)\,. \tag{6.88}$$

Proposition 6.23. *The sesquilinear form $\mathcal{B}[\cdot,\cdot]$ is coercive in $W\times H^1_\dagger(\Omega_C)$, provided that the maximum value σ_2^* of the scalar function σ^* is small enough. In particular, if one has chosen $\sigma^* = \frac{\kappa}{3}\operatorname{trace}(\boldsymbol{\sigma})$, $\mathcal{B}[\cdot,\cdot]$ is coercive provided that κ is small enough.*

Proof. From (6.86) we have

$$\begin{aligned}\mathcal{B}[(\mathbf{w},Q_C),(\mathbf{w},Q_C)] &= \int_\Omega \boldsymbol{\mu}^{-1}\,\mathrm{curl}\,\mathbf{w}\cdot\mathrm{curl}\,\overline{\mathbf{w}} + \mu_*^{-1}\int_{\Omega_I}|\,\mathrm{div}\,\mathbf{w}_I|^2\\ &+\mu_*^{-1}\int_{\Omega_C}|\,\mathrm{div}\,\mathbf{w}_C+\mu_*\sigma^* Q_C|^2\\ &+i\omega^{-1}\int_{\Omega_C}\boldsymbol{\sigma}(i\omega\mathbf{w}_C+\mathrm{grad}\,Q_C)\cdot(-i\omega\overline{\mathbf{w}_C}+\mathrm{grad}\,\overline{Q_C})\,.\end{aligned}$$

Let us denote by $\mu_{\max}$ a uniform bound in Ω for the maximum eigenvalues of $\boldsymbol{\mu}(\mathbf{x})$, and by $\sigma_{\min}$ a uniform lower bound in Ω_C for the minimum eigenvalues of $\boldsymbol{\sigma}(\mathbf{x})$. Remember also that the auxiliary function σ^* is assumed to satisfy $0<\sigma_1^*\le\sigma^*(\mathbf{x})\le\sigma_2^*$ in Ω_C. Proceeding as in the proof of the coerciveness of the sesquilinear form $\mathcal{A}[\cdot,\cdot]$ in Section 6.1.2, we have

$$\begin{aligned}&|\mathrm{Re}\,\mathcal{B}[(\mathbf{w},Q_C),(\mathbf{w},Q_C)]|\\ &\quad\ge \mu_{\max}^{-1}\int_\Omega|\,\mathrm{curl}\,\mathbf{w}|^2+\mu_*^{-1}\int_{\Omega_I}|\,\mathrm{div}\,\mathbf{w}_I|^2\\ &\qquad+\tfrac{1}{2}\mu_*^{-1}\int_{\Omega_C}|\,\mathrm{div}\,\mathbf{w}_C|^2-\mu_*(\sigma_2^*)^2\int_{\Omega_C}|Q_C|^2\end{aligned}$$

and, for each $0<\delta<1$,

$$\begin{aligned}&|\mathrm{Im}\,\mathcal{B}[(\mathbf{w},Q_C),(\mathbf{w},Q_C)]|\\ &\quad\ge|\omega|^{-1}\sigma_{\min}(1-\delta)\int_{\Omega_C}|\,\mathrm{grad}\,Q_C|^2-|\omega|\sigma_{\min}(1-\delta)\delta^{-1}\int_{\Omega_C}|\mathbf{w}_C|^2\,.\end{aligned}$$

Taking into account the Poincaré inequalities (6.39) and (6.88), the coerciveness of $\mathcal{B}[\cdot,\cdot]$ easily follows by choosing at first $(1-\delta)$ small enough and then σ_2^* small enough. □

Remark 6.24. The existence of a unique solution to the Lorenz gauged problem (6.58) can be proved also in a more general geometrical setting, namely, without assuming that Ω is simply-connected. To be precise, we suppose that (6.2) is satisfied and that $n_{\partial\Omega}=n_\Omega$, which means that we can choose the surfaces $\hat{\Sigma}_k$ "cutting" the non-bounding cycles on $\partial\Omega$ without intersecting Ω_C.

In this situation, we proceed as for the Coulomb gauged problem, considering a variational formulation in the space $W_\sharp$ (see (6.13)). Since the Poincaré-like inequality (6.39) holds, the coerciveness of the sesquilinear form $\mathcal{B}[\cdot,\cdot]$ in $W_\sharp\times H_\dagger^1(\Omega_C)$ follows as before, provided that the maximum value σ_2^* of the scalar function σ^* is small enough. Hence, there exists a unique solution of the weak problem

$$\begin{aligned}(\mathbf{A},V_C)\in W_\sharp\times H_\dagger^1(\Omega_C)\ :\ \mathcal{B}[(\mathbf{A},V_C),(\mathbf{w},Q_C)]&=\mathcal{F}(\mathbf{w},Q_C)\\ \forall\,(\mathbf{w},Q_C)&\in W_\sharp\times H_\dagger^1(\Omega_C)\,.\end{aligned}$$

In Section 6.1.1 we have considered three possible choices of the linear contraints $\mathcal{G}_k(\cdot)$ appearing in the definition of $W_\sharp$: since $\mathrm{div}\,\mathbf{A}$ is not vanishing in Ω, these choices are not equivalent and do not furnish the same solution $\mathbf{A}$. The proof that the weak solution is indeed a solution of the strong problem (6.58) is easily done only when $\mathcal{G}_k(\mathbf{w})=\int_\Omega\mathbf{w}\cdot\hat{\boldsymbol{\pi}}_k$, where $\hat{\boldsymbol{\pi}}_k$ are the basis functions of $\mathcal{H}(m;\Omega)$. As already noted, this choice is not the best suited for numerical approximation.

On the other hand, concerning the assumption on the geometry of the conductor, it can be proved that if Ω_C is not connected than the solution to (6.58) is not unique, even for a simply-connected domain Ω. More precisely, there exist solutions $(\mathbf{A},V_C)$ of (6.58) with $\mathbf{J}_e=\mathbf{0}$ satisfying $\mathbf{A}\ne\mathbf{0}$ and $\mathrm{grad}\,V_C\ne\mathbf{0}$ (the proof is left to the reader). □

6.2.4 Numerical approximation

Assume that Ω, Ω_C and Ω_I are Lipschitz polyhedra, and that $\mathcal{T}_{I,h}$ and $\mathcal{T}_{C,h}$ are two regular families of triangulations of Ω_I and Ω_C, respectively. For the sake of simplicity, we suppose that each element K of $\mathcal{T}_{I,h}$ and $\mathcal{T}_{C,h}$ is a tetrahedron. We also assume that these triangulations match on Γ, so that they furnish a family of triangulations $\mathcal{T}_h$ of Ω.

Numerical approximation of problem (6.87) via conforming finite elements can be easily devised. In fact, it is enough to choose suitable finite element subspaces $W_h^r \subset W$ and $(L^s_{C,h})_\dagger \subset H^1_\dagger(\Omega_C)$, and rewrite (6.87) in $W_h^r \times (L^s_{C,h})_\dagger$. The uniqueness of the discrete solution follows as in Theorem 6.22; its existence is then a consequence of the uniqueness result.

Moreover, in Section 6.2.3 we have proved that, assuming that the maximum value σ_2^* of the scalar function σ^* is small enough, the sesquilinear form $\mathcal{B}[\cdot,\cdot]$ is continuous and coercive. Therefore, the convergence analysis is easily performed as follows.

Denoting by $\mathbb{P}_k$, $k \geq 1$, the space of polynomials of degree less than or equal to k, for $r \geq 1$ and $s \geq 1$ we choose the discrete spaces of Lagrange nodal elements

$$W_h^r := \left\{\mathbf{w}_h \in (C^0(\Omega))^3 \mid \mathbf{w}_{h|K} \in (\mathbb{P}_r)^3 \;\forall\, K \in \mathcal{T}_h \;,\; \mathbf{w}_h \cdot \mathbf{n} = 0 \text{ on } \partial\Omega\right\},$$

and

$$(L^s_{C,h})_\dagger := \left\{Q_{C,h} \in C^0(\Omega_C) \mid Q_{C,h|K} \in \mathbb{P}_s \;\forall\, K \in \mathcal{T}_{C,h} \;,\; \textstyle\int_{\Omega_C} \sigma^* Q_{C,h} = 0\right\}$$

(for this last space, the construction of the basis functions satisfying the average constraint on Ω_C can be done as at the beginning of Section 6.1.4, where the basis functions of the constrained space $(W_h^r)_\sharp$ have been determined).

Via Céa lemma for each $\mathbf{w}_h \in W_h^r$ and $Q_{C,h} \in (L^s_{C,h})_\dagger$ we have

$$\begin{aligned}
&\Big(\textstyle\int_\Omega (|\mathbf{A} - \mathbf{A}_h|^2 + |\operatorname{curl}(\mathbf{A} - \mathbf{A}_h)|^2 + |\operatorname{div}(\mathbf{A} - \mathbf{A}_h)|^2) \\
&\qquad\qquad + \textstyle\int_{\Omega_C} (|V_C - V_{C,h}|^2 + |\operatorname{grad}(V_C - V_{C,h})|^2) \Big)^{1/2} \\
&\quad \leq \tfrac{C_0}{\kappa_0} \Big(\textstyle\int_\Omega (|\mathbf{A} - \mathbf{w}_h|^2 + |\operatorname{curl}(\mathbf{A} - \mathbf{w}_h)|^2 + |\operatorname{div}(\mathbf{A} - \mathbf{w}_h)|^2) \\
&\qquad\qquad + \textstyle\int_{\Omega_C} (|V_C - Q_{C,h}|^2 + |\operatorname{grad}(V_C - Q_{C,h})|^2) \Big)^{1/2} ,
\end{aligned}$$

where $\kappa_0 > 0$ and $C_0 > 0$ are the coerciveness and the continuity constant of $\mathcal{B}[\cdot,\cdot]$, respectively. Therefore, provided that Ω has no reentrant corners or edges and that the solutions $\mathbf{A}$ and V_C are regular enough, by means of well-known interpolation results we find the error estimate

$$\begin{aligned}\Big(\int_\Omega (|\mathbf{A} - \mathbf{A}_h|^2 + |\operatorname{curl}(\mathbf{A} - \mathbf{A}_h)|^2 + |\operatorname{div}(\mathbf{A} - \mathbf{A}_h)|^2) \\ + \int_{\Omega_C} (|V_C - V_{C,h}|^2 + |\operatorname{grad}(V_C - V_{C,h})|^2)\Big)^{1/2} \\ \leq C h^{\min(r,s)} .\end{aligned} \quad (6.89)$$

Remark 6.25. As for the Coulomb gauged problem (see Remark 6.6), convergence is questionable if Ω is a non-convex polyhedron. In any case, the speed of convergence of the finite element approximation depends on the smoothness of $\mathbf{A}$ and V_C, and the smoothness of $\mathbf{A}$ cannot be high, as, due to the particular structure of the Lorenz gauge, $\operatorname{div} \mathbf{A}$ has a jump on Γ (unless the electric potential satisfies $V_{C|\Gamma} = 0$). □

6.3 Other potential formulations

In this section we briefly present some other potential formulations that have been often used in engineering applications (again, for the sake of definiteness, let us consider the magnetic boundary value problem (1.22)).

Let us start from two that are strictly related: the first, proposed by Pillsbury [193] (see also Rodger and Eastham [211], Emson and Simkin [100], Silvester and Ferrari [227]) is based on the potentials $\mathbf{A}_C$ and V_C in Ω_C, combined with a scalar magnetic potential ψ_I in Ω_I (and, when the topology of Ω_I is not simple, with an additional harmonic field in Ω_I, as seen in Chapter 5); the second, proposed by Leonard and Rodger [166] (for a similar approach see also Kameari [143]), is based on $\mathbf{A}_C$ and V_C in Ω_C, on a vector magnetic potential $\mathbf{A}_I$ in a suitable subset $\Omega_A \setminus \overline{\Omega_C}$ of the insulator Ω_I, and on the scalar magnetic potential ψ_I in $\Omega \setminus \overline{\Omega_A}$ (here, we have assumed that the open connected set Ω_A satisfies $\overline{\Omega_C} \subset \Omega_A$ and $\overline{\Omega_A} \subset \Omega$).

We refer to Section 7.1 for a presentation of the first method: there, to avoid the technicalities required by the determination of the harmonic field, it is assumed that the conductor Ω_C is simply-connected, the interface Γ is connected and $\Omega_I = \mathbb{R}^3 \setminus \overline{\Omega_C}$.

Instead, we describe here the second formulation, assuming that the auxiliary domain Ω_A is the union of a finite number of disjoint simply-connected domains with connected boundary, and that the physical domain Ω is simply-connected with connected boundary $\partial\Omega$. Note that these geometrical assumptions are not concerned with the conductor Ω_C, therefore they are not very restrictive for engineering applications.

In particular, as a consequence of these geometrical assumptions, in $\Omega \setminus \overline{\Omega_A}$ we can write $\mathbf{H}_I - \mathbf{H}_{e,I} = \operatorname{grad} \psi_I$, the vector field $\mathbf{H}_{e,I}$ being defined in (3.3). We thus have

$$\begin{aligned} \mathbf{E}_C &= -i\omega \mathbf{A}_C - \operatorname{grad} V_C \quad && \text{in } \Omega_C \\ \mu \mathbf{H} &= \operatorname{curl} \mathbf{A} \quad && \text{in } \Omega_A \\ \mathbf{H}_I &= \operatorname{grad} \psi_I + \mathbf{H}_{e,I} \quad && \text{in } \Omega \setminus \overline{\Omega_A} . \end{aligned} \quad (6.90)$$

Setting $\Gamma_A := \partial\Omega_A$, we have $\partial(\Omega \setminus \overline{\Omega_A}) = \Gamma_A \cup \partial\Omega$. Inserting the Coulomb gauge in the Ampère equation by means of the usual penalization argument, it is easily seen that the eddy current problem reads

$$
\begin{cases}
\operatorname{curl}(\boldsymbol{\mu}^{-1}\operatorname{curl}\mathbf{A}) - \mu_*^{-1}\operatorname{grad}\operatorname{div}\mathbf{A} & \\
\qquad\qquad + i\omega\boldsymbol{\sigma}\mathbf{A} + \boldsymbol{\sigma}\operatorname{grad}V_C = \mathbf{J}_e & \text{in } \Omega_A \\
\operatorname{div}(i\omega\boldsymbol{\sigma}\mathbf{A}_C + \boldsymbol{\sigma}\operatorname{grad}V_C) = \operatorname{div}\mathbf{J}_{e,C} & \text{in } \Omega_C \\
\operatorname{div}(\boldsymbol{\mu}_I\operatorname{grad}\psi_I) = -\operatorname{div}(\boldsymbol{\mu}_I\mathbf{H}_{e,I}) & \text{in } \Omega\setminus\overline{\Omega_A} \\
(i\omega\boldsymbol{\sigma}\mathbf{A}_C + \boldsymbol{\sigma}\operatorname{grad}V_C)\cdot\mathbf{n}_C = \mathbf{J}_{e,C}\cdot\mathbf{n}_C + \mathbf{J}_{e,I}\cdot\mathbf{n}_I & \text{on } \Gamma \\
\mathbf{A}\cdot\mathbf{n}_A = 0 & \text{on } \Gamma_A \\
(\boldsymbol{\mu}^{-1}\operatorname{curl}\mathbf{A})\times\mathbf{n}_A - \operatorname{grad}\psi_I\times\mathbf{n}_A = \mathbf{H}_{e,I}\times\mathbf{n}_A & \text{on } \Gamma_A \\
\operatorname{curl}\mathbf{A}\cdot\mathbf{n}_A - \boldsymbol{\mu}_I\operatorname{grad}\psi_I\cdot\mathbf{n}_A = \boldsymbol{\mu}_I\mathbf{H}_{e,I}\cdot\mathbf{n}_A & \text{on } \Gamma_A \\
\psi_I = 0 & \text{on } \partial\Omega\ ,
\end{cases}
\tag{6.91}
$$

where V_C is determined up to an additive constant in each connected component $\Omega_{C,j}$ of Ω_C, $j = 1, \ldots, p_\Gamma + 1$.

The analysis of this problem has been performed by Acevedo and Rodríguez [1]. We briefly present here the main points of their proof of the existence and uniqueness of the solution, and of its numerical approximation.

A couple of remarks are in order. First of all, the topological conditions $(3.25)_6$ are satisfied: in fact, since $\operatorname{div}(\boldsymbol{\mu}\mathbf{H}) = 0$ in Ω and Ω has a connected boundary, there exists a vector field $\mathbf{P}$ such that $\operatorname{curl}\mathbf{P} = \boldsymbol{\mu}\mathbf{H}$ in Ω. Therefore, since Ω_A is the union of a finite number of disjoint simply-connected domains, we have $\mathbf{A} - \mathbf{P} = \operatorname{grad}\xi$ in Ω_A for a suitable function ξ, and also

$$\mathbf{E}_C = -i\omega\mathbf{A}_C - \operatorname{grad}V_C = -i\omega\mathbf{P}_C - i\omega\operatorname{grad}\xi_C - \operatorname{grad}V_C \quad \text{in } \Omega_C\,.$$

We thus have

$$
\begin{aligned}
\int_{\Omega_I} i\omega\boldsymbol{\mu}_I\mathbf{H}_I\cdot\boldsymbol{\rho}_{l,I} &= \int_{\Omega_I} i\omega\operatorname{curl}\mathbf{P}_I\cdot\boldsymbol{\rho}_{l,I} = -\int_\Gamma i\omega\,\mathbf{n}_C\times\mathbf{P}_C\cdot\boldsymbol{\rho}_{l,I} \\
&= \int_\Gamma \mathbf{n}_C\times\mathbf{E}_C\cdot\boldsymbol{\rho}_{l,I} + \int_\Gamma[\mathbf{n}_C\times\operatorname{grad}(i\omega\xi_C + V_C)]\cdot\boldsymbol{\rho}_{l,I} \\
&= -\int_\Gamma \mathbf{E}_C\times\mathbf{n}_C\cdot\boldsymbol{\rho}_{l,I} + \int_\Gamma(\boldsymbol{\rho}_{l,I}\times\mathbf{n}_C)\cdot\operatorname{grad}(i\omega\xi_C + V_C) \\
&= -\int_\Gamma \mathbf{E}_C\times\mathbf{n}_C\cdot\boldsymbol{\rho}_{l,I} - \int_\Gamma \operatorname{div}_\tau(\boldsymbol{\rho}_{l,I}\times\mathbf{n}_C)\,(i\omega\xi_C + V_C) \\
&= -\int_\Gamma \mathbf{E}_C\times\mathbf{n}_C\cdot\boldsymbol{\rho}_{l,I} \quad \forall\, l = 1, \ldots, n_\Gamma\,,
\end{aligned}
$$

as $\operatorname{div}_\tau(\boldsymbol{\rho}_{l,I}\times\mathbf{n}_C) = \operatorname{curl}\boldsymbol{\rho}_{l,I}\cdot\mathbf{n}_C = 0$ on Γ.

As a second remark, by adapting the arguments in the proof of Lemma 6.1 it can be shown that $\operatorname{div}\mathbf{A} = 0$ in Ω_A and therefore that, through (6.90), a solution to (6.91) indeed provides a solution of the eddy current problem.

By integration by parts, and using the interface conditions in (6.91), it is easily seen that the weak formulation corresponding to (6.91) is

$$
\begin{array}{l}
\text{Find } (\mathbf{A}, V_C, \psi_I) \in W_A \times H^1_\sharp(\Omega_C) \times H^1_{0,\partial\Omega}(\Omega \setminus \overline{\Omega_A}) \text{ such that} \\
\quad \int_{\Omega_A} (\boldsymbol{\mu}^{-1} \operatorname{curl} \mathbf{A} \cdot \operatorname{curl} \overline{\mathbf{w}} + \mu_*^{-1} \operatorname{div} \mathbf{A} \operatorname{div} \overline{\mathbf{w}}) \\
\qquad\quad + \int_{\Omega_C} (i\omega\boldsymbol{\sigma} \mathbf{A}_C \cdot \overline{\mathbf{w}_C} + \boldsymbol{\sigma} \operatorname{grad} V_C \cdot \overline{\mathbf{w}_C}) \\
\qquad\quad + \int_{\Gamma_A} \mathbf{n}_A \times \operatorname{grad} \psi_I \cdot \overline{\mathbf{w}} \\
\qquad = \int_{\Omega_A} \mathbf{J}_e \cdot \overline{\mathbf{w}} - \int_{\Gamma_A} \mathbf{n}_A \times \mathbf{H}_{e,I} \cdot \overline{\mathbf{w}} \\
\quad \int_{\Omega_C} (-\boldsymbol{\sigma} \mathbf{A}_C \cdot \operatorname{grad} \overline{Q_C} + i\omega^{-1} \boldsymbol{\sigma} \operatorname{grad} V_C \cdot \operatorname{grad} \overline{Q_C}) \\
\qquad = i\omega^{-1} \int_{\Omega_C} \mathbf{J}_{e,C} \cdot \operatorname{grad} \overline{Q_C} + i\omega^{-1} \int_{\Gamma} \mathbf{J}_{e,I} \cdot \mathbf{n}_I \overline{Q_C} \\
\quad \int_{\Omega \setminus \overline{\Omega_A}} \boldsymbol{\mu}_I \operatorname{grad} \psi_I \cdot \operatorname{grad} \overline{\chi_I} - \int_{\Gamma_A} \mathbf{n}_A \times \operatorname{grad} \overline{\chi_I} \cdot \mathbf{A} \\
\qquad = - \int_{\Omega \setminus \overline{\Omega_A}} \boldsymbol{\mu}_I \mathbf{H}_{e,I} \cdot \operatorname{grad} \overline{\chi_I} \\
\text{for all } (\mathbf{w}, Q_C, \chi_I) \in W_A \times H^1_\sharp(\Omega_C) \times H^1_{0,\partial\Omega}(\Omega \setminus \overline{\Omega_A}) \, ,
\end{array}
\tag{6.92}
$$

where

$$
W_A := H(\operatorname{curl}; \Omega_A) \cap H_0(\operatorname{div}; \Omega_A) \, ,
$$

and, as in (6.14),

$$
H^1_\sharp(\Omega_C) := \prod_{j=1}^{p_\Gamma + 1} H^1(\Omega_{C,j})/\mathbb{C} \, .
$$

Let us denote by $\mathcal{A}_\dagger[\cdot,\cdot]$ the sesquilinear form at the left hand side in (6.92). The proof that it is coercive has been given by Acevedo and Rodríguez [1], and can be also obtained by adapting the proof of the coerciveness of the sesquilinear form $\mathcal{A}[\cdot,\cdot]$ presented in Section 6.1.2. In this respect, it is useful to note that

$$
\begin{array}{l}
\operatorname{Re} \mathcal{A}_\dagger[(\mathbf{w}, Q_C, \chi_I), (\mathbf{w}, Q_C, \chi_I)] \\
\quad = \int_{\Omega_A} (\boldsymbol{\mu}^{-1} \operatorname{curl} \mathbf{w} \cdot \operatorname{curl} \overline{\mathbf{w}} + \mu_*^{-1} |\operatorname{div} \mathbf{w}|^2) + \int_{\Omega \setminus \overline{\Omega_A}} \boldsymbol{\mu}_I \operatorname{grad} \chi_I \cdot \operatorname{grad} \overline{\chi_I}
\end{array}
$$

and

$$
\begin{array}{l}
\operatorname{Im} \mathcal{A}_\dagger[(\mathbf{w}, Q_C, \chi_I), (\mathbf{w}, Q_C, \chi_I)] \\
\quad = \omega^{-1} \int_{\Omega_C} \boldsymbol{\sigma}(i\omega \mathbf{w}_C + \operatorname{grad} Q_C) \cdot (-i\omega \overline{\mathbf{w}_C} + \operatorname{grad} \overline{Q_C}) \\
\qquad + 2 \operatorname{Im}\big(\int_{\Gamma_A} \mathbf{n}_A \times \operatorname{grad} \chi_I \cdot \overline{\mathbf{w}} \big) \, .
\end{array}
$$

In Acevedo and Rodríguez [1] one can also find a detailed analysis of the numerical approximation of (6.92) via Lagrange nodal elements. The convergence of the approximation scheme suffers as usual if at least one of the connected components of the domain Ω_A is a non-convex polyhedron (for similar remarks, see Section 6.1.3). However, since Ω_A is an auxiliary domain that has not a precise physical meaning, in numerical computations we can assume that it is the union of a finite number of disjoint convex polyhedral domains, assuring in this way the convergence of the Galerkin finite element approximation scheme.

In is also worth noting that some numerical experiments described by Leonard and Rodger [168] show that the computational efficiency of this approach is less than that of the schemes based on the introduction of "cutting" surfaces, like the methods introduced in Sections 5.1 and 5.3.

Another potential formulation is the so-called $(\mathbf{T}_C, \psi_C) - \psi_I$ formulation (see, for instance, Carpenter [74], Preston and Reece [197]). Let us describe it for the following geometrical situation: the physical domain Ω is a "box" (namely, a bounded simply-connected domain with a connected boundary $\partial\Omega$), while the conductor Ω_C is a torus. Consequently, the space of harmonic fields $\mathcal{H}_{\mu_I}(\partial\Omega, \Gamma; \Omega_I)$ has dimension 1, and as usual we denote its basis function by $\boldsymbol{\rho}_{1,I}$. Let us also introduce a function $\mathbf{R}_{1,C} \in H(\mathrm{curl}; \Omega_C)$ that satisfies $\mathbf{R}_{1,C} \times \mathbf{n}_C + \boldsymbol{\rho}_{1,I} \times \mathbf{n}_I = \mathbf{0}$ on Γ.

When considering the $(\mathbf{T}_C, \psi_C) - \psi_I$ formulation, we are looking for $\mathbf{T}_C$, ψ_C, ψ_I and $\beta_{I,1} \in \mathbb{C}$ such that

$$
\begin{array}{ll}
\mathbf{H}_I - \mathbf{H}_{e,I} = \mathrm{grad}\,\psi_I + \beta_{I,1}\boldsymbol{\rho}_{1,I} & \text{in } \Omega_I \\
\mathbf{H}_C - \mathbf{H}_{e,C} = \mathbf{T}_C + \mathrm{grad}\,\psi_C + \beta_{I,1}\mathbf{R}_{1,C} & \text{in } \Omega_C \ ,
\end{array}
\tag{6.93}
$$

with the interface conditions on Γ

$$
\begin{array}{l}
\mathbf{T}_C \times \mathbf{n}_C = \mathbf{0} \\
\psi_C - \psi_I = 0 \ ,
\end{array}
\tag{6.94}
$$

the vector field $\mathbf{H}_e$ being defined in (3.5).

Let us first show that it is possible to satisfy the relations (6.93) and (6.94). From the results in Section 5.1 we know that ψ_I and $\beta_{I,1}$ satisfying $(6.93)_1$ are straightforwardly determined from $\mathbf{H}_I$ and $\mathbf{H}_{e,I}$. Imposing the Coulomb-like gauge condition $\mathrm{div}\,\mathbf{T}_C = 0$ in Ω_C we also find $\mathbf{T}_C$ as the solution to

$$
\begin{cases}
\mathrm{curl}\,\mathbf{T}_C = \mathrm{curl}(\mathbf{H}_C - \mathbf{H}_{e,C} - \beta_{I,1}\mathbf{R}_{1,C}) & \text{in } \Omega_C \\
\mathrm{div}\,\mathbf{T}_C = 0 & \text{in } \Omega_C \\
\mathbf{T}_C \times \mathbf{n}_C = \mathbf{0} & \text{on } \Gamma \ .
\end{cases}
\tag{6.95}
$$

Note that the solvability conditions for this problem are satisfied: in fact, first we have

$$
\mathrm{div}\,\mathrm{curl}(\mathbf{H}_C - \mathbf{H}_{e,C} - \beta_{I,1}\mathbf{R}_{1,C}) = 0 \qquad \text{in } \Omega_C \ ,
$$

and

$$
\begin{array}{l}
\mathrm{curl}(\mathbf{H}_C - \mathbf{H}_{e,C} - \beta_{I,1}\mathbf{R}_{1,C}) \cdot \mathbf{n}_C \\
\quad = -\,\mathrm{curl}(\mathbf{H}_I - \mathbf{H}_{e,I} - \beta_{I,1}\boldsymbol{\rho}_{1,I}) \cdot \mathbf{n}_I = 0 \qquad \text{on } \Gamma \ .
\end{array}
$$

Moreover, denote by $\boldsymbol{\rho}^*_{1,C}$ the basis function of the space of harmonic fields $\mathcal{H}(m; \Omega_C)$; referring to the related results presented in the Appendix, we know that it can be written as the $(L^2(\Omega_C))^3$-extension of $\mathrm{grad}\,p^*_{1,C}$, where $p^*_{1,C}$ is the harmonic function, determined up to an additive constant, satisfying $\mathrm{grad}\,p^*_{1,C} \cdot \mathbf{n}_C = 0$ on Γ and having a jump equal to 1 through a section S of the torus Ω_C. Then one also has

$$
\begin{array}{l}
\int_{\Omega_C} \mathrm{curl}(\mathbf{H}_C - \mathbf{H}_{e,C} - \beta_{I,1}\mathbf{R}_{1,C}) \cdot \boldsymbol{\rho}^*_{1,C} \\
\quad = \int_{\Omega_C \setminus S} \mathrm{curl}(\mathbf{H}_C - \mathbf{H}_{e,C} - \beta_{I,1}\mathbf{R}_{1,C}) \cdot \mathrm{grad}\,p^*_{1,C} \\
\quad = \int_S \mathrm{curl}(\mathbf{H}_C - \mathbf{H}_{e,C} - \beta_{I,1}\mathbf{R}_{1,C}) \cdot \mathbf{n}_S \\
\quad = \int_{\partial S} (\mathbf{H}_C - \mathbf{H}_{e,C} - \beta_{I,1}\mathbf{R}_{1,C}) \cdot d\boldsymbol{\tau} \\
\quad = \int_{\partial S} (\mathbf{H}_I - \mathbf{H}_{e,I} - \beta_{I,1}\boldsymbol{\rho}_{1,I}) \cdot d\boldsymbol{\tau} \\
\quad = \int_{\partial S} \mathrm{grad}\,\psi_I \cdot d\boldsymbol{\tau} = 0 \ ,
\end{array}
$$

namely, the last compatibility condition on the data that has to be satisfied for solving (6.95).

Having determined $\mathbf{T}_C$, we find ψ_C as the solution to

$$\begin{cases} \operatorname{div}(\boldsymbol{\mu}_C \operatorname{grad} \psi_C) \\ \qquad = \operatorname{div}[\boldsymbol{\mu}_C(\mathbf{H}_C - \mathbf{H}_{e,C} - \beta_{I,1}\mathbf{R}_{1,C} - \mathbf{T}_C)] & \text{in } \Omega_C \\ \psi_C = \psi_I & \text{on } \Gamma . \end{cases} \tag{6.96}$$

It is easily shown that $\mathbf{H}_C - \mathbf{H}_{e,C} = \mathbf{T}_C + \operatorname{grad} \psi_C + \beta_{I,1}\mathbf{R}_{1,C}$ in Ω_C, as the left hand side and the right hand side of this formula have the same curl, $\boldsymbol{\mu}_C$-divergence and tangential component; in conclusion, (6.93) and (6.94) are verified.

Let us write now the eddy current problem in terms of $\mathbf{T}_C$, ψ_C, ψ_I and $\beta_{I,1}$. Setting

$$\mathbf{E}_C = \boldsymbol{\sigma}^{-1}[\operatorname{curl}(\mathbf{H}_{e,C} + \mathbf{T}_C + \beta_{I,1}\mathbf{R}_{1,C}) - \mathbf{J}_{e,C}] \quad \text{in } \Omega_C ,$$

the Ampère equation in Ω is clearly satisfied, hence we only need to impose the Faraday equation in Ω_C and the Gauss magnetic equation in Ω, plus the topological condition $(3.25)_6$.

Take $\chi \in H_0^1(\Omega)$; since $\operatorname{div}(\boldsymbol{\mu}_I \boldsymbol{\rho}_{1,I}) = 0$ in Ω_I and $\boldsymbol{\mu}_I \boldsymbol{\rho}_{1,I} \cdot \mathbf{n}_I = 0$ on Γ, we have

$$\int_{\Omega_I} \boldsymbol{\mu}_I \boldsymbol{\rho}_{1,I} \cdot \operatorname{grad} \overline{\chi_I} = 0 .$$

Therefore the weak form of the Gauss magnetic equation reads (having multiplied by $i\omega$)

$$\begin{aligned} &\textstyle\int_\Omega i\omega\boldsymbol{\mu} \operatorname{grad} \psi \cdot \operatorname{grad} \overline{\chi} + \int_{\Omega_C} i\omega\boldsymbol{\mu}_C \mathbf{T}_C \cdot \operatorname{grad} \overline{\chi_C} \\ &\qquad = - \textstyle\int_\Omega i\omega\boldsymbol{\mu}\mathbf{H}_e \cdot \operatorname{grad} \overline{\chi} - \beta_{I,1} \int_{\Omega_C} i\omega\boldsymbol{\mu}_C \mathbf{R}_{1,C} \cdot \operatorname{grad} \overline{\chi_C} . \end{aligned} \tag{6.97}$$

Instead, the weak form of the Faraday equation in Ω_C, where we have already inserted the penalization term associated to the divergence-free condition for $\mathbf{T}_C$, is given by

$$\begin{aligned} &\textstyle\int_{\Omega_C} \boldsymbol{\sigma}^{-1} \operatorname{curl} \mathbf{T}_C \cdot \operatorname{curl} \overline{\mathbf{v}_C} + \sigma_*^{-1} \int_{\Omega_C} \operatorname{div} \mathbf{T}_C \operatorname{div} \overline{\mathbf{v}_C} \\ &\quad + \textstyle\int_{\Omega_C} i\omega\boldsymbol{\mu}_C \mathbf{T}_C \cdot \overline{\mathbf{v}_C} + \int_{\Omega_C} i\omega\boldsymbol{\mu}_C \operatorname{grad} \psi_C \cdot \overline{\mathbf{v}_C} \\ &= - \textstyle\int_{\Omega_C} \boldsymbol{\sigma}^{-1} \operatorname{curl} \mathbf{H}_{e,C} \cdot \operatorname{curl} \overline{\mathbf{v}_C} - \beta_{I,1} \int_{\Omega_C} \boldsymbol{\sigma}^{-1} \operatorname{curl} \mathbf{R}_{1,C} \cdot \operatorname{curl} \overline{\mathbf{v}_C} \\ &\quad - \textstyle\int_{\Omega_C} i\omega\boldsymbol{\mu}_C \mathbf{H}_{e,C} \cdot \overline{\mathbf{v}_C} - \beta_{I,1} \int_{\Omega_C} i\omega\boldsymbol{\mu}_C \mathbf{R}_{1,C} \cdot \overline{\mathbf{v}_C} \\ &\quad + \textstyle\int_{\Omega_C} \boldsymbol{\sigma}^{-1} \mathbf{J}_{e,C} \cdot \operatorname{curl} \overline{\mathbf{v}_C} , \end{aligned} \tag{6.98}$$

where $\mathbf{v}_C \in H_0(\operatorname{curl}; \Omega_C) \cap H(\operatorname{div}; \Omega_C)$ and $\sigma_* > 0$ is a dimensional constant (say, a suitable average in Ω_C of the entries of the matrix $\boldsymbol{\sigma}(\mathbf{x})$).

Let us assume, for the time being, that the conductor Ω_C, instead of being a torus, is a simply-connected domain with a connected boundary. This means that in (6.93) the terms $\boldsymbol{\rho}_{1,I}$ and $\mathbf{R}_{1,C}$ disappear: formally, we can suppose that $\beta_{I,1} = 0$. Moreover, also the topological condition $(3.25)_6$ has to be discarded. Hence, in this case we are only looking for ψ and $\mathbf{T}_C$, solutions to (6.97) and (6.98), where we have set $\beta_{I,1} = 0$.

For a domain Ω_C as above, the Poincaré-like inequality

$$\begin{aligned} &\textstyle\int_{\Omega_C}(|\operatorname{curl}\mathbf{v}_C|^2 + |\operatorname{div}\mathbf{v}_C|^2) \\ &\qquad \geq K_0 \textstyle\int_{\Omega_C}(|\operatorname{curl}\mathbf{v}_C|^2 + |\operatorname{div}\mathbf{v}_C|^2 + |\mathbf{v}_C|^2) \end{aligned} \tag{6.99}$$

is satisfied for any $\mathbf{v}_C \in H_0(\text{curl}; \Omega_C) \cap H(\text{div}; \Omega_C)$ (see, e.g., Girault and Raviart [111], Chap. I, Lemma 3.4). Then, by adapting the proof of the coerciveness of the sesquilinear form $\mathcal{A}[\cdot,\cdot]$ presented in Section 6.1.2, it is not difficult to show that the sesquilinear form at the left hand sides of (6.97) and (6.98) is coercive in $H_0^1(\Omega) \times [H_0(\text{curl}; \Omega_C) \cap H(\text{div}; \Omega_C)]$. Therefore the solution to the weak problem (6.97), (6.98) is unique, and it is the right physical solution $(\psi, \mathbf{T}_C)$ obtained from the eddy current solution through (6.93) and (6.94).

This fact has led some researchers to use the $(\mathbf{T}_C, \psi_C) - \psi_I$ formulation for solving the eddy current problem with assigned total current intensity: coming back to the case in which Ω_C is a torus, we know that the current intensity I^0 through a section S of Ω_C is given by $\int_S \operatorname{curl}\mathbf{H}_C \cdot \mathbf{n}_S$, therefore, by the Stokes theorem, by $\int_{\partial S} \mathbf{H}_I \cdot d\boldsymbol{\tau}$. Using $(6.93)_1$, we can write

$$\begin{aligned} I^0 &= \textstyle\int_{\partial S} \mathbf{H}_I \cdot d\boldsymbol{\tau} \\ &= \textstyle\int_{\partial S} \mathbf{H}_{e,I} \cdot d\boldsymbol{\tau} + \beta_{I,1} \int_{\partial S} \boldsymbol{\rho}_{1,I} \cdot d\boldsymbol{\tau} \\ &= \textstyle\int_{\partial S} \mathbf{H}_{e,I} \cdot d\boldsymbol{\tau} \pm \beta_{I,1}\,, \end{aligned} \tag{6.100}$$

where the sign depends on the orientation of ∂S and on the choice of the basis function $\boldsymbol{\rho}_{1,I}$. Hence it seems enough to insert this value of $\beta_{I,1}$ in the right hand sides of (6.97) and (6.98): as before, the problem has a unique solution, and it would appear that we are done (in this regard, see also the formulation proposed by Reissel [205], reported and commented in Section 3.3.2).

Instead, what is wrong is that the right value of $\beta_{I,1}$ is not the one determined in (6.100), but the one that allows us to solve $(3.25)_6$, namely, as we have seen in Section 3.3.1, the Faraday equation on the surface, contained in $\overline{\Omega_I}$, that "cuts" the non-bounding cycle ∂S.

Indeed, we know that the solution of the eddy current problem (1.22) or (1.20) is unique, therefore it is not possible to impose additional conditions, like, for instance, the total current intensity through S. We will see in Chapter 8 which type of boundary conditions and which type of geometrical configuration allow us to impose the current intensity or the voltage (in particular, in Section 8.1.3 we will adopt the $(\mathbf{T}_C, \psi_C) - \psi_I$ formulation).

Let us now analyze the correct formulation of the eddy current problem in the case in which Ω_C is a torus. The unknowns are $\mathbf{T}_C$, ψ_C, ψ_I and $\beta_{I,1}$, and, beside equations (6.97) and (6.98), one has also to impose the Faraday equation on the surface "cutting"

∂S, namely equation $(3.25)_6$. We have

$$
\begin{aligned}
\int_\Gamma \mathbf{E}_C \times \mathbf{n}_C \cdot \boldsymbol{\rho}_{1,I} &= \int_\Gamma \mathbf{E}_C \times \mathbf{n}_C \cdot \mathbf{R}_{1,C} \\
&= -\int_{\Omega_C} \operatorname{curl} \mathbf{E}_C \cdot \mathbf{R}_{1,C} + \int_{\Omega_C} \mathbf{E}_C \cdot \operatorname{curl} \mathbf{R}_{1,C} \\
&= \int_{\Omega_C} i\omega \boldsymbol{\mu}_C \mathbf{H}_C \cdot \mathbf{R}_{1,C} + \int_{\Omega_C} \boldsymbol{\sigma}^{-1}(\operatorname{curl} \mathbf{H}_C - \mathbf{J}_{e,C}) \cdot \operatorname{curl} \mathbf{R}_{1,C} \\
&= \int_{\Omega_C} i\omega \boldsymbol{\mu}_C \mathbf{T}_C \cdot \mathbf{R}_{1,C} + \int_{\Omega_C} i\omega \boldsymbol{\mu}_C \operatorname{grad} \psi_C \cdot \mathbf{R}_{1,C} \\
&\quad + \beta_{I,1} \int_{\Omega_C} i\omega \boldsymbol{\mu}_C \mathbf{R}_{1,C} \cdot \mathbf{R}_{1,C} + \int_{\Omega_C} i\omega \boldsymbol{\mu}_C \mathbf{H}_{e,C} \cdot \mathbf{R}_{1,C} \\
&\quad + \int_{\Omega_C} \boldsymbol{\sigma}^{-1} \operatorname{curl} \mathbf{T}_C \cdot \operatorname{curl} \mathbf{R}_{1,C} + \beta_{I,1} \int_{\Omega_C} \boldsymbol{\sigma}^{-1} \operatorname{curl} \mathbf{R}_{1,C} \cdot \operatorname{curl} \mathbf{R}_{1,C} \\
&\quad + \int_{\Omega_C} \boldsymbol{\sigma}^{-1} \operatorname{curl} \mathbf{H}_{e,C} \cdot \operatorname{curl} \mathbf{R}_{1,C} - \int_{\Omega_C} \boldsymbol{\sigma}^{-1} \mathbf{J}_{e,C} \cdot \operatorname{curl} \mathbf{R}_{1,C} .
\end{aligned}
$$

Hence $(3.25)_6$ reads

$$
\begin{aligned}
&\overline{\theta_I} \int_{\Omega_C} i\omega \boldsymbol{\mu}_C \mathbf{T}_C \cdot \mathbf{R}_{1,C} + \overline{\theta_I} \int_{\Omega_C} i\omega \boldsymbol{\mu}_C \operatorname{grad} \psi_C \cdot \mathbf{R}_{1,C} \\
&\quad + \beta_{I,1} \overline{\theta_I} \int_{\Omega_C} i\omega \boldsymbol{\mu}_C \mathbf{R}_{1,C} \cdot \mathbf{R}_{1,C} + \beta_{I,1} \overline{\theta_I} \int_{\Omega_I} i\omega \boldsymbol{\mu}_I \boldsymbol{\rho}_{1,I} \cdot \boldsymbol{\rho}_{1,I} \\
&\quad + \overline{\theta_I} \int_{\Omega_C} \boldsymbol{\sigma}^{-1} \operatorname{curl} \mathbf{T}_C \cdot \operatorname{curl} \mathbf{R}_{1,C} \\
&\quad + \beta_{I,1} \overline{\theta_I} \int_{\Omega_C} \boldsymbol{\sigma}^{-1} \operatorname{curl} \mathbf{R}_{1,C} \cdot \operatorname{curl} \mathbf{R}_{1,C} \\
&= \overline{\theta_I} \int_{\Omega_C} \boldsymbol{\sigma}^{-1} \mathbf{J}_{e,C} \cdot \operatorname{curl} \mathbf{R}_{1,C} - \overline{\theta_I} \int_{\Omega_C} i\omega \boldsymbol{\mu}_C \mathbf{H}_{e,C} \cdot \mathbf{R}_{1,C} \\
&\quad - \overline{\theta_I} \int_{\Omega_C} \boldsymbol{\sigma}^{-1} \operatorname{curl} \mathbf{H}_{e,C} \cdot \operatorname{curl} \mathbf{R}_{1,C} - \overline{\theta_I} \int_{\Omega_I} i\omega \boldsymbol{\mu}_I \mathbf{H}_{e,I} \cdot \boldsymbol{\rho}_{1,I} ,
\end{aligned}
\tag{6.101}
$$

for each $\theta_I \in \mathbb{C}$.

Thus the global problem can be written as

$$
\begin{aligned}
&\int_{\Omega_C} \boldsymbol{\sigma}^{-1} \operatorname{curl}(\mathbf{T}_C + \beta_{I,1} \mathbf{R}_{1,C}) \cdot \operatorname{curl}(\overline{\mathbf{v}_C} + \overline{\theta_I} \mathbf{R}_{1,C}) \\
&\qquad + \sigma_*^{-1} \int_{\Omega_C} \operatorname{div} \mathbf{T}_C \operatorname{div} \overline{\mathbf{v}_C} + i\omega \int_{\Omega_I} \boldsymbol{\mu}_I \operatorname{grad} \psi_I \cdot \operatorname{grad} \overline{\chi_I} \\
&\qquad + i\omega \int_{\Omega_C} \boldsymbol{\mu}_C (\mathbf{T}_C + \operatorname{grad} \psi_C + \beta_{I,1} \mathbf{R}_{1,C}) \\
&\qquad\qquad\qquad \cdot (\overline{\mathbf{v}_C} + \operatorname{grad} \overline{\chi_C} + \overline{\theta_I} \mathbf{R}_{1,C}) \\
&\qquad + i\omega\, \beta_{I,1}\, \overline{\theta_I} \int_{\Omega_I} \boldsymbol{\mu}_I \boldsymbol{\rho}_{1,I} \cdot \boldsymbol{\rho}_{1,I} \\
&\quad - \int_{\Omega_C} \boldsymbol{\sigma}^{-1} \mathbf{J}_{e,C} \cdot \operatorname{curl}(\overline{\mathbf{v}_C} + \overline{\theta_I} \mathbf{R}_{1,C}) \\
&\qquad - \int_{\Omega_C} \boldsymbol{\sigma}^{-1} \operatorname{curl} \mathbf{H}_{e,C} \cdot \operatorname{curl}(\overline{\mathbf{v}_C} + \overline{\theta_I} \mathbf{R}_{1,C}) \\
&\qquad - i\omega \int_{\Omega_C} \boldsymbol{\mu}_C \mathbf{H}_{e,C} \cdot (\overline{\mathbf{v}_C} + \overline{\theta_I} \mathbf{R}_{1,C}) \\
&\qquad - i\omega \int_{\Omega} \boldsymbol{\mu} \mathbf{H}_e \cdot \operatorname{grad} \overline{\chi} - i\omega \overline{\theta_I} \int_{\Omega_I} \boldsymbol{\mu}_I \mathbf{H}_{e,I} \cdot \boldsymbol{\rho}_{1,I} ,
\end{aligned}
\tag{6.102}
$$

and we are looking for $\mathbf{T}_C \in H_0(\operatorname{curl}; \Omega_C) \cap H(\operatorname{div}; \Omega_C)$, $\psi \in H_0^1(\Omega)$, and also $\beta_{I,1} \in \mathbb{C}$, taking the test functions $\mathbf{v}_C$, χ in the same spaces and θ_I in $\mathbb{C}$.

Let us denote by $\mathcal{S}(\cdot,\cdot)$ the sesquilinear form at the left hand side of (6.102). We have

$$
\begin{aligned}
&|\operatorname{Re} \mathcal{S}((\mathbf{v}_C, \chi, \theta_I), (\mathbf{v}_C, \chi, \theta_I))| \\
&\qquad \geq \sigma_{\max}^{-1} \int_{\Omega_C} |\operatorname{curl}(\mathbf{v}_C + \theta_I \mathbf{R}_{1,C})|^2 + \sigma_*^{-1} \int_{\Omega_C} |\operatorname{div} \mathbf{v}_C|^2,
\end{aligned}
$$

and

$$
\begin{aligned}
&|\operatorname{Im} \mathcal{S}((\mathbf{v}_C, \chi, \theta_I), (\mathbf{v}_C, \chi, \theta_I))| \\
&\quad \geq |\omega| \mu_{\min} \int_{\Omega_I} |\operatorname{grad} \chi_I|^2 + |\omega| \mu_{\min} \int_{\Omega_C} |\mathbf{v}_C + \operatorname{grad} \chi_C + \theta_I \mathbf{R}_{1,C}|^2 \\
&\qquad + |\omega| |\theta_I|^2 \int_{\Omega_I} \boldsymbol{\mu}_I \boldsymbol{\rho}_{1,I} \cdot \boldsymbol{\rho}_{1,I} ,
\end{aligned}
$$

where $\sigma_{\max}$ is a uniform upper bound for the maximum eigenvalues of $\boldsymbol{\sigma}(\mathbf{x})$ in Ω_C and $\mu_{\min}$ is a uniform lower bound for the minimum eigenvalues of $\boldsymbol{\mu}(\mathbf{x})$ in Ω.

Proceeding as in the proof of the coerciveness of the sesquilinear form $\mathcal{A}[\cdot,\cdot]$ in Section 6.1.2, we see that for each $0 < \delta < 1$ we have

$$\begin{aligned}&\int_{\Omega_C} |\operatorname{curl}(\mathbf{v}_C + \theta_I \mathbf{R}_{1,C})|^2 \\ &\quad \geq (1-\delta) \int_{\Omega_C} |\operatorname{curl} \mathbf{v}_C|^2 - (1-\delta)\delta^{-1}|\theta_I|^2 \int_{\Omega_C} |\operatorname{curl} \mathbf{R}_{1,C}|^2 .\end{aligned}$$

Similarly, for each $0 < \eta < 1$ we find

$$\begin{aligned}&\int_{\Omega_C} |\mathbf{v}_C + \operatorname{grad} \chi_C + \theta_I \mathbf{R}_{1,C}|^2 \\ &\quad \geq (1-\eta) \int_{\Omega_C} |\operatorname{grad} \chi_C|^2 - 2(1-\eta)\eta^{-1} \int_{\Omega_C} |\mathbf{v}_C|^2 \\ &\qquad -2(1-\eta)\eta^{-1}|\theta_I|^2 \int_{\Omega_C} |\mathbf{R}_{1,C}|^2 .\end{aligned}$$

Since the Poincaré-like inequality (6.99) holds, choosing $1-\delta = \tau$, $1-\eta = \tau^2$ and τ small enough it is now easy to prove that $\mathcal{S}(\cdot,\cdot)$ is coercive in

$$[H_0(\operatorname{curl};\Omega_C) \cap H(\operatorname{div};\Omega_C)] \times H_0^1(\Omega) \times \mathbb{C},$$

and therefore problem (6.102) has a unique solution via the Lax–Milgram lemma.

Concerning the numerical approximation, the use of Lagrange nodal elements can be a viable option. As usual for nodal finite elements (see Section 6.1.3), the convergence of the Galerkin finite element approximation scheme can be ensured only if Ω_C is a convex polyhedral domain (which is never the case if Ω_C is a torus). However, we want to mention that the analysis of the convergence in the case of smooth boundaries $\partial\Omega$ and Γ can be found in Tsukerman [236].

In conclusion, in a general topological case (say, if Ω_C is a polyhedral torus) the $(\mathbf{T}_C, \psi_C) - \psi_I$ approach has some defects: if used as described in (6.102), the convergence of the associated nodal finite element scheme is not ensured; if used prescribing the total current intensity through a suitable section S of the torus Ω_C, it leads to a wrong result since, as we have clarified before, the Faraday equation is violated on the surface which "cuts" the non-bounding cycle ∂S. This remark should sound interesting, as sometimes in the engineering literature the use the $(\mathbf{T}_C, \psi_C) - \psi_I$ approach has been proposed particularly for the case in which Ω_C is a torus and the total current intensity is assigned.

Remark 6.26. As usual, for numerical implementation it is better to replace the basis function $\boldsymbol{\rho}_{1,I}$ with the function $\boldsymbol{\lambda}_1$ introduced in (5.15). The corresponding variational formulation is easily devised by proceeding as was done in obtaining (6.102). □

Remark 6.27. Instead of the Coulomb-like gauge condition $\operatorname{div} \mathbf{T}_C = 0$ in Ω_C, a Lorenz-like gauge condition $\operatorname{div} \mathbf{T}_C + i\omega\mu_C\sigma\psi_C = 0$ in Ω_C has been also proposed (see Tang et al. [232]). As for the $(\mathbf{A}_C, V_C) - \mathbf{A}_I$ formulation, in general this approach leads to a problem which is not positive definite, hence it does not seem to be the best choice for numerical approximation. □

Remark 6.28. For a non-simply connected conductor, Ren [207] has proposed a ungauged $\mathbf{T} - \psi$ formulation where the vector potential $\mathbf{T}$ is approximated by edge elements in Ω_C and by curl-free edge elements in a one-layer domain around the "cutting" surface Ξ_1. The presented numerical results are in good agreement with the physical problem; however, for ungauged vector potential formulations a complete analysis of the convergence is not available (see also Remark 6.10). □

We conclude this section presenting another vector potential formulation, the so-called $(\mathbf{T}^*_C, \Phi_C) - \mathbf{A}_I$ formulation, proposed by Bíró and Preis [50], [51]. We assume again, for simplicity, that Ω is a "box" and that Ω_C is a torus. The starting point is to decompose the magnetic field $\mathbf{H}$, known in Ω, as

$$\mathbf{H}_C = \mathbf{T}^*_C + \operatorname{grad} \Phi_C \quad \text{in } \Omega_C \, , \; \mathbf{H}_I = \boldsymbol{\mu}_I^{-1} \operatorname{curl} \mathbf{A}_I \quad \text{in } \Omega_I \, , \tag{6.103}$$

where $\mathbf{T}^*_C$, Φ_C and $\mathbf{A}_I$ satisfy the interface conditions on Γ

$$\begin{aligned} &\mathbf{T}^*_C \cdot \mathbf{n}_C = 0 \\ &[\boldsymbol{\sigma}^{-1}(\operatorname{curl} \mathbf{T}^*_C - \mathbf{J}_{e,C})] \times \mathbf{n}_C - i\omega \mathbf{A}_I \times \mathbf{n}_I = \mathbf{0} \, , \end{aligned} \tag{6.104}$$

and the gauge conditions

$$\operatorname{div} \mathbf{T}^*_C = 0 \quad \text{in } \Omega_C \, , \; \operatorname{div} \mathbf{A}_I = 0 \quad \text{in } \Omega_I \, , \; \mathbf{A}_I \cdot \mathbf{n} = 0 \quad \text{on } \partial\Omega \, . \tag{6.105}$$

The existence of $\mathbf{T}^*_C$ and Φ_C satisfying $(6.103)_1$, $(6.104)_1$, and $(6.105)_1$ follows from Theorem A.8, having replaced Ω_I with Ω_C and $\boldsymbol{\mu}_I$ with Id. Then, proceeding as in Section 3.2, in the present geometrical configuration it is easily seen that the solvability conditions for determining $\mathbf{A}_I$ are satisfied, and we conclude that there exists a unique solution $\mathbf{A}_I$ of $(6.103)_2$, $(6.104)_2$, $(6.105)_2$, and $(6.105)_3$.

Clearly, from the Ampère equation the electric field in the conductor can be written as

$$\mathbf{E}_C = \boldsymbol{\sigma}^{-1}(\operatorname{curl} \mathbf{T}^*_C - \mathbf{J}_{e,C}) \quad \text{in } \Omega_C \, , \tag{6.106}$$

and the following interface conditions on Γ are also satisfied

$$\begin{aligned} &\mathbf{T}^*_C \times \mathbf{n}_C + \operatorname{grad} \Phi_C \times \mathbf{n}_C + (\boldsymbol{\mu}_I^{-1} \operatorname{curl} \mathbf{A}_I) \times \mathbf{n}_I = \mathbf{0} \\ &\boldsymbol{\mu}_C \mathbf{T}^*_C \cdot \mathbf{n}_C + \boldsymbol{\mu}_C \operatorname{grad} \Phi_C \cdot \mathbf{n}_C + \operatorname{curl} \mathbf{A}_I \cdot \mathbf{n}_I = 0 \, . \end{aligned} \tag{6.107}$$

Taking into account the Ampère equation in Ω_I, the Faraday equation in Ω_C, the Gauss magnetic equation in Ω_C, and using the interface conditions it is easily seen that for the functions $\mathbf{T}^*_C$, Φ_C and $\mathbf{A}_I$ thus determined the following variational formulation holds

$$\begin{aligned} &\int_{\Omega_C} \boldsymbol{\sigma}^{-1} \operatorname{curl} \mathbf{T}^*_C \cdot \operatorname{curl} \overline{\mathbf{v}_C} + \sigma_*^{-1} \int_{\Omega_C} \operatorname{div} \mathbf{T}^*_C \operatorname{div} \overline{\mathbf{v}_C} \\ &\quad + i\omega \int_{\Omega_C} \boldsymbol{\mu}_C (\mathbf{T}^*_C + \operatorname{grad} \Phi_C) \cdot (\overline{\mathbf{v}_C} + \operatorname{grad} \overline{\eta_C}) \\ &\quad + i\omega \int_{\Omega_I} \boldsymbol{\mu}_I^{-1} \operatorname{curl} \mathbf{A}_I \cdot \operatorname{curl} \overline{\mathbf{w}_I} + i\omega \mu_*^{-1} \int_{\Omega_I} \operatorname{div} \mathbf{A}_I \operatorname{div} \overline{\mathbf{w}_I} \\ &\quad + i\omega \int_\Gamma (\mathbf{T}^*_C \cdot \overline{\mathbf{w}_I} \times \mathbf{n}_I - \mathbf{A}_I \times \mathbf{n}_I \cdot \overline{\mathbf{v}_C}) \\ &\quad + i\omega \int_\Gamma (\operatorname{curl} \mathbf{A}_I \cdot \mathbf{n}_I \, \overline{\eta_C} - \Phi_C \operatorname{curl} \overline{\mathbf{w}_I} \cdot \mathbf{n}_I) \\ &= \int_{\Omega_C} \boldsymbol{\sigma}^{-1} \mathbf{J}_{e,C} \cdot \operatorname{curl} \overline{\mathbf{v}_C} + i\omega \int_{\Omega_I} \mathbf{J}_{e,I} \cdot \overline{\mathbf{w}_I} \, , \end{aligned} \tag{6.108}$$

with $\mathbf{v}_C \in H(\text{curl}; \Omega_C) \cap H_0(\text{div}; \Omega_C)$, $\eta_C \in H^1(\Omega_C)/\mathbb{C}$, $\mathbf{w}_I \in H(\text{curl}; \Omega_I) \cap H_{0,\partial\Omega}(\text{div}; \Omega_I)$. Here we have made use of the facts that $\mathbf{T}_C^* \times \mathbf{n}_C \cdot \overline{\mathbf{w}_I} = \mathbf{T}_C^* \cdot \overline{\mathbf{w}_I} \times \mathbf{n}_I$ and that

$$\int_\Gamma \text{grad}\, \Phi_C \times \mathbf{n}_C \cdot \overline{\mathbf{w}_I} = -\int_\Gamma \text{div}_\tau(\overline{\mathbf{w}_I} \times \mathbf{n}_I)\, \Phi_C = -\int_\Gamma \text{curl}\, \overline{\mathbf{w}_I} \cdot \mathbf{n}_I\, \Phi_C \,.$$

To our knowledge, it is not known whether the sesquilinear form $\mathcal{Q}(\cdot,\cdot)$ at the left hand side of (6.108) is coercive or not. However, the solution can be shown to be unique (Φ_C up to an additive constant).

In fact, by choosing the test functions $\mathbf{v}_C = \text{grad}\, u_C$, $\eta_C = -u_C$ and $\mathbf{w}_I = \mathbf{0}$, where $\Delta u_C = \text{div}\, \mathbf{T}_C^*$ in Ω_C and $\text{grad}\, u_C \cdot \mathbf{n}_C = 0$ on Γ, from (6.108) it follows at once that $\text{div}\, \mathbf{T}_C^* = 0$ in Ω_C. Similarly, choosing $\mathbf{v}_C = \mathbf{0}$, $\eta_C = 0$ and $\mathbf{w}_I = \text{grad}\, u_I$, where $\Delta u_I = \text{div}\, \mathbf{A}_I$ in Ω_I, $\text{grad}\, u_I \cdot \mathbf{n} = 0$ on $\partial\Omega$, and $u_I = 0$ on Γ, we find $\text{div}\, \mathbf{A}_I = 0$ in Ω_I.

Then one verifies that the terms

$$i\omega \int_\Gamma (\mathbf{T}_C^* \cdot \overline{\mathbf{A}_I} \times \mathbf{n}_I - \mathbf{A}_I \times \mathbf{n}_I \cdot \overline{\mathbf{T}_C^*}),$$

and

$$i\omega \int_\Gamma (\text{curl}\, \mathbf{A}_I \cdot \mathbf{n}_I\, \overline{\Phi_C} - \Phi_C\, \text{curl}\, \overline{\mathbf{A}_I} \cdot \mathbf{n}_I)$$

are real numbers, therefore setting $\mathbf{J}_e = \mathbf{0}$ in (6.108) gives at once $\mathbf{T}_C^* + \text{grad}\, \Phi_C = \mathbf{0}$ in Ω_C, and $\text{curl}\, \mathbf{A}_I = \mathbf{0}$ in Ω_I. From $\text{div}\, \mathbf{T}_C^* = 0$ in Ω_C and $\mathbf{T}_C^* \cdot \mathbf{n}_C = 0$ on Γ we obtain that $\text{grad}\, \Phi_C = \mathbf{0}$ in Ω_C, hence $\mathbf{T}_C^* = \mathbf{0}$.

To conclude the proof of the uniqueness it is enough to note that problem (6.108) has become

$$\int_\Gamma \mathbf{A}_I \times \mathbf{n}_I \cdot \overline{\mathbf{v}_C} = 0$$

for each test function $\mathbf{v}_C$, therefore $\mathbf{A}_I \times \mathbf{n}_I = \mathbf{0}$ on Γ and thus $\mathbf{A}_I = \mathbf{0}$ in Ω_I.

Though a finite element discretization based on this method has not been completely analyzed, this scheme has been used and has performed well in several numerical computations. The same can be said for a modified version, the $(\mathbf{T}_C^*, \Phi_C) - \mathbf{A}_I - \Phi_I$ approach, in which the magnetic field is written as $\mathbf{H}_I = \mathbf{H}_{e,I} + \text{grad}\, \Phi_I$ in a simply-connected domain contained in Ω_I (see Bíró and Preis [50], [51], Bíró et al. [53]).

7

Coupled FEM–BEM approaches

In this chapter we focus on some procedures for solving eddy current problems that are based on a strategy which couples the finite element method (FEM) and the boundary element method (BEM). This kind of coupling allows the numerical approximation of the solution in unbounded domains, a typical situation in electromagnetism. The boundary element method is used for the approximation in the complement of a bounded domain: either the conductor Ω_C or else an artificial computational domain Ω, containing Ω_C but in general not very large. Instead, in the bounded domain the solution is approximated using the finite element method. Compared with the formulations presented in the previous chapters, the coupled FEM–BEM approaches compute the FEM approximation of the solution in a smaller region (say, the conductor), not required to be so large that the use of homogeneous boundary conditions is justified. This can be done because the BEM method takes into account the behaviour of the solution in the external region.

The idea of coupling a variational approach in one region with a potential approach in another region of the computational domain has been first proposed by engineers for the Laplace operator (see, e.g., Zienkiewicz et al. [248], Jami and Lenoir [138]), and then widely analyzed from the mathematical point of view, starting from the pioneering works of Brezzi and Johnson [66] and Johnson and Nédélec [141]. An important improvement has been furnished by the papers of Costabel [85], [86], that, for elliptic boundary value problems, show how to obtain a symmetric (or else to a positive definite) problem. Extensions to the full Maxwell equations are due to Ammari and Nédélec [25], [26].

Coming to the eddy current problem, the first FEM–BEM couplings have been proposed by Bossavit and Vérité [62], [63] (for the magnetic field, and using the Steklov–Poincaré operator) and Mayergoyz et al. [174] (for the electric field, and using special basis functions near Γ). A more recent result in FEM–BEM coupling, for axisymmetric problems associated to the modeling of induction furnaces, is due to Bermúdez et al. [40].

The approach of Bossavit and Vérité [62], [63] has led them to devise a very popular numerical code, named TRIFOU, that has been often used in engineering computations. A complete presentation of this approach can be found in Bossavit [59], Sect. 8.2; we describe its basic ideas in Section 7.6.1.

Symmetric formulations à la Costabel have been proposed for eddy current problems by Hiptmair [127] (unknowns: $\mathbf{E}_C$ in Ω_C, $\mathbf{H} \times \mathbf{n}$ on Γ) and Meddahi and Selgas [176] (unknowns: $\mathbf{H}_C$ in Ω_C, $\boldsymbol{\mu}\mathbf{H} \cdot \mathbf{n}$ on Γ), and are briefly presented in Sections 7.6.3 and 7.6.2, respectively.

The chapter begins with Sections 7.1–7.5, where we describe a FEM–BEM formulation proposed by Alonso Rodríguez and Valli [19], based on a vector magnetic potential and a scalar electric potential in the conductor, and on a scalar magnetic potential in the external part. An approach in terms of magnetic vector potentials has been also proposed for magnetostatics by Kuhn et al. [159] and Kuhn and Steinbach [160]; with respect to the choice of potentials, the presentation in Sections 7.1–7.5 is close to these last ones.

The reader mainly interested in numerical approximation and implementation can focus on problems (7.12), (7.30) and (7.31) ($(\mathbf{A}_C, V_C, q)$ formulation), on problem (7.36) (TRIFOU formulation), on problem (7.42) ($(\mathbf{H}_C, \lambda)$ formulation), and on problem (7.52) ($(\mathbf{E}_C, \mathbf{p}_\Gamma)$ formulation).

Let us focus now on a different aspect: not all the known methods devised for studying the Maxwell equations are robust enough to be used, without any modification, for both the time-harmonic case and the static case (namely, the case in which the electric and magnetic inductions are assumed to be time-independent; in other words, in the equations one has to set $\omega = 0$). In Sections 7.1–7.5 we show how one can treat without distinction the cases $\omega \neq 0$ and $\omega = 0$. Moreover, the numerical approximation there proposed is quite simple, since we use standard Lagrange nodal finite elements in the conductor, while a cheap formulation based on boundary elements is proposed in the external insulator.

Being simple, robust and cheap, this method can be therefore a suitable direct solver for some inverse problems in electromagnetism, for instance in electroencephalography (EEG) or magnetoencephalography (MEG) (see Section 9.2). In this respect, though in many papers devoted to these topics only the static case is considered (see, e.g., Sarvas [220], Hämäläinen et al. [117]), recently some researchers have focused on the time-harmonic case, which is a more precise model for describing the electric and magnetic activities in the brain (see Ammari et al. [22]). Clearly, the static case is much easier to solve, as, due to the irrotationality condition, one can reduce the problem to the sole determination of a scalar potential of the electric field in Ω_C (a suitable Neumann condition on Γ is the correct boundary condition to add). However, in no way that simple approach can be extended to the time-harmonic case, as irrotationality no longer holds.

In this chapter the geometrical assumptions on the conductor Ω_C are more restrictive than in the preceding chapters. In fact, we consider a bounded simply-connected open set $\Omega_C \subset \mathbb{R}^3$, with a Lipschitz boundary Γ (for EEG and MEG applications, Ω_C represents the human head). For simplicity, as in the preceding chapters we also assume that $\Omega_I := \mathbb{R}^3 \setminus \overline{\Omega_C}$ is connected, so that Γ is connected, too. The unit outward normal vector on Γ will be denoted by $\mathbf{n}_C = -\mathbf{n}_I$.

As usual, we assume that the electric conductivity $\boldsymbol{\sigma}$ and the magnetic permeability $\boldsymbol{\mu}_C$ are uniformly positive definite symmetric matrices in Ω_C, with entries belonging to $L^\infty(\Omega_C)$. The electric conductivity $\boldsymbol{\sigma}$ and the applied current density $\mathbf{J}_e \in (L^2(\mathbb{R}^3))^3$ are vanishing in Ω_I. Moreover, the magnetic permeability $\boldsymbol{\mu}_I$ and the electric permittivity ε_I are assumed to be a positive constant in Ω_I, say $\mu_0 > 0$ and $\varepsilon_0 > 0$.

7.1 The $(\mathbf{A}_C, V_C) - \psi_I$ formulation

In the present situation the eddy current problem in terms of the magnetic field $\mathbf{H}$ and the electric field $\mathbf{E}_C$ reads (see (3.25))

$$\begin{cases} \operatorname{curl} \mathbf{E}_C + i\omega\boldsymbol{\mu}_C\mathbf{H}_C = \mathbf{0} & \text{in } \Omega_C \\ \operatorname{curl} \mathbf{H}_C - \boldsymbol{\sigma}\mathbf{E}_C = \mathbf{J}_{e,C} & \text{in } \Omega_C \\ \operatorname{curl} \mathbf{H}_I = \mathbf{0} & \text{in } \Omega_I \\ \operatorname{div}(\mu_0\mathbf{H}_I) = 0 & \text{in } \Omega_I \\ \boldsymbol{\mu}_C\mathbf{H}_C \cdot \mathbf{n}_C + \mu_0\mathbf{H}_I \cdot \mathbf{n}_I = 0 & \text{on } \Gamma \\ \mathbf{H}_C \times \mathbf{n}_C + \mathbf{H}_I \times \mathbf{n}_I = \mathbf{0} & \text{on } \Gamma \\ \mathbf{H}_I(\mathbf{x}) = O(|\mathbf{x}|^{-1}) & \text{as } |\mathbf{x}| \to \infty\,. \end{cases} \tag{7.1}$$

If needed, but here we are not dealing with this aspect, the electric field $\mathbf{E}_I$ can be computed after having determined $\mathbf{H}_I$ and $\mathbf{E}_C$ in (7.1), by solving

$$\begin{cases} \operatorname{curl} \mathbf{E}_I = -i\omega\mu_0\mathbf{H}_I & \text{in } \Omega_I \\ \operatorname{div}(\varepsilon_0\mathbf{E}_I) = 0 & \text{in } \Omega_I \\ \mathbf{E}_I \times \mathbf{n}_I = -\mathbf{E}_C \times \mathbf{n}_C & \text{on } \Gamma \\ \int_\Gamma \varepsilon_0\mathbf{E}_I \cdot \mathbf{n}_I = 0 & \\ \mathbf{E}_I(\mathbf{x}) = O(|\mathbf{x}|^{-1}) & \text{as } |\mathbf{x}| \to \infty\,. \end{cases} \tag{7.2}$$

Since Ω_I is unbounded, note that we have to impose the no-flux condition on Γ, though it is a connected surface.

As proposed by Pillsbury [193], Rodger and Eastham [211], Emson and Simkin [100], we look for a vector magnetic potential $\mathbf{A}_C$, a scalar electric potential V_C and a scalar magnetic potential ψ_I such that

$$\boldsymbol{\mu}_C\mathbf{H}_C = \operatorname{curl} \mathbf{A}_C \;\;,\;\; \mathbf{E}_C = -i\omega\mathbf{A}_C - \operatorname{grad} V_C \;\;,\;\; \mathbf{H}_I = \operatorname{grad} \psi_I\,. \tag{7.3}$$

In this way one has $\operatorname{curl} \mathbf{E}_C = -i\omega \operatorname{curl} \mathbf{A}_C = -i\omega\boldsymbol{\mu}_C\mathbf{H}_C$, and therefore the Faraday equation in Ω_C is satisfied. Note that, in particular, when $\omega = 0$ one finds $\mathbf{E}_C = -\operatorname{grad} V_C$, therefore for the static case the usual formulation in terms of a scalar electric potential is recovered.

As usual, in order to have a unique vector potential $\mathbf{A}_C$, it is necessary to impose some gauge conditions: here we are considering the Coulomb gauge $\operatorname{div} \mathbf{A}_C = 0$ in Ω_C, with $\mathbf{A}_C \cdot \mathbf{n}_C = 0$ on Γ. Moreover, we also impose that

$$|\psi_I(\mathbf{x})| = O(|\mathbf{x}|^{-1}) \quad \text{as } |\mathbf{x}| \to \infty\,.$$

In conclusion, we are left with the problem

$$
\begin{cases}
\operatorname{curl}(\boldsymbol{\mu}_C^{-1} \operatorname{curl} \mathbf{A}_C) & \\
\qquad + i\omega\boldsymbol{\sigma}\mathbf{A}_C + \boldsymbol{\sigma} \operatorname{grad} V_C = \mathbf{J}_{e,C} & \text{in } \Omega_C \\
\Delta\psi_I = 0 & \text{in } \Omega_I \\
\operatorname{div} \mathbf{A}_C = 0 & \text{in } \Omega_C \\
\mathbf{A}_C \cdot \mathbf{n}_C = 0 & \text{on } \Gamma \\
\operatorname{curl} \mathbf{A}_C \cdot \mathbf{n}_C + \mu_0 \operatorname{grad} \psi_I \cdot \mathbf{n}_I = 0 & \text{on } \Gamma \\
(\boldsymbol{\mu}_C^{-1} \operatorname{curl} \mathbf{A}_C) \times \mathbf{n}_C + \operatorname{grad} \psi_I \times \mathbf{n}_I = \mathbf{0} & \text{on } \Gamma \\
|\psi_I(\mathbf{x})| + |\operatorname{grad} \psi_I(\mathbf{x})| = O(|\mathbf{x}|^{-1}) & \text{as } |\mathbf{x}| \to \infty ,
\end{cases}
\tag{7.4}
$$

where V_C is determined up to an additive constant.

In order to obtain a problem which is stable also in the case $\omega = 0$, and for which Lagrange nodal finite elements can be used for approximation, it is well-known (see, e.g., Coulomb [91], Morisue [180], Bìrò and Preis [49] and Chapter 6) that the Coulomb gauge condition $\operatorname{div} \mathbf{A}_C = 0$ in Ω_C can be incorporated as a penalization term in the Ampère equation. Introducing the constant $\mu_* > 0$, that for physical consistency can be chosen, for example, as a suitable average in Ω_C of the entries of the matrix $\boldsymbol{\mu}_C$, one considers

$$
\begin{cases}
\operatorname{curl}(\boldsymbol{\mu}_C^{-1} \operatorname{curl} \mathbf{A}_C) - \mu_*^{-1} \operatorname{grad} \operatorname{div} \mathbf{A}_C & \\
\qquad + i\omega\boldsymbol{\sigma}\mathbf{A}_C + \boldsymbol{\sigma} \operatorname{grad} V_C = \mathbf{J}_{e,C} & \text{in } \Omega_C \\
\Delta\psi_I = 0 & \text{in } \Omega_I \\
\operatorname{div}(i\omega\boldsymbol{\sigma}\mathbf{A}_C + \boldsymbol{\sigma} \operatorname{grad} V_C) = \operatorname{div} \mathbf{J}_{e,C} & \text{in } \Omega_C \\
(i\omega\boldsymbol{\sigma}\mathbf{A}_C + \boldsymbol{\sigma} \operatorname{grad} V_C) \cdot \mathbf{n}_C = \mathbf{J}_{e,C} \cdot \mathbf{n}_C & \text{on } \Gamma \\
\mathbf{A}_C \cdot \mathbf{n}_C = 0 & \text{on } \Gamma \\
\operatorname{curl} \mathbf{A}_C \cdot \mathbf{n}_C + \mu_0 \operatorname{grad} \psi_I \cdot \mathbf{n}_I = 0 & \text{on } \Gamma \\
(\boldsymbol{\mu}_C^{-1} \operatorname{curl} \mathbf{A}_C) \times \mathbf{n}_C + \operatorname{grad} \psi_I \times \mathbf{n}_I = \mathbf{0} & \text{on } \Gamma \\
|\psi_I(\mathbf{x})| + |\operatorname{grad} \psi_I(\mathbf{x})| = O(|\mathbf{x}|^{-1}) & \text{as } |\mathbf{x}| \to \infty ,
\end{cases}
\tag{7.5}
$$

the two additional equations appearing in (7.5) being necessary as the modification in the Ampère equation does not ensure now that $\mathbf{E}_C = -i\omega\mathbf{A}_C - \operatorname{grad} V_C$ satisfies the necessary conditions $\operatorname{div}(\boldsymbol{\sigma}\mathbf{E}_C) = -\operatorname{div} \mathbf{J}_{e,C}$ in Ω_C and $\boldsymbol{\sigma}\mathbf{E}_C \cdot \mathbf{n}_C = -\mathbf{J}_{e,C} \cdot \mathbf{n}_C$ on Γ.

Moreover, taking the divergence of $(7.5)_1$ and using $(7.5)_3$, we have $\Delta \operatorname{div} \mathbf{A}_C = 0$ in Ω_C, and, taking the scalar product of $(7.5)_1$ by $\mathbf{n}_C$, using $(7.5)_4$ and $(7.5)_7$, we find

$$
\begin{aligned}
\mu_*^{-1} \operatorname{grad} \operatorname{div} \mathbf{A}_C \cdot \mathbf{n}_C &= \operatorname{curl}(\boldsymbol{\mu}^{-1} \operatorname{curl} \mathbf{A}_C) \cdot \mathbf{n}_C \\
&= \operatorname{div}_\tau[(\boldsymbol{\mu}^{-1} \operatorname{curl} \mathbf{A}_C) \times \mathbf{n}_C] = -\operatorname{div}_\tau(\operatorname{grad} \psi_I \times \mathbf{n}_I) \\
&= -\operatorname{curl} \operatorname{grad} \psi_I \cdot \mathbf{n}_I = 0 \quad \text{on } \Gamma .
\end{aligned}
$$

Therefore $\operatorname{div} \mathbf{A}_C$ is constant in Ω_C, and this constant is 0 as a consequence of $(7.5)_5$. In conclusion, any solution to (7.5) satisfies $\operatorname{div} \mathbf{A}_C = 0$ in Ω_C, and thus (7.4) and (7.5) are equivalent.

7.2 The $(\mathbf{A}_C, V_C) - \psi_\Gamma$ weak formulation

In this chapter we have assumed that $\boldsymbol{\mu}_I$ is a positive constant μ_0 and we are looking for a scalar magnetic potential ψ_I. Therefore for determining this potential we have to solve the Laplace equation in Ω_I. This allows us to use potential theory, transforming the problem for ψ_I into a problem on the interface Γ and reducing in a significative way the number of unknowns in numerical computations.

Referring for notation to Section A.1, it is well-known from potential theory (see, e.g., McLean [175], Nédélec [187]) that we can introduce on Γ the single layer and double layer potentials

$$\mathcal{S} : H^{-1/2}(\Gamma) \to H^{1/2}(\Gamma) \ , \ \ \mathcal{S}(\xi)(\mathbf{x}) := \int_\Gamma \frac{1}{4\pi|\mathbf{x}-\mathbf{y}|}\, \xi(\mathbf{y}) dS_y \tag{7.6}$$

$$\mathcal{D} : H^{1/2}(\Gamma) \to H^{1/2}(\Gamma) \ , \ \ \mathcal{D}(\eta)(\mathbf{x}) := \int_\Gamma \frac{\mathbf{x}-\mathbf{y}}{4\pi|\mathbf{x}-\mathbf{y}|^3} \cdot \eta(\mathbf{y})\mathbf{n}_C(\mathbf{y}) dS_y, \tag{7.7}$$

and the hypersingular integral operator

$$\begin{aligned} &\mathcal{H} : H^{1/2}(\Gamma) \to H^{-1/2}(\Gamma) \ , \\ &\quad \mathcal{H}(\eta)(\mathbf{x}) := -\,\mathrm{grad} \left(\int_\Gamma \tfrac{\mathbf{x}-\mathbf{y}}{4\pi|\mathbf{x}-\mathbf{y}|^3} \cdot \eta(\mathbf{y})\mathbf{n}_C(\mathbf{y}) dS_y \right) \cdot \mathbf{n}_C(\mathbf{x}) \ . \end{aligned} \tag{7.8}$$

We also recall that the adjoint operator $\mathcal{D}' : H^{-1/2}(\Gamma) \to H^{-1/2}(\Gamma)$ reads

$$\mathcal{D}'(\xi)(\mathbf{x}) = \left(\int_\Gamma \frac{\mathbf{y}-\mathbf{x}}{4\pi|\mathbf{x}-\mathbf{y}|^3}\, \xi(\mathbf{y}) dS_y \right) \cdot \mathbf{n}_C(\mathbf{x}) \ . \tag{7.9}$$

Since we have that $\Delta\psi_I = 0$ in Ω_I and $\mathrm{grad}\, \psi_I \cdot \mathbf{n}_I = -\frac{1}{\mu_0} \mathrm{curl}\, \mathbf{A}_C \cdot \mathbf{n}_C$ on Γ, a first result is that the trace $\psi_\Gamma := \psi_{I|\Gamma}$ satisfies

$$\frac{1}{2}\psi_\Gamma - \mathcal{D}(\psi_\Gamma) + \frac{1}{\mu_0}\mathcal{S}(\mathrm{curl}\, \mathbf{A}_C \cdot \mathbf{n}_C) = 0 \ \text{ on } \Gamma \tag{7.10}$$

$$\frac{1}{2\mu_0}\, \mathrm{curl}\, \mathbf{A}_C \cdot \mathbf{n}_C + \frac{1}{\mu_0}\mathcal{D}'(\mathrm{curl}\, \mathbf{A}_C \cdot \mathbf{n}_C) + \mathcal{H}(\psi_\Gamma) = 0 \ \text{ on } \Gamma \tag{7.11}$$

(see, e.g., McLean [175], Nédélec [187]).

As a second step, we can devise a weak formulation in terms of $(\mathbf{A}_C, V_C) - \psi_\Gamma$. A standard integration by parts yields

$$\int_\Gamma \mathbf{n}_I \times \mathrm{grad}\, \psi_I \cdot \overline{\mathbf{w}_C} = \int_\Gamma \psi_\Gamma\, \mathrm{curl}\, \overline{\mathbf{w}_C} \cdot \mathbf{n}_C \ .$$

Moreover, multiplying $(7.5)_1$, $(7.5)_3$ and (7.11) by suitable test functions $(\mathbf{w}_C, Q_C, \eta)$ with $\mathbf{w}_C \cdot \mathbf{n}_C = 0$ on Γ, integrating in Ω_C and Γ, and integrating by parts, from the other matching condition

$$\boldsymbol{\mu}_C^{-1}\, \mathrm{curl}\, \mathbf{A}_C \times \mathbf{n}_C + \mathrm{grad}\, \psi_I \times \mathbf{n}_I = \mathbf{0} \text{ on } \Gamma,$$

and the interface equation (7.10) we end up with the following weak problem

$$\begin{aligned}
&\int_{\Omega_C}(\boldsymbol{\mu}_C^{-1}\,\mathrm{curl}\,\mathbf{A}_C\cdot\mathrm{curl}\,\overline{\mathbf{w}_C}+\mu_*^{-1}\,\mathrm{div}\,\mathbf{A}_C\,\mathrm{div}\,\overline{\mathbf{w}_C})\\
&\qquad+\int_{\Omega_C}(i\omega\boldsymbol{\sigma}\mathbf{A}_C\cdot\overline{\mathbf{w}_C}+\boldsymbol{\sigma}\,\mathrm{grad}\,V_C\cdot\overline{\mathbf{w}_C})\\
&\qquad+\int_\Gamma[-\tfrac{1}{2}\psi_\Gamma-\mathcal{D}(\psi_\Gamma)+\tfrac{1}{\mu_0}\mathcal{S}(\mathrm{curl}\,\mathbf{A}_C\cdot\mathbf{n}_C)]\,\mathrm{curl}\,\overline{\mathbf{w}_C}\cdot\mathbf{n}_C\\
&\quad=\int_{\Omega_C}\mathbf{J}_{e,C}\cdot\overline{\mathbf{w}_C}\\
&\int_{\Omega_C}(i\omega\boldsymbol{\sigma}\mathbf{A}_C\cdot\mathrm{grad}\,\overline{Q_C}+\boldsymbol{\sigma}\,\mathrm{grad}\,V_C\cdot\mathrm{grad}\,\overline{Q_C})\\
&\quad=\int_{\Omega_C}\mathbf{J}_{e,C}\cdot\mathrm{grad}\,\overline{Q_C}\\
&\int_\Gamma[\tfrac{1}{2}\,\mathrm{curl}\,\mathbf{A}_C\cdot\mathbf{n}_C+\mathcal{D}'(\mathrm{curl}\,\mathbf{A}_C\cdot\mathbf{n}_C)+\mu_0\mathcal{H}(\psi_\Gamma)]\overline{\eta}=0\,.
\end{aligned}$$

We note that, for the ease of notation, as usual here above we have written the integration symbol on Γ instead of the pairing between $H^{-1/2}(\Gamma)$ and $H^{1/2}(\Gamma)$; the same notation will be used in the sequel.

Since the hypersingular operator $\mathcal{H}(\cdot)$ is coercive in the constrained space $H^{1/2}(\Gamma)/\mathbb{C}$, it is convenient to rewrite the preceding problem for test functions $\eta\in H^{1/2}(\Gamma)/\mathbb{C}$, looking for $q\in H^{1/2}(\Gamma)/\mathbb{C}$, which differs from ψ_Γ by an additive constant.

We know that $\mathcal{H}(1)=0$ and $\mathcal{D}(1)=-\frac{1}{2}$ (see, e.g., McLean [175], Nédélec [187]), and that $\int_\Gamma\mathcal{H}(\eta)=0$ for each $\eta\in H^{1/2}(\Gamma)$ (see, e.g., Nédélec [187], Theor. 3.3.2). Hence $\mathcal{H}(\psi_\Gamma+c_0)=\mathcal{H}(\psi_\Gamma)$,

$$-\frac{1}{2}(\psi_\Gamma+c_0)-\mathcal{D}(\psi_\Gamma+c_0)=-\frac{1}{2}\psi_\Gamma-\mathcal{D}(\psi_\Gamma),$$

and

$$\begin{aligned}
&\int_\Gamma[\tfrac{1}{2}\,\mathrm{curl}\,\mathbf{A}_C\cdot\mathbf{n}_C+\mathcal{D}'(\mathrm{curl}\,\mathbf{A}_C\cdot\mathbf{n}_C)+\mu_0\mathcal{H}(\psi_\Gamma)]\\
&\quad=\int_\Gamma[\tfrac{1}{2}\,\mathrm{curl}\,\mathbf{A}_C\cdot\mathbf{n}_C+\mathrm{curl}\,\mathbf{A}_C\cdot\mathbf{n}_C\,\mathcal{D}(1)+\mu_0\mathcal{H}(\psi_\Gamma)]=0\,.
\end{aligned}$$

In conclusion, introducing the space

$$W_C:=H(\mathrm{curl};\Omega_C)\cap H_0(\mathrm{div};\Omega_C)$$

we are looking for the solution of the following coupled problem

$$\begin{aligned}
&\text{Find } (\mathbf{A}_C,V_C,q)\in W_C\times H^1(\Omega_C)/\mathbb{C}\times H^{1/2}(\Gamma)/\mathbb{C}\text{ such that}\\
&\int_{\Omega_C}(\boldsymbol{\mu}_C^{-1}\,\mathrm{curl}\,\mathbf{A}_C\cdot\mathrm{curl}\,\overline{\mathbf{w}_C}+\mu_*^{-1}\,\mathrm{div}\,\mathbf{A}_C\,\mathrm{div}\,\overline{\mathbf{w}_C})\\
&\qquad+\int_{\Omega_C}(i\omega\boldsymbol{\sigma}\mathbf{A}_C\cdot\overline{\mathbf{w}_C}+\boldsymbol{\sigma}\,\mathrm{grad}\,V_C\cdot\overline{\mathbf{w}_C})\\
&\qquad+\int_\Gamma[-\tfrac{1}{2}q-\mathcal{D}(q)+\tfrac{1}{\mu_0}\mathcal{S}(\mathrm{curl}\,\mathbf{A}_C\cdot\mathbf{n}_C)]\,\mathrm{curl}\,\overline{\mathbf{w}_C}\cdot\mathbf{n}_C\\
&\quad=\int_{\Omega_C}\mathbf{J}_{e,C}\cdot\overline{\mathbf{w}_C}\\
&\int_{\Omega_C}(i\omega\boldsymbol{\sigma}\mathbf{A}_C\cdot\mathrm{grad}\,\overline{Q_C}+\boldsymbol{\sigma}\,\mathrm{grad}\,V_C\cdot\mathrm{grad}\,\overline{Q_C})\\
&\quad=\int_{\Omega_C}\mathbf{J}_{e,C}\cdot\mathrm{grad}\,\overline{Q_C}\\
&\int_\Gamma[\tfrac{1}{2}\,\mathrm{curl}\,\mathbf{A}_C\cdot\mathbf{n}_C+\mathcal{D}'(\mathrm{curl}\,\mathbf{A}_C\cdot\mathbf{n}_C)+\mu_0\mathcal{H}(q)]\overline{\eta}=0\\
&\text{for all } (\mathbf{w}_C,Q_C,\eta)\in W_C\times H^1(\Omega_C)/\mathbb{C}\times H^{1/2}(\Gamma)/\mathbb{C}\,.
\end{aligned}\tag{7.12}$$

Now we want to prove that from a solution to (7.12) we can construct a solution to the strong problem (7.4). Let us note that the condition $\mathbf{A}_C \cdot \mathbf{n}_C = 0$ on Γ in (7.4) is already contained in the definition of the space W_C.

Lemma 7.1. *Suppose that* $(\mathbf{A}_C, V_C, q)$ *is a solution to* (7.12). *Then* $\operatorname{div} \mathbf{A}_C = 0$ *in* Ω_C.

Proof. Since $\int_{\Omega_C} \operatorname{div} \mathbf{A}_C = \int_\Gamma \mathbf{A}_C \cdot \mathbf{n}_C = 0$, we can consider the solution $v_C \in H^1(\Omega_C)/\mathbb{C}$ to the Neumann problem

$$\begin{cases} \Delta v_C = \operatorname{div} \mathbf{A}_C & \text{in } \Omega_C \\ \operatorname{grad} v_C \cdot \mathbf{n}_C = 0 & \text{on } \Gamma \, . \end{cases}$$

Setting $\mathbf{w}_C = \operatorname{grad} v_C$, clearly we have $\mathbf{w}_C \in W_C$. Using in $(7.12)_1$ and $(7.12)_2$ the test function $(\mathbf{w}_C, v_C)$ we find $\int_{\Omega_C} |\operatorname{div} \mathbf{A}_C|^2 = 0$, therefore $\operatorname{div} \mathbf{A}_C = 0$ in Ω_C. □

Concerning the interface equations (7.10) and (7.11) we have:

Lemma 7.2. *Suppose that* $(\mathbf{A}_C, V_C, q)$ *is a solution to* (7.12). *Then*

$$\frac{1}{2} q - \mathcal{D}(q) + \frac{1}{\mu_0} \mathcal{S}(\operatorname{curl} \mathbf{A}_C \cdot \mathbf{n}_C) = \text{const} \ \ \textit{on } \Gamma \tag{7.13}$$

$$\frac{1}{2\mu_0} \operatorname{curl} \mathbf{A}_C \cdot \mathbf{n}_C + \frac{1}{\mu_0} \mathcal{D}'(\operatorname{curl} \mathbf{A}_C \cdot \mathbf{n}_C) + \mathcal{H}(q) = 0 \ \ \textit{on } \Gamma \, . \tag{7.14}$$

Proof. As already seen, we have $\int_\Gamma \mathcal{H}(\eta) = 0$ for each $\eta \in H^{1/2}(\Gamma)$ and $\mathcal{D}(1) = -\frac{1}{2}$, thus

$$\begin{aligned} &\textstyle\int_\Gamma [\frac{1}{2} \operatorname{curl} \mathbf{A}_C \cdot \mathbf{n}_C + \mathcal{D}'(\operatorname{curl} \mathbf{A}_C \cdot \mathbf{n}_C) + \mu_0 \mathcal{H}(q)] \\ &\quad = \textstyle\int_\Gamma [\frac{1}{2} \operatorname{curl} \mathbf{A}_C \cdot \mathbf{n}_C + \operatorname{curl} \mathbf{A}_C \cdot \mathbf{n}_C \, \mathcal{D}(1) + \mu_0 \mathcal{H}(q)] = 0 \, . \end{aligned}$$

Therefore equation $(7.12)_3$ is satisfied not only for all $\eta \in H^{1/2}(\Gamma)/\mathbb{C}$, but also for all $\eta \in H^{1/2}(\Gamma)$, and equation (7.14) follows at once.

Consequently, it is well-known from potential theory that we also obtain (7.13). □

We need now to introduce the single layer and double layer operators in the interior of Ω_I (namely, the exterior of Ω_C). For $\mathbf{x} \in \Omega_I$ we can define (see, e.g., McLean [175], Nédélec [187])

$$\mathcal{S}_I : H^{-1/2}(\Gamma) \to W^1(\Omega_I) \ , \ \ \mathcal{S}_I(\xi)(\mathbf{x}) := \int_\Gamma \frac{1}{4\pi |\mathbf{x} - \mathbf{y}|} \xi(\mathbf{y}) dS_y \tag{7.15}$$

$$\begin{aligned} &\mathcal{D}_I : H^{1/2}(\Gamma)/\mathbb{C} \to W^1(\Omega_I) \ , \\ &\qquad \mathcal{D}_I(\eta)(\mathbf{x}) := \textstyle\int_\Gamma \frac{(\mathbf{x}-\mathbf{y})}{4\pi|\mathbf{x}-\mathbf{y}|^3} \cdot \eta(\mathbf{y}) \mathbf{n}_C(\mathbf{y}) dS_y \, , \end{aligned} \tag{7.16}$$

where

$$\begin{aligned} W^1(\Omega_I) := \{ &\chi_I \in (C_0^\infty(\Omega_I))' \mid \\ &(1 + |\mathbf{x}|^2)^{-1/2} \chi_I \in L^2(\Omega_I), \operatorname{grad} \chi_I \in (L^2(\Omega_I))^3 \} \, . \end{aligned} \tag{7.17}$$

We conclude our argument by showing that:

Lemma 7.3. *Suppose that* $(\mathbf{A}_C, V_C, q)$ *is a solution to* (7.12)*. In the domain* Ω_I *define the function* $\psi_I := \mathcal{D}_I(q) - \frac{1}{\mu_0}\mathcal{S}_I(\operatorname{curl}\mathbf{A}_C \cdot \mathbf{n}_C)$*. Then*

$$\begin{aligned}&\int_{\Omega_C}(\boldsymbol{\mu}_C^{-1}\operatorname{curl}\mathbf{A}_C \cdot \operatorname{curl}\overline{\mathbf{w}_C^*} + i\omega\boldsymbol{\sigma}\mathbf{A}_C \cdot \overline{\mathbf{w}_C^*} + \boldsymbol{\sigma}\operatorname{grad}V_C \cdot \overline{\mathbf{w}_C^*})\\&\quad + \int_\Gamma \mathbf{n}_C \times \operatorname{grad}\psi_I \cdot \overline{\mathbf{w}_C^*} = \int_{\Omega_C}\mathbf{J}_{e,C} \cdot \overline{\mathbf{w}_C^*}\end{aligned} \tag{7.18}$$

for all $\mathbf{w}_C^* \in H(\operatorname{curl};\Omega_C)$*. Therefore,*

$$\begin{cases}\operatorname{curl}(\boldsymbol{\mu}_C^{-1}\operatorname{curl}\mathbf{A}_C) + i\omega\boldsymbol{\sigma}\mathbf{A}_C + \boldsymbol{\sigma}\operatorname{grad}V_C = \mathbf{J}_{e,C} & \text{in } \Omega_C\\ \Delta\psi_I = 0 & \text{in } \Omega_I\\ \operatorname{curl}\mathbf{A}_C \cdot \mathbf{n}_C + \mu_0 \operatorname{grad}\psi_I \cdot \mathbf{n}_I = 0 & \text{on } \Gamma\\ (\boldsymbol{\mu}_C^{-1}\operatorname{curl}\mathbf{A}_C) \times \mathbf{n}_C + \operatorname{grad}\psi_I \times \mathbf{n}_I = \mathbf{0} & \text{on } \Gamma\\ |\psi_I(\mathbf{x})| + |\operatorname{grad}\psi_I(\mathbf{x})| = O(|\mathbf{x}|^{-1}) & \text{as } |\mathbf{x}| \to \infty\,.\end{cases} \tag{7.19}$$

Proof. Well-known results of potential theory imply that ψ_I is a harmonic function with $|\psi_I(\mathbf{x})|$ and $|\operatorname{grad}\psi_I(\mathbf{x})|$ decaying at infinity as $O(|\mathbf{x}|^{-1})$. Moreover, ψ_I satisfies the trace relations

$$\psi_{I|\Gamma} = \frac{1}{2}q + \mathcal{D}(q) - \frac{1}{\mu_0}\mathcal{S}(\operatorname{curl}\mathbf{A}_C \cdot \mathbf{n}_C), \tag{7.20}$$

and

$$\operatorname{grad}\psi_I \cdot \mathbf{n}_I = \mathcal{H}(q) + \mu_0^{-1}\Big[-\frac{1}{2}\operatorname{curl}\mathbf{A}_C \cdot \mathbf{n}_C + \mathcal{D}'(\operatorname{curl}\mathbf{A}_C \cdot \mathbf{n}_C)\Big] \tag{7.21}$$

(see, e.g., McLean [175], Nédélec [187]).

From (7.14) and (7.21) we see that the interface condition $(7.19)_3$ is satisfied. Moreover, from Lemma 7.1, $(7.12)_1$ and (7.20) we find that

$$\begin{aligned}&\int_{\Omega_C}(\boldsymbol{\mu}_C^{-1}\operatorname{curl}\mathbf{A}_C \cdot \operatorname{curl}\overline{\mathbf{w}_C} + i\omega\boldsymbol{\sigma}\mathbf{A}_C \cdot \overline{\mathbf{w}_C} + \boldsymbol{\sigma}\operatorname{grad}V_C \cdot \overline{\mathbf{w}_C})\\&\quad - \int_\Gamma \psi_{I|\Gamma}\operatorname{curl}\overline{\mathbf{w}_C} \cdot \mathbf{n}_C = \int_{\Omega_C}\mathbf{J}_{e,C} \cdot \overline{\mathbf{w}_C}\,.\end{aligned}$$

Since we have $-\int_\Gamma \psi_{I|\Gamma}\operatorname{curl}\overline{\mathbf{w}_C} \cdot \mathbf{n}_C = \int_\Gamma \mathbf{n}_C \times \operatorname{grad}\psi_I \cdot \overline{\mathbf{w}_C}$, for each $\mathbf{w}_C \in W_C$ we have obtained

$$\begin{aligned}&\int_{\Omega_C}(\boldsymbol{\mu}_C^{-1}\operatorname{curl}\mathbf{A}_C \cdot \operatorname{curl}\overline{\mathbf{w}_C} + i\omega\boldsymbol{\sigma}\mathbf{A}_C \cdot \overline{\mathbf{w}_C} + \boldsymbol{\sigma}\operatorname{grad}V_C \cdot \overline{\mathbf{w}_C})\\&\quad + \int_\Gamma \mathbf{n}_C \times \operatorname{grad}\psi_I \cdot \overline{\mathbf{w}_C} = \int_{\Omega_C}\mathbf{J}_{e,C} \cdot \overline{\mathbf{w}_C}\,.\end{aligned} \tag{7.22}$$

If $\mathbf{w}_C^* \in H(\operatorname{curl};\Omega_C)$, consider the solution $v_C^* \in H^1(\Omega_C)/\mathbb{C}$ of the Neumann problem $\Delta v_C^* = \operatorname{div}\mathbf{w}_C^*$ in Ω_C with $\operatorname{grad}v_C^* \cdot \mathbf{n}_C = \mathbf{w}_C^* \cdot \mathbf{n}_C$ on Γ. Setting $\mathbf{w}_C = \mathbf{w}_C^* - \operatorname{grad}v_C^*$, we have $\mathbf{w}_C \in W_C$, and using it in (7.22) we obtain

$$\begin{aligned}&\int_{\Omega_C}(\boldsymbol{\mu}_C^{-1}\operatorname{curl}\mathbf{A}_C \cdot \operatorname{curl}\overline{\mathbf{w}_C^*} + i\omega\boldsymbol{\sigma}\mathbf{A}_C \cdot \overline{\mathbf{w}_C^*} + \boldsymbol{\sigma}\operatorname{grad}V_C \cdot \overline{\mathbf{w}_C^*})\\&\qquad + \int_\Gamma \mathbf{n}_C \times \operatorname{grad}\psi_I \cdot \overline{\mathbf{w}_C^*}\\&\quad = \int_{\Omega_C}(\boldsymbol{\mu}_C^{-1}\operatorname{curl}\mathbf{A}_C \cdot \operatorname{curl}\overline{\mathbf{w}_C} + i\omega\boldsymbol{\sigma}\mathbf{A}_C \cdot \overline{\mathbf{w}_C} + \boldsymbol{\sigma}\operatorname{grad}V_C \cdot \overline{\mathbf{w}_C})\\&\qquad + \int_\Gamma \mathbf{n}_C \times \operatorname{grad}\psi_I \cdot \overline{\mathbf{w}_C}\\&\qquad + \int_{\Omega_C}(i\omega\boldsymbol{\sigma}\mathbf{A}_C \cdot \operatorname{grad}\overline{v_C^*} + \boldsymbol{\sigma}\operatorname{grad}V_C \cdot \operatorname{grad}\overline{v_C^*})\\&\qquad + \int_\Gamma \mathbf{n}_C \times \operatorname{grad}\psi_I \cdot \operatorname{grad}\overline{v_C^*}\\&\quad = \int_{\Omega_C}\mathbf{J}_{e,C} \cdot \overline{\mathbf{w}_C} + \int_{\Omega_C}\mathbf{J}_{e,C} \cdot \operatorname{grad}\overline{v_C^*} = \int_{\Omega_C}\mathbf{J}_{e,C} \cdot \overline{\mathbf{w}_C^*}\,,\end{aligned}$$

having used $(7.12)_2$ and the fact that

$$\begin{aligned}&\int_\Gamma \mathbf{n}_C \times \operatorname{grad} \psi_I \cdot \operatorname{grad} \overline{v_C^*} = -\int_\Gamma \operatorname{div}_\tau(\mathbf{n}_C \times \operatorname{grad} \psi_I)\, \overline{v_C^*}\\ &\quad = -\int_\Gamma \operatorname{curl} \operatorname{grad} \psi_I \cdot \mathbf{n}_I\, \overline{v_C^*} = 0\,.\end{aligned}$$

Taking a test function $\mathbf{w}_C^* \in (C_0^\infty(\Omega_C))^3$ and integrating by parts we verify that $(7.19)_1$ is satisfied; repeating the same argument for $\mathbf{w}_C^* \in H(\operatorname{curl}; \Omega_C)$, we see that the interface condition $(7.19)_4$ is satisfied as well. □

Remark 7.4. The function $q \in H^{1/2}(\Gamma)/\mathbb{C}$ determined in (7.12) is defined up to an additive constant. It is easily seen that, as functions in $H^{1/2}(\Gamma)/\mathbb{C}$, q and the trace on Γ of the harmonic scalar potential ψ_I, namely, what we have called ψ_Γ, are the same function. Indeed, from (7.13) and (7.20) we see that $\psi_\Gamma + \text{const} = q$. □

7.3 Existence and uniqueness of the weak solution

In order to prove the existence and uniqueness of the solution to (7.12), let us introduce the following sesquilinear forms: for $\omega \neq 0$

$$\begin{aligned}&\mathcal{A}_{(\omega\neq 0)}[(\mathbf{A}_C, V_C, q), (\mathbf{w}_C, Q_C, \eta)]\\ &\quad = \int_{\Omega_C} (\boldsymbol{\mu}_C^{-1} \operatorname{curl} \mathbf{A}_C \cdot \operatorname{curl} \overline{\mathbf{w}_C} + \mu_*^{-1} \operatorname{div} \mathbf{A}_C \operatorname{div} \overline{\mathbf{w}_C})\\ &\qquad + i\omega^{-1} \int_{\Omega_C} \boldsymbol{\sigma}(i\omega \mathbf{A}_C + \operatorname{grad} V_C) \cdot (-i\omega \overline{\mathbf{w}_C} + \operatorname{grad} \overline{Q_C})\\ &\qquad + \int_\Gamma [-\tfrac{1}{2} q - \mathcal{D}(q)] \operatorname{curl} \overline{\mathbf{w}_C} \cdot \mathbf{n}_C\\ &\qquad + \int_\Gamma [\tfrac{1}{2} \operatorname{curl} \mathbf{A}_C \cdot \mathbf{n}_C + \mathcal{D}'(\operatorname{curl} \mathbf{A}_C \cdot \mathbf{n}_C)] \overline{\eta}\\ &\qquad + \int_\Gamma [\tfrac{1}{\mu_0} \mathcal{S}(\operatorname{curl} \mathbf{A}_C \cdot \mathbf{n}_C) \operatorname{curl} \overline{\mathbf{w}_C} \cdot \mathbf{n}_C + \mu_0 \mathcal{H}(q)\, \overline{\eta}]\,,\end{aligned} \tag{7.23}$$

and for $\omega = 0$

$$\begin{aligned}&\mathcal{A}_{(\omega=0)}[(\mathbf{A}_C, V_C, q), (\mathbf{w}_C, Q_C, \eta)]\\ &\quad = \int_{\Omega_C} (\boldsymbol{\mu}_C^{-1} \operatorname{curl} \mathbf{A}_C \cdot \operatorname{curl} \overline{\mathbf{w}_C} + \mu_*^{-1} \operatorname{div} \mathbf{A}_C \operatorname{div} \overline{\mathbf{w}_C})\\ &\qquad + \int_{\Omega_C} (\boldsymbol{\sigma} \operatorname{grad} V_C \cdot \overline{\mathbf{w}_C} + \beta \boldsymbol{\sigma} \operatorname{grad} V_C \cdot \operatorname{grad} \overline{Q_C})\\ &\qquad + \int_\Gamma [-\tfrac{1}{2} q - \mathcal{D}(q)] \operatorname{curl} \overline{\mathbf{w}_C} \cdot \mathbf{n}_C\\ &\qquad + \int_\Gamma [\tfrac{1}{2} \operatorname{curl} \mathbf{A}_C \cdot \mathbf{n}_C + \mathcal{D}'(\operatorname{curl} \mathbf{A}_C \cdot \mathbf{n}_C)] \overline{\eta}\\ &\qquad + \int_\Gamma [\tfrac{1}{\mu_0} \mathcal{S}(\operatorname{curl} \mathbf{A}_C \cdot \mathbf{n}_C) \operatorname{curl} \overline{\mathbf{w}_C} \cdot \mathbf{n}_C + \mu_0 \mathcal{H}(q)\, \overline{\eta}]\,.\end{aligned} \tag{7.24}$$

These forms are obtained by adding the left hand sides in (7.12): however, in the case $\omega \neq 0$ we have multiplied the second equation by $i\omega^{-1}$, obtaining $\mathcal{A}_{(\omega\neq 0)}[\cdot,\cdot]$, while in the case $\omega = 0$ we have multiplied the second equation by $\beta > 0$, to be chosen in the sequel, obtaining $\mathcal{A}_{(\omega=0)}[\cdot,\cdot]$.

The main result of this section is:

Theorem 7.5. *The sesquilinear form $\mathcal{A}_{(\omega\neq 0)}[\cdot,\cdot]$ is coercive in the space $W_C \times H^1(\Omega_C)/\mathbb{C} \times H^{1/2}(\Gamma)/\mathbb{C}$, uniformly as $\omega \to 0$; namely, there exists a constant $\kappa > 0$, independent of ω, such that for each $(\mathbf{w}_C, Q_C, \eta) \in W_C \times H^1(\Omega_C) \times H^{1/2}(\Gamma)$ with*

$\int_{\Omega_C} Q_C = 0$ *and* $\int_\Gamma \eta = 0$ *one has*

$$
\begin{aligned}
&|\mathcal{A}_{(\omega\neq 0)}[(\mathbf{w}_C, Q_C, \eta), (\mathbf{w}_C, Q_C, \eta)]| \\
&\quad \geq \kappa\Big(\textstyle\int_{\Omega_C} (|\mathbf{w}_C|^2 + |\operatorname{curl} \mathbf{w}_C|^2 + |\operatorname{div} \mathbf{w}_C|^2) \\
&\qquad + ||\eta||^2_{1/2,\Gamma} + \chi(\omega) \textstyle\int_{\Omega_C} (|Q_C|^2 + |\operatorname{grad} Q_C|^2) \Big) \,,
\end{aligned} \tag{7.25}
$$

where the constant $\chi(\omega) > 0$ *is equal to* $|\omega|^{-1}$ *in the case* $0 < |\omega| < 1$ *and is equal to* ω^{-2} *in the case* $|\omega| \geq 1$.

Moreover, the sesquilinear form $\mathcal{A}_{(\omega=0)}[\cdot,\cdot]$ *is coercive in the space* $W_C \times H^1(\Omega_C)/\mathbb{C} \times H^{1/2}(\Gamma)/\mathbb{C}$, *namely, there exists a constant* $\kappa_0 > 0$ *such that for each* $(\mathbf{w}_C, Q_C, \eta) \in W_C \times H^1(\Omega_C) \times H^{1/2}(\Gamma)$ *with* $\int_{\Omega_C} Q_C = 0$ *and* $\int_\Gamma \eta = 0$ *one has*

$$
\begin{aligned}
&|\mathcal{A}_{(\omega=0)}[(\mathbf{w}_C, Q_C, \eta), (\mathbf{w}_C, Q_C, \eta)]| \\
&\quad \geq \kappa_0\Big(\textstyle\int_{\Omega_C} (|\mathbf{w}_C|^2 + |\operatorname{curl} \mathbf{w}_C|^2 + |\operatorname{div} \mathbf{w}_C|^2) \\
&\qquad + ||\eta||^2_{1/2,\Gamma} + \textstyle\int_{\Omega_C} (|Q_C|^2 + |\operatorname{grad} Q_C|^2) \Big) \,.
\end{aligned} \tag{7.26}
$$

As a consequence, for each $\mathbf{J}_{e,C} \in (L^2(\Omega_C))^3$, *existence and uniqueness of the solution to* (7.12) *follow from the Lax–Milgram lemma.*

Proof. First of all, let us recall that the operators $\mathcal{S}$ and $\mathcal{H}$ are continuous from $H^{-1/2}(\Gamma)$ into $H^{1/2}(\Gamma)$ and from $H^{1/2}(\Gamma)$ into $H^{-1/2}(\Gamma)$, respectively, and satisfy

$$
\int_\Gamma \mathcal{S}(\xi)\,\overline{\xi} \geq \kappa_1 ||\xi||^2_{-1/2,\Gamma} \,, \quad \int_\Gamma \mathcal{H}(\eta)\,\overline{\eta} \geq \kappa_2 ||\eta||^2_{1/2,\Gamma} \tag{7.27}
$$

for each $\xi \in H^{-1/2}(\Gamma)$ and $\eta \in H^{1/2}(\Gamma)$ with $\int_\Gamma \eta = 0$, and moreover that the operator $\mathcal{D}$ is continuous from $H^{1/2}(\Gamma)$ into itself (see, e.g., McLean [175], Nédélec [187]).

The sesquilinear form $\mathcal{A}_{(\omega\neq 0)}[\cdot,\cdot]$ satisfies

$$
\begin{aligned}
&\mathcal{A}_{(\omega\neq 0)}[(\mathbf{w}_C, Q_C, \eta), (\mathbf{w}_C, Q_C, \eta)] \\
&\quad = \textstyle\int_{\Omega_C} (\boldsymbol{\mu}_C^{-1} \operatorname{curl} \mathbf{w}_C \cdot \operatorname{curl} \overline{\mathbf{w}_C} + \mu_*^{-1} |\operatorname{div} \mathbf{w}_C|^2) \\
&\qquad + i\omega^{-1} \textstyle\int_{\Omega_C} \boldsymbol{\sigma}(i\omega \mathbf{w}_C + \operatorname{grad} Q_C) \cdot (-i\omega \overline{\mathbf{w}_C} + \operatorname{grad} \overline{Q_C}) \\
&\qquad + \textstyle\int_\Gamma [-\frac{1}{2}\eta - \mathcal{D}(\eta)] \operatorname{curl} \overline{\mathbf{w}_C} \cdot \mathbf{n}_C \\
&\qquad + \textstyle\int_\Gamma [\frac{1}{2} \operatorname{curl} \mathbf{w}_C \cdot \mathbf{n}_C + \mathcal{D}'(\operatorname{curl} \mathbf{w}_C \cdot \mathbf{n}_C)]\overline{\eta} \\
&\qquad + \textstyle\int_\Gamma [\frac{1}{\mu_0} \mathcal{S}(\operatorname{curl} \mathbf{w}_C \cdot \mathbf{n}_C) \operatorname{curl} \overline{\mathbf{w}_C} \cdot \mathbf{n}_C + \mu_0 \mathcal{H}(\eta)\,\overline{\eta}] \,.
\end{aligned}
$$

Since

$$
\int_\Gamma \mathcal{D}'(\operatorname{curl} \mathbf{w}_C \cdot \mathbf{n}_C)]\overline{\eta} = \int_\Gamma \mathcal{D}(\overline{\eta}) \operatorname{curl} \mathbf{w}_C \cdot \mathbf{n}_C \,,
$$

and

$$
\begin{aligned}
&[-\tfrac{1}{2}\eta - \mathcal{D}(\eta)] \operatorname{curl} \overline{\mathbf{w}_C} \cdot \mathbf{n}_C + \operatorname{curl} \mathbf{w}_C \cdot \mathbf{n}_C\, [\tfrac{1}{2}\overline{\eta} + \mathcal{D}(\overline{\eta})] \\
&\qquad = -2\,i \operatorname{Im}\Big([\tfrac{1}{2}\eta + \mathcal{D}(\eta)] \operatorname{curl} \overline{\mathbf{w}_C} \cdot \mathbf{n}_C \Big) \,,
\end{aligned}
$$

we have

$$
\begin{aligned}
&\mathrm{Re}\,\mathcal{A}_{(\omega\neq 0)}[(\mathbf{w}_C,Q_C,\eta),(\mathbf{w}_C,Q_C,\eta)]\\
&\quad=\int_{\Omega_C}(\boldsymbol{\mu}_C^{-1}\,\mathrm{curl}\,\mathbf{w}_C\cdot\mathrm{curl}\,\overline{\mathbf{w}_C}+\mu_*^{-1}|\,\mathrm{div}\,\mathbf{w}_C|^2)\\
&\qquad+\int_\Gamma[\tfrac{1}{\mu_0}\mathcal{S}(\mathrm{curl}\,\mathbf{w}_C\cdot\mathbf{n}_C)\;\mathrm{curl}\,\overline{\mathbf{w}_C}\cdot\mathbf{n}_C+\mu_0\mathcal{H}(\eta)\,\overline{\eta}],
\end{aligned}
$$

and

$$
\begin{aligned}
&\mathrm{Im}\,\mathcal{A}_{(\omega\neq 0)}[(\mathbf{w}_C,Q_C,\eta),(\mathbf{w}_C,Q_C,\eta)]\\
&\quad=\omega^{-1}\int_{\Omega_C}\boldsymbol{\sigma}(i\omega\mathbf{w}_C+\mathrm{grad}\,Q_C)\cdot(-i\omega\overline{\mathbf{w}_C}+\mathrm{grad}\,\overline{Q_C})\\
&\qquad-2\,\mathrm{Im}\,\int_\Gamma[\tfrac{1}{2}\eta+\mathcal{D}(\eta)]\;\mathrm{curl}\,\overline{\mathbf{w}_C}\cdot\mathbf{n}_C\,.
\end{aligned}
$$

Hence, for a suitable constant $\kappa_3>0$, independent of ω, we find

$$
\begin{aligned}
&\mathrm{Re}\,\mathcal{A}_{(\omega\neq 0)}[(\mathbf{w}_C,Q_C,\eta),(\mathbf{w}_C,Q_C,\eta)]\\
&\quad\geq\kappa_3\Big(\int_{\Omega_C}(|\,\mathrm{curl}\,\mathbf{w}_C|^2+|\,\mathrm{div}\,\mathbf{w}_C|^2)+||\,\mathrm{curl}\,\mathbf{w}_C\cdot\mathbf{n}_C||^2_{-1/2,\Gamma}+||\eta||^2_{1/2,\Gamma}\Big)\,,
\end{aligned}
$$

and moreover, taking into account that the operator $\mathcal{D}$ is continuous from $H^{1/2}(\Gamma)$ into itself, for a suitable constant $C_1>0$, independent of ω, we obtain

$$
\begin{aligned}
&\Big|2\,\mathrm{Im}\,\int_\Gamma[\tfrac{1}{2}\eta+\mathcal{D}(\eta)]\;\mathrm{curl}\,\overline{\mathbf{w}_C}\cdot\mathbf{n}_C\Big|\leq C_1\,||\eta||_{1/2,\Gamma}\,||\,\mathrm{curl}\,\mathbf{w}_C\cdot\mathbf{n}_C||_{-1/2,\Gamma}\\
&\quad\leq\tfrac{C_1}{2}||\eta||^2_{1/2,\Gamma}+\tfrac{C_1}{2}||\,\mathrm{curl}\,\mathbf{w}_C\cdot\mathbf{n}_C||^2_{-1/2,\Gamma}\,.
\end{aligned}
$$

Hence, proceeding as in the proof of the coerciveness of the sesquilinear form $\mathcal{A}[\cdot,\cdot]$ in Section 6.1.2, we find, for each $0<\tau\leq 1$,

$$
\begin{aligned}
&|\mathrm{Im}\,\mathcal{A}_{(\omega\neq 0)}[(\mathbf{w}_C,Q_C,\eta),(\mathbf{w}_C,Q_C,\eta)]|\\
&\quad\geq\tau|\mathrm{Im}\,\mathcal{A}_{(\omega\neq 0)}[(\mathbf{w}_C,Q_C,\eta),(\mathbf{w}_C,Q_C,\eta)]|\\
&\quad\geq\tfrac{1}{2}\tau|\omega|^{-1}\sigma_{\min}\int_{\Omega_C}|\,\mathrm{grad}\,Q_C|^2-\tau|\omega|\sigma_{\min}\int_{\Omega_C}|\mathbf{w}_C|^2\\
&\qquad-\tau\tfrac{C_1}{2}||\eta||^2_{1/2,\Gamma}-\tau\tfrac{C_1}{2}||\,\mathrm{curl}\,\mathbf{w}_C\cdot\mathbf{n}_C||^2_{-1/2,\Gamma}\,,
\end{aligned}
$$

where $\sigma_{\min}$ is a uniform lower bound in Ω_C for the minimum eigenvalues of $\boldsymbol{\sigma}(\mathbf{x})$.

Let us recall now the Poincaré inequalities (6.38) and (6.39): there exist constants $\kappa_5>0$ and $\kappa_6>0$ such that

$$
\int_{\Omega_C}|\,\mathrm{grad}\,Q_C|^2\geq\kappa_5\int_{\Omega_C}(|\,\mathrm{grad}\,Q_C|^2+|Q_C|^2)
$$

for all $Q_C\in H^1(\Omega_C)$ with $\int_{\Omega_C}Q_C=0$, and

$$
\begin{aligned}
&\int_{\Omega_C}(|\,\mathrm{curl}\,\mathbf{w}_C|^2+|\,\mathrm{div}\,\mathbf{w}_C|^2)\\
&\qquad\geq\kappa_6\int_{\Omega_C}(|\,\mathrm{curl}\,\mathbf{w}_C|^2+|\,\mathrm{div}\,\mathbf{w}_C|^2+|\mathbf{w}_C|^2)
\end{aligned}
$$

for all $\mathbf{w}_C\in W_C$. Coerciveness follows by choosing τ small enough to have $\tau|\omega|\sigma_{\min}<\kappa_3\kappa_6$ and $\tau\frac{C_1}{2}<\kappa_3$. In particular, we have $\tau=O(1)$ for $0<|\omega|<1$ and $\tau=O(|\omega|^{-1})$ for $|\omega|\geq 1$. Thus the constant κ in (7.25) can be clearly chosen independent of ω, and the constant $\chi(\omega)$ is $O(|\omega|^{-1})$ for $0<|\omega|<1$ and $O(\omega^{-2})$ for $|\omega|\geq 1$.

In the case $\omega = 0$, the sesquilinear form satisfies

$$
\begin{aligned}
&\mathcal{A}_{(\omega=0)}[(\mathbf{w}_C, Q_C, \eta), (\mathbf{w}_C, Q_C, \eta)] \\
&\quad = \int_{\Omega_C} (\boldsymbol{\mu}_C^{-1} \operatorname{curl} \mathbf{w}_C \cdot \operatorname{curl} \overline{\mathbf{w}_C} + \mu_*^{-1} |\operatorname{div} \mathbf{w}_C|^2 \\
&\qquad\quad + \boldsymbol{\sigma} \operatorname{grad} Q_C \cdot \overline{\mathbf{w}_C} + \beta \boldsymbol{\sigma} \operatorname{grad} Q_C \cdot \operatorname{grad} \overline{Q_C}) \\
&\qquad + \int_\Gamma [-\tfrac{1}{2}\eta - \mathcal{D}(\eta)] \operatorname{curl} \overline{\mathbf{w}_C} \cdot \mathbf{n}_C \\
&\qquad + \int_\Gamma [\tfrac{1}{2} \operatorname{curl} \mathbf{w}_C \cdot \mathbf{n}_C + \mathcal{D}'(\operatorname{curl} \mathbf{w}_C \cdot \mathbf{n}_C)] \overline{\eta} \\
&\qquad + \int_\Gamma [\tfrac{1}{\mu_0} \mathcal{S}(\operatorname{curl} \mathbf{w}_C \cdot \mathbf{n}_C) \operatorname{curl} \overline{\mathbf{w}_C} \cdot \mathbf{n}_C + \mu_0 \mathcal{H}(\eta)\, \overline{\eta}] \,.
\end{aligned}
$$

We split $\int_{\Omega_C} \boldsymbol{\sigma} \operatorname{grad} Q_C \cdot \overline{\mathbf{w}_C}$ into its real and imaginary part, and, for each $\delta > 0$ and suitable constants $\kappa_7 > 0$ and $C_2 > 0$, we end up with

$$
\begin{aligned}
&\left|\operatorname{Re} \mathcal{A}_{(\omega=0)}[(\mathbf{w}_C, Q_C, \eta), (\mathbf{w}_C, Q_C, \eta)]\right| \\
&\quad \geq \kappa_7 \big(\int_{\Omega_C} (|\operatorname{curl} \mathbf{w}_C|^2 + |\operatorname{div} \mathbf{w}_C|^2 + \beta |\operatorname{grad} Q_C|^2) \\
&\qquad\qquad + ||\operatorname{curl} \mathbf{w}_C \cdot \mathbf{n}_C||^2_{-1/2,\Gamma} + ||\eta||^2_{1/2,\Gamma} \big) \\
&\qquad - C_2 \delta^{-1} \int_{\Omega_C} |\operatorname{grad} Q_C|^2 - \delta \int_{\Omega_C} |\mathbf{w}_C|^2 \,,
\end{aligned}
$$

thus the conclusion follows by choosing δ so small that $\kappa_7 \kappa_6 - \delta > 0$, and then β large enough to have $\kappa_7 \beta - C_2 \delta^{-1} > 0$. □

7.4 Stability as ω goes to 0

We are now interested in showing that the solution to problem (7.12) is stable with respect to ω, namely, if we denote by $(\mathbf{A}_C^\omega, V_C^\omega, q^\omega)$ the solution to (7.12) corresponding to the angular frequency ω, we have $(\mathbf{A}_C^\omega, V_C^\omega, q^\omega) \to (\mathbf{A}_C^0, V_C^0, q^0)$ as $\omega \to 0$.

Theorem 7.6. *There exists a constant $K > 0$, independent of ω, such that for each ω with $0 < |\omega| < 1$, the solutions to (7.12) satisfy*

$$
\begin{aligned}
&\int_{\Omega_C} (|\mathbf{A}_C^\omega - \mathbf{A}_C^0|^2 + |\operatorname{curl} \mathbf{A}_C^\omega - \operatorname{curl} \mathbf{A}_C^0|^2) \leq K\, \omega^2 \\
&\int_{\Omega_C} (|V_C^\omega - V_C^0|^2 + |\operatorname{grad} V_C^\omega - \operatorname{grad} V_C^0|^2) \leq K\, \omega^2 \\
&||q^\omega - q^0||^2_{1/2,\Gamma} \leq K\, \omega^2 \,,
\end{aligned}
$$

having chosen V_C^ω, V_C^0, q^ω and q^0 such that $\int_{\Omega_C} V_C^\omega = \int_{\Omega_C} V_C^0 = 0$ and $\int_\Gamma q^\omega = \int_\Gamma q^0 = 0$.

Proof. By linearity, the difference $(\mathbf{Z}_C, N_C, p) := (\mathbf{A}_C^\omega, V_C^\omega, q^\omega) - (\mathbf{A}_C^0, V_C^0, q^0)$ satisfies

$$
\begin{aligned}
&\int_{\Omega_C} (\boldsymbol{\mu}_C^{-1} \operatorname{curl} \mathbf{Z}_C \cdot \operatorname{curl} \overline{\mathbf{w}_C} + \mu_*^{-1} \operatorname{div} \mathbf{Z}_C \operatorname{div} \overline{\mathbf{w}_C}) \\
&\qquad + \int_{\Omega_C} (i\omega\boldsymbol{\sigma} \mathbf{Z}_C \cdot \overline{\mathbf{w}_C} + \boldsymbol{\sigma} \operatorname{grad} N_C \cdot \overline{\mathbf{w}_C}) \\
&\qquad + \int_\Gamma [-\tfrac{1}{2} p - \mathcal{D}(p) + \tfrac{1}{\mu_0} \mathcal{S}(\operatorname{curl} \mathbf{Z}_C \cdot \mathbf{n}_C)] \operatorname{curl} \overline{\mathbf{w}_C} \cdot \mathbf{n}_C \\
&\quad = -\int_{\Omega_C} i\omega\boldsymbol{\sigma} \mathbf{A}_C^0 \cdot \overline{\mathbf{w}_C} \\
&\int_{\Omega_C} (-\boldsymbol{\sigma} \mathbf{Z}_C \cdot \operatorname{grad} \overline{Q_C} + i\omega^{-1} \boldsymbol{\sigma} \operatorname{grad} N_C \cdot \operatorname{grad} \overline{Q_C}) \\
&\quad = \int_{\Omega_C} \boldsymbol{\sigma} \mathbf{A}_C^0 \cdot \operatorname{grad} \overline{Q_C} \\
&\int_\Gamma [\tfrac{1}{2} \operatorname{curl} \mathbf{Z}_C \cdot \mathbf{n}_C + \mathcal{D}'(\operatorname{curl} \mathbf{Z}_C \cdot \mathbf{n}_C) + \mu_0 \mathcal{H}(p)] \overline{\eta} = 0
\end{aligned}
\tag{7.28}
$$

(here, we have $\operatorname{div} \mathbf{Z}_C = 0$ in Ω_C by Lemma 7.1; however, we prefer to write everything in terms of the sesquilinear form $\mathcal{A}_{(\omega \neq 0)}[\cdot, \cdot]$).

Therefore, from the coerciveness of $\mathcal{A}_{(\omega \neq 0)}[\cdot, \cdot]$ and taking into account that $0 < |\omega| < 1$, from (7.25) we obtain at once that

$$
\begin{aligned}
&\int_{\Omega_C} (|\mathbf{Z}_C|^2 + |\operatorname{curl} \mathbf{Z}_C|^2 + |\operatorname{div} \mathbf{Z}_C|^2) \\
&\qquad + ||p||_{1/2,\Gamma}^2 + \chi(\omega) \int_{\Omega_C} (|N_C|^2 + |\operatorname{grad} N_C|^2) \\
&\quad \leq \kappa^{-1} c_1 \Big[|\omega| \Big(\int_{\Omega_C} |\mathbf{A}_C^0|^2 \Big)^{1/2} \Big(\int_{\Omega_C} |\mathbf{Z}_C|^2 \Big)^{1/2} \\
&\qquad\qquad + \Big(\int_{\Omega_C} |\mathbf{A}_C^0|^2 \Big)^{1/2} \Big(\int_{\Omega_C} |\operatorname{grad} N_C|^2 \Big)^{1/2} \Big] \\
&\quad \leq \kappa^{-1} c_2 |\omega|^2 \alpha_1^{-1} \int_{\Omega_C} |\mathbf{A}_C^0|^2 + \kappa^{-1} c_2 \alpha_2^{-1} \int_{\Omega_C} |\mathbf{A}_C^0|^2 \\
&\qquad + \alpha_1 \int_{\Omega_C} |\mathbf{Z}_C|^2 + \alpha_2 \int_{\Omega_C} |\operatorname{grad} N_C|^2
\end{aligned}
\tag{7.29}
$$

for each $\alpha_1 > 0$ and $\alpha_2 > 0$. Choosing $\alpha_1 = 1/2$ and $\alpha_2 = \chi(\omega)/2 = O(|\omega|^{-1})$ (see Theorem 7.5), we have that the left hand side in (7.29) is $O(|\omega|)$. In particular,

$$
\int_{\Omega_C} (|N_C|^2 + |\operatorname{grad} N_C|^2) = [\chi(\omega)]^{-1} O(|\omega|) = O(\omega^2),
$$

and

$$
\int_{\Omega_C} (|\mathbf{Z}_C|^2 + |\operatorname{curl} \mathbf{Z}_C|^2 + |\operatorname{div} \mathbf{Z}_C|^2) + ||p||_{1/2,\Gamma}^2 = O(|\omega|) .
$$

Rewriting the first equation in (7.28) as

$$
\begin{aligned}
&\int_{\Omega_C} (\boldsymbol{\mu}_C^{-1} \operatorname{curl} \mathbf{Z}_C \cdot \operatorname{curl} \overline{\mathbf{w}_C} + \mu_*^{-1} \operatorname{div} \mathbf{Z}_C \operatorname{div} \overline{\mathbf{w}_C}) \\
&\qquad + \int_\Gamma [-\tfrac{1}{2} p - \mathcal{D}(p) + \tfrac{1}{\mu_0} \mathcal{S}(\operatorname{curl} \mathbf{Z}_C \cdot \mathbf{n}_C)] \operatorname{curl} \overline{\mathbf{w}_C} \cdot \mathbf{n}_C \\
&\quad = -\int_{\Omega_C} (i\omega\boldsymbol{\sigma} \mathbf{Z}_C \cdot \overline{\mathbf{w}_C} + \boldsymbol{\sigma} \operatorname{grad} N_C \cdot \overline{\mathbf{w}_C}) - \int_{\Omega_C} i\omega\boldsymbol{\sigma} \mathbf{A}_C^0 \cdot \overline{\mathbf{w}_C} ,
\end{aligned}
$$

using $(7.28)_3$ and proceeding as in the proof of Theorem 7.5, we obtain that the sesquilinear form at the left hand side is coercive (with coerciveness constant $K_0 > 0$ independent of ω). Hence we have

$$
\begin{aligned}
&\int_{\Omega_C} (|\mathbf{Z}_C|^2 + |\operatorname{curl} \mathbf{Z}_C|^2 + |\operatorname{div} \mathbf{Z}_C|^2) + ||\operatorname{curl} \mathbf{Z}_C \cdot \mathbf{n}_C||_{-1/2,\Gamma}^2 + ||p||_{1/2,\Gamma}^2 \\
&\quad \leq K_0^{-1} c_3 \Big[|\omega| \int_{\Omega_C} |\mathbf{Z}_C|^2 + \Big(\int_{\Omega_C} |\operatorname{grad} N_C|^2 \Big)^{1/2} \Big(\int_{\Omega_C} |\mathbf{Z}_C|^2 \Big)^{1/2} \\
&\qquad\qquad + |\omega| \Big(\int_{\Omega_C} |\mathbf{A}_C^0|^2 \Big)^{1/2} \Big(\int_{\Omega_C} |\mathbf{Z}_C|^2 \Big)^{1/2} \Big] \\
&\quad = O(\omega^2) + O(|\omega|) \Big(\int_{\Omega_C} |\mathbf{Z}_C|^2 \Big)^{1/2} \leq O(\omega^2) + \tfrac{1}{2} \int_{\Omega_C} |\mathbf{Z}_C|^2 .
\end{aligned}
$$

The result thus follows. □

Remark 7.7. As we recalled in Section 2.3.1, an analysis of the asymptotic behaviour of the solution of the eddy current model with respect to $\omega \to 0$ has been presented in Ammari et al. [23]. In particular they prove, by a formal asymptotic expansion, that the electric and the magnetic fields solutions of the eddy current problem converge linearly to the corresponding solutions of the static problem in the L^2-norm. Expressing the electric and magnetic fields in terms of $\mathbf{A}_C$, V_C and ψ_I, it can be easily checked that Theorem 7.6 is in agreement with their result. □

7.5 Numerical approximation

In this section we deal with the numerical approximation of problem (7.12). In the sequel we assume that Ω_C is a Lipschitz polyhedral domain, and that $\mathcal{T}_{C,h}$ and $\mathcal{T}_{\Gamma,h}$ are two regular families of triangulations of Ω_C and Γ, respectively. For the sake of simplicity, we suppose that each element K of $\mathcal{T}_{C,h}$ is a tetrahedron and each element T of $\mathcal{T}_{\Gamma,h}$ is a triangle; however, the results below also hold for hexahedral and rectangular elements, respectively. Let us note that the mesh induced on Γ by $\mathcal{T}_{C,h}$ is not assumed to coincide with $\mathcal{T}_{\Gamma,h}$.

Let $\mathbb{P}_k$, $k \geq 1$, be the space of polynomials of degree less than or equal to k. For $r \geq 1$, $s \geq 1$ and $t \geq 1$ we employ the discrete spaces given by Lagrange nodal elements

$$\begin{aligned} W^r_{C,h} := \{\mathbf{w}_{C,h} \in (C^0(\Omega_C))^3 \\ \mid \mathbf{w}_{C,h|K} \in (\mathbb{P}_r)^3 \ \forall\, K \in \mathcal{T}_h \ , \ \mathbf{w}_{C,h} \cdot \mathbf{n}_C = 0 \text{ on } \Gamma\} \ , \end{aligned}$$

$$L^s_{C,h} := \{Q_{C,h} \in C^0(\Omega_C) \mid Q_{C,h|K} \in \mathbb{P}_s \ \forall\, K \in \mathcal{T}_{C,h}\} \ ,$$

and

$$L^t_{\Gamma,h} := \{\eta_h \in C^0(\Gamma) \mid \eta_{h|T} \in \mathbb{P}_t \ \forall\, T \in \mathcal{T}_{\Gamma,h}\} \ .$$

Clearly, we have $W^r_{C,h} \subset W_C$, $L^s_{C,h} \subset H^1(\Omega_C)$ and $L^t_{\Gamma,h} \subset H^{1/2}(\Gamma)$, therefore we are ready to consider a conforming finite element approximation.

The discrete problem is given by

$$\begin{aligned}
&\text{Find } (\mathbf{A}_{C,h}, V_{C,h}, q_h) \in W^r_{C,h} \times L^s_{C,h}/\mathbb{C} \times L^t_{\Gamma,h}/\mathbb{C} \text{ such that} \\
&\quad \textstyle\int_{\Omega_C} (\boldsymbol{\mu}_C^{-1} \operatorname{curl} \mathbf{A}_{C,h} \cdot \operatorname{curl} \overline{\mathbf{w}_{C,h}} + \mu_*^{-1} \operatorname{div} \mathbf{A}_{C,h} \operatorname{div} \overline{\mathbf{w}_{C,h}} \\
&\qquad\qquad + i\omega\boldsymbol{\sigma} \mathbf{A}_{C,h} \cdot \overline{\mathbf{w}_{C,h}} + \boldsymbol{\sigma} \operatorname{grad} V_{C,h} \cdot \overline{\mathbf{w}_{C,h}}) \\
&\qquad + \textstyle\int_\Gamma [-\frac{1}{2} q_h - \mathcal{D}(q_h) + \frac{1}{\mu_0} \mathcal{S}(\operatorname{curl} \mathbf{A}_{C,h} \cdot \mathbf{n}_C)] \operatorname{curl} \overline{\mathbf{w}_{C,h}} \cdot \mathbf{n}_C \\
&\qquad = \textstyle\int_{\Omega_C} \mathbf{J}_{e,C} \cdot \overline{\mathbf{w}_{C,h}} \\
&\quad \textstyle\int_{\Omega_C} (i\omega\boldsymbol{\sigma} \mathbf{A}_{C,h} \cdot \operatorname{grad} \overline{Q_{C,h}} + \boldsymbol{\sigma} \operatorname{grad} V_{C,h} \cdot \operatorname{grad} \overline{Q_{C,h}}) \\
&\qquad = \textstyle\int_{\Omega_C} \mathbf{J}_{e,C} \cdot \operatorname{grad} \overline{Q_{C,h}} \\
&\quad \textstyle\int_\Gamma [\frac{1}{2} \operatorname{curl} \mathbf{A}_{C,h} \cdot \mathbf{n}_C + \mathcal{D}'(\operatorname{curl} \mathbf{A}_{C,h} \cdot \mathbf{n}_C) + \mu_0 \mathcal{H}(q_h)] \overline{\eta_h} = 0 \\
&\text{for all } (\mathbf{w}_{C,h}, Q_{C,h}, \eta_h) \in W^r_{C,h} \times L^s_{C,h}/\mathbb{C} \times L^t_{\Gamma,h}/\mathbb{C} \ .
\end{aligned} \tag{7.30}$$

Existence and uniqueness of the discrete solution follow by the Lax–Milgram lemma, applied in $W^r_{C,h} \times L^s_{C,h}/\mathbb{C} \times L^t_{\Gamma,h}/\mathbb{C}$.

We also have

Theorem 7.8. *Assume that Ω_C is a convex polyhedron, or else that the solution $(\mathbf{A}_C, V_C, q)$ is smooth enough. Then the discrete solution $(\mathbf{A}_{C,h}, V_{C,h}, q_h)$ converges in $W_C \times H^1(\Omega_C)/\mathbb{C} \times H^{1/2}(\Gamma)/\mathbb{C}$ to the exact solution $(\mathbf{A}_C, V_C, q)$.*

Proof. Let us start noting that, as proved in Lemma 7.2, $(\mathbf{A}_C, V_C, q)$ and $(\mathbf{A}_{C,h}, V_{C,h}, q_h)$ are solutions to problems (7.12) and (7.30), respectively, also for all test functions $\eta \in H^{1/2}(\Gamma)$ and $\eta_h \in L^t_{\Gamma,h}$. Similarly, it is obvious that (7.12) and (7.30) also hold for all test functions $Q_C \in H^1(\Omega_C)$ and $Q_{C,h} \in L^s_{C,h}$.

Therefore, finite element interpolants can be used as test functions, and, if the solution $(\mathbf{A}, V_C, q)$ is smooth enough, the convergence follows by applying Céa lemma and standard interpolation results.

If the domain Ω_C is convex, it is known (see Costabel et al. [89]) that smooth functions with vanishing normal component are dense in W_C, and the same arguments can be applied. □

Remark 7.9. As noted in Remark 6.6, if Ω_C is a non-convex polyhedral domain it can happen that the solution $\mathbf{A}_C$ is non-smooth, namely, not even an element of $(H^1(\Omega_C))^3$, and that $H^1_\tau(\Omega_C) := (H^1(\Omega_C))^3 \cap H_0(\mathrm{div}; \Omega_C)$ is a closed proper subspace of W_C. Since the finite element space $W^r_{C,h}$ is contained in $H^1_\tau(\Omega_C)$, in that case a convergence result in W_C cannot hold. For non-convex domains, an alternative approch is presented in Section 7.5.1. □

The determination of a precise order of convergence requires the knowledge of the regularity of the solution: as usual, if $(\mathbf{A}_C, V_C, q) \in H^{k+1}(\Omega_C) \times H^{k+1}(\Omega_C) \times H^{k+1/2}(\Gamma)$, where the integer $k \geq 1$ is equal to $r = s = t$, the degree of polynomial approximation, we have

$$\Big(\int_{\Omega_C} (|\mathbf{A}_C - \mathbf{A}_{C,h}|^2 + |\,\mathrm{curl}(\mathbf{A}_C - \mathbf{A}_{C,h})|^2 + |\,\mathrm{div}(\mathbf{A}_C - \mathbf{A}_{C,h})|^2) \\ + \int_{\Omega_C} (|V_C - V_{C,h}|^2 + |\,\mathrm{grad}(V_C - V_{C,h})|^2) + ||q - q_h||^2_{1/2,\Gamma} \Big)^{1/2} = O(h^k)\,,$$

having chosen V_C, $V_{C,h}$, q and q_h such that $\int_{\Omega_C} V_C = \int_{\Omega_C} V_{C,h} = 0$ and $\int_\Gamma q = \int_\Gamma q_h = 0$.

On the other hand, in EEG and MEG applications a typical assumption for $\boldsymbol{\sigma}$, the human head conductivity, is that it is a piecewise-smooth (but not globally continuous) positive definite symmetric matrix. In this case, it is not clear if the solution is regular as required above. In general, one could expect that the solution belongs to $H^{1+\gamma}(\Omega_C) \times H^{1+\gamma}(\Omega_C) \times H^{1/2+\gamma}(\Gamma)$ for some γ with $0 < \gamma < 1/2$; however, we do not know a proof of this result.

It is worth noting that the same difficulty arises if one assumes $\omega = 0$, namely, one just considers the problem of electrostatics. In this case one has to approximate the solution V_C (determined up to an additive constant) of

$$\begin{cases} \mathrm{div}(\boldsymbol{\sigma}\,\mathrm{grad}\, V_C) = \mathrm{div}\,\mathbf{J}_{e,C} & \text{in } \Omega_C \\ \boldsymbol{\sigma}\,\mathrm{grad}\, V_C \cdot \mathbf{n}_C = \mathbf{J}_{e,C} \cdot \mathbf{n}_C & \text{on } \Gamma\,, \end{cases}$$

and the regularity of V_C is not easily determined for a piecewise-smooth positive definite symmetric matrix $\boldsymbol{\sigma}$. Therefore, also in this case the rate of convergence of a finite element approximation scheme is not easily determined.

Concerning the behaviour with respect to the angular frequency ω, in the discrete case we can repeat the proof of Theorem 7.6 and obtain (with obvious notation):

Theorem 7.10. *There exists a constant* $K > 0$, *independent of* ω *and* h, *such that for each* ω *with* $0 < |\omega| < 1$, *the solutions to* (7.30) *satisfy*

$$\int_{\Omega_C} (|\mathbf{A}^{\omega}_{C,h} - \mathbf{A}^{0}_{C,h}|^2 + |\operatorname{curl} \mathbf{A}^{\omega}_{C,h} - \operatorname{curl} \mathbf{A}^{0}_{C,h}|^2 \\ + |\operatorname{div} \mathbf{A}^{\omega}_{C,h} - \operatorname{div} \mathbf{A}^{0}_{C,h}|^2) \leq K\,\omega^2$$

$$\int_{\Omega_C} (|V^{\omega}_{C,h} - V^{0}_{C,h}|^2 + |\operatorname{grad} V^{\omega}_{C,h} - \operatorname{grad} V^{0}_{C,h}|^2) \leq K\,\omega^2$$

$$||q^{\omega}_h - q^0_h||^2_{1/2,\Gamma} \leq K\,\omega^2 \,,$$

having chosen $V^{\omega}_{C,h}, V^{0}_{C,h}, q^{\omega}_h$ and q^0_h such that $\int_{\Omega_C} V^{\omega}_{C,h} = \int_{\Omega_C} V^{0}_{C,h} = 0$ and $\int_\Gamma q^{\omega}_h = \int_\Gamma q^0_h = 0$.

An important point of the above result is that the behaviour in ω is uniform with respect to h; it is not evident that this is true for other finite element approximation schemes, as it is not always possible to show that the associated sesquilinear form is coercive uniformly with respect to ω (for our approach, this has been proved in Theorem 7.5).

Remark 7.11. A delicate point of the discretization is the efficient computation of the terms involving the single layer and double layer potentials and the hypersingular integral operator: an extensive literature is devoted to analyze this problem.

By integration by parts it is possible to restrict the problem to the computation of terms of the form

$$\int_{T\times T'} \frac{1}{|\mathbf{x}-\mathbf{y}|}\, p(\mathbf{y})q(\mathbf{x})\; dS_y dS_x$$

or

$$\int_{T\times T'} \frac{\mathbf{x}-\mathbf{y}}{|\mathbf{x}-\mathbf{y}|^3} \cdot \mathbf{n}(\mathbf{y})\; p(\mathbf{y})q(\mathbf{x})\; dS_y dS_x \,,$$

where p, q are polynomials and T, T' are triangles of the mesh $\mathcal{T}_{\Gamma,h}$. If $T \cap T' = \emptyset$ the integrands are regular functions and standard cubature methods can be used. On the other hand, if $T = T'$ or $T \cap T'$ is an edge or a vertex the integrands have a singular behavior. As indicated in Börm and Hackbusch [56], different techniques can be applied to evaluate these terms. One possibility is to use quadrature rules adapted to the singularity of the kernel (see Schwab and Wendland [224]). Another possibility is to apply a suitable regularizing coordinate transformation that renders regular the integrand, and then to use standard cubature formulas (see Duffy [98], Erichsen and Sauter [101], Sauter and Lage [221]). Finally, semi-analytical approaches apply an exact integration at least for the inner integral (see Sauter and Schwab [222], Gray et al. [113]). □

7.5.1 The non-convex case

As we noted in Remark 7.9, if the conductor Ω_C is a polyhedral non-convex set it can happen that the convergence of the finite element approximation does not hold. Therefore, it is suitable to follow an alternative approach.

We start by recalling that, when the conductor has a complex geometry, it is usual to enclose it into a "simpler" set, and in this new region to look for a vector potential of the magnetic induction. This procedure, that is generally called the $(\mathbf{A}_C, V_C)-\mathbf{A}_I-\psi_I$ formulation, has been described in Section 6.3.

In our case, we assume that the conductor $\overline{\Omega_C}$ is included into a *polyhedral convex* bounded open set Ω_A, as small as possible. Setting now $\Omega_I := \mathbb{R}^3 \setminus \overline{\Omega_A}$, $\Gamma_A := \partial\Omega_A$,

$$W_A := H(\mathrm{curl}; \Omega_A) \cap H_0(\mathrm{div}; \Omega_A) ,$$

and denoting by $\mathbf{n}_A$ the unit outward normal vector on Γ_A, the weak formulation reads

$$\begin{cases} \text{Find } (\mathbf{A}, V_C, q) \in W_A \times H^1(\Omega_C)/\mathbb{C} \times H^{1/2}(\Gamma_A)/\mathbb{C} \text{ such that} \\ \int_{\Omega_A} (\boldsymbol{\mu}^{-1} \,\mathrm{curl}\, \mathbf{A} \cdot \mathrm{curl}\, \overline{\mathbf{w}} + \mu_*^{-1} \,\mathrm{div}\, \mathbf{A} \,\mathrm{div}\, \overline{\mathbf{w}}) \\ \quad + \int_{\Omega_C} (i\omega\boldsymbol{\sigma} \mathbf{A}_C \cdot \overline{\mathbf{w}_C} + \boldsymbol{\sigma} \,\mathrm{grad}\, V_C \cdot \overline{\mathbf{w}_C}) \\ \quad + \int_{\Gamma_A} [-\frac{1}{2} q - \mathcal{D}(q) + \frac{1}{\mu_0} \mathcal{S}(\mathrm{curl}\, \mathbf{A} \cdot \mathbf{n}_A)] \,\mathrm{curl}\, \overline{\mathbf{w}} \cdot \mathbf{n}_A \\ \quad = \int_{\Omega_C} \mathbf{J}_{e,C} \cdot \overline{\mathbf{w}_C} \\ \int_{\Omega_C} (i\omega\boldsymbol{\sigma} \mathbf{A}_C \cdot \mathrm{grad}\, \overline{Q_C} + \boldsymbol{\sigma} \,\mathrm{grad}\, V_C \cdot \mathrm{grad}\, \overline{Q_C}) \\ \quad = \int_{\Omega_C} \mathbf{J}_{e,C} \cdot \mathrm{grad}\, \overline{Q_C} \\ \int_{\Gamma_A} [\frac{1}{2} \mathrm{curl}\, \mathbf{A} \cdot \mathbf{n}_A + \mathcal{D}'(\mathrm{curl}\, \mathbf{A} \cdot \mathbf{n}_A) + \mu_0 \mathcal{H}(q)] \overline{\eta} = 0 \\ \text{for all } (\mathbf{w}, Q_C, \eta) \in W_A \times H^1(\Omega_C)/\mathbb{C} \times H^{1/2}(\Gamma_A)/\mathbb{C} . \end{cases} \tag{7.31}$$

The results presented in Section 7.3, as well as those in Sections 7.4 and 7.5, can be easily obtained also for this formulation, with essentially the same proofs. In particular, the finite element approximation scheme converges, as stated in Theorem 7.8, since the domain Ω_A is convex. All the details concerning this approach have been given in Alonso Rodríguez and Valli [19].

7.6 Other FEM–BEM approaches

Among the FEM–BEM formulations that we mentioned at the beginning of this chapter, in this section we briefly present those due to Bossavit and Vérité [62], [63], Meddahi and Selgas [176] and Hiptmair [127].

7.6.1 The code TRIFOU

The first authors who proposed a coupled FEM–BEM formulation of the eddy current problem are Bossavit and Vérité [62], [63]. Based on this coupled approach, they have also developed a popular numerical code, named TRIFOU, widely used at Electricité de France since 1980.

We recall that, for the sake of simplicity, we are assuming that Ω_C is simply-connected and that $\Omega_I = \mathbb{R}^3 \setminus \overline{\Omega_C}$ is connected, so that the boundary $\Gamma = \partial\Omega_C$ is connected. As a consequence, as in Chapter 5 we can write $\mathbf{H}_I = \operatorname{grad}\psi_I$.

As in (3.9), for each test function $\mathbf{v} \in H(\text{curl}; \mathbb{R}^3)$ with $\operatorname{curl}\mathbf{v}_I = \mathbf{0}$ in Ω_I we have

$$\int_{\Omega_C} \boldsymbol{\sigma}^{-1} \operatorname{curl}\mathbf{H}_C \cdot \operatorname{curl}\overline{\mathbf{v}_C} + \int_{\mathbb{R}^3} i\omega\boldsymbol{\mu}\mathbf{H}\cdot\overline{\mathbf{v}} = \int_{\Omega_C} \boldsymbol{\sigma}^{-1}\mathbf{J}_{e,C}\cdot\operatorname{curl}\overline{\mathbf{v}_C}\,. \tag{7.32}$$

On the other hand, writing $\mathbf{v}_I = \operatorname{grad}\chi_I$ and remembering that ψ_I is a harmonic function vanishing at infinity, we find by integration by parts

$$\begin{aligned}\textstyle\int_{\Omega_I} i\omega\mu_0\mathbf{H}_I\cdot\overline{\mathbf{v}_I} &= \textstyle\int_{\Omega_I} i\omega\mu_0 \operatorname{grad}\psi_I\cdot\operatorname{grad}\overline{\chi_I}\\ &= \textstyle\int_\Gamma i\omega\mu_0 \operatorname{grad}\psi_I\cdot\mathbf{n}_I\,\overline{\chi_I}\,.\end{aligned} \tag{7.33}$$

We introduce the linear and continuous Steklov–Poincaré operator $\mathcal{R}$ as

$$\mathcal{R}: H^{1/2}(\Gamma) \to H^{-1/2}(\Gamma)\ ,\ \ \mathcal{R}(\chi_\Gamma) := \operatorname{grad}\chi_I\cdot\mathbf{n}_I \quad \text{on } \Gamma\,, \tag{7.34}$$

where χ_I belongs to the space $W^1(\Omega_I)$ introduced in (7.17) and satisfies $\Delta\chi_I = 0$ in Ω_I and $\chi_{I|\Gamma} = \chi_\Gamma$. We also set

$$\begin{aligned}\widetilde{W} := \{(\mathbf{v}_C, \chi_\Gamma) \in H(\text{curl}; \Omega_C) \times H^{1/2}(\Gamma)&\\ \,|\, \mathbf{v}_C\times\mathbf{n}_C + \operatorname{grad}_\tau\chi_\Gamma\times\mathbf{n}_I = \mathbf{0} \text{ on } \Gamma\}&\,.\end{aligned} \tag{7.35}$$

We can thus rewrite the eddy current problem as

$$\begin{aligned}&\text{Find } (\mathbf{H}_C, \psi_\Gamma) \in \widetilde{W} \text{ such that}\\ &\quad \textstyle\int_{\Omega_C}\boldsymbol{\sigma}^{-1}\operatorname{curl}\mathbf{H}_C\cdot\operatorname{curl}\overline{\mathbf{v}_C} + \int_{\Omega_C} i\omega\boldsymbol{\mu}_C\mathbf{H}_C\cdot\overline{\mathbf{v}_C}\\ &\qquad\qquad \textstyle +i\omega\mu_0\int_\Gamma \mathcal{R}(\psi_\Gamma)\,\overline{\chi_\Gamma}\\ &\qquad \textstyle = \int_{\Omega_C}\boldsymbol{\sigma}^{-1}\mathbf{J}_{e,C}\cdot\operatorname{curl}\overline{\mathbf{v}_C}\\ &\text{for each } (\mathbf{v}_C, \chi_\Gamma) \in \widetilde{W}\,.\end{aligned} \tag{7.36}$$

By the trace inequality (A.8) and the Poincaré inequality in $W^1(\Omega_I)$ (see, e.g., Nédélec [187], Theor. 2.5.13) we have

$$\int_\Gamma \mathcal{R}(\chi_\Gamma)\,\overline{\chi_\Gamma} = \int_{\Omega_I} \operatorname{grad}\chi_I\cdot\operatorname{grad}\overline{\chi_I} \ge \kappa_0\|\chi_\Gamma\|_{1/2,\Gamma}\,,$$

hence the sesquilinear form $a_T(\cdot,\cdot)$ at the left hand side of (7.36) is clearly coercive in $\widetilde{W}$, and the problem is well-posed.

If one is interested in finding also the magnetic field in Ω_I, one has to set

$$\psi_I = \mathcal{D}_I(\psi_\Gamma) - \frac{1}{\mu_0}\mathcal{S}_I(\boldsymbol{\mu}_C\mathbf{H}_C\cdot\mathbf{n}_C)\,,$$

where the operators $\mathcal{S}_I$ and $\mathcal{D}_I$ have been introduced in (7.15) and (7.16).

When considering numerical approximation we assume that Ω_C is a Lipschitz polyhedral domain, and we denote by $\mathcal{T}_{C,h}$ a regular family of triangulations of Ω_C and by $\mathcal{T}_{\partial,h}$ the mesh induced on Γ by $\mathcal{T}_{C,h}$. We also suppose that each element K of $\mathcal{T}_{C,h}$ is a tetrahedron. We consider

$$\begin{aligned}\widetilde{W}_h := \{(\mathbf{v}_{C,h}, \chi_{\Gamma,h}) \in N^1_{C,h} \times C^0(\Gamma) \,|\, \chi_{\Gamma,h|T} \in \mathbb{P}_1 \; \forall\, T \in \mathcal{T}_{\partial,h}, \\ \mathbf{v}_{C,h} \times \mathbf{n}_C + \operatorname{grad}_\tau \chi_{\Gamma,h} \times \mathbf{n}_I = \mathbf{0} \text{ on } \Gamma\} \,,\end{aligned}$$

where $N^1_{C,h}$ is the space of Nédélec curl-conforming edge elements of the lowest order in Ω_C (see Section A.2).

Due to the constraint on Γ, any function $(\mathbf{v}_{C,h}, \chi_{\Gamma,h})$ in $\widetilde{W}_h$ can be clearly written as

$$\mathbf{v}_{C,h} = \sum_{e \in \mathcal{E}^0_{C,h}} \alpha_e \mathbf{q}_e + \sum_{v \in \mathcal{V}_{\Gamma,h}} \alpha_v \operatorname{grad} \varphi_v \;\; , \;\; \chi_{\Gamma,h} = \sum_{v \in \mathcal{V}_{\Gamma,h}} \alpha_v \varphi_v \;,$$

where $\mathcal{E}^0_{C,h}$ is the set of edges $e \in \mathcal{T}_{C,h}$ that are internal to Ω_C, $\mathcal{V}_{\Gamma,h}$ is the set of vertices $v \in \mathcal{T}_{\partial,h}$, and we have denoted by $\mathbf{q}_e$ the edge basis function defined in Ω_C and associated to the edge e, and by φ_v the nodal basis function defined in Ω_C and associated to the vertex v.

For a suitable implementation it is necessary to find a sound and computationally cheap approximation of the Steklov–Poincaré operator $\mathcal{R}$. Recalling the definition of the operators $\mathcal{S}$ in (7.6) and $\mathcal{S}_I$ in (7.15), we can write $\psi_I = \mathcal{S}_I(\lambda_\Gamma)$ and $\psi_\Gamma = \mathcal{S}(\lambda_\Gamma)$, where, as a consequence of well-known results in potential theory, $\lambda_\Gamma \in H^{-1/2}(\Gamma)$ satisfies

$$\operatorname{grad} \psi_I \cdot \mathbf{n}_I = \frac{1}{2}\lambda_\Gamma - \mathcal{D}'(\lambda_\Gamma) \qquad \text{on } \Gamma$$

(see, e.g., McLean [175], Nédélec [187]). Passing to a variational formulation, we are looking for $\psi_\Gamma \in H^{1/2}(\Gamma)$ and $\lambda_\Gamma \in H^{-1/2}(\Gamma)$ such that

$$\begin{aligned} &\textstyle\int_\Gamma \mathcal{S}(\lambda_\Gamma)\,\overline{\xi_\Gamma} = \int_\Gamma \psi_\Gamma\,\overline{\xi_\Gamma} \\ &\textstyle\int_\Gamma \mathcal{R}(\psi_\Gamma)\,\overline{\chi_\Gamma} - \int_\Gamma [\frac{1}{2}\lambda_\Gamma - \mathcal{D}'(\lambda_\Gamma)]\,\overline{\chi_\Gamma} \end{aligned}$$

for all $\chi_\Gamma \in H^{1/2}(\Gamma)$ and $\xi_\Gamma \in H^{-1/2}(\Gamma)$.

In matrix form we can write, with obvious notation,

$$\begin{aligned} &S\boldsymbol{\lambda}_\Gamma = B_\Gamma^T \boldsymbol{\psi}_\Gamma \\ &R\boldsymbol{\psi}_\Gamma = \tfrac{1}{2} B_\Gamma \boldsymbol{\lambda}_\Gamma - D'\boldsymbol{\lambda}_\Gamma \;, \end{aligned}$$

where the vector unknowns are complex-valued, while the matrices are real-valued (as we can choose real-valued finite element basis functions). We also see at once that the matrix S is symmetric and positive definite, hence we can rewrite

$$R = \Big(\frac{1}{2}B_\Gamma - D'\Big) S^{-1} B_\Gamma^T \,.$$

Unfortunately, this matrix is not symmetric, though the Steklov–Poincaré operator $\mathcal{R}$ is hermitian. Therefore, in the TRIFOU code the following symmetric matrix

$$R_\star := \frac{1}{2}(R + R^T)$$

has been proposed as an approximation of the operator $\mathcal{R}$.

Though a complete analysis of the convergence of the method is not available, the TRIFOU code has been used in many engineering applications with satisfactory results (for a deeper insight and additional comments, see Bossavit and Vérité [62], [63], Bossavit [57]; in particular, note that the matrix $R_\star$ may even happen to be singular: see Bossavit [59], p. 214).

7.6.2 An approach based on the magnetic field $\mathbf{H}_C$

In Meddahi and Selgas [176], following an approach that is close to that presented in the preceding section, the authors choose as unknowns $\mathbf{H}_C$ in Ω_C and $\boldsymbol{\mu}_C\mathbf{H}_C \cdot \mathbf{n}_C$ on Γ, and derive a symmetric formulation. Again we assume, for the sake of simplicity, that Ω_C is simply-connected and that the boundary $\Gamma = \partial\Omega_C$ is connected (for the general not simply-connected case, see Meddahi and Selgas [176]). Consequently, we can write $\mathbf{H}_I = \operatorname{grad}\psi_I$.

As before, we obtain (7.32) and (7.33), and, using the interface condition $\boldsymbol{\mu}_C\mathbf{H}_C \cdot \mathbf{n}_C + \mu_0 \operatorname{grad}\psi_I \cdot \mathbf{n}_I = 0$ on Γ, we also find

$$\begin{aligned}\int_{\Omega_I} i\omega\mu_0\mathbf{H}_I \cdot \overline{\mathbf{v}_I} &= \int_\Gamma i\omega\mu_0 \operatorname{grad}\psi_I \cdot \mathbf{n}_I \overline{\chi_I} \\ &= -\int_\Gamma i\omega\boldsymbol{\mu}_C\mathbf{H}_C \cdot \mathbf{n}_C \overline{\chi_I} \,.\end{aligned}$$

Furthermore, we can rewrite (7.10) and (7.11) as

$$\frac{1}{2}\psi_\Gamma - \mathcal{D}(\psi_\Gamma) + \frac{1}{\mu_0}\mathcal{S}(\boldsymbol{\mu}_C\mathbf{H}_C \cdot \mathbf{n}_C) = 0 \text{ on } \Gamma \tag{7.37}$$

$$\frac{1}{2\mu_0}\boldsymbol{\mu}_C\mathbf{H}_C \cdot \mathbf{n}_C + \frac{1}{\mu_0}\mathcal{D}'(\boldsymbol{\mu}_C\mathbf{H}_C \cdot \mathbf{n}_C) + \mathcal{H}(\psi_\Gamma) = 0 \text{ on } \Gamma \,. \tag{7.38}$$

Thus, setting $\chi_\Gamma := \chi_{I|\Gamma}$, we easily find

$$\begin{aligned}&\int_{\Omega_C} \boldsymbol{\sigma}^{-1} \operatorname{curl}\mathbf{H}_C \cdot \operatorname{curl}\overline{\mathbf{v}_C} + \int_{\Omega_C} i\omega\boldsymbol{\mu}_C\mathbf{H}_C \cdot \overline{\mathbf{v}_C} \\ &\qquad + i\omega \int_\Gamma [-\tfrac{1}{2}\boldsymbol{\mu}_C\mathbf{H}_C \cdot \mathbf{n}_C + \mathcal{D}'(\boldsymbol{\mu}_C\mathbf{H}_C \cdot \mathbf{n}_C) + \mu_0\mathcal{H}(\psi_\Gamma)]\,\overline{\chi_\Gamma} \\ &= \int_{\Omega_C} \boldsymbol{\sigma}^{-1}\mathbf{J}_{e,C} \cdot \operatorname{curl}\overline{\mathbf{v}_C} \,,\end{aligned} \tag{7.39}$$

and for any test function ξ on Γ we also have

$$\int_\Gamma [\frac{1}{2}\psi_\Gamma - \mathcal{D}(\psi_\Gamma) + \frac{1}{\mu_0}\mathcal{S}(\boldsymbol{\mu}_C\mathbf{H}_C \cdot \mathbf{n}_C)]\,\overline{\xi} = 0 \,. \tag{7.40}$$

Let us set $\lambda := \mu_0^{-1}\boldsymbol{\mu}_C\mathbf{H}_C \cdot \mathbf{n}_C$. From the Stokes theorem for closed surfaces we have $\lambda \in H_\sharp^{-1/2}(\Gamma)$, where

$$H_\sharp^{-1/2}(\Gamma) := \left\{\xi \in H^{-1/2}(\Gamma) \mid \int_\Gamma \xi = 0\right\} .$$

Moreover, as in Section 4.4, define

$$\widetilde{X}_C := \{\mathbf{v}_C \in H(\mathrm{curl}; \Omega_C) \,|\, \mathrm{div}_\tau(\mathbf{v}_C \times \mathbf{n}_C) = 0 \text{ on } \Gamma\} ,$$

and set

$$\widetilde{X}_\Gamma := \{(\mathbf{v}_C \times \mathbf{n}_C)_{|\Gamma} \,|\, \mathbf{v}_C \in \widetilde{X}_C\} . \tag{7.41}$$

Introducing the operator

$$\mathrm{Curl}_\tau \chi_\Gamma := \mathrm{grad}\, \chi_I \times \mathbf{n}_I$$

(see also Section A.1), it is straightforward to verify that $\mathrm{Curl}_\tau \chi_\Gamma \in \widetilde{X}_\Gamma$.

We have seen in Section 7.2 that $\mathcal{D}(1) = -\frac{1}{2}$, so that

$$\begin{aligned}\textstyle\int_\Gamma [-\frac{1}{2}\boldsymbol{\mu}_C \mathbf{H}_C \cdot \mathbf{n}_C + \mathcal{D}'(\boldsymbol{\mu}_C \mathbf{H}_C \cdot \mathbf{n}_C)] &= \textstyle\mu_0 \int_\Gamma [-\frac{1}{2}\lambda + \mathcal{D}'(\lambda)] \\ &= \textstyle\mu_0 \int_\Gamma [-\frac{1}{2}\lambda + \lambda \mathcal{D}(1)] = -\mu_0 \int_\Gamma \lambda = 0 .\end{aligned}$$

Moreover, it holds $\int_\Gamma \mathcal{H}(\eta) = 0$ for each $\eta \in H^{1/2}(\Gamma)$ and $\mathcal{H}(1) = 0$, hence equation (7.39) does not change if we add a constant to ψ_Γ and χ_Γ. Instead, adding a constant to ψ_Γ we have

$$\frac{1}{2}(\psi_\Gamma + c_0) - \mathcal{D}(\psi_\Gamma + c_0) = \frac{1}{2}\psi_\Gamma - \mathcal{D}(\psi_\Gamma) + c_0 ,$$

therefore equation (7.40) does not change if we choose the test function $\xi \in H^{-1/2}_\sharp(\Gamma)$.

Meddahi and Selgas [176] proved that Curl_τ is an isomorphism from $H^{1/2}(\Gamma)/\mathbb{C}$ onto $\widetilde{X}_\Gamma$. Since $\mathbf{H}_C \times \mathbf{n}_C = -\,\mathrm{grad}\, \psi_I \times \mathbf{n}_I = -\mathrm{Curl}_\tau \psi_\Gamma$ on Γ, in (7.39) we can replace ψ_Γ and the test function χ_Γ with $\mathrm{Curl}_\tau^{-1}(\mathbf{H}_C \times \mathbf{n}_C)$ and $\mathrm{Curl}_\tau^{-1}(\mathbf{v}_C \times \mathbf{n}_C)$, respectively, and we finally obtain that the eddy current problem can be rewritten as

$$\begin{cases} \text{Find } (\mathbf{H}_C, \lambda) \in \widetilde{X}_C \times H^{-1/2}_\sharp(\Gamma) \text{ such that} \\ \quad \int_{\Omega_C} \boldsymbol{\sigma}^{-1} \mathrm{curl}\, \mathbf{H}_C \cdot \mathrm{curl}\, \overline{\mathbf{v}_C} + \int_{\Omega_C} i\omega \boldsymbol{\mu}_C \mathbf{H}_C \cdot \overline{\mathbf{v}_C} \\ \qquad\qquad + i\omega\mu_0 \int_\Gamma [\frac{1}{2}\lambda - \mathcal{D}'(\lambda)]\, \mathrm{Curl}_\tau^{-1}(\overline{\mathbf{v}_C} \times \mathbf{n}_C) \\ \qquad\qquad + i\omega\mu_0 \int_\Gamma \mathcal{H}(\mathrm{Curl}_\tau^{-1}(\mathbf{H}_C \times \mathbf{n}_C))\, \mathrm{Curl}_\tau^{-1}(\overline{\mathbf{v}_C} \times \mathbf{n}_C) \\ \qquad = \int_{\Omega_C} \boldsymbol{\sigma}^{-1} \mathbf{J}_{e,C} \cdot \mathrm{curl}\, \overline{\mathbf{v}_C} \\ \quad i\omega\mu_0 \int_\Gamma [-\frac{1}{2}\mathrm{Curl}_\tau^{-1}(\mathbf{H}_C \times \mathbf{n}_C) + \mathcal{D}(\mathrm{Curl}_\tau^{-1}(\mathbf{H}_C \times \mathbf{n}_C))]\, \overline{\xi} \\ \qquad\qquad + i\omega\mu_0 \int_\Gamma \mathcal{S}(\lambda)\, \overline{\xi} = 0 \\ \text{for each } (\mathbf{v}_C, \xi) \in \widetilde{X}_C \times H^{-1/2}_\sharp(\Gamma) . \end{cases} \tag{7.42}$$

Note that

$$\begin{aligned} & \textstyle i\omega\mu_0 \int_\Gamma [\frac{1}{2}\xi - \mathcal{D}'(\xi)]\, \mathrm{Curl}_\tau^{-1}(\overline{\mathbf{v}_C} \times \mathbf{n}_C) \\ & \qquad \textstyle + i\omega\mu_0 \int_\Gamma [-\frac{1}{2}\mathrm{Curl}_\tau^{-1}(\mathbf{v}_C \times \mathbf{n}_C) + \mathcal{D}(\mathrm{Curl}_\tau^{-1}(\mathbf{v}_C \times \mathbf{n}_C))]\, \overline{\xi} \end{aligned}$$

is a real number, namely, it is equal to

$$-2\omega\mu_0 \operatorname{Im}\left(\int_\Gamma [\frac{1}{2}\xi - \mathcal{D}'(\xi)] \operatorname{Curl}_\tau^{-1}(\overline{\mathbf{v}_C} \times \mathbf{n}_C)\right).$$

Moreover, taking into account that the operator $\mathcal{D}$ is continuous from $H^{1/2}(\Gamma)$ into itself and that the operator $\operatorname{Curl}_\tau^{-1}$ is continuous from $\widetilde{X}_\Gamma$ into $H^{1/2}(\Gamma)/\mathbb{C}$, it follows that for each $0 < \delta < 1$ one has

$$\begin{aligned}
&2|\omega|\mu_0 \left|\int_\Gamma [\tfrac{1}{2}\xi - \mathcal{D}'(\xi)] \operatorname{Curl}_\tau^{-1}(\overline{\mathbf{v}_C} \times \mathbf{n}_C)\right| \\
&\quad \leq c_* \|\xi\|_{-1/2,\Gamma}(\|\mathbf{v}_C\|_{0,\Omega_C} + \|\operatorname{curl} \mathbf{v}_C\|_{0,\Omega_C}) \\
&\quad \leq \delta \|\operatorname{curl} \mathbf{v}_C\|^2_{0,\Omega_C} + C_* \delta^{-1} \|\xi\|^2_{-1/2,\Gamma} + C_* \|\mathbf{v}_C\|^2_{0,\Omega_C} .
\end{aligned}$$

Then, recalling (7.27) and adapting the proof of Theorem 7.5, by choosing δ small enough it is not difficult to show that the sesquilinear form $a_C^\Gamma(\cdot,\cdot)$ at the left hand side of (7.42) is coercive in $\widetilde{X}_C \times H_\sharp^{-1/2}(\Gamma)$. Problem (7.42) is therefore well-posed.

Having solved (7.42), one can determine ψ_I in Ω_I by setting

$$\psi_I = -\mathcal{D}_I(\operatorname{Curl}_\tau^{-1}(\mathbf{H}_C \times \mathbf{n}_C)) - \mathcal{S}_I(\lambda) .$$

The numerical approximation needs some remarks, as a conforming discretization requires that the finite element functions $\mathbf{v}_{C,h}$ satisfy the constraint $\operatorname{div}_\tau(\mathbf{v}_{C,h} \times \mathbf{n}_C) = \mathbf{0}$ on Γ. Instead of introducing a Lagrange multiplier, as done in Section 4.5, here we present an alternative approach, based on the explicit construction of a basis for the space

$$\widetilde{X}_{C,h} := \{\mathbf{v}_{C,h} \in N^1_{C,h} \mid \operatorname{div}_\tau(\mathbf{v}_{C,h} \times \mathbf{n}_C) = 0 \text{ on } \Gamma\} ,$$

where $N^1_{C,h}$ is the space of Nédélec curl-conforming edge elements of the lowest order (see Section A.2). Note that this construction could be used also for the approach presented in Section 4.5.

As in the preceding section, we assume that Ω_C is a Lipschitz polyhedral domain, and we denote by $\mathcal{T}_{C,h}$ and $\mathcal{T}_{\Gamma,h}$ two regular families of triangulations of Ω_C and Γ, respectively. We suppose that each element K of $\mathcal{T}_{C,h}$ is a tetrahedron and that each element T of $\mathcal{T}_{\Gamma,h}$ is a triangle. Let us also denote by $\mathcal{T}_{\partial,h}$ the mesh induced on Γ by $\mathcal{T}_{C,h}$; it is not assumed to coincide with $\mathcal{T}_{\Gamma,h}$. Finally, $\mathcal{E}^0_{C,h}$ denotes the set of edges $e \in \mathcal{T}_{C,h}$ that are internal to Ω_C, $\mathcal{V}_{\Gamma,h}$ the set of vertices $v \in \mathcal{T}_{\partial,h}$, $\mathbf{q}_e$ the lowest-order edge basis function defined in Ω_C and associated to the edge e, and φ_v the piecewise-linear nodal basis function defined in Ω_C and associated to the vertex v.

Proposition 7.12. *Let $v_0 \in \Gamma$ be a fixed vertex of $\mathcal{V}_{\Gamma,h}$. The set*

$$\widetilde{\mathcal{B}}_h := \{\mathbf{q}_e \mid e \in \mathcal{E}^0_{C,h}\} \cup \{\operatorname{grad} \varphi_v \mid v \in \mathcal{V}_{\Gamma,h}, v \neq v_0\}$$

is a basis of $\widetilde{X}_{C,h}$.

Proof. Let us start by showing that the elements of $\widetilde{\mathcal{B}}_h$ are linearly independent. Suppose that

$$\sum_{e \in \mathcal{E}^0_{C,h}} \alpha_e \mathbf{q}_e + \sum_{\substack{v \in \mathcal{V}_{\Gamma,h} \\ v \neq v_0}} \alpha_v \operatorname{grad} \varphi_v = \mathbf{0} \, .$$

Then on Γ we have

$$\mathbf{0} = \Big(\sum_{e \in \mathcal{E}^0_{C,h}} \alpha_e \mathbf{q}_e + \sum_{\substack{v \in \mathcal{V}_{\Gamma,h} \\ v \neq v_0}} \alpha_v \operatorname{grad} \varphi_v \Big) \times \mathbf{n}_C = \operatorname{grad} \Big(\sum_{\substack{v \in \mathcal{V}_{\Gamma,h} \\ v \neq v_0}} \alpha_v \varphi_v \Big) \times \mathbf{n}_C \, ,$$

so that

$$\sum_{\substack{v \in \mathcal{V}_{\Gamma,h} \\ v \neq v_0}} \alpha_v \varphi_v = k_0 \text{ on } \Gamma \, ,$$

where k_0 is a constant. Since $\varphi_v(v_0) = 0$ for each $v \neq v_0$, we have $k_0 = 0$ and therefore $\alpha_v = 0$ for each $v \in \mathcal{V}_{\Gamma,h}, v \neq v_0$. Then we are left with $\sum_{e \in \mathcal{E}^0_{C,h}} \alpha_e \mathbf{q}_e = \mathbf{0}$, which gives $\alpha_e = 0$ for each $e \in \mathcal{E}^0_{C,h}$.

On the other hand, the inclusion $\widetilde{\mathcal{B}}_h \subset \widetilde{X}_{C,h}$ is clearly true. Moreover, take $\mathbf{v}_{C,h} \in \widetilde{X}_{C,h}$: since $\operatorname{div}_\tau(\mathbf{v}_{C,h} \times \mathbf{n}_C) = \mathbf{0}$ on Γ, recalling that Γ is simply-connected it is possible to find a piecewise-linear function $\varphi_{\Gamma,h}$, defined on Γ, such that $\operatorname{grad}_\tau \varphi_{\Gamma,h} \times \mathbf{n}_C = \mathbf{v}_{C,h} \times \mathbf{n}_C$ on Γ. The function $\varphi_{\Gamma,h}$ is uniquely determined by requiring $\varphi_{\Gamma,h}(v_0) = 0$. The extension of $\varphi_{\Gamma,h}$ in Ω_C, obtained by setting all its internal nodal values equal to 0, will be denoted by φ_h. Clearly, $\operatorname{grad} \varphi_h$ belongs to the space spanned by the set of functions $\{\operatorname{grad} \varphi_v \,|\, v \in \mathcal{V}_{\Gamma,h}, v \neq v_0\}$. Since $\mathbf{v}_{C,h} \times \mathbf{n}_C = \operatorname{grad} \varphi_h \times \mathbf{n}_C$ on Γ, it follows that $(\mathbf{v}_{C,h} - \operatorname{grad} \varphi_h)$ is an edge element belonging to the space spanned by the set of functions $\{\mathbf{q}_e \,|\, e \in \mathcal{E}^0_{C,h}\}$, and the thesis follows. □

Thus we have a viable description of the finite element space $X_{C,h}$, and, since we know that $\widetilde{X}_{C,h} \subset \widetilde{X}_C$, a conforming approximation scheme is readily devised. The finite element space used for approximating functions in $H^{-1/2}_\sharp(\Gamma)$ is typically

$$M_{\Gamma,h} := \{\xi_h \in L^2(\Gamma) \,|\, \xi_{h|T} \in \mathbb{P}_0 \; \forall \, T \in \mathcal{T}_{\Gamma,h} \, , \; \textstyle\int_\Gamma \xi_h = 0\} \, ,$$

and the convergence of the scheme is a straightforward consequence of Céa lemma.

It should also be noted that, in the implementation of the finite element scheme, the inverse of the tangential operator Curl_τ does not appear. In fact, let $\mathbf{b}_C \in \widetilde{\mathcal{B}}_h$ be a basis function. If $\mathbf{b}_C = \mathbf{q}_e$, one has $\mathbf{b}_C \times \mathbf{n}_C = \mathbf{0}$ on Γ; if $\mathbf{b}_C = \operatorname{grad} \varphi_v$, it holds $\operatorname{Curl}_\tau \varphi_v = -\operatorname{grad} \varphi_v \times \mathbf{n}_C = -\mathbf{b}_C \times \mathbf{n}_C$ on Γ, thus in the finite element approximation of (7.42) we can replace $\operatorname{Curl}^{-1}_\tau(\mathbf{b}_C \times \mathbf{n}_C)$ with $-\varphi_{v|\Gamma}$.

To illustrate the performance of this method, let us exhibit some numerical results presented in Selgas Buznego [226] for a couple of academic problems. In the first one the computational domain is the cube $\Omega_C = (-1, 1)^3$, all the physical parameters are set equal to 1, and the current density $\mathbf{J}_e$ is computed starting from the exact solution

Table 7.1. Absolute errors for $\mathbf{H}_C$ and λ in the first example (courtesy of V. Selgas)

h	$\|\mathbf{H}_C - \mathbf{H}_{C,h}\|_{H(\text{curl};\Omega_C)}$	$\|\lambda - \lambda_h\|_{0,\Gamma}$	α
0.7733	34.8247	0.3136	-
0.5330	25.0762	0.1634	0.8825
0.2989	14.0539	0.0257	1.0010
0.2337	11.4203	0.0131	0.8433

$\mathbf{E} = \text{curl}(f, f, f)$, where

$$f(\mathbf{x}) := \begin{cases} (1-x_1^2)^4(1-x_2^2)^4(1-x_3^2)^4 & \text{in } \overline{\Omega_C} \\ 0 & \text{in } \Omega_I = \mathbb{R}^3 \setminus \overline{\Omega_C}. \end{cases}$$

In Table 7.1 and Figure 7.1 we report the absolute error and the convergence rate for $\mathbf{H}_C$ and λ for different value of the mesh size h. We have defined

$$\alpha := \frac{\log(\|\mathbf{H}_C - \mathbf{H}_{C,h_i}\|_{H(\text{curl};\Omega_C)}/\|\mathbf{H}_C - \mathbf{H}_{C,h_{i+1}}\|_{H(\text{curl};\Omega_C)})}{\log(h_i/h_{i+1})},$$

h_i and h_{i+1} being the mesh sizes of two consecutive computations.

In the second example the conductor Ω_C is the torus given by

$$\Omega_C := [(-1,1)\times(-1,1)\times(-1/2,1/2)] \setminus (-1/2,1/2)^3,$$

contained in the computational domain

$$\Omega := (-3/2,3/2)\times(-3/2,3/2)\times(-1,1)$$

(see Figure 7.2).

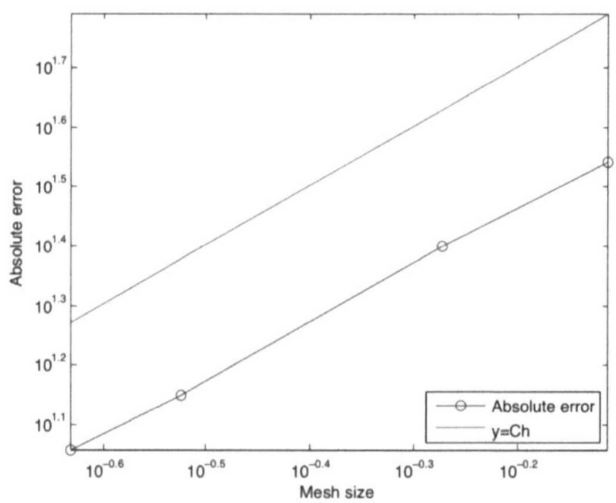

Fig. 7.1. Convergence rate for $\mathbf{H}_C$ in the first example (courtesy of V. Selgas)

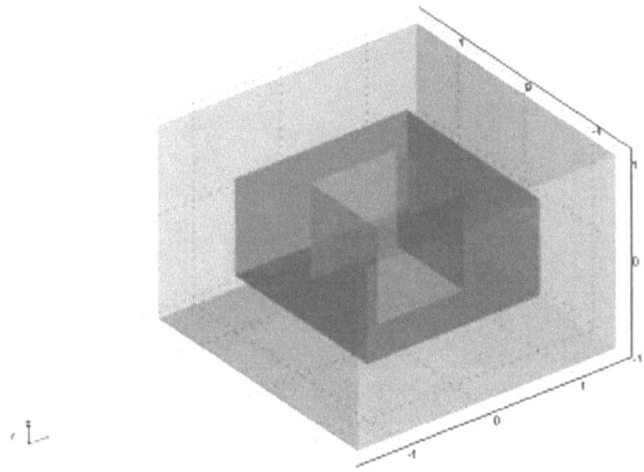

Fig. 7.2. The computational domain Ω for the second example (courtesy of V. Selgas)

In order to avoid the technical difficulties arising from the fact that Ω_C is not simply-connected, the variational formulation has been modified: to be precise, in $\mathbb{R}^3 \setminus \overline{\Omega}$ the usual approach based on the potential theory is used for reducing the contribution of the magnetic field $\mathbf{H}_I = \operatorname{grad} \psi_I$ to suitable integrals on the boundary $\partial\Omega$, while the eddy current problem is solved in Ω by adopting the $\mathbf{H}$-based formulation described in Section 4.3.

Again, all the physical parameters are equal to 1, and the current density $\mathbf{J}_e$ is computed starting from the exact solution $\mathbf{E} = \operatorname{curl}(g, g, g)$, where

$$g(\mathbf{x}) := \begin{cases} (1 - |\mathbf{x}|^2)^4 & \text{in } \overline{B(0,1)} \\ 0 & \text{in } \Omega_I = \mathbb{R}^3 \setminus \overline{B(0,1)} \end{cases} ,$$

where $B(0,1)$ is the ball of center 0 and radius 1. Note that in $\Omega \setminus \overline{\Omega_C}$ the curl of the magnetic field is not vanishing and consequently $\mathbf{H}_I$ is not the gradient of a scalar potential, while this is true outside Ω.

In Table 7.2 and Figure 7.3 the absolute error and the convergence rate for $\mathbf{H}_C$ and λ are presented for different value of the mesh size h.

Table 7.2. Absolute errors for $\mathbf{H}_C$ and λ in the second example (courtesy of V. Selgas)

h	$\|\mathbf{H}_C - \mathbf{H}_{C,h}\|_{H(\operatorname{curl};\Omega_C)}$	$\|\lambda - \lambda_h\|_{0,\Gamma}$	α
0.9299	56.1051	0.7107	-
0.6601	39.3998	0.1413	1.0315
0.4572	28.6757	0.0620	0.8651
0.3653	22.0725	0.0173	1.1663

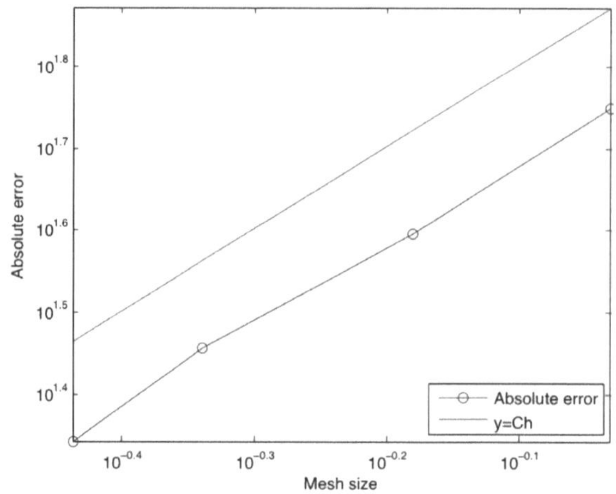

Fig. 7.3. Convergence rate for $\mathbf{H}_C$ in the second example (courtesy of V. Selgas)

7.6.3 An approach based on the electric field $\mathbf{E}_C$

Another FEM–BEM approach can be devised if one keeps the electric field $\mathbf{E}_C$ as principal unknown. Again, for the sake of simplicity, we assume that Ω_C is simply-connected and that the boundary $\Gamma = \partial\Omega_C$ is connected. Starting from the Ampère equation and inserting in it the Faraday equation, for a test function $\mathbf{z}$ that decays sufficiently fast at infinity we find

$$\begin{aligned}-i\omega \int_{\Omega_C} \boldsymbol{\sigma}\mathbf{E}_C \cdot \overline{\mathbf{z}_C} - i\omega \int_{\Omega_C} \mathbf{J}_{e,C} \cdot \overline{\mathbf{z}_C} &= -i\omega \int_{\mathbb{R}^3} \operatorname{curl} \mathbf{H} \cdot \overline{\mathbf{z}} \\ &= -i\omega \int_{\mathbb{R}^3} \mathbf{H} \cdot \operatorname{curl} \overline{\mathbf{z}} = \int_{\mathbb{R}^3} \boldsymbol{\mu}^{-1} \operatorname{curl} \mathbf{E} \cdot \operatorname{curl} \overline{\mathbf{z}} \,.\end{aligned}$$

Since we know that $\mu_0^{-1} \operatorname{curl} \operatorname{curl} \mathbf{E}_I = -i\omega \operatorname{curl} \mathbf{H}_I = \mathbf{0}$ in Ω_I, we also have

$$\int_{\Omega_I} \mu_0^{-1} \operatorname{curl} \mathbf{E}_I \cdot \operatorname{curl} \overline{\mathbf{z}_I} = \int_{\Gamma} \mu_0^{-1} \operatorname{curl} \mathbf{E}_I \times \mathbf{n}_I \cdot \overline{\mathbf{z}_I} \,,$$

therefore we are left with

$$\begin{aligned}&\int_{\Omega_C} \boldsymbol{\mu}_C^{-1} \operatorname{curl} \mathbf{E}_C \cdot \operatorname{curl} \overline{\mathbf{z}_C} + i\omega \int_{\Omega_C} \boldsymbol{\sigma}\mathbf{E}_C \cdot \overline{\mathbf{z}_C} \\ &\qquad + \mu_0^{-1} \int_{\Gamma} \operatorname{curl} \mathbf{E}_I \times \mathbf{n}_I \cdot \overline{\mathbf{z}_I} \\ &\quad = -i\omega \int_{\Omega_C} \mathbf{J}_{e,C} \cdot \overline{\mathbf{z}_C} \,.\end{aligned} \tag{7.43}$$

Let us go on without giving all the details about the functional framework, but just presenting the main idea. Denote by R the vectorial Steklov–Poincaré operator given by

$$\mathsf{R}(\mathbf{q}) := \operatorname{curl} \mathbf{e}_I \times \mathbf{n}_I \qquad \text{on } \Gamma \,,$$

where $\mathbf{q} \cdot \mathbf{n}_I = 0$ on Γ and $\mathbf{e}_I$ is the solution to

$$\begin{cases} \operatorname{curl}\operatorname{curl}\mathbf{e}_I = \mathbf{0} & \text{in } \Omega_I \\ \operatorname{div}(\varepsilon_0 \mathbf{e}_I) = 0 & \text{in } \Omega_I \\ \mathbf{n}_I \times \mathbf{e}_I \times \mathbf{n}_I = \mathbf{q} & \text{on } \Gamma \\ \int_\Gamma \varepsilon_0 \mathbf{e}_I \cdot \mathbf{n}_I = 0 & \\ \mathbf{e}_I(\mathbf{x}) = O(|\mathbf{x}|^{-1}) & \text{as } |\mathbf{x}| \to \infty \, . \end{cases} \tag{7.44}$$

We thus have $\operatorname{curl}\mathbf{E}_I \times \mathbf{n}_I = \mathsf{R}(\mathbf{n}_C \times \mathbf{E}_C \times \mathbf{n}_C)$ on Γ, and we can rewrite (7.43) as

$$\begin{aligned} &\int_{\Omega_C} \boldsymbol{\mu}_C^{-1} \operatorname{curl}\mathbf{E}_C \cdot \operatorname{curl}\overline{\mathbf{z}_C} + i\omega \int_{\Omega_C} \boldsymbol{\sigma}\mathbf{E}_C \cdot \overline{\mathbf{z}_C} \\ &\qquad + \mu_0^{-1} \int_\Gamma \mathsf{R}(\mathbf{n}_C \times \mathbf{E}_C \times \mathbf{n}_C) \cdot \overline{\mathbf{z}_I} \\ &= -i\omega \int_{\Omega_C} \mathbf{J}_{e,C} \cdot \overline{\mathbf{z}_C} \, . \end{aligned} \tag{7.45}$$

Bossavit [57] has extended the TRIFOU approach to this formulation, which is based on the electric field. We do not dwell on this here, referring the interested reader to the paper just quoted (see also Ren et al. [208]).

Instead, we present an alternative approach, proposed and analyzed by Hiptmair [127], which leads to a symmetric formulation (we also note that in that paper no restrictive assumption on the geometrical shape of the conducting domain Ω_C is imposed). First of all, as in (A.1), (A.3) and (A.4) introduce the trace spaces

$$H_T^{1/2}(\Gamma) := \{(\mathbf{n} \times \mathbf{v} \times \mathbf{n})_{|\Gamma} \,|\, \mathbf{v} \in (H^1(\Omega))^3\}$$

$$H^{-1/2}(\operatorname{div}_\tau; \Gamma) = \{(\mathbf{v}_C \times \mathbf{n}_C)_{|\Gamma} \,|\, \mathbf{v}_C \in H(\operatorname{curl}; \Omega_C)\},$$

and

$$H^{-1/2}(\operatorname{curl}_\tau; \Gamma) = \{(\mathbf{n}_C \times \mathbf{v}_C \times \mathbf{n}_C)_{|\Gamma} \,|\, \mathbf{v}_C \in H(\operatorname{curl}; \Omega_C)\} \, ;$$

note that the last two spaces are one the dual space of the other.

Let us define now on Γ the vectorial single layer and double layer potentials

$$\begin{aligned} &\mathsf{S} : (H_T^{1/2}(\Gamma))' \to H_T^{1/2}(\Gamma) \\ &\mathsf{S}(\mathbf{p})(\mathbf{x}) := \int_\Gamma \frac{1}{4\pi|\mathbf{x}-\mathbf{y}|} \, \mathbf{p}(\mathbf{y}) dS_y \end{aligned} \tag{7.46}$$

$$\begin{aligned} &\mathsf{D} : H^{-1/2}(\operatorname{curl}_\tau; \Gamma) \to H^{-1/2}(\operatorname{curl}_\tau; \Gamma) \ , \\ &\mathsf{D}(\mathbf{q})(\mathbf{x}) := \int_\Gamma \frac{\mathbf{x}-\mathbf{y}}{4\pi|\mathbf{x}-\mathbf{y}|^3} \times [\mathbf{q}(\mathbf{y}) \times \mathbf{n}_C(\mathbf{y})] dS_y \ , \end{aligned} \tag{7.47}$$

the hypersingular integral operator

$$\begin{aligned} &\mathsf{H} : H^{-1/2}(\operatorname{curl}_\tau; \Gamma) \to H^{-1/2}(\operatorname{div}_\tau; \Gamma) \ , \\ &\mathsf{H}(\mathbf{q})(\mathbf{x}) := -\operatorname{curl}\left(\int_\Gamma \frac{\mathbf{x}-\mathbf{y}}{4\pi|\mathbf{x}-\mathbf{y}|^3} \times [\mathbf{q}(\mathbf{y}) \times \mathbf{n}_C(\mathbf{y})] dS_y \right) \times \mathbf{n}_C(\mathbf{x}), \end{aligned} \tag{7.48}$$

and the adjoint operator

$$\begin{aligned} &\mathsf{D}' : \widetilde{X}_\Gamma \to \widetilde{X}_\Gamma \ , \\ &\mathsf{D}'(\mathbf{p})(\mathbf{x}) := \left(\int_\Gamma \frac{\mathbf{y}-\mathbf{x}}{4\pi|\mathbf{x}-\mathbf{y}|^3} \times \mathbf{p}(\mathbf{y}) dS_y \right) \times \mathbf{n}_C(\mathbf{x}) \ , \end{aligned} \tag{7.49}$$

where the space $\widetilde{X}_\Gamma$ has been introduced in (7.41), and is given by the vector functions $\mathbf{p}$ belonging to $H^{-1/2}(\mathrm{div}_\tau;\Gamma)$ and such that $\mathrm{div}_\tau\,\mathbf{p}=0$ on Γ.

In Hiptmair [127] (see also Reissel [205], Hiptmair and Ostrowski [129]) it has been shown that these operators are continuous, and moreover that the solution $\mathbf{E}_I$ satisfies $\mathbf{n}_C\times\mathbf{E}_I\times\mathbf{n}_C=\mathbf{n}_C\times\mathbf{E}_C\times\mathbf{n}_C\in H^{-1/2}(\mathrm{curl}_\tau;\Gamma)$, $\mathrm{curl}\,\mathbf{E}_I\times\mathbf{n}_C\in\widetilde{X}_\Gamma$ and also

$$\begin{aligned}&\tfrac{1}{2}\textstyle\int_\Gamma \mathbf{n}_C\times\mathbf{E}_C\times\mathbf{n}_C\cdot\overline{\mathbf{p}'}-\int_\Gamma \mathsf{D}(\mathbf{n}_C\times\mathbf{E}_C\times\mathbf{n}_C)\cdot\overline{\mathbf{p}'}\\&\qquad+\textstyle\int_\Gamma \mathsf{S}(\mathrm{curl}\,\mathbf{E}_I\times\mathbf{n}_C)\cdot\overline{\mathbf{p}'}=0\qquad\forall\,\mathbf{p}'\in\widetilde{X}_\Gamma,\end{aligned}\tag{7.50}$$

and

$$\begin{aligned}&\tfrac{1}{2}\textstyle\int_\Gamma \mathrm{curl}\,\mathbf{E}_I\times\mathbf{n}_C\cdot\overline{\mathbf{q}'}+\int_\Gamma \mathsf{D}'(\mathrm{curl}\,\mathbf{E}_I\times\mathbf{n}_C)\cdot\overline{\mathbf{q}'}\\&\qquad+\textstyle\int_\Gamma \mathsf{H}(\mathbf{n}_C\times\mathbf{E}_C\times\mathbf{n}_C)\cdot\overline{\mathbf{q}'}=0\qquad\forall\,\mathbf{q}'\in H^{-1/2}(\mathrm{curl}_\tau;\Gamma)\,.\end{aligned}\tag{7.51}$$

Setting $\mathbf{p}_\Gamma:=\mathrm{curl}\,\mathbf{E}_I\times\mathbf{n}_C$, the term on Γ in equation (7.43) can be written as

$$\mu_0^{-1}\int_\Gamma \mathrm{curl}\,\mathbf{E}_I\times\mathbf{n}_I\cdot\overline{\mathbf{z}_I}=-\mu_0^{-1}\int_\Gamma \mathbf{p}_\Gamma\cdot\overline{\mathbf{z}_I}=-\mu_0^{-1}\int_\Gamma \mathbf{p}_\Gamma\cdot\mathbf{n}_C\times\overline{\mathbf{z}_C}\times\mathbf{n}_C\,.$$

Therefore, inserting (7.51) in (7.43), we see that the eddy current problem can be formulated as follows

Find $(\mathbf{E}_C,\mathbf{p}_\Gamma)\in H(\mathrm{curl};\Omega_C)\times\widetilde{X}_\Gamma$ such that

$$\begin{aligned}&\textstyle\int_{\Omega_C}\boldsymbol{\mu}_C^{-1}\,\mathrm{curl}\,\mathbf{E}_C\cdot\mathrm{curl}\,\overline{\mathbf{z}_C}+i\omega\int_{\Omega_C}\boldsymbol{\sigma}\mathbf{E}_C\cdot\overline{\mathbf{z}_C}\\&\qquad-\tfrac{1}{2}\mu_0^{-1}\textstyle\int_\Gamma \mathbf{p}_\Gamma\cdot\mathbf{n}_C\times\overline{\mathbf{z}_C}\times\mathbf{n}_C\\&\qquad+\mu_0^{-1}\textstyle\int_\Gamma \mathsf{D}'(\mathbf{p}_\Gamma)\cdot\mathbf{n}_C\times\overline{\mathbf{z}_C}\times\mathbf{n}_C\\&\qquad+\mu_0^{-1}\textstyle\int_\Gamma \mathsf{H}(\mathbf{n}_C\times\mathbf{E}_C\times\mathbf{n}_C)\cdot\mathbf{n}_C\times\overline{\mathbf{z}_C}\times\mathbf{n}_C\\&\quad=-i\omega\textstyle\int_{\Omega_C}\mathbf{J}_{e,C}\cdot\overline{\mathbf{z}_C}\\[1ex]&\tfrac{1}{2}\mu_0^{-1}\textstyle\int_\Gamma \mathbf{n}_C\times\mathbf{E}_C\times\mathbf{n}_C\cdot\overline{\mathbf{p}'}-\mu_0^{-1}\int_\Gamma \mathsf{D}(\mathbf{n}_C\times\mathbf{E}_C\times\mathbf{n}_C)\cdot\overline{\mathbf{p}'}\\&\qquad+\mu_0^{-1}\textstyle\int_\Gamma \mathsf{S}(\mathbf{p}_\Gamma)\cdot\overline{\mathbf{p}'}=0\end{aligned}\tag{7.52}$$

for each $(\mathbf{z}_C,\mathbf{p}')\in H(\mathrm{curl};\Omega_C)\times\widetilde{X}_\Gamma$.

Note that

$$\begin{aligned}&-\tfrac{1}{2}\textstyle\int_\Gamma \mathbf{p}'\cdot\mathbf{n}_C\times\overline{\mathbf{z}_C}\times\mathbf{n}_C+\int_\Gamma \mathsf{D}'(\mathbf{p}')\cdot\mathbf{n}_C\times\overline{\mathbf{z}_C}\times\mathbf{n}_C\\&\qquad+\tfrac{1}{2}\textstyle\int_\Gamma \mathbf{n}_C\times\mathbf{z}_C\times\mathbf{n}_C\cdot\overline{\mathbf{p}'}-\int_\Gamma \mathsf{D}(\mathbf{n}_C\times\mathbf{z}_C\times\mathbf{n}_C)\cdot\overline{\mathbf{p}'}\end{aligned}$$

is a purely imaginary number, and it is equal to

$$2\,i\,\mathrm{Im}\left(\frac{1}{2}\int_\Gamma \mathbf{n}_C\times\mathbf{z}_C\times\mathbf{n}_C\cdot\overline{\mathbf{p}'}-\int_\Gamma \mathsf{D}(\mathbf{n}_C\times\mathbf{z}_C\times\mathbf{n}_C)\cdot\overline{\mathbf{p}'}\right).$$

Moreover, the boundedness of the operator D and the trace inequality for $H(\text{curl}; \Omega_C)$ (see (A.11)) give that, for each $0 < \delta < 1$,

$$\begin{aligned}
&\left|\tfrac{1}{2}\int_\Gamma \mathbf{n}_C \times \mathbf{z}_C \times \mathbf{n}_C \cdot \overline{\mathbf{p}}' - \int_\Gamma \mathsf{D}(\mathbf{n}_C \times \mathbf{z}_C \times \mathbf{n}_C) \cdot \overline{\mathbf{p}}'\right| \\
&\quad \leq c_* \|\mathbf{p}'\|_{H^{-1/2}(\text{div}_\tau;\Sigma)} (\|\mathbf{z}_C\|_{0,\Omega_C} + \|\text{curl}\, \mathbf{z}_C\|_{0,\Omega_C}) \\
&\quad \leq \delta \|\mathbf{z}_C\|^2_{0,\Omega_C} + C_* \delta^{-1} \|\mathbf{p}'\|^2_{H^{-1/2}(\text{div}_\tau;\Sigma)} + C_* \|\text{curl}\, \mathbf{z}_C\|^2_{0,\Omega_C} \, .
\end{aligned}$$

Finally, in Hiptmair [127] it is shown that the operators S and H satisfies

$$\int_\Gamma \mathsf{H}(\mathbf{n}_C \times \mathbf{z}_C \times \mathbf{n}_C) \cdot \mathbf{n}_C \times \overline{\mathbf{z}_C} \times \mathbf{n}_C \geq 0 \ , \quad \int_\Gamma \mathsf{S}(\mathbf{p}') \cdot \overline{\mathbf{p}}' \geq \kappa_0 \|\mathbf{p}'\|^2_{H^{-1/2}(\text{div}_\tau;\Sigma)} \, .$$

Thus, adapting the proof of Theorem 7.5, by choosing δ small enough it is not difficult to prove that the sesquilinear form $a^\Gamma_{e,C}(\cdot,\cdot)$ at the left hand side of (7.52) is coercive in $H(\text{curl}; \Omega_C) \times \widetilde{X}_\Gamma$, and we conclude that problem (7.52) is well-posed.

Having determined $\mathbf{E}_C$ and $\mathbf{p}_\Gamma = \text{curl}\, \mathbf{E}_I \times \mathbf{n}_C = i\omega\mu_0 \mathbf{H}_I \times \mathbf{n}_I$, one can also find the magnetic field in Ω_I. In fact, setting

$$\mathsf{S}_I(\mathbf{p})(\mathbf{x}) := \int_\Gamma \frac{1}{4\pi|\mathbf{x} - \mathbf{y}|} \mathbf{p}(\mathbf{y}) dS_y \ , \quad \mathbf{x} \in \Omega_I \ ,$$

$$\mathsf{D}_I(\mathbf{q})(\mathbf{x}) := \int_\Gamma \frac{\mathbf{x} - \mathbf{y}}{4\pi|\mathbf{x} - \mathbf{y}|^3} \times [\mathbf{q}(\mathbf{y}) \times \mathbf{n}_C(\mathbf{y})] dS_y \ , \quad \mathbf{x} \in \Omega_I \ ,$$

from well-known results of potential theory we easily obtain by integration by parts the representation formula

$$\mathbf{E}_I(\mathbf{x}) = \mathsf{D}_I(\mathbf{E}_C) - \mathsf{S}_I(\mathbf{p}_\Gamma) - \text{grad}\, \mathcal{S}_I(\mathbf{E}_I \cdot \mathbf{n}_C) \ , \tag{7.53}$$

where the operator $\mathcal{S}_I$ has been introduced in (7.15). Then the magnetic field $\mathbf{H}_I = -(i\omega\mu_0)^{-1} \text{curl}\, \mathbf{E}_I$ can be written as

$$\mathbf{H}_I = -(i\omega\mu_0)^{-1} (\text{curl}\, \mathsf{D}_I(\mathbf{E}_C) - \text{curl}\, \mathsf{S}_I(\mathbf{p}_\Gamma)) \ .$$

Instead, since we do not know the value of the normal component $\mathbf{E}_I \cdot \mathbf{n}_C$ on Γ, the electric field $\mathbf{E}_I$ cannot be computed through (7.53), and one has to solve (7.2).

The numerical approximation is quite similar to that presented in Section 7.6.2. In fact, Nédélec curl-conforming edge element of the lowest order can be used in Ω_C; instead, the conforming approximation of $\widetilde{X}_\Gamma$ is given by the space spanned by

$$\{\text{Curl}_\tau \varphi_v \,|\, v \in \mathcal{V}_{\Gamma,h}, v \neq v_0\} \ ,$$

where $\mathcal{V}_{\Gamma,h}$ is the set of vertices $v \in \mathcal{T}_{\partial,h}$, the mesh induced on Γ by $\mathcal{T}_{C,h}$, φ_v is the piecewise-linear nodal basis function defined on Γ and associated to the vertex v, and $v_0 \in \Gamma$ is a fixed vertex of $\mathcal{V}_{\Gamma,h}$.

In Hiptmair [127] the convergence of the approximation scheme, based on Céa lemma and suitable interpolation estimates, is completely proved. Moreover, the discrete problem is analyzed also when Ω_C is not simply-connected: this geometric situation has the drawback that the boundary element space for approximating $\widetilde{X}_\Gamma$ is more complicated. Finally, some remarks on implementation are also added: in particular, it is shown that the operators S, D and H can be expressed in terms of the analogous operators constructed for the Laplace operator. For example, one has

$$\int_\Gamma \mathsf{S}(\mathrm{Curl}_\tau \psi_{\Gamma,h}) \cdot \mathrm{Curl}_\tau \overline{\chi_{\Gamma,h}} = \int_\Gamma \mathcal{H}(\psi_{\Gamma,h})\, \overline{\chi_{\Gamma,h}} \,,$$

$$\begin{aligned} \textstyle\int_\Gamma \mathsf{D}'(\mathbf{p}_{\Gamma,h}) \cdot \mathbf{n}_C \times \overline{\mathbf{z}_{C,h}} \times \mathbf{n}_C &= \textstyle\int_\Gamma \mathbf{p}_{\Gamma,h} \cdot \mathcal{D}(\mathbf{n}_C \times \overline{\mathbf{z}_{C,h}} \times \mathbf{n}_C) \\ &+ \textstyle\int_\Gamma \int_\Gamma \frac{\mathbf{x}-\mathbf{y}}{4\pi|\mathbf{x}-\mathbf{y}|^3} \cdot [\mathbf{n}_C(\mathbf{x}) \times \overline{\mathbf{z}_{C,h}}(\mathbf{x}) \times \mathbf{n}_C(\mathbf{x})](\mathbf{p}_{\Gamma,h}(\mathbf{y}) \cdot \mathbf{n}_C(\mathbf{x}))dS_x dS_y \,, \end{aligned}$$

$$\int_\Gamma \mathsf{H}(\mathbf{n}_C \times \mathbf{E}_{C,h} \times \mathbf{n}_C) \cdot \mathbf{n}_C \times \overline{\mathbf{z}_{C,h}} \times \mathbf{n}_C = \int_\Gamma \mathcal{S}(\mathrm{div}_\tau(\mathbf{E}_{C,h} \times \mathbf{n}_C))\, \mathrm{div}_\tau(\overline{\mathbf{z}_{C,h}} \times \mathbf{n}_C) \,,$$

where $\mathcal{S}$, $\mathcal{D}$ and $\mathcal{H}$ are the operators introduced in (7.6), (7.7) and (7.8), respectively. Therefore, the techniques developed for Galerkin boundary element methods for the Laplace operator can be used in this framework (in this respect, see also Remark 7.11).

8

Voltage and current intensity excitation

In many electromagnetic phenomena it is useful to couple formulations in terms of electrical circuits with more general formulations based on Maxwell equations or on some reduced model like the eddy current system. On the common interface between the two models, the boundary data for the domain where the eddy current model is considered are current intensities or voltages.

In its simplest configuration, in these cases the computational domain is a simply-connected bounded open set Ω. The conducting region Ω_C, unlike in the previous chapters, is not strictly contained in the computational domain and the region where it touches the boundary of Ω has at least two disjoint connected components, the electric ports. Our aim is to model the electromagnetic problem in the case of an electric current passing inside the conductor connecting the electric ports; this electric current is imposed by assigning its intensity or a potential difference.

A critical point is which kind of boundary conditions can be considered outside the electric ports. In fact, for certain boundary conditions the eddy current problem is well-posed even if no additional condition like voltage or current intensity is imposed. The same occurs, for any type of boundary conditions, in the case of a device where the conductor is strictly contained in the computational domain. To overcome this apparent contradiction it is necessary to focus on the modeling of the problem, and devise a formulation for which it becomes possible to impose the voltage or current intensity, but without giving up Maxwell equations.

In this chapter we propose a systematic approach to eddy current problems driven by voltage or current intensity. In Section 8.1 we start by making precise for which set of boundary conditions the voltage excitation and the current intensity excitation problems in the presence of electric ports are well-posed. Then, following Alonso Rodríguez et al. [20], we present a hybrid approach in terms of the electric field in Ω_C and the scalar magnetic potential in Ω_I. We also briefly describe two other approaches that have been proposed for addressing the eddy current problem in the presence of electric ports (see Meunier et al. [177], Bíró et al. [52], Bermúdez et al. [45]). Finally, we propose a finite element method for the approximation of the hybrid problem, obtain the error estimates and illustrate the performance of the methods with some numerical results. In Section 8.2, as in Alonso Rodríguez and Valli [18], we discuss how to take

A. Alonso Rodríguez, A. Valli: Eddy Current Approximation of Maxwell Equations.

into account voltage and current intensity excitation in the case of a conductor strictly contained in the computational domain (a similar point of view can also be adopted for the problem with electric ports, when subjected to electric or magnetic boundary conditions on $\partial\Omega \cap \partial\Omega_I$).

To a reader interested in numerical approximation and implementation we suggest to concentrate on problems (8.38) and (8.40) (voltage excitation), on problems (8.39) and (8.42) (current intensity excitation), on Section 8.1.2 ($(\mathbf{H}_C, \psi_I^*)$ formulation), on Section 8.1.3 ($(\mathbf{T}_C, \psi^*)$ formulation) and on Section 8.1.5.

8.1 The eddy current problem in the presence of electric ports

The computational domain is a simply-connected bounded open set $\Omega \subset \mathbb{R}^3$, with a connected and Lipschitz boundary $\partial\Omega$. It is split into two Lipschitz subdomains, a conducting region Ω_C and a non-conducting region $\Omega_I = \Omega \setminus \overline{\Omega_C}$; the latter is assumed to be connected. The conducting region Ω_C is not strictly contained in Ω, i.e., $\partial\Omega \cap \partial\Omega_C \neq \emptyset$. For the sake of simplicity, we assume that it is simply-connected; for more general geometrical situations, which could be more interesting for applications, see Remark 8.8. As usual, we denote the interface between the two regions by Γ, while the different parts of the boundary $\partial\Omega$ are indicated by $\Gamma_C = \partial\Omega \cap \partial\Omega_C$ and $\Gamma_I = \partial\Omega \cap \partial\Omega_I$. Moreover, we suppose that $\Gamma_C = \Gamma_E \cup \Gamma_J$, where Γ_E and Γ_J are two disjoint and connected surfaces on Γ_C ('electric ports'). Therefore, with these notations we have $\partial\Omega_C = \Gamma_E \cup \Gamma_J \cup \Gamma$, $\partial\Omega_I = \Gamma_I \cup \Gamma$ (see Figure 8.1).

Concerning the material coefficients, as usual we assume that the matrix $\boldsymbol{\mu}$ is symmetric and uniformly positive definite in Ω, with entries belonging to $L^\infty(\Omega)$, the matrix ε_I is symmetric and uniformly positive definite in Ω_I, with entries belonging to $L^\infty(\Omega_I)$, and the matrix $\boldsymbol{\sigma}$ is symmetric and uniformly positive definite in Ω_C, with entries belonging to $L^\infty(\Omega_C)$, whereas it vanishes in Ω_I.

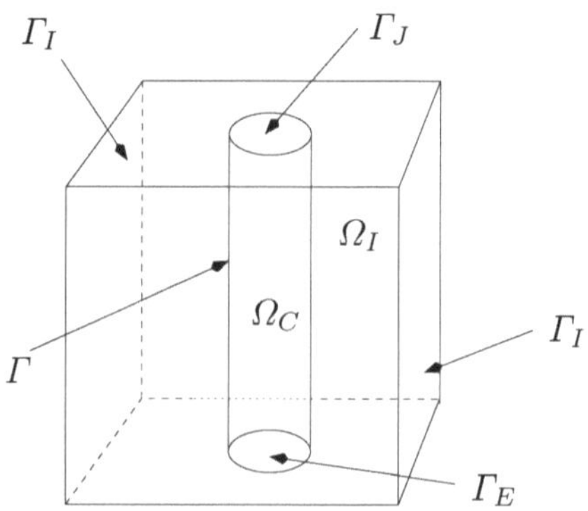

Fig. 8.1. The computational domain

The eddy current equations as usual read

$$\begin{aligned} &\operatorname{curl} \mathbf{H} - \boldsymbol{\sigma} \mathbf{E} = \mathbf{J}_e && \text{in } \Omega \\ &\operatorname{curl} \mathbf{E} + i\omega \boldsymbol{\mu} \mathbf{H} = \mathbf{0} && \text{in } \Omega \\ &\operatorname{div}(\varepsilon_I \mathbf{E}_I) = 0 && \text{in } \Omega_I \,. \end{aligned} \tag{8.1}$$

We want to model the electromagnetic problem in the case of an electric current passing along the 'cylinder' Ω_C, and to impose this electric current as a potential difference between Γ_E and Γ_J, or as a certain given intensity through Γ_J.

A first point in the modelization is to require that the electric field is normal to the boundary on the two electric ports, namely, $\mathbf{E}_C \times \mathbf{n}_C = \mathbf{0}$ on $\Gamma_E \cup \Gamma_J$. Then, following Bossavit [61], we consider the no-flux boundary conditions

$$\begin{aligned} &\mathbf{E}_C \times \mathbf{n}_C = \mathbf{0} && \text{on } \Gamma_E \cup \Gamma_J \\ &\varepsilon_I \mathbf{E}_I \cdot \mathbf{n}_I = 0 && \text{on } \Gamma_I \\ &\boldsymbol{\mu} \mathbf{H} \cdot \mathbf{n} = 0 && \text{on } \partial\Omega \,. \end{aligned} \tag{8.2}$$

Since $\boldsymbol{\mu}\mathbf{H} \cdot \mathbf{n} = 0$ on $\partial\Omega$, from the Faraday law one has $\operatorname{div}_\tau(\mathbf{E} \times \mathbf{n}) = 0$ on $\partial\Omega$, which is a simply-connected surface; hence there exists a surface potential $v \in H^1(\Omega)$ such that $\mathbf{E} \times \mathbf{n} = \operatorname{grad} v \times \mathbf{n}$ on $\partial\Omega$. Moreover $\mathbf{E}_C \times \mathbf{n}_C = \mathbf{0}$ on $\Gamma_E \cup \Gamma_J$, then the function v must be constant on Γ_E and Γ_J. Since v is defined up to a constant, we can take it equal to 0 on Γ_E.

In the voltage excitation problem, given $V \in \mathbb{C}$ we look for a solution of the eddy current problem with the boundary condition

$$\mathbf{E} \times \mathbf{n} = \operatorname{grad} v \times \mathbf{n} \text{ on } \partial\Omega \,, \text{ with } v_{|\Gamma_J} = V \text{ and } v_{|\Gamma_E} = 0 \,, \tag{8.3}$$

plus $\varepsilon_I \mathbf{E}_I \cdot \mathbf{n}_I = 0$ on Γ_I (in particular, all the conditions (8.2) are satisfied).

Remark 8.1. It is worth noting that the quantity V is given by

$$V = \int_{\widehat{\gamma}} \mathbf{E} \cdot d\boldsymbol{\tau} \,, \tag{8.4}$$

where $\widehat{\gamma}$ is any curve lying on Γ_I and joining the electric port Γ_E to the electric port Γ_J. Hence the voltage excitation problem can also be written as the eddy current problem with the boundary conditions (8.2) and the additional condition (8.4). □

In the current intensity excitation problem, given $I^0 \in \mathbb{C}$ we impose the eddy current equations, the boundary conditions (8.2) and moreover

$$\int_{\Gamma_J} \operatorname{curl} \mathbf{H}_C \cdot \mathbf{n}_C = I^0 \,. \tag{8.5}$$

We notice that the set of boundary conditions (8.2) allows us to assign the voltage or the current intensity. This is not the case for other boundary conditions such as the electric boundary condition

$$\mathbf{E} \times \mathbf{n} = \mathbf{0} \quad \text{on } \partial\Omega \tag{8.6}$$

or the magnetic boundary conditions

$$\begin{aligned} &\mathbf{E}_C \times \mathbf{n}_C = \mathbf{0} \quad \text{on } \Gamma_E \cup \Gamma_J , \\ &\varepsilon_I \mathbf{E}_I \cdot \mathbf{n}_I = 0 \quad \text{on } \Gamma_I , \\ &\mathbf{H}_I \times \mathbf{n}_I = \mathbf{0} \quad \text{on } \Gamma_I . \end{aligned} \tag{8.7}$$

In fact, the following result holds:

Proposition 8.2. *Let us consider the solutions* $\mathbf{H}$ *and* $\mathbf{E}$ *of the eddy current problem*

$$\begin{aligned} &\operatorname{curl} \mathbf{H} - \boldsymbol{\sigma}\mathbf{E} = \mathbf{J}_e \quad \text{in } \Omega \\ &\operatorname{curl} \mathbf{E} + i\omega\boldsymbol{\mu}\mathbf{H} = \mathbf{0} \text{ in } \Omega . \end{aligned} \tag{8.8}$$

The magnetic field $\mathbf{H}$ *in* Ω *and the electric field* $\mathbf{E}_C$ *in* Ω_C *are uniquely determined for each one of the boundary condition* (8.6) *and* (8.7).

Proof. Assume that $\mathbf{J}_e = \mathbf{0}$ in Ω. Multiply the Faraday equation by $\overline{\mathbf{H}}$ and integrate in Ω. Integration by parts gives

$$\begin{aligned} 0 &= \int_\Omega \operatorname{curl} \mathbf{E} \cdot \overline{\mathbf{H}} + \int_\Omega i\omega\boldsymbol{\mu}\mathbf{H} \cdot \overline{\mathbf{H}} \\ &= \int_\Omega \mathbf{E} \cdot \operatorname{curl} \overline{\mathbf{H}} + \int_\Omega i\omega\boldsymbol{\mu}\mathbf{H} \cdot \overline{\mathbf{H}} - \int_{\partial\Omega} \mathbf{E} \times \mathbf{n} \cdot \overline{\mathbf{H}} . \end{aligned}$$

Recalling that $\operatorname{curl} \mathbf{H}_I = \mathbf{0}$ in Ω_I and replacing $\operatorname{curl} \mathbf{H}_C$ by $\boldsymbol{\sigma}\mathbf{E}_C$ we have

$$0 = \int_{\Omega_C} \boldsymbol{\sigma}\mathbf{E}_C \cdot \overline{\mathbf{E}_C} + \int_\Omega i\omega\boldsymbol{\mu}\mathbf{H} \cdot \overline{\mathbf{H}} - \int_{\partial\Omega} \mathbf{E} \times \mathbf{n} \cdot \overline{\mathbf{H}} .$$

It is clear that the boundary integral vanishes for both choices (8.6) and (8.7), and then it follows that $\mathbf{H} = \mathbf{0}$ and $\mathbf{E}_C = \mathbf{0}$. □

If we repeat the computation here above with the set of boundary conditions (8.2), writing $\mathbf{E}_C = \boldsymbol{\sigma}^{-1} \operatorname{curl} \mathbf{H}_C$ we find

$$\begin{aligned} &\int_{\Omega_C} \boldsymbol{\sigma}^{-1} \operatorname{curl} \mathbf{H}_C \cdot \operatorname{curl} \overline{\mathbf{H}_C} + \int_\Omega i\omega\boldsymbol{\mu}\mathbf{H} \cdot \overline{\mathbf{H}} \\ &\quad = \int_{\partial\Omega} \operatorname{grad} v \times \mathbf{n} \cdot \overline{\mathbf{H}} = -\int_{\partial\Omega} \overline{\mathbf{H}} \times \mathbf{n} \cdot \operatorname{grad} v \\ &\quad = \int_{\partial\Omega} \operatorname{curl} \overline{\mathbf{H}} \cdot \mathbf{n}\, v = v_{|\Gamma_J} \int_{\Gamma_J} \operatorname{curl} \overline{\mathbf{H}_C} \cdot \mathbf{n} . \end{aligned} \tag{8.9}$$

In this case a degree of freedom is indeed still free, either the voltage $v_{|\Gamma_J}$ or else the current intensity $\int_{\Gamma_J} \operatorname{curl} \mathbf{H}_C \cdot \mathbf{n}$.

8.1.1 Hybrid formulations in term of $\mathbf{E}_C$ and ψ_I^*

In this section, following Alonso Rodríguez et al. [20], we introduce a weak formulation of problem (8.8) with boundary conditions (8.2) and assigned voltage (8.3) or assigned current intensity (8.5). It is a hybrid formulation since the main unknowns are the electric field in the conductor and the magnetic field in the insulator (see Chapter 4). The latter is decomposed as the sum of the gradient of a function in $H^1(\Omega_I)$ plus a harmonic field (see Chapter 5, in particular Section 5.3, where a formulation in terms of the electric field in Ω_C and of a scalar magnetic potential in Ω_I is devised for the internal conductor case). When the input current intensity is given, this harmonic field is uniquely determined, hence the unknowns of the problem reduce to the electric field in the conductor and a scalar magnetic potential in the insulator. On the other hand, when the voltage is given the unknowns of the problem are the electric field in the conductor, a scalar magnetic potential in the insulator and the current intensity.

We recall the following orthogonal decomposition of $(L^2(\Omega_I))^3$ presented in Theorem A.8: any given vector function $\mathbf{v}_I \in (L^2(\Omega_I))^3$ can be decomposed as

$$\mathbf{v}_I = \boldsymbol{\mu}_I^{-1} \operatorname{curl} \mathbf{Q}_I^* + \operatorname{grad} \chi_I^* + \mathbf{k}_I^* ,$$

with $\mathbf{Q}_I^* \in H(\operatorname{curl}; \Omega_I)$, $\chi_I^* \in H^1(\Omega_I)/\mathbb{C}$ and $\mathbf{k}_I^* \in \mathcal{H}_{\mu_I}(m; \Omega_I)$; in particular, $\mathbf{Q}_I^* = \mathbf{0}$ if $\operatorname{curl} \mathbf{v}_I = \mathbf{0}$ in Ω_I.

Since the conductor Ω_C touches the boundary of the computational domain in the two electric ports, the non-conducting region Ω_I is not simply-connected; its first Betti number, which coincides with the dimension of the space $\mathcal{H}_{\mu_I}(m; \Omega_I)$, is equal to one. Moreover, there exists one "cutting" surface Ξ_I^* such that $\Xi_I^* \subset \Omega_I$, $\partial \Xi_I^* \subset \partial \Omega_I$ and the open set $\Omega_I \setminus \Xi_I^*$ is simply-connected (see Figure 8.2). Let p_I^* denote the solution (determined up to an additive constant) of the following elliptic problem

$$\begin{cases} \operatorname{div}(\boldsymbol{\mu}_I \operatorname{grad} p_I^*) = 0 & \text{in } \Omega_I \setminus \Xi_I^* \\ \boldsymbol{\mu}_I \operatorname{grad} p_I^* \cdot \mathbf{n}_I = 0 & \text{on } \partial\Omega_I \setminus \partial\Xi_I^* \\ \left[\boldsymbol{\mu}_I \operatorname{grad} p_I^* \cdot \mathbf{n}_{\Xi_I^*}\right]_{\Xi_I^*} = 0 & \\ [p_I^*]_{\Xi_I^*} = 1 \, , & \end{cases} \tag{8.10}$$

where $[p_I^*]_{\Xi_I^*}$ denotes the jump of p_I^* across Ξ_I^* and $\mathbf{n}_{\Xi_I^*}$ denotes the unit normal vector on Ξ_I^*. Then $\boldsymbol{\rho}_I^* := \widetilde{\operatorname{grad} p_I^*}$ is a basis function of $\mathcal{H}_{\mu_I}(m; \Omega_I)$, $\widetilde{\operatorname{grad} p_I^*}$ denoting the $(L^2(\Omega_I))^3$-extension of $\operatorname{grad} p_I^*$ computed in $\Omega_I \setminus \Xi_I^*$ (see Section A.4), and we can choose the definition of the jump of p_I^* on Ξ_I^* in such a way that $\int_{\partial \Gamma_J} \boldsymbol{\rho}_I^* \cdot \mathbf{t} = 1$, where $\mathbf{t}$ is the tangential vector counterclockwise oriented with respect to $\mathbf{n}_C$ on Γ_J. We also choose, for the sake of definiteness, the orientation of $\mathbf{n}_{\Xi_I^*}$ on Ξ_I^* equal to that of $\mathbf{t}$.

We assume that the current density $\mathbf{J}_e \in (L^2(\Omega))^3$ and, for the sake of simplicity, in the sequel we also assume that $\mathbf{J}_{e,I} = \mathbf{0}$ in Ω_I (the general case is considered in Remark 8.3). Therefore, it follows that $\operatorname{curl} \mathbf{H}_I = \mathbf{0}$ in Ω_I and as a consequence $\mathbf{H}_I = \operatorname{grad} \psi_I^* + K \boldsymbol{\rho}_I^*$ for some $\psi_I^* \in H^1(\Omega_I)$ and $K \in \mathbb{C}$.

From the Stokes theorem

$$I^0 = \int_{\Gamma_J} \operatorname{curl} \mathbf{H}_C \cdot \mathbf{n}_C = \int_{\partial \Gamma_J} \mathbf{H}_C \cdot \mathbf{t} = \int_{\partial \Gamma_J} \mathbf{H}_I \cdot \mathbf{t} = K \int_{\partial \Gamma_J} \boldsymbol{\rho}_I^* \cdot \mathbf{t} = K \, .$$

Hence, when the current intensity is assigned, the main unknowns in our formulation are in fact $\mathbf{E}_C$ and the magnetic scalar potential ψ_I^* only.

Computing the magnetic field from the Faraday equation and inserting it in the Ampère law, we obtain

$$\operatorname{curl}(\boldsymbol{\mu}_C^{-1} \operatorname{curl} \mathbf{E}_C) + i\omega\boldsymbol{\sigma}\mathbf{E}_C = -i\omega\mathbf{J}_{e,C} .$$

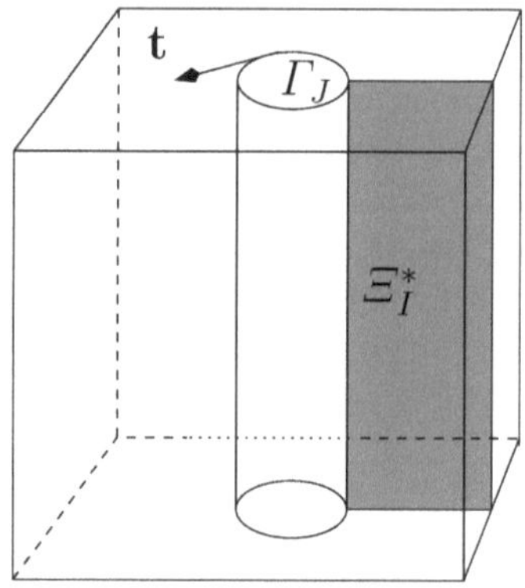

Fig. 8.2. The "cutting" surface

For each $\mathbf{z}_C \in H(\mathrm{curl}; \Omega_C)$, by integration by parts one finds

$$\int_{\Omega_C} \boldsymbol{\mu}_C^{-1} \,\mathrm{curl}\, \mathbf{E}_C \cdot \mathrm{curl}\, \overline{\mathbf{z}_C} + i\omega \int_{\Omega_C} \boldsymbol{\sigma} \mathbf{E}_C \cdot \overline{\mathbf{z}_C} - \int_\Gamma \boldsymbol{\mu}_C^{-1} \,\mathrm{curl}\, \mathbf{E}_C \times \mathbf{n}_C \cdot \overline{\mathbf{z}_C} \\ = -i\omega \int_{\Omega_C} \mathbf{J}_{e,C} \cdot \overline{\mathbf{z}_C} \,.$$

From the Faraday equation and the interface condition

$$\mathbf{H}_C \times \mathbf{n}_C + \mathbf{H}_I \times \mathbf{n}_I = \mathbf{0} \quad \text{on } \Gamma$$

one has that

$$\boldsymbol{\mu}_C^{-1} \,\mathrm{curl}\, \mathbf{E}_C \times \mathbf{n}_C = i\omega \mathbf{H}_I \times \mathbf{n}_I \quad \text{on } \Gamma \,,$$

therefore,

$$\int_{\Omega_C} \boldsymbol{\mu}_C^{-1} \,\mathrm{curl}\, \mathbf{E}_C \cdot \mathrm{curl}\, \overline{\mathbf{z}_C} + i\omega \int_{\Omega_C} \boldsymbol{\sigma} \mathbf{E}_C \cdot \overline{\mathbf{z}_C} - i\omega \int_\Gamma \overline{\mathbf{z}_C} \times \mathbf{n}_C \cdot \mathbf{H}_I \\ = -i\omega \int_{\Omega_C} \mathbf{J}_C \cdot \overline{\mathbf{z}_C} \,. \qquad (8.11)$$

On the other hand, multiplying the Faraday equation by a test function $\mathbf{v}_I = \mathrm{grad}\, \chi_I^*$, with $\chi_I^* \in H^1(\Omega_I)$, by integration by parts one has

$$i\omega \int_{\Omega_I} \boldsymbol{\mu}_I \mathbf{H}_I \cdot \mathrm{grad}\, \overline{\chi_I^*} = -\int_{\Omega_I} \mathrm{curl}\, \mathbf{E}_I \cdot \mathrm{grad}\, \overline{\chi_I^*} = \int_{\partial\Omega_I} \mathbf{E}_I \times \mathbf{n}_I \cdot \mathrm{grad}\, \overline{\chi_I^*} \,.$$

Denoting by χ^* any extension of χ_I^* in $H^1(\Omega)$, from the interface condition

$$\mathbf{E}_C \times \mathbf{n}_C + \mathbf{E}_I \times \mathbf{n}_I = \mathbf{0} \quad \text{on } \Gamma$$

we have

$$\int_{\partial\Omega_I} \mathbf{E}_I \times \mathbf{n}_I \cdot \mathrm{grad}\, \overline{\chi_I^*} \\ = \int_{\partial\Omega} \mathbf{E} \times \mathbf{n} \cdot \mathrm{grad}\, \overline{\chi^*} + \int_\Gamma \mathbf{E}_I \times \mathbf{n}_I \cdot \mathrm{grad}\, \overline{\chi_I^*} - \int_{\Gamma_E \cup \Gamma_J} \mathbf{E}_C \times \mathbf{n}_C \cdot \mathrm{grad}\, \overline{\chi^*} \\ = -\int_\Gamma \mathbf{E}_C \times \mathbf{n}_C \cdot \mathrm{grad}\, \overline{\chi_I^*} \,,$$

because $\mathrm{div}_\tau(\mathbf{E}\times\mathbf{n}) = 0$ on $\partial\Omega$ and $\mathbf{E}_C \times \mathbf{n}_C = \mathbf{0}$ on $\Gamma_E \cup \Gamma_J$. Hence

$$i\omega \int_{\Omega_I} \boldsymbol{\mu}\mathbf{H}_I \cdot \mathrm{grad}\,\overline{\chi_I^*} = -\int_\Gamma \mathbf{E}_C \times \mathbf{n}_C \cdot \mathrm{grad}\,\overline{\chi_I^*}\,. \tag{8.12}$$

In a similar way, taking $\boldsymbol{\rho}_I^*$ as test function one obtains

$$i\omega \int_{\Omega_I} \boldsymbol{\mu}_I\mathbf{H}_I \cdot \boldsymbol{\rho}_I^* = -\int_{\Omega_I} \mathrm{curl}\,\mathbf{E}_I \cdot \boldsymbol{\rho}_I^* = \int_{\partial\Omega_I} \mathbf{E}_I \times \mathbf{n}_I \cdot \boldsymbol{\rho}_I^*\,,$$

and denoting by $\mathbf{R}^*$ any extension of $\boldsymbol{\rho}_I^*$ in $H(\mathrm{curl};\Omega)$ it follows

$$\int_{\partial\Omega_I} \mathbf{E}_I \times \mathbf{n}_I \cdot \boldsymbol{\rho}_I^* = \int_{\partial\Omega} \mathbf{E}\times\mathbf{n}\cdot\mathbf{R}^* - \int_\Gamma \mathbf{E}_C \times \mathbf{n}_C \cdot \boldsymbol{\rho}_I^*\,.$$

Using that $\mathbf{E}\times\mathbf{n} = \mathrm{grad}\,v \times \mathbf{n}$ on $\partial\Omega$ we have

$$\int_{\partial\Omega} \mathbf{E}\times\mathbf{n}\cdot\mathbf{R}^* = \int_{\partial\Omega} \mathrm{grad}\,v\times\mathbf{n}\cdot\mathbf{R}^* = -\int_{\partial\Omega} \mathbf{R}^*\times\mathbf{n}\cdot\mathrm{grad}\,v = \int_{\partial\Omega} \mathrm{curl}\,\mathbf{R}^*\cdot\mathbf{n}\,v\,.$$

Since $\mathrm{curl}\,\mathbf{R}_I^* = \mathrm{curl}\,\boldsymbol{\rho}_I^* = \mathbf{0}$ in Ω_I, $v = V$ on Γ_J and $v = 0$ on Γ_E, using the Stokes theorem on Γ_J we obtain that

$$\int_{\partial\Omega} \mathrm{curl}\,\mathbf{R}^*\cdot\mathbf{n}\,v = V\int_{\Gamma_J} \mathrm{curl}\,\mathbf{R}^*\cdot\mathbf{n}_C = V\int_{\partial\Gamma_J} \boldsymbol{\rho}_I^*\cdot\mathbf{t} = V\,.$$

Hence

$$i\omega \int_{\Omega_I} \boldsymbol{\mu}_I\mathbf{H}_I \cdot \boldsymbol{\rho}_I^* = V - \int_\Gamma \mathbf{E}_C \times \mathbf{n}_C \cdot \boldsymbol{\rho}_I^*\,. \tag{8.13}$$

As we noticed before, $\mathbf{H}_I = \mathrm{grad}\,\psi_I^* + I^0\boldsymbol{\rho}_I^*$ where $\psi_I^* \in H^1(\Omega_I)$ and $I^0 \in \mathbb{C}$ is the current intensity. Moreover, this decomposition is orthogonal in the sense that

$$\begin{aligned}&\textstyle\int_{\Omega_I} \boldsymbol{\mu}_I(\mathrm{grad}\,\psi_I^* + K\boldsymbol{\rho}_I^*)\cdot(\mathrm{grad}\,\overline{\chi_I^*} + \overline{Q}\boldsymbol{\rho}_I^*)\\ &\qquad= \textstyle\int_{\Omega_I} \boldsymbol{\mu}_I\ \mathrm{grad}\,\psi_I^*\cdot\mathrm{grad}\,\overline{\chi_I^*} + K\overline{Q}\int_{\Omega_I} \boldsymbol{\mu}_I\boldsymbol{\rho}_I^*\cdot\boldsymbol{\rho}_I^*\end{aligned}$$

for all ψ_I^*, $\chi_I^* \in H^1(\Omega_I)$ and K, $Q \in \mathbb{C}$. Hence, from (8.11), (8.12) and (8.13), multiplying these two last equations by $-i\omega$, we have that $\mathbf{E}_C$ and $\mathbf{H}_I = \mathrm{grad}\,\psi_I^* + I^0\boldsymbol{\rho}_I^*$ are such that for each $\mathbf{z}_C \in H_{0,\Gamma_C}(\mathrm{curl};\Omega_C)$, $\chi_I^* \in H^1(\Omega_I)$ and $Q \in \mathbb{C}$ it holds

$$\begin{cases}\textstyle\int_{\Omega_C}(\boldsymbol{\mu}_C^{-1}\,\mathrm{curl}\,\mathbf{E}_C\cdot\mathrm{curl}\,\overline{\mathbf{z}_C} + i\omega\boldsymbol{\sigma}\mathbf{E}_C\cdot\overline{\mathbf{z}_C})\\ \qquad\textstyle -i\omega\int_\Gamma \overline{\mathbf{z}_C}\times\mathbf{n}_C\cdot\mathrm{grad}\,\psi_I^* - i\omega I^0\int_\Gamma \overline{\mathbf{z}_C}\times\mathbf{n}_C\cdot\boldsymbol{\rho}_I^*\\ \quad\textstyle = -i\omega\int_{\Omega_C}\mathbf{J}_{e,C}\cdot\overline{\mathbf{z}_C}\\ \textstyle -i\omega\int_\Gamma \mathbf{E}_C\times\mathbf{n}_C\cdot\mathrm{grad}\,\overline{\chi_I^*} + \omega^2\int_{\Omega_I}\boldsymbol{\mu}_I\ \mathrm{grad}\,\psi_I^*\cdot\mathrm{grad}\,\overline{\chi_I^*} = 0\\ \textstyle -i\omega\overline{Q}\int_\Gamma \mathbf{E}_C\times\mathbf{n}_C\cdot\boldsymbol{\rho}_I^* + \omega^2 I^0\overline{Q}\int_{\Omega_I}\boldsymbol{\mu}_I\,\boldsymbol{\rho}_I^*\cdot\boldsymbol{\rho}_I^* = -i\omega V\overline{Q}\,.\end{cases} \tag{8.14}$$

When the voltage V is given and the current intensity I^0 is unknown, these three equations determine $\mathbf{E}_C$, ψ_I^* and I^0. On the other hand, when the current intensity

I^0 is given, the first two equations are enough to determine the two unknowns of the problem $\mathbf{E}_C$ and ψ_I^*. The voltage V can be computed using the third equation.

In conclusion, the *voltage excitation* problem reads

$$\begin{aligned}
&\text{Find } (\mathbf{E}_C, \psi_I^*, I^0) \in H_{0,\Gamma_C}(\text{curl}; \Omega_C) \times H^1(\Omega_I)/\mathbb{C} \times \mathbb{C} : \\
&\int_{\Omega_C} (\boldsymbol{\mu}_C^{-1} \,\text{curl}\, \mathbf{E}_C \cdot \text{curl}\, \overline{\mathbf{z}_C} + i\omega\boldsymbol{\sigma}\mathbf{E}_C \cdot \overline{\mathbf{z}_C}) \\
&\qquad - i\omega \int_\Gamma \overline{\mathbf{z}_C} \times \mathbf{n}_C \cdot \text{grad}\, \psi_I^* - i\omega I^0 \int_\Gamma \overline{\mathbf{z}_C} \times \mathbf{n}_C \cdot \boldsymbol{\rho}_I^* \\
&\quad = -i\omega \int_{\Omega_C} \mathbf{J}_{e,C} \cdot \overline{\mathbf{z}_C} \\
&-i\omega \int_\Gamma \mathbf{E}_C \times \mathbf{n}_C \cdot \text{grad}\, \overline{\chi_I^*} + \omega^2 \int_{\Omega_I} \boldsymbol{\mu}_I \,\text{grad}\, \psi_I^* \cdot \text{grad}\, \overline{\chi_I^*} = 0 \\
&-i\omega\overline{Q} \int_\Gamma \mathbf{E}_C \times \mathbf{n}_C \cdot \boldsymbol{\rho}_I^* + \omega^2 I^0 \overline{Q} \int_{\Omega_I} \boldsymbol{\mu}_I \boldsymbol{\rho}_I^* \cdot \boldsymbol{\rho}_I^* = -i\omega V \overline{Q} \\
&\text{for all } (\mathbf{z}_C, \chi_I^*, Q) \in H_{0,\Gamma_C}(\text{curl}; \Omega_C) \times H^1(\Omega_I)/\mathbb{C} \times \mathbb{C} ,
\end{aligned} \tag{8.15}$$

while the *current intensity excitation* problem is given by

$$\begin{aligned}
&\text{Find } (\mathbf{E}_C, \psi_I^*) \in H_{0,\Gamma_C}(\text{curl}; \Omega_C) \times H^1(\Omega_I)/\mathbb{C} : \\
&\int_{\Omega_C} (\boldsymbol{\mu}_C^{-1} \,\text{curl}\, \mathbf{E}_C \cdot \text{curl}\, \overline{\mathbf{z}_C} + i\omega\boldsymbol{\sigma}\mathbf{E}_C \cdot \overline{\mathbf{z}_C}) \\
&\qquad - i\omega \int_\Gamma \overline{\mathbf{z}_C} \times \mathbf{n}_C \cdot \text{grad}\, \psi_I^* \\
&\quad = -i\omega \int_{\Omega_C} \mathbf{J}_{e,C} \cdot \overline{\mathbf{z}_C} + i\omega I^0 \int_\Gamma \overline{\mathbf{z}_C} \times \mathbf{n}_C \cdot \boldsymbol{\rho}_I^* \\
&-i\omega \int_\Gamma \mathbf{E}_C \times \mathbf{n}_C \cdot \text{grad}\, \overline{\chi_I^*} + \omega^2 \int_{\Omega_I} \boldsymbol{\mu}_I \,\text{grad}\, \psi_I^* \cdot \text{grad}\, \overline{\chi_I^*} = 0 \\
&\text{for all } (\mathbf{z}_C, \chi_I^*) \in H_{0,\Gamma_C}(\text{curl}; \Omega_C) \times H^1(\Omega_I)/\mathbb{C} .
\end{aligned} \tag{8.16}$$

If $(\mathbf{E}_C, \psi_I^*)$ is the solution of the current intensity excitation problem then

$$V = \int_\Gamma \mathbf{E}_C \times \mathbf{n}_C \cdot \boldsymbol{\rho}_I^* + i\omega I^0 \int_{\Omega_I} \boldsymbol{\mu}_I \boldsymbol{\rho}_I^* \cdot \boldsymbol{\rho}_I^* . \tag{8.17}$$

Remark 8.3. When $\mathbf{J}_{e,I}$ is not vanishing, it has to satisfy the necessary condition $\text{div}\, \mathbf{J}_{e,I} = 0$ in Ω_I. Therefore, since the boundary of Ω_I is connected, there exists a vector field $\mathbf{H}_{e,I}^* \in H(\text{curl}; \Omega_I)$ such that $\text{curl}\, \mathbf{H}_{e,I}^* = \mathbf{J}_{e,I}$ in Ω_I (for its explicit construction, see Section 5.4.1). We can thus repeat the arguments of this section for $\mathbf{H}_I - \mathbf{H}_{e,I}^* = \text{grad}\, \psi_I^* + I^0 \boldsymbol{\rho}_I^*$.

In particular, it can be easily checked that we have only to modify the right hand sides of (8.15) or (8.16), (8.17). To be more specific, at the right hand side of $(8.15)_1$, $(8.15)_2$ and $(8.15)_3$ we have to add $i\omega \int_\Gamma \overline{\mathbf{w}_C} \times \mathbf{n}_C \cdot \mathbf{H}_I^*$, $-\omega^2 \int_{\Omega_I} \boldsymbol{\mu}_I \mathbf{H}_I^* \cdot \text{grad}\, \overline{\chi_I^*}$ and $-\omega^2 \overline{Q} \int_{\Omega_I} \boldsymbol{\mu}_I \mathbf{H}_I^* \cdot \boldsymbol{\rho}_I^*$, respectively. In a similar way one has to proceed for equations (8.16), (8.17). □

In order to analyze these two problems, let us recall the sesquilinear form and the anti-linear functional introduced in (4.3) and (4.5), respectively

$$\begin{aligned}
\mathcal{C}((\mathbf{w}_C, \mathbf{u}_I), (\mathbf{z}_C, \mathbf{v}_I)) := &\int_{\Omega_C} (\boldsymbol{\mu}_C^{-1} \,\text{curl}\, \mathbf{w}_C \cdot \text{curl}\, \overline{\mathbf{z}_C} + i\omega\boldsymbol{\sigma}\mathbf{w}_C \cdot \overline{\mathbf{z}_C}) \\
&+ \omega^2 \int_{\Omega_I} \boldsymbol{\mu}_I \mathbf{u}_I \cdot \overline{\mathbf{v}_I} - i\omega \big[\int_\Gamma \overline{\mathbf{z}_C} \times \mathbf{n}_C \cdot \mathbf{u}_I + \int_\Gamma \mathbf{w}_C \times \mathbf{n}_C \cdot \overline{\mathbf{v}_I} \big]
\end{aligned}$$

$$L(\mathbf{z}_C) := -i\omega \int_{\Omega_C} \mathbf{J}_{e,C} \cdot \overline{\mathbf{z}_C}\,;$$

moreover, define the antilinear functionals

$$\begin{aligned} S_V(\mathbf{v}_I) &:= -i\omega c_0 V \int_{\Omega_I} \boldsymbol{\mu}_I \boldsymbol{\rho}_I^* \cdot \overline{\mathbf{v}_I} \\ S_{I^0}(\mathbf{z}_C) &:= i\omega I^0 \int_\Gamma \overline{\mathbf{z}_C} \times \mathbf{n}_C \cdot \boldsymbol{\rho}_I^*\,, \end{aligned}$$

where V and I^0 are given complex constants and $c_0 := (\int_{\Omega_I} \boldsymbol{\mu}_I \boldsymbol{\rho}_I^* \cdot \boldsymbol{\rho}_I^*)^{-1}$. Recall that any $\mathbf{v}_I \in H^0(\text{curl}; \Omega_I)$ can be univocally decomposed as $\mathbf{v}_I = \text{grad}\,\chi_I^* + Q\boldsymbol{\rho}_I^*$, with $\chi_I^* \in H^1(\Omega_I)/\mathbb{C}$ and $Q \in \mathbb{C}$, and that $\text{grad}\,\chi_I^*$ and $\boldsymbol{\rho}_I^*$ are orthogonal with respect to the scalar product $(\mathbf{u}_I, \mathbf{v}_I)_{\mu_I,\Omega_I} := \int_{\Omega_I} \boldsymbol{\mu}_I \mathbf{u}_I \cdot \mathbf{v}_I$. Then for those $\mathbf{v}_I$ we have $S_V(\mathbf{v}_I) = -i\omega V\overline{Q}$.

Using this orthogonal decomposition of $H^0(\text{curl}; \Omega_I)$, it is easy to see that problem (8.15) is equivalent to the following one

$$\begin{aligned} &\text{Find } (\mathbf{E}_C, \mathbf{H}_I) \in H_{0,\Gamma_C}(\text{curl}; \Omega_C) \times H^0(\text{curl}; \Omega_I) \;: \\ &\quad \mathcal{C}((\mathbf{E}_C, \mathbf{H}_I), (\mathbf{z}_C, \mathbf{v}_I)) = L(\mathbf{z}_C) + S_V(\mathbf{v}_I) \\ &\text{for all } (\mathbf{z}_C, \mathbf{v}_I) \in H_{0,\Gamma_C}(\text{curl}; \Omega_C) \times H^0(\text{curl}; \Omega_I)\,, \end{aligned} \tag{8.18}$$

whereas problem (8.16) is equivalent to

$$\begin{aligned} &\text{Find } (\mathbf{E}_C, \psi_I^*) \in H_{0,\Gamma_C}(\text{curl}; \Omega_C) \times H^1(\Omega_I)/\mathbb{C} \;: \\ &\quad \mathcal{C}((\mathbf{E}_C, \text{grad}\,\psi_I^*), (\mathbf{z}_C, \text{grad}\,\chi_I^*)) = L(\mathbf{z}_C) + S_{I^0}(\mathbf{z}_C) \\ &\text{for all } (\mathbf{z}_C, \chi_I^*) \in H_{0,\Gamma_C}(\text{curl}; \Omega_C) \times H^1(\Omega_I)/\mathbb{C}\,. \end{aligned} \tag{8.19}$$

The antilinear functionals $L(\cdot)$ and $S_{I^0}(\cdot)$ are clearly continuous in the space $H(\text{curl}; \Omega_C)$, whereas $S_V(\cdot)$ is continuous in $H^0(\text{curl}; \Omega_I)$. The sesquilinear form $\mathcal{C}(\cdot,\cdot)$ is continuous in $H_{0,\Gamma_C}(\text{curl}; \Omega_C) \times H^0(\text{curl}; \Omega_I)$; moreover in Theorem 4.1 we proved that it is coercive, hence from the Lax–Milgram lemma we have obtained:

Theorem 8.4. *There exists a unique solution of both problems* (8.18) *and* (8.19).

Proceeding as in Section 3.2, it is easy to see that the weak solutions $\mathbf{E}_C$ and $\mathbf{H}_I = \text{grad}\,\psi_I^* + I^0\boldsymbol{\rho}_I^*$ of problems (8.18) or (8.19), together with

$$\mathbf{H}_C = -(i\omega)^{-1}\boldsymbol{\mu}_C^{-1}\,\text{curl}\,\mathbf{E}_C\,, \tag{8.20}$$

furnish indeed a solution of the strong eddy current problem

$$\begin{cases} \text{curl}\,\mathbf{E}_C + i\omega\boldsymbol{\mu}_C\mathbf{H}_C = \mathbf{0} & \text{in } \Omega_C \\ \text{curl}\,\mathbf{H} - \boldsymbol{\sigma}\mathbf{E}_C = \mathbf{J}_e & \text{in } \Omega \\ \text{div}(\boldsymbol{\mu}_I\mathbf{H}_I) = 0 & \text{in } \Omega_I \\ \mathbf{E}_C \times \mathbf{n}_C = \mathbf{0} & \text{on } \Gamma_E \cup \Gamma_J \\ \boldsymbol{\mu}\mathbf{H} \cdot \mathbf{n} = 0 & \text{on } \partial\Omega \\ \boldsymbol{\mu}_I\mathbf{H}_I \cdot \mathbf{n}_I + \boldsymbol{\mu}_C\mathbf{H}_C \cdot \mathbf{n}_C = 0 & \text{on } \Gamma\,, \end{cases} \tag{8.21}$$

with in addition, for the assigned voltage problem,

$$\int_{\Gamma} \mathbf{E}_C \times \mathbf{n}_C \cdot \boldsymbol{\rho}_I^* + i\omega \int_{\Omega_I} \boldsymbol{\mu}_I \mathbf{H}_I \cdot \boldsymbol{\rho}_I^* = V \,, \tag{8.22}$$

or, for the assigned current intensity,

$$\int_{\Gamma_J} \operatorname{curl} \mathbf{H}_C \cdot \mathbf{n}_C = I^0 \,. \tag{8.23}$$

Remark 8.5. Taking into account that $\boldsymbol{\rho}_I^* := \widetilde{\operatorname{grad} p_I^*}$, where p_I^* is defined in (8.10), it is easy to see that (8.22) formally corresponds to

$$\int_{\widetilde{\gamma}} \mathbf{E}_C \cdot d\boldsymbol{\tau} + i\omega \int_{\Xi_I^*} \boldsymbol{\mu}_I \mathbf{H}_I \cdot \mathbf{n}_{\Xi_I^*} = V \,,$$

where $\widetilde{\gamma} = \partial \Xi_I^* \cap \Gamma$, oriented from Γ_E to Γ_J. Hence, if we find an electric field $\mathbf{E}_I$ satisfying the Faraday equation on the surface Ξ_I^*, we have

$$\int_{\widehat{\gamma}} \mathbf{E}_I \cdot d\boldsymbol{\tau} = V \,,$$

where $\widehat{\gamma} = \partial \Xi_I^* \cap \Gamma_I$, oriented from the electric port Γ_E to the electric port Γ_J. □

Now we need to find the electric field $\mathbf{E}_I$ in Ω_I. Clearly we have to solve the Faraday equation $\operatorname{curl} \mathbf{E}_I + i\omega \boldsymbol{\mu}_I \mathbf{H}_I = \mathbf{0}$ in Ω_I and the interface condition $\mathbf{E}_I \times \mathbf{n}_I + \mathbf{E}_C \times \mathbf{n}_C = \mathbf{0}$ on Γ, where $\mathbf{H}_I$ and $\mathbf{E}_C$ can be seen as given data. This problem has infinitely many solutions, depending on the type of "gauge" conditions we choose for $\mathbf{E}_I$ (but none of these solutions has a vanishing tangential component on $\partial\Omega$: see Proposition 8.2). For instance, as we see here below, it is possible to solve the following problem:

$$\begin{cases} \operatorname{curl} \mathbf{E}_I = -i\omega \boldsymbol{\mu}_I \mathbf{H}_I & \text{in } \Omega_I \\ \operatorname{div}(\varepsilon_I \mathbf{E}_I) = 0 & \text{in } \Omega_I \\ \mathbf{E}_I \times \mathbf{n}_I = -\mathbf{E}_C \times \mathbf{n}_C & \text{on } \Gamma \\ \varepsilon_I \mathbf{E}_I \cdot \mathbf{n}_I = 0 & \text{on } \Gamma_I \,, \end{cases} \tag{8.24}$$

that is similar to (3.13).

Theorem 8.6. *There exists a unique solution* $\mathbf{E}_I$ *of problem* (8.24).

Proof. It is easy to see that the solution of (8.24) is unique since the space

$$\begin{aligned} \mathcal{H}_{\varepsilon_I}(\Gamma, \Gamma_I; \Omega_I) := \{ \mathbf{v}_I \in (L^2(\Omega_I))^3 |\ & \operatorname{curl} \mathbf{v}_I = \mathbf{0}, \operatorname{div}(\varepsilon_I \mathbf{v}_I) = 0, \\ & \varepsilon_I \mathbf{v}_I \cdot \mathbf{n}_I = 0 \text{ on } \Gamma_I, \ \mathbf{v}_I \times \mathbf{n}_I = \mathbf{0} \text{ on } \Gamma \} \end{aligned}$$

is trivial in the geometrical situation under consideration. In fact, given a function $\mathbf{v}_I \in \mathcal{H}_{\varepsilon_I}(\Gamma, \Gamma_I; \Omega_I)$, one has $\operatorname{curl} \mathbf{v}_I = \mathbf{0}$ in $\Omega_I \setminus \Xi_I^*$, that is a simply-connected subset. Hence there exists $U_I \in H^1(\Omega_I \backslash \Xi_I^*)$ such that $\operatorname{grad} U_I = \mathbf{v}_I$ and

$$\begin{cases} \operatorname{div}(\varepsilon_I \operatorname{grad} U_I) = 0 & \text{in } \Omega_I \backslash \Xi_I^* , \\ \varepsilon_I \operatorname{grad} U_I \cdot \mathbf{n}_I = 0 & \text{on } \Gamma_I \setminus \partial \Xi_I^* , \\ U_I = \kappa^* & \text{on } \Gamma \setminus \partial \Xi_I^* , \\ [U_I]_{\Xi_I^*} = c^* , & \\ [\varepsilon_I \operatorname{grad} U_I \cdot \mathbf{n}_I]_{\Xi_I^*} = 0 , & \end{cases} \tag{8.25}$$

κ^* and c^* being constants. Since $\Gamma \cap \Xi_I^* \neq \emptyset$, it follows that the constant c^* must be zero; therefore the unique solution of (8.25) is $U_I = \kappa^*$ and consequently $\mathbf{v}_I = \mathbf{0}$.

The existence of the solution to (8.24) can be proved as in Theorem 3.3, noting that, similarly to what shown here above, the space of harmonic fields

$$\mathcal{H}(\Gamma_I, \Gamma; \Omega_I) := \{\mathbf{v}_I \in (L^2(\Omega_I))^3 | \operatorname{curl} \mathbf{v}_I = \mathbf{0}, \operatorname{div} \mathbf{v}_I = 0, \\ \mathbf{v}_I \cdot \mathbf{n}_I = 0 \text{ on } \Gamma, \mathbf{v}_I \times \mathbf{n}_I = \mathbf{0} \text{ on } \Gamma_I\}$$

is trivial. □

This ends the proof of the existence of a solution to the current intensity excitation problem (8.1), (8.2) and (8.5): it is enough to take the solution $(\mathbf{E}_C, \psi_I^*)$ of problem (8.19), $\mathbf{H}_I = \operatorname{grad} \psi_I^* + I^0 \boldsymbol{\rho}_I^*$, $\mathbf{H}_C = -(i\omega)^{-1} \boldsymbol{\mu}_C^{-1} \operatorname{curl} \mathbf{E}_C$ and the solution $\mathbf{E}_I$ of problem (8.24).

Concerning the voltage excitation problem, one takes the solution $(\mathbf{E}_C, \mathbf{H}_I)$ of problem (8.18) (or, equivalently, the solution $(\mathbf{E}_C, \psi_I^*, I^0)$ of problem (8.15), with $\mathbf{H}_I = \operatorname{grad} \psi_I^* + I^0 \boldsymbol{\rho}_I^*$), $\mathbf{H}_C = -(i\omega)^{-1} \boldsymbol{\mu}_C^{-1} \operatorname{curl} \mathbf{E}_C$ and the solution $\mathbf{E}_I$ of problem (8.24). Since with this choice the Faraday equation is satisfied in all of Ω, we obtain that

$$\operatorname{div}_\tau(\mathbf{E} \times \mathbf{n}) = \operatorname{curl} \mathbf{E} \cdot \mathbf{n} = -i\omega \boldsymbol{\mu} \mathbf{H} \cdot \mathbf{n} = 0 \quad \text{on } \partial\Omega ,$$

hence $\mathbf{E} \times \mathbf{n} = \operatorname{grad} U \times \mathbf{n}$ on $\partial\Omega$, where the boundary potential U is constant on Γ_J and vanishes on Γ_E. By proceeding as was done for obtaining (8.13) it is readily seen that $\mathbf{E}_C$ and $\mathbf{H}_I$ satisfies (8.13) with V replaced by $U_{|\Gamma_J}$. Comparing this last equation with $(8.15)_3$ it follows $U_{|\Gamma_J} = V$, hence $(\mathbf{E}, \mathbf{H})$ satisfies (8.1), (8.2) and (8.3).

Remark 8.7. Let us underline an important difference between the eddy current problem for an internal conductor (with magnetic or electric boundary conditions) and the eddy current problem with the electric port boundary conditions (8.2) and assigned voltage (8.3) or current intensity (8.5). In the former case the spaces of harmonic fields $\mathcal{H}_{\mu_I}(\partial\Omega, \Gamma; \Omega_I)$ (relevant for the magnetic boundary conditions) or $\mathcal{H}_{\mu_I}(m; \Omega_I)$ (relevant for the electric boundary condition) are non-trivial, therefore the strong form of the eddy current problem in terms of the magnetic field $\mathbf{H}$ contains the topological conditions $(3.23)_4$, $(3.23)_5$ (for the magnetic boundary conditions) or $(3.42)_4$ (for the electric boundary condition). In the electric port case with the no-flux boundary conditions the space of harmonic fields $\mathcal{H}_{\mu_I}(\Gamma_I, \Gamma; \Omega_I)$ is trivial, therefore no topological conditions are present in the strong formulation in terms of $\mathbf{H}$: one has only to impose the Faraday equation in Ω_C, the Ampère equation in Ω and the Gauss magnetic equation in Ω. □

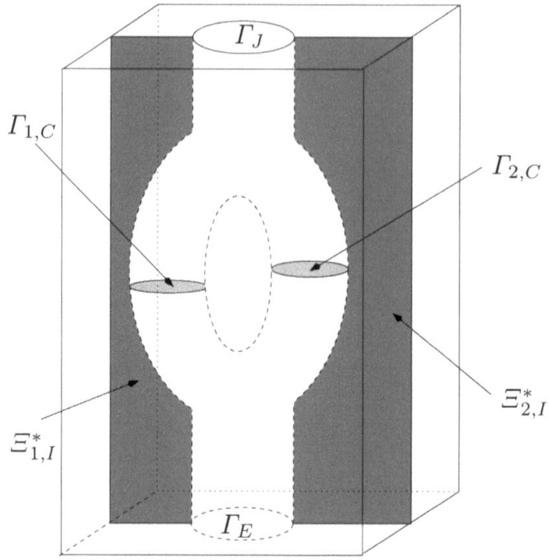

Fig. 8.3. A conductor which is not simply-connected

Remark 8.8. The two formulations (8.18) and (8.19) can be adapted to the case of a connected but not simply-connected conductor Ω_C with two electric ports $\partial\Omega_C \cap \partial\Omega = \Gamma_E \cup \Gamma_J$ (see Figure 8.3). In this case the space $\mathcal{H}_{\mu_I}(m;\Omega_I)$ has dimension $n_{\Omega_I} > 1$, the first Betti number of Ω_I, or, equivalently, the number of independent non-bounding cycles in Ω_I. Given a basis of $\mathcal{H}_{\mu_I}(m;\Omega_I)$, $\{\boldsymbol{\rho}^*_{\alpha,I}\}_{\alpha=1}^{n_{\Omega_I}}$, any function $\mathbf{v}_I \in H^0(\mathrm{curl};\Omega_I)$ can be written as $\mathbf{v}_I = \mathrm{grad}\,\chi^*_I + \sum_{\alpha=1}^{n_{\Omega_I}} Q_\alpha \boldsymbol{\rho}^*_{\alpha,I}$ for some $\chi^*_I \in H^1(\Omega_I)/\mathbb{C}$ and $Q_\alpha \in \mathbb{C}$, $\alpha = 1,\ldots,n_{\Omega_I}$. If the basis functions $\boldsymbol{\rho}^*_{\alpha,I}$ are suitably chosen, we can find n_{Ω_I} connected orientable Lipschitz surfaces $\{\Gamma_{1,C},\ldots,\Gamma_{n_{\Omega_I},C}\}$ contained in $\overline{\Omega_C}$ such that $\gamma_\alpha := \partial\Gamma_{\alpha,C}$ are independent non-bounding cycles in $\overline{\Omega_I}$ for which

$$\int_{\gamma_\beta} \boldsymbol{\rho}^*_{\alpha,I}\cdot\mathbf{t} = \delta_{\alpha\beta}\,, \quad \alpha,\beta = 1,\ldots,n_{\Omega_I}$$

(see, e.g., Hiptmair and Ostrowski [128]). Since the magnetic field $\mathbf{H}_I$ can be written as $\mathbf{H}_I = \mathrm{grad}\,\psi^*_I + \sum_{\alpha=1}^{n_{\Omega_I}} K_\alpha \boldsymbol{\rho}^*_{\alpha,I}$, from the Stokes theorem we have

$$K_\alpha = \int_{\gamma_\alpha} \mathbf{H}_I\cdot\mathbf{t} = \int_{\gamma_\alpha} \mathbf{H}_C\cdot\mathbf{t} = \int_{\Gamma_{\alpha,C}} \mathrm{curl}\,\mathbf{H}_C\cdot\mathbf{n}_\alpha = I^0_\alpha\,,$$

where $\mathbf{n}_\alpha$ is the unit normal vector to $\Gamma_{\alpha,C}$ such that $\mathbf{t}$ is counterclockwise oriented with respect to $\mathbf{n}_\alpha$ on $\Gamma_{\alpha,C}$. The quantity I^0_α is the current intensity through the surface $\Gamma_{\alpha,C}$.

Multiplying the Faraday equation by the function $\boldsymbol{\rho}^*_{\beta,I}$ and proceeding as in the case of a simply-connected conductor, we obtain

$$i\omega\int_{\Omega_I} \boldsymbol{\mu}_I\mathbf{H}_I\cdot\boldsymbol{\rho}^*_{\beta,I} = V\int_{\partial\Gamma_J} \boldsymbol{\rho}^*_{\beta,I}\cdot\mathbf{t} - \int_{\Gamma} \mathbf{E}_C\times\mathbf{n}_C\cdot\boldsymbol{\rho}^*_{\beta,I}\,.$$

So we have obtained that $\mathbf{E}_C$ and $\mathbf{H}_I = \operatorname{grad}\psi_I^* + \sum_{\alpha=1}^{n_{\Omega_I}} I_\alpha^0 \boldsymbol{\rho}_{\alpha,I}^*$ are such that for each $\mathbf{z}_C \in H_{0,\Gamma_C}(\operatorname{curl};\Omega_C)$, $\chi_I^* \in H^1(\Omega_I)$ and $\mathbf{Q} \in \mathbb{C}^{n_{\Omega_I}}$ it holds

$$
\begin{aligned}
&\int_{\Omega_C}(\boldsymbol{\mu}_C^{-1}\operatorname{curl}\mathbf{E}_C\cdot\operatorname{curl}\overline{\mathbf{z}_C} + i\omega\boldsymbol{\sigma}\mathbf{E}_C\cdot\overline{\mathbf{z}_C})\\
&\qquad -i\omega\int_\Gamma \overline{\mathbf{z}_C}\times\mathbf{n}_C\cdot\operatorname{grad}\psi_I^* - i\omega\sum_{\alpha=1}^{n_{\Omega_I}} I_\alpha^0\int_\Gamma\overline{\mathbf{z}}_C\times\mathbf{n}_C\cdot\boldsymbol{\rho}_{\alpha,I}^*\\
&\qquad\qquad = -i\omega\int_{\Omega_C}\mathbf{J}_{e,C}\cdot\overline{\mathbf{z}_C}\\
&-i\omega\int_\Gamma\mathbf{E}_C\times\mathbf{n}_C\cdot\operatorname{grad}\overline{\chi_I^*} + \omega^2\int_{\Omega_I}\boldsymbol{\mu}_I\operatorname{grad}\psi_I^*\cdot\operatorname{grad}\overline{\chi_I^*} = 0\\
&-i\omega\overline{Q}_\beta\int_\Gamma\mathbf{E}_C\times\mathbf{n}_C\cdot\boldsymbol{\rho}_{\beta,I}^* + \omega^2\overline{Q}_\beta\sum_{\alpha=1}^{n_{\Omega_I}} I_\alpha^0\int_{\Omega_I}\boldsymbol{\mu}_I\boldsymbol{\rho}_{\alpha,I}^*\cdot\boldsymbol{\rho}_{\beta,I}^*\\
&\qquad\qquad = -i\omega\overline{Q}_\beta V\int_{\partial\Gamma_J}\boldsymbol{\rho}_{\beta,I}^*\cdot\mathbf{t}\ ,\ \forall\beta = 1,\ldots,n_{\Omega_I}\,.
\end{aligned}
\tag{8.26}
$$

On the other hand, if Ω_C has q connected components $\Omega_{C,j}$, $j = 1,\ldots,q$, each one with two electric ports, then there are q different voltages V_j. In fact, on $\partial\Omega$ we have $\mathbf{E}\times\mathbf{n} = \operatorname{grad} v\times\mathbf{n}$, and, setting $\partial\Omega_{C,j}\cap\partial\Omega = \Gamma_{J,j}\cup\Gamma_{E,j}$, with $\Gamma_{J,j}$ and $\Gamma_{E,j}$ disjoint and connected surfaces, we have $v_{|\Gamma_{J,j}} = V_j^1$ and $v_{|\Gamma_{E,j}} = V_j^0$, where V_j^1 and V_j^0 are complex constants; then the voltages are defined as $V_j = V_j^1 - V_j^0$.

Multiplying the Faraday equation by $\boldsymbol{\rho}_{\beta,I}^*$, a basis function of the space $\mathcal{H}_{\mu_I}(m;\Omega_I)$, by integration by parts one has

$$
i\omega\int_{\Omega_I}\boldsymbol{\mu}_I\mathbf{H}_I\cdot\boldsymbol{\rho}_{\beta,I}^* = \int_{\partial\Omega_I}\mathbf{E}_I\times\mathbf{n}_I\cdot\boldsymbol{\rho}_{\beta,I}^* = \int_{\partial\Omega}\mathbf{E}\times\mathbf{n}\cdot\mathbf{R}_\beta^* - \int_\Gamma\mathbf{E}_C\times\mathbf{n}_C\cdot\boldsymbol{\rho}_{\beta,I}^*\,,
$$

where $\mathbf{R}_\beta^*$ is any extension of $\boldsymbol{\rho}_{\beta,I}^*$ in $H(\operatorname{curl};\Omega)$. Moreover

$$
\begin{aligned}
\int_{\partial\Omega}\mathbf{E}\times\mathbf{n}\cdot\mathbf{R}_\beta^* &= \int_{\partial\Omega}\operatorname{curl}\mathbf{R}_\beta^*\cdot\mathbf{n}\,v\\
&= \sum_{j=1}^q\left(V_j^1\int_{\Gamma_{J,j}}\operatorname{curl}\mathbf{R}_\beta^*\cdot\mathbf{n}_C + V_j^0\int_{\Gamma_{E,j}}\operatorname{curl}\mathbf{R}_\beta^*\cdot\mathbf{n}_C\right)\\
&= \sum_{j=1}^q V_j\int_{\Gamma_{J,j}}\operatorname{curl}\mathbf{R}_\beta^*\cdot\mathbf{n}_C\\
&= \sum_{j=1}^q V_j\int_{\partial\Gamma_{J,j}}\boldsymbol{\rho}_{\beta,I}^*\cdot\mathbf{t}\,,
\end{aligned}
$$

since, denoting by $\Gamma_j = \partial\Omega_{C,j}\setminus(\Gamma_{J,j}\cup\Gamma_{E,j})$, from the Stokes theorem for closed surfaces we have

$$
\begin{aligned}
&\int_{\Gamma_{E,j}}\operatorname{curl}\mathbf{R}_\beta^*\cdot\mathbf{n}_C\\
&\quad = \int_{\partial\Omega_{C,j}}\operatorname{curl}\mathbf{R}_\beta^*\cdot\mathbf{n}_C - \int_{\Gamma_{J,j}}\operatorname{curl}\mathbf{R}_\beta^*\cdot\mathbf{n}_C - \int_{\Gamma_j}\operatorname{curl}\mathbf{R}_\beta^*\cdot\mathbf{n}_C\\
&\quad = -\int_{\Gamma_{J,j}}\operatorname{curl}\mathbf{R}_\beta^*\cdot\mathbf{n}_C - \int_{\Gamma_j}\operatorname{curl}\boldsymbol{\rho}_{\beta,I}^*\cdot\mathbf{n}_C = -\int_{\Gamma_{J,j}}\operatorname{curl}\mathbf{R}_\beta^*\cdot\mathbf{n}_C\,.
\end{aligned}
$$

So, the third equation in (8.26) becomes

$$
\begin{aligned}
-i\omega\overline{Q}_\beta\int_\Gamma\mathbf{E}_C\times\mathbf{n}_C\cdot\boldsymbol{\rho}_{\beta,I}^* &+ \omega^2\overline{Q}_\beta\sum_{\alpha=1}^{n_{\Omega_I}} I_\alpha^0\int_{\Omega_I}\boldsymbol{\mu}_I\boldsymbol{\rho}_{\alpha,I}^*\cdot\boldsymbol{\rho}_{\beta,I}^*\\
&= -i\omega\overline{Q}_\beta\sum_{j=1}^q V_j\int_{\partial\Gamma_{J,j}}\boldsymbol{\rho}_{\beta,I}^*\cdot\mathbf{t}\ ,\ \forall\beta = 1,\ldots,n_{\Omega_I}\,.
\end{aligned}
$$

In the voltage excitation problem the q voltages V_j are given; the unknowns of the problem are the electric field in the conductor, the function ψ_I^* appearing in the orthogonal decomposition of $\mathbf{H}_I$ and the n_{Ω_I} intensities I_α^0.

In the current intensity problem the n_{Ω_I} current intensities I_α^0 are given and therefore the unknowns of the problem are the electric field in the conductor and the function

ψ_I^*. For this problem the q voltages V_j can be then computed in the following way: for each $j = 1, \ldots, q$, let $\boldsymbol{\rho}^*_{\beta(j),I}$ be a basis function of $\mathcal{H}_{\mu_I}(m;\Omega_I)$ corresponding to a non-bounding cycle $\gamma_{\beta(j)} = \partial\Gamma_{\beta(j),C}$ such that $\Gamma_{\beta(j),C} \subset \overline{\Omega_{C,j}}$. For this basis function one has in particular that $\int_{\partial\Gamma_{J,s}} \boldsymbol{\rho}^*_{\beta(j),I} \cdot \mathbf{t} = 0$ for $s = 1, \ldots, q$, $s \neq j$. Then

$$V_j = \frac{\int_\Gamma \mathbf{E}_C \times \mathbf{n}_C \cdot \boldsymbol{\rho}^*_{\beta(j),I} + i\omega \sum_{\alpha=1}^{n_{\Omega_I}} I^0_\alpha \int_{\Omega_I} \boldsymbol{\mu}_I\, \boldsymbol{\rho}^*_{\alpha,I} \cdot \boldsymbol{\rho}^*_{\beta(j),I}}{\int_{\partial\Gamma_{J,j}} \boldsymbol{\rho}^*_{\beta(j),I} \cdot \mathbf{t}},$$

and this value depends on j but not on the choice of $\beta(j)$, as it can be checked that it is equal to the line integral of $\mathbf{E}$ on a curve lying on $\partial\Omega$ and connecting $\Gamma_{E,j}$ to $\Gamma_{J,j}$.

Let us also note that, when Ω_C has $q > 1$ connected components, the electric field $\mathbf{E}_I$ determined in (8.24) is not unique, unless one also requires that $\int_{\Gamma_j} \varepsilon_I \mathbf{E}_I \cdot \mathbf{n}_I = 0$ for each $j = 1, \ldots, q-1$.

For the sake of simplicity, in the following we limit ourselves to the case of a simply-connected conductor. □

8.1.2 Formulations in terms of $\mathbf{H}_C$ and ψ_I^*

In this section we follow Bermúdez et al. [45]. We again assume that $\mathbf{J}_{e,I} = \mathbf{0}$ in Ω_I (otherwise, we can proceed as in Remark 8.3). If $\mathbf{E}$ and $\mathbf{H}$ are the solutions of the eddy current problem subject to voltage or current intensity excitation, whose existence and uniqueness have been proved in the preceding section, multiplying the Faraday equation by a test function $\overline{\mathbf{v}^*}$ satisfying $\operatorname{curl} \mathbf{v}_I^* = \mathbf{0}$ in Ω_I and integrating by parts we find

$$\int_\Omega \mathbf{E}_C \cdot \operatorname{curl} \overline{\mathbf{v}_C^*} + \int_\Omega i\omega\boldsymbol{\mu}\mathbf{H} \cdot \overline{\mathbf{v}^*} - \int_{\partial\Omega} \mathbf{E} \times \mathbf{n} \cdot \overline{\mathbf{v}^*} = 0\,.$$

Knowing that $\mathbf{E} \times \mathbf{n} = \operatorname{grad} v \times \mathbf{n}$ on $\partial\Omega$ we have

$$\begin{aligned}\int_{\partial\Omega} \mathbf{E} \times \mathbf{n} \cdot \overline{\mathbf{v}^*} &= \int_{\partial\Omega} \operatorname{grad} v \times \mathbf{n} \cdot \overline{\mathbf{v}^*} = -\int_{\partial\Omega} \operatorname{grad} v \cdot \overline{\mathbf{v}^*} \times \mathbf{n}\\ &= \int_{\partial\Omega} v \operatorname{div}_\tau(\overline{\mathbf{v}^*} \times \mathbf{n}) = \int_{\partial\Omega} v \operatorname{curl} \overline{\mathbf{v}^*} \cdot \mathbf{n} = V \int_{\Gamma_J} \operatorname{curl} \overline{\mathbf{v}_C^*} \cdot \mathbf{n}\,.\end{aligned}$$

Moreover, we can write $\mathbf{v}_I^* = \operatorname{grad} \chi_I^* + Q\boldsymbol{\rho}_I^*$ in Ω_I, where $\chi_I^* \in H^1(\Omega_I)/\mathbb{C}$ and $Q \in \mathbb{C}$, hence from the Stokes theorem we obtain

$$\begin{aligned}\int_{\Gamma_J} \operatorname{curl} \overline{\mathbf{v}_C^*} \cdot \mathbf{n} &= \int_{\partial\Gamma_J} \overline{\mathbf{v}_C^*} \cdot \mathbf{t} = \int_{\partial\Gamma_J} \overline{\mathbf{v}_I^*} \cdot \mathbf{t}\\ &= \int_{\partial\Gamma_J} \operatorname{grad} \overline{\chi_I^*} \cdot \mathbf{t} + \overline{Q} \int_{\partial\Gamma_J} \boldsymbol{\rho}_I^* \cdot \mathbf{t} = \overline{Q}\,.\end{aligned}$$

In conclusion, recalling that $\mathbf{H}_I = \operatorname{grad} \psi_I^* + I^0 \boldsymbol{\rho}_I^*$ in Ω_I and that

$$\mathbf{E}_C = \boldsymbol{\sigma}^{-1}(\operatorname{curl} \mathbf{H}_C - \mathbf{J}_{e,C})$$

in Ω_C, we have seen that $\mathbf{H}_C$, ψ_I^*, I^0 and V satisfy

$$\begin{aligned}&\int_{\Omega_C} (\boldsymbol{\sigma}^{-1} \operatorname{curl} \mathbf{H}_C \cdot \operatorname{curl} \overline{\mathbf{v}_C^*} + i\omega\boldsymbol{\mu}_C \mathbf{H}_C \cdot \overline{\mathbf{v}_C^*})\\ &\quad + i\omega \int_{\Omega_I} \boldsymbol{\mu}_I \operatorname{grad} \psi_I^* \cdot \operatorname{grad} \overline{\chi_I^*} + i\omega I^0\, \overline{Q} \int_{\Omega_I} \boldsymbol{\mu}_I \boldsymbol{\rho}_I^* \cdot \boldsymbol{\rho}_I^* - V\overline{Q}\\ &= \int_{\Omega_C} \boldsymbol{\sigma}^{-1} \mathbf{J}_{e,C} \cdot \operatorname{curl} \overline{\mathbf{v}_C^*}\end{aligned} \tag{8.27}$$

for each $\mathbf{v}_C^* \in H(\mathrm{curl}; \Omega_C)$, $\chi_I^* \in H^1(\Omega_I)$ and $Q \in \mathbb{C}$ satisfying

$$\mathbf{v}_C^* \times \mathbf{n}_C + \mathrm{grad}\, \chi_I^* \times \mathbf{n}_I + Q \boldsymbol{\rho}_I^* \times \mathbf{n}_I = \mathbf{0} \quad \text{on } \Gamma \,.$$

When the current intensity I^0 is given, one has to solve

$$\begin{array}{l} \text{Find } (\mathbf{H}_C, \psi_I^*) \in W^*(I^0) \text{ such that} \\ \int_{\Omega_C} (\boldsymbol{\sigma}^{-1} \mathrm{curl}\, \mathbf{H}_C \cdot \mathrm{curl}\, \overline{\mathbf{v}_C^*} + i\omega \boldsymbol{\mu}_C \mathbf{H}_C \cdot \overline{\mathbf{v}_C^*}) \\ \qquad + i\omega \int_{\Omega_I} \boldsymbol{\mu}_I \,\mathrm{grad}\, \psi_I^* \cdot \mathrm{grad}\, \overline{\chi_I^*} \\ \quad = \int_{\Omega_C} \boldsymbol{\sigma}^{-1} \mathbf{J}_{e,C} \cdot \mathrm{curl}\, \overline{\mathbf{v}_C^*} \\ \text{for all } (\mathbf{v}_C^*, \chi_I^*) \in W^*(0) \,, \end{array} \tag{8.28}$$

where

$$\begin{aligned} W^*(Q) := \{ (\mathbf{v}_C^*, \chi_I^*) \in H(\mathrm{curl}; \Omega_C) \times H^1(\Omega_I)/\mathbb{C} \,|\, \\ \mathbf{v}_C^* \times \mathbf{n}_C + \mathrm{grad}\, \chi_I^* \times \mathbf{n}_I = -Q \boldsymbol{\rho}_I^* \times \mathbf{n}_I \text{ on } \Gamma \} \,. \end{aligned}$$

It is easily seen that for each $I^0 \in \mathbb{C}$ the set $W^*(I^0)$ is non-empty: in fact, it is enough to set $\chi_I^* = 0$ and choose for $\mathbf{v}_C^*$ a vector field in $H(\mathrm{curl}; \Omega_C)$, continuously dependent on I^0, with $\mathbf{v}_C^* \times \mathbf{n}_C = -I^0 \boldsymbol{\rho}_I^* \times \mathbf{n}_I$ on Γ. Moreover, the sesquilinear form at the left hand side of (8.28) is clearly coercive in $W^*(0)$, thus there exists a unique solution to (8.28).

Having determined $\mathbf{H}_C$ and ψ_I^*, the voltage V is obtained by setting $\mathbf{v}_C^* = \mathbf{H}_C$, $\chi_I^* = \psi_I^*$ and $Q = I^0$ in (8.27), namely,

$$\begin{aligned} V = \overline{I^0}^{-1} \int_{\Omega_C} (\boldsymbol{\sigma}^{-1} \mathrm{curl}\, \mathbf{H}_C \cdot \mathrm{curl}\, \overline{\mathbf{H}_C} + i\omega \boldsymbol{\mu}_C \mathbf{H}_C \cdot \overline{\mathbf{H}_C}) \\ + i\omega \overline{I^0}^{-1} \int_{\Omega_I} \boldsymbol{\mu}_I \,\mathrm{grad}\, \psi_I^* \cdot \mathrm{grad}\, \overline{\psi_I^*} + i\omega I^0 \int_{\Omega_I} \boldsymbol{\mu}_I \boldsymbol{\rho}_I^* \cdot \boldsymbol{\rho}_I^* \\ - \overline{I^0}^{-1} \int_{\Omega_C} \boldsymbol{\sigma}^{-1} \mathbf{J}_{e,C} \cdot \mathrm{curl}\, \overline{\mathbf{H}_C} \,. \end{aligned}$$

Instead, when the voltage V is assigned one has to consider

$$\begin{array}{l} \text{Find } (\mathbf{H}_C, \psi_I^*, I^0) \in W^{**} \text{ such that} \\ \int_{\Omega_C} (\boldsymbol{\sigma}^{-1} \mathrm{curl}\, \mathbf{H}_C \cdot \mathrm{curl}\, \overline{\mathbf{v}_C^*} + i\omega \boldsymbol{\mu}_C \mathbf{H}_C \cdot \overline{\mathbf{v}_C^*}) \\ \qquad + i\omega \int_{\Omega_I} \boldsymbol{\mu}_I \,\mathrm{grad}\, \psi_I^* \cdot \mathrm{grad}\, \overline{\chi_I^*} + i\omega I^0\, \overline{Q} \int_{\Omega_I} \boldsymbol{\mu}_I \boldsymbol{\rho}_I^* \cdot \boldsymbol{\rho}_I^* \\ \quad = \int_{\Omega_C} \boldsymbol{\sigma}^{-1} \mathbf{J}_{e,C} \cdot \mathrm{curl}\, \overline{\mathbf{v}_C^*} + V \overline{Q} \\ \text{for all } (\mathbf{v}_C^*, \chi_I^*, Q) \in W^{**} \,, \end{array} \tag{8.29}$$

where

$$\begin{aligned} W^{**} := \{ (\mathbf{v}_C^*, \chi_I^*, Q) \in H(\mathrm{curl}; \Omega_C) \times H^1(\Omega_I)/\mathbb{C} \times \mathbb{C} \,|\, \\ \mathbf{v}_C^* \times \mathbf{n}_C + \mathrm{grad}\, \chi_I^* \times \mathbf{n}_I + Q \boldsymbol{\rho}_I^* \times \mathbf{n}_I = \mathbf{0} \text{ on } \Gamma \} \,. \end{aligned}$$

Again, the sesquilinear form at the left hand side of (8.29) is coercive in W^{**}, thus ensuring existence and uniqueness of the solution.

For a more detailed presentation of the $(\mathbf{H}_C, \psi_I^*)$ approach and of its finite element approximation we refer to Bermúdez et al. [43], [45]. Here, we only note that the constraint on Γ that is present in the definition of $W^*(I^0)$, $W^*(0)$ and W^{**} has to be kept at the finite dimensional level, and this leads to the requirement that the meshes in Ω_C and Ω_I match on Γ, and that the finite elements satisfy the constraint. These technical aspects are also described in Section 5.4.2.

Remark 8.9. The voltage excitation problem is also addressed in Hiptmair and Sterz [130]. As in Bermúdez et al. [43], [45], the problem is written in terms of the magnetic field, and then the electric field is computed by imposing a different boundary condition; more precisely, it is assumed to satisfy

$$\mathbf{E} \times \mathbf{n} = V \operatorname{grad} v \times \mathbf{n} \quad \text{on } \partial\Omega ,$$

where $v \in H^{1/2}(\partial\Omega)$, $v = 1$ on Γ_J, $v = 0$ on $\partial\Omega \setminus (\Gamma_J \cup \Theta)$, and Θ is a transition zone around Γ_J. In particular, $\mathbf{E} \times \mathbf{n} = \mathbf{0}$ on $\partial\Omega \setminus \Theta$ and $\mathbf{E} \times \mathbf{n} \neq \mathbf{0}$ on Θ. Thus the model depends on the choices of the set Θ and the function v, and it can be seen that, though the magnetic field does not depend on Θ and v, the electric field does depend on them.

The current intensity excitation problem is also considered in Hiptmair and Sterz [130], and it is formulated in terms of the electric field; again, the solution depends on the choice of Θ and v, and, moreover, a complete analysis of well-posedness is not furnished. □

8.1.3 Formulations in terms of $\mathbf{T}_C$ and ψ^*

In this section we present a formulation proposed by Meunier et al. [177], Bíró et al. [52]. We assume as before that $\mathbf{J}_{e,I} = \mathbf{0}$ in Ω_I (otherwise, one can proceed as in Remark 8.3); therefore, we have $\mathbf{H}_I = \operatorname{grad} \psi_I^* + I^0 \boldsymbol{\rho}_I^*$ in Ω_I. Let us start by denoting by $\mathbf{R}_C^*$ a vector field that belongs to $H(\operatorname{curl}; \Omega_C)$ and satisfies $\mathbf{R}_C^* \times \mathbf{n}_C + \boldsymbol{\rho}_I^* \times \mathbf{n}_I = \mathbf{0}$ on Γ. Following the same arguments used in Section 6.3 for a similar situation, the first step is to see that we can formulate the eddy current problem by means of a couple $(\mathbf{T}_C, \psi^*)$ such that

$$\begin{aligned} \mathbf{H}_I &= \operatorname{grad} \psi_I^* + I^0 \boldsymbol{\rho}_I^* && \text{in } \Omega_I \\ \mathbf{H}_C &= \mathbf{T}_C + \operatorname{grad} \psi_C^* + I^0 \mathbf{R}_C^* && \text{in } \Omega_C , \end{aligned} \tag{8.30}$$

with the interface conditions

$$\begin{aligned} \mathbf{T}_C \times \mathbf{n}_C = \mathbf{0} \quad & \text{on } \Gamma \\ \psi_C^* - \psi_I^* = 0 \quad & \text{on } \Gamma . \end{aligned} \tag{8.31}$$

Let us verify that $(8.30)_2$ and (8.31) can be satisfied. First, the vector field $\mathbf{T}_C$ can be determined as the solution to

$$\begin{cases} \operatorname{curl} \mathbf{T}_C = \operatorname{curl}(\mathbf{H}_C - I^0 \mathbf{R}_C^*) & \text{in } \Omega_C \\ \operatorname{div} \mathbf{T}_C = 0 & \text{in } \Omega_C \\ \mathbf{T}_C \times \mathbf{n}_C = \mathbf{0} & \text{on } \Gamma \\ \mathbf{T}_C \cdot \mathbf{n}_C = 0 & \text{on } \Gamma_E \cup \Gamma_J . \end{cases} \tag{8.32}$$

This problem is well-posed. Indeed, uniqueness is straightforward, as, in the geometrical situation under consideration, the space of harmonic fields

$$\mathcal{H}(\Gamma,\Gamma_C;\Omega_C) := \{\mathbf{v}_C \in (L^2(\Omega_C))^3 \mid \operatorname{curl}\mathbf{v}_C = \mathbf{0}, \operatorname{div}\mathbf{v}_C = 0,\\ \mathbf{v}_C\cdot\mathbf{n}_C = 0 \text{ on } \Gamma_C,\ \mathbf{v}_C\times\mathbf{n}_C = \mathbf{0} \text{ on } \Gamma\}\,.$$

is trivial. Concerning existence, we can easily show that the right hand side $\mathbf{G}_C := \operatorname{curl}(\mathbf{H}_C - I^0\mathbf{R}_C^*)$ satisfies the necessary solvability conditions. In fact, we clearly have $\operatorname{div}\mathbf{G}_C = 0$ in Ω_C, and also $\mathbf{G}_C\cdot\mathbf{n}_C = 0$ on Γ, as $\mathbf{G}_C\cdot\mathbf{n}_C = \operatorname{curl}(\mathbf{H}_I - I^0\boldsymbol{\rho}_I^*)\cdot\mathbf{n}_C$ on Γ. Finally, let us define φ_C to be the solution to

$$\begin{cases} \Delta\varphi_C = 0 & \text{in } \Omega_C\\ \varphi_C = 1 & \text{on } \Gamma_J\\ \varphi_C = 0 & \text{on } \Gamma_E\\ \operatorname{grad}\varphi_C\cdot\mathbf{n}_C = 0 & \text{on } \Gamma\,, \end{cases}$$

so that $\operatorname{grad}\varphi_C$ is a basis function of the (one-dimensional) space of harmonic fields

$$\mathcal{H}(\Gamma_C,\Gamma;\Omega_C) := \{\mathbf{v}_C \in (L^2(\Omega_C))^3 \mid \operatorname{curl}\mathbf{v}_C = \mathbf{0}, \operatorname{div}\mathbf{v}_C = 0,\\ \mathbf{v}_C\cdot\mathbf{n}_C = 0 \text{ on } \Gamma,\ \mathbf{v}_C\times\mathbf{n}_C = \mathbf{0} \text{ on } \Gamma_C\}\,.$$

Integrating by parts and using the Stokes theorem we have that the last solvability condition is satisfied

$$\begin{aligned}\int_{\Omega_C}\mathbf{G}_C\cdot\operatorname{grad}\varphi_C &= \int_{\Gamma_J}\mathbf{G}_C\cdot\mathbf{n}_C = \int_{\Gamma_J}\operatorname{curl}(\mathbf{H}_C - I^0\mathbf{R}_C^*)\cdot\mathbf{n}_C\\ &= \int_{\partial\Gamma_J}(\mathbf{H}_C - I^0\mathbf{R}_C^*)\cdot d\boldsymbol{\tau} = \int_{\partial\Gamma_J}(\mathbf{H}_I - I^0\boldsymbol{\rho}_I^*)\cdot d\boldsymbol{\tau} = 0\,.\end{aligned}$$

Having determined $\mathbf{T}_C$, we find ψ_C^* as the solution to

$$\begin{cases} \operatorname{div}(\boldsymbol{\mu}_C\operatorname{grad}\psi_C^*) = \operatorname{div}[\boldsymbol{\mu}_C(\mathbf{H}_C - I^0\mathbf{R}_C^* - \mathbf{T}_C)] & \text{in } \Omega_C\\ \psi_C^* = {\psi_I}^* & \text{on } \Gamma\\ \boldsymbol{\mu}_C\operatorname{grad}\psi_C^*\cdot\mathbf{n}_C = -\boldsymbol{\mu}_C(I^0\mathbf{R}_C^* + \mathbf{T}_C)\cdot\mathbf{n}_C & \text{on } \Gamma_E\cup\Gamma_J\,. \end{cases} \tag{8.33}$$

Now it is easily checked that $\mathbf{H}_C - \mathbf{T}_C - \operatorname{grad}\psi_C^* - I^0\mathbf{R}_C^*$ belongs to the space

$$\mathcal{H}_{\mu_C}(\Gamma,\Gamma_C;\Omega_C) := \{\mathbf{v}_C \in (L^2(\Omega_C))^3 \mid \operatorname{curl}\mathbf{v}_C = \mathbf{0}, \operatorname{div}(\boldsymbol{\mu}_C\mathbf{v}_C) = 0,\\ \boldsymbol{\mu}_C\mathbf{v}_C\cdot\mathbf{n}_C = 0 \text{ on } \Gamma_C,\ \mathbf{v}_C\times\mathbf{n}_C = \mathbf{0} \text{ on } \Gamma\}\,,$$

and this space is trivial, so that $(8.30)_2$ is satisfied.

We can thus write the eddy current problem in terms of $\mathbf{T}_C$ and ψ^*. First of all we set

$$\mathbf{E}_C = \boldsymbol{\sigma}^{-1}[\operatorname{curl}(\mathbf{T}_C + I^0\mathbf{R}_C^*) - \mathbf{J}_{e,C}] \quad \text{in } \Omega_C\,,$$

so that the Ampère equation in Ω is clearly satisfied. Hence, as explained in Remark 8.7, we only need to impose the Faraday equation in Ω_C and the Gauss magnetic equation in Ω. Proceeding as in Section 6.3, we find

$$\begin{aligned}&\int_\Omega i\omega\boldsymbol{\mu}\operatorname{grad}\psi^*\cdot\operatorname{grad}\overline{\chi^*} + \int_{\Omega_C} i\omega\boldsymbol{\mu}_C\mathbf{T}_C\cdot\operatorname{grad}\overline{\chi_C^*}\\ &+I^0\int_{\Omega_C} i\omega\boldsymbol{\mu}_C\mathbf{R}_C^*\cdot\operatorname{grad}\overline{\chi_C^*} = 0\end{aligned} \tag{8.34}$$

for each $\chi^* \in H^1(\Omega)$, and moreover (having already inserted the penalization term associated to the divergence-free condition for $\mathbf{T}_C$)

$$\begin{aligned}
&\int_{\Omega_C} \boldsymbol{\sigma}^{-1} \operatorname{curl} \mathbf{T}_C \cdot \operatorname{curl} \overline{\mathbf{v}_C} + \sigma_*^{-1} \int_{\Omega_C} \operatorname{div} \mathbf{T}_C \operatorname{div} \overline{\mathbf{v}_C} \\
&\quad + \int_{\Omega_C} i\omega \boldsymbol{\mu}_C \mathbf{T}_C \cdot \overline{\mathbf{v}_C} + \int_{\Omega_C} i\omega \boldsymbol{\mu}_C \operatorname{grad} \psi_C^* \cdot \overline{\mathbf{v}_C} \\
&\quad + I^0 \int_{\Omega_C} \boldsymbol{\sigma}^{-1} \operatorname{curl} \mathbf{R}_C^* \cdot \operatorname{curl} \overline{\mathbf{v}_C} + I^0 \int_{\Omega_C} i\omega \boldsymbol{\mu}_C \mathbf{R}_C^* \cdot \overline{\mathbf{v}_C} \\
&= \int_{\Omega_C} \boldsymbol{\sigma}^{-1} \mathbf{J}_{e,C} \cdot \operatorname{curl} \overline{\mathbf{v}_C} ,
\end{aligned} \tag{8.35}$$

where $\mathbf{v}_C \in H_{0,\Gamma}(\mathrm{curl}; \Omega_C) \cap H_{0,\Gamma_C}(\mathrm{div}; \Omega_C)$ and $\sigma_* > 0$ is a dimensional constant (say, a suitable average in Ω_C of the entries of the matrix $\boldsymbol{\sigma}$).

When the current intensity I^0 is assigned, moving the terms containing it to the right hand side in (8.34) and (8.35) leaves on the left hand side the sesquilinear form

$$\begin{aligned}
&\int_{\Omega_C} \boldsymbol{\sigma}^{-1} \operatorname{curl} \mathbf{T}_C \cdot \operatorname{curl} \overline{\mathbf{v}_C} + \sigma_*^{-1} \int_{\Omega_C} \operatorname{div} \mathbf{T}_C \operatorname{div} \overline{\mathbf{v}_C} \\
&\quad + \int_{\Omega_C} i\omega \boldsymbol{\mu}_C \mathbf{T}_C \cdot \overline{\mathbf{v}_C} + \int_{\Omega_C} i\omega \boldsymbol{\mu}_C \operatorname{grad} \psi_C^* \cdot \overline{\mathbf{v}_C} \\
&\quad + \int_{\Omega} i\omega \boldsymbol{\mu} \operatorname{grad} \psi^* \cdot \operatorname{grad} \overline{\chi^*} + \int_{\Omega_C} i\omega \boldsymbol{\mu}_C \mathbf{T}_C \cdot \operatorname{grad} \overline{\chi_C^*} .
\end{aligned}$$

As in Section 6.3, it can be proved that it is coercive in

$$(H_{0,\Gamma}(\mathrm{curl}; \Omega_C) \cap H_{0,\Gamma_C}(\mathrm{div}; \Omega_C)) \times H^1(\Omega)/\mathbb{C} ,$$

therefore problem (8.34), (8.35) has a unique solution $(\mathbf{T}_C, \psi^*)$ for each assigned I^0 and $\mathbf{J}_{e,C}$.

Instead, when the voltage V is prescribed another suitable equation comes from (8.17). We have

$$\begin{aligned}
&\int_\Gamma \mathbf{E}_C \times \mathbf{n}_C \cdot \boldsymbol{\rho}_I^* = \int_\Gamma \mathbf{E}_C \times \mathbf{n}_C \cdot \mathbf{R}_C^* \\
&\quad = - \int_{\Omega_C} \operatorname{curl} \mathbf{E}_C \cdot \mathbf{R}_C^* + \int_{\Omega_C} \mathbf{E}_C \cdot \operatorname{curl} \mathbf{R}_C^* \\
&\quad = \int_{\Omega_C} i\omega \boldsymbol{\mu}_C \mathbf{H}_C \cdot \mathbf{R}_C^* + \int_{\Omega_C} \boldsymbol{\sigma}^{-1} (\operatorname{curl} \mathbf{H}_C - \mathbf{J}_{e,C}) \cdot \operatorname{curl} \mathbf{R}_C^* \\
&\quad = \int_{\Omega_C} i\omega \boldsymbol{\mu}_C \mathbf{T}_C \cdot \mathbf{R}_C^* + \int_{\Omega_C} i\omega \boldsymbol{\mu}_C \operatorname{grad} \psi_C^* \cdot \mathbf{R}_C^* + I^0 \int_{\Omega_C} i\omega \boldsymbol{\mu}_C \mathbf{R}_C^* \cdot \mathbf{R}_C^* \\
&\qquad + \int_{\Omega_C} \boldsymbol{\sigma}^{-1} \operatorname{curl} \mathbf{T}_C \cdot \operatorname{curl} \mathbf{R}_C^* + I^0 \int_{\Omega_C} \boldsymbol{\sigma}^{-1} \operatorname{curl} \mathbf{R}_C^* \cdot \operatorname{curl} \mathbf{R}_C^* \\
&\qquad - \int_{\Omega_C} \boldsymbol{\sigma}^{-1} \mathbf{J}_{e,C} \cdot \operatorname{curl} \mathbf{R}_C^* .
\end{aligned}$$

Therefore (8.17) can be rewritten as

$$\begin{aligned}
&\overline{Q} \int_{\Omega_C} i\omega \boldsymbol{\mu}_C \mathbf{T}_C \cdot \mathbf{R}_C^* + \overline{Q} \int_{\Omega_C} i\omega \boldsymbol{\mu}_C \operatorname{grad} \psi_C^* \cdot \mathbf{R}_C^* \\
&\quad + I^0 \overline{Q} \int_{\Omega_C} i\omega \boldsymbol{\mu}_C \mathbf{R}_C^* \cdot \mathbf{R}_C^* + I^0 \overline{Q} \int_{\Omega_I} i\omega \boldsymbol{\mu}_I \boldsymbol{\rho}_I^* \cdot \boldsymbol{\rho}_I^* \\
&\quad + \overline{Q} \int_{\Omega_C} \boldsymbol{\sigma}^{-1} \operatorname{curl} \mathbf{T}_C \cdot \operatorname{curl} \mathbf{R}_C^* \\
&\quad + I^0 \overline{Q} \int_{\Omega_C} \boldsymbol{\sigma}^{-1} \operatorname{curl} \mathbf{R}_C^* \cdot \operatorname{curl} \mathbf{R}_C^* \\
&= V \overline{Q} + \overline{Q} \int_{\Omega_C} \boldsymbol{\sigma}^{-1} \mathbf{J}_{e,C} \cdot \operatorname{curl} \mathbf{R}_C^*
\end{aligned} \tag{8.36}$$

for each $Q \in \mathbb{C}$.

Putting together (8.34), (8.35) and (8.36), we end up with the final variational formulation for the voltage problem in terms of $\mathbf{T}_C$, ψ^* and I^0 (that now is an unknown to be determined).

This variational problem is well-posed. In fact, the existence of the solution follows from our procedure, as we have shown how to construct $\mathbf{T}_C$ and ψ_C^* starting from the solutions $\mathbf{H}_C = (-i\omega\boldsymbol{\mu}_C)^{-1} \operatorname{curl} \mathbf{E}_C$, ψ_I^* and I^0 of the eddy current problem (8.15). Concerning the uniqueness of the solution, the first step consists in rewriting the problem as

$$\begin{aligned}
&\int_{\Omega_C} \boldsymbol{\sigma}^{-1} \operatorname{curl}(\mathbf{T}_C + I^0\mathbf{R}_C^*) \cdot \operatorname{curl}(\overline{\mathbf{v}_C} + \overline{Q}\mathbf{R}_C^*) \\
&\quad +\sigma_*^{-1} \int_{\Omega_C} \operatorname{div} \mathbf{T}_C \operatorname{div} \overline{\mathbf{v}_C} + i\omega \int_{\Omega_I} \boldsymbol{\mu}_I \operatorname{grad} \psi_I^* \cdot \operatorname{grad} \overline{\chi_I^*} \\
&\quad +i\omega \int_{\Omega_C} \boldsymbol{\mu}_C(\mathbf{T}_C + \operatorname{grad} \psi_C^* + I^0\mathbf{R}_C^*) \cdot (\overline{\mathbf{v}_C} + \operatorname{grad} \overline{\chi_C^*} + \overline{Q}\mathbf{R}_C^*) \\
&\quad +i\omega I^0 \overline{Q} \int_{\Omega_I} \boldsymbol{\mu}_I \boldsymbol{\rho}_I^* \cdot \boldsymbol{\rho}_I^* \\
&= \int_{\Omega_C} \boldsymbol{\sigma}^{-1} \mathbf{J}_{e,C} \cdot \operatorname{curl} \overline{\mathbf{v}_C} + V\overline{Q} + \overline{Q} \int_{\Omega_C} \boldsymbol{\sigma}^{-1} \mathbf{J}_{e,C} \cdot \operatorname{curl} \mathbf{R}_C^* \,.
\end{aligned} \tag{8.37}$$

Let us denote by $\mathcal{S}_*(\cdot,\cdot)$ the sesquilinear form at the left hand side in (8.37). Similarly to what proved in Section 6.3, we have

$$\begin{aligned}
&|\operatorname{Re} \mathcal{S}_*((\mathbf{v}_C, \chi^*, Q), (\mathbf{v}_C, \chi^*, Q))| \\
&\qquad \geq \sigma_{\max}^{-1} \int_{\Omega_C} |\operatorname{curl}(\mathbf{v}_C + Q\mathbf{R}_C^*)|^2 + \sigma_*^{-1} \int_{\Omega_C} |\operatorname{div} \mathbf{v}_C|^2 ,
\end{aligned}$$

and

$$\begin{aligned}
&|\operatorname{Im} \mathcal{S}_*((\mathbf{v}_C, \chi^*, Q), (\mathbf{v}_C, \chi^*, Q))| \\
&\qquad \geq |\omega| \mu_{\min} \int_{\Omega_I} |\operatorname{grad} \chi_I^*|^2 + |\omega| \mu_{\min} \int_{\Omega_C} |\mathbf{v}_C + \operatorname{grad} \chi_C^* + Q\mathbf{R}_C^*|^2 \\
&\qquad\quad +|\omega||Q|^2 \int_{\Omega_I} \boldsymbol{\mu}_I \boldsymbol{\rho}_I^* \cdot \boldsymbol{\rho}_I^* ,
\end{aligned}$$

where $\sigma_{\max}$ is an uniform upper bound for the maximum eigenvalues of $\boldsymbol{\sigma}(\mathbf{x})$ in Ω_C and $\mu_{\min}$ is an uniform lower bound for the minimum eigenvalues of $\boldsymbol{\mu}(\mathbf{x})$ in Ω. Moreover, for each $0 < \delta < 1$ we obtain

$$\begin{aligned}
&\int_{\Omega_C} |\operatorname{curl}(\mathbf{v}_C + Q\mathbf{R}_C^*)|^2 \\
&\qquad \geq (1-\delta) \int_{\Omega_C} |\operatorname{curl} \mathbf{v}_C|^2 - (1-\delta)\delta^{-1}|Q|^2 \int_{\Omega_C} |\mathbf{R}_C^*|^2 ,
\end{aligned}$$

and for each $0 < \eta < 1$

$$\begin{aligned}
&\int_{\Omega_C} |\mathbf{v}_C + \operatorname{grad} \chi_C^* + Q\mathbf{R}_C^*|^2 \\
&\qquad \geq (1-\eta) \int_{\Omega_C} |\operatorname{grad} \chi_C^*|^2 - 2(1-\eta)\eta^{-1} \int_{\Omega_C} |\mathbf{v}_C|^2 \\
&\qquad\quad -2(1-\eta)\eta^{-1}|Q|^2 \int_{\Omega_C} |\mathbf{R}_C^*|^2 .
\end{aligned}$$

Since in the present topological situation the Poincaré-like inequality

$$\int_{\Omega_C} (|\operatorname{curl} \mathbf{v}_C|^2 + |\operatorname{div} \mathbf{v}_C|^2) \geq K_0 \int_{\Omega_C} (|\operatorname{curl} \mathbf{v}_C|^2 + |\operatorname{div} \mathbf{v}_C|^2 + |\mathbf{v}_C|^2)$$

holds true in $H_{0,\Gamma}(\operatorname{curl}; \Omega_C) \cap H_{0,\Gamma_C}(\operatorname{div}; \Omega_C)$ (see, e.g., Fernandes and Gilardi [104]), choosing $1-\delta = \tau$, $1-\eta = \tau^2$ and τ small enough it is now easy to prove that $\mathcal{S}_*(\cdot,\cdot)$ is coercive in $[H_{0,\Gamma}(\operatorname{curl}; \Omega_C) \cap H_{0,\Gamma_C}(\operatorname{div}; \Omega_C)] \times H^1(\Omega)/\mathbb{C} \times \mathbb{C}$.

Remark 8.10. The use of the vector potential $\mathbf{T}_C$ and the scalar magnetic potential ψ^* for solving eddy current problems coupled with circuits has been also proposed by Specogna et al. [228]. They consider an approach based on the integral form of the Faraday, Gauss magnetic and Ampère equations, and homology theory is deeply used in order to devise the complete formulation of the problem. □

8.1.4 Finite element approximation

In this section we present some finite element approximation schemes based on the hybrid approaches described in Section 8.1.1.

Let us start noting that the variational formulations (8.15) and (8.16) are not suitable for numerical approximation. In fact, a conforming finite element discretization based directly on them requires that $\boldsymbol{\rho}_I^*$ is explicitly known. An alternative approach, that overcomes this difficulty, is based on a different decomposition of $\mathbf{H}_I$.

Let $\boldsymbol{\zeta}_I$ be the generalized gradient of a function $\eta \in H^1(\Omega_I \setminus \Xi_I^*)$ such that $[\eta]_{\Xi_I^*} = 1$. Then $\operatorname{curl} \boldsymbol{\zeta}_I = \mathbf{0}$ and $\int_{\partial \Gamma_J} \boldsymbol{\zeta}_I \cdot \mathbf{t} = 1$, but in general $\boldsymbol{\zeta}_I \notin H(\operatorname{div}; \Omega_I)$. However, since $\operatorname{curl}(\boldsymbol{\rho}_I^* - \boldsymbol{\zeta}_I) = \mathbf{0}$ in Ω_I and $\int_{\partial \Gamma_J} (\boldsymbol{\rho}_I^* - \boldsymbol{\zeta}_I) \cdot \mathbf{t} = 0$, one easily verifies that $\boldsymbol{\rho}_I^* = \boldsymbol{\zeta}_I + \operatorname{grad} g_{\zeta_I}$ for some $g_{\zeta_I} \in H^1(\Omega_I)$. Hence

$$\mathbf{H}_I = \operatorname{grad} \psi_I^* + I^0 \boldsymbol{\rho}_I^* = \operatorname{grad} \psi_I^* + I^0 (\boldsymbol{\zeta}_I + \operatorname{grad} g_{\zeta_I}) = \operatorname{grad} \widehat{\psi}_I^* + I^0 \boldsymbol{\zeta}_I \,,$$

with $\widehat{\psi}_I^* \in H^1(\Omega_I)/\mathbb{C}$ that depends of the choice of $\boldsymbol{\zeta}_I$. This alternative decomposition is not orthogonal with respect to the scalar product $(\cdot,\cdot)_{\mu_I, \Omega_I}$, and this has as a consequence that the corresponding weak formulation has some additional terms. In fact the voltage excitation problem now reads

$$\begin{aligned}
&\text{Find } (\mathbf{E}_C, \widehat{\psi}_I^*, I^0) \in H_{0,\Gamma_C}(\operatorname{curl}; \Omega_C) \times H^1(\Omega_I)/\mathbb{C} \times \mathbb{C} : \\
&\int_{\Omega_C} (\boldsymbol{\mu}_C^{-1} \operatorname{curl} \mathbf{E}_C \cdot \operatorname{curl} \overline{\mathbf{z}_C} + i\omega \boldsymbol{\sigma} \mathbf{E}_C \cdot \overline{\mathbf{z}_C}) \\
&\qquad -i\omega \int_\Gamma \overline{\mathbf{z}_C} \times \mathbf{n}_C \cdot \operatorname{grad} \widehat{\psi}_I^* - i\omega I^0 \int_\Gamma \overline{\mathbf{z}_C} \times \mathbf{n}_C \cdot \boldsymbol{\zeta}_I \\
&\quad = -i\omega \int_{\Omega_C} \mathbf{J}_{e,C} \cdot \overline{\mathbf{z}_C} \\
&-i\omega \int_\Gamma \mathbf{E}_C \times \mathbf{n}_C \cdot \operatorname{grad} \overline{\chi_I^*} + \omega^2 \int_{\Omega_I} \boldsymbol{\mu}_I \operatorname{grad} \widehat{\psi}_I^* \cdot \operatorname{grad} \overline{\chi_I^*} \\
&\qquad + \omega^2 I^0 \int_{\Omega_I} \boldsymbol{\mu}_I \boldsymbol{\zeta}_I \cdot \operatorname{grad} \overline{\chi_I^*} = 0 \\
&-i\omega \overline{Q} \int_\Gamma \mathbf{E}_C \times \mathbf{n}_C \cdot \boldsymbol{\zeta}_I + \omega^2 \overline{Q} \int_{\Omega_I} \boldsymbol{\mu}_I \operatorname{grad} \widehat{\psi}_I^* \cdot \boldsymbol{\zeta}_I \\
&\qquad + \omega^2 I^0 \overline{Q} \int_{\Omega_I} \boldsymbol{\mu}_I \boldsymbol{\zeta}_I \cdot \boldsymbol{\zeta}_I = -i\omega V \overline{Q} \\
&\text{for all } (\mathbf{z}_C, \chi_I^*, Q) \in H_{0,\Gamma_C}(\operatorname{curl}; \Omega_C) \times H^1(\Omega_I)/\mathbb{C} \times \mathbb{C} \,,
\end{aligned} \tag{8.38}$$

while the current intensity excitation problem reads

$$\begin{aligned}
&\text{Find } (\mathbf{E}_C, \widehat{\psi}_I^*) \in H_{0,\Gamma_C}(\operatorname{curl}; \Omega_C) \times H^1(\Omega_I)/\mathbb{C} : \\
&\int_{\Omega_C} (\boldsymbol{\mu}_C^{-1} \operatorname{curl} \mathbf{E}_C \cdot \operatorname{curl} \overline{\mathbf{z}_C} + i\omega \boldsymbol{\sigma} \mathbf{E}_C \cdot \overline{\mathbf{z}_C}) \\
&\qquad -i\omega \int_\Gamma \overline{\mathbf{z}_C} \times \mathbf{n}_C \cdot \operatorname{grad} \widehat{\psi}_I^* \\
&\quad = -i\omega \int_{\Omega_C} \mathbf{J}_{e,C} \cdot \overline{\mathbf{z}_C} + i\omega I^0 \int_\Gamma \overline{\mathbf{z}_C} \times \mathbf{n}_C \cdot \boldsymbol{\zeta}_I \\
&-i\omega \int_\Gamma \mathbf{E}_C \times \mathbf{n}_C \cdot \operatorname{grad} \overline{\chi_I^*} + \omega^2 \int_{\Omega_I} \boldsymbol{\mu}_I \operatorname{grad} \widehat{\psi}_I^* \cdot \operatorname{grad} \overline{\chi_I^*} \\
&\quad = -\omega^2 I^0 \int_{\Omega_I} \boldsymbol{\mu}_I \boldsymbol{\zeta}_I \cdot \operatorname{grad} \overline{\chi_I^*} \\
&\text{for all } (\mathbf{z}_C, \chi_I^*) \in H_{0,\Gamma_C}(\operatorname{curl}; \Omega_C) \times H^1(\Omega_I)/\mathbb{C} \,.
\end{aligned} \tag{8.39}$$

Here below, in the framework of finite element approximation, we present a possible choice of $\boldsymbol{\zeta}_I$, depending on the mesh size h. An alternative choice, for which $\boldsymbol{\zeta}_I$ is not dependent on h, is proposed in Alonso Rodríguez et al. [20].

Let us now describe our finite element approximation schemes. We assume that Ω_C and Ω_I are Lipschitz polyhedral domains, and that $\mathcal{T}_{C,h}$ and $\mathcal{T}_{I,h}$ are two families of tetrahedral meshes of Ω_C and Ω_I, respectively. In particular, let us underline that we do not need that the finite elements match on Γ, and therefore the compatibility of the meshes is not necessary. We employ $N^k_{C,h}$, the Nédélec curl-conforming edge elements of degree k, to approximate the functions in $H(\mathrm{curl};\Omega_C)$ and $L^k_{I,h}$, the Lagrange nodal elements of degree k, to approximate the functions in $H^1(\Omega_I)$ (see Section A.2). Let us also set $Y^k_{C,h} := N^k_{C,h} \cap H_{0,\Gamma_C}(\mathrm{curl};\Omega_C)$.

The choice of the function $\boldsymbol{\zeta}_I$ is carried out as follows. Let us start denoting by $\Xi^*_{I,h}$ a discrete "cutting" surface that depends on the mesh $\mathcal{T}_{I,h}$ and by $\eta_{I,h}$ the piecewise-linear function taking value 1 at the nodes on one side of $\Xi^*_{I,h}$, say $\Xi^{*+}_{I,h}$, and 0 at all the other nodes including those on $\Xi^{*-}_{I,h}$, the other side of $\Xi^*_{I,h}$. Then define $\boldsymbol{\lambda}^h_I$ the $(L^2(\Omega_I))^3$-extension of the gradient of $\eta_{I,h}$ computed in $\Omega_I \setminus \Xi^*_{I,h}$, and choose $\boldsymbol{\zeta}_I = \boldsymbol{\lambda}^h_I$. This approach is similar to the one analyzed in Bermúdez et al. [45] for the current intensity excitation problem.

The sesquilinear form associated to problem (8.38) depends on h, and it is given by

$$\begin{aligned}
&\mathcal{C}^*_h((\mathbf{w}_C,\psi^*_I,K),(\mathbf{z}_C,\chi^*_I,Q)) \\
&\quad := \int_{\Omega_C} (\boldsymbol{\mu}_C^{-1}\,\mathrm{curl}\,\mathbf{w}_C \cdot \mathrm{curl}\,\overline{\mathbf{z}_C} + i\omega\boldsymbol{\sigma}\mathbf{w}_C \cdot \overline{\mathbf{z}_C}) \\
&\qquad +\omega^2 \int_{\Omega_I} \boldsymbol{\mu}_I\,\mathrm{grad}\,\psi^*_I \cdot \mathrm{grad}\,\overline{\chi^*_I} + \omega^2 K\overline{Q}\int_{\Omega_I} \boldsymbol{\mu}_I \boldsymbol{\lambda}^h_I \cdot \boldsymbol{\lambda}^h_I \\
&\qquad -i\omega\big[\int_\Gamma \overline{\mathbf{z}_C} \times \mathbf{n}_C \cdot \mathrm{grad}\,\psi^*_I + \int_\Gamma \mathbf{w}_C \times \mathbf{n}_C \cdot \mathrm{grad}\,\overline{\chi^*_I}\big] \\
&\qquad -i\omega\big[K\int_\Gamma \overline{\mathbf{z}_C} \times \mathbf{n}_C \cdot \boldsymbol{\lambda}^h_I + \overline{Q}\int_\Gamma \mathbf{w}_C \times \mathbf{n}_C \cdot \boldsymbol{\lambda}^h_I\big] \\
&\qquad +\omega^2\big[K\int_{\Omega_I} \boldsymbol{\mu}_I\,\mathrm{grad}\,\overline{\chi^*_I} \cdot \boldsymbol{\lambda}^h_I + \overline{Q}\int_{\Omega_I} \boldsymbol{\mu}_I\,\mathrm{grad}\,\psi^*_I \cdot \boldsymbol{\lambda}^h_I\big]\,.
\end{aligned}$$

However

$$\mathcal{C}^*_h((\mathbf{w}_C,\psi^*_I,K),(\mathbf{z}_C,\chi^*_I,Q)) = \mathcal{C}((\mathbf{w}_C,\mathrm{grad}\,\psi^*_I + K\boldsymbol{\lambda}^h_I),(\mathbf{z}_C,\mathrm{grad}\,\chi^*_I + Q\boldsymbol{\lambda}^h_I))\,,$$

where $\mathcal{C}(\cdot,\cdot)$ has been introduced in (4.3). Hence the finite element approximation of the voltage excitation problem (8.38) reads

$$\begin{aligned}
&\text{Find } (\mathbf{E}_{C,h},\widehat{\psi}^*_{I,h},I^0_h) \in Y^k_{C,h} \times L^k_{I,h}/\mathbb{C} \times \mathbb{C} : \\
&\quad \mathcal{C}((\mathbf{E}_{C,h},\mathrm{grad}\,\widehat{\psi}^*_{I,h} + I^0_h\boldsymbol{\lambda}^h_I),(\mathbf{z}_{C,h},\mathrm{grad}\,\chi^*_{I,h} + Q\boldsymbol{\lambda}^h_I)) \\
&\qquad\qquad = -i\omega\int_{\Omega_C} \mathbf{J}_{e,C} \cdot \overline{\mathbf{z}_{C,h}} - i\omega V\overline{Q} \\
&\text{for all } (\mathbf{z}_{C,h},\chi^*_{I,h},Q) \in Y^k_{C,h} \times L^k_{I,h}/\mathbb{C} \times \mathbb{C}\,.
\end{aligned} \tag{8.40}$$

Let us consider the error estimate. Setting

$$\mathbf{H}_{I,h} := \mathrm{grad}\,\widehat{\psi}^*_{I,h} + I^0_h\boldsymbol{\lambda}^h_I \in H^0(\mathrm{curl};\Omega_I)\,,$$

from (8.38) and (8.40) we have the following equation for the error

$$\mathcal{C}((\mathbf{E}_C - \mathbf{E}_{C,h},\mathbf{H}_I - \mathbf{H}_{I,h}),(\mathbf{z}_{C,h},\mathrm{grad}\,\chi^*_{I,h} + Q\boldsymbol{\lambda}^h_I)) = 0$$

for all $\mathbf{z}_{C,h} \in Y^k_{C,h}$, $\chi^*_{I,h} \in L^k_{I,h}/\mathbb{C}$ and $Q \in \mathbb{C}$. Hence

$$\begin{aligned} &\|\mathbf{E}_C - \mathbf{E}_{C,h}\|_{H(\mathrm{curl};\Omega_C)} + \|\mathbf{H}_I - \mathbf{H}_{I,h}\|_{0,\Omega_I} \\ &\quad = \|\mathbf{E}_C - \mathbf{E}_{C,h}\|_{H(\mathrm{curl};\Omega_C)} + \|\mathbf{H}_I - \operatorname{grad}\widehat{\psi}^*_{I,h} - I^0_h \boldsymbol{\lambda}^h_I\|_{0,\Omega_I} \\ &\quad \le C \inf_{(\mathbf{z}_{C,h},\mathbf{v}_{I,h})\in Y^k_{C,h}\times Z^k_{I,h}} \big(\|\mathbf{E}_C - \mathbf{z}_{C,h}\|_{H(\mathrm{curl};\Omega_C)} \\ &\qquad\qquad + \|\mathbf{H}_I - \mathbf{v}_{I,h}\|_{0,\Omega_I}\big) , \end{aligned} \tag{8.41}$$

where

$$Z^k_{I,h} := \operatorname{grad} L^k_{I,h} \oplus \operatorname{span}\{\boldsymbol{\lambda}^h_I\} .$$

An error estimate for the intensity is obtained by noting that

$$\int_{\Omega_I} \boldsymbol{\mu}_I(\mathbf{H}_I - \mathbf{H}_{I,h}) \cdot \boldsymbol{\rho}^*_I = (I^0 - I^0_h) \int_{\Omega_I} \boldsymbol{\mu}_I \boldsymbol{\rho}^*_I \cdot \boldsymbol{\rho}^*_I ,$$

as $\mathbf{H}_I = \operatorname{grad}\psi^*_I + I^0 \boldsymbol{\rho}^*_I$, $\mathbf{H}_{I,h} = \operatorname{grad}\widehat{\psi}^*_{I,h} + I^0_h \boldsymbol{\lambda}^h_I$ and $\boldsymbol{\lambda}^h_I = \boldsymbol{\rho}^*_I - \operatorname{grad} g_{\lambda^h_I}$. Since from the assumptions on $\boldsymbol{\mu}$ there exist two positive constants μ_* and μ^* such that $\mu_* \|\mathbf{v}_I\|^2_{0,\Omega_I} \le \int_{\Omega_I} \boldsymbol{\mu}_I \mathbf{v}_I \cdot \overline{\mathbf{v}_I} \le \mu^* \|\mathbf{v}_I\|^2_{0,\Omega_I}$ for all $\mathbf{v}_I \in (L^2(\Omega_I))^3$, it follows at once

$$|I^0 - I^0_h| \le (\mu^*/c_0)^{1/2} \|\mathbf{H}_I - \mathbf{H}_{I,h}\|_{0,\Omega_I} ,$$

where $c_0 := (\int_{\Omega_I} \boldsymbol{\mu}_I \boldsymbol{\rho}^*_I \cdot \boldsymbol{\rho}^*_I)^{-1}$.

The convergence of the approximation scheme is an easy consequence of (8.41). Let us denote by $N^k_{I,h}$ the space of Nédélec curl-conforming edge elements of degree k in $\mathcal{T}_{I,h}$ and by $\pi_{I,h}$ the interpolation operator. If $\mathbf{H}_I$ is so regular that $\pi_{I,h}\mathbf{H}_I$ is well-defined, then in (8.41) we can choose $\mathbf{v}_{I,h} = \pi_{I,h}\mathbf{H}_I$, as indeed $\pi_{I,h}\mathbf{H}_I \in Z^k_{I,h}$. This assertion can be demonstrated as follows: since $\operatorname{curl}(\pi_{I,h}\mathbf{H}_I - I^0\boldsymbol{\lambda}^h_I) = \mathbf{0}$ in Ω_I and $\int_{\partial\Gamma_J}(\pi_{I,h}\mathbf{H}_I - I^0\boldsymbol{\lambda}^h_I)\cdot\mathbf{t} = 0$, then $\pi_{I,h}\mathbf{H}_I - I^0\boldsymbol{\lambda}^h_I = \operatorname{grad}\xi_I$ for some $\xi_I \in H^1(\Omega_I)$. Knowing that $\pi_{I,h}\mathbf{H}_I - I^0\boldsymbol{\lambda}^h_I \in N^k_{I,h}$, from Girault and Raviart [111], Chap. III, Lemma 5.3, it follows that $\xi_{I|K}$ is a polynomial of degree k for each $K \in \mathcal{T}_{I,h}$, therefore $\xi_I \in L^k_{I,h}$.

As a consequence, from standard interpolation estimates, for a regular solution $(\mathbf{E}_C, \mathbf{H}_I)$ it is straightforward to specify the order of convergence of the approximation method.

Instead, if one has no information about the regularity of the solution, by a density argument it is possible to prove the convergence of the finite element scheme if the permeability coefficient $\boldsymbol{\mu}_I$ is regular enough in Ω_I (say, a constant as in the usual physical case) or if the family of meshes $\mathcal{T}_{I,h}$ is obtained by refining a coarse mesh $\mathcal{T}_{I,h^0}$.

In fact, when $\boldsymbol{\mu}_I$ is constant we know that the harmonic field $\boldsymbol{\rho}^*_I$ is regular enough to define the interpolant $\pi_{I,h}\boldsymbol{\rho}^*_I$ (see Amrouche et al. [27], Alonso and Valli [9]). Since $\mathbf{H}_I = \operatorname{grad}\psi^*_I + I^0\boldsymbol{\rho}^*_I$, a density argument applied to ψ^*_I permits to conclude the proof.

In the other case, first we note that we can write $\boldsymbol{\rho}^*_I = \operatorname{grad} g_{\lambda_I} + \boldsymbol{\lambda}_I$, where $\boldsymbol{\lambda}_I$ is defined as $\boldsymbol{\lambda}^h_I$, but on the fixed coarse mesh $\mathcal{T}_{I,h^0}$. Then, knowing that $\mathcal{T}_{I,h}$ is a refinement of $\mathcal{T}_{I,h^0}$, it follows that $\boldsymbol{\lambda}_I \in N^k_{I,h}$, hence $\boldsymbol{\lambda}_I = \pi_{I,h}\boldsymbol{\lambda}_I$, and a density argument for $\psi_I + g_{\lambda_I}$ gives the result.

For the current intensity excitation problem (8.39) the finite element approach reads

$$
\begin{aligned}
&\text{Find } (\mathbf{E}_{C,h}, \widehat{\psi}^*_{I,h}) \in Y^k_{C,h} \times L^k_{I,h}/\mathbb{C} : \\
&\quad \mathcal{C}((\mathbf{E}_{C,h}, \operatorname{grad} \widehat{\psi}^*_{I,h}), (\mathbf{z}_{C,h}, \operatorname{grad} \chi^*_{I,h})) \\
&\qquad = -i\omega \textstyle\int_{\Omega_C} \mathbf{J}_{e,C} \cdot \overline{\mathbf{z}_{C,h}} + i\omega I^0 \int_\Gamma \overline{\mathbf{z}_{C,h}} \times \mathbf{n}_C \cdot \boldsymbol{\lambda}^h_I \\
&\qquad\quad -\omega^2 I^0 \textstyle\int_{\Omega_I} \boldsymbol{\mu}_I \boldsymbol{\lambda}^h_I \cdot \operatorname{grad} \overline{\chi^*_{I,h}} \\
&\text{for all } (\mathbf{z}_{C,h}, \chi^*_{I,h}) \in Y^k_{C,h} \times L^k_{I,h}/\mathbb{C} .
\end{aligned}
\tag{8.42}
$$

Recalling that $\mathbf{H}_I = \operatorname{grad} \widehat{\psi}^*_I + I^0 \boldsymbol{\lambda}^h_I$ and setting $\mathbf{H}_{I,h} := \operatorname{grad} \widehat{\psi}^*_{I,h} + I^0 \boldsymbol{\lambda}^h_I$, from (8.39) and (8.42) we have the following equation for the error

$$
\begin{aligned}
&\mathcal{C}((\mathbf{E}_C - \mathbf{E}_{C,h}, \mathbf{H}_I - \mathbf{H}_{I,h}), (\mathbf{z}_{C,h}, \operatorname{grad} \chi^*_{I,h})) \\
&\quad = \mathcal{C}((\mathbf{E}_C - \mathbf{E}_{C,h}, \operatorname{grad} \widehat{\psi}^*_I - \operatorname{grad} \widehat{\psi}^*_{I,h}), (\mathbf{z}_{C,h}, \operatorname{grad} \chi^*_{I,h})) = 0
\end{aligned}
$$

for each $(\mathbf{z}_{C,h}, \chi^*_{I,h}) \in Y^k_{C,h} \times L^k_{I,h}/\mathbb{C}$. Therefore, the coerciveness of $\mathcal{C}(\cdot,\cdot)$ gives

$$
\begin{aligned}
&\|\mathbf{E}_C - \mathbf{E}_{C,h}\|_{H(\mathrm{curl};\Omega_C)} + \|\mathbf{H}_I - \mathbf{H}_{I,h}\|_{0;\Omega_I} \\
&\quad = \|\mathbf{E}_C - \mathbf{E}_{C,h}\|_{H(\mathrm{curl};\Omega_C)} + \|\operatorname{grad} \widehat{\psi}^*_I - \operatorname{grad} \widehat{\psi}^*_{I,h}\|_{0,\Omega_I} \\
&\quad \le C \big(\|\mathbf{E}_C - \mathbf{z}_{C,h}\|_{H(\mathrm{curl};\Omega_C)} + \|\operatorname{grad} \widehat{\psi}^*_I - \operatorname{grad} \chi^*_{I,h}\|_{0,\Omega_I} \big)
\end{aligned}
$$

for each $(\mathbf{z}_{C,h}, \chi^*_{I,h}) \in Y^k_{C,h} \times L^k_{I,h}/\mathbb{C}$. Consequently, we find the error estimate

$$
\begin{aligned}
&\|\mathbf{E}_C - \mathbf{E}_{C,h}\|_{H(\mathrm{curl};\Omega_C)} + \|\mathbf{H}_I - \mathbf{H}_{I,h}\|_{0,\Omega_I} \\
&\quad \le C \inf_{(\mathbf{z}_{C,h}, \mathbf{v}_{I,h}) \in Y^k_{C,h} \times Z^k_{I,h}(I^0)} \big(\|\mathbf{E}_C - \mathbf{z}_{C,h}\|_{H(\mathrm{curl};\Omega_C)} + \|\mathbf{H}_I - \mathbf{v}_{I,h}\|_{0,\Omega_I} \big) ,
\end{aligned}
$$

where

$$
Z^k_{I,h}(I^0) := \operatorname{grad} L^k_{I,h} + I^0 \boldsymbol{\lambda}^h_I .
$$

The convergence of the approximation scheme can be proved following the arguments presented for the voltage problem (the only difference is that now we work with the space $Z^k_{I,h}(I^0)$ instead of $Z^k_{I,h}$, and this fact gives no problems to the procedure).

Once we have obtained $\mathbf{E}_{C,h}$ and $\widehat{\psi}^*_{I,h}$ from (8.39), we can compute

$$
V_h := \int_\Gamma \mathbf{E}_{C,h} \times \mathbf{n}_C \cdot \boldsymbol{\lambda}^h_I + i\omega \int_{\Omega_I} \boldsymbol{\mu}_I \mathbf{H}_{I,h} \cdot \boldsymbol{\lambda}^h_I .
$$

We can show that this quantity is an approximation of the voltage, that, from (8.13), can be written as

$$
V = \int_\Gamma \mathbf{E}_C \times \mathbf{n}_C \cdot \boldsymbol{\rho}^*_I + i\omega \int_{\Omega_I} \boldsymbol{\mu}_I \mathbf{H}_I \cdot \boldsymbol{\rho}^*_I .
$$

In fact, let us introduce the auxiliary quantity

$$
\widehat{V}_h := \int_\Gamma \mathbf{E}_{C,h} \times \mathbf{n}_C \cdot \boldsymbol{\rho}^*_I + i\omega \int_{\Omega_I} \boldsymbol{\mu}_I \mathbf{H}_{I,h} \cdot \boldsymbol{\rho}^*_I .
$$

We easily have

$$|V - \widehat{V}_h| \leq C_1(\|\mathbf{E}_C - \mathbf{E}_{C,h}\|_{H(\mathrm{curl};\Omega_C)} + \|\mathbf{H}_I - \mathbf{H}_{I,h}\|_{0,\Omega_I}) .$$

On the other hand, taking $\mathbf{z}_{C,h} = \mathbf{0}$ in (8.42), it is easy to see that

$$V_h = \int_\Gamma \mathbf{E}_{C,h} \times \mathbf{n}_C \,(\mathrm{grad}\, \chi^*_{I,h} + \boldsymbol{\lambda}^h_I) + i\omega \int_{\Omega_I} \boldsymbol{\mu}_I \mathbf{H}_{I,h} \cdot (\mathrm{grad}\, \chi^*_{I,h} + \boldsymbol{\lambda}^h_I)$$

for all $\chi^*_{I,h} \in L^k_{I,h}/\mathbb{C}$. Thus

$$|\widehat{V}_h - V_h| \leq C_2 \left(\|\mathbf{E}_{C,h}\|_{H(\mathrm{curl};\Omega_C)} + \|\mathbf{H}_{I,h}\|_{0,\Omega_I}\right) \|\boldsymbol{\rho}^*_I - (\mathrm{grad}\, \chi^*_{I,h} + \boldsymbol{\lambda}^h_I)\|_{0,\Omega_I} ,$$

for all $\chi^*_{I,h} \in L^k_{I,h}/\mathbb{C}$. Therefore, since $(\mathrm{grad}\, \chi^*_{I,h} + \boldsymbol{\lambda}^h_I) \in \mathbf{Z}^k_{I,h}(1)$ (namely, the line integral of this vector function along $\partial\Gamma_J$ is equal to 1), we have

$$\begin{aligned} |V - V_h| &\leq C_1 \left(\|\mathbf{E}_C - \mathbf{E}_{C,h}\|_{H(\mathrm{curl};\Omega_C)} + \|\mathbf{H}_I - \mathbf{H}_{I,h}\|_{0,\Omega_I}\right) \\ &\quad + C_2 \left(\|\mathbf{E}_{C,h}\|_{H(\mathrm{curl};\Omega_C)} + \|\mathbf{H}_{I,h}\|_{0,\Omega_I}\right) \inf_{\mathbf{v}_{I,h} \in Z^k_{I,h}(1)} \|\boldsymbol{\rho}^*_I - \mathbf{v}_{I,h}\|_{0,\Omega_I} \\ &\leq \left(C_1 + C_2 \inf_{\mathbf{v}_{I,h} \in Z^k_{I,h}(1)} \|\boldsymbol{\rho}^*_I - \mathbf{v}_{I,h}\|_{0,\Omega_I}\right) \\ &\qquad \times \left(\|\mathbf{E}_C - \mathbf{E}_{C,h}\|_{H(\mathrm{curl};\Omega_C)} + \|\mathbf{H}_I - \mathbf{H}_{I,h}\|_{0,\Omega_I}\right) \\ &\quad + C_2 \left(\|\mathbf{E}_C\|_{H(\mathrm{curl};\Omega_C)} + \|\mathbf{H}_I\|_{0,\Omega_I}\right) \inf_{\mathbf{v}_{I,h} \in Z^k_{I,h}(1)} \|\boldsymbol{\rho}^*_I - \mathbf{v}_{I,h}\|_{0,\Omega_I} . \end{aligned}$$

If the permeability coefficient $\boldsymbol{\mu}_I$ is smooth enough or if the family of meshes $\mathcal{T}_{I,h}$ is a refinement of a coarse mesh $\mathcal{T}_{I,h^0}$, the convergence can be proved as in the preceding cases.

Remark 8.11. Suitable finite element approximation schemes can be devised starting from the variational formulations (8.28) and (8.29), related to the $(\mathbf{H}_C, \psi^*_I)$-approach, or (8.34), (8.35), (8.36), related to the $(\mathbf{T}_C, \psi^*)$-approach.

In the former case, we have already noted that the meshes in Ω_C and Ω_I must be compatible on Γ, while this is not the case for the hybrid $(\mathbf{E}_C, \psi^*_I)$ formulations.

In the latter case, the use of Lagrange nodal elements is the natural choice, but in Section 6.1.3 we have seen that the convergence of the Galerkin finite element approximation scheme can be ensured only if Ω_C is a convex polyhedral domain (which is not the most interesting case in real-life applications). Moreover, since the scalar potential ψ^* is present also in Ω_C, the total number of degrees of freedom is higher than in the hybrid formulations. □

8.1.5 Numerical results

The finite element method presented in the preceding section has been implemented in MATLAB, using Nédélec edge elements of first order $Y^1_{C,h}$ for the electric field in the conductor, and scalar Lagrange nodal elements of first order for the magnetic potential in the insulator.

The method has been tested by solving a problem with a known analytical solution presented in Bermúdez et al. [42]. The conducting domain Ω_C and the whole domain Ω are two coaxial cylinders of radii R_C and R_I, respectively, with height L. An alternating current of intensity $\mathbb{I}(t) = I^0 \cos(\omega t)$ is traversing the conductor in the axial direction. Supposing that the physical parameters $\boldsymbol{\sigma}$ and $\boldsymbol{\mu}$ are constant scalars, the solution of the problem in cylindrical coordinates is given by

$$
\begin{aligned}
\mathbf{E}_C(r,\theta,z) &= \frac{\kappa}{2\pi R_C \sigma} \frac{\mathcal{I}_0(\kappa r)}{\mathcal{I}_1(\kappa R_C)} \mathbf{e}_z \quad \text{in } \Omega_C \\
\mathbf{H}_C(r,\theta,z) &= \frac{I^0}{2\pi R_C} \frac{\mathcal{I}_1(\kappa r)}{\mathcal{I}_1(\kappa R_C)} \mathbf{e}_\theta \quad \text{in } \Omega_C \\
\mathbf{H}_I(r,\theta,z) &= \frac{I^0}{2\pi r} \mathbf{e}_\theta \quad \text{in } \Omega_I \,,
\end{aligned}
$$

where $\mathcal{I}_0$ and $\mathcal{I}_1$ denote the modified Bessel functions of the first kind and order 0 and 1, respectively, and $\kappa = \sqrt{i\omega\mu\sigma}$. Moreover, for this particular geometry it holds $\boldsymbol{\rho}_I^* = \frac{1}{2\pi r}\,\mathbf{e}_\theta$, thus $\mathbf{H}_I = I^0 \boldsymbol{\rho}_I^*$.

Once the fields and the function $\boldsymbol{\rho}_I^*$ are known, the value of V is computed from the expression (8.17), obtaining

$$
V = \frac{\kappa L I^0}{2\pi\sigma R_C} \frac{\mathcal{I}_0(\kappa R_C)}{\mathcal{I}_1(\kappa R_C)} + i\omega\mu \frac{L I^0}{2\pi} \ln \frac{R_I}{R_C} \,.
$$

For our particular case we have used the following geometry and data

$$
\begin{aligned}
R_C &= 0.25 \text{ m}, \\
R_I &= 0.5 \text{ m}, \\
L &= 0.25 \text{ m}, \\
\sigma &= 151565.8 \text{ S/m}, \\
\mu &= 4\pi \times 10^{-7} \text{ H/m}, \\
\omega &= 2\pi \times 50 \text{ rad/s},
\end{aligned}
$$

and either assigned current intensity or voltage,

$$
I^0 = 10^4 \text{ A}, \quad \text{or} \quad V = 0.08979 + 0.14680i,
$$

where the value of V has been computed for an intensity of 10^4 A.

To test the order of convergence, the problem has been solved in four successively refined meshes, for either assigned current intensity or voltage. We present in Tables 8.1 and 8.2 the relative errors of our numerical solutions against the analytical solution, that have been set as follows

$$
\begin{aligned}
e_E &= \frac{\|\mathbf{E}_C - \mathbf{E}_{C,h}\|_{H(\mathrm{curl};\Omega_C)}}{\|\mathbf{E}_C\|_{H(\mathrm{curl};\Omega_C)}}, \quad e_V = \frac{|V - V_h|}{|V|} \\
e_H &= \frac{\|\mathbf{H}_I - \mathbf{H}_{I,h}\|_{0,\Omega_I}}{\|\mathbf{H}_I\|_{0,\Omega_I}}, \quad e_{I^0} = \frac{|I^0 - I^0_h|}{|I^0|}.
\end{aligned}
$$

Table 8.1. Relative errors for assigned intensity

Elements	*DoF*	e_E	e_H	e_V
2304	1684	0.2341	0.1693	0.0312
18432	11240	0.1132	0.0847	0.0089
62208	35580	0.0750	0.0567	0.0048
147456	81616	0.0561	0.0425	0.0018

Table 8.2. Relative errors for assigned voltage

Elements	*DoF*	e_E	e_H	e_{I^0}
2304	1685	0.2336	0.1685	0.0274
18432	11241	0.1132	0.0847	0.0085
62208	35581	0.0750	0.0566	0.0041
147456	81617	0.0561	0.0425	0.0024

Finally, Figures 8.4 and 8.5 show the plots in a log-log scale of the relative errors versus the degrees of freedom. A linear dependence on the mesh size is obtained for the errors of electric and magnetic fields, either for assigned intensity or voltage. The voltage and intensity errors turn out to be quadratic with respect to h; to our knowledge, this superconvergence result has not been theoretically proved, but it looks quite evident from these numerical computations.

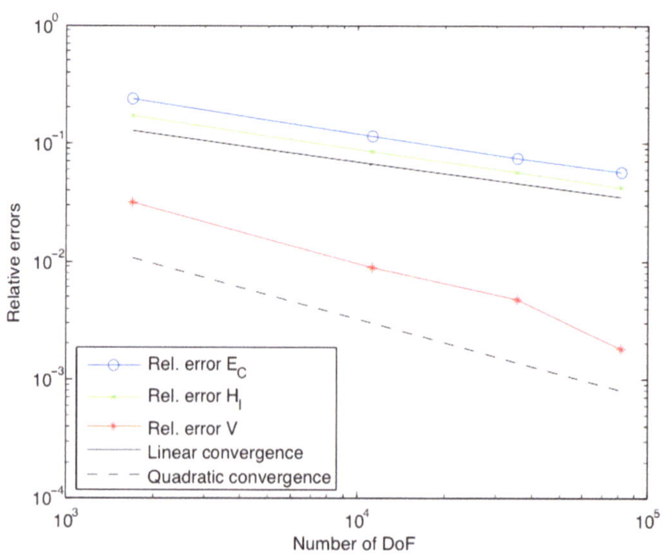

Fig. 8.4. Relative errors versus number of DoF (assigned intensity)

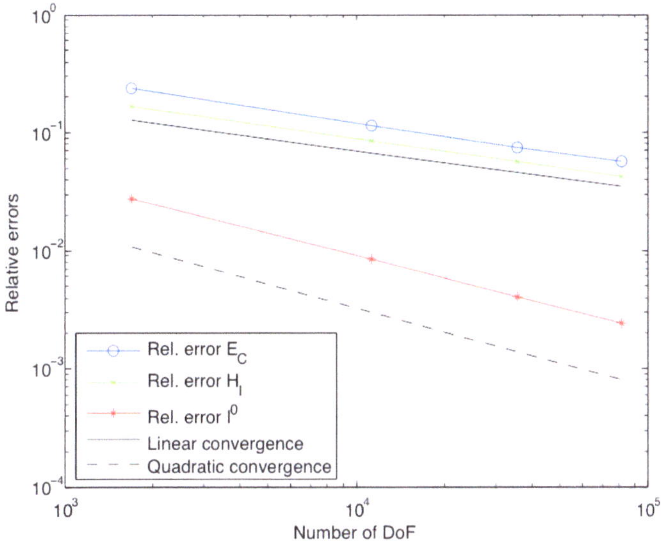

Fig. 8.5. Relative errors versus number of DoF (assigned voltage)

The method has been also applied to a more realistic problem which was presented in Bermúdez et al. [43]. In this case the domain is a cylindrical electric furnace with three ELSA electrodes equally distanced. The dimensions of the furnace are the following: furnace height: 2 m; furnace diameter: 8.88 m; electrodes height: 1.25 m; electrodes diameter: 1 m; distance from the center of the electrodes to the wall: 3 m.

The three ELSA electrodes inside the furnace are formed by a graphite core of 0.4 m of diameter, and an outer part of Søderberg paste. The electric current enters the electrodes through horizontal copper bars of rectangular section (0.07 m × 0.25 m), connecting the top of the electrode with the external boundary.

For the simulation we have considered the angular frequency $\omega = 2\pi \times 50$ rad/s, the magnetic permeability $\mu = 4\pi \times 10^{-7}$ H/m, and the electric conductivities $\sigma = 10^6$ S/m for graphite, $\sigma = 10^4$ S/m for Søderberg paste, and $\sigma = 5 \times 10^6$ S/m for copper. We have imposed an intensity of $I_j^0 = 7 \times 10^4 A$ for each electrode, using the approach that has been explained in Remark 8.8 for the case of a non-connected conductor. With the same notation used there, the boundaries $\Gamma_{E,j}$ correspond to the contacts of the copper bars on the boundary of the furnace, and $\Gamma_{J,j}$ to the bottom of the electrodes.

In Figure 8.6 we present the absolute value of the magnetic potential, i.e., $|\widehat{\psi}^*_{I,h} + \sum_{j=1}^3 I_j^0 \eta_{I,j,h}|$, where $\eta_{I,j,h}$ are the piecewise-linear functions with a jump of height 1 on the “cutting” surfaces $\Xi^*_{j,h}$. In Figures 8.7 and 8.8 the magnitude of the current density $\mathbf{J}_h = \sigma \mathbf{E}_{C,h}$ on a horizontal and a vertical section of one electrode is shown.

Fig. 8.6. Magnetic potential in the dielectric

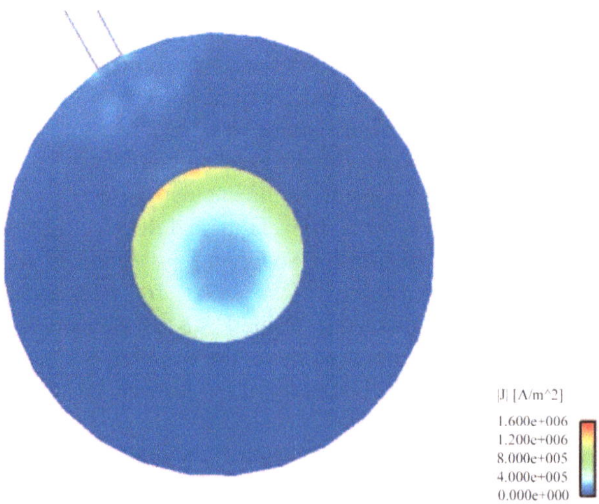

Fig. 8.7. Magnitude of the current density $\mathbf{J}_h$ (A/m^2) on a horizontal section of one electrode

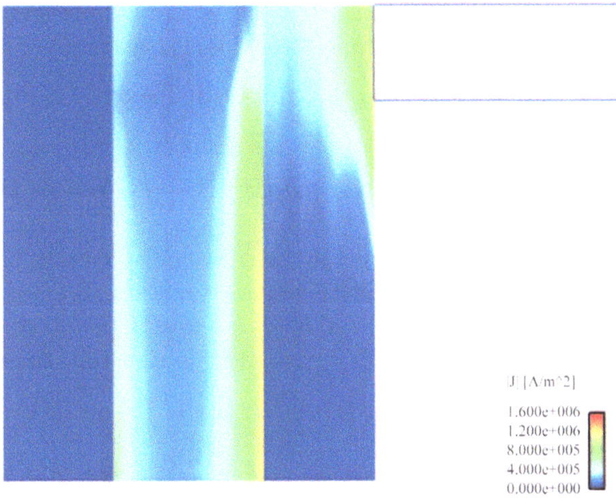

Fig. 8.8. Magnitude of the current density $\mathbf{J}_h$ (A/m^2) on a vertical section of one electrode

8.2 Voltage and current intensity excitation for an internal conductor

In this section, following Alonso Rodríguez and Valli [18], where a more complete analysis is presented, we consider the question of how to impose a current intensity or a voltage when the conductor is strictly contained in the computational domain. For the sake of simplicity we assume that Ω is a simply-connected bounded open set, with a connected boundary $\partial\Omega$, and that the conductor Ω_C is a torus-like domain. We consider three different kinds of boundary conditions: the electric boundary condition

$$\mathbf{E} \times \mathbf{n} = \mathbf{0} \ \text{ on } \partial\Omega \,, \tag{8.43}$$

the magnetic boundary conditions

$$\begin{array}{ll} \mathbf{H} \times \mathbf{n} = \mathbf{0} & \text{on } \partial\Omega \\ \varepsilon\mathbf{E} \cdot \mathbf{n} & \text{on } \partial\Omega, \end{array} \tag{8.44}$$

and the no-flux boundary conditions

$$\begin{array}{ll} \mu\mathbf{H} \cdot \mathbf{n} = 0 & \text{on } \partial\Omega \\ \varepsilon\mathbf{E} \cdot \mathbf{n} & \text{on } \partial\Omega \,. \end{array} \tag{8.45}$$

First we notice that for each one of these boundary conditions the solution of the eddy current problem is unique. In fact we have the following result (similar to Proposition 8.2).

Proposition 8.12. *Let us consider the solutions* $\mathbf{H}$ *and* $\mathbf{E}$ *of the eddy current problem*

$$\begin{array}{l} \operatorname{curl} \mathbf{H} - \sigma\mathbf{E} = \mathbf{J}_e \quad \text{in } \Omega \\ \operatorname{curl} \mathbf{E} + i\omega\mu\mathbf{H} = \mathbf{0} \ \text{in } \Omega \,. \end{array}$$

The magnetic field $\mathbf{H}$ *in* Ω *and the electric field* $\mathbf{E}_C$ *in* Ω_C *are uniquely determined for each one of the boundary condition* (8.43), (8.44) *and* (8.45).

Proof. Assuming that $\mathbf{J}_e = \mathbf{0}$ and proceeding as in the proof of Proposition 8.2 we obtain that

$$0 = \int_{\Omega_C} \boldsymbol{\sigma} \mathbf{E}_C \cdot \overline{\mathbf{E}_C} + \int_{\Omega} i\omega\boldsymbol{\mu} \mathbf{H} \cdot \overline{\mathbf{H}} - \int_{\partial\Omega} \mathbf{E} \times \mathbf{n} \cdot \overline{\mathbf{H}} \,.$$

The uniqueness of $\mathbf{H}$ and $\mathbf{E}_C$ follows at once if we prove that the boundary integral vanishes. This is clear in the case of the boundary conditions (8.43) and (8.44). For the case (8.45) we have $\operatorname{div}_\tau(\mathbf{E} \times \mathbf{n}) = \operatorname{curl} \mathbf{E} \cdot \mathbf{n} = -i\omega\boldsymbol{\mu}\mathbf{H} \cdot \mathbf{n} = 0$ on $\partial\Omega$, hence there exists a scalar function v such that $\mathbf{E} \times \mathbf{n} = \operatorname{grad} v \times \mathbf{n}$ on $\partial\Omega$. Therefore

$$\begin{aligned} \textstyle\int_{\partial\Omega} \mathbf{E} \times \mathbf{n} \cdot \overline{\mathbf{H}} &= -\textstyle\int_{\partial\Omega} \overline{\mathbf{H}} \times \mathbf{n} \cdot \operatorname{grad} v = \textstyle\int_{\partial\Omega} \operatorname{div}_\tau(\overline{\mathbf{H}} \times \mathbf{n})\, v \\ &= \textstyle\int_{\partial\Omega} \operatorname{curl} \overline{\mathbf{H}} \cdot \mathbf{n}\, v = 0 \,, \end{aligned}$$

as $\operatorname{curl} \mathbf{H}_I = \mathbf{0}$ in Ω_I and $\partial\Omega \subset \partial\Omega_I$. □

Since the eddy current problem has a unique solution for each of the boundary conditions described in (8.43), (8.44) and (8.45), it is not possible to impose an additional condition, say, voltage or current intensity, if we do not relax some of the other equations. However we cannot renounce to Maxwell equations, namely, to Faraday and Ampère equations.

The point is therefore to devise a different interpretation, and this will be the object of our presentation in the sequel. We revisit what was done in Section 8.1, and analyze again the electric port case with boundary conditions (8.2). Indeed, this is the case in which it is clear how to impose voltage or current intensity. First of all, it is worthwhile to recall that, assuming $\mathbf{J}_e = \mathbf{0}$, as in (8.9) one has

$$\int_{\Omega_C} \boldsymbol{\sigma}^{-1} \operatorname{curl} \mathbf{H}_C \cdot \operatorname{curl} \overline{\mathbf{H}_C} + \int_{\Omega} i\omega\boldsymbol{\mu} \mathbf{H} \cdot \overline{\mathbf{H}} = V\, \overline{I^0} \,.$$

On the other hand, given a complex number $q \in \mathbb{C}$ take $\mathbf{J}_{e,I} = \mathbf{0}$ and $\mathbf{J}_{e,C} = q\boldsymbol{\sigma} \operatorname{grad} \phi_C$, where ϕ_C is the unique solution to

$$\begin{cases} \operatorname{div}(\boldsymbol{\sigma} \operatorname{grad} \phi_C) = 0 & \text{in } \Omega_C \\ \phi_C = 1 & \text{on } \Gamma_J \\ \phi_C = 0 & \text{on } \Gamma_E \\ \boldsymbol{\sigma} \operatorname{grad} \phi_C \cdot \mathbf{n} = 0 & \text{on } \Gamma \,. \end{cases} \tag{8.46}$$

Then

$$\begin{aligned} &\textstyle\int_{\Omega_C} \boldsymbol{\sigma}^{-1} \operatorname{curl} \mathbf{H}_C \cdot \operatorname{curl} \overline{\mathbf{H}_C} + i\omega \textstyle\int_{\Omega} \boldsymbol{\mu} \mathbf{H} \cdot \overline{\mathbf{H}} \\ &\quad = \textstyle\int_{\Omega_C} \boldsymbol{\sigma}^{-1} \mathbf{J}_{e,C} \cdot \operatorname{curl} \overline{\mathbf{H}_C} + V\, \overline{I^0} \\ &\quad = q \textstyle\int_{\Omega_C} \operatorname{grad} \phi_C \cdot \operatorname{curl} \overline{\mathbf{H}_C} + V\, \overline{I^0} \,. \end{aligned}$$

Moreover,

$$\begin{aligned} &\textstyle\int_{\Omega_C} \operatorname{grad} \phi_C \cdot \operatorname{curl} \overline{\mathbf{H}_C} \\ &\quad = -\textstyle\int_{\Omega_C} \phi_C \operatorname{div} \operatorname{curl} \overline{\mathbf{H}_C} + \textstyle\int_{\Gamma_E \cup \Gamma_J \cup \Gamma} \phi_C \operatorname{curl} \overline{\mathbf{H}_C} \cdot \mathbf{n}_C \\ &\quad = \textstyle\int_{\Gamma_J} \operatorname{curl} \overline{\mathbf{H}_C} \cdot \mathbf{n} \,, \end{aligned} \tag{8.47}$$

as $\phi_C = 0$ on Γ_E, $\phi_C = 1$ on Γ_J and $\operatorname{curl} \mathbf{H}_C \cdot \mathbf{n}_C = \operatorname{curl} \mathbf{H}_I \cdot \mathbf{n}_C = 0$ on Γ. Hence

$$\int_{\Omega_C} \boldsymbol{\sigma}^{-1} \operatorname{curl} \mathbf{H}_C \cdot \operatorname{curl} \overline{\mathbf{H}_C} + i\omega \int_{\Omega} \boldsymbol{\mu} \mathbf{H} \cdot \overline{\mathbf{H}} = (q + V)\, \overline{I^0} \,. \tag{8.48}$$

This is telling us that, when considering the electric port case with boundary conditions (8.2), assigning a voltage V is in some sense equivalent to impose a current density $\mathbf{J}_{e,C} = V\boldsymbol{\sigma} \operatorname{grad} \phi_C$ in Ω_C.

More precisely, let us consider the electric port eddy current problem with $\mathbf{J}_{e,C} = -V\boldsymbol{\sigma} \operatorname{grad} \phi_C$, $\mathbf{J}_{e,I} = \mathbf{0}$ and assigned voltage equal to V, namely,

$$\begin{cases} \operatorname{curl} \widehat{\mathbf{H}} - \boldsymbol{\sigma}\widehat{\mathbf{E}} = -V\boldsymbol{\sigma} \operatorname{grad} \phi_C & \text{in } \Omega \\ \operatorname{curl} \widehat{\mathbf{E}} + i\omega\boldsymbol{\mu}\widehat{\mathbf{H}} = \mathbf{0} & \text{in } \Omega \\ \operatorname{div}(\boldsymbol{\varepsilon}_I \widehat{\mathbf{E}}_I) = 0 & \text{in } \Omega_I \\ \boldsymbol{\varepsilon}_I \widehat{\mathbf{E}}_I \cdot \mathbf{n}_I = 0 & \text{on } \Gamma_I \\ \widehat{\mathbf{E}} \times \mathbf{n} = \operatorname{grad} v \times \mathbf{n} & \text{on } \partial\Omega \\ v_{|\Gamma_J} = V & \text{on } \Gamma_J \\ v_{|\Gamma_E} = 0 & \text{on } \Gamma_E \,. \end{cases}$$

From (8.48) it follows at once that $\widehat{\mathbf{H}} = \mathbf{0}$, and it is also easily seen that $\widehat{\mathbf{E}} = V \operatorname{grad} v$, where v is equal to ϕ_C in Ω_C and to v_I in Ω_I, v_I being the unique solution to

$$\begin{cases} \operatorname{div}(\boldsymbol{\varepsilon}_I \operatorname{grad} v_I) = 0 & \text{in } \Omega_I \\ v_I = \phi_C & \text{on } \Gamma \\ \boldsymbol{\varepsilon}_I \operatorname{grad} v_I \cdot \mathbf{n} = 0 & \text{on } \Gamma_I \,. \end{cases} \tag{8.49}$$

This will lead us to propose a suitable formulation for the eddy current problem with an internal conductor subjected to a given voltage or current intensity excitation: the key point will be that these excitations have to be interpreted as a particular applied current density.

Note that the function $\operatorname{grad} \phi_C$ is the basis function of the space of harmonic fields

$$\begin{aligned} \mathcal{H}_\sigma(\Gamma_C, \Gamma; \Omega_C) := \{ \mathbf{v}_C \in (L^2(\Omega_C))^3 \mid \operatorname{curl} \mathbf{v}_C = \mathbf{0}, \operatorname{div}(\boldsymbol{\sigma}\mathbf{v}_C) = 0, \\ \boldsymbol{\sigma}\mathbf{v}_C \cdot \mathbf{n}_C = 0 \text{ on } \Gamma,\ \mathbf{v}_C \times \mathbf{n}_C = \mathbf{0} \text{ on } \Gamma_C \} \,, \end{aligned}$$

normalized with the condition

$$\int_{\widetilde{\gamma}} \operatorname{grad} \phi_C \cdot d\boldsymbol{\tau} = 1 \,,$$

where $\widetilde{\gamma}$ is a curve lying in $\overline{\Omega_C}$ and joining Γ_E to Γ_J. Thus, for the internal conductor case we are led to introduce the space of harmonic fields

$$\begin{aligned} \mathcal{H}_\sigma(m; \Omega_C) := \{ \mathbf{v}_C \in (L^2(\Omega_C))^3 \mid \operatorname{curl} \mathbf{v}_C = \mathbf{0}, \operatorname{div}(\boldsymbol{\sigma}\mathbf{v}_C) = 0, \\ \boldsymbol{\sigma}\mathbf{v}_C \cdot \mathbf{n}_C = 0 \text{ on } \Gamma \} \,, \end{aligned}$$

denoting by $\boldsymbol{\rho}_C^*$ its basis function normalized with the condition

$$\int_{\gamma_*} \boldsymbol{\rho}_C^* \cdot d\boldsymbol{\tau} = 1 \,,$$

where γ_* is a non-bounding cycle internal to Ω_C (whose orientation has been freely chosen).

The voltage and current intensity excitation problems are therefore formulated as follows.

Voltage rule. *When the voltage V is imposed, modify the Ohm law in Ω_C by adding to the current density $\boldsymbol{\sigma}\mathbf{E}_C$ the "applied" current density $\mathbf{J}_{e,C} = V\boldsymbol{\sigma}\boldsymbol{\rho}_C^*$. Thus the Ampère equation reads*

$$\operatorname{curl}\mathbf{H}_C - \boldsymbol{\sigma}\mathbf{E}_C = V\boldsymbol{\sigma}\boldsymbol{\rho}_C^* \ .$$

Our convention is that the voltage passes from 0 to V along the basic cycle γ_*.

Current intensity rule. *When the current intensity I^0 is imposed, modify the Ohm law in Ω_C by adding to the current density $\boldsymbol{\sigma}\mathbf{E}_C$ the "applied" current density $\mathbf{J}_{e,C} = V\boldsymbol{\sigma}\boldsymbol{\rho}_C^*$. Thus the Ampère equation reads*

$$\operatorname{curl}\mathbf{H}_C - \boldsymbol{\sigma}\mathbf{E}_C - V\boldsymbol{\sigma}\boldsymbol{\rho}_C^* = 0 \ ,$$

where V has to be determined by imposing the additional constraint

$$\int_{\Xi_C^*} \operatorname{curl}\mathbf{H}_C \cdot \mathbf{n}_* = I^0 \ .$$

Here Ξ_C^* is a section of Ω_C and the unit normal vector $\mathbf{n}_*$ has the same orientation of the basic cycle γ_*.

Let us show that, when adopting these two rules, we are respecting the following power law

$$P := \int_{\Omega_C} \boldsymbol{\sigma}^{-1} \operatorname{curl}\mathbf{H}_C \cdot \operatorname{curl}\overline{\mathbf{H}_C} + i\omega \int_\Omega \mu\mathbf{H}\cdot\overline{\mathbf{H}} = V\overline{I^0} \ .$$

In fact, for $\mathbf{J}_{e,C} = V\boldsymbol{\sigma}\boldsymbol{\rho}_C^*$ and $\mathbf{J}_{e,I} = \mathbf{0}$, we have as usual

$$\int_{\Omega_C} \boldsymbol{\sigma}^{-1} \operatorname{curl}\mathbf{H}_C \cdot \operatorname{curl}\overline{\mathbf{H}_C} + i\omega \int_\Omega \mu\mathbf{H}\cdot\overline{\mathbf{H}} = V \int_{\Omega_C} \boldsymbol{\rho}_C^* \cdot \operatorname{curl}\overline{\mathbf{H}_C} \ .$$

Recalling that the basis function $\boldsymbol{\rho}_C^*$ is the $L^2(\Omega_C)$-extension of the gradient of a suitable scalar function p_C^*, defined in $\Omega_C \setminus \Xi_C^*$ and having a jump equal to 1 across Ξ_C^*, we obtain,

$$\begin{aligned}
\textstyle\int_{\Omega_C} \boldsymbol{\rho}_C^* \cdot \operatorname{curl}\overline{\mathbf{H}_C} &= \textstyle\int_{\Omega_C\setminus\Xi_C^*} \operatorname{grad} p_C^* \cdot \operatorname{curl}\overline{\mathbf{H}_C} \\
&= -\textstyle\int_{\Omega_C\setminus\Xi_C^*} p_C^* \operatorname{div}\operatorname{curl}\overline{\mathbf{H}_C} + \int_\Gamma p_C^* \operatorname{curl}\overline{\mathbf{H}_C}\cdot\mathbf{n}_C \\
&\quad + \textstyle\int_{\Xi_C^*} \operatorname{curl}\overline{\mathbf{H}_C}\cdot\mathbf{n}_* \\
&= \textstyle\int_{\Xi_C^*} \operatorname{curl}\overline{\mathbf{H}_C}\cdot\mathbf{n}_* \ ,
\end{aligned} \tag{8.50}$$

as $\operatorname{curl}\mathbf{H}_C\cdot\mathbf{n}_C = \operatorname{curl}\mathbf{H}_I\cdot\mathbf{n}_C = 0$ on Γ and the jump of p_C^* on Ξ_C^* is equal to 1. Hence we end up with

$$\begin{aligned}
P &= \textstyle\int_{\Omega_C} \boldsymbol{\sigma}^{-1} \operatorname{curl}\mathbf{H}_C \cdot \operatorname{curl}\overline{\mathbf{H}_C} + i\omega \int_\Omega \mu\mathbf{H}\cdot\overline{\mathbf{H}} \\
&= V \textstyle\int_{\Omega_C} \boldsymbol{\rho}_C^* \cdot \operatorname{curl}\overline{\mathbf{H}_C} = V\,\overline{I^0} \ .
\end{aligned}$$

8.2.1 Variational formulations

We can consider $\mathbf{H}$-based formulations, $\mathbf{E}$-based formulations or hybrid formulations. In our opinion, the simplest approach is in terms of $\mathbf{H}$, thus we start using the latter.

Concerning the voltage excitation problem we have already made explicit the following rule: applying a voltage is equivalent to consider a current density $\mathbf{J}_{e,C} = V\boldsymbol{\sigma}\boldsymbol{\rho}_C^*$. Hence the weak $\mathbf{H}$-based formulation of the voltage excitation problem is similar to the formulation considered in Chapter 3: given $V \in \mathbb{C}$,

Find $\mathbf{H} \in X$ such that:

$$\int_{\Omega_C} \boldsymbol{\sigma}^{-1} \operatorname{curl} \mathbf{H}_C \cdot \operatorname{curl} \overline{\mathbf{v}_C} + \int_\Omega i\omega\boldsymbol{\mu}\mathbf{H} \cdot \overline{\mathbf{v}} = V \int_{\Omega_C} \boldsymbol{\rho}_C^* \cdot \operatorname{curl} \overline{\mathbf{v}_C} \tag{8.51}$$

for each $\mathbf{v} \in X$,

where

$$X := \{\mathbf{v} \in H(\operatorname{curl}; \Omega) \mid \operatorname{curl} \mathbf{v}_I = \mathbf{0} \text{ in } \Omega_I\} \tag{8.52}$$

in the case of the electric boundary condition or the no-flux boundary conditions, while

$$X := \{\mathbf{v} \in H_0(\operatorname{curl}; \Omega) \mid \operatorname{curl} \mathbf{v}_I = \mathbf{0} \text{ in } \Omega_I\} \tag{8.53}$$

in the case of the magnetic boundary conditions.

Then set $\mathbf{E}_C := \boldsymbol{\sigma}^{-1} \operatorname{curl} \mathbf{H}_C - V\boldsymbol{\rho}_C^*$ in Ω_C, and in Ω_I define $\mathbf{E}_I$ to be the solution to

$$\begin{cases} \operatorname{curl} \mathbf{E}_I = -i\omega\boldsymbol{\mu}_I\mathbf{H}_I & \text{in } \Omega_I \\ \operatorname{div}(\boldsymbol{\varepsilon}_I\mathbf{E}_I) = 0 & \text{in } \Omega_I \\ \mathbf{E}_I \times \mathbf{n}_I = -\mathbf{E}_C \times \mathbf{n}_C & \text{on } \Gamma \\ \int_{\partial\Omega} \boldsymbol{\varepsilon}_I\mathbf{E}_I \cdot \mathbf{n} = 0 & \\ \mathbf{E}_I \times \mathbf{n} = \mathbf{0} & \text{on } \partial\Omega \end{cases} \tag{8.54}$$

when considering the electric boundary condition, or to

$$\begin{cases} \operatorname{curl} \mathbf{E}_I = -i\omega\boldsymbol{\mu}_I\mathbf{H}_I & \text{in } \Omega_I \\ \operatorname{div}(\boldsymbol{\varepsilon}_I\mathbf{E}_I) = 0 & \text{in } \Omega_I \\ \mathbf{E}_I \times \mathbf{n}_I = -\mathbf{E}_C \times \mathbf{n}_C & \text{on } \Gamma \\ \boldsymbol{\varepsilon}_I\mathbf{E}_I \cdot \mathbf{n} = 0 & \text{on } \partial\Omega \end{cases} \tag{8.55}$$

when considering the magnetic or the no-flux boundary conditions.

The well-posedness of problem (8.51) comes from the coerciveness in X of the sesquilinear form $a(\mathbf{u}, \mathbf{v}) := \int_{\Omega_C} \boldsymbol{\sigma}^{-1} \operatorname{curl} \mathbf{u}_C \cdot \operatorname{curl} \overline{\mathbf{v}_C} + \int_\Omega i\omega\boldsymbol{\mu}\mathbf{u} \cdot \overline{\mathbf{v}}$. Instead, a delicate point here is the unique solvability of problem (8.54) and problem (8.55). In fact, as it is well-known, boundary-value problems for the curl-div system in general require that some compatibility conditions be satisfied in order to ensure the existence of a solution, and that suitable additional conditions are imposed to guarantee its uniqueness; some of these conditions are related to the non-trivial topology of Ω_I. We refer the reader to Chapter 3, where this analysis has been performed for both the electric and the magnetic boundary conditions. In the case of no-flux boundary conditions the verification of the compatibility conditions for the well-posedness of problem (8.55) is similar to the magnetic boundary condition case.

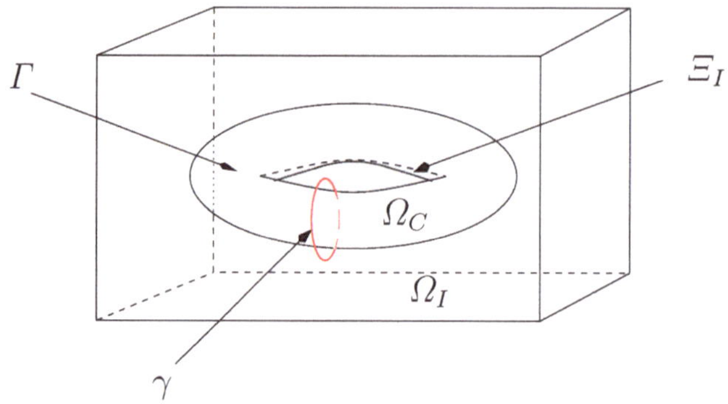

Fig. 8.9. The "cutting" surface Ξ_I for the internal conductor case

Remark 8.13. Let us also recall that the same variational formulation (8.51) has been proposed in Dular et al. [99], Rappaz et al. [202] and Hiptmair and Sterz [130]. However, there it has been set $\mathbf{E}_C := \boldsymbol{\sigma}^{-1} \operatorname{curl} \mathbf{H}_C$, leading to the violation of the Faraday equation on the surface Ξ_I "cutting" the basic non-bounding cycle γ of Ω_I (see Figure 8.9).

Let us show that this is in fact the case, and that all the other equations are verified. Since $V \int_{\Omega_C} \boldsymbol{\rho}_C^* \cdot \operatorname{curl} \overline{\mathbf{v}_C} = V \int_\Gamma \boldsymbol{\rho}_C^* \times \mathbf{n}_C \cdot \overline{\mathbf{v}_C}$ and this term is vanishing for a test function $\mathbf{v}_C$ with compact support in Ω_C, from (8.51) one verifies that the Faraday equation in Ω_C is satisfied. Defining $\mathbf{E}_C = \boldsymbol{\sigma}^{-1} \operatorname{curl} \mathbf{H}_C$, the Ampère equation without sources is clearly satisfied in the whole Ω. However, in Section 3.3.1 we have shown that the Faraday law on Ξ_I can be written as

$$\int_{\Omega_I} i\omega \boldsymbol{\mu}_I \mathbf{H}_I \cdot \boldsymbol{\rho}_I = - \int_\Gamma (\mathbf{E}_C \times \mathbf{n}_C) \cdot \boldsymbol{\rho}_I \,, \tag{8.56}$$

where $\boldsymbol{\rho}_I$ is the $(L^2(\Omega_I))^3$-extension of the gradient of a function $p_I \in H^1(\Omega_I \setminus \Xi_I)$ such that

$$\begin{cases} \operatorname{div}(\boldsymbol{\mu}_I \operatorname{grad} p_I) = 0 & \text{in } \Omega_I \setminus \Xi_I \\ \boldsymbol{\mu}_I \operatorname{grad} p_I \cdot \mathbf{n}_I = 0 & \text{on } \Gamma \setminus \partial \Xi_I \\ [\boldsymbol{\mu}_I \operatorname{grad} p_I \cdot \mathbf{n}_I]_{\Xi_I} = 0 & \\ [p_I]_{\Xi_I} = 1 \,, & \end{cases}$$

plus a boundary condition on $\partial\Omega$ that depends on the boundary conditions for $\mathbf{H}$ and $\mathbf{E}$ under consideration. Precisely, for the magnetic boundary conditions one has $p_I = 0$ on $\partial\Omega$, while for the electric and the no-flux boundary conditions $\boldsymbol{\mu}_I \operatorname{grad} p_I \cdot \mathbf{n}_I = 0$ on $\partial\Omega$ (and in this case p_I is defined up to an additive constant).

From (8.51) we find at once

$$\begin{aligned} \textstyle\int_{\Omega_I} i\omega \boldsymbol{\mu} \mathbf{H}_I \cdot \boldsymbol{\rho}_I = - \textstyle\int_{\Omega_C} i\omega \boldsymbol{\mu} \mathbf{H}_C \cdot \mathbf{R}_C - \textstyle\int_{\Omega_C} \boldsymbol{\sigma}^{-1} \operatorname{curl} \mathbf{H}_C \cdot \operatorname{curl} \mathbf{R}_C \\ + V \textstyle\int_{\Omega_C} \boldsymbol{\rho}_C^* \cdot \operatorname{curl} \mathbf{R}_C \,, \end{aligned}$$

where $\mathbf{R}_C$ is any extension of $\boldsymbol{\rho}_I$ in Ω_C giving a global function belonging to X.

Setting $\mathbf{E}_C = \boldsymbol{\sigma}^{-1} \operatorname{curl} \mathbf{H}_C$ and integrating by parts one has

$$\begin{aligned}\int_{\Omega_I} i\omega\boldsymbol{\mu}\mathbf{H}_I \cdot \boldsymbol{\rho}_I &= -\int_{\Omega_C} i\omega\boldsymbol{\mu}\mathbf{H}_C \cdot \mathbf{R}_C - \int_{\Omega_C} \operatorname{curl} \mathbf{E}_C \cdot \mathbf{R}_C \\ &\quad - \int_\Gamma (\mathbf{E}_C \times \mathbf{n}_C) \cdot \boldsymbol{\rho}_I + V \int_\Gamma (\boldsymbol{\rho}_C^* \times \mathbf{n}_C) \cdot \boldsymbol{\rho}_I \\ &= -\int_\Gamma (\mathbf{E}_C \times \mathbf{n}_C) \cdot \boldsymbol{\rho}_I + V \,,\end{aligned}$$

having used the Faraday equation in Ω_C and the result $\int_\Gamma (\boldsymbol{\rho}_C^* \times \mathbf{n}_C) \cdot \boldsymbol{\rho}_I = 1$. Since $V \neq 0$, the Faraday equation on the surface Ξ_I does not hold.

In other words, taking $\mathbf{E}_C = \boldsymbol{\sigma}^{-1} \operatorname{curl} \mathbf{H}_C$ has as a consequence that there is no electric field in Ω_I solving the Faraday equation $\operatorname{curl} \mathbf{E}_I = -i\omega\boldsymbol{\mu}_I \mathbf{H}_I$ with $\mathbf{E}_I \times \mathbf{n}_I = -\mathbf{E}_C \times \mathbf{n}_C$ on Γ: the necessary compatibility condition on the data

$$\int_{\Omega_I} i\omega\boldsymbol{\mu}_I \mathbf{H}_I \cdot \boldsymbol{\rho}_I = -\int_\Gamma (\mathbf{E}_C \times \mathbf{n}_C) \cdot \boldsymbol{\rho}_I$$

is not satisfied. □

Now we consider the current intensity excitation problem. It can be expressed in this way: given $I^0 \in \mathbb{C}$,

Find $(\mathbf{H}, V) \in X \times \mathbb{C}$ such that

$$\begin{cases} \int_{\Omega_C} \boldsymbol{\sigma}^{-1} \operatorname{curl} \mathbf{H}_C \cdot \operatorname{curl} \overline{\mathbf{v}_C} + \int_\Omega i\omega\boldsymbol{\mu}\mathbf{H} \cdot \overline{\mathbf{v}} - V \int_{\Omega_C} \boldsymbol{\rho}_C^* \cdot \operatorname{curl} \overline{\mathbf{v}_C} = 0 \\ \int_{\Omega_C} \boldsymbol{\rho}_C^* \cdot \operatorname{curl} \mathbf{H}_C = I^0 \end{cases} \tag{8.57}$$

for each $\mathbf{v} \in X$,

where X is as in (8.52) or (8.53). Then one sets $\mathbf{E}_C := \boldsymbol{\sigma}^{-1} \operatorname{curl} \mathbf{H}_C - V\boldsymbol{\rho}_C^*$ in Ω_C and determines $\mathbf{E}_I$ as in (8.54) or (8.55).

The well-posedness of problem (8.57) comes from the theory of saddle-point problems. In fact, we have already noted that the sesquilinear form $a(\mathbf{u}, \mathbf{v}) := \int_{\Omega_C} \boldsymbol{\sigma}^{-1} \operatorname{curl} \mathbf{u}_C \cdot \operatorname{curl} \overline{\mathbf{v}_C} + \int_\Omega i\omega\boldsymbol{\mu}\mathbf{u} \cdot \overline{\mathbf{v}}$ is coercive in X; moreover, since the unknown $V \in \mathbb{C}$ is a number, for showing that the inf–sup condition is satisfied it is enough to find $\mathbf{w}^* \in X$ such that

$$\left| \int_{\Omega_C} \boldsymbol{\rho}_C^* \cdot \operatorname{curl} \mathbf{w}_C^* \right| > 0 \,.$$

This can be achieved as follows: $\mathbf{w}^* \in X$ is any extension of $\mathbf{w}_C^*$, the solution to

$$\begin{cases} \operatorname{curl} \mathbf{w}_C^* = \boldsymbol{\sigma}\boldsymbol{\rho}_C^* & \text{in } \Omega_C \\ \operatorname{div} \mathbf{w}_C^* = 0 & \text{in } \Omega_C \\ \mathbf{w}_C^* \times \mathbf{n}_C = c_0^* \boldsymbol{\rho}_I \times \mathbf{n}_C & \text{on } \Gamma \,, \end{cases}$$

where $c_0^* := \int_{\Omega_C} \boldsymbol{\sigma}\boldsymbol{\rho}_C^* \cdot \boldsymbol{\rho}_C^*$. Note that the existence of the solution $\mathbf{w}_C^*$ is a consequence of the relation $\int_\Gamma (\boldsymbol{\rho}_C^* \times \mathbf{n}_C) \cdot \boldsymbol{\rho}_I = 1$.

The variational formulations (8.51) and (8.57) can be used as starting points for devising finite element methods for approximating the solution.

In fact, the voltage excitation reduces to a standard problem with a given current density $V\boldsymbol{\sigma}\boldsymbol{\rho}^*_C$, therefore any method used for eddy current problems can be applied.

It is worth noting that the construction of the function $\boldsymbol{\rho}^*_C$ is not required. Indeed, one can proceed as in Section 5.1: consider a fixed (and coarse) mesh in Ω_C that induces a triangulation of the "cutting" surface Ξ^*_C. The first step is to denote by Π_C the piecewise-linear function taking value 1 at the nodes on one side of Ξ^*_C, say Ξ^{*+}_C, and 0 at all the other nodes (including those on Ξ^{*-}_C). Then define $g_C \in H^1(\Omega_C)$ to be the solution (determined up to an additive constant) to

$$\begin{cases} \operatorname{div}(\boldsymbol{\sigma} \operatorname{grad} g_C) = -\operatorname{div}(\boldsymbol{\sigma} \operatorname{grad} \Pi_C) & \text{in } \Omega_C \\ \boldsymbol{\sigma} \operatorname{grad} g_C \cdot \mathbf{n} = -\boldsymbol{\sigma} \operatorname{grad} \Pi_C \cdot \mathbf{n} & \text{on } \Gamma \, . \end{cases}$$

Thus $\boldsymbol{\rho}^*_C = \widetilde{\operatorname{grad}} \, \Pi_C + \operatorname{grad} g_C$, where $\widetilde{\operatorname{grad}} \, \Pi_C$ denotes the $(L^2(\Omega_C))^3$-extension of $\operatorname{grad} \Pi_C$ computed in $\Omega_C \setminus \Xi^*_C$. If $\mathbf{v} \in X$ one has

$$\begin{aligned} \int_{\Omega_C} \boldsymbol{\rho}^*_C \cdot \operatorname{curl} \overline{\mathbf{v}_C} &= \int_{\Omega_C \setminus \Xi^*_C} \operatorname{grad} \Pi_C \cdot \operatorname{curl} \overline{\mathbf{v}_C} + \int_{\Omega_C} \operatorname{grad} g_C \cdot \operatorname{curl} \overline{\mathbf{v}_C} \\ &= \int_{\Omega_C \setminus \Xi^*_C} \operatorname{grad} \Pi_C \cdot \operatorname{curl} \overline{\mathbf{v}}_C \, , \end{aligned}$$

as $\operatorname{div} \operatorname{curl} \overline{\mathbf{v}_C} = 0$ and $\operatorname{curl} \overline{\mathbf{v}_C} \cdot \mathbf{n} = 0$ on Γ.

Therefore, we have verified that in (8.51) one can substitute $\boldsymbol{\rho}^*_C$ by the easily computable $\operatorname{grad} \Pi_C$, and the solution $\mathbf{H}$ remains the same. Clearly, the need to compute $\boldsymbol{\rho}^*_C$ comes again into play if one wants to recover $\mathbf{E}_C$, which is given by

$$\mathbf{E}_C = \boldsymbol{\sigma}^{-1} \operatorname{curl} \mathbf{H}_C - V \boldsymbol{\rho}^*_C = \boldsymbol{\sigma}^{-1} \operatorname{curl} \mathbf{H}_C - V \widetilde{\operatorname{grad}} \, \Pi_C - V \operatorname{grad} g_C \, .$$

If the current intensity is given, the constraint $\int_{\Omega_C} \boldsymbol{\rho}^*_C \cdot \operatorname{curl} \mathbf{H}_C = I^0$ has to be added. In (8.57), the voltage V plays the role of a Lagrange multiplier associated to this constraint, and the global problem is a saddle-point problem. For any type of conforming finite element discretization using edge elements in Ω_C, the presence of this Lagrange multiplier requires that an inf–sup condition like

$$\left| \int_{\Omega_C} \boldsymbol{\rho}^*_C \cdot \operatorname{curl} \mathbf{w}^*_{C,h} \right| \geq \beta ||\mathbf{w}^*_h||_X$$

be satisfied for a constant $\beta > 0$, independent of h, and a suitable discrete vector function $\mathbf{w}^*_h$.

This can be achieved as follows: expressing $\boldsymbol{\rho}^*_C$ in terms of $\widetilde{\operatorname{grad}} \, \Pi_C$, as done before, we have by integration by parts and the Stokes theorem

$$\begin{aligned} \int_{\Omega_C} \boldsymbol{\rho}^*_C \cdot \operatorname{curl} \mathbf{w}^*_{C,h} &= \int_{\Omega_C \setminus \Xi^*_C} \operatorname{grad} \Pi_C \cdot \operatorname{curl} \mathbf{w}^*_{C,h} \\ = \int_{\partial(\Omega_C \setminus \Xi^*_C)} \Pi_C \operatorname{curl} \mathbf{w}^*_{C,h} \cdot \mathbf{n} &= \int_{\Xi^*_C} \operatorname{curl} \mathbf{w}^*_{C,h} \cdot \mathbf{n} \\ &= \int_{\partial \Xi^*_C} \mathbf{w}^*_{C,h} \cdot d\boldsymbol{\tau} = \int_{\partial \Xi^*_C} \mathbf{w}^*_{I,h} \cdot d\boldsymbol{\tau} \, . \end{aligned}$$

Now let us consider a fixed (and coarse) mesh in Ω_I that induces a triangulation of the "cutting" surface Ξ_I (see Figure 8.10). Proceeding as in the conductor region we denote by Π_I the piecewise-linear function taking value 1 at the nodes on one side of

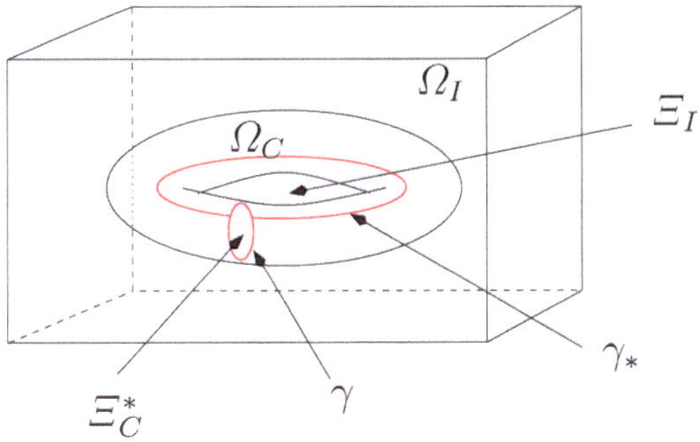

Fig. 8.10. The "cutting" surfaces Ξ_C^* and Ξ_I for the internal conductor case

Ξ_I and 0 at all the other nodes. From now on we consider triangulations that are all obtained as a refinement of the basic coarse mesh, in such a way that a discrete function on the coarse mesh is also a discrete function on all the other meshes. Then choose as $\mathbf{w}_{I,h}^*$ the $(L^2(\Omega_I))^3$-extension of grad Π_I, computed in $\Omega_I \setminus \Xi_I$. Finally, take as $\mathbf{w}_{C,h}^*$ the edge element interpolant, on the coarse mesh in Ω_C, of the value $\mathbf{w}_{I,h}^* \times \mathbf{n}_I$ on Γ. Clearly $\int_{\partial \Xi_C^*} \mathbf{w}_{I,h}^* \cdot d\boldsymbol{\tau} = 1$ and the norm $||\mathbf{w}_h^*||_X$ does not depend on h, therefore the uniform inf–sup condition is satisfied.

Up to now, in this section we have focused on the $\mathbf{H}$-based formulation; however, the $\mathbf{E}$-based formulation of voltage and current intensity excitation problems can also be considered.

The "voltage rule" is telling us that we have just to consider a current density $\mathbf{J}_{e,C} = V\boldsymbol{\sigma}\boldsymbol{\rho}_C^*$, hence the $\mathbf{E}$-based formulation is devised proceeding as in Section 4.6, where the magnetic boundary conditions have been considered.

The variational formulation is that described in (4.75), and the only point that needs to be clarified is the choice of the variational space Z in which the problem is formulated.

In the case of the magnetic boundary conditions one takes Z as in (4.74), which in the present geometrical situation reads

$$\begin{aligned} Z := \{\mathbf{z} \in H(\mathrm{curl}; \Omega) \mid \mathrm{div}(\varepsilon_I \mathbf{z}_I) = 0 \text{ in } \Omega_I, \\ \varepsilon_I \mathbf{z}_I \cdot \mathbf{n} = 0 \text{ on } \partial\Omega\}\,; \end{aligned} \tag{8.58}$$

for the electric boundary condition one has

$$\begin{aligned} Z := \{\mathbf{z} \in H(\mathrm{curl}; \Omega) \mid \mathrm{div}(\varepsilon_I \mathbf{z}_I) = 0 \text{ in } \Omega_I, \\ \mathbf{z} \times \mathbf{n} = \mathbf{0} \text{ on } \partial\Omega, \textstyle\int_{\partial\Omega} \varepsilon_I \mathbf{z}_I \cdot \mathbf{n} = 0\}\,; \end{aligned} \tag{8.59}$$

for the no-flux boundary conditions one chooses

$$\begin{aligned} Z := \{\mathbf{z} \in H(\mathrm{curl}; \Omega) \mid \mathrm{div}(\varepsilon_I \mathbf{z}_I) = 0 \text{ in } \Omega_I, \\ \varepsilon_I \mathbf{z}_I \cdot \mathbf{n} = 0 \text{ on } \partial\Omega, \mathrm{div}_\tau(\mathbf{z} \times \mathbf{n}) = \mathbf{0} \text{ on } \partial\Omega\}\,. \end{aligned} \tag{8.60}$$

Passing to the "current intensity rule", it says that the given current intensity I^0 is generating not only the electric field but also a current density $V\boldsymbol{\sigma}\boldsymbol{\rho}_C^*$. Moreover, we have $\operatorname{curl}\mathbf{H}_C = \boldsymbol{\sigma}\mathbf{E}_C + V\boldsymbol{\sigma}\boldsymbol{\rho}_C^*$. Then, the problem is: for each given $I^0 \in \mathbb{C}$,

$$\begin{array}{l}\text{Find } (\mathbf{E}, V) \in Z \times \mathbb{C} \text{ such that} \\ \quad \int_\Omega \boldsymbol{\mu}^{-1} \operatorname{curl}\mathbf{E} \cdot \operatorname{curl}\overline{\mathbf{z}} + \int_{\Omega_C} i\omega\boldsymbol{\sigma}\mathbf{E}_C \cdot \overline{\mathbf{z}_C} + i\omega V \int_{\Omega_C} \boldsymbol{\sigma}\boldsymbol{\rho}_C^* \cdot \overline{\mathbf{z}_C} = 0 \\ \quad \int_{\Omega_C} \boldsymbol{\rho}_C^* \cdot \boldsymbol{\sigma}\mathbf{E}_C + V \int_{\Omega_C} \boldsymbol{\sigma}\boldsymbol{\rho}_C^* \cdot \boldsymbol{\rho}_C^* = I^0 \\ \text{for each } \mathbf{z} \in Z\,,\end{array} \qquad (8.61)$$

where Z is as in (8.58) for the magnetic boundary conditions, as in (8.59) for the electric boundary condition, and as in (8.60) for the no-flux boundary conditions.

The existence of solution is a consequence of what was already proved for the **H**-based formulation. On the other hand, uniqueness needs some work. First of all, multiplying $(8.61)_2$ by $i\omega\overline{Q}$, where $Q \in \mathbb{C}$, we find

$$\begin{array}{l}\int_\Omega \boldsymbol{\mu}^{-1} \operatorname{curl}\mathbf{E} \cdot \operatorname{curl}\overline{\mathbf{z}} \\ \qquad + i\omega \int_{\Omega_C} \boldsymbol{\sigma}(\mathbf{E}_C + V\boldsymbol{\rho}_C^*) \cdot (\overline{\mathbf{z}_C} + \overline{Q}\boldsymbol{\rho}_C^*) = i\omega I^0\, \overline{Q}\,.\end{array}$$

Thus, putting $I^0 = 0$ and choosing $\mathbf{z} = \mathbf{E}$ and $Q = V$, we obtain $\operatorname{curl}\mathbf{E} = \mathbf{0}$ in Ω and $\mathbf{E}_C + V\boldsymbol{\rho}_C^* = \mathbf{0}$ in Ω_C. Since Ω is simply-connected, we also have $\mathbf{E} = \operatorname{grad} U$ in Ω. Therefore, integrating $\mathbf{E}_C$ on the cycle γ_* we find

$$0 = \int_{\gamma_*} \operatorname{grad} U_C \cdot d\boldsymbol{\tau} = \int_{\gamma_*} \mathbf{E}_C \cdot d\boldsymbol{\tau} = -\int_{\gamma_*} V\boldsymbol{\rho}_C^* \cdot d\boldsymbol{\tau} = -V\,,$$

thus $V = 0$, and consequently $\mathbf{E}_C = \mathbf{0}$ in Ω_C. Finally, the interface condition $\mathbf{E}_I \times \mathbf{n}_I = -\mathbf{E}_C \times \mathbf{n}_C$ on Γ is sufficient to conclude that $\mathbf{E}_I = \mathbf{0}$ in Ω_I.

Solved (8.61), the magnetic field in Ω is as usual defined as $\mathbf{H} = -\frac{1}{i\omega}\boldsymbol{\mu}^{-1} \operatorname{curl}\mathbf{E}$.

Remark 8.14. A formulation similar to (8.61) has been proposed in Hiptmair and Sterz [130], Bermúdez et al. [40] (in the former paper, replacing the source $V\boldsymbol{\rho}_C^*$ by $V\widetilde{\operatorname{grad}\widetilde{\Phi}_C}$, where $\widetilde{\Phi}_C$ is a function jumping by 1 on the "cutting" surface Ξ_C^* and $\widetilde{\operatorname{grad}\widetilde{\Phi}_C}$ denotes the $(L^2(\Omega_C))^3$-extension of $\operatorname{grad}\widetilde{\Phi}_C$ computed in $\Omega_C \setminus \Xi_C^*$). However, in these papers the electric field is not the solution $\mathbf{E}_C$ to (8.61), but it is corrected in Ω_C by adding the source term. In this way the Faraday law is no longer verified across the interface Γ. □

Remark 8.15. We note that the **E**-based formulation (8.61) takes a non-standard form: in fact, it is questionable if the sesquilinear forms at the left hand sides of (8.61) are coercive, and, on the other hand, the current intensity condition is not a pure constraint, so that a formulation using Lagrange multipliers is not suitable. Therefore, a complete analysis of a finite element approximation method could be a delicate task. However, this approach has been used in Bermúdez et al. [40] for an axisymmetric problem, with good numerical performances. □

Remark 8.16. The two rules presented at the beginning of this section can also be used to derive a model for the voltage and current intensity excitation problems in the presence of electric ports, in the case in which the electric boundary condition $\mathbf{E} \times \mathbf{n} = \mathbf{0}$ on $\partial\Omega$ is imposed.

The procedure is quite simple: in fact, to adapt the voltage and the current intensity rules one has only to replace $\boldsymbol{\rho}_C^*$ with $\operatorname{grad}\phi_C$, with ϕ_C defined in (8.46).

Thus, the voltage excitation problem reads: given $V \in \mathbb{C}$,

$$
\begin{aligned}
&\text{Find } \mathbf{H} \in X \text{ such that} \\
&\int_{\Omega_C} \boldsymbol{\sigma}^{-1} \operatorname{curl} \mathbf{H}_C \cdot \operatorname{curl} \overline{\mathbf{v}_C} + \int_\Omega i\omega\boldsymbol{\mu}\mathbf{H} \cdot \overline{\mathbf{v}} \\
&\qquad = V \int_{\Omega_C} \operatorname{grad}\phi_C \cdot \operatorname{curl} \overline{\mathbf{v}_C} \\
&\text{for each } \mathbf{v} \in X ,
\end{aligned}
\tag{8.62}
$$

where $X := \{\mathbf{v} \in H(\operatorname{curl};\Omega) \mid \operatorname{curl} \mathbf{v}_I = \mathbf{0} \text{ in } \Omega_I\}$. Then set

$$
\mathbf{E}_C := \boldsymbol{\sigma}^{-1} \operatorname{curl} \mathbf{H}_C - V \operatorname{grad}\phi_C \quad \text{in } \Omega_C ,
$$

and in Ω_I define $\mathbf{E}_I$ to be the solution to

$$
\begin{cases}
\operatorname{curl} \mathbf{E}_I = -i\omega\boldsymbol{\mu}_I \mathbf{H}_I & \text{in } \Omega_I \\
\operatorname{div}(\boldsymbol{\varepsilon}_I \mathbf{E}_I) = 0 & \text{in } \Omega_I \\
\mathbf{E}_I \times \mathbf{n}_I = -\mathbf{E}_C \times \mathbf{n}_C & \text{on } \Gamma \\
\mathbf{E}_I \times \mathbf{n} = \mathbf{0} & \text{on } \Gamma_I .
\end{cases}
\tag{8.63}
$$

The current intensity excitation problem reads: given $I^0 \in \mathbb{C}$,

$$
\begin{aligned}
&\text{Find } (\mathbf{H}, V) \in X \times \mathbb{C} \text{ such that} \\
&\int_{\Omega_C} \boldsymbol{\sigma}^{-1} \operatorname{curl} \mathbf{H}_C \cdot \operatorname{curl} \overline{\mathbf{v}_C} + \int_\Omega i\omega\boldsymbol{\mu}\mathbf{H} \cdot \overline{\mathbf{v}} \\
&\qquad - V \int_{\Omega_C} \operatorname{grad}\phi_C \cdot \operatorname{curl} \overline{\mathbf{v}_C} = 0 \\
&\int_{\Omega_C} \operatorname{grad}\phi_C \cdot \operatorname{curl} \mathbf{H}_C = I^0 \\
&\text{for each } \mathbf{v} \in X .
\end{aligned}
\tag{8.64}
$$

Then $\mathbf{E}_C$ and $\mathbf{E}_I$ are determined in the same way as before.

The corresponding $\mathbf{E}$-based formulations can be found in Alonso Rodríguez and Valli [18]. □

Remark 8.17. Let us also discuss the case of electric ports with the magnetic boundary conditions $\mathbf{E}_C \times \mathbf{n}_C = \mathbf{0}$ on $\Gamma_E \cup \Gamma_J$, $\boldsymbol{\varepsilon}_I \mathbf{E}_I \cdot \mathbf{n}_I = 0$ and $\mathbf{H}_I \times \mathbf{n}_I = \mathbf{0}$ on Γ_I. The crucial remark is that in this case the current intensity cannot be assigned freely, because

$$
I^0 = \int_{\Gamma_J} \operatorname{curl} \mathbf{H}_C \cdot \mathbf{n} = \int_{\partial\Gamma_J} \mathbf{H}_C \cdot d\boldsymbol{\tau} = \int_{\partial\Gamma_J} \mathbf{H}_I \cdot d\boldsymbol{\tau} = 0 ,
$$

as $\mathbf{H}_I \times \mathbf{n}_I = \mathbf{0}$ on Γ_I and thus on $\partial\Gamma_J$.

Imposing $I^0 = 0$ is thus the only possible case. Though it does not seem very interesting, let us have a deeper look at the problem.

First of all, for each $\mathbf{v} \in H(\operatorname{curl};\Omega)$ satisfying $\operatorname{curl} \mathbf{v}_I = \mathbf{0}$ in Ω_I and $\mathbf{v}_I \times \mathbf{n}_I = \mathbf{0}$ on Γ_I it holds

$$
\begin{aligned}
&\int_{\Omega_C} \operatorname{grad}\phi_C \cdot \operatorname{curl} \overline{\mathbf{v}_C} \\
&\quad = \int_{\Gamma \cup \Gamma_E \cup \Gamma_J} \phi_C \operatorname{curl} \overline{\mathbf{v}_C} \cdot \mathbf{n}_C = \int_{\Gamma_J} \operatorname{curl} \overline{\mathbf{v}_C} \cdot \mathbf{n}_C \\
&\quad = \int_{\partial\Gamma_J} \overline{\mathbf{v}_C} \cdot d\boldsymbol{\tau} = \int_{\partial\Gamma_J} \overline{\mathbf{v}_I} \cdot d\boldsymbol{\tau} = 0 .
\end{aligned}
\tag{8.65}
$$

Considering now the Faraday equation, integration by parts gives

$$-i\omega \int_{\Omega} \boldsymbol{\mu}\mathbf{H} \cdot \overline{\mathbf{H}} = \int_{\Omega} \operatorname{curl} \mathbf{E} \cdot \overline{\mathbf{H}} = \int_{\Omega} \mathbf{E} \cdot \operatorname{curl} \overline{\mathbf{H}} - \int_{\partial\Omega} \mathbf{E} \times \mathbf{n} \cdot \overline{\mathbf{H}} .$$

Since $\mathbf{E} \times \mathbf{n} = \mathbf{0}$ on $\Gamma_E \cup \Gamma_J$ and $\mathbf{H} \times \mathbf{n} = \mathbf{0}$ on Γ_I, using the Ampère equation we obtain

$$\begin{aligned} &\textstyle\int_{\Omega} \mathbf{E} \cdot \operatorname{curl} \overline{\mathbf{H}} - \int_{\partial\Omega} \mathbf{E} \times \mathbf{n} \cdot \overline{\mathbf{H}} \\ &\quad = \textstyle\int_{\Omega_C} \mathbf{E}_C \cdot \operatorname{curl} \overline{\mathbf{H}_C} = \int_{\Omega_C} (\boldsymbol{\sigma}^{-1} \operatorname{curl} \mathbf{H}_C - V \operatorname{grad} \phi_C) \cdot \operatorname{curl} \overline{\mathbf{H}_C} . \end{aligned}$$

From (8.65) we have $\int_{\Omega_C} \operatorname{grad} \phi_C \cdot \operatorname{curl} \overline{\mathbf{H}_C} = 0$, thus we conclude that

$$\int_{\Omega_C} \boldsymbol{\sigma}^{-1} \operatorname{curl} \mathbf{H}_C \cdot \operatorname{curl} \overline{\mathbf{H}_C} + i\omega \int_{\Omega} \boldsymbol{\mu}\mathbf{H} \cdot \overline{\mathbf{H}} = 0 , \tag{8.66}$$

hence $\mathbf{H} = \mathbf{0}$ in Ω. Moreover, setting $\mathbf{E}_C = -V \operatorname{grad} \phi_C$ in Ω_C and $\mathbf{E}_I = -V \operatorname{grad} v_I$ in Ω_I, where v_I is the solution to (8.49), we find infinitely many electric fields solution of the eddy current problem with vanishing current intensity: one for each choice of the voltage $V \in \mathbb{C}$.

On the other hand, following the voltage rule for a given $V \in \mathbb{C}$ we have to consider the problem

$$\begin{cases} \operatorname{curl} \mathbf{H} - \boldsymbol{\sigma}\mathbf{E} = V \boldsymbol{\sigma} \operatorname{grad} \phi_C & \text{in } \Omega \\ \operatorname{curl} \mathbf{E} + i\omega\boldsymbol{\mu}\mathbf{H} = \mathbf{0} & \text{in } \Omega . \end{cases}$$

Repeating the same arguments described above, one arrives to (8.66). This means that, for any assigned voltage $V \in \mathbb{C}$, the voltage excitation problem with the magnetic boundary conditions leads to $\mathbf{H} = \mathbf{0}$ in Ω, and, moreover, one also finds that the electric field in Ω is given by $\mathbf{E} = -V \operatorname{grad} v$, where $v = \phi_C$ in Ω_C and $v = v_I$ in Ω_I. □

9

Selected applications

In this chapter we present some real-life problems that can be modeled by the eddy current equations. In some of these examples the time-harmonic eddy current system is used for numerical simulations, and a rich bibliography on the subject is available. However, we also include some applications where, to our knowledge, the eddy current model has not yet been used. We believe that it could be a more accurate description than the ones actually employed, and, using the method proposed in this book, it should be suitable for numerical simulations.

In the following we focus on the illustration of the physical phenomena; the descriptions do not pretend to be complete and fully detailed, but just to give a flavour of different technological problems that are related to low-frequency electromagnetism.

9.1 Metallurgical thermoelectrical problems

We consider in this section two kind of electromagnetic furnaces used in the metallurgical industry: induction heating systems and electric reduction furnaces. There is an increasing interest in numerical simulations as means to optimize the design and to improve the performances of these kind of electromagnetic devices. In an induction furnace the eddy currents generated within conductors and resistances lead to Joule heating; in an electric reduction furnace the charged material is directly exposed to an electric arc. In both cases the mathematical model for the behaviour of the furnace involves thermal and electromagnetic phenomena, that can be described through the coupling of the Maxwell equations and the heat transfer equation.

Normally the electromagnetic submodel is solved in the frequency domain and the effect of displacement currents can be neglected, thus leading to the time-harmonic eddy current problem analyzed in this book. The electromagnetic and the thermal problem are coupled for two reasons: the electromagnetic properties of the different materials, in particular the electric conductivity, depend on the temperature, and the Joule effect is one of the source terms in the heat transfer equation. Other phenomena can be also taken into account; for instance, hydrodynamic phenomena must be consid-

A. Alonso Rodríguez, A. Valli: Eddy Current Approximation of Maxwell Equations.

ered when melting a metal in an induction furnace, while mechanical effects play an important role on the design of the electrodes of a reduction furnace.

In the following we present two different industrial applications. The first is the modelization of a coreless induction furnace designed for melting and stirring, and the second the modelization of an electric reduction furnace for the production of silicon.

9.1.1 Induction furnaces

In this section we follow the presentation of the problem given by Bermúdez et al. [41] and Vázquez [240]. Induction furnaces are widely used in metallurgical industry for hardening, melting or casting. An induction heating system is basically composed by an inductor, fed by an alternating electrical current, and a conducting object that has to be heated. More precisely, a coreless induction furnace for melting consists of a helical copper coil, connected to a power supply, and a workpiece formed by the crucible and the load within (see Figure 9.1). The alternating current traversing the inductor produces an oscillating magnetic field, which generates eddy currents. These currents, due to the Joule effect, produce heat in the conducting crucible, and the metal inside is also heated until it melts. The crucible is surrounded by refractory and insulating materials, and the inductor coil is water-cooled to avoid overheating due to Ohmic losses. The operating frequencies of the supplied alternating current may vary from utility frequency (50 or 60 Hz) to few kHz.

Numerical simulations are a valuable help in the shape optimization of this kind of system. There are many different aspects that must be taken into account for the design: the frequency and intensity of the applied current affect the temperature profile in the furnace and the stirring action within the molten metal, thus influencing the quality of the final product; ohmic losses could generate very high temperatures in the crucible, damaging it and reducing its lifetime; some physical parameters, such as the thermal and electrical conductivity of the refractory layer, and some geometrical parameters, as the crucible thickness or its distance from the coil, are also important for the performance of the device.

Fig. 9.1. An induction furnace (left, courtesy of V. Valcarcel, Ceramic Institute, Universidade de Santiago de Compostela) and a sketch of the computational geometry (right)

Melting systems were probably the first industrial application of eddy currents. Their modelization involves three main coupled phenomena: the electromagnetic field provides Joule heating and give rise to Lorentz forces that act on the molten metal.

Since the inductor is fed by an alternating current and the effect of the displacement current can be disregarded, the problem is modeled by the time-harmonic eddy current system

$$\begin{aligned} \operatorname{curl} \mathbf{H} &= \mathbf{J} \\ \operatorname{curl} \mathbf{E} + i\omega\boldsymbol{\mu}\mathbf{H} &= \mathbf{0} \,. \end{aligned}$$

The heating of the conductor due to the Joule effect is governed by the transient heat transfer equation with change of phase

$$\rho\Big(\frac{\partial e}{\partial t} + \mathbf{u}\cdot \operatorname{grad} e\Big) - \operatorname{div}(k \operatorname{grad} T) = \mathcal{J}\cdot\mathcal{E} \,,$$

where the heating due to viscous terms has been neglected, and e is the energy per unit mass, T the temperature, ρ the density, and k the thermal conductivity. The energy can be expressed as a multivalued function of the temperature, depending on different physical parameters. The right-hand side $\mathcal{J}(t,\mathbf{x})\cdot\mathcal{E}(t,\mathbf{x})$ is the heat generated by eddy currents ($\mathcal{J}(t,\mathbf{x})$ and $\mathcal{E}(t,\mathbf{x})$ are the time-dependent total current density and electric field, respectively). The term $\mathbf{u}\cdot \operatorname{grad} e$ corresponds to the convective heat transfer; $\mathbf{u}$ is the velocity of the molten metal and it is given by the Navier-Stokes equations

$$\begin{aligned} &\rho\left(\frac{\partial \mathbf{u}}{\partial t} + (\mathbf{u}\cdot \operatorname{grad})\mathbf{u}\right) - \operatorname{div}(2\eta D(\mathbf{u})) + \operatorname{grad} p = \mathbf{f}_g + \mathbf{f}_l \\ &\operatorname{div} \mathbf{u} = 0 \,, \end{aligned}$$

where η is the viscosity, p the pressure of the molten metal, and $D(\mathbf{u})$ the symmetric part of grad $\mathbf{u}$, namely,

$$D(\mathbf{u}) := \frac{\operatorname{grad} \mathbf{u} + (\operatorname{grad} \mathbf{u})^T}{2} \,.$$

The forces at the right-hand side of the Navier-Stokes equations are the buoyancy force $\mathbf{f}_g$, given by

$$\mathbf{f}_g = \rho\mathbf{g} \,,$$

where $\mathbf{g}$ is the acceleration of gravity, and the Lorentz force $\mathbf{f}_l$, given by

$$\mathbf{f}_l = \mathcal{J}\times\mathcal{B} \,,$$

where $\mathcal{B}(t,\mathbf{x})$ is the time dependent magnetic induction.

The heat source and the Lorentz force are determined taking the mean value on a cycle. Taking into account that $\mathcal{J}(t,\mathbf{x}) = \operatorname{Re}[e^{i\omega t}\mathbf{J}(\mathbf{x})]$ and analogously $\mathcal{E}(t,\mathbf{x}) = \operatorname{Re}[e^{i\omega t}\mathbf{E}(\mathbf{x})]$ and $\mathcal{B}(t,\mathbf{x}) = \operatorname{Re}[e^{i\omega t}\mathbf{B}(\mathbf{x})]$, an easy computation gives

$$\begin{aligned} &\tfrac{\omega}{2\pi}\textstyle\int_0^{2\pi} \mathcal{J}(t,\mathbf{x})\cdot\mathcal{E}(t,\mathbf{x})\,dt \\ &\quad = \tfrac{1}{2}\left(\operatorname{Re}\mathbf{J}(\mathbf{x})\cdot\operatorname{Re}\mathbf{E}(\mathbf{x}) + \operatorname{Im}\mathbf{J}(\mathbf{x})\cdot\operatorname{Im}\mathbf{E}(\mathbf{x})\right) , \end{aligned}$$

and

$$\frac{\omega}{2\pi}\int_0^{2\pi} \mathcal{J}(t,\mathbf{x}) \times \mathcal{B}(t,\mathbf{x})\,dt \\ = \frac{1}{2}\left(\mathrm{Re}\,\mathbf{J}(\mathbf{x}) \times \mathrm{Re}\,\mathbf{B}(\mathbf{x}) + \mathrm{Im}\,\mathbf{J}(\mathbf{x}) \times \mathrm{Im}\,\mathbf{B}(\mathbf{x})\right) .$$

It should be noted that the Ohm law for a moving conductor reads

$$\mathbf{J} = \boldsymbol{\sigma}(\mathbf{E} + \mathbf{u} \times \mathbf{B}) .$$

However, when working with molten metals on a laboratory scale, the term $\boldsymbol{\sigma}(\mathbf{u} \times \mathbf{B})$ can be neglected.

In most numerical schemes the coupled problem is solved using an iterative time stepping procedure, in which the electromagnetic field is first determined for temperature dependent conductivity and permeability, then the momentum and the temperature equations are advanced using the resulting Lorentz force and Joule loss distribution, and finally the material properties are updated and another step can be applied. Proceeding in this way, at each time step it is necessary to solve an eddy current problem like the one analyzed in the previous chapters of this book.

There is a very rich literature on numerical modeling of induction heating, and we refer to Lavers [163] for an extensive bibliographic review on this subject. More often, taking advantage of some geometrical symmetries, in many works concerning the coupling of electromagnetic and thermal problems the computational domain is two-dimensional. For instance, Chaboudez et al. [78] consider a two-dimensional problem involved in induction heating of long workpieces; Chaboudez et al. [77] do the same for an axisymmetric configuration; Bermúdez et al. [40] study the thermo-electromagnetic problem in induction furnaces used for melting, proposing and analyzing a FEM/BEM method for the approximation of the electromagnetic subproblem; Bay et al. [35] consider a model which couples electromagnetic, thermal and mechanical effects in axisymmetrical induction heating processes; Henneberger et al. [124], Natarajan and El-Kaddah [183] deal with the magneto-hydrodynamic problem in the context of induction melting systems with axisymmetric geometry, but they do not take explicitly into account thermal effects. Let us also mention Rappaz and Touzani [203] for the numerical analysis of a two-dimensional magneto-hydrodynamic problem.

By contrast, there are few works concerning the numerical approximation of the thermal-magneto-hydrodynamic problem: we mention the results by Henneberger and Obrecht [123], Katsumura et al. [149], and in particular the more recent paper by Bermúdez et al. [41] (see also Vázquez [240]); in all these works the axisymmetric geometry is assumed. More specifically, in Bermúdez et al. [41] and Vázquez [240] a BEM/FEM method is used for the approximation of the solution of the electromagnetic problem. The problem is formulated in terms of a magnetic vector potential and the input data of the problem are the current intensities through the inductor coils. Some numerical simulations for an industrial furnace are presented, and show a good agreement with experimental data. In this modelization, the induction coil has been replaced by a suitable set of rings, each one having toroidal geometry (see Figure 9.2).

As far as we know, there are no three-dimensional simulations of the thermal-magneto-hydrodynamic problem for the more realistic situation in which the coil is a

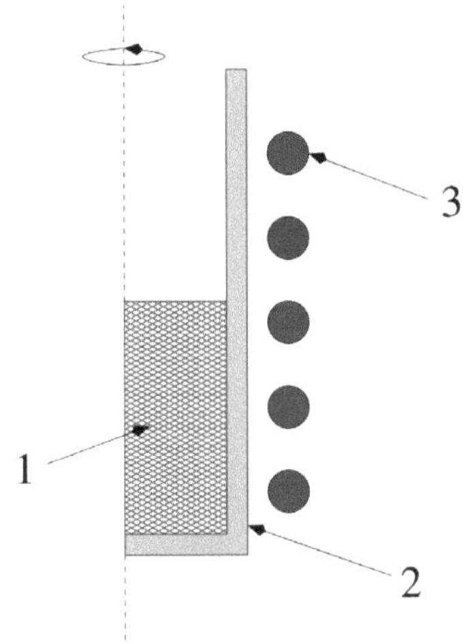

Fig. 9.2. Axisymmetric induction furnace: 1 metal, 2 crucible, 3 coils

simply-connected helix with two electric ports (for this type of coils in Section 8.1 we have presented a complete analysis of the eddy current problem).

9.1.2 Metallurgical electrodes

The time-harmonic eddy current model is also used in modeling the thermoelectrical behaviour of electrodes for electric reduction furnaces. Although the model is rather general, we focus on metallurgical electric furnaces for silicon metal and ferro-silicon production, following the presentation in Bermúdez et al. [43] (see also Bermúdez et al. [45], Salgado [217]). This kind of furnace basically consists of a cylindrical pot containing charged material and three electrodes symmetrically disposed. The pot is a steel cylinder charged with quartz and quartzite, as silicon oxide source, and carbonaceous substance, as coal and coke. At temperatures over 1900 degrees the carbon reduces the silica to silicon by the chemical reaction $SiO_2 + 2C \rightarrow Si + 2CO$. The electrodes are made of carbon materials and they serve to conduct the electric current to the center of the furnace. Different transformers change the high-voltage current usually supplied into the low-voltage high-intensity current suitable for the process. The electric current enters each electrode through a metal ring, which completely embraces the electrode above the charge level. The ring is composed by several copper sections, called contact clamps; bus bars connect the transformers to the contact clamps. At the tip of each electrode an electric arc is produced, generating the high temperatures that activate the chemical reaction. In Figure 9.3 we give a sketch of a reduction furnace.

The electrodes can be of different kinds, depending on the type of production, namely, silicon metal or ferro-silicon. Traditionally, in furnaces for silicon metal production two types of electrodes are mainly used: the pure graphite electrodes, composed by graphite bars joined by threated graphite connecting pieces, called nipples;

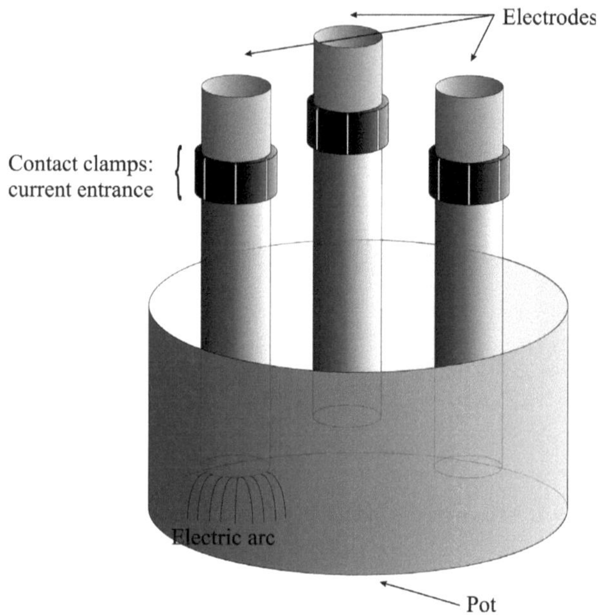

Fig. 9.3. Sketch of a reduction furnace (courtesy of A. Bermúdez, R. Rodríguez and P. Salgado)

the prebaked electrodes, composed by a mixture of carbonaceous substance known as paste, which has been previously baked to eliminate volatile substances.

Instead, the most used electrode in ferro-silicon industry is the Søderberg one, composed by a mixture of petroleum coke and coal-tar pitch contained into a steel cylinder. This paste is put in the cylinder at the top of the electrode, and it bakes in the zone of the contact clamps, employing the heat generated by the Joule effect. In this process the initially non-conductive paste at the top of the electrode becomes a solid carbon conductor. The baked electrode is consumed during the reaction that takes place at the tip of the pot, and has to be continuously replaced by pushing the carbon body down. This is done by moving the casing, but this procedure has the drawback that the steel melts and pollutes silicon. For this reason the Søderberg electrodes, that can be built in larger size and cost less than pure graphite or prebaked electrodes, are only used in the production of ferro-silicon, which can contain a large percentage of iron, but cannot be used to obtain pure silicon metal.

For many years graphite or prebaked electrodes have been the only kind of electrodes used in silicon metal production. In the early 1990s, the Spanish company Ferroatlántica S.L. built a new type of electrode named ELSA, that serves for the production of silicon metal at a lower cost. It consists of a central column of baked carbonaceous material, graphite or similar, surrounded by a Søderberg-like paste. There is a steel casing that contains the paste until it is baked, but the carbon core is responsible for slipping, so the casing does not move with the baked electrode and it does not melt. In this way it is possible to produce silicon with metallurgical quality. In Figure 9.4 we see a sketch of the ELSA electrode.

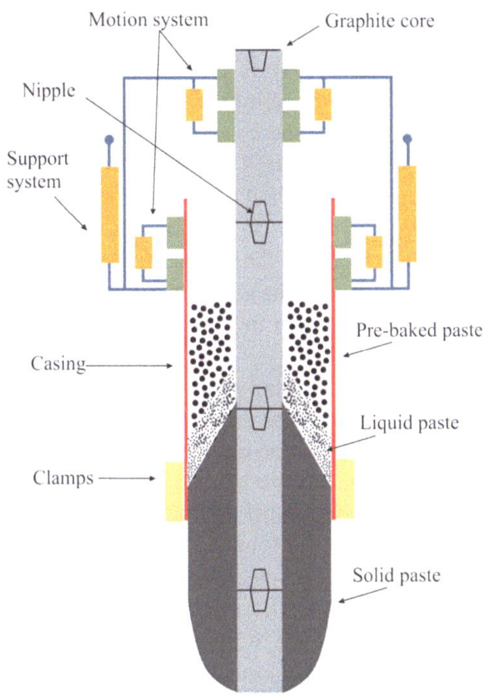

Fig. 9.4. Sketch of an ELSA electrode (courtesy of A. Bermúdez, R. Rodríguez and P. Salgado)

During the last decades many models and codes have been developed to simulate the working conditions of electric reduction furnaces. They compute the temperature distribution, the electromagnetic fields and the stress distribution inside the electrodes by solving the heat equation, the Maxwell equations and the elasticity equations. The system is coupled since the heat source depends on the electromagnetic fields, and the conductivity and stresses depend on the temperature. The alternating current and the low frequency (50 Hz) used make the eddy current model a good approximation for the electromagnetic submodel.

The early works concerning the modeling of a single electrode assume cylindrical symmetry (see, for instance, Bermúdez et al. [39]). The problem is solved in a vertical section of the electrode, writing the equations in cylindrical coordinates. Clearly, the two-dimensional model reduces the computational cost, but introduces some simplifications. Axisymmetric boundary conditions are assumed, but these are not realistic in industrial applications, for which the current enters the electrode through the contact clamps, and in each electrode half of the clamps is connected to one transformer while the others are connected to a second transformer. Moreover, the conductivity is not axisymmetric in the electrode, since it depends on the temperature, which is greater in the central part of the furnace containing the electrode.

There are few works concerning three dimensional simulation of metallurgical electrodes. The mathematical analysis of the continuous and the discrete problems in the case of a single electrode can be found in Bermúdez et al. [46] for the $\mathbf{H}$-based formulation, and Bermúdez et al. [44] for the $\mathbf{E}$-based one. Notice that, in this case, the conductivity is assumed to be uniformly positive definite in the whole domain, and an insulating region is not present: the computational domain simply corresponds to the electrode and to the contact clamps. The problem is not axisymmetric, because it takes into account the non-symmetric boundary conditions that are typical in industrial applications.

Here we show some numerical simulations due to Bermúdez et al. [43] for an ELSA electrode. In Figures 9.5 we describe the geometrical configuration; in Figures 9.6, 9.7, 9.8 and 9.9 we show the magnitude of the current density in different sections of the electrode.

A delicate issue of the model that only considers one single electrode is the determination of the boundary conditions, as explained in Bermúdez et al. [44]. On the tip of the electrode, where the electric arc arises, the current exits freely, hence $\mathbf{E} \times \mathbf{n} = \mathbf{0}$. Also on the contacts, namely, the cross-sections of the bus bars through which the electric current enters the domain, the condition $\mathbf{E} \times \mathbf{n} = \mathbf{0}$ is imposed, and moreover the current intensity through each bus bar is known. Then one has $\mathbf{J} \cdot \mathbf{n} = 0$ outside the tip of the electrode and the contacts, since there is no current flux through this part of the boundary. Finally, $\mu\mathbf{H} \cdot \mathbf{n}$ is set equal to 0 on the whole boundary, though this assumption is not valid in general: for instance, it is exactly true in the axisymmetric case, and it is admissible when the number of external bus bars feeding the electrode is large and they are arranged radially, because in this case the normal magnetic fluxes that they generate tend to cancel out. In more general situations one could take a larger domain around the electrode and the bus bars, and assign the boundary condition $\mu\mathbf{H} \cdot \mathbf{n} = 0$ on

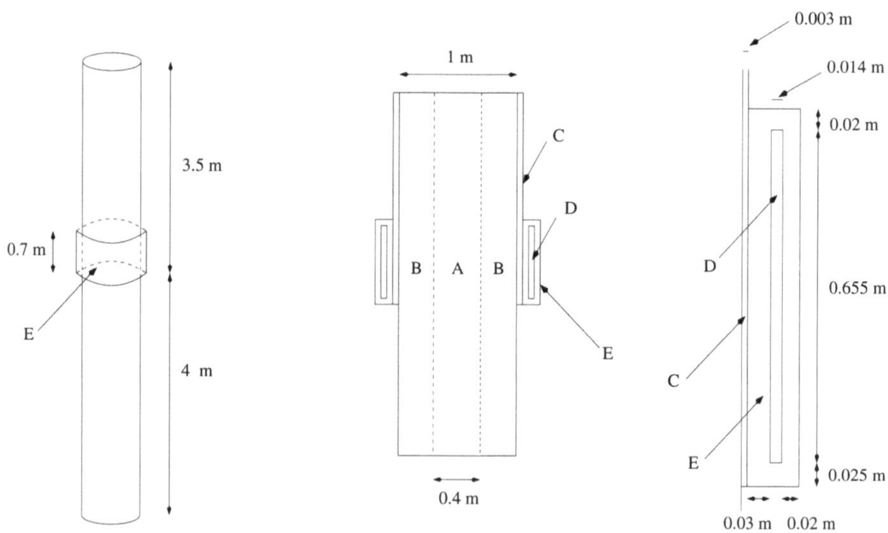

Fig. 9.5. The geometric configuration of the ELSA electrode: A graphite, B paste, C casing, D water, E contact clamp

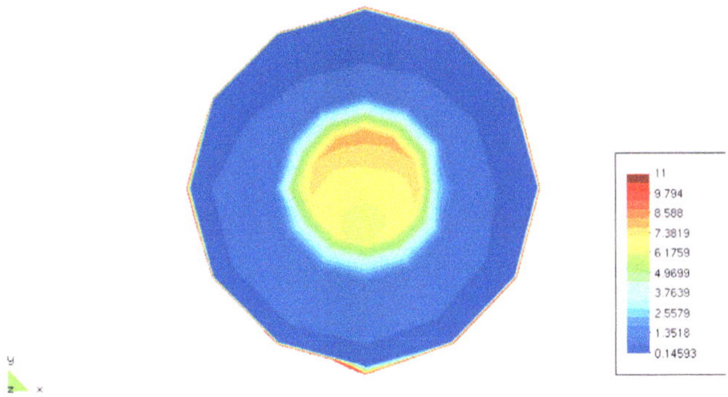

Fig. 9.6. Magnitude of the current density (A/cm^2): horizontal section at the top of contact clamps (courtesy of A. Bermúdez, R. Rodríguez and P. Salgado)

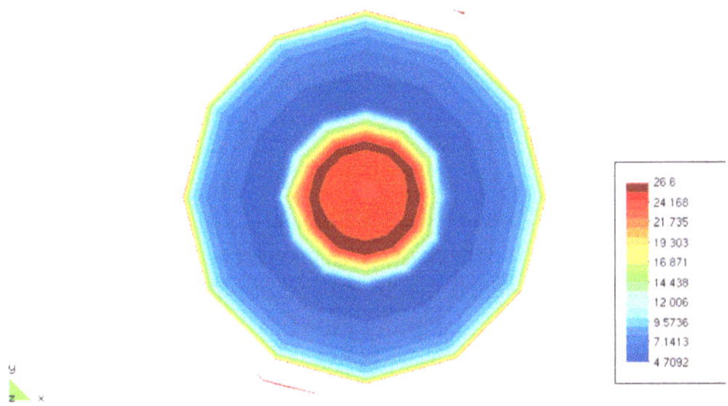

Fig. 9.7. Magnitude of the current density (A/cm^2): horizontal section at the bottom of contact clamps (courtesy of A. Bermúdez, R. Rodríguez and P. Salgado)

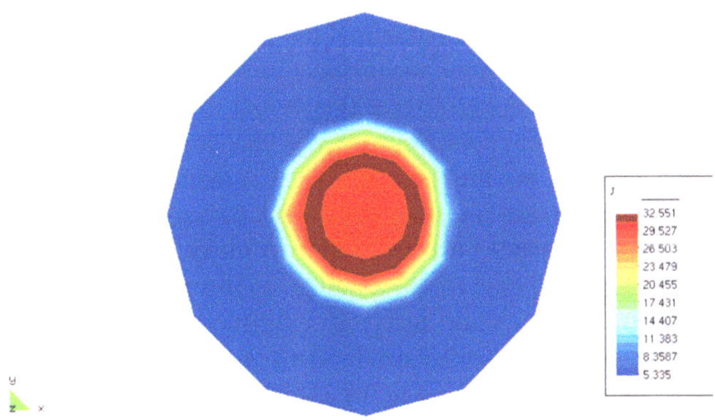

Fig. 9.8. Magnitude of the current density (A/cm^2): horizontal section 25 cm below the contact clamps (courtesy of A. Bermúdez, R. Rodríguez and P. Salgado)

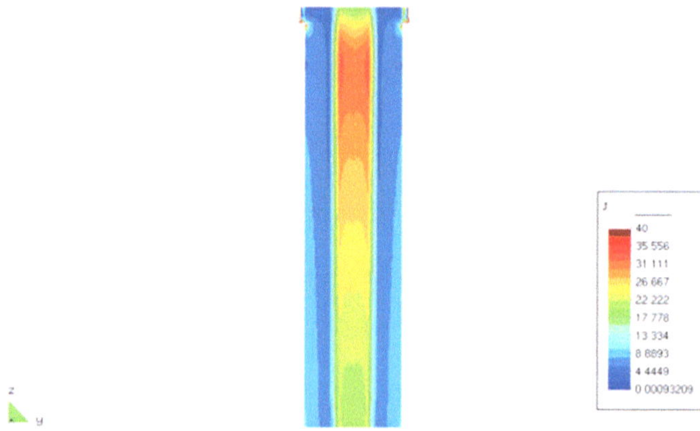

Fig. 9.9. Magnitude of the current density (A/cm^2): vertical section (courtesy of A. Bermúdez, R. Rodríguez and P. Salgado)

the external boundary: however, in this way a non-conductive air region is introduced (see Bermúdez et al. [45], Alonso Rodríguez et al. [20]).

When considering a single electrode the proximity effect is neglected: though the magnetic field generated by each electrode induces eddy currents in the other electrodes, this is not considered in the simulations. A first attempt at taking into account this effect has been carried out by Bermúdez et al. [38] for ELSA electrodes, solving numerically the electromagnetic problem on a horizontal section of the three electrodes. A drawback is that these two-dimensional models are valid only in the lower part of the electrode, where it can be assumed that the electric current is orthogonal to the considered two-dimensional section.

The more realistic modeling of the reduction furnace requires to consider a three-dimensional non-symmetric computational domain, formed by a conducting region and an insulating region. We conclude this section by presenting some numerical simulations due to Bermúdez et al. [43] for this model of the furnace, with three ELSA compound electrodes. The contact clamps and the casing are not explicitly considered in the modelization, and the Søderberg paste is assumed to be baked in the whole domain. The electric current enters the electrodes through copper bars of rectangular section. In Figures 9.10 and 9.11 we show the geometrical data of the problem.

The numerical method used for the simulations illustrated in Figures 9.12 and 9.13 is the finite element discretization analyzed in Bermúdez et al. [45], and presented in Section 5.4.2. The problem is formulated in terms of the magnetic field in the conductor and of the scalar magnetic potential in the insulator, and the finite elements used are first order edge elements in the conductor and first order nodal elements in the insulator.

Numerical results for the same problem, but formulated in terms of the electric field in the conductor and of the scalar magnetic potential in the insulator, have been also presented in Section 8.1.5.

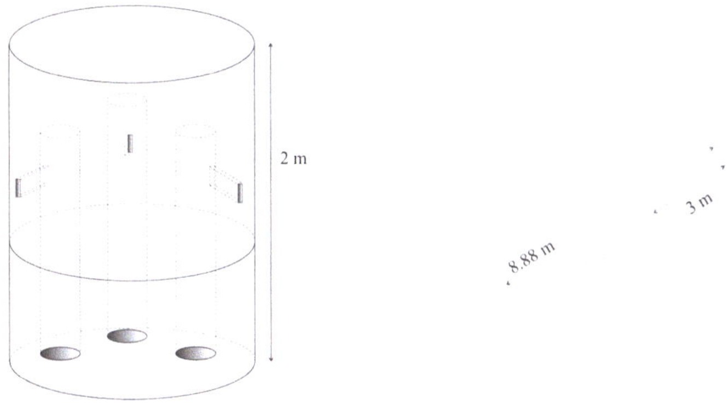

Fig. 9.10. Sketch of the model domain (courtesy of A. Bermúdez, R. Rodríguez and P. Salgado)

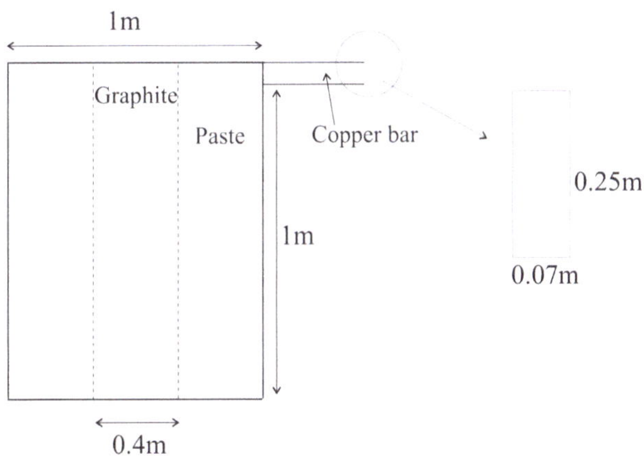

Fig. 9.11. Geometrical data corresponding to a vertical section of each electrode (courtesy of A. Bermúdez, R. Rodríguez and P. Salgado)

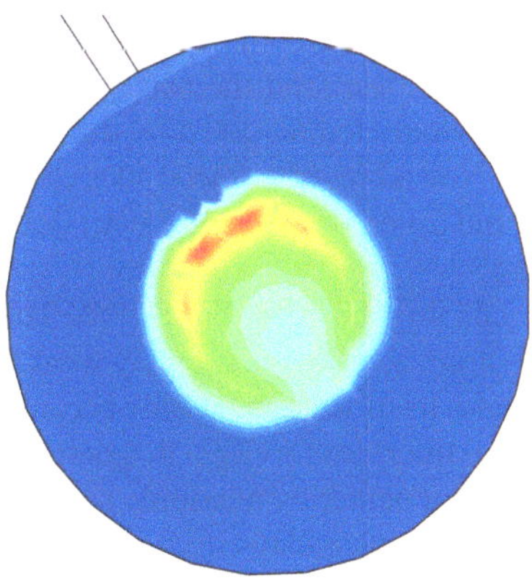

Fig. 9.12. Magnitude of the current density (A/cm^2): horizontal section (courtesy of A. Bermúdez, R. Rodríguez and P. Salgado)

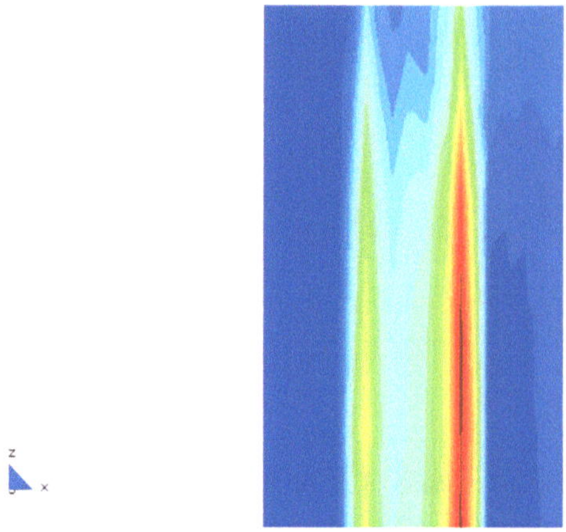

Fig. 9.13. Magnitude of the current density (A/cm^2): vertical section (courtesy of A. Bermúdez, R. Rodríguez and P. Salgado)

9.2 Bioelectromagnetism: EEG and MEG

Electroencephalography (EEG) and magnetoencephalography (MEG) are two non-invasive techniques used to localize electric activity in the brain from measurements of external electromagnetic signals. Electroencephalography measures the scalp electric potential, while magnetoencephalography measures the external magnetic flux.

The electromagnetic activity of the brain is due to the movements of ions within activated regions of the cortex sheet, the so-called impressed currents (or primary currents). In addition, Ohmic currents are generated in the surrounding medium, the so-called return currents. The measures of EEG and MEG correspond to both impressed and return currents, but the source of interest are the impressed currents, as they represent the area of neural activity associated to a sensory stimulus.

The first EEG recording in man (and the name Electroenkephalogram) is due to H. Berger in 1924. He measured electric potential differences between pairs of electrodes placed on the scalp. Nowadays these electrodes can be directly glued to the skin or fitted in an elastic cap, and tipically up to 256 electrodes are used (see in Figure 9.14 a cap with 128 electrodes).

The first magnetoencephalograms date back at the late 1960s by D. Cohen. The magnetic signal related to brain activity is extremely weak, about 10^8 times lower than the earth's geomagnetic field. Its measurement only becames possible with the SQUID (Superconducting QUantum Interface Devices) magnetometer introduced by Zimmerman [249]. This kind of instrumentation measures some component of the magnetic induction on different locations, nowadays up to 100, close but external to the head (see Figure 9.15).

For a comprehensive introduction to theory and instrumentation in MEG see Hämäläinen et al. [117]. A complete description of the models used in EEG/MEG

Fig. 9.14. The distribution of the sensors for EEG

Fig. 9.15. The distribution of the sensors for MEG (courtesy of Elekta)

source localization is provided in Baillet et al. [34] (see also Mosher et al. [182]). Let us give here a concise presentation of the topic.

Source localization is an inverse problem: knowing the value of the magnetic field or of the electric field on the surface of the head (or, possibly, external to the head, but close to its surface), the aim is to determine the position and some physical characteristics of the current density that has given rise to that value.

Since the current distribution inside a conductor cannot be retrieved uniquely from knowledge of the electromagnetic field outside the conductor, the mathematical problem does not have a unique solution if some additional conditions on the source model are not assumed (see Sarvas [220]). Two different approaches are mainly used to reconstruct the brain neural sources: equivalent dipole and distributed source models.

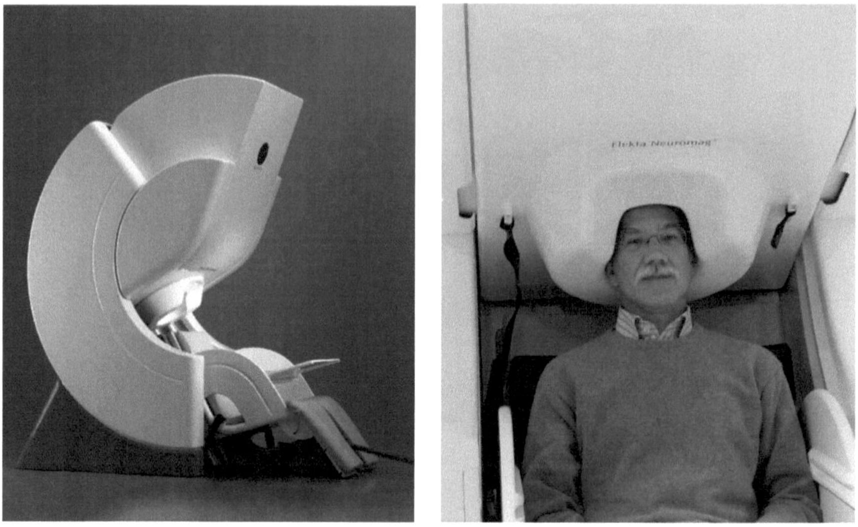

Fig. 9.16. The Elekta Neuromag MEG system (left, courtesy of Elekta) and the second author in the CIMeC Laboratory, University of Trento (right)

Moreover, see for instance Kaipio and Somersalo [142] for statistical approaches that we do not consider here.

In the dipolar model the primary current distribution is represented as a point source located at $\mathbf{x}_q$ with moment $\mathbf{q}$, namely,

$$\mathbf{J}_e(\mathbf{x}) = \mathbf{q}\,\delta(\mathbf{x} - \mathbf{x}_q)\,,$$

where $\delta(\cdot)$ is the Dirac delta distribution. The dipole is a convenient representation for a unidirectional impressed current due to the activation of a large number of pyramidal cells, that in real situations may indeed extend over several square centimeters of the cortex. More generally, it is assumed that a primary current source can be decomposed as the sum of (few) current dipoles. In the standard dipolar method the parameters of the dipoles (location, amplitude and orientation) are found using a nonlinear least-squares search.

The distributed source model (also called imaging approach) assumes that a lot of dipoles are located perpendicularly to the cortical surface. The geometry of the cortical surface can be extracted from brain magnetic resonance imaging (MRI) data. A tessellation of this surface is constructed and a current dipole is placed on each element with itsorientation normal to the surface. The inverse problem in this case turns out to be linear: only the magnitudes of the dipole moments have to be reconstructed, and not the location nor the orientation. Proceeding in this way the number of unknowns is typically greater than the number of measured data and the inverse problem is solved using regularization schemes, such as a truncated singular value decomposition of the Tikhonov regularization.

In both cases, a preliminary step for the solution of the inverse problem is an efficient resolution of the forward problem. In fact, the procedure is essentially the following: given a source $\mathbf{J}_e$, solve the forward problem, thus determining the electric and magnetic fields generated by $\mathbf{J}_e$, and then minimize in a suitable way the difference between the computed and the measured data. The current density $\mathbf{J}_e^*$ which achieves the minimum is the source we are trying to determine.

Let us focus now on the forward problem. For biological tissues, the linear constitutive equations $\mathcal{D} = \varepsilon\mathcal{E}$ and $\mathcal{B} = \mu\mathcal{H}$ can be assumed (see Plonsey and Heppner [194]). Due to its complicated detailed structure, the human brain must be modeled as a heterogeneous anisotropic medium, with physical parameters that depend on the spatial variable and that may be tensors. The frequency spectrum for electrophysiological signals in MEG is tipically below 1000 Hz, and most studies deal with frequency between 0.1 and 100 Hz.

As far as we know, in almost all the studies concerning the neural generation of electromagnetic fields the static approximation of Maxwell equations is considered

$$\begin{aligned} &\operatorname{curl}\mathbf{H} = \mathbf{J} \\ &\operatorname{div}\mathbf{B} = 0 \\ &\operatorname{curl}\mathbf{E} = \mathbf{0}\,, \end{aligned} \tag{9.1}$$

neglecting not only the displacement current but also the electromagnetic diffusion.

From Ohm law the total current density $\mathbf{J}$ is the sum of the impressed currents plus the return currents

$$\mathbf{J} = \mathbf{J}_e + \sigma\mathbf{E} = \mathbf{J}_e - \sigma \operatorname{grad} V\,,$$

where V is the electric scalar potential. From the first equation in (9.1) it follows that

$$0 = \operatorname{div}\mathbf{J} = \operatorname{div}(\mathbf{J}_e - \sigma \operatorname{grad} V)\,.$$

Hence V can be obtained by solving the Poisson equation

$$\operatorname{div}(\sigma \operatorname{grad} V) = \operatorname{div}\mathbf{J}_e\,, \tag{9.2}$$

usually with the boundary condition $\sigma \operatorname{grad} V \cdot \mathbf{n} = \mathbf{J}_e \cdot \mathbf{n}$, which is a consequence of the fact that outside the head the magnetic field is supposed to be curl-free (the source $\mathbf{J}_e$ is located inside the head, and the conductivity is vanishing outside the head, so that $\mathbf{J} = \mathbf{0}$).

For EEG this is the point: solving this elliptic problem gives the electric field, and the inverse problem of source localization can be dealt with.

For MEG, one has to go further. Since the magnetic permeability can be assumed to be homogeneous and equal to μ_0, the free-space permeability, $\mathbf{B}$ is given by the Biot-Savart law

$$\mathbf{B}(\mathbf{x}) = \frac{\mu_0}{4\pi}\int_{\mathbb{R}^3} \mathbf{J}(\mathbf{y}) \times \frac{\mathbf{x}-\mathbf{y}}{|\mathbf{x}-\mathbf{y}|^3}\, d\mathbf{y}\,. \tag{9.3}$$

Here the integration is indeed carried out on a bounded domain Ω, representing the head, as $\mathbf{J}$ is vanishing outside Ω. Note that this formula furnishes a direct way to compute the magnetic induction $\mathbf{B}$, but only after we have determined the electric scalar potential V through (9.2).

However, in some cases solving the elliptic problem (9.2) can be avoided. In fact, the typical (though simplified) head model assumes that the head can be described by three (scalp, skull and brain) to five (scalp, skull, cerebrospinal fluid, gray matter and white matter) contiguous layers Ω_j, $j = 1, \ldots, n$. The different layers of the head and the air region are separated by the surfaces S_j, $j = 1, \ldots, n$, S_1 being the outermost one. Assuming that the conductivity of each layer is a scalar constant, by employing classical results of potential theory it is possible to derive a surface integral equation for $V_k := V_{|S_k}$, $k = 1, \ldots, n$,

$$\begin{aligned}\frac{\sigma_k^- + \sigma_k^+}{2} V_k(\mathbf{x}) \\ = V_\infty(\mathbf{x}) - \frac{1}{4\pi} \sum_{j=1}^{n} (\sigma_j^- - \sigma_j^+) \int_{S_j} V_j(\mathbf{y})\, \mathbf{n}_j(\mathbf{y}) \cdot \frac{\mathbf{x}-\mathbf{y}}{|\mathbf{x}-\mathbf{y}|^3}\, dS_y\end{aligned} \tag{9.4}$$

(see Sarvas [220]), where

$$V_\infty(\mathbf{x}) := \frac{1}{4\pi} \int_\Omega \mathbf{J}_e(\mathbf{y}) \cdot \frac{\mathbf{x}-\mathbf{y}}{|\mathbf{x}-\mathbf{y}|^3}\, d\mathbf{y}\,,$$

$\mathbf{n}_j$ is the unit outward normal vector to S_j, σ_j^- is the inside conductivity and σ_j^+ is the outside conductivity, with $\sigma_1^+ = 0$ and, clearly, $\sigma_j^- = \sigma_{j+1}^+$, $j = 1, \ldots, n-1$. Note that, in the particular case of a current dipole, one has

$$V_\infty(\mathbf{x}) = \frac{1}{4\pi} \mathbf{q} \cdot \frac{\mathbf{x}-\mathbf{x}_q}{|\mathbf{x}-\mathbf{x}_q|^3}\,.$$

For constant conductivities integration by parts in (9.3) shows that also the Biot–Savart law can be written as a sum of surface integrals on the interfaces between layers, obtaining the formula due to Geselowitz [109]

$$\mathbf{B}(\mathbf{x}) = \mathbf{B}_\infty(\mathbf{x}) - \frac{\mu_0}{4\pi} \sum_{j=1}^{n} (\sigma_j^- - \sigma_j^+) \int_{S_j} V_j(\mathbf{y})\, \mathbf{n}_j(\mathbf{y}) \times \frac{\mathbf{x}-\mathbf{y}}{|\mathbf{x}-\mathbf{y}|^3}\, dS_y\,, \tag{9.5}$$

where the vector field

$$\mathbf{B}_\infty(\mathbf{x}) := \frac{\mu_0}{4\pi} \int_\Omega \mathbf{J}_e(\mathbf{y}) \times \frac{\mathbf{x}-\mathbf{y}}{|\mathbf{x}-\mathbf{y}|^3}\, d\mathbf{y}$$

is called the primary field. In the case of a current dipole it becomes

$$\mathbf{B}_\infty(\mathbf{x}) = \frac{\mu_0}{4\pi} \mathbf{q} \times \frac{\mathbf{x}-\mathbf{x}_q}{|\mathbf{x}-\mathbf{x}_q|^3}\,.$$

At this stage, for MEG the main point turns out to be the determination of the functions V_j on the surfaces S_j, which furnish the magnetic induction $\mathbf{B}$ via the explicit formula

(9.5). Hence a boundary element approach can be introduced, with the aim of finding a solution to the discrete approximation of (9.4), then inserting the obtained results in (9.5).

In some particular cases one can even avoid solving (9.4). Indeed, a simplified model assumes that the head consists of a set of nested concentric spheres, each layer with a scalar constant conductivity. In the special case of a current dipole and of a MEG system that measures only the radial component of the magnetic induction $\mathbf{B}$, the contribution of the return currents vanishes, as for $\mathbf{y} \in S_j$ and $\mathbf{x} \neq \mathbf{y}$ one has

$$\begin{aligned}&\left(\mathbf{n}_j(\mathbf{y}) \times \tfrac{\mathbf{x}-\mathbf{y}}{|\mathbf{x}-\mathbf{y}|^3}\right) \cdot \tfrac{\mathbf{x}}{|\mathbf{x}|} \\ &\quad = \left(\tfrac{\mathbf{y}}{|\mathbf{y}|} \times \tfrac{\mathbf{x}}{|\mathbf{x}-\mathbf{y}|^3}\right) \cdot \tfrac{\mathbf{x}}{|\mathbf{x}|} - \left(\tfrac{\mathbf{y}}{|\mathbf{y}|} \times \tfrac{\mathbf{y}}{|\mathbf{x}-\mathbf{y}|^3}\right) \cdot \tfrac{\mathbf{x}}{|\mathbf{x}|} = 0 \,.\end{aligned}$$

Therefore, the radial component of $\mathbf{B}(\mathbf{x})$ reduces to

$$B_r(\mathbf{x}) = \frac{\mathbf{x}}{|\mathbf{x}|} \cdot \mathbf{B}(\mathbf{x}) = \frac{\mathbf{x}}{|\mathbf{x}|} \cdot \mathbf{B}_\infty(\mathbf{x}) = \frac{\mu_0}{4\pi} \frac{\mathbf{x} \times \mathbf{x}_q}{|\mathbf{x}||\mathbf{x} - \mathbf{x}_q|^3} \cdot \mathbf{q} \tag{9.6}$$

(note the linear dependence of B_r on the moment $\mathbf{q}$ and the nonlinear dependence on the position $\mathbf{x}_q$). Hence, the radial component of $\mathbf{B}$ turns out to be independent of the potential V, and in this case the solution of the inverse MEG problem does not require the previous computation of V, and simply uses the explicit formula (9.6).

These spherical models work reasonably well and are routinely used in most applications of EEG and MEG source localization. However, it seems clear that, in order to improve the source reconstruction, more accurate solutions to the forward problem are needed, and a more realistic model must be considered. Anatomical information can be obtained from brain magnetic resonance imaging or X-ray computed tomography imaging (see, for instance, Khan et al. [151]). From these images it is possible to construct a realistic head model (see, e.g., Van Uiter et al. [239], Kybic et al. [162], Wolters et al. [245]) and extract precise informations about surface boundaries for scalp, skull and brain. On the other hand, recent studies of Marin et al. [172], Wolters et al. [244], and Haueisen et al. [118] show that the anisotropy of the conductivity in the skull and brain must be taken into account and in particular the conductivity cannot be assumed to be piecewise-constant. From a numerical point of view this means that one has to go back to the numerical solution of (9.2), and this can be done by using a finite element scheme.

However, a modelization through the elliptic equation (9.2) is not completely satisfactory. In fact, as already remarked, the physiological frequency involved in the problem ranges between 0.1 and 100 Hz, and in general cannot be assumed to vanish. Therefore the static model (9.1) has to be replaced by the eddy current model. To the best of our knowledge, the latter has not been used yet for brain activity reconstruction from MEG data, but this could be an important direction for further researches.

In this respect, since it is necessary to reduce as far as possible the computational cost of the forward solver, the approach presented in Sections 7.1–7.5 could be a useful tool.

Remark 9.1. The necessity of taking into account a non-vanishing frequency has been underlined by He and Romanov [122], Ammari et al. [22], who use the full Maxwell system as forward problem. They consider the inverse source problem that arises in determining the location of an epileptic focus, i.e., the localization of a single dipole in a homogeneous or even heterogeneous medium. Unlike many inverse methods, the proposed algorithm is non-iterative. Following Ammari et al. [22], in this case the considered forward problem is the time-harmonic full Maxwell system in $\mathbb{R}^3$

$$\begin{aligned} &\operatorname{curl} \mathbf{H} - i\omega\varepsilon\mathbf{E} = \boldsymbol{\sigma}\mathbf{E} + \mathbf{q}\,\delta(\mathbf{x} - \mathbf{x}_q) \\ &\operatorname{curl} \mathbf{E} + i\omega\boldsymbol{\mu}\mathbf{H} = \mathbf{0}\,, \end{aligned} \tag{9.7}$$

with the Silver–Müller radiation condition

$$\lim_{|\mathbf{x}|\to+\infty} |\mathbf{x}| \left(\sqrt{\mu_0}\,\mathbf{H} \times \frac{\mathbf{x}}{|\mathbf{x}|} - \sqrt{\varepsilon_0}\,\mathbf{E} \right) = \mathbf{0}\,,$$

ε_0 being the free-space electric permittivity.

As usual, let Ω_C denote the conductor (the human head). If φ is a scalar harmonic function in Ω_C and $\mathbf{u}$ is a solution to

$$\operatorname{curl}(\mu_0^{-1} \operatorname{curl} \mathbf{u}) = i\sigma \operatorname{grad} \varphi\,,$$

then it can be proved that

$$\mathbf{q} \cdot \operatorname{grad} \varphi(\mathbf{x}_q) = \int_{\partial\Omega_C} \mathbf{H} \times \mathbf{n} \cdot \operatorname{grad} \varphi + i \int_{\partial\Omega_C} \mu_0^{-1} \operatorname{curl} \mathbf{u} \cdot \mathbf{E} \times \mathbf{n} + O(\omega)\,.$$

Choosing in this formula six particular harmonic functions ($\varphi_k = x_k$ for $k = 1, 2, 3$ and $\varphi_k = e^{i\boldsymbol{\xi}_{k-3}\cdot\mathbf{x}}$ for $k = 4, 5, 6$, where $\boldsymbol{\xi}_j \in \mathbb{C}^3$, $j = 1, 2, 3$, are such that $\sum_{i=1}^3 \xi_{j,i}^2 = 0$), the six components of $\mathbf{q}$ and $\mathbf{x}_q$ can be approximated. It is worth noting that this reconstruction is carried out without a priori knowledge of the angular frequency ω. □

Remark 9.2. A related forward problem, where the eddy current system or else the full Maxwell equations have been adopted, is the numerical simulation of transcranial magnetic stimulation (TMS): see, e.g., Ueno et al. [237], Sekino et al. [225]. This is a non-invasive method for stimulating neurons in the brain, and it is widely used in neuroscience, in order to study the functional organization of human brain, and in the diagnosis and the treatement of neurological diseases. A transcranial magnetic stimulation system consists in a coil placed on the scalp, that produces a time-harmonic magnetic field which induces eddy currents in the brain. The operating frequency ranges from 1 to 4 kHz. It is also possible to use more than one coil to stimulate different parts of the brain simultaneously: this is the so-called multichannel transcranial magnetic stimulation, that has recently attracted particular interest (see Lu et al. [170] and references therein).

Accurate numerical simulations of the induced fields inside the brain are necessary for optimizing the design of the coils that have to generate the desired stimulation. □

9.3 Magnetic levitation

Due to the relatively low frequencies involved, magnetic levitation problems are an interesting field of application for eddy current models: in fact, “the magnetic energy storage is dominant (as compared to energy stored in the electric field) and wave phenomena are small enough to be ignored” (Thompson [234]; see also, e.g., Kriezis et al. [157]).

Let us start this section with a brief presentation of problem 28 of the TEAM workshop, a simple electrodynamic levitation problem which can serve as a model problem for more complex computations in moving domains. It is an axisymmetric transient problem with electromechanical coupling (see Karl et al. [148] and Kurz et al. [161]). The device is described as follows: a cylindrical aluminium plate is located above two cylindrical coils, formed by electric wires, all the parts with the same axis (see Figures 9.17 and 9.18). At the initial instant the plate is above the coils at a certain distance, then an applied current density is imposed. Both coils are connected in series, with different sense of winding.

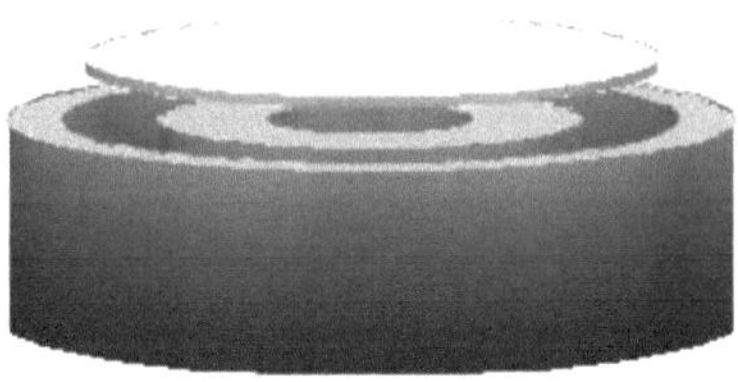

Fig. 9.17. The geometry in problem 28 of the TEAM workshop

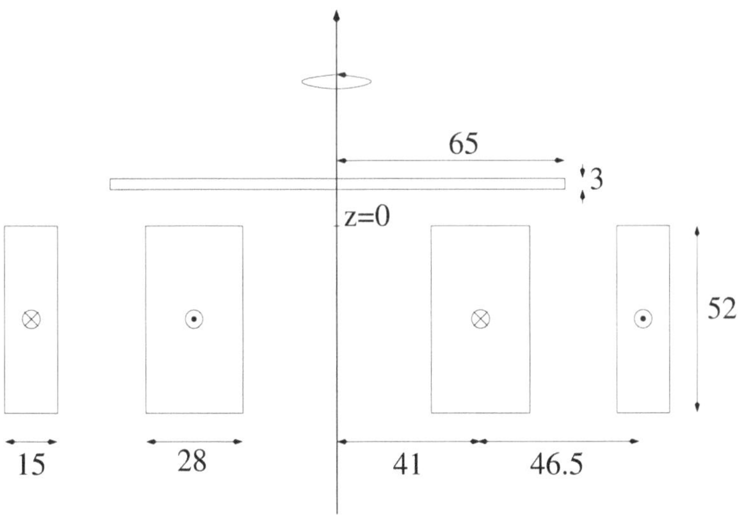

Fig. 9.18. The dimensions in problem 28 of the TEAM workshop

Due to the induced eddy currents, a repulsive Lorentz force

$$\mathbf{F}_L(\mathbf{u},\mathbf{E},\mathbf{B}) := \int_{\Omega_C} [\boldsymbol{\sigma}(\mathbf{E}+\mathbf{u}\times\mathbf{B})]\times\mathbf{B}$$

acts on the plate Ω_C, which reaches, after a transient time, a stationary levitation height. Here $\mathbf{u}$ denotes the velocity of the plate, which depends on mechanical as well as electromagnetic forces, in particular on $\mathbf{B}$ and $\mathbf{E}$.

From the mathematical point of view, the problem is described by the time-dependent eddy current equations

$$\begin{cases} \operatorname{curl}\mathbf{H} - \boldsymbol{\sigma}\mathbf{E} = \mathbf{J}_e + \boldsymbol{\sigma}(\mathbf{u}\times\mathbf{B}) & \text{in } \Omega \\ \operatorname{curl}\mathbf{E} + \frac{\partial\mathbf{B}}{\partial t} = \mathbf{0} & \text{in } \Omega \\ \operatorname{div}(\varepsilon_I\mathbf{E}_I) = 0 & \text{in } \Omega_I\ , \end{cases} \tag{9.8}$$

where Ω is a "box" containing the plate and the support of the coils, and $\mathbf{J}_e$ is supported only in the coils.

The constitutive relation between $\mathbf{B}$ and $\mathbf{H}$ in general is given by $\mathbf{B} = \boldsymbol{\mu}\mathbf{H} + \mathbf{M}$, where $\mathbf{M}$ is the magnetization; however, as always done in this book, here below we assume that $\mathbf{M} = \mathbf{0}$.

Employing an implicit time-discretization scheme and computing the nonlinear term $\mathbf{u}\times\mathbf{B}$ at the previous time level leads at each time step to the solution of

$$\begin{cases} \operatorname{curl}\mathbf{H}^{n+1} - \boldsymbol{\sigma}\mathbf{E}^{n+1} = \mathbf{J}_e^{n+1} + \boldsymbol{\sigma}(\mathbf{u}^n\times\mathbf{B}^n) & \text{in } \Omega \\ \operatorname{curl}\mathbf{E}^{n+1} + \mathbf{B}^{n+1}/\Delta t = \mathbf{B}^n/\Delta t & \text{in } \Omega \\ \operatorname{div}(\varepsilon_I\mathbf{E}_I^{n+1}) = 0 & \text{in } \Omega_I\ . \end{cases} \tag{9.9}$$

Then at the time step $n+1$ the velocity $\mathbf{u}$ and the position $\mathbf{r}$ of the center of gravity of the plate are obtained by setting

$$\mathbf{u}^{n+1} = \mathbf{u}^n + \mathbf{g} + m^{-1}\Delta t\,\mathbf{F}_L(\mathbf{u}^n,\mathbf{E}^{n+1},\mathbf{B}^{n+1}),$$

and

$$\mathbf{r}^{n+1} = \mathbf{r}^n + \Delta t\,\mathbf{u}^{n+1}\,,$$

where $\mathbf{g}$ is the acceleration of gravity, m the mass of the plate and $\mathbf{F}_L$ the Lorentz force.

Most of the results that we have presented for time-harmonic eddy current problems can be adapted to the system of equations (9.9). For instance, Kurz et al. [161] have computed the levitation height by using a FEM–BEM approach for the $(\mathbf{A}_C, V_C) - \mathbf{A}_I$ vector potential formulation. Moreover, Rapetti [201] has used an approach based on the vector potential $\mathbf{A}$ and on an overlapping mortar element technique for taking into account the movement of the plate.

In particular, in the TEAM workshop problem 28 the data of the problem are as follows: the mass of the plate is $m = 0.107$ kg, the initial distance of the plate from the coils is 3.8 mm, the applied current density is given by $\mathbf{J}_e(t) = (-1)^k J_k I^0 \sin(\omega t)\mathbf{e}_\phi$, where $\mathbf{e}_\phi$ is the (counterclockwise) azymuthal unit vector in the cylindrical system,

$k = 1$ refers to the outer coil and $k = 2$ to the inner coil, $I^0 = 20$ A, $\omega = 2\pi \times 50$ rad/s, $J_1 = N_1/S_1$ and $J_2 = N_2/S_2$, where $N_1 = 576$ and $N_2 = 960$ are the number of turns of the electric wires in the coils, and S_1 and S_2 are the cross sections. Finally, the conductivity and the permeability are given by $\sigma = 3.4 \times 10^7$ S/m, $\mu = 4\pi \times 10^{-7}$ H/m.

The results obtained by Kurz et al. [161] and Rapetti [201] are in very good agreement with the experimental data: after a transient time of about 1.6 ms, a stationary levitation height of about 11.3 mm is reached.

TEAM workshop problem 28 is clearly a very simplified model for realistic phenomena based on magnetic levitation. In order to give a more detailed description of the effective technological problems related to this topic, below we briefly outline a presentation of magnetic levitation trains.

Since the 1960s some industrial companies attempted to design a train without wheels, suspended over a specialized track by magnetic levitation and with a propulsion system based on magnetic forces (for more details about these early projects, see, e.g., the review papers by Thornton [235], Yamamura [246], Rogg [213], Powell and Danby [195]).

Two related but different techniques have been mainly used to reach this goal: electrodynamic levitation with superconducting magnets and electromagnetic levitation with normal conductive magnets (for an up-to-date presentation of these technologies, see, e.g., Cassat and Jufer [75], Lee et al. [164], Yan [247]).

In electrodynamic levitation the train is lifted and guided by means of repulsive forces between superconducting coils placed on the vehicle and coils inserted in the guideway (see Figure 9.19). The repulsive forces are produced only when the magnets are moving, hence the train does not levitate at low speed and it still needs wheels for "take-off and landing". The air gap (the distance between the vehicle and the ground) can be larger that 10 cm, and the system turns out to be magnetically stable: if the levitation height becomes lower than the equilibrium position, the magnetic repulsion

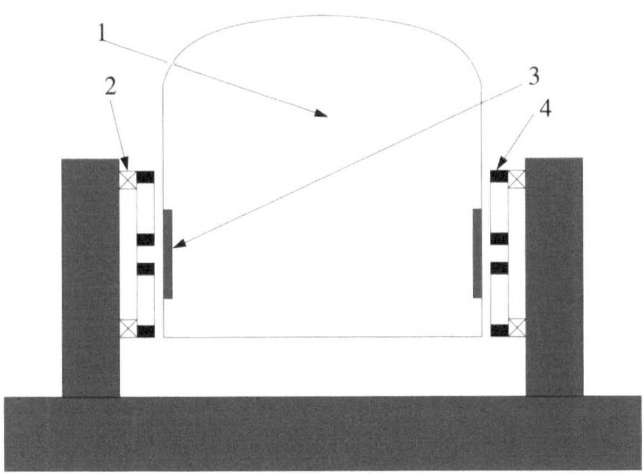

Fig. 9.19. Scheme of the electrodynamic levitation system: 1 vehicle, 2 propulsion windings, 3 superconducting magnets, 4 levitation and guidance windings

force increases, restoring equilibrium; if the levitation height turns out to be too large, then gravity prevails and the air gap is reduced. For this stability property, this system is the most indicated for high speeds and for use in regions that can be subjected to strong earthquakes. The electrodynamic levitation system, originally proposed by Danby and Powell [92], has been adopted by Japanese National Railways, that since the 1970s have produced the series of MLU trains and more recently the last prototype MLX, that in 2003 obtained the train speed record of 581 km/h.

By contrast, electromagnetic levitation makes use of attracting forces between normal conducting electromagnets situated on board and an iron-core armature winding on the rail. The attracting forces produces an inherently unstable levitation system, and the air gap, which is about 1 cm and is nearly velocity-independent, has to be controlled via a high-precision device. Sensors measure the air gap and accelerometers measure the acceleration of the magnets, and information about both of these are passed to the control system. Levitation and guidance can be either integrated in a single system (see Figure 9.20), or else separated (see Figure 9.21). The first choice has been adopted for the Japanese HSST train, operating in public service since 2005 in Nagoya (short distance and medium speed), the second one for the German Transrapid train, operating since 2004 in Shanghai (long distance and high speed).

Concerning propulsion, in both levitation systems the power to the coils at the guideway is supplied by a linear synchronous motor, whose structure is simpler than that of a standard rotating electric motor, not requiring the use of mechanical coupling: plainly-speaking, it is like a conventional rotating motor in which stator, rotor and windings have been unrolled and stretched along the guideway. The working principle is the same: an alternating current inside the motor windings on the guideway generates a space–time depending magnetic field in the air gap, and it induces an electromotive force in the secondary part, a conducting sheet with (standard or superconducting)

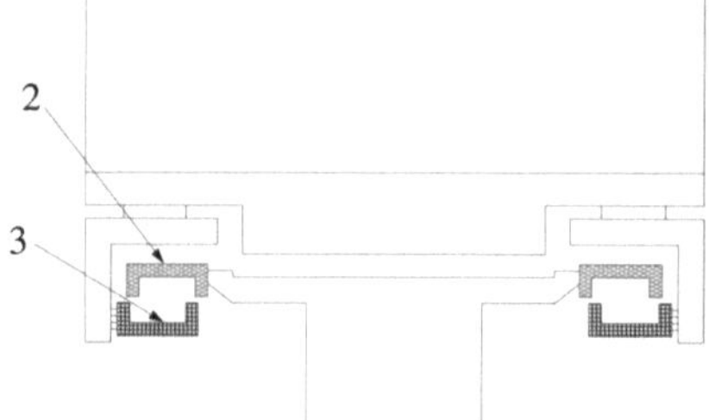

Fig. 9.20. Scheme of the electromagnetic levitation system (levitation and guidance integrated): 1 vehicle, 2 iron-core rail windings, 3 levitation and guidance magnets

magnets placed on the vehicle. This electromotive force generates the eddy currents, whose interaction with the flux in the air gap produces the thrust force. The speed is regulated by varying the frequency of the alternating current, and, if the direction of the traveling field is reversed, the motor becomes a generator and the the train is braked, without contact. The advantage given by a linear motor is that, in the case of a (more or less) rectilinear motion, its efficiency is higher than that of a rotating motor, because of the minor amount of vibration and noise.

Also guidance is based on magnetic forces. In the MLX prototype, the levitation coils on the sideways are connected in such a way that, if a train is closer to one side, then induced currents are produced and this generates a guiding force (in other words, the coils work as a guide system, based on a repulsive force). For the Transrapid train, electromagnets are placed on both sides of the vehicle, and reaction rails on the guideway interact with them maintaining the train suitably centered on the track.

Summing up, the magnetic levitation train is a technological problem of low-frequency electromagnetism coupled with dynamics. To our knowledge, a complete modeling of the whole process has not been performed, due to its high complexity. However, some of its parts have been considered in detail and analyzed by means of the finite element method, though mainly for simplified mathematical models derived from the eddy current equations: as examples we recall the calculation of the magnetic field around the HSST train magnet (see Aoki [28]) or that of induced currents and forces for an hybrid levitation magnet (see Albertz et al. [4]), the investigation of the stability of repulsive forces (see He et al. [121]), the analysis of the heating problem arising in superconducting magnets (see Saito et al. [216]), the design of high temperature superconducting coils (see Jenkins et al. [139]). Finally, the analysis of the MLX train levitation system and some of its variants has been investigated only by means of the dynamic circuit theory (see He et al. [120], Davey [96]), and would give more precise results if the analysis relied on the complete eddy current model.

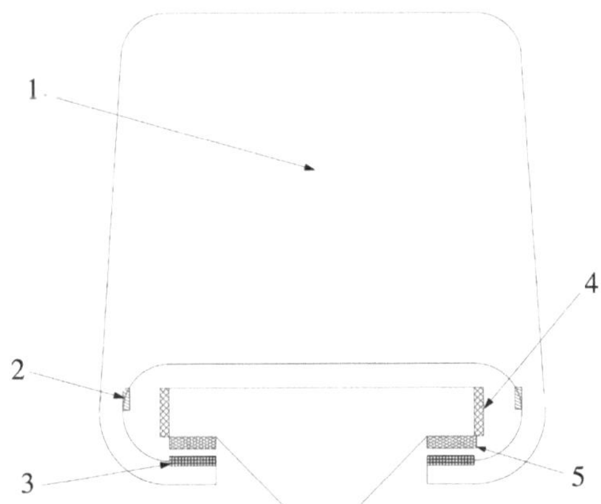

Fig. 9.21. Scheme of the electromagnetic levitation system (levitation and guidance separated): 1 vehicle, 2 guidance magnets, 3 levitation and propulsion magnets, 4 rail guidance windings, 5 iron-core rail windings

Fig. 9.22. The Transrapid magnetic levitation train in Shanghai (left) and the first author before take-off (right)

9.4 Power transformers

As it is well-known, power transformers are used to produce an alternating current with low intensity and high voltage starting from an alternating current with high intensity and low voltage, and viceversa.

In its most common form a transformer is constituted by two windings, wrapped around an iron core. A time-dependent current through the primary winding generates a time-varying magnetic field in the core, and by mutual induction this field induces a voltage in the secondary winding.

The ratio between the voltage in the secondary winding and the voltage in the primary winding is proportional to the ratio between their respective winding numbers. Therefore, tweaking on the number of turns makes it possible to tune the electromotive force at the exit of a transformer, hence to reduce the resistive loss in the conducting wires employed for the transmission of electric power from the power plant to the user.

In 1831 M. Faraday was the first to discover the electromagnetic induction between coils, but the first transformer for commercial use, based on an alternating current for creating the flux variations necessary for induction, was designed by W. Stanley in 1886 (see Figure 9.23).

Nowadays, transformers are in general polyphase; the most common used in electric power distributions are three-phase transformers (see Figures 9.24 and 9.25).

Looking in more detail, their structure can be rather complicated, as they include the coils and the core, an oil tank for refrigeration and insulation, pressing and clamping plates around the coils, shields at the walls of the tank.

With the increase of the power, stray losses become more and more important, lowering the efficiency of transformers and producing significative local overheating in the metallic components: by Joule effect the current in the windings generates a resistive heating, eddy currents are responsible of the increasing of the heat in the core, and the rise of the temperature also occurs in plates and shields. The reduction of these unintended effects, which influences the reliability and decreases the operating life of transformers, is one of the most important points for optimal design.

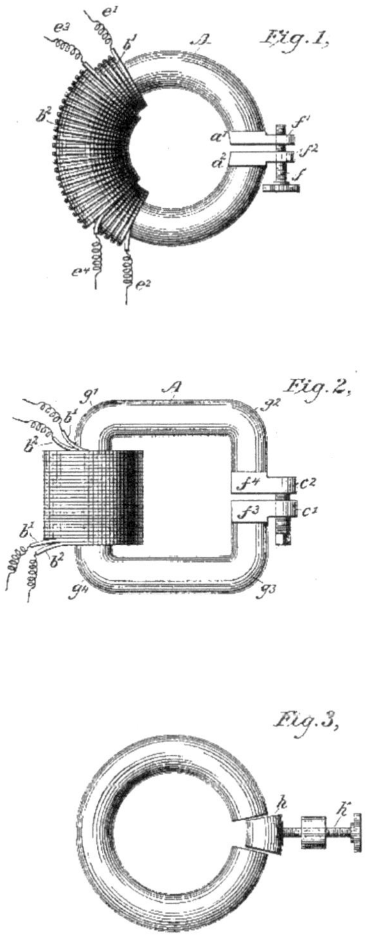

Fig. 9.23. The Stanley transformer (1886, U.S. Patent and Trademark Office)

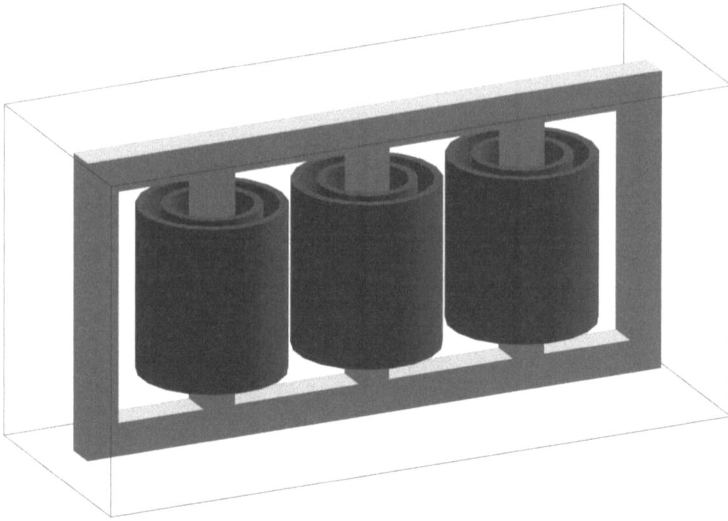

Fig. 9.24. A three-phase, five-leg transformer

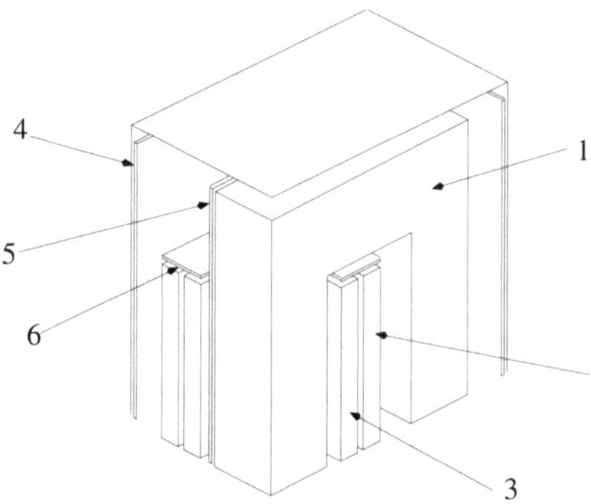

Fig. 9.25. A detail of a transformer: 1 iron core, 2 high winding, 3 low winding, 4 shielding, 5 clamping plate, 6 pressing plate

A thorough modeling of a power transformer based on the eddy current equations, possibly coupled with the heat equation, has been proposed by many authors.

Chen et al. [80] have considered the finite element approximation of a three-phase, five-leg transformer, using an approach based on a suitably modified $\mathbf{T}_C - \psi$ method, preferred to others since the current potential $\mathbf{T}_C$ only appears in the conductor, a rather small region in power transformers. Considering, for the sake of definiteness, the magnetic boundary value problem $\mathbf{H} \times \mathbf{n} = \mathbf{0}$ on $\partial\Omega$, the magnetic field is represented as

$$\mathbf{H} = \begin{cases} \operatorname{grad} \psi_I + \mathbf{K}_I & \text{in } \Omega_I \\ \mathbf{T}_C + \operatorname{grad} \psi_C + \mathbf{K}_C & \text{in } \Omega_C \,, \end{cases}$$

where $\mathbf{K}_I \in \mathcal{H}_{\mu_I}(\partial\Omega, \Gamma; \Omega_I)$, $\mathbf{K}_C \in H(\text{curl}; \Omega_C)$ and satisfies $\mathbf{K}_C \times \mathbf{n}_C + \mathbf{K}_I \times \mathbf{n}_I = \mathbf{0}$ on Γ, and $\psi_I = \psi_C$ and $\mathbf{T}_C \times \mathbf{n}_C = \mathbf{0}$ on Γ. Note that the terms $\mathbf{K}_I$ and $\mathbf{K}_C$ are related to the topology of the domain, and cannot be discarded if the conductor Ω_C is not simply-connected.

The windings are modeled as a conducting region, and the conductivity σ assumes two different constant values in the windings and in the core. Therefore, Ω_C has several connected components, and at least one of them (the core) is not simply-connected.

The magnetic permeability μ is assumed to be a positive constant in the whole transformer, and, moreover, a Lorenz-like gauge

$$\operatorname{div} \mathbf{T}_C = -i\omega\sigma\mu_C\psi_C \qquad \text{in } \Omega_C$$

is used, in order to resort to the problem

$$-\Delta\mathbf{T}_C + i\omega\mu_C\sigma\mathbf{T}_C = \operatorname{curl} \mathbf{J}_{e,C} - \operatorname{curl} \operatorname{curl} \mathbf{K}_C - i\omega\mu_C\sigma\mathbf{K}_C \,.$$

Chen et al. [80] have computed the distribution of the eddy current density on the metallic parts of the tank, with the aim of determining the best design for their shape. In particular, they have shown that the maximum reduction of stray losses is obtained with the use of vertical magnetic shunts instead of aluminium screens.

The same approach has been proposed by Tang et al. [230] for computing the magnetic field on pressing plates, yoke-clamps and the tank wall when both windings and heavy current leads are taken into account, with the aim of optimizing the shape and dimension of a copper shield employed for minimizing overheating in the wall.

The $\mathbf{T}_C - \psi$ method has been also used by Preis et al. [196], for computing the electromagnetic field and the temperature rise in bushing adapters carrying eddy currents due to high-current low-voltage leads. In that paper the coupling with the heat equation has been taken into account, considering the Joule effect due to the eddy currents and assuming that the conductivity σ is a positive scalar function depending in a nonlinear way on the temperature. An iterative coupling strategy has been proposed, with the choice of recalculating the magnetic field in the conductor only, in order to avoid the heavy computations related to the solution of the complete electromagnetic problem.

However, as explained in Chapter 6, the $\mathbf{T}_C - \psi$ approach has various flaws, as it does not include the Faraday equation on the "cutting" surfaces, that are indeed present in multi-phase power transformers, as the conductor is not simply-connected. Moreover, the use of nodal elements, which is natural for this formulation, is not the best choice for the approximation of a problem where the conductor is a polyhedral non-convex domain, as the convergence of the approximate solutions could fail. Therefore, a better modeling of power transformers could be achieved by resorting to one of the formulations that are more suitable for problems with complex topology (for instance, those presented in Chapters 4 and 5).

Tang et al. [231] and Ho et al. [131], employing the $(\mathbf{A}, V_C) - \mathbf{A}_I$ formulation with Coulomb gauge, has considered the numerical simulation of transient eddy current fields in power transformers connected to voltage source through electric circuits. The windings are modeled as coils included in the insulator Ω_I, and there the current density is written in the form $\mathbf{J}_{e,I}(t) = \frac{N}{S}\,\mathbb{I}(t)\,\mathbf{t}$, where $\mathbf{t}$ is the unit coil direction vector, tangential to the windings, N is the number of turns in the coil, S is its cross section and $\mathbb{I}(t)$ is the unknown current intensity. Outside the windings and in the conductor the applied current density is assumed to vanish.

Since the total induced electromotive force in windings can be expressed in term of the vector magnetic potential as follows

$$\text{emf} = \frac{N}{S}\frac{d}{dt}\int_{\Omega_w} \mathbf{A}\cdot\mathbf{t}\,,$$

where Ω_w is the space filled by windings, the problem is closed by adding a suitable equation, representing a circuit model of power transformers. Summing up, for the magnetic boundary value problem $\mathbf{H}\times\mathbf{n} = \mathbf{0}$ on $\partial\Omega$ and a domain Ω of simple shape, the global problem reads

$$\begin{cases} \operatorname{curl}(\boldsymbol{\mu}^{-1}\operatorname{curl}\mathbf{A}) - \mu_*^{-1}\operatorname{grad}\operatorname{div}\mathbf{A} & \\ \qquad +\boldsymbol{\sigma}\frac{\partial}{\partial t}\mathbf{A} + \boldsymbol{\sigma}\operatorname{grad}V_C - \frac{N}{S}\,\mathbb{I}\,\mathbf{t} = \mathbf{0} & \text{in } \Omega\times(0,T) \\ \operatorname{div}(\boldsymbol{\sigma}\frac{\partial}{\partial t}\mathbf{A}_C + \boldsymbol{\sigma}\operatorname{grad}V_C) = 0 & \text{in } \Omega_C\times(0,T) \\ (\boldsymbol{\sigma}\frac{\partial}{\partial t}\mathbf{A}_C + \boldsymbol{\sigma}\operatorname{grad}V_C)\cdot\mathbf{n}_C = 0 & \text{on } \Gamma\times(0,T) \\ \mathbf{A}\cdot\mathbf{n} = 0 & \text{on } \partial\Omega\times(0,T) \\ (\boldsymbol{\mu}^{-1}\operatorname{curl}\mathbf{A})\times\mathbf{n} = \mathbf{0} & \text{on } \partial\Omega\times(0,T) \\ \frac{N}{S}\frac{d}{dt}\int_{\Omega_w}\mathbf{A}\cdot\mathbf{t} + L\frac{d\mathbb{I}}{dt} + R\mathbb{I} = V & \text{on } (0,T), \end{cases} \tag{9.10}$$

plus suitable initial conditions for $\mathbf{A}$ and $\mathbb{I}$, where L is the inductance, R the resistance, and V the voltage source of the circuit modeling the transformer. Here, the current density $\frac{N}{S}\,\mathbb{I}\,\mathbf{t}$ is intended to vanish outside the windings Ω_w.

In particular, Tang et al. [231] and Ho et al. [131] have computed the transient performance of a single-phase, three-leg power transformer, focusing in particular on the magnetic flux density on the surface of the iron core and in the windings. For a three-phase, five-leg transformer they have determined the distribution of torsional forces acting on the coils, in order to control their robustness and stability, as well as the eddy current losses in the clamping plates, checking in this way the efficiency of a magnetic by-pass plate designed for reducing overheating.

Alternative approaches to the simulation of power transformers coupled with circuits are those described in Chapter 8. In these cases, the windings are modeled as conductors, each one with two electric ports, where the voltage drop can be assigned. The topology of the insulator Ω_I becomes more complex, but the total number of degrees of freedom in formulation (8.15) is much less than that in (9.10), therefore its numerical accuracy and efficiency should be better.

9.5 Defect detection

In this section we present non-destructive evaluation (NDE) techniques based on electromagnetic methods. The aim of these techniques is to detect and characterize defects in conducting materials without causing damage.

In eddy current non-destructive testing, a coil supplied with alternating current is placed near the conductive object being inspected. Thus eddy currents are induced and generate a secondary magnetic field. Flaws are detected by monitoring changes in this magnetic field. The measured quantity is usually the impedance of the exciting coil or of a receipt coil. One can distinguish between absolute probes, where the same coil is source and receiver, and differential probes with source coils and receptive coils. This kind of techniques is widely employed in aerospace, transportation energy, nuclear and other industries. It is used, for instance, for the in-service inspection of steam generator tubes in power plants, or for the verification of aging aircraft structures. For instance, problem 27 of the TEAM workshop concerns the detection of deep flaws in a riveted assembly of aluminum sheets with a filler between the sheets, held together by titanium fasteners (see Figure 9.26). This is an example of the kind of structures that are subjected to control in aeronautical industry.

Numerical simulations are needed for the design of the probe coil and for the qualification of monitoring device. In order to develop more reliable instruments it is important to clarify the correlation between the flaws and the changes in the generated eddy currents, and numerical simulations can be used in place of more expensive experiments. From the numerical point of view a great effort has been made in the last years to obtain efficient computational schemes to simulate probe-defect interactions: see, for instance, the pioneering works of Lord [169] and Ida and Lord [136], Rasolonjanahary et al. [204], Badics et al. [33], Sabariego and Dular [214], [215], Henneron

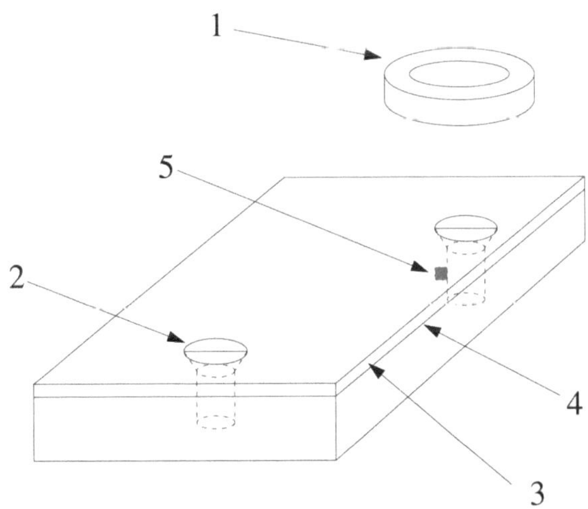

Fig. 9.26. A typical screwed assembled structure that needs to be to controlled (from problem 27 of the TEAM workshop): 1 sensor, 2 rivets, 3 sheets, 4 filler, 5 flaw to detect

et al. [125], Krebs et al. [156]; see also the review by Auld and Moulder [32] and the references therein.

Let us consider a computational domain Ω that contains the conductor Ω_C to be inspected. The electromagnetic fields generated in the case of no-flaw can be assumed to be known; in particular, they satisfy the eddy current approximation of Maxwell equations

$$\begin{aligned} \operatorname{curl} \mathbf{H}^u - \boldsymbol{\sigma}^u \mathbf{E}^u &= \mathbf{J}_e && \text{in } \Omega \\ \operatorname{curl} \mathbf{E}^u + i\omega \boldsymbol{\mu}^u \mathbf{H}^u &= \mathbf{0} && \text{in } \Omega\,, \end{aligned} \tag{9.11}$$

where the superscript u denotes unflawed quantities.

The impedance of the unflawed configuration is given by

$$Z^u = \frac{1}{|I^0|^2} \left(\int_{\Omega_C} \boldsymbol{\sigma}^u \mathbf{E}^u \cdot \overline{\mathbf{E}^u} + i\omega \int_{\Omega} \boldsymbol{\mu}^u \mathbf{H}^u \cdot \overline{\mathbf{H}^u} \right) , \tag{9.12}$$

where I^0 is the applied current intensity (see, e.g., Jackson [137], p. 266).

Let us now assume that a flaw Ω_f (typically, a non-conducting region) is present in Ω_C, the object to be inspected. The conductivity and permeability of the flaw are different than those of the host material, thus the electromagnetic fields in the flawed arrangement satisfy

$$\begin{aligned} \operatorname{curl} \mathbf{H}^f - \boldsymbol{\sigma}^f \mathbf{E}^f &= \mathbf{J}_e && \text{in } \Omega \\ \operatorname{curl} \mathbf{E}^f + i\omega \boldsymbol{\mu}^f \mathbf{H}^f &= \mathbf{0} && \text{in } \Omega\,, \end{aligned}$$

where the superscript f denotes the quantities when the flaw is present. For the sake of simplicity, in the following we assume that the permeability is the same in the unflawed and flawed arrangements, and that the conductivity of the flaw is equal to 0, while outside it coincides with $\boldsymbol{\sigma}^u$, namely,

$$\boldsymbol{\sigma}^f = \begin{cases} \boldsymbol{\sigma}^u & \text{in } \Omega_C \setminus \Omega_f \\ \mathbf{0} & \text{in } \Omega_f\,. \end{cases}$$

We also set $\boldsymbol{\mu} := \boldsymbol{\mu}^f = \boldsymbol{\mu}^u$ and $\boldsymbol{\sigma} := \boldsymbol{\sigma}^u$. Hence when the flaw is present the impedance is given by

$$Z^f = \frac{1}{|I^0|^2} \left(\int_{\Omega_C \setminus \Omega_f} \boldsymbol{\sigma} \mathbf{E}^f \cdot \overline{\mathbf{E}^f} + i\omega \int_{\Omega} \boldsymbol{\mu} \mathbf{H}^f \cdot \overline{\mathbf{H}^f} \right) . \tag{9.13}$$

The direct approach computes the difference between the impedance values with and without flaws, determined as in (9.12) and (9.13). The change of the observed quantity is very small, usually under 1% of the unflawed impedance value, so very high accuracy is needed in the finite element approximation of the fields. Sometimes the unperturbed configuration has symmetries that make it possible to simplify the computation, however the approximation of the perturbed problem requires a very fine three-dimensional mesh and can be extremely expensive for complicated geometrical situations.

In order to minimize this computational cost a different approach is based on perturbation techniques, that lead to the computation of the impedance variation as an integral on the flaw, thus making it possible to obtain sufficiently high accuracy by refining the mesh only in the suspect region. These techniques can be described as follows. The computational domain Ω is assumed to have a simply-connected boundary $\partial\Omega$, and to be given by $\Omega = \Omega_C \cup \Omega_{\rm coil} \cup \Omega_I$, where the conductive coil $\Omega_{\rm coil}$ is an absolute probe, with two contacts $\partial\Omega_{\rm coil} \cap \partial\Omega = \Gamma_E \cup \Gamma_J$. The excitation is given by a current intensity I^0 (and not by a current density as in (9.11)). From the Ampère law, this assigned current intensity can be expressed as $I^0 = \int_{\Gamma_J} {\rm curl}\, \mathbf{H}^f \cdot \mathbf{n} = \int_{\Gamma_J} {\rm curl}\, \mathbf{H}^u \cdot \mathbf{n}$. Finally, let us also assume that the no-flux boundary conditions

$$
\begin{aligned}
&\mathbf{E}^u \times \mathbf{n} = \mathbf{0} \;\text{ on } \Gamma_E \cup \Gamma_J \\
&\boldsymbol{\mu}\mathbf{H}^u \cdot \mathbf{n} = 0 \;\text{ on } \partial\Omega
\end{aligned}
$$

are satisfied (see (8.2)).

For each $\mathbf{v} \in H({\rm curl}; \Omega)$ such that ${\rm curl}\, \mathbf{v} = 0$ in $\Omega \setminus (\overline{\Omega_C} \cup \overline{\Omega_{\rm coil}})$ one has, as in Section 8.1,

$$
\begin{aligned}
-i\omega \int_\Omega \boldsymbol{\mu}\mathbf{H}^u \cdot \overline{\mathbf{v}} &= \int_\Omega {\rm curl}\, \mathbf{E}^u \cdot \overline{\mathbf{v}} \\
&= \int_\Omega \mathbf{E}^u \cdot {\rm curl}\, \overline{\mathbf{v}} - \int_{\partial\Omega} \mathbf{E}^u \times \mathbf{n} \cdot \overline{\mathbf{v}} \\
&= \int_{\Omega_C \cup \Omega_{\rm coil}} \boldsymbol{\sigma}^{-1} {\rm curl}\, \mathbf{H}^u \cdot {\rm curl}\, \overline{\mathbf{v}} - V^u \int_{\Gamma_J} {\rm curl}\, \overline{\mathbf{v}} \cdot \mathbf{n}\,.
\end{aligned}
\tag{9.14}
$$

Analogously, assuming that also when the flaw is present the electromagnetic fields satisfy no-flux boundary conditions, one finds

$$
\begin{aligned}
-i\omega \int_\Omega \boldsymbol{\mu}\mathbf{H}^f \cdot \overline{\mathbf{v}} &= \int_\Omega {\rm curl}\, \mathbf{E}^f \cdot \overline{\mathbf{v}} \\
&= \int_\Omega \mathbf{E}^f \cdot {\rm curl}\, \overline{\mathbf{v}} - \int_{\partial\Omega} \mathbf{E}^f \times \mathbf{n} \cdot \overline{\mathbf{v}} \\
&= \int_{(\Omega_C \setminus \Omega_f) \cup \Omega_{\rm coil}} \boldsymbol{\sigma}^{-1} {\rm curl}\, \mathbf{H}^f \cdot {\rm curl}\, \overline{\mathbf{v}} \\
&\qquad + \int_{\Omega_f} \mathbf{E}^f \cdot {\rm curl}\, \overline{\mathbf{v}} - V^f \int_{\Gamma_J} {\rm curl}\, \overline{\mathbf{v}} \cdot \mathbf{n}\,.
\end{aligned}
\tag{9.15}
$$

Taking $\mathbf{v} = \overline{\mathbf{H}^f}$ in (9.14) and $\mathbf{v} = \overline{\mathbf{H}^u}$ in (9.15), and recalling that $I^0 = \int_{\Gamma_J} {\rm curl}\, \mathbf{H}^f \cdot \mathbf{n} = \int_{\Gamma_J} {\rm curl}\, \mathbf{H}^u \cdot \mathbf{n}$, one has

$$
V^u I^0 = \int_{\Omega_C \cup \Omega_{\rm coil}} \boldsymbol{\sigma}^{-1} {\rm curl}\, \mathbf{H}^u \cdot {\rm curl}\, \mathbf{H}^f + i\omega \int_\Omega \boldsymbol{\mu}\mathbf{H}^u \cdot \mathbf{H}^f
$$

and

$$
\begin{aligned}
V^f I^0 = &\int_{\Omega_f} \mathbf{E}^f \cdot {\rm curl}\, \mathbf{H}^u + \int_{(\Omega_C \setminus \Omega_f) \cup \Omega_{\rm coil}} \boldsymbol{\sigma}^{-1} {\rm curl}\, \mathbf{H}^f \cdot {\rm curl}\, \mathbf{H}^u \\
&+ i\omega \int_\Omega \boldsymbol{\mu}\mathbf{H}^f \cdot \mathbf{H}^u\,,
\end{aligned}
$$

hence

$$
\begin{aligned}
(V^f - V^u) I^0 &= \int_{\Omega_f} \mathbf{E}^f \cdot {\rm curl}\, \mathbf{H}^u - \int_{\Omega_f} \boldsymbol{\sigma}^{-1} {\rm curl}\, \mathbf{H}^u \cdot {\rm curl}\, \mathbf{H}^f \\
&= \int_{\Omega_f} \mathbf{E}^f \cdot {\rm curl}\, \mathbf{H}^u\,,
\end{aligned}
$$

since ${\rm curl}\, \mathbf{H}^f = \mathbf{0}$ in Ω_f.

Proceeding as in (8.9) and taking into account the relations (9.12) and (9.13) we see that $Z^u = V^u/I^0$ and $Z^f = V^f/I^0$, hence the impedance variation is given by

$$(Z^f - Z^u) = \frac{V^f - V^u}{I^0} = \frac{1}{(I^0)^2} \int_{\Omega_f} \mathbf{E}^f \cdot \operatorname{curl} \mathbf{H}^u \,.$$

For the finite element approximation different formulations have been considered. More often the problem is formulated in terms of a magnetic vector potential $\mathbf{A}_C$ and an electric scalar potential V_C in the conductor, and a magnetic scalar potential ψ_I in the insulator (see Section 6.3). The first three-dimensional simulations, due to Ida and Lord [136], use this formulation and isoparametric hexahedral finite elements for the approximation of the impedance, given a source current density $\mathbf{J}_e$. They verify the validity of the formulation for a problem related to the non-destructive testing of a nuclear plant steam generator. In particular, the test problem consists of an Inconel 600 (a nickel-chromium allotrope of iron) tube and a carbon steel support plate; two conical defects are located on the outer surface of the tube.

Rasolonjanahary et al. [204] consider the problem of the inspection of flaws in a riveted aircraft structure. They use a $(\mathbf{A}_C, V_C)$ formulation in the conductor to be inspected and the magnetic scalar potential ψ_I in the surrounding non-conducting region. They compare the results obtained using in the flaw domain a formulation in terms of either the vector magnetic potential $\mathbf{A}$ or the scalar magnetic potential ψ, and obtain more accurate results with the former formulation.

Badics et al. [33] use perturbation techniques in terms of the vector potential $\mathbf{A}_C$ and the electric scalar potential V_C in the conductor, the vector magnetic potential $\mathbf{A}$ in the flaw and the magnetic scalar potential ψ_I in the air region. They compute the solution of the unperturbed model and the field distortion due to a flaw. Setting $\mathbf{H} := \mathbf{H}^f - \mathbf{H}^u$ and $\mathbf{E} := \mathbf{E}^f - \mathbf{E}^u$ in Ω and assuming as before that the permeability is the same in the unflawed and flawed arrangements and that the conductivity of the flaw is equal to 0, one has

$$\begin{aligned} \operatorname{curl} \mathbf{H} - \boldsymbol{\sigma}^f \mathbf{E} &= \mathbf{J}_* && \text{in } \Omega \\ \operatorname{curl} \mathbf{E} + i\omega\boldsymbol{\mu}\mathbf{H} &= \mathbf{0} && \text{in } \Omega \,, \end{aligned}$$

where

$$\mathbf{J}_* := (\boldsymbol{\sigma}^f - \boldsymbol{\sigma})\mathbf{E}^u = \begin{cases} \mathbf{0} & \text{in } \Omega \setminus \Omega_f \\ -\boldsymbol{\sigma}\mathbf{E}^u & \text{in } \Omega_f \,. \end{cases}$$

Hence, for calculating the impedance perturbation it is necessary to know the value of the unflawed electromagnetic fields only in the flaw.

The efficiency of the formulation is verified by solving the TEAM workshop problem 15 (see Figure 9.27 for a sketched description of the arrangement). The test specimen is an aluminum alloy plate of 260 mm of side (2d) and 12.22 mm of thickness (t). The defect is a parallelepipedal slot of length 12.60 mm (2c), depth 5 mm (h) and width 0.28 mm (w). The probe is a circular air-cored coil with inner radius 9.34 mm, outer radius 18.40 mm and 9 mm of length. The frequency of the applied current is equal to 7 kHz and the lift-off of the probe is 2.03 mm.

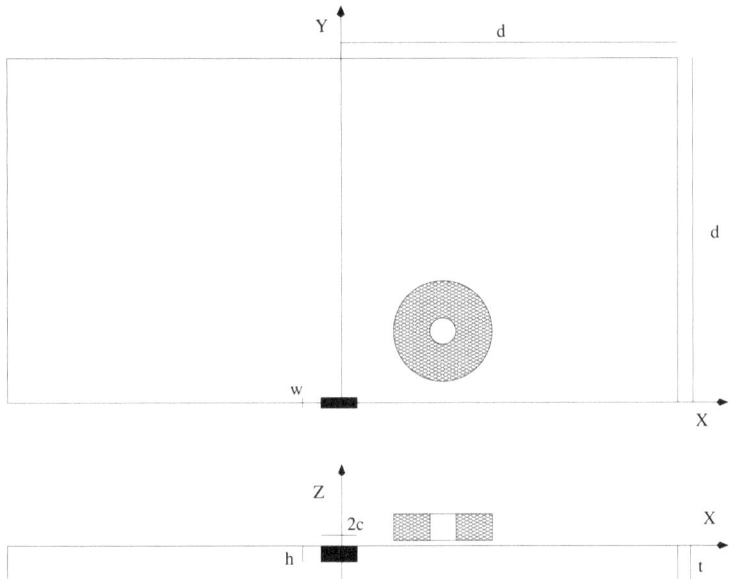

Fig. 9.27. Sketch of problem 15 of the TEAM workshop

They also solve two other test problems where the host specimen is a stainless steel tube and the probe is again a circular air-cored coil. The defect is a long axial slot in the first case and a short circumferential slot in the second case.

Sabariego and Dular [214], [215] propose a perturbation approach using the **H**-based formulation and edge finite elements for its numerical approximation. The perturbed field is not computed in the whole domain but only in a reduced domain surrounding the flaw. The mesh of this reduced subdomain is independent of the mesh used for the unflawed problem and can be adapted to the dimensions of the flaw. To demonstrate the performance of the proposed method they consider the second eddy current benchmark problem proposed by the World Federation of NDE Centers, an Inconel tube with a defect on the outer surface and a circular coil that scans the inner surface.

In Henneron et al. [125] the $(\mathbf{A}, V_C)$ formulation and the $(\mathbf{T}_C, \psi)$ formulation are considered, and are compared in terms of numerical results and computational time. The numerical experiments concern the qualification process of testing devices used in heat exchanger tubes. Two different probes are considered. The first one consists in two coils with a ferrite core used as source and receptive coils, simultaneously. The second one has a source coil and two different receptive coils. Also Krebs et al. [156] use these two formulations to obtain a-posteriori error estimators within an adaptive meshing procedure.

Appendix

A.1 Functional spaces and notation

In this section we introduce some definitions and notations which have been often used in the preceding chapters. For more detailed presentations and descriptions of the functional spaces useful in electromagnetism, see, e.g., Nečas [184], Adams [2], Adams and Fournier [3], Girault and Raviart [111], Dautray and Lions [94], Cessenat [76], Bossavit [59], Monk [179].

Let us consider an open, connected and bounded set Ω contained in $\mathbb{R}^3$, with a Lipschitz continuous boundary $\partial\Omega$, and let Σ be a Lipschitz continuous surface contained in $\partial\Omega$. The unit outward normal vector on $\partial\Omega$ is indicated by $\mathbf{n}$.

We denote by $C_0^\infty(\Omega)$ the space of infinitely differentiable functions having compact support in Ω, i.e., vanishing outside an open set $\Omega' \subset \Omega$ which has a positive distance from the boundary $\partial\Omega$ of Ω.

The space of functions that are bounded in Ω (with the possible exception of a subset of measure equal to 0) is denoted by $L^\infty(\Omega)$, with norm $\| \cdot \|_{L^\infty(\Omega)}$.

For a function defined in Ω, for any $s \in \mathbb{R}$ the Sobolev space of order s is denoted by $H^s(\Omega)$. The norm in this space is indicated by $\| \cdot \|_{s,\Omega}$. For functions defined on the surface Σ, for any $t \in [-1, 1]$ the Sobolev space of order t is denoted by $H^t(\Sigma)$, with norm $\| \cdot \|_{t,\Sigma}$. As usual, the space $H^0(\Omega)$ (respectively, $H^0(\Sigma)$) is always denoted by $L^2(\Omega)$ (respectively, $L^2(\Sigma)$). We also recall that the space $H^{1/2}(\Sigma)$ is the space of the values on Σ (or, equivalently, the traces on Σ) of functions belonging to $H^1(\Omega)$, and that $H^{-t}(\Sigma)$ is the dual space of $H^t(\Sigma)$, $t \in [0, 1]$.

The space $H^1_{0,\Sigma}(\Omega)$ consists of those $H^1(\Omega)$-functions that have a vanishing value on Σ. When $\Sigma = \partial\Omega$, we simply write $H^1_0(\Omega)$ instead of $H^1_{0,\partial\Omega}(\Omega)$.

For a real number s with $0 \leq s \leq 1$ and a domain $D = \Omega$ or $D = \Sigma$, Σ a closed surface, we are also interested in the space $H^s(D)/\mathbb{C}$, whose elements are identified

A. Alonso Rodríguez, A. Valli: Eddy Current Approximation of Maxwell Equations.

if they differ by a (complex) constant. This space is endowed with the following norm

$$\|v\|_{H^s(D)/\mathbb{C}} := \begin{cases} \left(\int_D |v - v_D|^2 \right)^{1/2} & \text{for } s = 0 \\ \left(\int_D |v - v_D|^2 + |v|^2_{s,D} \right)^{1/2} & \text{for } 0 < s \leq 1 \, , \end{cases}$$

where $v_D := (\text{meas}\, D)^{-1} \int_D v$ is the mean value of v and $|v|_{s,D}$ denotes the semi-norm of v in $H^s(D)$. In particular, if the function $v \in H^s(D)/\mathbb{C}$ is chosen with $v_D = 0$, we have $\|v\|_{H^s(D)/\mathbb{C}} = \|v\|_{s,D}$. Moreover, due to the Poincaré inequality (see, e.g., Dautray and Lions [94], Chap. IV, Sect. 7, Prop. 2), we have that in $H^1(\Omega)/\mathbb{C}$ the semi-norm $\|\operatorname{grad} v\|_{0,\Omega}$ is indeed an equivalent norm.

When considering vector-valued functions $\mathbf{v} : \Omega \to \mathbb{R}^3$, the space

$$H(\text{div}; \Omega) := \left\{ \mathbf{v} \in (L^2(\Omega))^3 \mid \operatorname{div} \mathbf{v} \in L^2(\Omega) \right\}$$

is often used. It is endowed with the graph norm, i.e.,

$$\|\mathbf{v}\|_{H(\text{div};\Omega)} := (\|\mathbf{v}\|^2_{0,\Omega} + \|\operatorname{div} \mathbf{v}\|^2_{0,\Omega})^{1/2} \, .$$

Similarly, we employ the space

$$H(\text{curl}; \Omega) := \left\{ \mathbf{v} \in (L^2(\Omega))^3 \mid \operatorname{curl} \mathbf{v} \in (L^2(\Omega))^3 \right\} ,$$

with the norm

$$\|\mathbf{v}\|_{H(\text{curl};\Omega)} := (\|\mathbf{v}\|^2_{0,\Omega} + \|\operatorname{curl} \mathbf{v}\|^2_{0,\Omega})^{1/2} \, .$$

Moreover, we set

$$\begin{aligned} H_{0,\Sigma}(\text{div}; \Omega) &:= \{ \mathbf{v} \in H(\text{div}; \Omega) \mid \mathbf{v} \cdot \mathbf{n} = 0 \text{ on } \Sigma \} \\ H_{0,\Sigma}(\text{curl}; \Omega) &:= \{ \mathbf{v} \in H(\text{curl}; \Omega) \mid \mathbf{v} \times \mathbf{n} = \mathbf{0} \text{ on } \Sigma \} \\ H^0(\text{div}; \Omega) &:= \{ \mathbf{v} \in H(\text{div}; \Omega) \mid \operatorname{div} \mathbf{v} = 0 \text{ in } \Omega \} \\ H^0(\text{curl}; \Omega) &:= \{ \mathbf{v} \in H(\text{curl}; \Omega) \mid \operatorname{curl} \mathbf{v} = \mathbf{0} \text{ in } \Omega \} \\ H^0_{0,\Sigma}(\text{div}; \Omega) &:= H_{0,\Sigma}(\text{div}; \Omega) \cap H^0(\text{div}; \Omega) \\ H^0_{0,\Sigma}(\text{curl}; \Omega) &:= H_{0,\Sigma}(\text{curl}; \Omega) \cap H^0(\text{curl}; \Omega) \, . \end{aligned}$$

When $\Sigma = \partial\Omega$, we simply write $H_0(\text{div}; \Omega)$ instead of $H_{0,\partial\Omega}(\text{div}; \Omega)$, and similarly for the other cases.

For a symmetric matrix $\boldsymbol{\eta} = \boldsymbol{\eta}(\mathbf{x})$, uniformly positive definite in Ω and with entries belonging to $L^\infty(\Omega)$, we also set

$$\begin{aligned} H(\boldsymbol{\eta}, \text{div}; \Omega) &:= \{ \mathbf{v} \in (L^2(\Omega))^3 \mid \operatorname{div}(\boldsymbol{\eta}\mathbf{v}) \in L^2(\Omega) \} \\ H_{0,\Sigma}(\boldsymbol{\eta}, \text{div}; \Omega) &:= \{ \mathbf{v} \in H(\boldsymbol{\eta}, \text{div}; \Omega) \mid \boldsymbol{\eta}\mathbf{v} \cdot \mathbf{n} = 0 \text{ on } \Sigma \} \, . \end{aligned}$$

To characterize the tangential boundary value of a vector belonging to $H(\text{curl}; \Omega)$ we need some preliminaries concerning tangential differential operators. The standard definition of the tangential gradient and the tangential curl on the flat surface $\{x_3 = 0\}$, having chosen the unit outward normal vector $\mathbf{n} = (0, 0, 1)$, is

$$\mathrm{grad}_\tau \phi = (\partial_1 \phi, \partial_2 \phi, 0) \ , \ \ \mathrm{Curl}_\tau \, \phi = \mathrm{grad}_\tau \phi \times \mathbf{n} = (\partial_2 \phi, -\partial_1 \phi, 0) \, .$$

Starting from this, if $\Sigma \subset \partial\Omega$ is a closed surface (namely, a surface without boundary), using local coordinates (see, e.g., Nečas [184]) it is possible to define the operators grad_τ and Curl_τ for functions belonging to $H^1(\Sigma)$, and one obtains $\mathrm{grad}_\tau \phi \in \mathcal{L}^2_t(\Sigma)$ and $\mathrm{Curl}_\tau \phi \in \mathcal{L}^2_t(\Sigma)$, where

$$\mathcal{L}^2_t(\Sigma) := \{\mathbf{v} \in (L^2(\Sigma))^3 \, | \, \mathbf{v} \cdot \mathbf{n} = 0\} \, .$$

By a duality argument, the adjoint operators

$$\mathrm{div}_\tau : \mathcal{L}^2_t(\Sigma) \to H^{-1}(\Sigma)$$

and

$$\mathrm{curl}_\tau : \mathcal{L}^2_t(\Sigma) \to H^{-1}(\Sigma)$$

are also introduced, and the Laplace–Beltrami operator

$$\Delta_\tau : H^1(\Sigma) \to H^{-1}(\Sigma)$$

is defined as $\Delta_\tau := \mathrm{div}_\tau \, \mathrm{grad}_\tau = - \mathrm{curl}_\tau \, \mathrm{Curl}_\tau$.

These operators can be restricted to other spaces: in particular, one can verify that the following relation holds

$$\mathrm{grad}_\tau \phi = (\mathbf{n} \times \mathrm{grad} \, \widetilde{\phi} \times \mathbf{n})_{|\Sigma} \ , \ \ \phi \in H^{3/2}(\Sigma) \, ,$$

where we have set $H^{3/2}(\Sigma) := \{\varphi_{|\Sigma} \, | \, \varphi \in H^2(\Omega)\}$ and $\widetilde{\phi}$ is any extension of ϕ to $H^2(\Omega)$. Similarly, it holds

$$\mathrm{Curl}_\tau \phi = \mathrm{grad}_\tau \phi \times \mathbf{n} \ , \ \ \phi \in H^{3/2}(\Sigma) \, .$$

Clearly, in this case we have $\mathrm{grad}_\tau \phi \in H^{1/2}_T(\Sigma)$, where

$$H^{1/2}_T(\Sigma) := \{(\mathbf{n} \times \mathbf{v} \times \mathbf{n})_{|\Sigma} \, | \, \mathbf{v} \in (H^1(\Omega))^3\} \, , \tag{A.1}$$

and $\mathrm{Curl}_\tau \phi \in H^{1/2}_\times(\Sigma)$, where

$$H^{1/2}_\times(\Sigma) := \{(\mathbf{v} \times \mathbf{n})_{|\Sigma} \, | \, \mathbf{v} \in (H^1(\Omega))^3\} \, , \tag{A.2}$$

and moreover for each $\boldsymbol{\lambda} \in H^{1/2}_T(\Sigma)$ we have $(\boldsymbol{\lambda} \times \mathbf{n}) \in H^{1/2}_\times(\Sigma)$, and viceversa for each $\boldsymbol{\lambda} \in H^{1/2}_\times(\Sigma)$ we have $(\boldsymbol{\lambda} \times \mathbf{n}) \in H^{1/2}_T(\Sigma)$. Let us also note that the spaces $H^{1/2}_T(\Sigma)$ and $H^{1/2}_\times(\Sigma)$ are both equal to the space

$$H^{1/2}_t(\Sigma) := \{\boldsymbol{\lambda} \in H^{1/2}(\Sigma) \, | \, \boldsymbol{\lambda} \cdot \mathbf{n} = 0\}$$

if Σ is a smooth surface, while a characterization of them for a polyhedral domain is given in Buffa and Ciarlet [70].

The dual operators div_τ and curl_τ now read (all the integrals should be intended as duality pairings)

$$\int_\Sigma (\mathrm{div}_\tau \boldsymbol{\lambda}) \, \phi := - \int_\Sigma \boldsymbol{\lambda} \cdot \mathrm{grad}_\tau \phi \ , \ \ \boldsymbol{\lambda} \in (H^{1/2}_T(\Sigma))' \ , \ \phi \in H^{3/2}(\Sigma)$$

$$\int_\Sigma (\mathrm{curl}_\tau \boldsymbol{\lambda})\, \phi := \int_\Sigma \boldsymbol{\lambda} \cdot \mathbf{Curl}_\tau \phi \ , \ \ \boldsymbol{\lambda} \in (H_\times^{1/2}(\Sigma))' \ , \ \phi \in H^{3/2}(\Sigma) \, ,$$

and clearly one has that $\mathrm{div}_\tau \boldsymbol{\lambda} \in (H^{3/2}(\Sigma))'$ and $\mathrm{curl}_\tau \boldsymbol{\lambda} \in (H^{3/2}(\Sigma))'$. It can also be checked that $\boldsymbol{\lambda} \in (H_\times^{1/2}(\Sigma))'$ if and only if $(\boldsymbol{\lambda} \times \mathbf{n}) \in (H_T^{1/2}(\Sigma))'$, and moreover that $\mathrm{div}_\tau(\boldsymbol{\lambda} \times \mathbf{n}) = \mathrm{curl}_\tau \boldsymbol{\lambda}$ for each $\boldsymbol{\lambda} \in (H_\times^{1/2}(\Sigma))'$.

We can now state the result concerning the characterization of the space of tangential traces on Σ or of tangential components on Σ of functions belonging to $H(\mathrm{curl};\Omega)$. In Buffa and Ciarlet [69], Buffa et al. [71] it has been proved that the space of tangential traces $(\mathbf{v} \times \mathbf{n})_{|\Sigma}$ on Σ for $\mathbf{v} \in H(\mathrm{curl};\Omega)$ is given by

$$H^{-1/2}(\mathrm{div}_\tau;\Sigma) := \{\boldsymbol{\lambda} \in (H_T^{1/2}(\Sigma))' \mid \mathrm{div}_\tau \boldsymbol{\lambda} \in H^{-1/2}(\Sigma)\} \, , \qquad \text{(A.3)}$$

with the graph norm

$$\|\boldsymbol{\lambda}\|_{H^{-1/2}(\mathrm{div}_\tau;\Sigma)} := (\|\boldsymbol{\lambda}\|^2_{(H_T^{1/2}(\Sigma))'} + \|\,\mathrm{div}_\tau \boldsymbol{\lambda}\|^2_{-1/2,\Sigma})^{1/2} \, ,$$

while the space of tangential components $(\mathbf{n} \times \mathbf{v} \times \mathbf{n})_{|\Sigma}$ on Σ for $\mathbf{v} \in H(\mathrm{curl};\Omega)$ is given by

$$H^{-1/2}(\mathrm{curl}_\tau;\Sigma) := \{\boldsymbol{\lambda} \in (H_\times^{1/2}(\Sigma))' \mid \mathrm{curl}_\tau \boldsymbol{\lambda} \in H^{-1/2}(\Sigma)\} \, , \qquad \text{(A.4)}$$

with the graph norm

$$\|\boldsymbol{\lambda}\|_{H^{-1/2}(\mathrm{curl}_\tau;\Sigma)} := (\|\boldsymbol{\lambda}\|^2_{(H_\times^{1/2}(\Sigma))'} + \|\,\mathrm{curl}_\tau \boldsymbol{\lambda}\|^2_{-1/2,\Sigma})^{1/2} \, .$$

It can be also shown that two these spaces are in duality, and that one has $\boldsymbol{\lambda} \in H^{-1/2}(\mathrm{curl}_\tau;\Sigma)$ if and only if $(\boldsymbol{\lambda} \times \mathbf{n}) \in H^{-1/2}(\mathrm{div}_\tau;\Sigma)$. Moreover, it holds $\mathrm{div}_\tau(\boldsymbol{\lambda} \times \mathbf{n}) = \mathrm{curl}_\tau \boldsymbol{\lambda}$ for each $\boldsymbol{\lambda} \in H^{-1/2}(\mathrm{curl}_\tau;\Sigma)$. Let us finally note that, when Σ is a smooth surface, these trace spaces can be described as

$$H^{-1/2}(\mathrm{div}_\tau;\Sigma) := \{\boldsymbol{\lambda} \in H^{-1/2}(\Sigma) \mid \boldsymbol{\lambda} \cdot \mathbf{n} = 0, \mathrm{div}_\tau \boldsymbol{\lambda} \in H^{-1/2}(\Sigma)\}$$
$$H^{-1/2}(\mathrm{curl}_\tau;\Sigma) := \{\boldsymbol{\lambda} \in H^{-1/2}(\Sigma) \mid \boldsymbol{\lambda} \cdot \mathbf{n} = 0, \mathrm{curl}_\tau \boldsymbol{\lambda} \in H^{-1/2}(\Sigma)\}$$

(see Paquet [190], Alonso and Valli [6], Cessenat [76]).

In this functional framework it is thus possible to extend the operators $\mathbf{grad}_\tau$ and $\mathbf{Curl}_\tau$ on $H^{1/2}(\Sigma)$ as

$$\int_\Sigma \mathbf{grad}_\tau \phi \cdot \boldsymbol{\lambda} = -\int_\Sigma (\mathrm{div}_\tau \boldsymbol{\lambda})\, \phi \ , \ \ \boldsymbol{\lambda} \in H^{-1/2}(\mathrm{div}_\tau;\Sigma),$$

and

$$\int_\Sigma \mathbf{Curl}_\tau \phi \cdot \boldsymbol{\lambda} = \int_\Sigma (\mathrm{curl}_\tau \boldsymbol{\lambda})\, \phi \ , \ \ \boldsymbol{\lambda} \in H^{-1/2}(\mathrm{curl}_\tau;\Sigma) \, ,$$

obtaining by duality $\mathbf{grad}_\tau \phi \in H^{-1/2}(\mathrm{curl}_\tau;\Sigma)$ and $\mathbf{Curl}_\tau \phi \in H^{-1/2}(\mathrm{div}_\tau;\Sigma)$. Again, $\mathbf{Curl}_\tau \phi = \mathbf{grad}_\tau \phi \times \mathbf{n}$ for each $\phi \in H^{1/2}(\Sigma)$. In particular, we have also obtained

$$\int_\Sigma \mathbf{grad}_\tau(\varphi_{|\Sigma}) \cdot \mathbf{u} \times \mathbf{n} = -\int_\Sigma \mathrm{div}_\tau(\mathbf{u} \times \mathbf{n})\, \varphi_{|\Sigma} \qquad \text{(A.5)}$$

for each $\mathbf{u} \in H(\mathrm{curl};\Omega)$ and $\varphi \in H^1(\Omega)$.

For each $\mathbf{u} \in H(\text{curl};\Omega)$, $\mathbf{v} \in H_{0,\partial\Omega\setminus\Sigma}(\text{curl};\Omega)$ it can be proved that the following formula of integration by parts holds true

$$\int_\Omega \text{curl}\,\mathbf{u}\cdot\mathbf{v} = \int_\Omega \mathbf{u}\cdot\text{curl}\,\mathbf{v} - \int_\Sigma (\mathbf{u}\times\mathbf{n})\cdot\mathbf{v} \qquad \text{(A.6)}$$

(where the last integral is indeed the duality pairing between $(\mathbf{u}\times\mathbf{n}) \in H^{-1/2}(\text{div}_\tau;\Sigma)$ and $(\mathbf{n}\times\mathbf{v}\times\mathbf{n}) \in H^{-1/2}(\text{curl}_\tau;\Sigma)$).

Taking $\mathbf{v} = \text{grad}\,\varphi$, with $\varphi \in H^1_{0,\partial\Omega\setminus\Sigma}(\Omega)$, from (A.6), (A.5) and the Gauss divergence theorem it follows

$$\begin{aligned}\textstyle\int_\Sigma \text{curl}\,\mathbf{u}\cdot\mathbf{n}\,\varphi_{|\Sigma} &= \textstyle -\int_\Omega (\text{div}\,\text{curl}\,\mathbf{u})\,\varphi + \int_\Sigma \text{curl}\,\mathbf{u}\cdot\mathbf{n}\,\varphi_{|\Sigma}\\ &= \textstyle\int_\Omega \text{curl}\,\mathbf{u}\cdot\text{grad}\,\varphi = \int_\Omega \mathbf{u}\cdot\text{curl}\,\text{grad}\,\varphi - \int_\Sigma (\mathbf{u}\times\mathbf{n})\cdot\text{grad}\,\varphi\\ &= \textstyle -\int_\Sigma (\mathbf{u}\times\mathbf{n})\cdot\text{grad}_\tau(\varphi_{|\Sigma}) = \int_\Sigma \text{div}_\tau(\mathbf{u}\times\mathbf{n})\,\varphi_{|\Sigma}\,,\end{aligned}$$

hence

$$\text{curl}\,\mathbf{u}\cdot\mathbf{n} = \text{div}_\tau(\mathbf{u}\times\mathbf{n}) \ \text{ on } \Sigma \qquad \text{(A.7)}$$

for each $\mathbf{u} \in H(\text{curl};\Omega)$.

We finally recall that the following trace inequalities hold true (in the second, third and fourth inequality we are assuming that Σ is a closed surface)

$$\|\phi_{|\Sigma}\|_{1/2,\Sigma} \le \kappa\,\|\phi\|_{1,\Omega} \quad \forall\,\phi \in H^1(\Omega) \qquad \text{(A.8)}$$

$$\|(\mathbf{v}\cdot\mathbf{n})_{|\Sigma}\|_{-1/2,\Sigma} \le \kappa\,\|\mathbf{v}\|_{H(\text{div};\Omega)} \quad \forall\,\mathbf{v} \in H(\text{div};\Omega) \qquad \text{(A.9)}$$

$$\|(\mathbf{v}\times\mathbf{n})_{|\Sigma}\|_{H^{-1/2}(\text{div}_\tau;\Sigma)} \le \kappa\,\|\mathbf{v}\|_{H(\text{curl};\Omega)} \quad \forall\,\mathbf{v} \in H(\text{curl};\Omega) \qquad \text{(A.10)}$$

$$\|(\mathbf{n}\times\mathbf{v}\times\mathbf{n})_{|\Sigma}\|_{H^{-1/2}(\text{curl}_\tau;\Sigma)} \le \kappa\,\|\mathbf{v}\|_{H(\text{curl};\Omega)} \quad \forall\,\mathbf{v} \in H(\text{curl};\Omega)\,, \qquad \text{(A.11)}$$

where $\kappa > 0$ is a suitable constant only depending on Ω and Σ. Moreover, in all these cases there exist linear continuous extension operators form the trace space to the corresponding space of functions defined in Ω.

A.2 Nodal and edge finite elements

We present in this section a brief description of the finite element spaces used for the approximation of the spaces $H^1(\Omega)$ and $H(\text{curl};\Omega)$. A more comprehensive presentation can be found, e.g., in Ciarlet [83], Quarteroni and Valli [199], Monk [179].

Let $\Omega \subset \mathbb{R}^3$ be a Lipschitz polyhedral domain and let us consider a finite decomposition of Ω given by

$$\overline{\Omega} = \bigcup_{K\in\mathcal{T}_h} K\,,$$

where, denoting by $\text{int}(K)$ and $\text{diam}(K)$ the internal part and the diameter of K, respectively,

- each K is a (closed) polyhedron with positive volume;
- if K_1 and K_2 are distinct elements in $\mathcal{T}_h$ then $\mathrm{int}(K_1) \cap \mathrm{int}(K_2) = \emptyset$;
- if K_1 and K_2 are distinct elements in $\mathcal{T}_h$ and $F = K_1 \cap K_2 \neq \emptyset$, then F is a common face, side, or vertex of K_1 and K_2;
- $\mathrm{diam}(K) \leq h$ for each $K \in \mathcal{T}_h$.

Under these conditions, $\mathcal{T}_h$ is called a triangulation of Ω. In the sequel we will consider triangulations where each element K can be obtained as an affine trasformation of a reference element $\hat{K}$, i.e., $K = T_K(\hat{K})$, where T_K is an invertible affine map $T_K(\hat{\mathbf{x}}) = B_K\hat{\mathbf{x}} + \mathbf{b}_K$, B_K being a non-singular matrix. The reference element can be the tetrahedron $\hat{K}$ of vertices $(0,0,0)$, $(1,0,0)$, $(0,1,0)$, $(0,0,1)$ or the cube $\hat{K} = [0,1]^3$.

Finite element spaces are based on piecewise-polynomial functions, hence some notations for polynomial spaces are necessary. Let us denote by $\mathbb{P}_k$, $k \geq 0$, the space of polynomials of degree less than or equal to k in the three variables x_1, x_2, x_3, and by $\widetilde{\mathbb{P}}_k$ the space of homogeneous polynomials of degree k. Let $\mathbb{Q}_{l,m,n}$ be the polynomial space given by polynomials of maximum degree l in x_1, m in x_2, and n in x_3. In particular $\mathbb{Q}_k$ denotes the space of polynomials that are of degree less than or equal to k with respect to each variable.

For the definition of finite elements in $H(\mathrm{curl};\Omega)$ the following space of vector polynomials is used

$$R_k := (\mathbb{P}_{k-1})^3 \oplus S_k \,,$$

where $k \geq 1$ and

$$S_k := \{\mathbf{q} \in (\widetilde{\mathbb{P}}_k)^3 \mid \mathbf{q}(\mathbf{x}) \cdot \mathbf{x} = 0\} \,.$$

Another space of incomplete vector polynomials is (for $k \geq 1$)

$$D_k := (\mathbb{P}_{k-1})^3 \oplus \widetilde{\mathbb{P}}_k \, \mathbf{x} \,.$$

We also use polynomial spaces defined on planes and lines. If e is a segment we denote by $\mathbb{P}_k(e)$ the space of polynomials of maximum degree k with respect to the arc length on e. If f is a plane subdomain in $\mathbb{R}^3$, $\mathbb{P}_k(f)$ denotes the space of polynomials of maximum degree k in two variables using an orthogonal coordinate system in the plane.

A.2.1 Grad-conforming finite elements

Let us first recall that a function $\phi : \Omega \to \mathbb{R}$ belongs to $H^1(\Omega)$ if and only if $\phi_{|K} \in H^1(K)$ for each $K \in \mathcal{T}_h$ and for each common face $f = K_1 \cap K_2$, $K_1, K_2 \in \mathcal{T}_h$, the value of $\phi_{|K_1}$ and $\phi_{|K_2}$ on f is the same.

Therefore, for any $k \geq 1$ the space

$$L_h^k := \{\phi_h \in C^0(\Omega) \mid \phi_{h|K} \in \mathbb{P}_k \quad \forall \, K \in \mathcal{T}_h\}$$

is a subspace of $H^1(\Omega)$.

In order to identify a basis of this finite dimensional subspace it is necessary to choose a set of degrees of freedom that are unisolvent on $\mathbb{P}_k$, namely, such that their values uniquely determine a polynomial belonging to $\mathbb{P}_k$.

Let us first assume that $\mathcal{T}_h$ is a triangulation of Ω composed by tetrahedra. Following Monk [179], for a regular enough function ϕ we consider the following set of degrees of freedom on a generic tetrahedron K:

- vertex degrees of freedom

$$m_v(\phi) := \{\phi(\mathbf{a}) \text{ for all vertices } \mathbf{a} \text{ of } K\} ;$$

- edge degrees of freedom (for $k \geq 2$)

$$m_e(\phi) := \left\{ \frac{1}{\text{length}(e)} \int_e \phi\, q\, ds \;\forall\, q \in \mathbb{P}_{k-2}(e) \text{ for all edges } e \text{ of } K \right\} ;$$

- face degrees of freedom (for $k \geq 3$)

$$m_f(\phi) := \left\{ \frac{1}{\text{area}(f)} \int_f \phi\, q\, dS \;\forall\, q \in \mathbb{P}_{k-3}(f) \text{ for all faces } f \text{ of } K \right\} ;$$

- volume degrees of freedom (for $k \geq 4$)

$$m_K(\phi) := \left\{ \frac{1}{\text{volume}(K)} \int_K \phi\, q\, dV \;\forall\, q \in \mathbb{P}_{k-4} \right\} .$$

It is easy to check that the total number of degrees of freedom in a tetrahedron concides with the dimension of $\mathbb{P}_k$; moreover it can be verified that a polynomial $\phi \in \mathbb{P}_k$ is vanishing in K provided that all its degrees of freedom are equal to 0. Hence these degrees of freedom are unisolvent on $\mathbb{P}_k$.

It can be also proved that, if all vertex, edge and face degrees of freedom of $\phi \in \mathbb{P}_k$ vanish for a particular face f of a tetrahedron, then $\phi = 0$ on that face. This means that, using these degrees of freedom for identifying a piecewise-polynomial functions that locally belongs to $\mathbb{P}_k$, we define a continuous function, hence an element of L_h^k. A basis of L_h^k is thus given by the collection of those functions that are locally in $\mathbb{P}_k$ and that have one degree of freedom equal to 1 and all the others equal to 0.

Remark A.1. A different and more often used set of degrees of freedom, consisting of the values of the function on different points of the tetrahedron, can be employed in order to describe these finite element spaces. Being expressed in terms of point values, this kind of finite dimensional spaces are often called *nodal* finite elements (see, e.g., Ciarlet [83], Quarteroni and Valli [199]). For instance, if $k = 2$ the values of the function ϕ at the vertices $\mathbf{a}_i$, $1 \leq i \leq 4$, and in the middle point of each edge constitutes another set of grad-conforming and unisolvent set of degrees of freedom; if $k = 3$ an analogous set of conditions is given by the values of ϕ at the 20 different points of the form $\frac{1}{3}\mathbf{a}_i + \frac{1}{3}\mathbf{a}_j + \frac{1}{3}\mathbf{a}_k$, with $1 \leq i, j, k \leq 4$.

Here we have preferred to adopt the vertex, edge, face and volume degrees of freedom for the sake of similarity with the curl-conforming finite elements introduced in Section A.2.2. □

For any $\phi \in H^{3/2+\delta}(K)$, $\delta > 0$, we can now define an interpolation operator π_K by requiring that

$$m_v(\phi - \pi_K\phi) = m_e(\phi - \pi_K\phi) = m_f(\phi - \pi_K\phi) = m_K(\phi - \pi_K\phi) = 0$$

for all the vertices, edges and faces of K; the corresponding global interpolation operator

$$\pi_h : H^{3/2+\delta}(\Omega) \to L_h^k$$

is defined by $(\pi_h\phi)_{|K} := \pi_K\phi_{|K}$ for each $K \in \mathcal{T}_h$. Note that the assumption on the regularity of ϕ ensures that ϕ is continuous in K, hence vertex values are well-defined.

We recall that a family of triangulations $\mathcal{T}_h$ is called regular if there exists a constant $\sigma > 0$ such that

$$\max_{K\in\mathcal{T}_h} \frac{h_K}{\rho_K} \leq \sigma \qquad \forall\, h > 0\,,$$

where

$$\rho_K := \sup\{\mathrm{diam}(R) \mid R \text{ is a ball contained in } K\}\,.$$

The following interpolation error estimate holds:

Theorem A.2. *Let $\mathcal{T}_h$ be a regular family of triangulations of Ω. Then if $\phi \in H^{s+1}(\Omega)$, $1/2 + \delta \leq s \leq k$, there exists a constant $C > 0$, independent of h, such that*

$$\|\phi - \pi_h\phi\|_{0,\Omega} + h\|\phi - \pi_h\phi\|_{1,\Omega} \leq Ch^{s+1}\|\phi\|_{s+1,\Omega}\,.$$

It is also possible to construct a finite element space analogous to L_h^k when considering a triangulation of Ω consisting of parallelepipeds. In this case one works with piecewise-polynomial functions ϕ_h such that $\phi_{h|K} \circ T_K \in \mathbb{Q}_k$. The space of nodal finite elements for a mesh composed by parallelepipeds and for $k \geq 1$ is

$$\widetilde{L}_h^k := \{\phi_h \in C^0(\Omega) \mid \phi_{h|K} \circ T_K \in \mathbb{Q}_k \quad \forall\, K \in \mathcal{T}_h\}\,.$$

On the reference element $\hat{K}$ the degrees of freedom, unisolvent on $\mathbb{Q}_k$, are:

- vertex degrees of freedom

$$m_v(\hat{\phi}) := \left\{\hat{\phi}(\hat{\mathbf{a}}) \text{ for all vertices } \hat{\mathbf{a}} \text{ of } \hat{K}\right\}\,;$$

- edge degrees of freedom (for $k \geq 2$)

$$m_e(\hat{\phi}) := \left\{\int_{\hat{e}} \hat{\phi}\,\hat{q}\,ds\; \forall\, \hat{q} \in \mathbb{P}_{k-2}(\hat{e}) \text{ for all edges } \hat{e} \text{ of } \hat{K}\right\}\,;$$

- face degrees of freedom (for $k \geq 2$)

$$m_f(\hat{\phi}) := \left\{\int_{\hat{f}} \hat{\phi}\,\hat{q}\,dS\; \forall\, \hat{q} \in \mathbb{Q}_{k-2}(\hat{f}) \text{ for all faces } \hat{f} \text{ of } \hat{K}\right\}\,;$$

- volume degrees of freedom (for $k \geq 2$)

$$m_K(\hat{\phi}) := \left\{ \int_{\hat{K}} \hat{\phi}\,\hat{q}\, dV \ \forall \hat{q} \in \mathbb{Q}_{k-2} \right\} .$$

The degrees of freedom on a general element K can be obtained from those on $\hat{K}$ using the transformation $\phi \circ T_K = \hat{\phi}$.

For these finite element spaces the interpolation error estimate described in Theorem A.2 still holds.

Remark A.3. Let us assume $k \geq 0$. A finite element subspace of $L^2(\Omega)$ is easily defined by

$$C_h^k := \{ q_h \in L^2(\Omega) \mid q_{h|K} \in \mathbb{P}_k \ \forall K \in \mathcal{T}_h \}$$

when the elements $K \in \mathcal{T}_h$ are tetrahedra, and by

$$\widetilde{C}_h^k := \{ q_h \in L^2(\Omega) \mid q_{h|K} \circ T_K \in \mathbb{Q}_k \ \forall K \in \mathcal{T}_h \}$$

when the elements $K \in \mathcal{T}_h$ are parallelepipeds.

If $P_{0,h} : L^2(\Omega) \to C_h^k$ denotes the $L^2(\Omega)$-projection, then one has

$$\| \phi - P_{0,h}\phi \|_{0,\Omega} \leq C h^{s+1} \| \phi \|_{s+1,\Omega} ,$$

for all $\phi \in H^{s+1}(\Omega)$, $0 \leq s \leq k$. The same holds true for the $L^2(\Omega)$-projection $\widetilde{P}_{0,h} : L^2(\Omega) \to \widetilde{C}_h^k$. □

A.2.2 Curl-conforming finite elements

Here we introduce the finite element spaces used for the approximation of the space $H(\mathrm{curl};\Omega)$. We present the two families of elements proposed by Nédélec in [185] and [186], which are also called *edge* elements.

We start by considering a triangulation of Ω composed by tetrahedra. For $k \geq 1$, the first family is defined as

$$N_h^k := \{ \mathbf{z}_h \in H(\mathrm{curl};\Omega) \mid \mathbf{z}_{h|K} \in R_k \quad \forall K \in \mathcal{T}_h \} .$$

We have the following set of degrees of freedom:

- edge degrees of freedom

$$m_e(\mathbf{z}) := \left\{ \int_e \mathbf{z} \cdot \boldsymbol{\tau}\, q\, ds \ \forall q \in \mathbb{P}_{k-1}(e) \text{ for all edges } e \text{ of } K \right\} ;$$

- face degrees of freedom (for $k \geq 2$)

$$m_f(\mathbf{z}) := \left\{ \int_f \mathbf{z} \times \boldsymbol{\nu} \cdot \mathbf{q}\, dS \ \forall \mathbf{q} \in (\mathbb{P}_{k-2}(f))^2 \text{ for all faces } f \text{ of } K \right\} ;$$

- volume degrees of freedom (for $k \geq 3$)

$$m_K(\mathbf{z}) := \left\{ \int_K \mathbf{z} \cdot \mathbf{q} \, dV \ \forall \mathbf{q} \in (\mathbb{P}_{k-3})^3 \right\} .$$

Here $\boldsymbol{\tau}$ denotes a unit vector with the direction of e, while $\boldsymbol{\nu}$ is the unit normal vector on f.

The total number of degrees of freedom on a tetrahedron K is equal to the dimension of R_k, and it can be shown that, if all the degrees of freedom are 0, then a polynomial $\mathbf{z} \in R_k$ is identically vanishing in K. Hence this set of degrees of freedom is unisolvent on R_k.

We recall that, if K_1, K_2 are two different elements of $\mathcal{T}_h$ with a common face $f = K_1 \cap K_2$, defining $\mathbf{z} \in L^2(K_1 \cup K_2)$ by

$$\mathbf{z} = \begin{cases} \mathbf{z}_1 \text{ in } K_1 \\ \mathbf{z}_2 \text{ in } K_2 , \end{cases}$$

where $\mathbf{z}_1 \in H(\mathrm{curl}; K_1)$ and $\mathbf{z}_2 \in H(\mathrm{curl}; K_2)$, it follows $\mathbf{z} \in H(\mathrm{curl}; K_1 \cup K_2)$ provided that $\mathbf{z}_1 \times \boldsymbol{\nu} = \mathbf{z}_2 \times \boldsymbol{\nu}$ on f.

It can also be proved that, if a vector function $\mathbf{z} \in R_k$ has all its degrees of freedom vanishing on a face f of K and on the three edges contained in f, then the tangential component of $\mathbf{z}$ vanishes on f. This means that, using these degrees of freedom for identifying a piecewise-polynomial functions that locally belongs to R_k, we obtain an element of $H(\mathrm{curl}; \Omega)$, hence an element of N_h^k.

We can introduce a natural interpolation operator. Assuming that $\mathbf{z}$ is sufficiently regular, the interpolant $\mathbf{r}_h\mathbf{z} \in N_h^k$ is the unique function in N_h^k that has the same degrees of freedom of $\mathbf{z}$, that is

$$m_e(\mathbf{z} - \mathbf{r}_h\mathbf{z}) = m_f(\mathbf{z} - \mathbf{r}_h\mathbf{z}) = m_h(\mathbf{z} - \mathbf{r}_h\mathbf{z}) = 0$$

for all the edges, faces and tetrahedra of $\mathcal{T}_h$.

We notice that the degrees of freedom $m_e(\mathbf{z})$ are not defined for a general function in $H(\mathrm{curl}; \Omega)$. However they are well-defined if $\mathbf{z} \in (H^s(\Omega))^3$ for some $s > 1/2$ and $\mathrm{curl}\, \mathbf{z} \in (L^p(\Omega))^3$ for some $p > 2$. For the proof of this result see Monk [179] (see also Amrouche et al. [27], where a more general result is proved).

In particular, the following interpolation error estimate holds (see Alonso and Valli [9]):

Theorem A.4. *Let $\mathcal{T}_h$ be a regular family of triangulations of Ω. If $\mathbf{z} \in (H^s(\Omega))^3$ and* $\mathrm{curl}\, \mathbf{z} \in (H^s(\Omega))^3$, $1/2 < s \leq k$, *then there exists a constant $C > 0$, independent of h, such that*

$$\|\mathbf{z} - \mathbf{r}_h\mathbf{z}\|_{0,\Omega} + \|\mathrm{curl}(\mathbf{z} - \mathbf{r}_h\mathbf{z})\|_{0,\Omega} \leq Ch^s(\|\mathbf{z}\|_{s,\Omega} + \|\mathrm{curl}\, \mathbf{z}\|_{s,\Omega}) .$$

It is also possible to define an analogous family of curl-conforming finite element spaces when considering a triangulation of Ω consisting of parallelepipeds. For the reference element $\hat{K} = [0,1]^3$ the polynomial space is $\mathbb{Q}_{k-1,k,k} \times \mathbb{Q}_{k,k-1,k} \times \mathbb{Q}_{k,k,k-1}$, with $k \geq 1$, and the degrees of freedom are given on edges $\hat{e}$ with unit tangent $\hat{\boldsymbol{\tau}}$, on faces $\hat{f}$ with unit normal $\hat{\boldsymbol{\nu}}$ and in the interior of $\hat{K}$. In particular we consider the following set of degrees of freedom, unisolvent on $\mathbb{Q}_{k-1,k,k} \times \mathbb{Q}_{k,k-1,k} \times \mathbb{Q}_{k,k,k-1}$:

- edge degrees of freedom

$$m_e(\hat{\mathbf{z}}) := \left\{ \int_{\hat{e}} \hat{\mathbf{z}} \cdot \hat{\boldsymbol{\tau}} \hat{q} \, ds \; \forall \, \hat{q} \in \mathbb{P}_{k-1}(\hat{e}) \text{ for all edges } \hat{e} \text{ of } \hat{K} \right\} ;$$

- face degrees of freedom (for $k \geq 2$)

$$m_f(\hat{\mathbf{z}}) := \Big\{ \textstyle\int_{\hat{f}} \hat{\mathbf{z}} \times \hat{\boldsymbol{\nu}} \cdot \hat{\mathbf{q}} \, dS \; \forall \, \hat{\mathbf{q}} \in \mathbb{Q}_{k-2,k-1}(\hat{f}) \times \mathbb{Q}_{k-1,k-2}(\hat{f}) \text{ for all faces } \hat{f} \text{ of } \hat{K} \Big\} ;$$

- volume degrees of freedom (for $k \geq 2$)

$$m_K(\hat{\mathbf{z}}) := \Big\{ \textstyle\int_{\hat{K}} \hat{\mathbf{z}} \cdot \hat{\mathbf{q}} \, dV \; \forall \, \hat{\mathbf{q}} \in \mathbb{Q}_{k-1,k-2,k-2} \times \mathbb{Q}_{k-2,k-1,k-2} \times \mathbb{Q}_{k-2,k-2,k-1} \Big\} .$$

They are well-defined if $\hat{\mathbf{z}} \in (H^s(\hat{K}))^3$ for some $s > 1/2$ and $\operatorname{curl} \hat{\mathbf{z}} \in (L^p(\hat{K}))^3$ for some $p > 2$. The basis functions on a general element K can be obtained from those on $\hat{K}$ using the transformation $\mathbf{z} \circ T_K = (B_K^T)^{-1} \hat{\mathbf{z}}$. In this way the curl of $\mathbf{z}$ is expressed in terms of the curl of $\hat{\mathbf{z}}$ by

$$\operatorname{curl} \mathbf{z} \circ T_K = \frac{1}{\det(B_K)} B_K \operatorname{curl} \hat{\mathbf{z}} .$$

For $k \geq 1$ we can thus consider the following curl-conforming finite element spaces defined on parallelepipeds

$$\widetilde{N}_h^k := \{ \mathbf{z}_h \in H(\operatorname{curl}; \Omega) \mid \mathbf{z}_{h|K} \circ T_K \in \mathbb{Q}_{k-1,k,k} \times \mathbb{Q}_{k,k-1,k} \times \mathbb{Q}_{k,k,k-1} \; \forall K \in \mathcal{T}_h \} .$$

The interpolation error estimate reported in Theorem A.4 still holds.

We notice that, for the curl-conforming finite elements presented here above, when using elements of degree k the interpolation error in the $L^2(\Omega)$-norm is $O(h^k)$. A second family of curl-conforming elements has been introduced by Nédélec in [186], in order to obtain an $O(h^{k+1})$-error estimate in $L^2(\Omega)$.

Let us first consider the case of a tetrahedral mesh. For $k \geq 1$, the discrete functions locally belong to the polynomial space $(\mathbb{P}_k)^3$, and the degrees of freedom, are the following:

- edge degrees of freedom

$$m_e(\mathbf{z}) := \left\{ \int_e \mathbf{z} \cdot \boldsymbol{\tau} \, q \, ds \; \forall \, q \in \mathbb{P}_k(e) \text{ for all edges } e \text{ of } K \right\} ;$$

- face degrees of freedom (for $k \geq 2$)

$$m_f(\mathbf{z}) := \left\{ \int_f \mathbf{z} \cdot \mathbf{q} \, dS \ \forall \mathbf{q} \in D_{k-1}(f) \text{ for all faces } f \text{ of } K \right\} ;$$

- volume degrees of freedom (for $k \geq 3$)

$$m_K(\mathbf{z}) := \left\{ \int_K \mathbf{z} \cdot \mathbf{q} \, dV \ \forall \mathbf{q} \in D_{k-2} \right\} .$$

Here $D_{k-1}(f)$ is the analogue of D_{k-1} in two dimensions.

This set of degrees of freedom has been proved to be curl-conforming and unisolvent on $(\mathbb{P}_k)^3$. Thus for $k \geq 1$ we can consider the discrete space

$$N^k_{*,h} := \{\mathbf{z}_h \in H(\mathrm{curl}; \Omega) \mid \mathbf{z}_{h|K} \in (\mathbb{P}_k)^3 \quad \forall K \in \mathcal{T}_h\} ,$$

and, for a function $\mathbf{z}$ which is regular enough, we can define the interpolant $\mathbf{r}_{*,h}\mathbf{z} \in N^k_{*,h}$.

Concerning the interpolation error the following estimate holds:

Theorem A.5. *Let $\mathcal{T}_h$ be a regular family of triangulations of Ω. If $\mathbf{z} \in (H^{s+1}(\Omega))^3$, $1 \leq s \leq k$, then there exists a constant $C > 0$, independent of h, such that*

$$\|\mathbf{z} - \mathbf{r}_{*,h}\mathbf{z}\|_{0,\Omega} + h\| \mathrm{curl}(\mathbf{z} - \mathbf{r}_{*,h}\mathbf{z})\|_{0,\Omega} \leq C h^{s+1} |\mathbf{z}|_{s+1,\Omega} .$$

Comparing the interpolation errors in N^k_h and $N^k_{*,h}$ we see that $L^2(\Omega)$-norms of the curl are of the same order with respect to h, while the $L^2(\Omega)$-norms of the fields are $O(h^s)$ for N^k_h and $O(h^{s+1})$ for $N^k_{*,h}$. On the other hand, the number of degrees of freedom of $N^k_{*,h}$ is greater than that of N^k_h.

It is also possible to define a second family of Nédélec curl-conforming finite elements when considering a triangulation of Ω consisting of parallelepipeds. For the reference element $\hat{K} = [0,1]^3$ and $k \geq 1$ the polynomial space is $(\mathbb{Q}_k)^3$ and the degrees of freedom, unisolvent on $(\mathbb{Q}_k)^3$, are given by:

- edge degrees of freedom

$$m_e(\hat{\mathbf{z}}) := \left\{ \int_{\hat{e}} \hat{\mathbf{z}} \cdot \hat{\boldsymbol{\tau}} \, \hat{q} \, ds \ \forall \hat{q} \in \mathbb{P}_k(\hat{e}) \text{ for all edges } \hat{e} \text{ of } \hat{K} \right\} ;$$

- face degrees of freedom (for $k \geq 2$)

$$m_f(\hat{\mathbf{z}}) := \left\{ \int_{\hat{f}} \hat{\mathbf{z}} \cdot \hat{\mathbf{q}} \, dS \ \forall \hat{\mathbf{q}} \in \mathbb{Q}_{k,k-2}(\hat{f}) \times \mathbb{Q}_{k-2,k}(\hat{f}) \text{ for all faces } \hat{f} \text{ of } \hat{K} \right\};$$

- volume degrees of freedom (for $k \geq 2$)

$$m_K(\hat{\mathbf{z}}) := \left\{ \int_{\hat{K}} \hat{\mathbf{z}} \cdot \hat{\mathbf{q}} \, dV \ \forall \hat{\mathbf{q}} \in \mathbb{Q}_{k,k-2,k-2} \times \mathbb{Q}_{k-2,k,k-2} \times \mathbb{Q}_{k-2,k-2,k} \right\} .$$

The corresponding discrete space is given by

$$\widetilde{N}^k_{*,h} := \{\mathbf{z}_h \in H(\mathrm{curl}; \Omega) \mid \mathbf{z}_{h|K} \circ T_K \in (\mathbb{Q}_k)^3 \quad \forall K \in \mathcal{T}_h\} .$$

A.3 Orthogonal decomposition results

We prove in this section some orthogonal decomposition results that are useful for splitting the magnetic field $\mathbf{H}_I$ or the electric field $\mathbf{E}_I$ into the sum of suitable terms (for a more detailed presentation, see also, e.g., Dautray and Lions [95], Saranen [218], [219], Auchmuty [29], Cantarella et al. [73]).

Here the geometrical assumptions on Ω, Ω_C and Ω_I are those of Section 1.3. Moreover, we again assume that the matrix $\boldsymbol{\mu}$ is symmetric and uniformly positive definite in Ω, with entries belonging to $L^\infty(\Omega)$ and that the matrix ε_I is symmetric and uniformly positive definite in Ω_I, with entries belonging to $L^\infty(\Omega_I)$. Finally, the spaces of harmonic fields are introduced in Section 1.4 (see also Section A.4).

A.3.1 First decomposition result

Let us start by introducing the scalar product

$$(\mathbf{w}_I, \mathbf{z}_I)_{\varepsilon_I, \Omega_I} := \int_{\Omega_I} \varepsilon_I \mathbf{w}_I \cdot \overline{\mathbf{z}_I} \,,$$

where $\overline{\mathbf{z}_I}$ indicates the complex conjugate of $\mathbf{z}_I$, and by denoting with the symbol $\perp^{\varepsilon_I}$ the orthogonality with respect to this scalar product. Instead, $\perp$ denotes the orthogonality with respect to the standard $L^2(\Omega_I)$-scalar product. We have the following theorem:

Theorem A.6. *Any vector function $\mathbf{z}_I \in (L^2(\Omega_I))^3$ can be written as*

$$\mathbf{z}_I = \varepsilon_I^{-1} \operatorname{curl} \mathbf{q}_I + \operatorname{grad} \varphi_I + \mathbf{h}_I \,, \tag{A.12}$$

where $\mathbf{q}_I \in H_{0,\partial\Omega}(\mathrm{curl}; \Omega_I) \cap H^0_{0,\Gamma}(\mathrm{div}; \Omega_I) \cap \mathcal{H}(\partial\Omega, \Gamma; \Omega_I)^\perp$, $\varphi_I \in H^1_{0,I}(\Omega_I)$ and $\mathbf{h}_I \in \mathcal{H}_{\varepsilon_I}(\Gamma, \partial\Omega; \Omega_I)$, and each term of the decomposition (A.12) *is orthogonal to the others, with respect to the scalar product $(\cdot,\cdot)_{\varepsilon_I,\Omega_I}$.*

Moreover, if $\operatorname{curl} \mathbf{z}_I = \mathbf{0}$ in Ω_I and $\mathbf{z}_I \times \mathbf{n}_I = \mathbf{0}$ on Γ it follows $\mathbf{q}_I = \mathbf{0}$, if $\mathrm{div}(\varepsilon_I \mathbf{z}_I) = 0$ in Ω_I and $\varepsilon_I \mathbf{z}_I \cdot \mathbf{n} = 0$ on $\partial\Omega$ one has $\varphi_I = 0$, and if $\mathbf{z}_I \perp^{\varepsilon_I} \mathcal{H}_{\varepsilon_I}(\Gamma, \partial\Omega; \Omega_I)$ one finds $\mathbf{h}_I = \mathbf{0}$.

Proof. To prove this result, let us start showing how $\mathbf{q}_I$, φ_I and $\mathbf{h}_I$ can be determined in terms of $\mathbf{z}_I$.

First of all, setting

$$Y_I := \left\{ \mathbf{p}_I \in H_{0,\partial\Omega}(\mathrm{curl}; \Omega_I) \cap H_{0,\Gamma}(\mathrm{div}; \Omega_I) \,|\, \mathbf{p}_I \perp \mathcal{H}(\partial\Omega, \Gamma; \Omega_I) \right\} , \tag{A.13}$$

the vector field $\mathbf{q}_I \in Y_I$ is the solution to

$$\int_{\Omega_I} (\varepsilon_I^{-1} \operatorname{curl} \mathbf{q}_I \cdot \operatorname{curl} \overline{\mathbf{p}_I} + \operatorname{div} \mathbf{q}_I \operatorname{div} \overline{\mathbf{p}_I}) = \int_{\Omega_I} \mathbf{z}_I \cdot \operatorname{curl} \overline{\mathbf{p}_I} \tag{A.14}$$

for all $\mathbf{p}_I \in Y_I$. Concerning the solvability of this problem, note that using the compactness of Y_I in $L^2(\Omega_I)$ (see, e.g., Fernandes and Gilardi [104]), the following Poincaré-like inequality is easily proved

$$\int_{\Omega_I} |\mathbf{p}_I|^2 \leq C \int_{\Omega_I} (|\operatorname{curl} \mathbf{p}_I|^2 + |\operatorname{div} \mathbf{p}_I|^2) \qquad \forall\, \mathbf{p}_I \in Y_I\,. \tag{A.15}$$

Therefore, the existence of a unique solution $\mathbf{q}_I$ to (A.14) is a consequence of the Lax–Milgram lemma. It can be also verified that, if $\operatorname{curl} \mathbf{z}_I = \mathbf{0}$ in Ω_I and $\mathbf{z}_I \times \mathbf{n}_I = \mathbf{0}$ on Γ, then the right-hand side of (A.14) vanishes, thus $\mathbf{q}_I = \mathbf{0}$.

Since equation (A.14) is trivially satisfied for each test function belonging to $\mathcal{H}(\partial\Omega, \Gamma; \Omega_I)$, there we can select the test functions $\mathbf{p}_I$ not only in Y_I but also in $H_{0,\partial\Omega}(\operatorname{curl}; \Omega_I) \cap H_{0,\Gamma}(\operatorname{div}; \Omega_I)$, namely, without imposing the orthogonality constraint.

We also see that $\operatorname{div} \mathbf{q}_I = 0$ in Ω_I, as in (A.14) we can choose $\mathbf{p}_I = \operatorname{grad} v_I$, where the function $v_I \in H^1(\Omega_I)$ satisfies $\Delta v_I = \operatorname{div} \mathbf{q}_I$ in Ω_I, $v_I = 0$ on $\partial\Omega$ and $\operatorname{grad} v_I \cdot \mathbf{n}_I = 0$ on Γ, thus obtaining $\int_{\Omega_I} |\operatorname{div} \mathbf{q}_I|^2 = 0$.

We have therefore shown that $\mathbf{q}_I$ satisfies

$$\int_{\Omega_I} \varepsilon_I^{-1} \operatorname{curl} \mathbf{q}_I \cdot \operatorname{curl} \overline{\mathbf{p}_I} = \int_{\Omega_I} \mathbf{z}_I \cdot \operatorname{curl} \overline{\mathbf{p}_I} \tag{A.16}$$

for all $\mathbf{p}_I \in H_{0,\partial\Omega}(\operatorname{curl}; \Omega_I) \cap H_{0,\Gamma}(\operatorname{div}; \Omega_I)$. We can indeed prove something more, namely, that equation (A.16) is satisfied for each test function $\mathbf{p}_I^* \in H_{0,\partial\Omega}(\operatorname{curl}; \Omega_I)$. In fact, denoting by $v_I^* \in H^1(\Omega_I)$ the solution of $\Delta v_I^* = \operatorname{div} \mathbf{p}_I^*$ in Ω_I, $v_I^* = 0$ on $\partial\Omega$ and $\operatorname{grad} v_I^* \cdot \mathbf{n}_I = \mathbf{p}_I^* \cdot \mathbf{n}_I$ on Γ, we have $\mathbf{p}_I = (\mathbf{p}_I^* - \operatorname{grad} v_I^*) \in H_{0,\partial\Omega}(\operatorname{curl}; \Omega_I) \cap H_{0,\Gamma}(\operatorname{div}; \Omega_I)$, and the result follows from the fact that $\operatorname{curl} \mathbf{p}_I^* = \operatorname{curl} \mathbf{p}_I$.

Hence, choosing in (A.16) a test function $\mathbf{p}_I^* \in (C_0^\infty(\Omega_I))^3$, we find by integration by parts that

$$\operatorname{curl}(\varepsilon_I^{-1} \operatorname{curl} \mathbf{q}_I - \mathbf{z}_I) = \mathbf{0} \qquad \text{in } \Omega_I\,;$$

finally, repeating the same computation for $\mathbf{p}_I^* \in H_{0,\partial\Omega}(\operatorname{curl}; \Omega_I)$ gives

$$(\varepsilon_I^{-1} \operatorname{curl} \mathbf{q}_I - \mathbf{z}_I) \times \mathbf{n}_I = \mathbf{0} \qquad \text{on } \Gamma\,.$$

The function $\varphi_I \in H^1_{0,\Gamma}(\Omega_I)$ is such that

$$\int_{\Omega_I} \varepsilon_I \operatorname{grad} \varphi_I \cdot \operatorname{grad} \overline{\eta_I} = \int_{\Omega_I} \varepsilon_I \mathbf{z}_I \cdot \operatorname{grad} \overline{\eta_I} \tag{A.17}$$

for all $\eta_I \in H^1_{0,\Gamma}(\Omega_I)$, and also this problem is uniquely solvable by the Lax–Milgram lemma, as the Poincaré inequality

$$\int_{\Omega_I} |\eta_I|^2 \leq C \int_{\Omega_I} |\operatorname{grad} \eta_I|^2 \tag{A.18}$$

holds in $H^1_{0,\Gamma}(\Omega_I)$ (see, e.g., Dautray and Lions [94], Chap. IV, Sect. 7, Rem. 4). It is readily seen that, if $\operatorname{div}(\varepsilon_I \mathbf{z}_I) = 0$ in Ω_I and $\varepsilon_I \mathbf{z}_I \cdot \mathbf{n} = 0$ on $\partial\Omega$, then the right-hand side of (A.17) vanishes, and consequently $\varphi_I = 0$.

Then, selecting $\eta_I \in C_0^\infty(\Omega_I)$, an integration by parts in (A.17) yields

$$\operatorname{div}[\varepsilon_I(\operatorname{grad}\varphi_I - \mathbf{z}_I)] = 0 \qquad \text{in } \Omega_I ,$$

and the choice $\eta_I \in H^1_{0,\Gamma}(\Omega_I)$ gives

$$\varepsilon_I(\operatorname{grad}\varphi_I - \mathbf{z}_I)\cdot \mathbf{n} = 0 \qquad \text{on } \partial\Omega .$$

Finally, $\mathbf{h}_I \in \mathcal{H}_{\varepsilon_I}(\Gamma, \partial\Omega; \Omega_I)$ satisfies

$$\int_{\Omega_I} \varepsilon_I \mathbf{h}_I \cdot \operatorname{grad} w_{j,I} = \int_{\Omega_I} \varepsilon_I \mathbf{z}_I \cdot \operatorname{grad} w_{j,I} \, , \; \int_{\Omega_I} \varepsilon_I \mathbf{h}_I \cdot \boldsymbol{\pi}_{k,I} = \int_{\Omega_I} \varepsilon_I \mathbf{z}_I \cdot \boldsymbol{\pi}_{k,I}$$

for each $j = 1, \ldots, p_\Gamma$ and $k = 1, \ldots, n_{\partial\Omega}$, the harmonic vector fields $\operatorname{grad} w_{j,I}$ and $\boldsymbol{\pi}_{k,I}$ being the basis functions of the space $\mathcal{H}_{\varepsilon_I}(\Gamma, \partial\Omega; \Omega_I)$. In other words, $\mathbf{h}_I$ can be written as

$$\mathbf{h}_I = \sum_{j=1}^{p_\Gamma} c_{I,j} \operatorname{grad} w_{j,I} + \sum_{k=1}^{n_{\partial\Omega}} d_{k,I} \boldsymbol{\pi}_{k,I} \, ,$$

where $(c_{I,j}, d_{k,I})$ are the solution of the linear system

$$A^\dagger \begin{pmatrix} c_{I,j} \\ d_{I,k} \end{pmatrix} = \begin{pmatrix} \int_{\Omega_I} \varepsilon_I \mathbf{z}_I \cdot \operatorname{grad} w_{g,I} \\ \int_{\Omega_I} \varepsilon_I \mathbf{z}_I \cdot \boldsymbol{\pi}_{i,I} \end{pmatrix} \, , \qquad \text{(A.19)}$$

$g = 1, \ldots, p_\Gamma$, $i = 1, \ldots, n_{\partial\Omega}$, where $A^\dagger := \begin{pmatrix} D^\dagger & B^\dagger \\ (B^\dagger)^T & C^\dagger \end{pmatrix}$ with

$$\begin{aligned} D^\dagger_{gj} &:= \textstyle\int_{\Omega_I} \varepsilon_I \operatorname{grad} w_{j,I} \cdot \operatorname{grad} w_{g,I} \\ B^\dagger_{gk} &:= \textstyle\int_{\Omega_I} \varepsilon_I \boldsymbol{\pi}_{k,I} \cdot \operatorname{grad} w_{g,I} \\ C^\dagger_{ik} &:= \textstyle\int_{\Omega_I} \varepsilon_I \boldsymbol{\pi}_{k,I} \cdot \boldsymbol{\pi}_{i,I} \, . \end{aligned}$$

It is easily proved that the matrix $A^\dagger$ is symmetric and positive definite, as the matrix $\varepsilon_I(\mathbf{x})$ is symmetric and positive definite, uniformly with respect to $\mathbf{x}$, and the functions $\boldsymbol{\pi}_{k,I}$ and $\operatorname{grad} w_{j,I}$ are linearly independent. Therefore (A.17) is uniquely solvable; its right-hand side vanishes if $\mathbf{z}_I \perp^{\varepsilon_I} \mathcal{H}_{\varepsilon_I}(\Gamma, \partial\Omega; \Omega_I)$, so that in that case one has $\mathbf{h}_I = \mathbf{0}$.

The three terms $\varepsilon_I^{-1} \operatorname{curl} \mathbf{q}_I$, $\operatorname{grad}\varphi_I$ and $\mathbf{h}_I$ are orthogonal with respect to the scalar product $(\cdot,\cdot)_{\varepsilon_I,\Omega_I}$: in fact

$$\int_{\Omega_I} \varepsilon_I(\varepsilon_I^{-1} \operatorname{curl} \mathbf{q}_I) \cdot \operatorname{grad} \overline{\varphi_I} = \int_{\partial\Omega\cup\Gamma} \operatorname{curl} \mathbf{q}_I \cdot \mathbf{n}_I \, \overline{\varphi_I} = 0 \, ,$$

as $\operatorname{curl} \mathbf{q}_I \cdot \mathbf{n} = \operatorname{div}_\tau(\mathbf{q}_I \times \mathbf{n}) = \mathbf{0}$ on $\partial\Omega$ and $\varphi_I = 0$ on Γ. Then

$$\int_{\Omega_I} \varepsilon_I \mathbf{h}_I \cdot \operatorname{grad} \overline{\varphi_I} = \int_{\partial\Omega\cup\Gamma} \varepsilon_I \mathbf{h}_I \cdot \mathbf{n}_I \, \overline{\varphi_I} = 0 \, ,$$

as $\operatorname{div}(\varepsilon_I \mathbf{h}_I) = 0$ in Ω_I, $\varepsilon_I \mathbf{h}_I \cdot \mathbf{n}_I = 0$ on $\partial\Omega$ and and $\varphi_I = 0$ on Γ. Finally,

$$\int_{\Omega_I} \varepsilon_I(\varepsilon_I^{-1} \operatorname{curl} \mathbf{q}_I) \cdot \overline{\mathbf{h}_I} = \int_{\partial\Omega\cup\Gamma} \mathbf{n}_I \times \mathbf{q}_I \cdot \overline{\mathbf{h}_I} = 0 \, ,$$

as $\operatorname{curl} \mathbf{h}_I = \mathbf{0}$ in Ω_I, $\mathbf{h}_I \times \mathbf{n}_I = \mathbf{0}$ on Γ and $\mathbf{q}_I \times \mathbf{n} = \mathbf{0}$ on $\partial\Omega$.

The decomposition result (A.12) is then straightforwardly verified. In fact, let us set $\mathbf{U}_I := \mathbf{z}_I - \varepsilon_I^{-1} \operatorname{curl} \mathbf{q}_I - \operatorname{grad} \varphi_I - \mathbf{h}_I$. Since $\operatorname{curl} \mathbf{q}_I \cdot \mathbf{n} = \operatorname{div}_\tau(\mathbf{q}_I \times \mathbf{n}) = 0$ on $\partial\Omega$ and $\operatorname{grad} \varphi_I \times \mathbf{n}_I = \mathbf{0}$ on Γ, from the results proved above we verify at once that $\mathbf{U}_I$ belongs to $\mathcal{H}_{\varepsilon_I}(\Gamma, \partial\Omega; \Omega_I)$. On the other hand, by construction $\mathbf{h}_I$ is the orthogonal projection of $\mathbf{z}_I$ on $\mathcal{H}_{\varepsilon_I}(\Gamma, \partial\Omega; \Omega_I)$ with respect to the scalar product $(\cdot,\cdot)_{\varepsilon_I,\Omega_I}$, hence $\mathbf{U}_I$ is orthogonal to $\mathcal{H}_{\varepsilon_I}(\Gamma, \partial\Omega; \Omega_I)$ with respect to the scalar product $(\cdot,\cdot)_{\varepsilon_I,\Omega_I}$, and the conclusion $\mathbf{U}_I = \mathbf{0}$ follows at once. □

A.3.2 Second decomposition result

Let us define the scalar product

$$(\mathbf{u}_I, \mathbf{v}_I)_{\mu_I,\Omega_I} := \int_{\Omega_I} \boldsymbol{\mu}_I \mathbf{u}_I \cdot \mathbf{v}_I \tag{A.20}$$

and denote by the symbol $\perp^{\mu_I}$ the orthogonality with respect to this scalar product. By interchanging the role of Γ and $\partial\Omega$ and by replacing ε_I with $\boldsymbol{\mu}_I$, by proceeding as in Section A.3.1 it is easy to obtain the following theorem, whose proof is presented for the ease of the reader.

Theorem A.7. *Any given vector function $\mathbf{v}_I \in (L^2(\Omega_I))^3$ can be decomposed into the following sum*

$$\mathbf{v}_I = \boldsymbol{\mu}_I^{-1} \operatorname{curl} \mathbf{Q}_I + \operatorname{grad} \chi_I + \sum_{r=1}^{p_{\partial\Omega}} a_{I,r} \operatorname{grad} z_{r,I} + \sum_{l=1}^{n_\Gamma} b_{I,l} \boldsymbol{\rho}_{l,I} \ , \tag{A.21}$$

where $\mathbf{Q}_I \in H_{0,\Gamma}(\mathrm{curl}; \Omega_I) \cap H^0_{0,\partial\Omega}(\mathrm{div}; \Omega_I) \cap \mathcal{H}(\Gamma, \partial\Omega; \Omega_I)^\perp$ are introduced in (A.22)*, $\chi_I \in H^1_{0,\partial\Omega}(\Omega_I)$ in* (A.23)*, and $a_{I,r}$, $b_{I,l}$, $r = 1, \ldots, p_{\partial\Omega}$, $l = 1, \ldots, n_\Gamma$, in* (A.24)*. Setting $\mathbf{k}_I := \sum_{r=1}^{p_{\partial\Omega}} a_{I,r} \operatorname{grad} z_{r,I} + \sum_{l=1}^{n_\Gamma} b_{I,l} \boldsymbol{\rho}_{l,I}$, each of the three terms $\boldsymbol{\mu}_I^{-1} \operatorname{curl} \mathbf{Q}_I$, $\operatorname{grad} \chi_I$ and $\mathbf{k}_I$ of the decomposition* (A.21) *is orthogonal to the others, with respect to the scalar product $(\cdot,\cdot)_{\mu_I,\Omega_I}$.*

Moreover, if $\operatorname{curl} \mathbf{v}_I = \mathbf{0}$ in Ω_I and $\mathbf{v}_I \times \mathbf{n} = \mathbf{0}$ on $\partial\Omega$ it follows $\mathbf{Q}_I = \mathbf{0}$, if $\operatorname{div}(\boldsymbol{\mu}_I \mathbf{v}_I) = 0$ in Ω_I and $\boldsymbol{\mu}_I \mathbf{v}_I \cdot \mathbf{n}_I = 0$ on Γ one has $\chi_I = 0$, and if $\mathbf{v}_I \perp^{\mu_I} \mathcal{H}_{\mu_I}(\partial\Omega, \Gamma; \Omega_I)$ one finds $a_{I,r} = 0$ and $b_{I,l} = 0$, $r = 1, \ldots, p_{\partial\Omega}$, $l = 1, \ldots, n_\Gamma$.

Proof. For the sake of variety with respect to the proof of Theorem A.6, let us present the result by resorting to the strong formulations. The vector function $\mathbf{Q}_I \in H(\mathrm{curl}; \Omega_I) \cap H(\mathrm{div}; \Omega_I)$ is the solution to

$$\begin{cases} \operatorname{curl}(\boldsymbol{\mu}_I^{-1} \operatorname{curl} \mathbf{Q}_I) = \operatorname{curl} \mathbf{v}_I & \text{in } \Omega_I \\ \operatorname{div} \mathbf{Q}_I = 0 & \text{in } \Omega_I \\ \mathbf{Q}_I \times \mathbf{n}_I = \mathbf{0} & \text{on } \Gamma \\ \mathbf{Q}_I \cdot \mathbf{n} = 0 & \text{on } \partial\Omega \\ (\boldsymbol{\mu}_I^{-1} \operatorname{curl} \mathbf{Q}_I) \times \mathbf{n} = \mathbf{v}_I \times \mathbf{n} & \text{on } \partial\Omega \\ \mathbf{Q}_I \perp \mathcal{H}(\Gamma, \partial\Omega; \Omega_I) \, . \end{cases} \tag{A.22}$$

The existence and uniqueness of the solution $\mathbf{Q}_I$ can be proved by proceeding as was done for problem (A.14). It is readily verified that one has $\mathbf{Q}_I = \mathbf{0}$ if $\operatorname{curl} \mathbf{v}_I = \mathbf{0}$ in Ω_I and $\mathbf{v}_I \times \mathbf{n} = \mathbf{0}$ on $\partial\Omega$.

The scalar function $\chi_I \in H^1(\Omega_I)$ is the solution to the elliptic mixed boundary value problem

$$\begin{cases} \operatorname{div}(\boldsymbol{\mu}_I \operatorname{grad} \chi_I) = \operatorname{div}(\boldsymbol{\mu}_I \mathbf{v}_I) & \text{in } \Omega_I \\ \boldsymbol{\mu}_I \operatorname{grad} \chi_I \cdot \mathbf{n}_I = \boldsymbol{\mu}_I \mathbf{v}_I \cdot \mathbf{n}_I & \text{on } \Gamma \\ \chi_I = 0 & \text{on } \partial\Omega \,. \end{cases} \tag{A.23}$$

The existence and uniqueness of the solution χ_I is well-known from the classical theory on elliptic boundary value problems. Clearly, if $\operatorname{div}(\boldsymbol{\mu}_I \mathbf{v}_I) = 0$ in Ω_I and $\boldsymbol{\mu}_I \mathbf{v}_I \cdot \mathbf{n}_I = 0$ on Γ it follows $\chi_I = 0$ in Ω_I.

Finally, the vector $(a_{I,r}, b_{I,l})$, $r = 1, \ldots, p_{\partial\Omega}$, $l = 1, \ldots, n_\Gamma$, is the solution of the linear system

$$A \begin{pmatrix} a_{I,r} \\ b_{I,l} \end{pmatrix} = \begin{pmatrix} \int_{\Omega_I} \boldsymbol{\mu}_I \mathbf{v}_I \cdot \operatorname{grad} z_{s,I} \\ \int_{\Omega_I} \boldsymbol{\mu}_I \mathbf{v}_I \cdot \boldsymbol{\rho}_{m,I} \end{pmatrix} , \tag{A.24}$$

$s = 1, \ldots, p_{\partial\Omega}$, $m = 1, \ldots, n_\Gamma$, where $A := \begin{pmatrix} D & B \\ B^T & C \end{pmatrix}$ with

$$\begin{aligned} D_{sr} &:= \textstyle\int_{\Omega_I} \boldsymbol{\mu}_I \operatorname{grad} z_{r,I} \cdot \operatorname{grad} z_{s,I} \\ B_{sl} &:= \textstyle\int_{\Omega_I} \boldsymbol{\mu}_I \boldsymbol{\rho}_{l,I} \cdot \operatorname{grad} z_{s,I} \\ C_{ml} &:= \textstyle\int_{\Omega_I} \boldsymbol{\mu}_I \boldsymbol{\rho}_{l,I} \cdot \boldsymbol{\rho}_{m,I} \,, \end{aligned} \tag{A.25}$$

and the harmonic vector fields $\operatorname{grad} z_{r,I}$ and $\boldsymbol{\rho}_{l,I}$ are the basis functions of the space $\mathcal{H}_{\mu_I}(\partial\Omega, \Gamma; \Omega_I)$. As in the preceeding Theorem A.6, it is easily proved that the matrix A is symmetric and positive definite, and that $a_{I,r} = 0$, $b_{I,l} = 0$ for $r = 1, \ldots, p_{\partial\Omega}$, $l = 1, \ldots, n_\Gamma$ if $\mathbf{v}_I \perp^{\mu_I} \mathcal{H}_{\mu_I}(\partial\Omega, \Gamma; \Omega_I)$.

The verification that the three terms in the decomposition are orthogonal with respect to the scalar product $(\cdot, \cdot)_{\mu_I, \Omega_I}$ is readily carried out by proceeding as in Theorem A.6. Moreover, defining

$$\mathbf{V}_I := \mathbf{v}_I - \boldsymbol{\mu}_I^{-1} \operatorname{curl} \mathbf{Q}_I - \operatorname{grad} \chi_I - \sum_{r=1}^{p_{\partial\Omega}} a_{I,r} \operatorname{grad} z_{r,I} - \sum_{l=1}^{n_\Gamma} b_{I,l} \boldsymbol{\rho}_{l,I} \,,$$

it can be directly verified that

$$\mathbf{V}_I \subset \mathcal{H}_{\mu_I}(\partial\Omega, \Gamma; \Omega_I) \,, \quad \mathbf{V}_I \perp^{\mu_I} \mathcal{H}_{\mu_I}(\partial\Omega, \Gamma; \Omega_I) \,,$$

and the thesis follows. □

A.3.3 Third decomposition result

Another decomposition result, based on different harmonic fields, is the following one

Theorem A.8. *Any vector function* $\mathbf{v}_I \in (L^2(\Omega_I))^3$ *can be decomposed into the following sum*

$$\mathbf{v}_I = \boldsymbol{\mu}_I^{-1} \operatorname{curl} \mathbf{Q}_I^* + \operatorname{grad} \chi_I^* + \sum_{\alpha=1}^{n_{\Omega_I}} \theta_{I,\alpha}^* \boldsymbol{\rho}_{\alpha,I}^* \,, \tag{A.26}$$

where $\mathbf{Q}_I^* \in H_0(\mathrm{curl};\Omega_I) \cap H^0(\mathrm{div};\Omega_I) \cap \mathcal{H}(e;\Omega_I)^\perp$ *is introduced in* (A.27), $\chi_I^* \in H^1(\Omega_I)/\mathbb{C}$ *in* (A.28) *and* $\theta_{I,\alpha}^*$, $\alpha = 1,\ldots,n_{\Omega_I}$, *in* (A.29), *and each term of the decomposition* (A.26) *is orthogonal to the others, with respect to the scalar product* $(\cdot,\cdot)_{\mu_I,\Omega_I}$.

Moreover, if $\operatorname{curl} \mathbf{v}_I = \mathbf{0}$ *in* Ω_I *it follows* $\mathbf{Q}_I^* = \mathbf{0}$, *if* $\operatorname{div}(\boldsymbol{\mu}_I \mathbf{v}_I) = 0$ *in* Ω_I *and* $\boldsymbol{\mu}_I \mathbf{v}_I \cdot \mathbf{n}_I = 0$ *on* $\Gamma \cup \partial\Omega$ *one has* $\operatorname{grad} \chi_I^* = \mathbf{0}$, *and if* $\mathbf{v}_I \perp^{\mu_I} \mathcal{H}_{\mu_I}(m;\Omega_I)$ *one finds* $\theta_{I,\alpha}^* = 0$, $\alpha = 1,\ldots,n_{\Omega_I}$.

Proof. Since the proof is similar to that of Theorems A.6 and A.7, let us just sketch it. For any vector function $\mathbf{v}_I \in (L^2(\Omega_I))^3$, construct the solution $\mathbf{Q}_I^* \in H(\mathrm{curl};\Omega_I) \cap H(\mathrm{div};\Omega_I)$ to

$$\begin{cases} \operatorname{curl}(\boldsymbol{\mu}_I^{-1} \operatorname{curl} \mathbf{Q}_I^*) = \operatorname{curl} \mathbf{v}_I & \text{in } \Omega_I \\ \operatorname{div} \mathbf{Q}_I^* = 0 & \text{in } \Omega_I \\ \mathbf{Q}_I^* \times \mathbf{n}_I = \mathbf{0} & \text{on } \Gamma \cup \partial\Omega \\ \mathbf{Q}_I^* \perp \mathcal{H}(e;\Omega_I) \,. \end{cases} \tag{A.27}$$

The existence and uniqueness of the solution $\mathbf{Q}_I^*$ can be proved as done for problem (A.14). Note that $\mathbf{Q}_I^* = \mathbf{0}$ if $\operatorname{curl} \mathbf{v}_I = \mathbf{0}$ in Ω_I.

The scalar function $\chi_I^* \in H^1(\Omega_I)/\mathbb{C}$ is the solution to the elliptic Neumann boundary value problem

$$\begin{cases} \operatorname{div}(\boldsymbol{\mu}_I \operatorname{grad} \chi_I^*) = \operatorname{div}(\boldsymbol{\mu}_I \mathbf{v}_I) & \text{in } \Omega_I \\ \boldsymbol{\mu}_I \operatorname{grad} \chi_I^* \cdot \mathbf{n}_I = \boldsymbol{\mu}_I \mathbf{v}_I \cdot \mathbf{n}_I & \text{on } \Gamma \cup \partial\Omega \,. \end{cases} \tag{A.28}$$

It is clear that $\operatorname{grad} \chi_I^* = \mathbf{0}$ in Ω_I provided that $\operatorname{div}(\boldsymbol{\mu}_I \mathbf{v}_I) = 0$ in Ω_I and $\boldsymbol{\mu}_I \mathbf{v}_I \cdot \mathbf{n}_I = 0$ on $\Gamma \cup \partial\Omega$.

Finally, the vector $\theta_{I,\alpha}^*$, $\alpha = 1,\ldots,n_{\Omega_I}$, is the solution of the linear system

$$\sum_{\alpha=1}^{n_{\Omega_I}} A_{\beta\alpha}^* \theta_{I,\alpha}^* = \int_{\Omega_I} \boldsymbol{\mu}_I \mathbf{v}_I \cdot \boldsymbol{\rho}_{\beta,I}^* \,, \quad \beta = 1,\ldots,n_{\Omega_I} \,, \tag{A.29}$$

where

$$A_{\beta\alpha}^* := \textstyle\int_{\Omega_I} \boldsymbol{\mu}_I \boldsymbol{\rho}_{\alpha,I}^* \cdot \boldsymbol{\rho}_{\beta,I}^* \,, \tag{A.30}$$

and the harmonic vector fields $\boldsymbol{\rho}_{\alpha,I}^*$ are the basis functions of the space $\mathcal{H}_{\mu_I}(m;\Omega_I)$. Hence $\theta_{I,\alpha}^* = 0$ for $\alpha = 1,\ldots,n_{\Omega_I}$ when $\mathbf{v}_I \perp^{\mu_I} \mathcal{H}_{\mu_I}(m;\Omega_I)$.

The proof of the theorem is now easily done by following the same procedure used in Theorems A.6 and A.7. □

A.4 More on harmonic fields

In this section we give an explicit construction of the basis functions of the spaces of harmonic fields presented in Section 1.4. Useful references about this topic are, e.g., the papers by Foias and Temam [106], Picard [192], Amrouche et al. [27], Fernandes and Gilardi [104], Hiptmair [126], Cantarella et al. [73], Auchmuty and Alexander [30], [31], and the books by Bossavit [59], Gross and Kotiuga [115]. The most complete results are in Ghiloni [110].

Let us start from the space $\mathcal{H}_{\varepsilon_I}(\Gamma, \partial\Omega; \Omega_I)$, defined as

$$\begin{aligned}\mathcal{H}_{\varepsilon_I}(\Gamma, \partial\Omega; \Omega_I) := \{\mathbf{v}_I \in (L^2(\Omega_I))^3 \mid \operatorname{curl} \mathbf{v}_I = \mathbf{0}, \operatorname{div}(\varepsilon_I \mathbf{v}_I) = 0,\\ \mathbf{v}_I \times \mathbf{n}_I = \mathbf{0} \text{ on } \Gamma, \varepsilon_I \mathbf{v}_I \cdot \mathbf{n} = 0 \text{ on } \partial\Omega\}\ .\end{aligned}$$

The determination of the basis functions $\operatorname{grad} w_{j,I}$, $j = 1, \ldots, p_\Gamma$, is easily done as follows: $w_{j,I} \in H^1(\Omega_I)$ is the solution of the elliptic problem

$$\begin{cases} \operatorname{div}(\varepsilon_I \operatorname{grad} w_{j,I}) = 0 & \text{in } \Omega_I \\ \varepsilon_I \operatorname{grad} w_{j,I} \cdot \mathbf{n} = 0 & \text{on } \partial\Omega \\ w_{j,I} = 0 & \text{on } \Gamma \setminus \Gamma_j \\ w_{j,I} = 1 & \text{on } \Gamma_j\ . \end{cases} \tag{A.31}$$

Instead, the basis functions $\boldsymbol{\pi}_{k,I}$, $k = 1, \ldots, n_{\partial\Omega}$, need some preliminary notation. It is known that in Ω_I there exist $n_{\partial\Omega}$ connected orientable Lipschitz surfaces Σ_k, with $\partial\Sigma_k \subset \partial\Omega$, such that every curl-free vector field in Ω_I with vanishing tangential component on Γ has a global potential in $\Omega_I \setminus \cup_k \Sigma_k$. These surfaces, usually called Seifert surfaces, are "cutting" surfaces: each one of them "cuts" a Γ-independent non-bounding cycle in Ω_I (for notation, see Section 1.4). We can now introduce the functions $q_{k,I} \in H^1(\Omega_I \setminus \Sigma_k)$, solutions to

$$\begin{cases} \operatorname{div}(\varepsilon_I \operatorname{grad} q_{k,I}) = 0 & \text{in } \Omega_I \setminus \Sigma_k \\ \varepsilon_I \operatorname{grad} q_{k,I} \cdot \mathbf{n} = 0 & \text{on } \partial\Omega \setminus \partial\Sigma_k \\ q_{k,I} = 0 & \text{on } \Gamma \\ [\varepsilon_I \operatorname{grad} q_{k,I} \cdot \mathbf{n}_\Sigma]_{\Sigma_k} = 0 & \\ [q_{k,I}]_{\Sigma_k} = 1\ , & \end{cases} \tag{A.32}$$

having denoted by $[\,\cdot\,]_{\Sigma_k}$ the jump across the surface Σ_k and by $\mathbf{n}_\Sigma$ the unit normal vector on Σ_k. We finally define $\boldsymbol{\pi}_{k,I}$ as the $(L^2(\Omega_I))^3$-extension of $\operatorname{grad} q_{k,I}$ (computed in $\Omega_I \setminus \Sigma_k$).

The basis functions for the other harmonic spaces can be defined in a similar way: let us go on with $\mathcal{H}_{\mu_I}(\partial\Omega, \Gamma; \Omega_I)$, defined as

$$\begin{aligned}\mathcal{H}_{\mu_I}(\partial\Omega, \Gamma; \Omega_I) := \{\mathbf{v}_I \in (L^2(\Omega_I))^3 \mid \operatorname{curl} \mathbf{v}_I = \mathbf{0}, \operatorname{div}(\boldsymbol{\mu}_I \mathbf{v}_I) = 0,\\ \mathbf{v}_I \times \mathbf{n} = \mathbf{0} \text{ on } \partial\Omega, \boldsymbol{\mu}_I \mathbf{v}_I \cdot \mathbf{n}_I = 0 \text{ on } \Gamma\}\ .\end{aligned}$$

The basis functions grad $z_{r,I}$, $r = 1, \dots, p_{\partial\Omega}$, are the gradients of the solutions $z_{r,I} \in H^1(\Omega_I)$ to

$$\begin{cases} \mathrm{div}(\boldsymbol{\mu}_I \,\mathrm{grad}\, z_{r,I}) = 0 & \text{in } \Omega_I \\ \boldsymbol{\mu}_I \,\mathrm{grad}\, z_{r,I} \cdot \mathbf{n}_I = 0 & \text{on } \Gamma \\ z_{r,I} = 0 & \text{on } \partial\Omega \setminus (\partial\Omega)_r \\ z_{r,I} = 1 & \text{on } (\partial\Omega)_r \,. \end{cases} \tag{A.33}$$

Moreover, similarly to the preceding case, in Ω_I there exist n_Γ connected orientable Lipschitz surfaces Ξ_l, $l = 1, \dots, n_\Gamma$, with $\partial\Xi_l \subset \Gamma$, such that every curl-free vector field in Ω_I with vanishing tangential component on $\partial\Omega$ has a global potential in $\Omega_I \setminus \cup_l \Xi_l$. These "cutting" surfaces "cuts" the $\partial\Omega$-independent non-bounding cycles in Ω_I. Then introduce the functions $p_{l,I} \in H^1(\Omega_I \setminus \Xi_l)$, solutions to

$$\begin{cases} \mathrm{div}(\boldsymbol{\mu}_I \,\mathrm{grad}\, p_{l,I}) = 0 & \text{in } \Omega_I \setminus \Xi_l \\ \boldsymbol{\mu}_I \,\mathrm{grad}\, p_{l,I} \cdot \mathbf{n}_I = 0 & \text{on } \Gamma \setminus \partial\Xi_l \\ p_{l,I} = 0 & \text{on } \partial\Omega \\ [\boldsymbol{\mu}_I \,\mathrm{grad}\, p_{l,I} \cdot \mathbf{n}_\Xi]_{\Xi_l} = 0 & \\ [p_{l,I}]_{\Xi_l} = 1 \,, & \end{cases} \tag{A.34}$$

having denoted by $[\,\cdot\,]_{\Xi_l}$ the jump across the surface Ξ_l and by $\mathbf{n}_\Xi$ the unit normal vector on Ξ_l. The basis functions $\boldsymbol{\rho}_{l,I}$ are the $(L^2(\Omega_I))^3$-extension of grad $p_{l,I}$ (computed in $\Omega_I \setminus \Xi_l$).

For the space $\mathcal{H}_{\varepsilon_I}(e; \Omega_I)$, defined as

$$\begin{aligned} \mathcal{H}_{\varepsilon_I}(e; \Omega_I) := \{\mathbf{v}_I \in (L^2(\Omega_I))^3 \mid \mathrm{curl}\, \mathbf{v}_I = \mathbf{0}, \mathrm{div}(\boldsymbol{\varepsilon}_I \mathbf{v}_I) = 0, \\ \mathbf{v}_I \times \mathbf{n}_I = \mathbf{0} \text{ on } \Gamma \cup \partial\Omega\} \,, \end{aligned}$$

the basis functions are grad $w^*_{\gamma,I}$, $\gamma = 0, \dots, p_{\partial\Omega} + p_\Gamma$, where $w^*_{\gamma,I} \in H^1(\Omega_I)$ is the solution to

$$\begin{cases} \mathrm{div}(\boldsymbol{\varepsilon}_I \,\mathrm{grad}\, w^*_{\gamma,I}) = 0 & \text{in } \Omega_I \\ w^*_{\gamma,I} = 0 & \text{on } (\partial\Omega \cup \Gamma) \setminus \Theta_\gamma \\ w^*_{\gamma,I} = 1 & \text{on } \Theta_\gamma \,. \end{cases} \tag{A.35}$$

Here the surfaces Θ_γ are defined as $\Theta_\gamma := (\partial\Omega)_\gamma$ for $\gamma = 0, \dots, p_{\partial\Omega}$ and $\Theta_\gamma := \Gamma_{\gamma - p_{\partial\Omega}}$ for $\gamma = p_{\partial\Omega} + 1, \dots, p_{\partial\Omega} + p_\Gamma$.

When considering the space $\mathcal{H}_{\mu_I}(m; \Omega_I)$, defined as

$$\begin{aligned} \mathcal{H}_{\mu_I}(m; \Omega_I) := \{\mathbf{v}_I \in (L^2(\Omega_I))^3 \mid \mathrm{curl}\, \mathbf{v}_I = \mathbf{0}, \mathrm{div}(\boldsymbol{\mu}_I \mathbf{v}_I) = 0, \\ \boldsymbol{\mu}_I \mathbf{v}_I \cdot \mathbf{n} = 0 \text{ on } \Gamma \cup \partial\Omega\} \,, \end{aligned}$$

one has first to introduce the "cutting" surfaces $\Xi^*_\alpha \subset \Omega_I$, $\alpha = 1, \dots, n_{\Omega_I}$, each one "cutting" an independent non-bounding cycle in Ω_I. They are connected orientable Lipschitz surfaces with $\partial\Xi^*_\alpha \subset \partial\Omega \cup \Gamma$, such that every curl-free vector field in Ω_I has a global potential in $\Omega_I \setminus \cup_\alpha \Xi^*_\alpha$. The basis functions $\boldsymbol{\rho}^*_{\alpha,I}$ are the $(L^2(\Omega_I))^3$-extension of grad $p^*_{\alpha,I}$, where $p^*_{\alpha,I} \in H^1(\Omega_I \setminus \Xi^*_\alpha)$ is the solution, determined up to an additive constant, to

$$\begin{cases} \operatorname{div}(\boldsymbol{\mu}_I \operatorname{grad} p^*_{\alpha,I}) = 0 & \text{in } \Omega_I \setminus \Xi^*_\alpha \\ \boldsymbol{\mu}_I \operatorname{grad} p^*_{\alpha,I} \cdot \mathbf{n}_I = 0 & \text{on } (\partial\Omega \cup \Gamma) \setminus \partial\Xi^*_\alpha \\ \left[\boldsymbol{\mu}_I \operatorname{grad} p^*_{\alpha,I} \cdot \mathbf{n}_{\Xi^*}\right]_{\Xi^*_\alpha} = 0 & \\ \left[p^*_{\alpha,I}\right]_{\Xi^*_\alpha} = 1 \, , & \end{cases} \tag{A.36}$$

having denoted by $[\,\cdot\,]_{\Xi^*_\alpha}$ the jump across the surface Ξ^*_α and by $\mathbf{n}_{\Xi^*}$ the unit normal vector on Ξ^*_α.

The basis functions $\operatorname{grad} \hat{z}_r$, $r = 1, \ldots, p_{\partial\Omega}$, of the space

$$\mathcal{H}(e;\Omega) := \{\mathbf{v} \in (L^2(\Omega))^3 \mid \operatorname{curl} \mathbf{v} = \mathbf{0}, \operatorname{div} \mathbf{v} = 0, \\ \mathbf{v} \times \mathbf{n} = \mathbf{0} \text{ on } \partial\Omega\} \, ,$$

and $\hat{\boldsymbol{\pi}}_t$, $t = 1, \ldots, n_\Omega$, of the space

$$\mathcal{H}(m;\Omega) := \{\mathbf{v} \in (L^2(\Omega))^3 \mid \operatorname{curl} \mathbf{v} = \mathbf{0}, \operatorname{div} \mathbf{v} = 0, \\ \mathbf{v} \cdot \mathbf{n} = 0 \text{ on } \partial\Omega\} \, ,$$

are determined in a similar way to those of the spaces $\mathcal{H}_{\varepsilon_I}(e;\Omega_I)$ and $\mathcal{H}_{\mu_I}(m;\Omega_I)$, respectively.

References

1. Acevedo, R., and Rodríguez, R. (2007) Analysis of the $A, V - A - \psi$ potential formulation for the eddy current problem in a bounded domain. *Electron. Trans. Numer. Anal.* 26, 270–284.
2. Adams, R. A. (1975) *Sobolev Spaces*. Academic Press, New York.
3. Adams, R. A., and Fournier, J. J. F. (2003) *Sobolev Spaces*, 2nd ed. Elsevier/Academic Press, Amsterdam.
4. Albertz, D., Dappen, S., and Henneberger, G. (1997) Calculation of the induced currents and forces for a hybrid magnetic levitation system. *IEEE Trans. Magn.* 33, 1263–1266.
5. Alonso, A. (1999) A mathematical justification of the low-frequency heterogeneous time-harmonic Maxwell equations. *Math. Models Methods Appl. Sci.* 9, 475–489.
6. Alonso, A., and Valli, A. (1996) Some remarks on the characterization of the space of tangential traces of $H(\mathrm{rot}; \Omega)$ and the construction of an extension operator. *Manuscripta Math.* 89, 159–178.
7. Alonso, A., and Valli, A. (1997) A domain decomposition approach for heterogeneous time-harmonic Maxwell equations. *Comput. Methods Appl. Mech. Engrg.* 143, 97–112.
8. Alonso, A., and Valli, A. (1998) Unique solvability for high-frequency heterogeneous time-harmonic Maxwell equations via the Fredholm alternative theory. *Math. Methods Appl. Sci.* 21, 463–477.
9. Alonso, A., and Valli, A. (1999) An optimal domain decomposition preconditioner for low-frequency time-harmonic Maxwell equations. *Math. Comp.* 68, 607–631.
10. Alonso Rodríguez, A., Fernandes, P., and Valli, A. (2003) The time-harmonic eddy-current problem in general domains: solvability via scalar potentials. In Carstensen, C., Funken, S., Hackbusch, W., Hoppe, R. H. W., and Monk, P. (eds.), *Computational Electromagnetics*. Springer, Berlin, 143–163.
11. Alonso Rodríguez, A., Fernandes, P., and Valli, A. (2003) Weak and strong formulations for the time-harmonic eddy-current problem in general multi-connected domains. *European J. Appl. Math.* 14, 387–406.
12. Alonso Rodríguez, A., Hiptmair, R., and Valli, A. (2004) Hybrid formulations of eddy current problems. Report UTM 663, Department of Mathematics, University of Trento.
13. Alonso Rodríguez, A., Hiptmair, R., and Valli, A. (2004) Mixed finite element approximation of eddy current problems. *IMA J. Numer. Anal.* 24, 255–271.
14. Alonso Rodríguez, A., Hiptmair, R., and Valli, A. (2005) A hybrid formulation of eddy current problems. *Numer. Methods Partial Differential Equations* 21, 742–763.

15. Alonso Rodríguez, A., and Raffetto, M. (2003) Unique solvability for electromagnetic boundary value problems in the presence of partly lossy inhomogeneous anisotropic media and mixed boundary conditions. *Math. Models Methods Appl. Sci.* 13, 597–611.
16. Alonso Rodríguez, A., and Valli, A. (2002) Domain decomposition methods for time-harmonic Maxwell equations: numerical results. In Pavarino, L. F. and Toselli, A. (eds.), *Recent Developments in Domain Decomposition Methods*. Springer, Berlin, 157–171.
17. Alonso Rodríguez, A., and Valli, A. (2004) Mixed finite element approximation of eddy current problems based on the electric field. In Neittaanmäki, P., Rossi, T., Majava, K., and Pironneau, O. (eds.), *ECCOMAS 2004: European Congress on Computational Methods in Applied Sciences and Engineering. Vol. 1,* University of Jyväskylä.
18. Alonso Rodríguez, A., and Valli, A. (2008) Voltage and current excitation for time-harmonic eddy-current problems. *SIAM J. Appl. Math.* 68, 1477–1494.
19. Alonso Rodríguez, A., and Valli, A. (2009) A FEM–BEM approach for electro–magnetostatics and time-harmonic eddy-current problems. *Appl. Numer. Math.* 59, 2036–2049.
20. Alonso Rodríguez, A., Valli, A., and Vázquez Hernández, R. (2009) A formulation of the eddy-current problem in the presence of electric ports. *Numer. Math.* 113, 643–672.
21. Alonso Rodríguez, A., and Vázquez Hernández, R. (2009) Iterative methods for the saddle-point problem arising from the $\mathbf{H}_C/\mathbf{E}_I$ formulation of the eddy current problem. *SIAM J. Sci. Comput.* 31, 3155–3178.
22. Ammari, H., Bao, G., and Fleming, J. L. (2002) An inverse source problem for Maxwell's equations in magnetoencephalography. *SIAM J. Appl. Math.* 62, 1369–1382.
23. Ammari, H., Buffa, A., and Nédélec, J.-C. (2000) A justification of eddy currents model for the Maxwell equations. *SIAM J. Appl. Math.* 60, 1805–1823.
24. Ammari, H., Latiri-Grouz, C., and Nédélec, J.-C. (1998) The Leontovich boundary value problem for the time-harmonic Maxwell equations. *Asymptot. Anal.* 18, 33–47.
25. Ammari, H., and Nédélec, J.-C. (1998) Couplage éléments finis/équations intégrales pour la résolution des équations de Maxwell en milieu hétérogène. In *Équations aux Dérivées Partielles et Applications*. Gauthier-Villars, Paris, 19–33.
26. Ammari, H., and Nédélec, J.-C. (1999) Coupling of finite and boundary element methods for the time-harmonic Maxwell equations. II. A symmetric formulation. In Rossmann, J., Takáč, P., and Wildenhain, G. (eds.), *The Maz′ya Anniversary Collection, Vol. 2*. Birkhäuser, Basel, 23–32.
27. Amrouche, C., Bernardi, C., Dauge, M., and Girault, V. (1998) Vector potentials in three–dimensional nonsmooth domains. *Math. Methods Appl. Sci.* 21, 823–864.
28. Aoki, S. (1980) 3-dimensional magnetic field calculation of the levitation magnet for HSST by the finite element method. *IEEE Trans. Magn.* MAG-16, 725–727.
29. Auchmuty, G. (1994) Orthogonal decompositions and bases for three-dimensional vector fields. *Numer. Funct. Anal. Optim.* 15, 455–488.
30. Auchmuty, G., and Alexander, J. C. (2005) L^2-well-posedness of 3D div-curl boundary value problems. *Quart. Appl. Math.* 63, 479–508.
31. Auchmuty, G., and Alexander, J. C. (2006) Finite-energy solutions of mixed 3D div-curl systems. *Quart. Appl. Math.* 64, 335–357.
32. Auld, B. A., and Moulder, J. C. (1999) Review of advances in quantitative eddy current nondestructive evaluation. *J. Nondestr. Eval.* 18, 3–36.
33. Badics, Z., Matsumoto, Y., Aoki, K., F. Nakayasu, F., Uesaka, M., and Miya, K. (1997) An effective 3-D finite element scheme for computing electromagnetic field distortions due to defects in eddy-current nondestructive evaluation. *IEEE Trans. Magn.* 33, 1012–1020.
34. Baillet, S., Mosher, J. C., and Leahy, R. M. (2001) Electromagnetic brain mapping. *IEEE Signal Process. Magaz.* 18, 14–30.

35. Bay, F., Labbe, V., Favennec, Y., and Chenot, J. L. (2003) A numerical model for induction heating processes coupling electromagnetism and thermomechanics. *Internat. J. Numer. Methods Engrg.* 58, 839–867.
36. Benedetti, R., Frigerio, R., and Ghiloni, R. (2010) The topology of Helmholtz domains. Report, arXiv:1001.4418v1 [math.GT].
37. Benzi, M., Golub, G. H., and Liesen, J. (2005) Numerical solution of saddle point problems. *Acta Numer.* 14, 1–137.
38. Bermúdez, A., Bullón, J., Muñiz, M. C., and Pena, F. (1999) Numerical computation of the electromagnetic field in the electrodes of a three-phase arc furnace. *Internat. J. Numer. Methods Engrg.* 46, 649–658.
39. Bermúdez, A., Bullón, J., and Pena, F. (1998) A finite element method for the themoelectrical modelling of electrodes. *Commun. Numer. Methods Engrg.* 14, 581–593.
40. Bermúdez, A., Gómez, D., Muñiz, M. C., and Salgado, P. (2007) A FEM/BEM for axisymmetric electromagnetic and thermal modelling of induction furnaces. *Internat. J. Numer. Methods Engrg.* 71, 856–878, 879–882.
41. Bermúdez, A., Gómez, D., Muñiz, M. C., Salgado, P., and Vázquez., R. (2009) Numerical simulation of a thermo-electromagneto-hydrodynamic problem in an induction heating furnace. *Appl. Numer. Math.* 59, 2082–2104.
42. Bermúdez, A., Rodríguez, R., and Salgado, P. (2002) A finite element method with Lagrange multipliers for low-frequency harmonic Maxwell equations. *SIAM J. Numer. Anal.* 40, 1823–1849.
43. Bermúdez, A., Rodríguez, R., and Salgado, P. (2005) FEM for 3D eddy current problems in bounded domains subject to realistic boundary conditions. An application to metallurgical electrodes. *Arch. Comput. Methods Engrg.* 12, 67–114.
44. Bermúdez, A., Rodríguez, R., and Salgado, P. (2005) Numerical analysis of electric field formulations of the eddy current model. *Numer. Math.* 102, 181–201.
45. Bermúdez, A., Rodríguez, R., and Salgado, P. (2005) Numerical solution of eddy-current problems in bounded domains using realistic boundary conditions. *Comput. Methods Appl. Mech. Engrg.* 194, 411–426.
46. Bermúdez, A., Rodríguez, R., and Salgado, P. (2005) Numerical treatment of realistic boundary conditions for the eddy current problem in an electrode via Lagrange multipliers. *Math. Comp.* 74, 123–151.
47. Biddlecombe, C. S., Heighway, E. A., Simkin, J., and Trowbridge, C. W. (1982) Methods for eddy current computation in three dimensions. *IEEE Trans. Magn.* MAG-18, 492–497.
48. Bíró, O. (1999) Edge element formulations of eddy current problems. *Comput. Methods Appl. Mech. Engrg.* 169, 391–405.
49. Bíró, O., and Preis, K. (1989) On the use of the magnetic vector potential in the finite element analysis of three-dimensional eddy currents. *IEEE Trans. Magn.* 25, 3145–3159.
50. Bíró, O., and Preis, K. (1990) Coupling electric and magnetic vector potentials in 3-D eddy current analysis. *COMPEL* 9, Supplement A, 45–48.
51. Bíró, O., and Preis, K. (1990) Finite element analysis of 3-D eddy currents. *IEEE Trans. Magn.* 26, 418–423.
52. Bíró, O., Preis, K., Buchgraber, G., and Tičar, I. (2004) Voltage-driven coils in finite-element formulations using a current vector and a magnetic scalar potential. *IEEE Trans. Magn.* 40, 1286–1289.
53. Bíró, O., Preis, K., Renhart, W., Richter, K. R., and Vrisk, G. (1990) Performance of different vector potential formulations in solving multiply connected 3-D eddy current problems. *IEEE Trans. Magn.* 26, 438–441.

54. Bíró, O., and Valli, A. (2007) The Coulomb gauged vector potential formulation for the eddy-current problem in general geometry: well-posedness and numerical approximation. *Comput. Methods Appl. Mech. Engrg.* 196, 1890–1904.
55. Bondeson, A., Rylander, T., and Ingelström, P. (2005) *Computational Electromagnetics*. Springer, New York.
56. Börm, S., and Hackbusch, W. (2005) Hierarchical quadrature for singular integrals. *Computing* 74, 75–100.
57. Bossavit, A. (1991) The computation of eddy-currents, in dimension 3, by using mixed finite elements and boundary elements in association. *Math. Comput. Modelling* 15, 33–42.
58. Bossavit, A. (1993) *Électromagnétisme, en Vue de la Modélisation*. Springer, Paris.
59. Bossavit, A. (1998) *Computational Electromagnetism. Variational Formulation, Complementarity, Edge Elements*. Academic Press, San Diego.
60. Bossavit, A. (1999) On the Lorenz gauge. *COMPEL* 18, 323–336.
61. Bossavit, A. (2000) Most general 'non-local' boundary conditions for the Maxwell equations in a bounded region. *COMPEL* 19, 239–245.
62. Bossavit, A., and Vérité, J. C. (1982) A mixed FEM/BIEM method to solve eddy-current problems. *IEEE Trans. Magn.* MAG-18, 431–435.
63. Bossavit, A., and Vérité, J. C. (1983) The TRIFOU code: solving the 3-D eddy-currents problem by using H as state variable. *IEEE Trans. Magn.* MAG-19, 2465–2470.
64. Bossavit, A., Vourdas, A., and Binns, K. J. (1989) Magnetostatics with scalar potentials in multiply connected regions. *IEE Proc. A* 136, 260–261.
65. Brezzi, F., and Fortin, M. (1991) *Mixed and Hybrid Finite Element Methods*. Springer, New York.
66. Brezzi, F., and Johnson, C. (1979) On the coupling of boundary integral and finite element methods. *Calcolo* 16, 189–201.
67. Bryant, C. F., Emson, C. R. I., Simkin, J., Trowbridge, C. W., and Nitta, K. (1998) A comparison of Lorentz and Coulomb gauges in the solution of Maxwell's equations. In El Dabaghi, F., Morgan, K., Parrott, K., and Périaux, J. (eds.), *Approximation and Numerical Methods for the Solution of Maxwell's Equations*. Clarendon Press, Oxford, 301–312.
68. Bryant, C. F., Emson, C. R. I., and Trowbridge, C. W. (1990) A comparison of Lorentz gauge formulations for eddy current computations. *IEEE Trans. Magn.* 26, 430–433.
69. Buffa, A., and Ciarlet, P. (2001) On traces for functional spaces related to Maxwell's equations. I. An integration by parts formula in Lipschitz polyhedra. *Math. Methods Appl. Sci.* 24, 9–30.
70. Buffa, A., and Ciarlet, P. (2001) On traces for functional spaces related to Maxwell's equations. II. Hodge decompositions on the boundary of Lipschitz polyhedra and applications. *Math. Methods Appl. Sci.* 24, 31–48.
71. Buffa, A., Costabel, M., and Sheen, D. (2002) On traces for $\mathbf{H}(\mathbf{curl}, \Omega)$ in Lipschitz domains. *J. Math. Anal. Appl.* 276, 845–867.
72. Buffa, A., Hiptmair, R., von Petersdorff, T., and Schwab, C. (2003) Boundary element methods for Maxwell equations on Lipschitz domains. *Numer. Math.* 95, 459–485.
73. Cantarella, J., DeTurck, D., and Gluck, H. (2002) Vector calculus and the topology of domains in 3-space. *Amer. Math. Monthly* 109, 409–442.
74. Carpenter, C. J. (1977) Comparison of alternative formulations of 3-dimensional magnetic-field and eddy-current problems at power frequencies. *Proc. IEE* 124, 1026–1034.
75. Cassat, A., and Jufer, M. (2002) MAGLEV projects technology aspects and choices. *IEEE Trans. Appl. Supercond.* 12, 915–925.
76. Cessenat, M. (1996) *Mathematical Methods in Electromagnetism*. World Scientific, Singapore.

77. Chaboudez, C., Clain, S., Glardon, R., Mari, D., Rappaz, J., and Swierkosz, M. (1997) Numerical modellng in induction heating for axisymmetric geometries. *IEEE Trans. Magn.* 33, 739–745.
78. Chaboudez, C., Clain, S., Glardon, R., Rappaz, J., Swierkosz, M., and Touzani, R. (1994) Numerical modeling of induction heating of long workpieces. *IEEE Trans. Magn.* 30, 5028–5037.
79. Chari, M. V. K., Konrad, A., Palmo, M. A., and D'Angelo, J. (1982) Three-dimensional vector potential analysis for machine field problems. *IEEE Trans. Magn.* MAG-18, 436–446.
80. Chen, Y. B., Yang, J. Y., Yu, H. I., and Tang, R. Y. (1994) Study on eddy current losses and shielding measures in large power transformers. *IEEE Trans. Magn.* 30, 3068–3071.
81. Chen, Z., Du, Q., and Zou, J. (2000) Finite element methods with matching and nonmatching meshes for Maxwell equations with discontinuous coefficients. *SIAM J. Numer. Anal.* 37, 1542–1570.
82. Christiansen, S. H., and Nédélec, J.-C. (2002) A preconditioner for the electric field integral equation based on Calderón formulas. *SIAM J. Numer. Anal.* 40, 1100–1135.
83. Ciarlet, P. G. (1978) *The Finite Element Method for Elliptic Problems*. North-Holland, Amsterdam.
84. Clemens, M., and Weiland, T. (1999) Transient eddy-current calculation with the FI-method. *IEEE Trans. Magn.* 35, 1163–1166.
85. Costabel, M. (1987) Symmetric methods for the coupling of finite elements and boundary elements. In Brebbia, C. A., Wendland, W. L., and Kuhn, G. (eds.), *Boundary Elements IX, Vol. 1 (Stuttgart, 1987)*. Springer, Berlin, 411–420.
86. Costabel, M. (1988) A symmetric method for the coupling of finite elements and boundary elements. In Whiteman, J. R. (ed.), *The Mathematics of Finite Elements and Applications VI (Uxbridge, 1987)*. Academic Press, London, 281–288.
87. Costabel, M., and Dauge, M. (2000) Singularities of electromagnetic fields in polyhedral domains. *Arch. Rational Mech. Anal.* 151, 221–276.
88. Costabel, M., and Dauge, M. (2002) Weighted regularization of Maxwell equations in polyhedral domains. A rehabilitation of nodal finite elements. *Numer. Math.* 93, 239–277.
89. Costabel, M., Dauge, M., and Nicaise, S. (1999) Singularities of Maxwell interface problems. *M2AN Math. Model. Numer. Anal.* 33, 627–649.
90. Costabel, M., Dauge, M., and Nicaise, S. (2003) Singularities of eddy current problems. *M2AN Math. Model. Numer. Anal.* 37, 807–831.
91. Coulomb, J. L. (1981) Finite element three dimensional magnetic field computation. *IEEE Trans. Magn.* MAG-17, 3241–3246.
92. Danby, G. T., and Powell, J. R. (1969) Levitation with null flux coils. *U.S. Patent 3 470 828.*
93. Dautray, R., and Lions, J.-L. (1992) *Mathematical Analysis and Numerical Methods for Science and Technology. Volume 1: Physical Origins and Classical Methods*. Springer, Berlin.
94. Dautray, R., and Lions, J.-L. (1992) *Mathematical Analysis and Numerical Methods for Science and Technology. Volume 2: Functional and Variational Methods*. Springer, Berlin.
95. Dautray, R., and Lions, J.-L. (1992) *Mathematical Analysis and Numerical Methods for Science and Technology. Volume 3: Spectral Theory and Applications*. Springer, Berlin.
96. Davey, K. R. (1997) Electrodynamic Maglev coil design and analysis. *IEEE Trans. Magn.* 33, 4227–4229.

97. Dłotko, P., Specogna, R., and Trevisan, F. (2009) Automatic generation of cuts on large-sized meshes for the T-Ω geometric eddy-current formulation. *Comput. Methods Appl. Mech. Engrg.* 198, 3765–3781.
98. Duffy, M. G. (1982) Quadrature over a pyramid or cube of integrands with a singularity at a vertex. *SIAM J. Numer. Anal.* 19, 1260–1262.
99. Dular, P., Geuzaine, C., and Legros, W. (1999) A natural method for coupling magnetodynamic **H**-formulations and circuits equations. *IEEE Trans. Magn.* 35, 1626–1629.
100. Emson, C. R. I., and Simkin, J. (1983) An optimal method for 3-D eddy currents. *IEEE Trans. Magn.* MAG-19, 2450–2452.
101. Erichsen, S., and Sauter, S. A. (1998) Efficient automatic quadrature in 3-d Galerkin BEM. *Comput. Methods Appl. Mech. Engrg.* 157, 215–224.
102. Eriksson, K., Estep, D., Hansbo, P., and Johnson, C. (1996) *Computational Differential Equations*. Cambridge University Press, Cambridge.
103. Fernandes, P. (1995) General approach to prove the existence and uniqueness of the solution in vector potential formulations of 3-D eddy current problems. *IEE Proc.-Sci. Meas. Technol.* 142, 299–306.
104. Fernandes, P., and Gilardi, G. (1997) Magnetostatic and electrostatic problems in inhomogeneous anisotropic media with irregular boundary and mixed boundary conditions. *Math. Models Methods Appl. Sci.* 7, 957–991.
105. Fernandes, P., and Valli, A. (2008) Lorenz gauged vector potential formulations for the time-harmonic eddy-current problem. *Math. Methods Appl. Sci.* 31, 71–98.
106. Foias, C., and Temam, R. (1978) Remarques sur les équations de Navier-Stokes stationnaires et les phénomènes successifs de bifurcations. *Ann. Scuola Norm. Sup. Pisa (IV)* 5, 29–63.
107. Fujiwara, K., and Nakata, T. (1990) Results for benchmark problem 7 (asymmetrical conductor with a hole). *COMPEL* 9, 137–154.
108. Fujiwara, K., Nakata, T., and Ohashi, H. (1996) Improvement of convergence characteristic of IICG method for the $\mathbf{A}$-ϕ method using edge elements. *IEEE Trans. Magn.* 32, 804–807.
109. Geselowitz, D. B. (1970) On the magnetic field generated outside an inhomogeneous volume conductor by internal current sources. *IEEE Trans. Magn.* MAG-6, 346–347.
110. Ghiloni, R. (2010) The Hodge decomposition theorem for general three-dimensional vector fields, without cuts. Report UTM 731, Department of Mathematics, University of Trento.
111. Girault, V., and Raviart, P.-A. (1986) *Finite Element Methods for Navier–Stokes Equations*. Springer, Berlin.
112. Golias, N. A., Antonopoulos, C. S., Tsiboukis, T. D., and Kriezis, E. E. (1998) 3D eddy current computation with edge elements in terms of the electric intensity. *COMPEL* 17, 667–673.
113. Gray, L. J., Glaeser, J. M., and Kaplan, T. (2004) Direct evaluation of hypersingular Galerkin surface integrals. *SIAM J. Sci. Comput.* 25, 1534–1556.
114. Gross, P. W., and Kotiuga, P. R. (2001) Finite element-based algorithms to make cuts for magnetic scalar potentials: topological constraints and computational complexity. In Teixeira, F. L. (ed.), *Geometric Methods for Computational Electromagnetics*. EMW Publishing, Cambridge, 207–245.
115. Gross, P. W., and Kotiuga, P. R. (2004) *Electromagnetic Theory and Computation. A Topological Approach*. Cambridge University Press, New York.
116. Guermond, J. L., and Minev, P. D. (2003) Mixed finite element approximation of an MHD problem involving conducting and insulating regions: the 3D case. *Numer. Methods Partial Differential Equations* 19, 709–731.

117. Hämäläinen, M., Hari, R., Ilmoniemi, R. J., Knuutila, J., and Lounasmaa, O. V. (1993) Magnetoencephalography – theory, instrumentation, and applications to noninvasive studies of the working human brain. *Rev. Mod. Phys.* 65, 413–497.
118. Haueisen, J., Tuch, D. S., Ramon, C., Schimpf, P. H., Wedeen, V. J., George, J. S., and Belliveau, J. W. (2002) The influence of brain anisotropy on human EEG and MEG. *NeuroImage* 15, 159–166.
119. Haus, H., and Melcher, J. (1989) *Electromagnetic Fields and Energy.* Prentice Hall, Englewood Cliffs.
120. He, J. L., Coffey, H. T., and Rote, D. M. (1995) Analysis of the combined maglev levitation, propulsion, and guidance system. *IEEE Trans. Magn.* 31, 981–987.
121. He, J. L., Wang, Z., Rote, D. M., and Winkleman, S. (1992) Investigation of the stability of AC repulsive-force levitation systems for low-speed maglev. *IEEE Trans. Magn.* 28, 3315–3317.
122. He, S., and Romanov, V. G. (1998) Identification of dipole sources in a bounded domain for Maxwell's equations. *Wave Motion* 28, 25–40.
123. Henneberger, G., and Obrecht, R. (1994) Numerical calculation of the temperature distribution in the melt of induction crucible furnaces. In *Second International Conference on Computation in Electromagnetics*, 16–19.
124. Henneberger, G., Sattler, P. K., Shen, D., and Hadrys, W. (1993) Coupling of magnetic and fluid flow problems and its application in induction melting apparatus. *IEEE Trans. Magn.* 29, 1589–1594.
125. Henneron, T., Menach, Y. L., Piriou, F., Moreau, O., Clenet, S., Ducreux, J.-P., and Verite, J.-C. (2007) Source field computation in NDT applications. *IEEE Trans. Magn.* 43, 1785–1788.
126. Hiptmair, R. (2002) Finite elements in computational electromagnetism. *Acta Numerica* 11, 237–339.
127. Hiptmair, R. (2002) Symmetric coupling for eddy current problems. *SIAM J. Numer. Anal.* 40, 41–65.
128. Hiptmair, R., and Ostrowski, J. (2002) Generators of $H_1(\Gamma_h, \mathbb{Z})$ for triangulated surfaces: construction and classification. *SIAM J. Computing* 31, 1405–1423.
129. Hiptmair, R., and Ostrowski, J. (2005) Coupled boundary-element scheme for eddy-current computation. *J. Engrg. Math.* 51, 231–250.
130. Hiptmair, R., and Sterz, O. (2005) Current and voltage excitations for the eddy current model. *Int. J. Numer. Modelling* 18, 1–21.
131. Ho, S. L., Li, Y., Wong, H. C., Wang, S. H., and Tang, R. Y. (2004) Numerical simulation of transient force and eddy current loss in a 720-MVA power transformer. *IEEE Trans. Magn.* 40, 687–690.
132. Hollaus, K., and Bíró, O. (2005) Comparison of tetrahedral edge finite-elements using different potential formulations. *IEEE Trans. Magn.* 41, 1676–1679.
133. Houston, P., Perugia, I., and Schötzau, D. (2004) Nonconforming mixed finite-element approximations to time-harmonic eddy current problems. *IEEE Trans. Magn.* 40, 1268–1273.
134. Hu, Q., and Zou, J. (2001) An iterative method with variable relaxation parameters for saddle-point problems. *SIAM J. Matrix Anal. Appl.* 23, 317–338.
135. Hu, Q., and Zou, J. (2004) Substructuring preconditioners for saddle-point problems arising from Maxwell's equations in three dimensions. *Math. Comp.* 73, 35–61.
136. Ida, N., and Lord, W. (1985) A finite element model for three-dimensional eddy current NDT phenomena. *IEEE Trans. Magn.* 21, 2490–2494.
137. Jackson, J. (1999) *Classical Electrodynamics*, 3rd ed. John Wiley & Sons, Hoboken.

138. Jami, A., and Lenoir, M. (1977) Formulation variationnelle pour le couplage entre une méthode d'éléments finis et une représentation intégrale. *C. R. Acad. Sci. Paris Sér. A-B* 285, A269–A272.
139. Jenkins, R. G., Jones, H., and Goodall, R. M. (1996) Designing HTS coils for magnetic circuits. *IEEE Trans. Magn.* 32, 2683–2686.
140. Jin, J. (2002) *The Finite Element Method in Electromagnetics*, 2nd ed. John Wiley & Sons, New York.
141. Johnson, C., and Nédélec, J.-C. (1980) On the coupling of boundary integral and finite element methods. *Math. Comp.* 35, 1063–1079.
142. Kaipio, J., and Somersalo, E. (2005) *Statistical and Computational Inverse Problems*. Springer, New York.
143. Kameari, A. (1988) Three dimensional eddy current calculation using finite element method with $\mathbf{A}$-V in conductor and Ω in vacuum. *IEEE Trans. Magn.* 24, 118–121.
144. Kameari, A. (1990) Calculation of transient 3D eddy current using edge-elements. *IEEE Trans. Magn.* 26, 466–469.
145. Kameari, A., and Koganezawa, K. (1997) Convergence of ICCG method in FEM using edge elements without gauge condition. *IEEE Trans. Magn.* 33, 1223–1226.
146. Kanayama, H., and Kikuchi, F. (1999) 3-D eddy current computation using the Nedelec elements. *Information* 2, 37–45.
147. Kanayama, H., Tagami, D., Saito, M., and Kikuchi, F. (2001) A numerical method for 3-D eddy current problems. *Japan J. Indust. Appl. Math.* 18, 603–612.
148. Karl, H., Fetzer, J., Kurz, S., Lehner, G., and Rucker, W. M. Description of TEAM Workshop problem 28: an electrodynamic levitation device. *www.compumag.co.uk/team.html*.
149. Katsumura, Y., Hashizume, H., and Toda, S. (1996) Numerical analysis of fluid flow with free surface and phase change under electromagnetic force. *IEEE Trans. Magn.* 32, 1002–1005.
150. Kettunen, L., Forsman, K., and Bossavit, A. (1998) Discrete spaces for div and curl-free fields. *IEEE Trans. Magn.* 34, 2551–2554.
151. Khan, S. H., Aristovich, K. Y., and Borovkov, A. I. (2009) Solution of the forward problem in magnetic-field tomography (MFT) based on magnetoencephalography (MEG). *IEEE Trans. Magn.* 45, 1416–1419.
152. Kotiuga, P. R. (1987) On making cuts for magnetic scalar potentials in multiply connected regions. *J. Appl. Phys.* 61, 3916–3918.
153. Kotiuga, P. R. (1988) Toward an algorithm to make cuts for magnetic scalar potentials in finite element meshes. *J. Appl. Phys.* 63, 3357–3359. Erratum: *J. Appl. Phys.* 64 (1988), 4257.
154. Kotiuga, P. R. (1989) An algorithm to make cuts for scalar potentials in tetrahedral meshes based on the finite element method. *IEEE Trans. Magn.* 25, 4129–4131.
155. Kotiuga, P. R., Vourdas, A., and Binns, K. J. (1990) Magnetostatics with scalar potentials in multiply connected regions. *IEE Proc. A* 137, 231–232.
156. Krebs, G., Abakar, A., and Clenet, S. (2009) Comparison of error estimators in eddy current testing. *IEEE Trans. Magn.* 45, 1420–1423.
157. Kriezis, E. E., Tsiboukis, T. D., Panas, S. M., and Tegopoulos, J. A. (1992) Eddy currents: theory and applications. *Proc. IEEE* 80, 1559–1589.
158. Křížek, M., and Neittaanmäki, P. (1996) *Mathematical and Numerical Modelling in Electrical Engineering*. Kluwer Academic Publishers, Dordrecht.
159. Kuhn, M., Langer, U., and Schöberl, J. (2000) Scientific computing tools for 3D magnetic field problems. In Whiteman, J. R. (ed.), *The Mathematics of Finite Elements and Applications X (Uxbridge 1999)*. Elsevier, Oxford, 239–258.

160. Kuhn, M., and Steinbach, O. (2002) Symmetric coupling of finite and boundary elements for exterior magnetic field problems. *Math. Methods Appl. Sci.* 25, 357–371.
161. Kurz, S., Fetzer, J., and Lehner, G. (1996) Three dimensional transient BEM–FEM coupled analysis of electrodynamic levitation problems. *IEEE Trans. Magn.* 32, 1062–1065.
162. Kybic, J., Clerc, M., Faugeras, O., Keriven, R., and Papadopoulo, T. (2006) Generalized head models for MEG/EEG: boundary element method beyond nested volumes. *Phys. Medicine Biol.* 51, 1333–1346.
163. Lavers, J. D. (2008) State of the art of numerical modeling for induction processes. *COMPEL* 27, 335–349.
164. Lee, H.-W., Kim, K.-C., and Lee, J. (2006) Review of Maglev train technologies. *IEEE Trans. Magn.* 42, 1917–1925.
165. Leonard, P. J., Lai, H. C., Hill-Cottingham, R. J., and Rodger, D. (1993) Automatic implementation of cuts in multiply connected magnetic scalar regions for 3D eddy current models. *IEEE Trans. Magn.* 29, 1368–1371.
166. Leonard, P. J., and Rodger, D. (1988) Finite element scheme for transient 3D eddy currents. *IEEE Trans. Magn.* 24, 90–93.
167. Leonard, P. J., and Rodger, D. (1989) A new method for cutting the magnetic scalar potential in multiply connected eddy current problems. *IEEE Trans. Magn.* 25, 4132–4134.
168. Leonard, P. J., and Rodger, D. (1990) Comparison of methods for dealing with multiply connected regions in 3D eddy current problems using the A-ψ method. *COMPEL* 9, Supplement A, 55–57.
169. Lord, W. (1983) Applications of numerical field modeling to electromagnetic methods of nondestructive testing. *IEEE Trans. Magn.* 19, 2437–2442.
170. Lu, M., Ueno, S., Thorlin, T., and Persson, M. (2009) Calculating the current density and electric field in human head by multichannel transcranial magnetic stimulation. *IEEE Trans. Magn.* 45, 1662–1665.
171. MacCamy, R. C., and Stephan, E. (1984) Solution procedures for three-dimensional eddy current problems. *J. Math. Anal. Appl.* 101, 348–379.
172. Marin, G., Guerin, C., Baillet, S., Garnero, L., and Meunier, G. (1998) Influence of skull anisotropy for the forward and inverse problem in EEG: simulation studies using FEM on realistic head models. *Hum. Brain Mapp.* 6, 250–269.
173. Mayergoyz, I. (1998) *Nonlinear Diffusion of Electromagnetic Fields.* Academic Press, San Diego.
174. Mayergoyz, I. D., Chari, M. V. K., and Konrad, A. (1983) Boundary Galerkin's method for three-dimensional finite element electromagnetic field computation. *IEEE Trans. Magn.* MAG-19, 2333–2336.
175. McLean, W. (2000) *Strongly Elliptic Systems and Boundary Integral Equations.* Cambridge University Press, Cambridge.
176. Meddahi, S., and Selgas, V. (2003) A mixed-FEM and BEM coupling for a three-dimensional eddy current problem. *M2AN Math. Model. Numer. Anal.* 37, 291–318.
177. Meunier, G., Le Floch, Y., and Guérin, C. (2003) A nonlinear circuit coupled $\mathbf{t} - \mathbf{t_0} - \phi$ formulation for solid conductors. *IEEE Trans. Magn.* 39, 1729–1732.
178. Monk, P. (1991) A mixed method for approximating Maxwell's equations. *SIAM J. Numer. Anal.* 28, 1610–1634.
179. Monk, P. (2003) *Finite Element Methods for Maxwell's Equations.* Oxford University Press, Oxford.
180. Morisue, T. (1982) Magnetic vector potential and electric scalar potential in three-dimensional eddy current problem. *IEEE Trans. Magn.* MAG-18, 531–535.

181. Morisue, T. (1993) A comparison of the Coulomb gauge and Lorentz gauge magnetic vector potential formulations for 3D eddy current calciulations. *IEEE Trans. Magn.* 29, 1372–1375.
182. Mosher, J. C., Leahy, R. M., and Lewis, P. S. (1999) EEG and MEG: forward solutions for inverse methods. *IEEE Trans. Biomed. Engrg.* 46, 245–259.
183. Natarajan, T., and El-Kaddah, N. (1999) A methodology for two-dimensional finite element analysis of electromagnetically driven flow in induction stirring systems. *IEEE Trans. Magn.* 35, 1773–1776.
184. Nečas, J. (1967) *Les Méthodes Directes en Théorie des Équations Elliptiques.* Masson, Paris.
185. Nédélec, J.-C. (1980) Mixed finite elements in $\mathbb{R}^3$. *Numer. Math.* 35, 315–341.
186. Nédélec, J.-C. (1986) A new family of mixed finite elements in $\mathbb{R}^3$. *Numer. Math.* 50, 57–81.
187. Nédélec, J.-C. (2001) *Acoustic and Electromagnetic Equations.* Springer, New York.
188. Nicaise, S. (2008) *Personal communication.*
189. Nicolet, A., and Delincé, F. (1996) Implicit Runge–Kutta methods for transient magnetic field computation. *IEEE Trans. Magn.* 32, 1405–1408.
190. Paquet, L. (1982) Problèmes mixtes pour le système de Maxwell. *Ann. Fac. Sci. Toulouse Math. (5)* 4, 103–141.
191. Perugia, I., and Schötzau, D. (2003) The *hp*-local discontinuous Galerkin method for low-frequency time-harmonic Maxwell equations. *Math. Comp.* 72, 1179–1214.
192. Picard, R. (1982) On the boundary value problems of electro- and magnetostatics. *Proc. Royal Soc. Edinburgh* 92 A, 165–174.
193. Pillsbury, R. D. (1983) A three dimensional eddy current formulation using two potentials: the magnetic vector potential and total magnetic scalar potential. *IEEE Trans. Magn.* MAG-19, 2284–2287.
194. Plonsey, R., and Heppner, D. B. (1967) Considerations of quasi-stationarity in electrophysiological systems. *Bull. Math. Biophys.* 29, 657–664.
195. Powell, J. R., and Danby, G. T. (1996) Maglev vehicles. Raising transportation advances off the ground. *IEEE Potentials* 15, 7–12.
196. Preis, K., Bíró, O., Buchgraber, G., and Tičar, I. (2006) Thermal-electromagnetic coupling in the finite-element simulation of power transformers. *IEEE Trans. Magn.* 42, 999–1002.
197. Preston, T. W., and Reece, A. B. J. (1982) Solution of 3-dimensional eddy current problems: the T-Ω method. *IEEE Trans. Magn.* MAG-18, 486–491.
198. Quarteroni, A. (2009) *Numerical Models for Differential Problems.* Springer, Milan.
199. Quarteroni, A., and Valli, A. (1994) *Numerical Approximation of Partial Differential Equations.* Springer, Berlin.
200. Quarteroni, A., and Valli, A. (1999) *Domain Decomposition Methods for Partial Differential Equations.* Oxford University Press, Oxford.
201. Rapetti, F. (2010) An overlapping mortar element approach to coupled magneto-mechanical problems. *Math. Comput. Simul.* 80, 1647–1656.
202. Rappaz, J., Swierkosz, M., and Trophime, C. (1999) Un modèle mathématique et numérique pour un logiciel de simulation tridimensionnelle d'induction électromagnétique. Tech. Rep. 05.99, Département de Mathématiques, École Polytechnique Fédérale de Lausanne.
203. Rappaz, J., and Touzani, R. (1996) On a two-dimensional magnetohydrodynamic problem. II. Numerical analysis. *RAIRO Modél. Math. Anal. Numér.* 30, 215–235.
204. Rasolonjanahary, J. L., Thollon, F., Burais, N., and Brunotte, X. (1996) Study of eddy currents non destructive testing system in riveted assemblies. *IEEE Trans. Magn.* 32, 1585–1588.

205. Reissel, M. (1993) On a transmission boundary-value problem for the time-harmonic Maxwell equations without displacement currents. *SIAM J. Math. Anal.* 24, 1440–1457.
206. Ren, Z. (1996) Influence of the R.H.S. on the convergence behaviour of the curl-curl equation. *IEEE Trans. Magn.* 32, 655–658.
207. Ren, Z. (2002) $\mathbf{T}$-Ω formulation for eddy-current problems in multiply connected regions. *IEEE Trans. Magn.* 38, 557–560.
208. Ren, Z., Bouillault, F., Razek, A., Bossavit, A., and Vérité, J. C. (1990) A new hybrid model using electric field formulation for 3-D eddy-current problems. *IEEE Trans. Magn.* 26, 470–473.
209. Ren, Z., and Ida, N. (2000) Solving 3D eddy current problems using second order nodal and edge elements. *IEEE Trans. Magn.* 36, 746–750.
210. Ren, Z., and Razek, A. (2000) Comparison of some 3D eddy current formulations in dual systems. *IEEE Trans. Magn.* 36, 751–755.
211. Rodger, D., and Eastham, J. F. (1983) A formulation for low frequency eddy current solutions. *IEEE Trans. Magn.* MAG-19, 2443–2446.
212. Rodger, D., and Eastham, J. F. (1985) The use of transformations in applying boundary conditions to three-dimensional vector field problems. *IEE Proc. A* 132, 165–170.
213. Rogg, D. (1984) General survey of the possible applications and development tendencies of magnetic levitation technology. *IEEE Trans. Magn.* MAG-20, 1696–1701.
214. Sabariego, R., and Dular, P. (2007) A perturbation approach for the modeling of eddy current nondestructive testing problems with differential probes. *IEEE Trans. Magn.* 43, 1289–1292.
215. Sabariego, R. V., and Dular, P. (2008) h- and b-conform finite element perturbation techniques for nondestructive eddy current testing. *COMPEL* 27, 319–327.
216. Saito, T., Maki, N., Kobayashi, T., and Shibata, M. (1993) Electromagnetic force and eddy current loss in dynamic behavior of a superconducting magnetically levitated vehicle. *IEEE Trans. Appl. Supercond.* 3, 417–420.
217. Salgado, P. (2002) Mathematical and numerical analysis of some electromagnetic problems. Application to the simulation of metallurgical electrodes. PhD thesis, Universidade de Santiago de Compostela.
218. Saranen, J. (1982) On generalized harmonic fields in domains with anisotropic nonhomogeneous media. *J. Math. Anal. Appl.* 88, 104–115. Erratum: *J. Math. Anal. Appl.* 91 (1983), 300.
219. Saranen, J. (1983) On electric and magnetic problems for vector fields in anisotropic nonhomogeneous media. *J. Math. Anal. Appl.* 91, 254–275.
220. Sarvas, J. (1987) Basic mathematical and electromagnetic concepts of the biomagnetic inverse problem. *Phys. Med. Biol.* 32, 11–22.
221. Sauter, S. A., and Lage, C. (2001) Transformation of hypersingular integrals and black-box cubature. *Math. Comp.* 70, 223–250.
222. Sauter, S. A., and Schwab, C. (1997) Quadrature for *hp*-Galerkin BEM in $\mathbf{R}^3$. *Numer. Math.* 78, 211–258.
223. Schmidt, K., Sterz, O., and Hiptmair, R. (2008) Estimating the eddy-current modeling error. *IEEE Trans. Magn.* 44, 686–689.
224. Schwab, C., and Wendland, W. L. (1992) On numerical cubatures of singular surface integrals in boundary element methods. *Numer. Math.* 62, 343–369.
225. Sekino, M., Hirata, M., Sakihara, K., Yorifuji, S., and Ueno, S. (2006) Intensity and localization of eddy currents in transcranial magnetic stimulation to the cerebellum. *IEEE Trans. Magn.* 42, 3575–3577.
226. Selgas Buznego, V. (2006) Análisis matemático y numérico de las ecuaciones de Maxwell cuasiestáticas. PhD thesis, Universidad de Oviedo.

227. Silvester, P. P., and Ferrari, R. L. (1996) *Finite Elements for Electrical Engineers*, 3rd ed. Cambridge University Press, Cambridge.
228. Specogna, R., Suuriniemi, S., and Trevisan, F. (2008) Geometric T-Ω approach to solve eddy currents coupled to electric circuits. *Internat. J. Numer. Methods Engrg.* 74, 101–115.
229. Sterz, O., and Schwab, C. (2001) A scalar boundary integrodifferential equation for eddy current problems using an impedance boundary condition. *Comput. Visul. Sci.* 3, 209–217.
230. Tang, R. Y., Li, Y., Lin, F., and Tian, L. J. (1996) Resultant magnetic fields due to both windings and heavy current leads in large power transformers. *IEEE Trans. Magn.* 32, 1641–1644.
231. Tang, R. Y., Wang, S. H., Li, Y., Wang., X. L., and Cui, X. (2000) Transient simulation of power transformers using 3-D finite element model coupled to electric circuit equations. *IEEE Trans. Magn.* 36, 1417–1420.
232. Tang, R. Y., Wu, Z. X., and Xie, S. M. (1990) 3D eddy current computation by a new FE method in large transformers. *COMPEL* 9, Supplement A, 63–66.
233. Tegopoulos, J. A., and Kriezis, E. E. (1985) *Eddy Currents in Linear Conducting Media*. Elsevier, New York.
234. Thompson, M. T. (2000) Eddy current magnetic levitation. Models and experiments. *IEEE Potentials* 19, 40–44.
235. Thornton, R. D. (1975) Magnetic levitation and propulsion, 1975. *IEEE Trans. Magn.* MAG-11, 981–995.
236. Tsukerman, I. A. (1990) Error estimation for finite element solutions of the eddy currents problem. *COMPEL* 9, 83–98.
237. Ueno, S., Tashiro, T., and Harada, K. (1988) Localized stimulation of neural tissues in the brain by means of a paired configuration of time-varying magnetic fields. *J. Appl. Phys.* 64, 5862–5864.
238. van Rienen, U. (2001) *Numerical Methods in Computational Electrodynamics. Linear Systems in Practical Applications*. Springer, Berlin.
239. van Uitert, R., Weinstein, D., and Johnson, C. (2003) Volume currents in forward and inverse magnetoencephalographic simulations using realistic head models. *Ann. Biomedical Engrg.* 31, 21–31.
240. Vázquez Hernández, R. (2008) Contributions to the mathematical study of some problems in magnetohydrodynamics and induction heating. PhD thesis, Universidade de Santiago de Compostela.
241. Webb, J. P., and Forghani, B. (1989) A single scalar potential method for 3D magnetostatics using edge elements. *IEEE Trans. Magn.* 25, 4126–4128.
242. Webb, J. P., and Forghani, B. (1990) A scalar-vector method for 3D eddy current problems using edge elements. *IEEE Trans. Magn.* 26, 2367–2369.
243. Weiland, T. (2003) Finite integration method and discrete electromagnetism. In Carstensen, C., Funken, S., Hackbusch, W., Hoppe, R. H. W., and Monk, P. (eds.), *Computational Electromagnetics*. Springer, Berlin, 183–198.
244. Wolters, C. H., Anwander, A., Tricoche, X., Weinstein, D., Koch, M. A., and MacLeod, R. S. (2006) Influence of tissue conductivity anisotropy on EEG/MEG field and return current computation in a realistic head model: a simulation and visualization study using high-resolution finite element modeling. *NeuroImage* 30, 813–826.
245. Wolters, C. H., Grasedyck, L., and Hackbusch, W. (2004) Efficient computation of lead field bases and influence matrix for the FEM-based EEG and MEG inverse problem. *Inverse Problems* 20, 1099–1116.
246. Yamamura, S. (1976) Magnetic levitation technology of tracked vehicles present status and prospects. *IEEE Trans. Magn.* MAG-12, 874–878.

247. Yan, L. (2008) Development and application of the Maglev transportation system. *IEEE Trans. Appl. Supercond.* 18, 92–99.
248. Zienkiewicz, O. C., Kelly, D. W., and Bettess, P. (1977) The coupling of the finite element method and boundary solution procedures. *Internat. J. Numer. Methods Engrg.* 11, 355–375.
249. Zimmerman, J. E., Thiene, P., and Harding, J. T. (1970) Design and operation of stable rf-biased superconducting point-contact quantum devices, and a note on the properties of perfectly clean metal contacts. *J. Appl. Phys.* 41, 1572–1580.

Index

MIX
Papier aus verantwortungsvollen Quellen
Paper from responsible sources
FSC® C105338

FSC
www.fsc.org

If you have any concerns about our products,
you can contact us on
ProductSafety@springernature.com

In case Publisher is established outside the EU,
the EU authorized representative is:
Springer Nature Customer Service Center GmbH
Europaplatz 3, 69115 Heidelberg, Germany

Printed by Libri Plureos GmbH
in Hamburg, Germany